Petr Hájek, Michal Johanis
Smooth Analysis in Banach Spaces

De Gruyter Series in Nonlinear Analysis and Applications

Volume 19

Petr Hájek, Michal Johanis

Smooth Analysis in Banach Spaces

DE GRUYTER

Mathematics Subject Classification 2010
Primary: 46B03, 46G25, 46T20; Secondary: 46B20, 46G05

Authors
Prof. Dr. Petr Hájek
Academy of Sciences of the Czech Republic
Institute of Mathematics
Žitná 25
11567, Prague
Czech Republic
hajek@math.cas.cz

and

Czech Technical University in Prague
Faculty of Electrical Engineering
Department of Mathematics
Zikova 4
16000, Prague
Czech Republic

Dr. Michal Johanis
Charles University
Faculty of Mathematics and Physics
Sokolovská 83
18675, Prague
Czech Republic
johanis@karlin.mff.cuni.cz

ISBN 978-3-11-025898-1
e-ISBN (PDF) 978-3-11-025899-8
e-ISBN (EPUB) 978-3-11-039199-2
Set-ISBN 978-3-11-220385-9
ISSN 0941-813X

Library of Congress Cataloging-in-Publication Data
A CIP catalog record for this book has been applied for at the Library of Congress.

Bibliographic information published by the Deutsche Nationalbibliothek
The Deutsche Nationalbibliothek lists this publication in the Deutsche Nationalbibliografie; detailed bibliographic data are available on the Internet at http://dnb.dnb.de.

Printing and binding: CPI books GmbH, Leck
♾Printed on acid-free paper
Printed in Germany

www.degruyter.com

Contents

Introduction

The purpose of this book is to lay down the foundations for the abstract theory of C^k-smoothness in infinite-dimensional real Banach spaces, and investigate its intimate connections with the structural properties of the underlying spaces.

The main objects of the theory are polynomials and C^k-smooth (including real analytic) mappings. In some sense, the most important result concerning C^k-smooth mappings is the Taylor formula, which takes the familiar form known from the theory of functions on $\mathbb{R}^n$. This formula unveils the prominent role played by polynomials in smoothness (especially higher smoothness) questions by way of approximating smooth functions in the neighbourhood of a point. In the infinite-dimensional setting this role is brought even further, as polynomials also provide the vital link with the structure of the underlying Banach space.

This explains why polynomials have received a great deal of attention in the present book. We have included plenty of results concerning polynomials with the intention to build up a supply of results and points of view that could be useful for the future development of the theory. The material is mostly organised according to the methods used. We study polynomials on $\mathbb{R}^n$ in connection with isometric theory of finite-dimensional spaces. It turns out that homogeneous polynomials are closely related to isometric subspaces of ℓ_p^n for p even. The theory of Chebyshev polynomials is used to obtain optimal estimates on the size of higher derivatives for a given polynomial. The theory of tensor products and (p, q)-summing operators is applied to obtain sharp estimates on the values of polynomial coefficients for polynomials from c_0 to ℓ_p. The concept of finite representability combined with powerful finite-dimensional results concerning type and cotype of Banach spaces, spreading models, and ultrapowers lead to strong structural results for Banach spaces admitting separating polynomials. All the above results and much more are covered in the first four chapters. The breadth of the above mentioned material precludes making our presentation completely self contained.

We decided, for the benefit of the reader, to include most of the needed auxiliary results (without proof) in the form of short survey paragraphs, or sometimes whole sections, which form an integral part of the text. This makes it possible for the reader to follow the text, without having to jump into an appendix or a specialised monograph, and keep track of the flow of ideas.

The remaining three chapters of the book are devoted to a detailed study of smooth mappings between Banach spaces. The properties being studied can be roughly divided into three main areas whose polynomial (and usually finite-dimensional) counterparts are covered in the first half of the book.

The first aspect are the structural properties implied by the existence of certain polynomials, resp. separating C^k-smooth functions (in the finite-dimensional setting this corresponds to the study of subspaces of ℓ_p^n). The second aspect is the supply of such mappings (this corresponds essentially to quantitative estimates on the coefficients), and finally the last aspect concerns approximation questions (this relates to the theory of algebras of sub-symmetric polynomials).

The Banach space c_0 plays a major role throughout the subject. From the technical point of view, this is due to the fact that its supply of polynomials (and uniformly smooth functions) is very small (they are weakly uniformly continuous), but on the other hand its supply of C^∞-smooth functions is very large (thanks to the phenomenon of functions that depend locally on finitely many coordinates). Moreover, it is a universal space for C^k-smooth embeddings. There are two important classes of spaces which share some of its important features. It is the class of polyhedral spaces, which are C^∞-smooth, c_0 saturated, and behave well with respect to taking subspaces. From the other side it is the class $\mathcal{W}$, which contains all $\mathcal{L}_\infty$-spaces and behaves rather well with respect to uniformly smooth mappings and quotients. The intersection of these two classes contains all isometric preduals of ℓ_1, in particular all $C(K)$ spaces, K countable compact. The examples of $\mathcal{L}_\infty$-spaces of Jean Bourgain and Freddy Delbaen do not contain c_0, so they are not polyhedral. On the other hand, Schreier's space is a polyhedral space (isomorphic to a subspaces of $C([0,\omega^\omega])$) which admits a non-compact linear operator into ℓ_2, and hence it does not belong to class $\mathcal{W}$. In fact, Ioannis Gasparis constructed polyhedral spaces with quotient ℓ_2.

The existence of a separating C^k-smooth function on a Banach space has strong structural consequences. For example a Banach space admitting a C^2-smooth bump either contains c_0 or it is super-reflexive. So the study of higher smoothness naturally splits into two rather distinct extreme situations, which also rely on distinct techniques. This feature is repeated with respect to smooth mappings, whose supply again depends heavily on the geometry of the underlying space. The former case corresponds to $\mathcal{W}$-spaces, which admit only few uniformly smooth mappings. More precisely, if Y^{**} has no subspace isomorphic to c_0, then any uniformly smooth mappings from a $\mathcal{W}$-space into Y is weakly compact. On the other hand, super-reflexive spaces admit surjective polynomials onto any separable Banach space.

In the questions on approximability of functions and mappings the space c_0 plays a fundamental role thanks to being a universal C^k-smooth embedding space. However, for real analytic approximations the available methods again seem to distinguish between spaces with a separating polynomial or merely a separating analytic function (the c_0-type). At present it is not clear if this distinction can be overcome by improving the methods of proof.

Our book is focusing on the case of real Banach spaces, using the complex case only as a tool for dealing with analytic functions. We feel that the theory has now reached a certain level of maturity, in the sense that extreme cases of higher smoothness, i.e. the

c_0-like, or finite rank phenomena, and on the other hand the separating polynomial-like phenomena have been well-understood. Our rendering of the state of the theory is rather complete and up to date, and it appears for the first time in a book form. A large part is devoted to very recent developments, sometimes in an outline form. We believe that our book may serve as a rather complete reference book for the results and techniques in the area of smoothness in separable real Banach spaces. It can also be used as a textbook by advanced graduate students and active researchers with solid background in Banach space theory, and for this purpose we pose a number of open problems for independent research. The future now lies in investigation of general polynomials, and higher order approximations. This will most probably require a new set of techniques.

Let us proceed with a more detailed description of some highlighted results in the respective chapters. The first chapter contains an abstract theory of C^k-smoothness in any Banach space. We introduce multilinear mappings and polynomials, as higher derivatives are defined to be these objects. The main result in this respect is the Taylor formula which provides a fundamental link between smoothness, polynomials, and the underlying Banach space structure. We also need to introduce the complexification of spaces and polynomials. This is indispensable for the development of the theory of analytic functions, but also very convenient in other situations thanks to efficient averaging methods for complex polynomials. Our treatment of (real) analytic functions was inspired by papers of Jacek Bochnak and Józef Siciak. Analytic functions are defined as functions locally admitting power series expansion. There are several important characterisations, in particular by using finite-dimensional restrictions, and by using the complexified series. These characterisations are also an important tool in developing the theory. The theory depends heavily on results from one or several complex variables, which are mentioned without proof and applied through the notion of higher Gâteaux smoothness.

In brief summary, the first chapter gives a complete and self contained introduction to the subject, including the case of real analytic functions which to the best of our knowledge is for the first time in a book form. The remaining six chapters are best read consecutively. Chapters 2–4, which focus on properties of polynomials, cover all the necessary background needed for the remaining Chapters 5–7, which are dealing with general smooth mappings. Most of the material contained therein is also new to book form.

In Chapter 2 we focus on the duality theory for spaces of polynomials on finite-dimensional Banach spaces, in particular the cubature formulae representing positive functionals. We present a very simple and original proof of the well-known Chakalov theorem. We outline the proofs of cubature formulae related to Chebyshev polynomials on $\mathbb{R}$ and $\mathbb{R}^2$. This topic seems to be rather special but it leads to the proof of the Skalyga-Markov-type inequality, an optimal estimate on the size of higher derivatives of real polynomials. We continue by showing that every homogeneous polynomial is a finite sum of powers of functionals, which leads to the theory of isometric embeddings of finite-dimensional spaces into ℓ_p-spaces where p is even. We present, without proof, the recent result of Vladimir Leonidovich Dol'nikov and Roman Nikolaevich Karasev, and its close (but easier) relative, the Birch theorem. According to these results, a k-homogeneous polynomial on $\mathbb{R}^N$ can be restricted to a suitable $n(N, k)$-dimensional subspace E so that this restriction is equivalent to a power of the Euclidean norm on E. We cover the basic theory of identities for polynomials, which implies that a continuous function is a polynomial of degree at most n provided that its restriction to every affine one-dimensional subspace is also a polynomial of degree at most n. The last section is devoted to a relatively simple but powerful averaging method for treating polynomials and multilinear mappings, which leads to important estimates of polynomial coefficients. More sophisticated methods which combine probabilistic averaging with other techniques are treated in the subsequent chapter.

Chapter 3 is devoted to the study of polynomials between Banach spaces and the way they act on sequences from the initial space. This is a classical theme in the theory of linear operators, represented by the Dunford-Pettis property and the theory of p-summing operators. We start the chapter by outlining the basics on tensor products as well as symmetric tensor products and their duality with polynomial spaces. We then outline the concepts of uniform spaces and uniform continuity which play a key role throughout the whole book. We proceed by developing in detail the basic theory of weakly, weakly sequentially, and weakly uniformly continuous mappings and in particular polynomials. This investigation was initiated in the early papers of Aleksander Pełczyński and more systematically developed by Richard Martin Aron and his coauthors. An important contribution of Raymond A. Ryan was the introduction of symmetric tensor products into the subject, as well as the proof that every weakly compact polynomial from a space with the Dunford-Pettis property maps weakly Cauchy sequences to norm convergent ones.

In Banach spaces not containing ℓ_1 the weak sequential continuity on bounded sets coincides with full (or even uniform) continuity. This important fact results from Rosenthal's ℓ_1 theorem and other related results of Edward Wilfred Odell and results of Jean Bourgain, David H. Fremlin, and Michel Talagrand.

We introduce the language of the theory of (p, q)-summing operators, mention some connections with tensor products, and formulate some of the fundamental results of the theory, introducing also the notions of type and cotype. Using the theory of multiple $(p; 1)$-summing operators, we give optimal estimates on the coefficients of polynomials in $\mathcal{P}(^n c_0; \ell_p)$, following the recent work of Andreas Defant and Pablo Sevilla-Peris.

In Chapter 4 we apply the concept of finite representability to the study of asymptotic behaviour of polynomials and the linear structure of the underlying Banach spaces. We describe the ultrapower construction for a Banach space X, which leads to a much larger Banach space $(X)_{\mathcal{U}}$ that is finitely representable in X and that is well-suited for constructions of uniformly continuous mappings. As an application we show that uniformly smooth mappings can be extended into the bidual. By using the spreading models of X, which capture the asymptotic behaviour of infinite sequences in X, we study the upper and lower estimates of sequences, the Banach-Saks and the weak p-Banach-Saks properties. We proceed by showing that if X has a sub-symmetric basis and a separating d-homogeneous polynomial, then it is isomorphic to some ℓ_p for p even, and $d = kp, k \in \mathbb{N}$. As a corollary we obtain a fundamental result of Robert Deville stating that every Banach space with a separating polynomial contains a subspace isomorphic to ℓ_p, p even. We study in detail the restrictions of polynomials on ℓ_p to subspaces generated by suitable subsequences of the canonical basis. The main result in this direction, which goes well beyond the asymptotic but finite-dimensional results based on spreading models, claims that these restrictions are almost sub-symmetric. As a corollary to all these results we show that for an arbitrary polynomial P on X there is an infinite-dimensional subspace Y of X such that the restriction of P to Y is either separating or asymptotically zero in a strong sense.

The last sections are devoted to the study of algebras $\mathcal{A}_n(X)$ of polynomials generated by polynomials of degree at most n on a Banach space X. The main technical tool is a finite-dimensional lemma which claims that the symmetric polynomial $s_n^N(x) = \sum_{j=1}^N x_j^n$ on $\mathbb{R}^N$ is not in the uniform closure of a suitably defined sub-symmetric subalgebra of $\mathcal{A}_{n-1}(\mathbb{R}^N)$, provided that N is large enough. We proceed by applying this result to infinite-dimensional spaces using spreading model techniques. The main result implies in particular that $\overline{\mathcal{A}_1(\ell_p)} = \cdots = \overline{\mathcal{A}_{n-1}(\ell_p)} \subsetneqq \overline{\mathcal{A}_n(\ell_p)} \subsetneqq \overline{\mathcal{A}_{n+1}(\ell_p)} \subsetneqq \cdots$, where $n = \lceil p \rceil$.

Chapter 5 is devoted to the detailed study of Banach spaces admitting smooth separating functions. An important set of tools for obtaining structural results are the variational principles. When applied to a given lower continuous and bounded below function f, they guarantee the existence of a point $x \in X$ and a smooth function g such that $f - g$ attains its minimum, sometimes in a strong sense. If f itself is a smooth function, then depending on the concrete conditions we may use the Taylor formula at x in order to obtain uniformly smooth separating functions on X, or at least some structural information about X. We describe two examples of this notion. Stegall's variational principle holds in every Banach space with the RNP and the function g can be chosen to be a functional from X^*. As a result, if a Banach space with the RNP admits twice Gâteaux smooth bump function, we conclude that X is super-reflexive and admits a norm with a power type 2 estimate on the modulus of smoothness.

The compact variational principle of Robert Deville and Marián Fabian, which was motivated by a paper of Jaroslav Pechanec, John H. M. Whitfield, and Václav Zizler, is a key tool for studying higher smoothness in Banach spaces. Its formulation is somewhat different from the general description given above, but it leads to similar applications. This principle (we prefer to avoid the precise formulation at this point) applies to Banach spaces which do not contain c_0, and it has lead to several strong structural results. The most important structural result is that if X has no subspace isomorphic to c_0 and has a C^k-smooth bump, for large enough k, then X is super-reflexive, admits a separating polynomial, and contains a subspace isomorphic to ℓ_p, p even.

Jaroslav Kurzweil has pioneered the field of higher smoothness by finding that the best order of smoothness of L_p-spaces is $C^{[p]}$, except in the case when p is even and the space admits a separating polynomial. In the latter case he constructed real analytic approximations for all continuous mappings from these spaces. In the rest of the chapter we compute the best order of smoothness for several classes of Banach spaces, notably the Orlicz spaces and the class of polyhedral Banach spaces.

In Chapter 6 we develop the theory of uniformly smooth mappings between Banach spaces. We study the relationship between uniform continuity and weak uniform continuity of the mapping and its higher derivatives. In particular, if a weakly uniformly continuous mapping has a uniformly continuous kth derivative, then this derivative is even weakly uniformly continuous. In Section 6.2 we introduce and study the important concept of bidual extension for $C^{k,+}$-smooth mappings from the unit ball of a Banach space X into Y to a $C^{k,+}$-smooth mapping from the unit ball of X^{**} into Y^{**}. This notion contains the classical bi-adjoint of a linear operator as a special case, but it is not completely canonical and in general depends on some parameters in the construction. This notion plays an important role later on in pushing some results for $C(K)$ spaces where K is scattered into the case of a general compact space K.

In Sections 6.3–6.6 we build the theory of $\mathcal{W}_\lambda$-spaces, i.e. all those X such that uniformly smooth functions in B_X map weakly Cauchy sequences in λB_X, for some $\lambda \in (0, 1]$, to convergent sequences. The main goal of this theory is to generalise some classical properties of linear operators from $C(K)$ spaces, described in Theorem 3.47, into the setting of uniformly smooth mappings. This objective is achieved in the main Theorem 6.57, which claims that weakly compact and uniformly smooth mappings from the unit ball of $C(K)$ spaces take weakly Cauchy sequences into norm convergent ones. The first step of the proof is to show the rigidity of ℓ_∞^n with respect to uniformly smooth functions. The second step consists of showing that $C(K)$, for K scattered, are $\mathcal{W}_1$-spaces. Next, we prove that the bidual extension of any uniformly smooth non-compact mapping in $\mathcal{C}^{1,+}(B_{c_0}; Y)$ has a point where the derivative is non-compact, and hence fixes a copy of c_0. This implies that Y^{**} has a subspace isomorphic to c_0. Moreover, non-compact uniformly smooth mappings from $C(K)$, K scattered, can always be reduced to a suitable subspace isomorphic to c_0 where the restriction remains non-compact. The case of a general $C(K)$ space requires another ingredient, Theorem 6.56, which claims that after passing to the bidual, weakly Cauchy sequences in $C(K)$ spaces are uniformly close to weakly Cauchy sequences in some $C([0, \alpha])$.

In Section 6.7 we give some rather general results on the ranges of smooth mappings (and derivatives of smooth functions) which illustrate that the theory of $\mathcal{W}$-spaces works under nearly optimal assumptions. Indeed, according to Theorem 6.69 no structural property of the initial space is generally preserved by surjective C^∞-smooth mappings from Banach spaces with property $\mathcal{B}$. In the rest of the chapter our attention shifts to smooth separating mappings from ℓ_p-spaces, proving strong structural result in this case, based on the notions of a separating mapping and harmonic behaviour.

Smooth approximations in Banach spaces are studied in the last chapter of our book. Unlike the finite-dimensional case, the unit ball of an infinite-dimensional Banach space X is not a compact set and hence it is easy to find uniformly continuous functions that are not uniformly approximable by polynomials on B_X. Interestingly enough, in the special case of $\mathcal{W}$-spaces X not containing ℓ_1, uniformly C^k-smooth functions can be uniformly approximated, together with their higher derivatives, by polynomials on B_X. This result underlines the extremely poor supply of uniformly smooth functions in these spaces, rather than the abundance of polynomials. Indeed, in this case all polynomials are weakly uniformly continuous on B_X. This is again a situation when smoothness properties take on very different shapes for c_0-like spaces (although X considered here need not even contain c_0!), and for super-reflexive spaces.

The main tool for studying C^k-smooth approximations of continuous mappings are C^k-smooth partitions of unity, or alternatively C^k-smooth embeddings into c_0. The problem becomes more challenging if additional conditions are put on the approximants, and the usual partitions of unity cannot be employed as they destroy the character of the approximating functions. This concerns for example the problem of smooth approximations preserving the Lipschitz constants, which is the first step for obtaining approximations together with higher derivatives, a problem that remains widely open. We finish the chapter by proving that if a separable Banach space admits a C^k-smooth equivalent norm, then every norm can be approximated on bounded sets by C^k-smooth renormings.

Let us make some remarks concerning the existing literature related to the subject of our book. Introduction to abstract smooth analysis (real or complex) forms part of Jean Dieudonné's book [Dieu], which was a great source of inspiration for us. The complex case, where all notions of smoothness coincide, has received more attention in monographs, e.g. in [Hill], [Din], [Mu], [Na]. We have relied much on Seán Dineen's book for its insights, historical comments, and ample references. The distinguishing feature of the real setting, as opposed to complex one, is the intricate role played by the Banach space structure and geometry for the supply of polynomials and C^k-smooth functions.

Concerning polynomials on finite dimension spaces, there are of course many sources, see e.g. [MMR], [BE]. Apparently the main interest in polynomials in the infinite-dimensional setting came from the study of analytic functions, see [Din] for historical comments.

Finally, it was the monograph of Robert Deville, Gilles Godefroy, and Václav Zizler [DGZ], which has undertaken a systematic study of smoothness (including higher smoothness) in the context of Banach space structure. Their book treats both the separable and non-separable situation (in the latter they are still relatively up to date, so we decided for the most part to omit it). Our book overlaps only to a small extent with [DGZ], but the problems posed therein played a decisive role for the subsequent development of the subject.

We mention that the first order smoothness theory is well covered (apart from [DGZ]) also in [Fab3], and the more introductory [FHHMZ]. The differentiability of Lipschitz mappings is treated in [LPT]. We are not covering any aspects of this direction of research. In a broader sense, our subject forms part of geometric non-linear analysis, a fast growing subject whose foundations are laid out in [BenLi].

Concerning prerequisites, most of the background material on the linear structural theory of Banach spaces can be found in [FHHMZ]. Deeper results from local theory, which play a crucial role throughout the book, can be found in [DJT]. These two references provide a solid background for the material presented in our book, but our theory draws on results from other areas as well. For the theory of Chebyshev polynomials we suggest [Ri], tensor products are treated in [DefFl] or [Ry3]. A detailed exposition of spreading models is in [BeaLa]. Structural theory of $C(K)$ and $\mathcal{L}_\infty$-spaces, which is used in Chapter 6, is covered in [LiTz1], and for $C(K)$ spaces also in [Ros3]. Structural theory of the classical Banach spaces is developed in more detail in [AK], [Dies2], [LiTz2], and [LiTz3]. For topological results we refer mostly to [Eng].

For an additional source of open problem we refer to [FMZ].

Acknowledgement

In the early 90's the first named author started his PhD research under the supervision of Václav Zizler, then at the University of Alberta in Edmonton. At that time Václav was finishing his book [DGZ], which has played a decisive role for the future development of this area, through its thorough exposition of the known results and techniques, but also through many attractive but approachable problems. Not only that. For the benefit of his students, Václav has compiled a list of additional 50 (later expanded to about 90) problems in the areas close to [DGZ], many of them related to higher smoothness. It was precisely this abundance of nice open problems and Václav's keen interest in any progress, that attracted the first named author to this field of research.

Some of the problems have been solved, and their solutions have lead to the theory presented in this book. Many remain still open, and have been again posed in the present work in the hope that they will do a similar service to the next generation of students.

We would like to thank Václav for all those years of support, interest, and friendship. We would also like to thank our closest colleagues and friends (in alphabetical order) Richard M. Aron, Robert Deville, Marián Fabian, Gilles Godefroy, Gilles Lancien, Vicente Montesinos, and Stanimir L. Troyanski for their important contribution to the present work in the form of open problems, shared knowledge, moral support, and opportunity for discussions and seminar presentations. Finally, we would like to thank our colleagues Michal Kraus and Luděk Zajíček for reading parts of the manuscript and supplying us with errata and critical comments.

Above all we thank our families for taking the burden of living with us through the long and painstaking process of writing this volume.

The work on this volume has been partially supported by grants GAČR 201/07/0394, GAČR 201/11/0345, GAAV A100190801, 7AMB12FR003, and RVO 67985840.

Notation

We fix some notation for objects and notions that the reader should be familiar with. By $\mathbb{N}$, $\mathbb{Z}$, $\mathbb{Q}$, $\mathbb{R}$, and $\mathbb{C}$ we denote the sets of natural numbers, integers, rational numbers, reals, and complex numbers respectively. We set $\mathbb{N}_0 = \mathbb{N} \cup \{0\}$. By $\mathbb{R}^+$ we denote the set of positive real numbers and by $\mathbb{R}_0^+$ the set of non-negative real numbers. By $\overline{\mathbb{R}}$ we denote the extended real line, i.e. $\overline{\mathbb{R}} = [-\infty, +\infty]$. By $\mathbb{K}$ we denote the scalar field $\mathbb{R}$ or $\mathbb{C}$. We use the convention that a sum over an empty set is zero and a product over an empty set is equal to 1. Further, $x^0 = 1$ for any $x \in \mathbb{K}$. For $x \in \mathbb{R}$ we denote by $[x]$ the integer part of x, i.e. the unique number $k \in \mathbb{Z}$ satisfying $k \leq x < k + 1$, by $\lceil x \rceil$ we denote the ceiling of x, i.e. the unique number $k \in \mathbb{Z}$ satisfying $k - 1 < x \leq k$.

For a set A we denote its cardinality by $|A|$ or card A. The cardinality of the continuum is denoted by $\mathfrak{c}$. By abusing the notation we write $\{x_\gamma\}_{\gamma \in \Gamma} \subset X$ meaning that $\{x_\gamma\}_{\gamma \in \Gamma}$ is a collection such that $x_\gamma \in X$ for each $\gamma \in \Gamma$.

Let (P, ρ) be a metric space. We denote $B(x, r) = \{y \in P;\ \rho(y, x) \leq r\}$ and $U(x, r) = \{y \in P;\ \rho(y, x) < r\}$ the closed, resp. open ball in P centred at $x \in P$ with radius $r > 0$. In case that it is necessary to distinguish the spaces in which the balls are taken, we will write $B_P(x, r)$, resp. $U_P(x, r)$. By B_X and U_X we denote the closed, resp. open unit ball of a normed linear space X. By S_X we denote the unit sphere of a normed linear space X. An interior of a set A in a topological space is denoted by Int A, its boundary is denoted by ∂A.

Throughout the book we use the following convention: In each statement involving multiple vector spaces we assume that all the spaces are over the same field $\mathbb{K}$ if not specified otherwise. Furthermore, if not specified explicitly or in the beginning of the chapter or section, then the statement holds both for $\mathbb{K} = \mathbb{R}$ and $\mathbb{K} = \mathbb{C}$. When we speak of a subspace of a Banach space, we always mean a closed subspace. General subspaces will be referred to as "linear subspaces". We define span $\emptyset = \{0\}$. If X is a normed linear space with a Schauder basis $\{e_n\}$ and $x = \sum_{n=1}^{\infty} x_n e_n \in X$, then $\operatorname{supp} x = \{n \in \mathbb{N};\ x_n \neq 0\}$ is called a support of x; a finitely supported vector is a vector with finite support. An algebraic dual of a vector space is denoted by $X^\#$, a topological dual of a topological vector space by X^*. Inner product is denoted by $\langle x, y \rangle$, and similarly we denote the evaluation in duality by $\langle f, x \rangle$. Let X, Y be normed linear spaces. For simplicity we say that X contains Y if X has a subspace isomorphic to Y.

By $C(X; Y)$ we denote the set of continuous mappings between topological spaces X, Y. If Y is a topological vector space, then $C(X; Y)$ is a vector space. For functions, i.e. mappings into the scalars, we use a shortened notation $C(X) = C(X; \mathbb{K})$; from the context it should always be clear whether $\mathbb{K} = \mathbb{R}$ or $\mathbb{K} = \mathbb{C}$. For a mapping $f\colon X \to Y$, where Y is a vector space, we denote $\operatorname{supp}_o f = f^{-1}(Y \setminus \{0\})$. If X is a topological space, then we denote $\operatorname{supp} f = \overline{\operatorname{supp}_o f}$. An L-Lipschitz mapping is a mapping that is Lipschitz with a constant L. By χ_A we denote the characteristic function of the set A.

If we say measure or Borel measure, we always mean a non-negative measure. On the other hand, Radon measure means scalar-valued Radon measure. By $\operatorname{supp}\mu$ we denote the support of a Borel measure μ. The n-dimensional Lebesgue measure will be denoted by λ_n, or just λ if the dimension is clear from the context.

All topological spaces in this volume are automatically and without mention assumed to be Hausdorff.

Chapter 1

Fundamental properties of smoothness

In this chapter we are going to develop the theory of C^k-smoothness on normed linear spaces and the theory of analytic mappings on Banach spaces. Completing this task requires a large number of conceptual and technical ingredients. The main concept here is the notion of a Fréchet derivative. Let $f: X \to Y$ be a mapping between normed linear spaces X and Y. The Fréchet derivative of f at $x \in X$ is a bounded linear operator from X to Y which approximates the mapping $h \mapsto f(x+h) - f(x)$ around 0. Higher order derivatives are defined by induction, so the second derivative is a linear operator from X into the space of bounded linear operators from X to Y. Things are getting more complicated as the order of the derivative grows and the higher derivatives at x are identified with bounded multilinear mappings from $X \times \cdots \times X$ to Y.

Higher derivatives turn out to be invariant with respect to permutations of the variables (we say that they are symmetric mappings), and this can be used to identify them with homogeneous polynomials from X into Y. This identification leads to no loss of information, but the benefit is enormous, as the domain of the differentials now remains the same space X, and we can formulate the fundamental Taylor theorem in the same way as for functions on the real line.

The program outlined above is carried out in the first four sections of the chapter, together with the additional construction of the complexification of a Banach space and the corresponding canonical extensions of multilinear mappings and polynomials to mappings between the complexifications. The remaining part of the chapter is devoted to building the theory of analytic mappings.

The formal definition of an analytic mapping looks rather simple, as we only require that on a neighbourhood of every point in the domain the mapping is the sum of a power series consisting of homogeneous polynomials. However, in order to establish the basic properties of analytic mappings we are forced to prove a number of non-trivial results and also work simultaneously with several equivalent conditions to being analytic. This program includes the detailed study of the modes and sets of convergence of the power series, the use of complex analytic results and methods (e.g. the Cauchy formula) from several complex variables, in tandem with the weaker notion of Gâteaux derivative and the fundamental estimate contained in the Polynomial lemma of Franciszek Leja.

In this chapter all the results hold both for real and complex spaces if not indicated otherwise. Sometimes, when the statement directly involves the field, we stress again that the spaces involved are over $\mathbb{K}$.

1. Multilinear mappings and polynomials

In this section we introduce the notions of multilinear mappings and polynomials between normed linear spaces, and prove some of their main properties. In particular, homogeneous polynomials are in a canonical one-to-one correspondence with the symmetric multilinear mappings via the Polarisation formula. Polynomials are continuous mappings whenever they have at least one point of continuity. We introduce non-homogeneous polynomials of degree at most n and prove that their space is a direct sum of the homogeneous summands.

Definition 1. Let $X_1, \dots, X_n$, and Y be vector spaces. We will say that a mapping $M\colon X_1 \times \dots \times X_n \to Y$ is n-linear if it is linear separately in each coordinate, i.e. if $x \mapsto M(x_1, \dots, x_{k-1}, x, x_{k+1}, \dots, x_n)$ is a linear mapping from X_k into Y for each $x_1 \in X_1, \dots, x_n \in X_n$ and each $k \in \{1, \dots, n\}$. By $L(X_1, \dots, X_n; Y)$ we denote the vector space of all n-linear mappings from $X_1 \times \dots \times X_n$ to Y. In the special case when $X_k = X$, $1 \le k \le n$, we use the short notation $L({}^nX; Y)$. A mapping is called multilinear if it is n-linear for some $n \in \mathbb{N}$. A 2-linear mapping will also be called bilinear.

We say that $M \in L({}^nX; Y)$ is symmetric if $M(x_1, \dots, x_n) = M(x_{\pi(1)}, \dots, x_{\pi(n)})$ for every permutation π of $\{1, \dots, n\}$ and every $x_1, \dots, x_n \in X$. By $L^{s}({}^nX; Y)$ we denote the vector space of all n-linear symmetric mappings from X^n to Y.

By repeated applications of the definition we can see that an n-linear mapping $M \in L(X_1, \dots, X_n; Y)$ satisfies the formula

$$M\left(\sum_{k_1=1}^{m_1} a^1_{k_1} x^1_{k_1}, \dots, \sum_{k_n=1}^{m_n} a^n_{k_n} x^n_{k_n}\right) = \sum_{\substack{1 \le k_j \le m_j \\ j=1,\dots,n}} \left(\prod_{j=1}^{n} a^j_{k_j}\right) M(x^1_{k_1}, \dots, x^n_{k_n}) \quad (1)$$

for an arbitrary choice of $m_j \in \mathbb{N}$, $a^j_{k_j} \in \mathbb{K}$, and $x^j_{k_j} \in X_j$, where $k_j = 1, \dots, m_j$ and $j = 1, \dots, n$. The following fact also follows easily from the definition.

Fact 2. *Let X, Y be vector spaces, $\dim X \ge n+1$, and $M\colon X^n \to Y$ a mapping such that $M\restriction_{E^n}$ is n-linear for every $(n+1)$-dimensional subspace $E \subset X$. Then $M \in L({}^nX; Y)$.*

Definition 3. Let $X_1, \dots, X_n$, and Y be normed linear spaces. We say that the mapping $M \in L(X_1, \dots, X_n; Y)$ is bounded if

$$\|M\| = \sup_{x_1 \in B_{X_1}, \dots, x_n \in B_{X_n}} \|M(x_1, \dots, x_n)\| < +\infty.$$

It is easily checked that this defines a norm on the subspace of $L(X_1, \dots, X_n; Y)$ consisting of bounded multilinear mappings. By $\big(\mathcal{L}(X_1, \dots, X_n; Y), \|\cdot\|\big)$, respectively $\big(\mathcal{L}({}^nX; Y), \|\cdot\|\big)$ and $\big(\mathcal{L}^{s}({}^nX; Y), \|\cdot\|\big)$, we denote the normed linear spaces of all respective n-linear bounded mappings. For bounded n-linear forms, i.e. when the target space is the scalar field, we use a shortened notation $\mathcal{L}({}^nX) = \mathcal{L}({}^nX; \mathbb{K})$.

One of the simplest examples of an n-linear mapping belonging to $\mathcal{L}(X_1,\dots,X_n;Y)$ is given by the formula $M(x_1,\dots,x_n)=\phi_1(x_1)\phi_2(x_2)\cdots\phi_n(x_n)y$, where $\phi_j\in X_j^*$, $j=1,\dots,n$, $y\in Y$. Note also that from (1) it follows that on finite-dimensional spaces all n-linear mappings are automatically bounded.

Let $M\in\mathcal{L}(X_1,\dots,X_n;Y)$. Then by the homogeneity we have

$$\|M(x_1,\dots,x_n)\|\le\|M\|\|x_1\|\cdots\|x_n\|\quad\text{for } x_j\in X_j,\ j=1,\dots,n. \tag{2}$$

Proposition 4. *Let $X_1,\dots,X_n,Y$ be normed linear spaces and $M\in L(X_1,\dots,X_n;Y)$. The following statements are equivalent:*

(i) *M is bounded.*
(ii) *M is Lipschitz on bounded sets.*
(iii) *M is continuous.*
(iv) *M is bounded on a neighbourhood of some point.*

Proof. We consider the maximum norm on the space $X_1\oplus\cdots\oplus X_n$. We write $x=(x_j)_{j=1}^n$ for $x\in X_1\oplus\cdots\oplus X_n$.

(i)$\Rightarrow$(ii) Let $S\subset X_1\oplus\cdots\oplus X_n$ be a given bounded set and let $c>0$ be such that $S\subset(cB_{X_1})\times\cdots\times(cB_{X_n})$. Using (1), for any $x,y\in S$ we obtain

$$M(y)=M(x+y-x)=M(x)+\sum_{A\subsetneqq\{1,\dots,n\}}M\Big(\big(\chi_A(j)x_j+(1-\chi_A(j))(y_j-x_j)\big)_{j=1}^n\Big).$$

Combining this with (2) we get

$$\|M(y)-M(x)\|\le 2^n\|M\|(2c)^{n-1}\max_{1\le j\le n}\|y_j-x_j\|. \tag{3}$$

(ii)$\Rightarrow$(iii)$\Rightarrow$(iv) is clear, so it remains to show that (iv)$\Rightarrow$(i). Using (1), for any $x,h\in X_1\oplus\cdots\oplus X_n$ we obtain

$$\begin{aligned}M(h)&=\sum_{A\subset\{1,\dots,n\}}M\Big(\big(\chi_A(j)(x_j+h_j)+(1-\chi_A(j))(-x_j)\big)_{j=1}^n\Big)\\&=\sum_{A\subset\{1,\dots,n\}}(-1)^{n-|A|}M\Big(\big(\chi_A(j)(x_j+h_j)+(1-\chi_A(j))x_j\big)_{j=1}^n\Big).\end{aligned}$$

It follows that if M is bounded on a neighbourhood of x, then it is also bounded on a neighbourhood of zero, and hence by the homogeneity also on the unit ball. □

Theorem 5 (Uniform boundedness principle; [MO2]). *Let $X_1,\dots,X_n$ be Banach spaces and Y a normed linear space. Let $\{M_\alpha\}_{\alpha\in\Gamma}\subset\mathcal{L}(X_1,\dots,X_n;Y)$ be a family such that $\sup_{\alpha\in\Gamma}\|M_\alpha(x_1,\dots,x_n)\|<+\infty$ for every $(x_1,\dots,x_n)\in X_1\times\cdots\times X_n$. Then $\sup_{\alpha\in\Gamma}\|M_\alpha\|<+\infty$.*

Proof. We use induction on n. For $n=1$ it is just the Banach-Steinhaus uniform boundedness principle. To prove the induction step from n to $n+1$ suppose that $\{M_\alpha\}_{\alpha\in\Gamma}\subset\mathcal{L}(X_1,\dots,X_n,X_{n+1};Y)$ satisfies the assumptions. By the inductive hypothesis for every $z\in X_{n+1}$ the family $\{(x_1,\dots,x_n)\mapsto M_\alpha(x_1,\dots,x_n,z)\}_{\alpha\in\Gamma}\subset\mathcal{L}(X_1,\dots,X_n;Y)$ is uniformly bounded. So for every $z\in X_{n+1}$ there exists $C_z>0$

such that $\|M_\alpha(x_1,\dots,x_n,z)\| \le C_z\|x_1\|\cdots\|x_n\|$ for every $x_1 \in X_1,\dots,x_n \in X_n$, and $\alpha \in \Gamma$. Since $X_{n+1} = \bigcup_{n=1}^{\infty}\{z \in X_{n+1};\ C_z \le n\}$, by the Baire category theorem there exist $C > 0$ and a set $V \subset X_{n+1}$ such that $\overline{V}$ has a non-empty interior and

$$\|M_\alpha(x_1,\dots,x_n,z)\| \le C\|x_1\|\cdots\|x_n\|$$

for every $x_j \in X_j$, $j = 1,\dots,n$, every $\alpha \in \Gamma$, and every $z \in V$. Because the mapping $z \mapsto M_\alpha(x_1,\dots,x_n,z)$ is continuous, the estimate above holds for every $z \in \overline{V}$ and hence for every z from some ball $B(w,r)$, where $w \in X_{n+1}$ and $r > 0$.

Choose arbitrary $x_j \in B_{X_j}$ for $j = 1,\dots,n+1$. We find $y \in X_{n+1}$ satisfying $\|y - w\| = r$ and such that $x_{n+1} = \lambda(y - w)$, where $\lambda = \|x_{n+1}\|/r$. Because $M_\alpha(x_1,\dots,x_{n+1}) = \lambda M_\alpha(x_1,\dots,x_n,y) - \lambda M_\alpha(x_1,\dots,x_n,w)$, we conclude that

$$\|M_\alpha(x_1,\dots,x_{n+1})\| \le \frac{2C}{r} \quad \text{for every } \alpha \in \Gamma.$$

This finishes the proof. □

The following result from elementary linear algebra gives a solution to an interpolation problem. It is of fundamental importance in the theory of polynomials.

Lemma 6. *Let $n \in \mathbb{N}_0$ and let $t_j \in \mathbb{K}$, $j = 0,\dots,n$, be distinct scalars. If Y is a vector space over $\mathbb{K}$ and $z_0,\dots,z_n \in Y$, then there are uniquely determined $y_0,\dots,y_n \in Y$ such that $z_j = t_j^0 y_0 + \cdots + t_j^n y_n$, $j = 0,\dots,n$. The vectors y_k are given by the formula $y_k = a_{k0}z_0 + \cdots + a_{kn}z_n$, $k = 0,\dots,n$, where a_{kj} are the elements of the inverse to the Vandermonde matrix*

$$\begin{pmatrix} 1 & t_0 & t_0^2 & \dots & t_0^n \\ 1 & t_1 & t_1^2 & \dots & t_1^n \\ & & \vdots & & \\ 1 & t_n & t_n^2 & \dots & t_n^n \end{pmatrix}.$$

In particular, the coefficients a_{kj} are independent of z_j.

We recall that a subset A of a topological space X is said to have the Baire property if A is a symmetric difference of an open set and a set of the first category. The family of all subsets of X with the Baire property is the smallest σ-algebra containing all open sets and all sets of the first category. In particular, every Borel set has the Baire property. The mapping $f\colon X \to Y$ between topological spaces X and Y is said to be Baire measurable if $f^{-1}(G) \subset X$ has the Baire property for every $G \subset Y$ open.

Theorem 7. *Let $X_1,\dots,X_n$ be Banach spaces, Y a normed linear space, and let $M \in L(X_1,\dots,X_n;Y)$. The following statements are equivalent:*

(i) *M is continuous.*
(ii) *M is separately continuous, i.e. $x \mapsto M(x_1,\dots,x_{k-1},x,x_{k+1},\dots,x_n) \in \mathcal{L}(X_k;Y)$ for each $x_1 \in X_1,\dots,x_n \in X_n$ and each $k \in \{1,\dots,n\}$.*
(iii) *M is Baire measurable.*
(iv) *$\phi \circ M$ is Baire measurable for every $\phi \in Y^*$.*
(v) *$\phi \circ M$ is separately Baire measurable for every $\phi \in Y^*$.*

In the proof we will use the following simple observation.

Lemma 8. *Let X be a topological vector space that is a Baire space, $G \subset X$ open, and $A \subset G$ residual in G. Then for each $y \in G$ and $n \in \mathbb{N}$ there is an open neighbourhood V of y such that for each $x \in V$ there is $h \in X$ satisfying $x + jh \in A$, $j = 1, \dots, n$.*

Proof. Without loss of generality we may assume that $y = 0$. Let W be a balanced open neighbourhood of 0 such that $W + W \subset G$ and put $V = \frac{1}{n}W$. Fix $x \in V$ and note that $V \subset \frac{1}{j}W$ for $j = 1, \dots, n$ and hence $V \subset \bigcap_{j=1}^{n} \frac{1}{j}(G - x)$. Each set $\frac{1}{j}(A - x)$ is residual in $\frac{1}{j}(G - x)$ and hence each $V \cap \frac{1}{j}(A - x)$, $j = 1, \dots, n$, is residual in V. Consequently, $V \cap \bigcap_{j=1}^{n} \frac{1}{j}(A - x)$ is residual in V and in particular it is non-empty and so it suffices to take any h from this set. □

Proof of Theorem 7. (i)⇒(iii)⇒(iv) is obvious.

To prove (iv)⇒(i) denote $X = X_1 \times \dots \times X_n$. Fix an arbitrary $\phi \in Y^*$. Let $A_k = \{x \in X;\ |\phi \circ M(x)| < k\}$, $k \in \mathbb{N}$. Then $X = \bigcup_{k \in \mathbb{N}} A_k$ and hence there is $m \in \mathbb{N}$ such that A_m is of the second category. Since by our assumption A_m has the Baire property, there is a non-empty open $G \subset X$ such that $A_m \cap G$ is residual in G. By Lemma 8 there is a non-empty open $V \subset X$ such that for each $x \in V$ there is $h \in X$ satisfying $x + jh \in A_m$, $j = 1, \dots, n + 1$. Let a_{kj} be the coefficients from Lemma 6 for $t_j = j + 1$, $j = 0, \dots, n$. For any $x \in V$ and $h \in X$ satisfying $x + jh \in A_m$, $j = 1, \dots, n + 1$ we have

$$M(x + jh) = M(x) + \sum_{k=1}^{n} j^k \sum_{\substack{C \subset \{1,\dots,n\} \\ |C| = k}} M\Big(\big((1 - \chi_C(l))x_l + \chi_C(l)h_l\big)_{l=1}^{n}\Big)$$

for $j = 1, \dots, n + 1$. Lemma 6 then implies

$$M(x) = \sum_{j=0}^{n} a_{0j} M\big(x + (j + 1)h\big)$$

and hence $|\phi \circ M(x)| \le m \sum_{j=0}^{n} |a_{0j}|$. This means that $\phi \circ M(V)$ is bounded and so $\phi \circ M(B_X)$ is bounded by Proposition 4. Therefore $M(B_X)$ is weakly bounded and hence it is bounded by the Banach-Steinhaus uniform boundedness principle. Thus M is continuous by Proposition 4.

(i)⇒(ii)⇒(v) is obvious and (v)⇒(ii) follows from the already proved (iv)⇒(i).

(ii)⇒(i) We use induction on n. For $n = 1$ there is nothing to prove. Let $n > 1$ and assume the statement holds for $(n-1)$-linear mappings. For each $x_1 \in X_1, \dots, x_n \in X_n$ we put $M_{x_2,\dots,x_n}(x_1) = M(x_1, \dots, x_n)$ and $M_{x_1}(x_2, \dots, x_n) = M(x_1, \dots, x_n)$. Then $M_{x_2,\dots,x_n} \in \mathcal{L}(X_1; Y)$ by the assumption, while $M_{x_1} \in \mathcal{L}(X_2, \dots, X_n; Y)$ by the inductive hypothesis. Now for each $x_1 \in X_1$ we have

$$\sup_{\substack{x_j \in B_{X_j} \\ 2 \le j \le n}} \|M_{x_2,\dots,x_n}(x_1)\| = \sup_{\substack{x_j \in B_{X_j} \\ 2 \le j \le n}} \|M(x_1, \dots, x_n)\| = \|M_{x_1}\| < +\infty,$$

so by the Uniform boundedness principle $\sup_{\substack{x_j\in B_{X_j}\\ 2\le j\le n}} \|M_{x_2,\dots,x_n}\| < +\infty$. Thus

$$\begin{aligned}\sup_{\substack{x_j\in B_{X_j}\\ 1\le j\le n}} \|M(x_1,\dots,x_n)\| &= \sup_{\substack{x_j\in B_{X_j}\\ 2\le j\le n}} \Big(\sup_{x_1\in B_{X_1}} \|M(x_1,\dots,x_n)\|\Big)\\ &= \sup_{\substack{x_j\in B_{X_j}\\ 2\le j\le n}} \|M_{x_2,\dots,x_n}\| < +\infty,\end{aligned}$$

which proves that M is bounded and we can use Proposition 4. □

The following fact is easy but important.

Fact 9. *Let $X_1,\dots,X_n,Y$ be normed linear spaces. Then $\mathcal{L}(X_1;\mathcal{L}(X_2,\dots,X_n;Y))$ is canonically (linearly) isometric to the space $\mathcal{L}(X_1,\dots,X_n;Y)$ in the following way: We identify the mapping $T\in\mathcal{L}(X_1;\mathcal{L}(X_2,\dots,X_n;Y))$ with $S\in\mathcal{L}(X_1,\dots,X_n;Y)$ if and only if $T(x_1)(x_2,\dots,x_n)=S(x_1,\dots,x_n)$ for all $x_j\in X_j$, $j=1,\dots,n$.*

Let X, Y be vector spaces and $M\in L(^nX;Y)$. We define the symmetrisation of M by

$$M^{s}(x_1,\dots,x_n)=\frac{1}{n!}\sum_{\pi\in S_n} M(x_{\pi(1)},\dots,x_{\pi(n)}),$$

where S_n is the set of all permutations of $\{1,\dots,n\}$. Then $M^s\in L^s(^nX;Y)$. Furthermore, if X and Y are normed linear spaces and M is bounded, then so is M^s and obviously $\|M^s\|\le\|M\|$. If M is symmetric, then clearly $M^s=M$.

Proposition 10. *Let $X_1,\dots,X_n$, and X be normed linear spaces, and let Y be a Banach space. Then $\mathcal{L}(X_1,\dots,X_n;Y)$ is a Banach space. Moreover, $\mathcal{L}^s(^nX;Y)$ is a closed and 1-complemented subspace of $\mathcal{L}(^nX;Y)$.*

Proof. Suppose that $\{M_k\}_{k=1}^{\infty}\subset\mathcal{L}(X_1,\dots,X_n;Y)$ is a Cauchy sequence. Then $\{M_k(x_1,\dots,x_n)\}_{k=1}^{\infty}$ is a Cauchy sequence in Y for every $x_1\in X_1,\dots,x_n\in X_n$ and so we may define the mapping $M(x_1,\dots,x_n)=\lim_{k\to\infty}M_k(x_1,\dots,x_n)$. Obviously M is n-linear. Since the sequence $\{\|M_k\|\}$ is bounded, we infer that the mapping M is bounded. Now choose any $\varepsilon>0$. There is $k_0\in\mathbb{N}$ such that

$$\sup_{x_1\in B_{X_1},\dots,x_n\in B_{X_n}} \|M_k(x_1,\dots,x_n)-M_m(x_1,\dots,x_n)\|\le\varepsilon$$

for all $k,m\ge k_0$. By passing to a limit as $m\to\infty$ we obtain

$$\sup_{x_1\in B_{X_1},\dots,x_n\in B_{X_n}} \|M_k(x_1,\dots,x_n)-M(x_1,\dots,x_n)\|\le\varepsilon$$

for all $k\ge k_0$, i.e. $\|M_k-M\|\le\varepsilon$. Thus $\lim_{k\to\infty}M_k=M$ in $\mathcal{L}(X_1,\dots,X_n;Y)$. This proves that $\mathcal{L}(X_1,\dots,X_n;Y)$ is indeed a Banach space.

The symmetrisation gives rise to a canonical projection $Q:\mathcal{L}(^nX;Y)\to\mathcal{L}^s(^nX;Y)$, $Q(M)=M^s$. Clearly, $\|Q\|=1$. □

Proposition 11 (Polarisation formula). *Let X, Y be vector spaces and $M \in L(^nX;Y)$. Then*

$$M^{s}(x_1,\ldots,x_n) = \frac{1}{2^n n!} \sum_{\varepsilon_j=\pm 1} \varepsilon_1\cdots\varepsilon_n M\left(a + \sum_{j=1}^{n} \varepsilon_j x_j, \ldots, a + \sum_{j=1}^{n} \varepsilon_j x_j\right)$$

for every $a, x_1, \ldots, x_n \in X$. In particular, if M is symmetric, then it is uniquely determined by its values $M(x,\ldots,x)$, $x \in X$, along the diagonal.

Proof. For convenience we put $x_0 = a$ and $\varepsilon_0 = 1$. Using (1) we obtain

$$\begin{aligned}
&\frac{1}{2^n n!} \sum_{\substack{\varepsilon_j=\pm 1\\ j=1,\ldots,n}} \varepsilon_1\cdots\varepsilon_n M\left(\sum_{j=0}^{n} \varepsilon_j x_j, \ldots, \sum_{j=0}^{n} \varepsilon_j x_j\right)\\
&\qquad= \frac{1}{2^n n!} \sum_{\substack{\varepsilon_j=\pm 1\\ j=1,\ldots,n}} \varepsilon_1\cdots\varepsilon_n \sum_{j_1,\ldots,j_n\in\{0,\ldots,n\}} \varepsilon_{j_1}\cdots\varepsilon_{j_n} M(x_{j_1},\ldots,x_{j_n})\\
&\qquad= \frac{1}{2^n n!} \sum_{j_1,\ldots,j_n\in\{0,\ldots,n\}} \left(\sum_{\substack{\varepsilon_j=\pm 1\\ j=1,\ldots,n}} \varepsilon_1\cdots\varepsilon_n\varepsilon_{j_1}\cdots\varepsilon_{j_n}\right) M(x_{j_1},\ldots,x_{j_n}).
\end{aligned}$$

If there is $k \in \{1,\ldots,n\} \setminus \{j_1,\ldots,j_n\}$, then

$$\begin{aligned}
&\sum_{\substack{\varepsilon_j=\pm 1\\ j=1,\ldots,n}} \varepsilon_1\cdots\varepsilon_n\varepsilon_{j_1}\cdots\varepsilon_{j_n}\\
&\qquad= \sum_{\substack{\varepsilon_j=\pm 1\\ j\neq k}} \varepsilon_{j_1}\cdots\varepsilon_{j_n} \prod_{j\neq k} \varepsilon_j + (-1)\cdot \sum_{\substack{\varepsilon_j=\pm 1\\ j\neq k}} \varepsilon_{j_1}\cdots\varepsilon_{j_n} \prod_{j\neq k} \varepsilon_j = 0.
\end{aligned} \tag{4}$$

It follows that

$$\begin{aligned}
&\frac{1}{2^n n!} \sum_{j_1,\ldots,j_n\in\{0,\ldots,n\}} \left(\sum_{\substack{\varepsilon_j=\pm 1\\ j=1,\ldots,n}} \varepsilon_1\cdots\varepsilon_n\varepsilon_{j_1}\cdots\varepsilon_{j_n}\right) M(x_{j_1},\ldots,x_{j_n})\\
&\qquad= \frac{1}{2^n n!} \sum_{\pi\in S_n} \left(\sum_{\varepsilon_j=\pm 1} \varepsilon_1^2\cdots\varepsilon_n^2\right) M(x_{\pi(1)},\ldots,x_{\pi(n)})\\
&\qquad= \frac{1}{n!} \sum_{\pi\in S_n} M(x_{\pi(1)},\ldots,x_{\pi(n)}) = M^{s}(x_1,\ldots,x_n),
\end{aligned}$$

where S_n is the set of all permutations of $\{1,\ldots,n\}$. This finishes the proof. □

Definition 12. Let X, Y be vector spaces and $n \in \mathbb{N}$. A mapping $P\colon X \to Y$ is called an n-homogeneous polynomial if there is an n-linear mapping $M \in L(^nX;Y)$ such that $P(x) = M(x,\ldots,x)$. We use the notation $P = \widehat{M}$. For convenience we also define 0-homogeneous polynomials as constant mappings from X to Y. We denote by $P({}^nX;Y)$, $n \in \mathbb{N}_0$, the vector space of all n-homogeneous polynomials from X into Y.

Suppose X, Y are normed linear spaces, $n \in \mathbb{N}_0$. We say that $P \in P({}^nX;Y)$ is a bounded polynomial, if

$$\|P\| = \sup_{x \in B_X} \|P(x)\| < +\infty.$$

We denote by $\left(\mathcal{P}({}^nX;Y), \|\cdot\|\right)$ the normed linear space of all n-homogeneous bounded polynomials from X into Y. When the target space is the scalar field, we use a shortened notation $\mathcal{P}({}^nX) = \mathcal{P}({}^nX;\mathbb{K})$.

Let $P \in P({}^nX;Y)$. From the definition it immediately follows that

$$P(cx) = c^n P(x)$$

for every $x \in X$ and $c \in \mathbb{K}$. This implies that for $P \in \mathcal{P}({}^nX;Y)$

$$\|P(x)\| \le \|P\|\|x\|^n$$

for every $x \in X$. This in turn gives

$$P(x) = o(\|x\|^p), \ x \to 0$$

for every $p \in \mathbb{R}$, $p < n$.

For a given n-homogeneous polynomial P the n-linear mapping M that gives rise to P is not determined uniquely. In particular, the symmetrised n-linear mapping leads to the same polynomial: for every $M \in L({}^nX;Y)$ we have $\widehat{M} = \widehat{M^s}$. However, the following fundamental result holds.

Proposition 13 (Polarisation formula; [BH], [MO1]). *Let X, Y be vector spaces and $n \in \mathbb{N}$. For every $P \in P({}^nX;Y)$ there exists a unique symmetric n-linear mapping $\check{P} \in L^s({}^nX;Y)$ such that $P(x) = \check{P}(x,\dots,x)$. It satisfies the formula*

$$\check{P}(x_1,\dots,x_n) = \frac{1}{2^n n!} \sum_{\varepsilon_j=\pm 1} \varepsilon_1 \cdots \varepsilon_n P\left(a + \sum_{j=1}^{n} \varepsilon_j x_j\right), \tag{5}$$

where $a \in X$ can be chosen arbitrarily. Moreover, if X, Y are normed linear spaces and P is bounded, then $\check{P}$ is also bounded and we have

$$\|P\| \le \|\check{P}\| \le \frac{n^n}{n!}\|P\|. \tag{6}$$

On the other hand, for every $m > n$ and $a, x_1, \dots, x_m \in X$ the following holds:

$$\sum_{\varepsilon_j=\pm 1} \varepsilon_1 \cdots \varepsilon_m P\left(a + \sum_{j=1}^{m} \varepsilon_j x_j\right) = 0. \tag{7}$$

Proof. Let $M \in L(^nX;Y)$ be such that $P = \widehat{M}$. Put $\check{P} = M^s$. Obviously, $P = \widehat{M^s}$ and so using Proposition 11 we obtain

$$\begin{aligned}\check{P}(x_1,\dots,x_n) &= M^s(x_1,\dots,x_n)\\ &= \frac{1}{2^n n!}\sum_{\varepsilon_j=\pm 1}\varepsilon_1\cdots\varepsilon_n M^s\left(a+\sum_{j=1}^n \varepsilon_j x_j,\dots,a+\sum_{j=1}^n \varepsilon_j x_j\right)\\ &= \frac{1}{2^n n!}\sum_{\varepsilon_j=\pm 1}\varepsilon_1\cdots\varepsilon_n P\left(a+\sum_{j=1}^n \varepsilon_j x_j\right).\end{aligned}$$

The uniqueness follows from the fact that a symmetric n-linear mapping is uniquely determined by its values along the diagonal (Proposition 11). The estimate (6) follows readily from the formula (5).

To prove (7) it is enough to follow the proof of Proposition 11 and notice that for each summand there is always $k \in \{1,\dots,m\} \setminus \{j_1,\dots,j_n\}$ and so (4) holds. □

Consequently, a homogeneous polynomial is continuous if and only if it is bounded. As another immediate corollary we obtain the next proposition.

Proposition 14. *Let X, Y be normed linear spaces and $n \in \mathbb{N}$. Then $\mathcal{P}(^nX;Y)$ is canonically isomorphic to $\mathcal{L}^s(^nX;Y)$. In particular, if Y is a Banach space, then so is $\mathcal{P}(^nX;Y)$.*

The following example shows that in general the right-hand side estimate in (6) cannot be improved.

Example 15. Consider $X = \ell_1$ with the canonical basis $\{(e_j; f_j)\}_{j=1}^{\infty}$. Choose $n \in \mathbb{N}$ and define $P \in \mathcal{P}(^nX)$ by $P(x) = f_1(x)\cdots f_n(x)$. It is indeed an n-homogeneous polynomial, since $P = \widehat{M}$, where $M(x_1,\dots,x_n) = f_1(x_1)\cdots f_n(x_n)$. Then $\|\check{P}\| \geq \check{P}(e_1,\dots,e_n) = \frac{1}{n!}$, while $|P(x)|$ attains its maximum $\frac{1}{n^n}$ over B_X at the point $x = (\frac{1}{n},\dots,\frac{1}{n},0,\dots)$.

However, interestingly if X is an inner product space, then the spaces $\mathcal{P}(^nX;Y)$ and $\mathcal{L}^s(^nX;Y)$ are isometric:

Theorem 16 ([Kel], [Bana]). *Let X be an inner product space, Y a normed linear space, $n \in \mathbb{N}$, and $M \in \mathcal{L}^s(^nX;Y)$. Then $\|M\| = \|\widehat{M}\|$.*

Proof. (Stanisław Łojasiewicz) By considering the real versions of the spaces we may without loss of generality assume that $\mathbb{K} = \mathbb{R}$. First suppose that $\dim X < \infty$. Then there are $y_1,\dots,y_n \in B_X$ such that $\|M\| = \|M(y_1,\dots,y_n)\|$. Pick $a \in X$ such that $\langle a, y_j\rangle \neq 0$ for all $j \in \{1,\dots,n\}$. Clearly there is $\delta > 0$ such that the set $K = \{(x_1,\dots,x_n) \in (B_X)^n;\ \|M(x_1,\dots,x_n)\| = \|M\|,\ \langle a, x_j\rangle \geq \delta,\ j = 1,\dots,n\}$ is non-empty. Put $\varphi(x_1,\dots,x_n) = \sum_{j=1}^n \langle a, x_j\rangle$. As K is compact, the function φ attains its maximum on K, say at $(z_1,\dots,z_n) \in K$. We claim that $z_1 = z_2 = \cdots = z_n$, which means that $\|M\| = \|M(z_1,\dots,z_n)\| = \|\widehat{M}(z_1)\| \leq \|\widehat{M}\|$ and this finishes the proof in the finite-dimensional case.

Assuming the contrary, by the symmetry of M we may without loss of generality assume that $z_1 \neq z_2$. Since $\langle a, z_1\rangle \geq \delta$ and $\langle a, z_2\rangle \geq \delta$, it follows that also $z_1 \neq -z_2$. Put $u = \frac{z_1+z_2}{\|z_1+z_2\|}$ and $v = \frac{z_1-z_2}{\|z_1-z_2\|}$. We show that in this case $(u, u, z_3, \dots, z_n) \in K$. Clearly, $\langle a, u\rangle \geq \delta$, as $\|z_1 + z_2\| \leq 2$. Let us define the mapping $T\colon X^2 \to Y$ by $T(x, y) = M(x, y, z_3, \dots, z_n)$. Then T is bilinear and symmetric and therefore $T(z_1, z_2) = \frac{1}{4}\big(T(z_1+z_2, z_1+z_2) - T(z_1-z_2, z_1-z_2)\big)$. Now if $\|T(u, u)\| < \|M\|$, then

$$\begin{aligned}\|M\| = \|T(z_1, z_2)\| &\leq \frac{1}{4}\big(\|z_1 + z_2\|^2\|T(u, u)\| + \|z_1 - z_2\|^2\|T(v, v)\|\big)\\ &< \frac{\|M\|}{4}\big(\|z_1 + z_2\|^2 + \|z_1 - z_2\|^2\big) = \frac{\|M\|}{2}\big(\|z_1\|^2 + \|z_2\|^2\big) \leq \|M\|,\end{aligned}$$

a contradiction. Thus $(u, u, z_3, \dots, z_n) \in K$. But since $\|z_1 - z_2\| > 0$, by the rotundity $\|z_1 + z_2\| < 2$, and so

$$\varphi(u, u, z_3, \dots, z_n) = \frac{2}{\|z_1 + z_2\|}\big(\langle a, z_1\rangle + \langle a, z_2\rangle\big) + \sum_{j=3}^{n}\langle a, z_j\rangle > \varphi(z_1, \dots, z_n),$$

which contradicts the fact that φ attains its maximum on K at $(z_1, \dots, z_n)$.

Finally, we deal with the infinite-dimensional case. Fix $x_1, \dots, x_n \in B_X$ and denote $Z = \operatorname{span}\{x_1, \dots, x_n\}$. Then $\|M(x_1, \dots, x_n)\| \leq \|M\restriction_{Z^n}\| \leq \|\widehat{M}\restriction_Z\| \leq \|\widehat{M}\|$ by the first part of the proof. This means that $\|M\| \leq \|\widehat{M}\|$, while the reverse inequality is obvious. □

To shorten the notation we will use the following convention: for $k \in \mathbb{N}_0$ we denote

$${}^{k}x = \underbrace{x, \dots, x}_{k \text{ times}}.$$

The following result holds also for complex spaces, but for those we show much stronger (and optimal) estimate in Corollary 2.70.

Corollary 17 ([Harr2]). *Let X, Y be real normed linear spaces and $M \in \mathcal{L}^s({}^nX; Y)$. Then*

$$\big\|M({}^{n_1}x_1, \dots, {}^{n_k}x_k)\big\| \leq \sqrt{\frac{n^n}{n_1^{n_1}\cdots n_k^{n_k}}}\,\|\widehat{M}\|$$

for all $x_1, \dots, x_k \in B_X$, $n_1, \dots, n_k \in \mathbb{N}$ with $n_1 + \dots + n_k = n$, $1 \leq k \leq n$.

Proof. Let $1 \leq k \leq n$, $x_1, \dots, x_k \in B_X$, and $n_1, \dots, n_k \in \mathbb{N}$ with $n_1 + \dots + n_k = n$ be given. Fix for a while $r = (r_1, \dots, r_k) \in (\mathbb{R}^+)^k$. Define $T \in \mathcal{L}(\ell_2^k; X)$ by the formula $T\big((t_1, \dots, t_k)\big) = \sum_{j=1}^k t_j r_j x_j$. Note that the Cauchy inequality implies $\big\|T\big((t_1, \dots, t_k)\big)\big\| \leq \sum_{j=1}^k |t_j|r_j\|x_j\| \leq \|(t_1, \dots, t_k)\|_2\|r\|_2$ and hence $\|T\| \leq \|r\|_2$. Further, let $Q(y_1, \dots, y_n) = M\big(T(y_1), \dots, T(y_n)\big)$ for $y_1, \dots, y_n \in \ell_2^k$. Clearly,

$Q \in \mathcal{L}^s({}^n\ell_2^k; Y)$, $\widehat{Q} = \widehat{M} \circ T$, and $\|\widehat{Q}\| \le \|\widehat{M}\|\|r\|_2^n$. Denote by $e_1, \dots, e_k$ the canonical basis in ℓ_2^k. Then by Theorem 16

$$\begin{aligned} r_1^{n_1} \cdots r_k^{n_k} \left\|M({}^{n_1}x_1, \dots, {}^{n_k}x_k)\right\| &= \left\|M\left({}^{n_1}(r_1x_1), \dots, {}^{n_k}(r_kx_k)\right)\right\| \\ &= \left\|Q({}^{n_1}e_1, \dots, {}^{n_k}e_k)\right\| \\ &\le \|Q\| = \|\widehat{Q}\| \le \|\widehat{M}\|\|r\|_2^n. \end{aligned}$$

Since the function $r \mapsto \frac{\|r\|_2^n}{r_1^{n_1} \cdots r_k^{n_k}}$ attains its minimum on $(\mathbb{R}^+)^k$ for $r_j = \sqrt{\frac{n_j}{n}}$, $j = 1, \dots, k$, the estimate follows. □

Fact 18. *Let X, Y be vector spaces, $n \in \mathbb{N}_0$, $P \in P({}^nX; Y)$, and let $Z \subset X$ be a subspace. Then $P\restriction_Z \in P({}^nZ; Y)$.*

Proof. For $n = 0$ it is obvious. Otherwise clearly $P\restriction_Z = \widehat{\check{P}\restriction_{Z^n}}$. □

Fact 19. *Let X, Y be vector spaces, $n \in \mathbb{N}_0$, $\dim X \ge n + 1$, and let $P : X \to Y$ be a mapping such that $P\restriction_E \in P({}^nE; Y)$ for every subspace $E \subset X$ of dimension $n + 1$. Then $P \in P({}^nX; Y)$.*

Proof. For $n = 0$ it is obvious. Otherwise define the mapping $M : X^n \to Y$ formally by formula (5) with $a = 0$. By Proposition 13, $M\restriction_{E^n} \in L({}^nE; Y)$ for every subspace $E \subset X$ of dimension $n + 1$. Thus $M \in L({}^nX; Y)$ by Fact 2. Clearly, $P = \widehat{M}$, as $P\restriction_E = \widehat{M}\restriction_E$ for every subspace $E \subset X$ of dimension $n + 1$. □

Definition 20. Let $n \in \mathbb{N}$. Given a multi-index $\alpha \in \mathbb{N}_0^n$ we denote its order by $|\alpha| = \sum_{j=1}^n \alpha_j$. Further, we denote the set of multi-indices of order $d \in \mathbb{N}_0$ by

$$\mathcal{I}(n, d) = \left\{\alpha \in \{0, \dots, d\}^n;\ |\alpha| = d\right\}.$$

We extend the definition also to the case when $n = \infty$, setting

$$\mathcal{I}(\infty, d) = \left\{\alpha \in \{0, \dots, d\}^{\mathbb{N}};\ |\alpha| = \sum_{j=1}^{\infty} \alpha_j = d\right\}.$$

For $n \in \mathbb{N}$ we have $|\mathcal{I}(n, d)| = \binom{n+d-1}{n-1}$ (indeed, from the combinatorial point of view it represents the number of distributions of d identical balls into n distinct boxes). A given $(k_j)_{j=1}^d \in \{1, \dots, n\}^d$ determines a unique $\alpha \in \mathcal{I}(n, d)$ by the relation

$$\alpha = \left(|\{j;\ k_j = 1\}|, |\{j;\ k_j = 2\}|, \dots, |\{j;\ k_j = n\}|\right). \tag{8}$$

Conversely, a given $\alpha \in \mathcal{I}(n, d)$ determines a unique $k(\alpha) = \left(k_1(\alpha), \dots, k_d(\alpha)\right)$, $k_1(\alpha) \le \dots \le k_d(\alpha)$, such that (8) holds. Given $x = (x_1, \dots, x_n) \in \mathbb{K}^n$ and $\alpha = (\alpha_1, \dots, \alpha_n) \in \mathcal{I}(n, d)$ we use the standard multi-index notation

$$x^\alpha = \prod_{l=1}^n x_l^{\alpha_l} = \prod_{j=1}^d x_{k_j(\alpha)}.$$

The case $n = \infty$ is similar and corresponds to multi-indices whose domain is $\mathbb{N}$.

Note that $x \mapsto x^\alpha \in \mathcal{P}({}^d\mathbb{K}^n)$ for any $\alpha \in \mathcal{I}(n,d)$. Indeed, denote by $\phi_l \in (\mathbb{K}^n)^*$ the coordinate functional $\phi_l(x) = x_l$. Let $M \in \mathcal{L}({}^d\mathbb{K}^n;\mathbb{K})$ be given by the formula $M(y_1,\dots,y_d) = \prod_{j=1}^d \phi_{k_j(\alpha)}(y_j)$. Then we have $x^\alpha = \widehat{M}(x)$.

Given $\alpha = (\alpha_1,\dots,\alpha_n) \in \mathcal{I}(n,d)$, we denote the corresponding multinomial coefficient by

$$\binom{d}{\alpha} = \binom{d}{\alpha_1,\dots,\alpha_n} = \frac{d!}{\alpha!} = \frac{d!}{\alpha_1!\cdots\alpha_n!}.$$

Lemma 21. *Let X, Y be vector spaces, $n \in \mathbb{N}$, $P \in P({}^nX;Y)$, and $x, y \in X$. Then*

$$P(x+y) = \sum_{j=0}^{n} \binom{n}{j} \check{P}({}^jx, {}^{n-j}y) \quad \textit{and}$$

$$P(x) - P(y) = \sum_{j=0}^{n-1} \binom{n}{j} \check{P}({}^jy, {}^{n-j}(x-y)).$$

Proof. To obtain the first formula, it suffices to apply (1) and the symmetry of $\check{P}$. The second formula follows easily from the first one. □

Proposition 22 (Multinomial formula). *Let X and Y be vector spaces, $d \in \mathbb{N}$, let $P \in P({}^dX;Y)$, and $x_1,\dots,x_n \in X$. Then*

$$P(x_1+\dots+x_n) = \sum_{\alpha\in\mathcal{I}(n,d)} \binom{d}{\alpha} \check{P}({}^{\alpha_1}x_1,\dots,{}^{\alpha_n}x_n).$$

Proof. Using (1) and then collecting the terms according to the value of α using the symmetry of $\check{P}$ we obtain

$$P\left(\sum_{j=1}^n x_j\right) = \sum_{\substack{1\le k_j\le n\\ j=1,\dots,d}} \check{P}(x_{k_1},\dots,x_{k_d}) = \sum_{\alpha\in\mathcal{I}(n,d)} \binom{d}{\alpha} \check{P}({}^{\alpha_1}x_1,\dots,{}^{\alpha_n}x_n).$$

□

The next proposition asserts that the abstract definition of homogeneous polynomials coincides on $\mathbb{K}^n$ with the classical definition that uses coordinates. Note that in this case all homogeneous polynomials are automatically bounded.

Proposition 23. *Let $n, d \in \mathbb{N}$ and let Y be a vector space over $\mathbb{K}$. A mapping $P\colon \mathbb{K}^n \to Y$ is a d-homogeneous polynomial if and only if there exists a collection $\{y_\alpha\}_{\alpha\in\mathcal{I}(n,d)} \subset Y$ such that $P(x) = \sum_{\alpha\in\mathcal{I}(n,d)} x^\alpha y_\alpha$. Moreover, each y_α is uniquely determined by*

$$y_\alpha = \binom{d}{\alpha} \check{P}({}^{\alpha_1}e_1,\dots,{}^{\alpha_n}e_n),$$

where $\{e_j\}_{j=1}^n$ is the canonical basis of $\mathbb{K}^n$.

Proof. Let $P \in \mathcal{P}(^d\mathbb{K}^n;Y)$. By Proposition 22 we have

$$P\left(\sum_{j=1}^{n} x_j e_j\right) = \sum_{\alpha\in\mathcal{J}(n,d)} x^\alpha \binom{d}{\alpha} \check{P}(^{\alpha_1}e_1,\dots,{}^{\alpha_n}e_n).$$

The converse is clear from the fact that $x \mapsto x^\alpha \in \mathcal{P}(^d\mathbb{K}^n)$.

To prove the uniqueness we show a bit more: if $Q(x_1,\dots,x_n) = \sum_{\alpha\in A} x^\alpha z_\alpha = 0$ for each $x \in \mathbb{K}^n$, then $z_\alpha = 0$ for each $\alpha \in A$, where $A = \{\alpha \in \mathbb{N}_0^n;\ |\alpha| \le d\}$. First notice that it suffices to show this only for $Y = \mathbb{K}$, the general case follows by using composition with linear functionals from Y^*. We use induction on n. For $n = 1$ it is a standard fact that can be proved for example by factorisation. To prove the induction step from $n-1$ to n put $A_k = \{\alpha \in A;\ \alpha_1 = k\}$, $k = 0,\dots,d$. Then $Q(x_1,\dots,x_n) = \sum_{k=0}^{d} x_1^k \sum_{\alpha\in A_k} x_2^{\alpha_2}\cdots x_n^{\alpha_n} z_\alpha = 0$ for each $x_1,\dots,x_n \in \mathbb{K}$. From the first step of the induction it follows that $\sum_{\alpha\in A_k} x_2^{\alpha_2}\cdots x_n^{\alpha_n} z_\alpha = 0$ for each $x_2,\dots,x_n \in \mathbb{K}$ and $k \in \{0,\dots,d\}$. Hence by the inductive hypothesis $z_\alpha = 0$ for every $\alpha \in \bigcup_{k=0}^{d} A_k = A$. □

In case that $Y = \mathbb{K}$ this reduces to the familiar formula $P(x) = \sum_{\alpha\in\mathcal{J}(n,d)} a_\alpha x^\alpha$, where the coefficients $a_\alpha \in \mathbb{K}$.

While working with polynomials $P \in P(^d\mathbb{K}^n;Y)$ we are often going to use freely the classical notation using the coordinates and multinomial expressions, e.g. we will sometimes write $P(x_1,\dots,x_n)$ instead of the formally correct $P\big((x_1,\dots,x_n)\big)$, $x_j \in \mathbb{K}$, $j = 1,\dots,n$.

Proposition 24. *Let X be a normed linear space with a Schauder basis $\{e_j\}_{j=1}^\infty$, Y a vector space, $d \in \mathbb{N}$, and $P \in P(^dX;Y)$. Denote $X_0 = \operatorname{span}\{e_j\}_{j=1}^\infty$. Then there is a unique collection of vectors $\{y_\alpha\}_{\alpha\in\mathcal{J}(\infty,d)} \subset Y$ such that the formula*

$$P(x) = \sum_{\alpha\in\mathcal{J}(\infty,d)} x^\alpha y_\alpha \tag{9}$$

holds for every $x = \sum x_j e_j \in X_0$. The coefficients y_α are given by

$$y_\alpha = \binom{d}{\alpha} \check{P}(^{\alpha_1}e_1, {}^{\alpha_2}e_2,\dots).$$

Conversely, any collection $\{y_\alpha\}_{\alpha\in\mathcal{J}(\infty,d)} \subset Y$ uniquely determines a polynomial $P \in P(^dX_0;Y)$ by the formula (9).

Proof. By Fact 18 the restrictions of P to $\operatorname{span}\{e_j\}_{j=1}^n$ are d-homogeneous polynomials, so we can use Proposition 23.

To prove the converse statement define $P\colon X_0 \to Y$ by (9). Each restriction of P to a finite-dimensional subspace of X_0 is a d-homogeneous polynomial, since it is a finite sum of d-homogeneous polynomials $x \mapsto x^\alpha y_\alpha$. Thus $P \in P(^dX_0;Y)$ by Fact 19. □

Even if X and Y are Banach spaces (or $Y = \mathbb{K}$) and $P \in \mathcal{P}(^dX;Y)$, the sum in (9) may not be convergent (absolutely) for infinitely supported vectors. However, the following statement holds.

Proposition 25. *Let X be a normed linear space with a Schauder basis $\{e_j\}_{j=1}^{\infty}$, Y a Banach space, and $d \in \mathbb{N}$. If there is a collection of vectors $\{y_\alpha\}_{\alpha\in\mathcal{I}(\infty,d)} \subset Y$ such that the expression*

$$\sum_{\alpha\in\mathcal{I}(\infty,d)} x^\alpha y_\alpha$$

is uniformly bounded for all finitely supported vectors $x \in B_X$, then there is a unique polynomial $P \in \mathcal{P}(^dX;Y)$ such that the formula (9) holds for every finitely supported vector $x \in X$.

Proof. By Proposition 24 the formula (9) defines a polynomial $P \in P(^dX_0;Y)$ on $X_0 = \operatorname{span}\{e_j\}_{j=1}^{\infty}$. Moreover, from our hypothesis it follows that P is bounded. Consequently, $\check{P}$ is bounded and hence Lipschitz on bounded sets by Proposition 4. Therefore $\check{P}$ can be uniquely extended to $X^n = (\overline{X_0})^n = \overline{X_0^n}$. Denote this extension by M. Using the continuity of M, the d-linearity of M passes from X_0^n to X^n. Since M is obviously also bounded, we conclude that $\widehat{M} \in \mathcal{P}(^dX;Y)$ is a desired extension of P. The uniqueness is clear from the density of X_0 in X. □

Lemma 26. *Let X, $Y_1,\dots,Y_k$, Z be vector spaces, let $M \in L(Y_1,\dots,Y_k;Z)$, and $P_j \in P(^{n_j}X;Y_j)$, $n_j \in \mathbb{N}_0$, for $j = 1,\dots,k$. Then the composite mapping $x \mapsto M\big(P_1(x),\dots,P_k(x)\big)$ belongs to $P(^{n_1+\dots+n_k}X;Z)$.*

Proof. It is clear that $M\big(\check{P}_1(\cdots),\dots,\check{P}_k(\cdots)\big) \in L(^{n_1+\dots+n_k}X;Z)$. □

As an immediate corollary we obtain the following two facts.

Fact 27. *Let X, Y be vector spaces over $\mathbb{K}$, $m,n \in \mathbb{N}_0$, and let $P \in P(^mX;\mathbb{K})$ and $Q \in P(^nX;Y)$. Then $P\cdot Q \in P(^{m+n}X;Y)$.*

Proof. The mapping $(t,y) \mapsto t\cdot y$ is a bilinear mapping from $\mathbb{K}\times Y$ to Y. □

Fact 28. *Let X, Y, Z be vector spaces, $m,n \in \mathbb{N}_0$, and let $P \in P(^mY;Z)$ and $Q \in P(^nX;Y)$. Then $P\circ Q \in P(^{mn}X;Z)$.*

Proof. We have $P\circ Q(x) = \check{P}\big(Q(x),\dots,Q(x)\big)$. □

Theorem 29. *Let X be a Banach space, Y a normed linear space, and $n \in \mathbb{N}_0$. If $\{P_k\}_{k=1}^{\infty} \subset \mathcal{P}(^nX;Y)$ is such that $\lim_{k\to\infty} P_k(x) = P(x)$ exists for every $x \in X$, then $P \in \mathcal{P}(^nX;Y)$.*

If $\{P_\gamma\}_{\gamma\in\Gamma}$ is a net in $\mathcal{P}(^nX;Y)$ such that $\lim_\gamma P_\gamma(x) = P(x)$ exists for every $x \in X$, then $P \in P(^nX;Y)$. If moreover $\{P_\gamma(x)\}_{\gamma\in\Gamma}$ is bounded for every $x \in X$, then $P \in \mathcal{P}(^nX;Y)$.

Proof. The statement is obvious for $n = 0$ so let $n > 0$. Then $\lim_\gamma \check{P}_\gamma(x_1,\dots,x_n)$ exists for every $x_1,\dots,x_n \in X$ by Proposition 13. Denote this limit by $M(x_1,\dots,x_n)$

and notice that $M \in L^s({}^nX;Y)$. It follows that $\lim_\gamma P_\gamma(x) = \widehat{M}(x)$ for all $x \in X$ and hence $P \in P({}^nX;Y)$. Note that in the case of the sequence, $\{P_k(x)\}_{k=1}^\infty$ is automatically bounded for each $x \in X$. Since again by the Polarisation formula $\{\check{P}_\gamma(x_1,\dots,x_n)\}_{\gamma\in\Gamma}$ is bounded for every $x_1,\dots,x_n \in X$, by Theorem 5 there is $C \in \mathbb{R}$ such that $\|\check{P}_\gamma\| \le C$ for every $\gamma \in \Gamma$. Thus $\|P(x)\| = \lim_\gamma \|P_\gamma(x)\| \le C$ for $x \in B_X$, which shows that $P \in \mathcal{P}({}^nX;Y)$. □

Definition 30. Let X, Y be vector spaces and $n \in \mathbb{N}_0$. A mapping $P: X \to Y$ is called a polynomial of degree at most n if there are $P_k \in P({}^kX;Y)$, $k = 0,\dots,n$, such that $P = \sum_{k=0}^n P_k$. If $P_n \neq 0$, we say that P has degree n and we use the notation $\deg P = n$. We denote by $P^n(X;Y)$ the space of all polynomials of degree at most n. We denote by $P(X;Y) = \bigcup_{n=0}^\infty P^n(X;Y)$ the space of all polynomials.

The fact that $\deg P$ is well-defined will be apparent from Corollary 32 below. The interpolation Lemma 6 has the following consequence:

Corollary 31. *For every $n \in \mathbb{N}_0$ and every $\delta > 0$ there exist numbers $a_{kj} \in \mathbb{R}$, $k, j = 0,\dots,n$, such that whenever X, Y are vector spaces over $\mathbb{K}$, $P \in P^n(X;Y)$, and $P_k \in P({}^kX;Y)$ satisfy $P = \sum_{k=0}^n P_k$, then $P_k(x) = \sum_{j=0}^n a_{kj} P\big((1-\frac{j}{n}\delta)x\big)$ for every $x \in X$.*

Proof. It suffices to notice that $P\big((1-\frac{j}{n}\delta)x\big) = \sum_{k=0}^n \big(1-\frac{j}{n}\delta\big)^k P_k(x)$ and apply Lemma 6 with $t_j = 1 - \frac{j}{n}\delta$. □

Corollary 32. *Let X, Y be vector spaces, $n \in \mathbb{N}_0$, and let $P \in P^n(X;Y)$ be such that $P = \sum_{k=0}^n P_k$, where $P_k \in P({}^kX;Y)$. Then the homogeneous summands P_k of P are uniquely determined.*

The next lemma shows that the Polarisation formula applied to a non-homogeneous polynomial extracts its "leading term".

Lemma 33. *Let X, Y be vector spaces, $n \in \mathbb{N}$, and let $P \in P^n(X;Y)$ be such that $P = \sum_{k=0}^n P_k$, $P_k \in P({}^kX;Y)$. Then*

$$\check{P}_n(x_1,\dots,x_n) = \frac{1}{2^n n!} \sum_{\varepsilon_j=\pm1} \varepsilon_1 \cdots \varepsilon_n P\left(a + \sum_{j=1}^n \varepsilon_j x_j\right)$$

for every $a, x_1,\dots,x_n \in X$.

Proof. By (7) in Proposition 13

$$\sum_{\varepsilon_j=\pm1} \varepsilon_1 \cdots \varepsilon_n P_k\left(a + \sum_{j-1}^n \varepsilon_j x_j\right) = 0$$

whenever $0 < k < n$, and it is of course true also for $k = 0$. Hence the result follows from the Polarisation formula (5). □

Lemma 34. *Let X, Y be vector spaces, $n \in \mathbb{N}_0$, $P \in P^n(X;Y)$, and $z \in X$. Then $R(x) = P(x+z) \in P^n(X;Y)$.*

Proof. Assume that $P = \sum_{k=0}^{n} P_k$, $P_k \in P({}^k X; Y)$. The statement follows from Lemma 21 and the obvious fact that $x \mapsto \check{P}_k({}^j x, {}^{k-j} z)$ is a j-homogeneous polynomial. □

The following fact immediately follows from Fact 18.

Fact 35. *Let X, Y be vector spaces, $n \in \mathbb{N}_0$, $P \in P^n(X; Y)$, and let $Z \subset X$ be a subspace. Then $P \restriction_Z \in P^n(Z; Y)$.*

Fact 36. *Let X, Y be vector spaces, $n \in \mathbb{N}_0$, $\dim X \geq n + 1$, and let $P : X \to Y$ be a mapping such that for every subspace $E \subset X$ of dimension $n + 1$ the mapping $P \restriction_E$ is a polynomial of degree at most n. Then $P \in P^n(X; Y)$.*

Proof. For every subspace $E \subset X$ of dimension $n + 1$ there are $P_k^E \in P({}^k E; Y)$, $0 \leq k \leq n$, such that $P \restriction_E = \sum_{k=0}^{n} P_k^E$. By Fact 35 and Corollary 32 for any subspaces $E \subset X$, $F \subset X$ of dimension $n + 1$ we have $P_k^E \restriction_{E \cap F} = P_k^F \restriction_{E \cap F}$, $0 \leq k \leq n$. So we may define $P_k(x) = P_k^E(x)$, $0 \leq k \leq n$, where $E \subset X$ is any subspace of dimension $n + 1$ containing x. Then $P_k \in P({}^k X; Y)$ by Fact 19. Clearly, $P = \sum_{k=0}^{n} P_k$. □

Later we show that it is actually enough to consider only restrictions to one-dimensional affine subspaces of X (Corollary 55, Theorem 2.49).

The polynomials behave as expected with regards to various operations:

Fact 37. *Let X, Y be vector spaces over $\mathbb{K}$, $m, n \in \mathbb{N}_0$, and let $P \in P^m(X; \mathbb{K})$ and $Q \in P^n(X; Y)$. Then $P \cdot Q \in P^{m+n}(X; Y)$.*

Fact 38. *Let X, Y, Z be vector spaces, $m, n \in \mathbb{N}_0$, and let $P \in P^m(Y; Z)$ and $Q \in P^n(X; Y)$. Then $P \circ Q \in P^{mn}(X; Z)$.*

The first fact follows from Fact 27 and the second one is proved using the Multinomial formula (Proposition 22) and Lemma 26.

Fact 39. *Let X, Y be normed linear spaces and let $P, Q \in P(X; Y)$ be such that P and Q are equal on some ball. Then $P = Q$.*

Proof. Clearly it suffices to prove that if a polynomial $P \in P(X; Y)$ vanishes on some ball $B(a, r)$, then $P = 0$. Assume that $P \neq 0$. Let $n = \deg P$ and denote the n-homogeneous summand of P by P_n. Lemma 33 implies that $P_n(x) = 0$ for $x \in B(0, \frac{r}{n})$, and so by homogeneity $P_n = 0$, a contradiction. □

Proposition 40. *Let X, Y be normed linear spaces, $n \in \mathbb{N}_0$, and $P \in P^n(X; Y)$, $P = \sum_{k=0}^{n} P_k$, where $P_k \in P({}^k X; Y)$. The following statements are equivalent:*

(i) *P_k, $k = 0, \ldots, n$, are Lipschitz on bounded sets.*
(ii) *P is Lipschitz on bounded sets.*
(iii) *P is continuous.*
(iv) *P is bounded on a neighbourhood of some point.*

Proof. (i)⇒(ii)⇒(iii)⇒(iv) are immediate, so it remains to prove (iv)⇒(i). Suppose that P is bounded on $B(a,2r)$ for some $a \in X$, $r > 0$. Then by Corollary 31 (letting $\delta = \frac{r}{r+\|a\|}$) each P_k, $k = 0,\dots,n$, is bounded on $B(a,r)$. Consequently, by the Polarisation formula (Proposition 13) each $\check{P}_k$ is bounded on $B(0,\frac{r}{k})^k$ and thus is Lipschitz on bounded sets by Proposition 4. Since the mapping $x \mapsto ({}^k x)$ is Lipschitz, the polynomials P_k are Lipschitz on bounded sets. □

Definition 41. Let X and Y be normed linear spaces over $\mathbb{K}$ and $n \in \mathbb{N}_0$. Set $\|P\| = \sup_{x\in B_X}\|P(x)\|$ whenever $P \in P^n(X;Y)$ is continuous. We denote by $\big(\mathcal{P}^n(X;Y),\|\cdot\|\big)$ (resp. $\big(\mathcal{P}(X;Y),\|\cdot\|\big)$) the normed linear space of all continuous polynomials of degree at most n (resp. all continuous polynomials) from X to Y. When the target space is the scalar field, we use a shortened notation $\mathcal{P}^n(X) = \mathcal{P}^n(X;\mathbb{K})$, resp. $\mathcal{P}(X) = \mathcal{P}(X;\mathbb{K})$.

The following fact follows at once from Corollary 31.

Fact 42. *Let X, Y be normed linear spaces and $n, j \in \mathbb{N}_0$, $j \le n$. The mapping $H^n_j \colon \mathcal{P}^n(X;Y) \to \mathcal{P}({}^jX;Y)$ given by $H^n_j(P) = P_j$ for $P = \sum_{k=0}^n P_k$, where $P_k \in \mathcal{P}({}^kX;Y)$, is a bounded linear projection. The norm of H^n_j depends on n and j, but not on X or Y.*

As an immediate consequence we obtain the next result.

Proposition 43. *Let X be a normed linear space, Y a Banach space, and $n \in \mathbb{N}_0$. Then the space $\mathcal{P}^n(X;Y)$ is a Banach space. Moreover, the canonical decomposition $P \mapsto \big(H^n_0(P),\dots,H^n_n(P)\big)$ for $P \in \mathcal{P}^n(X;Y)$ is a canonical isomorphism of Banach spaces $\mathcal{P}^n(X;Y)$ and $\mathcal{P}({}^0X;Y) \oplus \mathcal{P}({}^1X;Y) \oplus \dots \oplus \mathcal{P}({}^nX;Y)$.*

By combining Corollary 31, the Polarisation formula (Proposition 13), and condition (v) in Theorem 7 we obtain the following corollary:

Corollary 44. *Let X be a Banach space, Y a normed linear space, $n \in \mathbb{N}_0$, and $P \in P^n(X;Y)$. If $\phi \circ P$ is Baire measurable for every $\phi \in Y^*$, then P is continuous.*

Theorem 45 (Uniform boundedness principle; [MO2])**.** *Let X be a Banach space, Y a normed linear space, and $n \in \mathbb{N}_0$. Let $\{P_\alpha\}_{\alpha\in\Gamma} \subset \mathcal{P}^n(X;Y)$ be a family such that $\sup_{\alpha\in\Gamma}\|P_\alpha(x)\| < +\infty$ for every $x \in X$. Then $\sup_{\alpha\in\Gamma}\|P_\alpha\| < +\infty$.*

Proof. Corollary 31 entails that for a fixed $j \le n$, $\sup_{\alpha\in\Gamma}\|H^n_j(P_\alpha)(x)\| < +\infty$ for every $x \in X$. The result now follows from Proposition 13 and Theorem 5. □

Theorem 46. *Let X be a Banach space, Y a normed linear space, and $n \in \mathbb{N}_0$. If $\{P_k\}_{k=1}^\infty \subset \mathcal{P}^n(X;Y)$ is such that $\lim_{k\to\infty} P_k(x) = P(x)$ exists for every $x \in X$, then $P \in \mathcal{P}^n(X;Y)$.*

Proof. Fix $j \in \{0,\dots,n\}$ and denote $Q_k = H^n_j(P_k)$. Then $\lim_{k\to\infty} Q_k(x)$ exists for every $x \in X$ by Corollary 31. By Theorem 29, $\lim_{k\to\infty} Q_k \in \mathcal{P}({}^jX;Y)$. Therefore $P \in \mathcal{P}^n(X;Y)$. □

The following lemma is useful for estimating the norm of a homogeneous polynomial using its values on an arbitrary ball.

Lemma 47. *Let X, Y be normed linear spaces, $n \in \mathbb{N}$, let $P \in \mathcal{P}^n(X;Y)$ be such that $P = \sum_{k=0}^n P_k$, $P_k \in \mathcal{P}(^kX;Y)$, and let $a \in X$. Then*

$$\|P_n(x)\| \le \frac{n^n}{n!} \sup_{t\in[-1,1]} \|P(a+tx)\|$$

for every $x \in X$. In particular, for any $r > 0$

$$\sup_{x\in B(0,r)} \|P_n(x)\| \le \frac{n^n}{n!} \sup_{x\in B(a,r)} \|P(x)\|.$$

Proof. By Lemma 33,

$$\begin{aligned}\|P_n(x)\| = n^n \left\|P_n\left(\frac{x}{n}\right)\right\| &\le \frac{n^n}{2^n n!} \sum_{\varepsilon_j=\pm 1} \left\|P\left(a + \frac{x}{n}\sum_{j=1}^n \varepsilon_j\right)\right\| \\ &\le \frac{n^n}{n!} \sup_{t\in[-1,1]} \|P(a+tx)\|.\end{aligned}$$

□

Fact 48. *Let X, Y be normed linear spaces, $n \in \mathbb{N}_0$, and $P \in P^n(X;Y)$. Suppose that $P(x) = o(\|x\|^n)$, $x \to 0$. Then $P = 0$.*

Proof. Assume that $P = \sum_{j=k}^n P_j$, $P_j \in P(^jX;Y)$, $P_k \ne 0$. Then there is $y \in X$ such that $P_k(y) \ne 0$. Hence we have

$$0 = \lim_{t\to 0+} \frac{P(ty)}{\|ty\|^n} = \lim_{t\to 0+} \frac{P_k(y) + tP_{k+1}(y) + \cdots + t^{n-k}P_n(y)}{t^{n-k}\|y\|^n},$$

a contradiction. □

Fact 49. *Let X, Y be normed linear spaces, $n \in \mathbb{N}_0$, and let $P\colon X \to Y$. Then $P \in P^n(X;Y)$ if and only if $\phi \circ P$ is a polynomial of degree at most n for every $\phi \in \Phi$, where $\Phi \subset Y^*$ separates the points of Y. If moreover $\phi \circ P$ is continuous for every $\phi \in Y^*$, then so is P.*

Proof. $\Rightarrow$ This is straightforward.

$\Leftarrow$ We use induction on n. The case $n = 0$ follows easily from the fact that Φ separates the points of Y. Now we prove the induction step from $n-1$ to n. By Lemma 33 for every $\phi \in \Phi$ there exists an n-linear form $M_\phi \in L(^nX;\mathbb{K})$ such that

$$\begin{aligned}M_\phi(x_1,\dots,x_n) &= \frac{1}{2^n n!} \sum_{\varepsilon_j=\pm 1} \varepsilon_1\cdots\varepsilon_n (\phi\circ P)\left(\sum_{j=1}^n \varepsilon_j x_j\right) \\ &= \phi\left(\frac{1}{2^n n!} \sum_{\varepsilon_j=\pm 1} \varepsilon_1\cdots\varepsilon_n P\left(\sum_{j=1}^n \varepsilon_j x_j\right)\right).\end{aligned}$$

Set

$$M(x_1,\dots,x_n)=\frac{1}{2^n n!}\sum_{\varepsilon_j=\pm1}\varepsilon_1\cdots\varepsilon_n P\Bigl(\sum_{j=1}^n \varepsilon_j x_j\Bigr).$$

As

$$\begin{aligned}\phi\bigl(M(x_1,\dots,\alpha x+\beta y,\dots,x_n)\bigr)&=M_\phi(x_1,\dots,\alpha x+\beta y,\dots,x_n)\\&=\alpha M_\phi(x_1,\dots,x,\dots,x_n)+\beta M_\phi(x_1,\dots,y,\dots,x_n)\\&=\phi\bigl(\alpha M(x_1,\dots,x,\dots,x_n)+\beta M(x_1,\dots,y,\dots,x_n)\bigr)\end{aligned}$$

for any $\phi\in\Phi$ and Φ separates the points of Y, it follows that $M\in L(^nX;Y)$. Letting $P_n=\widehat{M}$, we see that by Lemma 33 the mapping $P-P_n$ satisfies the assumption of the theorem with $n-1$ instead of n, which finishes the induction step.

If $\phi\circ P$ is continuous for every $\phi\in Y^*$, then $P(B_X)$ is weakly bounded and hence it is bounded by the Banach-Steinhaus uniform boundedness principle. Thus P is continuous by Proposition 40. □

The above fact has the following immediate consequence (notice that the set of functionals $\{f\circ R_\gamma;\ f\in Y^*,\gamma\in\Gamma\}\subset\ell_\infty(\Gamma;Y)^*$ separates the points of $\ell_\infty(\Gamma;Y)$):

Fact 50. *Let X, Y be normed linear spaces, let Γ be an arbitrary set, $n\in\mathbb{N}_0$, and let $P\colon X\to\ell_\infty(\Gamma;Y)$. Denote by R_α, $\alpha\in\Gamma$ the canonical projections $R_\alpha\colon\ell_\infty(\Gamma;Y)\to Y$ given by $R_\alpha((y_\gamma))=y_\alpha$ and by $P_\gamma=R_\gamma\circ P$ the component mappings of P. Then $P\in\mathcal{P}^n(X;\ell_\infty(\Gamma;Y))$ if and only if $\{P_\gamma\}_{\gamma\in\Gamma}$ is a bounded set in $\mathcal{P}^n(X;Y)$.*

To prove that on real spaces it suffices to test the polynomiality of a mapping only on affine lines we need to introduce several technical tools.

Definition 51. Let X, Y be vector spaces, $U\subset X$, and $f\colon U\to Y$. For a fixed $h\in X$ we define the first difference by

$$\Delta^1 f(x;h)=f(x+h)-f(x)$$

for all $x\in U$ such that the right-hand side is defined. Further, we inductively define the differences of higher order by

$$\Delta^n f(x;h_1,\dots,h_n)=\Delta^{n-1}f(x+h_1;h_2,\dots,h_n)-\Delta^{n-1}f(x;h_2,\dots,h_n)$$

for all $x\in U$ such that the right-hand side is defined.

Lemma 52. *Let X, Y be vector spaces, $U\subset X$, $f\colon U\to Y$, and $n\in\mathbb{N}$. Then*

$$\Delta^n f(x;h_1,\dots,h_n)=\sum_{A\subset\{1,\dots,n\}}(-1)^{n-|A|}f\Bigl(x+\sum_{j\in A}h_j\Bigr)$$

if either of the sides is defined. Hence $\Delta^n f(x;h_1,\dots,h_n)=\Delta^n f(x;h_{\pi(1)},\dots,h_{\pi(n)})$ for every permutation π of $\{1,\dots,n\}$ and

$$\Delta^n f(x;h,\dots,h)=\sum_{k=0}^n(-1)^{n-k}\binom{n}{k}f(x+kh).$$

Proof. We use induction on n. For $n = 1$ it is just the definition. So let $n > 1$ and suppose that the formula holds for Δ^{n-1}. Then

$$\begin{aligned}
&\Delta^n f(x; h_1, \dots, h_n) \\
&= \sum_{A \subset \{2,\dots,n\}} (-1)^{n-1-|A|} f\Big(x + h_1 + \sum_{j \in A} h_j\Big) - \sum_{A \subset \{2,\dots,n\}} (-1)^{n-1-|A|} f\Big(x + \sum_{j \in A} h_j\Big) \\
&= \sum_{\substack{A \subset \{1,\dots,n\} \\ 1 \in A}} (-1)^{n-|A|} f\Big(x + \sum_{j \in A} h_j\Big) + \sum_{\substack{A \subset \{1,\dots,n\} \\ 1 \notin A}} (-1)^{n-|A|} f\Big(x + \sum_{j \in A} h_j\Big),
\end{aligned}$$

from which the formula follows. □

Lemma 53 ([Kuc]). *Let X, Y be vector spaces, $U \subset X$, $f : U \to Y$, and $n \in \mathbb{N}$. Then*

$$\Delta^n f(x; h_1, \dots, h_n) = \sum_{A \subset \{1,\dots,n\}} (-1)^{|A|} \Delta^n f\Big(x + \sum_{j \in A} h_j; -\sum_{j \in A} \tfrac{1}{j} h_j, \dots, -\sum_{j \in A} \tfrac{1}{j} h_j\Big)$$

for all $x, h_1, \dots, h_n \in X$ such that the right-hand side is defined.

Proof. First note that $\Delta^n f(x; u_1, \dots, u_n) = 0$ if $u_j = 0$ for some $j \in \{1, \dots, n\}$, which follows from the definition and the symmetry of Δ^n (Lemma 52). Therefore using Lemma 52 twice we obtain

$$\begin{aligned}
&\sum_{A \subset \{1,\dots,n\}} (-1)^{|A|} \Delta^n f\Big(x + \sum_{j \in A} h_j; -\sum_{j \in A} \tfrac{1}{j} h_j, \dots, -\sum_{j \in A} \tfrac{1}{j} h_j\Big) \\
&\quad = \sum_{A \subset \{1,\dots,n\}} (-1)^{|A|} \sum_{k=0}^{n} (-1)^{n-k} \binom{n}{k} f\Big(x + \sum_{j \in A} h_j - k \sum_{j \in A} \tfrac{1}{j} h_j\Big) \\
&\quad = \sum_{k=0}^{n} (-1)^{n-k} \binom{n}{k} \sum_{A \subset \{1,\dots,n\}} (-1)^{|A|} f\Big(x + \sum_{j \in A} (1 - \tfrac{k}{j}) h_j\Big) \\
&\quad = \sum_{k=0}^{n} (-1)^{n-k} \binom{n}{k} (-1)^n \Delta^n f\Big(x; (1-k) h_1, (1 - \tfrac{k}{2}) h_2, \dots, (1 - \tfrac{k}{n}) h_n\Big) \\
&\quad = \Delta^n f(x; h_1, \dots, h_n).
\end{aligned}$$

□

Let $X_1, \dots, X_n, Y$ be vector spaces. We say that a mapping $M : X_1 \times \dots \times X_n \to Y$ is n-additive if it is additive separately in each coordinate, i.e. if

$$\begin{aligned}
&M(x_1, \dots, x_{k-1}, x + y, x_{k+1}, \dots, x_n) \\
&\quad = M(x_1, \dots, x_{k-1}, x, x_{k+1}, \dots, x_n) + M(x_1, \dots, x_{k-1}, y, x_{k+1}, \dots, x_n)
\end{aligned}$$

for each $x_1 \in X_1, \dots, x_n \in X_n$, $x, y \in X_k$, and each $k \in \{1, \dots, n\}$. It is easy to see that if $M : X_1 \times \dots \times X_n \to Y$ is an n-additive mapping, then

$$M(x_1, \dots, x_{k-1}, t x_k, x_{k+1}, \dots, x_n) = t M(x_1, \dots, x_n)$$

for each $x_1 \in X_1, \dots, x_n \in X_n$, $t \in \mathbb{Q}$, and each $k \in \{1, \dots, n\}$.

By inspecting the proofs it is easy to see that Proposition 4, Theorem 7, and the Polarisation formula (Proposition 11, Proposition 13) hold also for n-additive mappings. In particular, if $X_1, \dots, X_n, Y$ are real normed linear spaces and $M : X_1 \times \cdots \times X_n \to Y$ is an n-additive mapping that is bounded on a neighbourhood of some point, then $M \in \mathcal{L}(X_1, \dots, X_n; Y)$.

Theorem 54 ([MO1]). *Let X, Y be vector spaces, $f : X \to Y$, and $n \in \mathbb{N}_0$. The following statements are equivalent:*

(i) $\Delta^{n+1} f(x; h, \dots, h) = 0$ *for all $x, h \in X$, i.e. f satisfies the Fréchet formula*

$$\sum_{k=0}^{n+1} (-1)^{n+1-k} \binom{n+1}{k} f(x + kh) = 0. \tag{10}$$

(ii) $\Delta^{n+1} f(x; h_1, \dots, h_{n+1}) = 0$ *for all $x, h_1, \dots, h_{n+1} \in X$.*

(iii) *There are symmetric k-additive mappings $M_k : X^k \to Y$, $k = 1, \dots, n$ such that $f(x) = f(0) + \sum_{k=1}^{n} \widehat{M_k}(x)$ for every $x \in X$, where $\widehat{M_k}(x) = M_k(x, \dots, x)$.*

Proof. (i)⇒(ii) follows from Lemma 53. (iii)⇒(i) follows from (7) in Proposition 13 by setting $a = x + (n+1)\frac{h}{2}$ and $x_j = \frac{h}{2}$.

(ii)⇒(iii) We use induction on n. For $n = 0$ it is clear, so assume that the implication holds for $n-1$ in place of n. Set $M_n(x_1, \dots, x_n) = \frac{1}{n!}\Delta^n f(0; x_1, \dots, x_n)$. It is easy to check using the definition that $\Delta^n f(0; x + y, x_2, \dots, x_n) = \Delta^n f(0; x, x_2, \dots, x_n) + \Delta^n f(x; y, x_2, \dots, x_n)$ for every $x, y, x_2, \dots, x_n \in X$. Hence

$$\begin{aligned} &M_n(x + y, x_2, \dots, x_n) - M_n(x, x_2, \dots, x_n) - M_n(y, x_2, \dots, x_n) \\ &= \frac{1}{n!}\big(\Delta^n f(0; x+y, x_2, \dots, x_n) - \Delta^n f(0; x, x_2, \dots, x_n) - \Delta^n f(0; y, x_2, \dots, x_n)\big) \\ &= \frac{1}{n!}\big(\Delta^n f(x; y, x_2, \dots, x_n) - \Delta^n f(0; y, x_2, \dots, x_n)\big) \\ &= \frac{1}{n!}\Delta^{n+1} f(0; x, y, x_2, \dots, x_n) = 0 \end{aligned}$$

by the assumption. So M_n is a symmetric n-additive mapping. Using the substitution $a = x + \sum_{j=1}^{n} \frac{h_j}{2}$ and $x_j = \frac{h_j}{2}$ the Polarisation formula (Proposition 11) yields $M_n(h_1, \dots, h_n) = \frac{1}{n!}\Delta^n \widehat{M_n}(x; h_1, \dots, h_n)$ for every $x, h_j \in X$. Finally, we put $g(x) = f(x) - \widehat{M_n}(x)$ for $x \in X$. Then

$$\begin{aligned} \Delta^n g(x; h_1, \dots, h_n) &= \Delta^n f(x; h_1, \dots, h_n) - n! M_n(h_1, \dots, h_n) \\ &= \Delta^n f(x; h_1, \dots, h_n) - \Delta^n f(0; h_1, \dots, h_n) \\ &= \Delta^{n+1} f(0; x, h_1, \dots, h_n) = 0. \end{aligned}$$

Thus we may use the inductive hypothesis on g and the proof is finished. □

Combining the previous theorem with Corollary 31, the Polarisation formula (Proposition 13), and condition (v) in Theorem 7 we obtain the following corollary:

Corollary 55. *Let X be a real Banach space, Y a real normed linear space, $n \in \mathbb{N}_0$, and let $P : X \to Y$ be such that $\phi \circ P$ is Baire measurable for each $\phi \in Y^*$. The following statements are equivalent:*

(i) $\Delta^{n+1} P(x; h, \dots, h) = 0$ *for all* $x, h \in X$.

(ii) $\Delta^{n+1} P(x; h_1, \dots, h_{n+1}) = 0$ *for all* $x, h_1, \dots, h_{n+1} \in X$.

(iii) *The restriction of P to each line in X is a polynomial of degree at most n.*

(iv) $P \in \mathcal{P}^n(X; Y)$.

We end this section by presenting the Lagrange interpolation formula for vector-valued polynomials on scalars which, although it belongs in spirit to Chapter 2, cannot be postponed, as it will be useful later on in this chapter.

Proposition 56 (Lagrange interpolation formula)**.** *Let Y be a normed linear space over $\mathbb{K}$, let $x_0, \dots, x_n$ be distinct points in $\mathbb{K}$, and let $y_0, \dots, y_n \in Y$. Then there exists a unique polynomial $P \in \mathcal{P}^n(\mathbb{K}; Y)$ satisfying $P(x_j) = y_j$, $j = 0, \dots, n$. The polynomial P can be expressed by the formula*

$$P(x) = \sum_{j=0}^{n} \frac{w_j(x)}{w_j(x_j)} y_j,$$

where

$$w_j(x) = \prod_{\substack{k=0 \\ k \neq j}}^{n} (x - x_k).$$

Proof. The formula for P of course gives the existence of such polynomial. The uniqueness follows from the fact that Y^* separates the points of Y, so we may apply the scalar case. □

2. Complexification

Complex Banach spaces are an indispensable tool in the theory of real analytic mappings. The complexification of a given real Banach space X can be described as $X + iX$, a space which is real-isomorphic to $X \oplus X$. The complex operations are obtained naturally. In general, the complex norm can be introduced in many non-isometric (but of course real-isomorphic) ways. This fact plays an inessential role for the most part of our theory, except perhaps for the precise estimates of norms obtained by passing to the complexification, and we choose to work with a complexification norm due to Angus Ellis Taylor. Here we focus on proving some basic extension inequalities to be used in later sections and chapters.

Definition 57. By a complexification of a real normed linear space $(X, \|\cdot\|)$ we mean the complex normed linear space $\tilde{X} = \{(x, y);\ x, y \in X\}$ with operations given by

$$(x, y) + (u, v) = (x + u, y + v),$$
$$(\alpha + i\beta)(x, y) = (\alpha x - \beta y, \alpha y + \beta x),$$

and a norm $\|\cdot\|_{\mathbb{C}}$ defined as

$$\|(x, y)\|_{\mathbb{C}} = \sup_{\phi \in B_{X^*}} \sqrt{\phi(x)^2 + \phi(y)^2}.$$

Checking that $(\tilde{X}, \|\cdot\|_{\mathbb{C}})$ is a complex normed linear space real-isomorphic to $X \oplus X$ is easy. The space X is canonically isometrically embedded in $\tilde{X}$ by $x \mapsto (x, 0)$ and if X is a real Banach space, then $\tilde{X}$ is also a Banach space. In some situations it is convenient to use this equivalent formula for the Taylor norm:

$$\begin{aligned} \|(x, y)\|_{\mathbb{C}} &= \sup_{\phi \in B_{X^*}} \sup_{t \in [0,2\pi]} \big(\cos(t)\phi(x) + \sin(t)\phi(y)\big) \\ &= \sup_{t \in [0,2\pi]} \|\cos(t)x + \sin(t)y\|. \end{aligned} \tag{11}$$

We will use an alternative notation for $\tilde{X} = X + iX$ using the correspondence $(x, y) = x + iy$ and assuming implicitly that $x, y \in X$. We will also use the usual projections $\operatorname{Re}(x + iy) = x$ and $\operatorname{Im}(x + iy) = y$. For $z = x + iy \in \tilde{X}$ we denote $\overline{z} = x - iy$. From the definition of the Taylor norm it is clear that $\|\overline{z}\|_{\mathbb{C}} = \|z\|_{\mathbb{C}}$ and $\|\operatorname{Re} z\| \le \|z\|_{\mathbb{C}}$, $\|\operatorname{Im} z\| \le \|z\|_{\mathbb{C}}$. We are going to omit the subscript $\mathbb{C}$ in order to simplify the notation later on in the text.

Let us make a few remarks. It is clear from the formula (11) that our complexification procedure respects subspaces, i.e. if $T\colon Z \to X$ is an isometric embedding, then $\|x + iy\|_{Z,\mathbb{C}} = \|T(x) + iT(y)\|_{X,\mathbb{C}}$ whenever $x, y \in Z$. Also, it is not difficult to check using (11) that the space $\widetilde{X \oplus Y}$ is isometric to the space $\tilde{X} \oplus \tilde{Y}$ if we take the maximum norm on both direct sums.

In case of the Banach spaces $C(K)$ Taylor's norm is natural and coincides with the usual definition of the norm of the respective complex spaces. However, for the spaces ℓ_p, $p \neq \infty$, the complexification norm does not coincide with the norm of the complex version of the sequence spaces. Taylor's norm also does not in general pass to quotients and it does not respect duality. We are not going to discuss the various possible alternative definitions of complexification, as they do not play a significant role for our purposes. Let us only remark that defining the complexification norm by the formula $\sqrt{\|x\|^2 + \|y\|^2}$ is not possible, as the complex homogeneity of the norm would be violated.

Given real normed linear spaces X, Y and $T \in \mathcal{L}(X; Y)$, it is easily checked that there is a uniquely determined (by complex linearity) complex extension $\tilde{T} \in \mathcal{L}(\tilde{X}; \tilde{Y})$, defined by $\tilde{T}(x + iy) = T(x) + iT(y)$.

Proposition 58. *Let X, Y be real normed linear spaces and $T \in \mathcal{L}(X; Y)$. Then* $\|\tilde{T}\|_{\mathbb{C}} = \|T\|$.

Proof. We only need to verify that $\|\tilde{T}\|_{\mathbb{C}} \le \|T\|$:

$$\begin{aligned}\|\tilde{T}(x+iy)\|_{\mathbb{C}} = \|T(x)+iT(y)\| &= \sup_{t\in[0,2\pi]} \|\cos(t)T(x)+\sin(t)T(y)\| \\ &= \sup_{t\in[0,2\pi]} \|T(\cos(t)x+\sin(t)y)\| \\ &\le \|T\| \sup_{t\in[0,2\pi]} \|\cos(t)x+\sin(t)y\| = \|T\|\|x+iy\|_{\mathbb{C}}.\end{aligned}$$ □

Obviously, T is one-to-one if and only if $\tilde{T}$ is one-to-one, and T is onto if and only if $\tilde{T}$ is onto.

Fact 59. *Let X be a real normed linear space and $z \in \tilde{X}$. Then there is $\phi \in S_{X^*}$ such that $|\tilde{\phi}(z)| = \|z\|$.*

Proof. Let $z = x + iy$. By the w^*-compactness of B_{X^*} there is $\phi \in S_{X^*}$ such that $\|z\| = \sqrt{\phi(x)^2 + \phi(y)^2} = |\tilde{\phi}(z)|$. □

Proposition 60 ([BoSi]). *Let X, Y be real normed linear spaces, $n \in \mathbb{N}$, and let $M \in L(^nX;Y)$. There is a unique extension $\widetilde{M} \in L(^n\tilde{X};\tilde{Y})$ such that $\widetilde{M}\restriction_{X^n} = M$. It is given by the formula*

$$\widetilde{M}(x_1^0 + ix_1^1, \dots, x_n^0 + ix_n^1) = \sum_{\varepsilon_1,\dots,\varepsilon_n\in\{0,1\}} i^{\sum_{j=1}^n \varepsilon_j} M(x_1^{\varepsilon_1}, \dots, x_n^{\varepsilon_n}). \tag{12}$$

Moreover, $\widetilde{M}$ is bounded if and only if M is bounded.

Proof. It is clear that $\widetilde{M}$ is n-linear as a real multilinear mapping and also that every complex extension must satisfy (12). It remains to prove that it is coordinatewise complex homogeneous. Let $a = \alpha + i\beta \in \mathbb{C}$. From (12) we obtain for every $u_2, \dots, u_n \in \tilde{X}$, $x_0, x_1 \in X$,

$$\widetilde{M}(x_0 + ix_1, u_2, \dots, u_n) = \widetilde{M}(x_0, u_2, \dots, u_n) + i\widetilde{M}(x_1, u_2, \dots, u_n).$$

So

$$\begin{aligned}\widetilde{M}\big(a(x_0+ix_1), u_2, \dots, u_n\big) &= \widetilde{M}\big((\alpha x_0 - \beta x_1) + i(\beta x_0 + \alpha x_1), u_2, \dots, u_n\big) \\ &= \alpha\widetilde{M}(x_0, u_2, \dots, u_n) - \beta\widetilde{M}(x_1, u_2, \dots, u_n) \\ &\qquad + i\beta\widetilde{M}(x_0, u_2, \dots, u_n) + i\alpha\widetilde{M}(x_1, u_2, \dots, u_n) \\ &= (\alpha + i\beta)\widetilde{M}(x_0, u_2, \dots, u_n) + (i\alpha - \beta)\widetilde{M}(x_1, u_2, \dots, u_n) \\ &= a\widetilde{M}(x_0 + ix_1, u_2, \dots, u_n).\end{aligned}$$

The assertion about boundedness is clear. □

Proposition 61 ([Tay1]). *Let X, Y be real normed linear spaces, $n \in \mathbb{N}$, and let $P \in P({}^nX;Y)$. Then there is a unique extension $\tilde{P} \in P({}^n\tilde{X};\tilde{Y})$ such that $\tilde{P}\restriction_X = P$. It is given by the formula*

$$\tilde{P}(x+iy) = \sum_{k=0}^{[\frac{n}{2}]}(-1)^k\binom{n}{2k}\check{P}({}^{n-2k}x, {}^{2k}y)$$
$$+\, i\sum_{k=0}^{[\frac{(n-1)}{2}]}(-1)^k\binom{n}{2k+1}\check{P}({}^{n-(2k+1)}x, {}^{2k+1}y).$$

Moreover, $\tilde{P}$ is bounded if and only if P is bounded.

Proof. Use Lemma 21 combined with Proposition 60. □

For $P \in \mathcal{P}({}^0X;Y)$ the extension is trivial. Naturally, if $P(x) = \sum_{k=0}^n P_k(x)$, $P_k \in P({}^kX;Y)$, then we denote $\tilde{P}(x) = \sum_{k=0}^n \tilde{P}_k(x)$.

The following is an immediate corollary of Proposition 61.

Fact 62. *Let X, Y be real normed linear spaces, $n \in \mathbb{N}_0$, and $P \in P^n(X;Y)$. Then $\tilde{P}(\overline{z}) = \overline{\tilde{P}(z)}$ for every $z \in \tilde{X}$.*

The next theorem gives in particular a norm estimate of the complexified homogeneous polynomials. For the general non-homogeneous case see Theorem 2.26.

Theorem 63 ([MST]). *Let X, Y be real normed linear spaces, $n \in \mathbb{N}$, and let $P \in \mathcal{P}^n(X;Y)$, $P = \sum_{k=0}^n P_k$, where $P_k \in \mathcal{P}({}^kX;Y)$. Then*

$$\|\tilde{P}_n\|_{\mathbb{C}} \le 2^{n-1}\|P\|,$$
$$\|\widetilde{P_{n-1}}\|_{\mathbb{C}} \le 2^{n-2}\|P\| \quad (\textit{for } n \ge 2).$$

Proof. Let $z = x+iy \in \tilde{X}$, $\|z\| = 1$ (hence also $\|\overline{z}\| = 1$). For $t \in \mathbb{R}$ define

$$f(t) = \tilde{P}\left(\frac{e^{-it}}{2}z + \frac{e^{it}}{2}\overline{z}\right) = P(\cos(t)x + \sin(t)y).$$

From (11) it follows that

$$\|f(t)\|_{\mathbb{C}} \le \|P\|, \quad t \in \mathbb{R}. \tag{13}$$

Using Lemma 21 we obtain

$$f(t) = P_0(0) + \sum_{k=1}^{n}\sum_{l=0}^{k}\binom{k}{l}\frac{e^{(k-2l)it}}{2^k}\check{\tilde{P}}_k({}^lz, {}^{k-l}\overline{z}) = \sum_{k=-n}^{n} e^{ikt}a_k$$

for suitable $a_k \in \tilde{Y}$. We can easily see that $a_{-n} = \frac{1}{2^n}\tilde{P}_n(z)$ and $a_n = \frac{1}{2^n}\tilde{P}_n(\overline{z})$. Furthermore, exploiting the fact that the inner sum for $k = n$ does not contain summands with factors $e^{(n-1)it}$, $e^{-(n-1)it}$, we obtain also $a_{-(n-1)} = \frac{1}{2^{n-1}}\widetilde{P_{n-1}}(z)$ and $a_{n-1} = \frac{1}{2^{n-1}}\widetilde{P_{n-1}}(\overline{z})$. Note also that the homogeneous summands of $\tilde{P}$ of lower degree are not expressed as easily.

Next, we use the following formula (a sum of a finite geometric series):

$$\frac{1}{2n}\sum_{p=0}^{2n-1}(-1)^p e^{\frac{ikp\pi}{n}} = \begin{cases} 1 & \text{for } k \equiv n \pmod{2n}, \\ 0 & \text{otherwise.} \end{cases} \tag{14}$$

Thus

$$\begin{aligned} \frac{1}{2n}\sum_{p=0}^{2n-1}(-1)^p f\Big(t+p\frac{\pi}{n}\Big) &= \frac{1}{2n}\sum_{p=0}^{2n-1}(-1)^p \sum_{k=-n}^{n} e^{ik(t+\frac{p\pi}{n})}a_k \\ &= \sum_{k=-n}^{n} e^{ikt}\Bigg(\frac{1}{2n}\sum_{p=0}^{2n-1}(-1)^p e^{\frac{ikp\pi}{n}}\Bigg)a_k \\ &= e^{-int}a_{-n} + e^{int}a_n = \frac{e^{-int}}{2^n}\widetilde{P}_n(z) + \frac{e^{int}}{2^n}\widetilde{P}_n(\overline{z}). \end{aligned}$$

Combining this with (13) we get

$$\sup_{t\in[0,2\pi]}\big\|e^{-int}\widetilde{P}_n(z)+e^{int}\widetilde{P}_n(\overline{z})\big\|_{\mathbb{C}} \le 2^n\|P\|.$$

Now let $\widetilde{P}_n(z) = u + iv$. By Fact 62 we have $\widetilde{P}_n(\overline{z}) = u - iv$ and so by (11)

$$\begin{aligned} \sup_{t\in[0,2\pi]}\big\|e^{-int}\widetilde{P}_n(z)+e^{int}\widetilde{P}_n(\overline{z})\big\|_{\mathbb{C}} &= \sup_{t\in[0,2\pi]}\|2\cos(nt)u + 2\sin(nt)v\|_{\mathbb{C}} \\ &= 2\|\widetilde{P}_n(z)\|_{\mathbb{C}}, \end{aligned}$$

which finishes the proof of the first estimate. The second estimate is proved analogously, using $n-1$ in place of n in (14). □

Corollary 64. *Let X, Y be real normed linear spaces, $n \in \mathbb{N}$, and $P = \sum_{k=0}^{n} P_k$, where $P_k \in \mathcal{P}(^kX;Y)$. Then $\|P_n\| \le 2^{n-1}\|P\|$ and $\|P_{n-1}\| \le 2^{n-2}\|P\|$ (for $n \ge 2$).*

3. Fréchet smoothness

We begin by introducing the directional and the Gâteaux derivative. The stronger concept of Fréchet derivative is by far the most important notion of differentiation. We develop its basic properties including the Implicit function theorem. We introduce higher derivatives inductively as multilinear mappings and prove some of their standard properties.

Definition 65. Let X, Y be normed linear spaces over $\mathbb{K}$, $A \subset X$, $f: A \to Y$, and $x \in A$. The directional derivative of f at x in the direction $h \in X$ is defined as

$$\frac{\partial f}{\partial h}(x) = \lim_{\substack{t \to 0 \\ t \in \mathbb{K}}} \frac{f(x+th) - f(x)}{t} \tag{15}$$

if the limit exists in Y. If $\frac{\partial f}{\partial h}(x)$ exists for all $h \in X$ and the mapping $L: X \to Y$, $L(h) = \frac{\partial f}{\partial h}(x)$, is a continuous linear operator, then we say that f is Gâteaux differentiable at x. The operator L is then called the Gâteaux derivative of f at x and is denoted by $\delta f(x)$. For the evaluation of the Gâteaux derivative we use the notation $\delta f(x)[h] = L(h)$.

Clearly, if the directional derivative $\frac{\partial f}{\partial h}(x)$ exists, then $\frac{\partial f}{\partial (Ch)}(x)$ exists for any $C \in \mathbb{K}$ and $\frac{\partial f}{\partial (Ch)}(x) = C\frac{\partial f}{\partial h}(x)$. For mappings defined in $\mathbb{R}^n$ the directional derivatives along the canonical basis vectors are just the usual partial derivatives. For a fixed $h \in X$ the natural domain of the mapping $\frac{\partial f}{\partial h}$ (i.e. of the mapping $x \mapsto \frac{\partial f}{\partial h}(x)$) is the set of all points $x \in A$ at which $\frac{\partial f}{\partial h}(x)$ exists. Similarly for the mapping δf.

Definition 66. Let X, Y be normed linear spaces over $\mathbb{K}$, $A \subset X$, $f: A \to Y$, and $x \in A$. We say that f is Fréchet differentiable at x if f is Gâteaux differentiable at x and the limit (15) is uniform with respect to $h \in S_X$. In this case we call the linear operator $\delta f(x)$ the Fréchet derivative of f at x and denote it by $Df(x)$. For the evaluation we again use the notation $Df(x)[h] = \delta f(x)[h]$.

Note that from the definition it follows that if f is Fréchet differentiable at x, then x is an interior point of A. If $f: A \to Y$ is Fréchet differentiable at every point of $B \subset A$, we say that f is Fréchet differentiable in B. Denoting by C the set of all $x \in A$ at which $Df(x)$ exists, $Df: C \to \mathcal{L}(X;Y)$ is then the mapping $x \mapsto Df(x)$. For convenience we also use the notation $D^1 f = Df$ and $D^0 f = f$, and analogously for the Gâteaux derivative.

Let us make a few remarks. It is clear that f is Fréchet differentiable at x if and only if there exists $L \in \mathcal{L}(X;Y)$ satisfying

$$\lim_{h \to 0} \frac{\|f(x+h) - f(x) - L(h)\|}{\|h\|} = 0. \tag{16}$$

The linear operator L is then the Fréchet derivative of f at x. It follows that if f is Fréchet differentiable at x, then it is continuous at x. This is not necessarily true for Gâteaux differentiable mappings, even for functions on $\mathbb{R}^2$ – consider the function $f(x,y) = \max\{(\frac{y}{x^2})^3(2 - \frac{y}{x^2})^3, 0\}$ for $x, y \in \mathbb{R}$, $x \neq 0$, and $f(0,y) = 0$, $y \in \mathbb{R}$, which is Gâteaux differentiable at every point of $\mathbb{R}^2$ but is discontinuous at the origin. It is however not difficult (using the compactness of the unit sphere) to show that for locally Lipschitz mappings defined on finite-dimensional spaces the notions of Fréchet and Gâteaux differentiability coincide. Note also that on one-dimensional spaces the existence of directional (for $h \neq 0$), Gâteaux, and Fréchet derivatives are all equivalent. Finally, unlike the finite-dimensional case, if X is infinite-dimensional, then even the equi-continuity of $\{x \mapsto \frac{\partial f}{\partial h}(x)\}_{h\in S_X}$ at $a \in X$ does not ensure the existence of the Gâteaux derivative $\delta f(a)$ (consider a discontinuous linear functional).

From (16) we immediately obtain the following two facts.

Fact 67. *Let X, Y be normed linear spaces and $L \in \mathcal{L}(X;Y)$. Then $DL(x) = L$ for every $x \in X$.*

Proposition 68. *Let X, Y be normed linear spaces over $\mathbb{K}$, $U \subset X$ an open set, let $f, g\colon U \to Y$ be Fréchet differentiable at $a \in U$, and $c \in \mathbb{K}$. Then $f + g$ is Fréchet differentiable at a and $D(f+g)(a) = Df(a) + Dg(a)$. Further, $c \cdot f$ is Fréchet differentiable at a and $D(cf)(a) = cDf(a)$.*

An equivalent reformulation of (16) is the following:

$$f(x+h) - f(x) - Df(x)[h] = o(\|h\|),\ h \to 0.$$

This form is useful for proving the Chain rule.

Theorem 69 (Chain rule)**.** *Let X, Y, Z be normed linear spaces, $U \subset X$ open, and let $g\colon U \to Y$ be Fréchet differentiable at $a \in U$. Suppose further that $g(U) \subset V$, where $V \subset Y$ is an open set, and $f\colon V \to Z$ is Fréchet differentiable at $g(a)$. Then $f \circ g\colon U \to Z$ is Fréchet differentiable at a and*

$$D(f \circ g)(a) = Df\big(g(a)\big) \circ Dg(a).$$

Proof. We denote $b = g(a)$ and put $\omega(h) = g(a+h) - g(a) - Dg(a)[h]$ and $\eta(k) = f(b+k) - f(b) - Df(b)[k]$. These mappings are defined on suitable neighbourhoods of the origins. Choose an arbitrary $\varepsilon \in (0,1)$. There is $\Delta > 0$ such that for $h \in X$, $\|h\| < \Delta$ and $k \in Y$, $\|k\| < \Delta$ we have $\|\omega(h)\| \le \varepsilon\|h\|$ and $\|\eta(k)\| \le \varepsilon\|k\|$. Put $\delta = \Delta/(1 + \|Dg(a)\|)$. Then for $h \in X$, $\|h\| < \delta$ we have $\|Dg(a)[h] + \omega(h)\| \le (\|Dg(a)\| + \varepsilon)\|h\| < \Delta$ and hence

$$\begin{aligned}
&\big\|f\big(g(a+h)\big) - f\big(g(a)\big) - Df\big(g(a)\big)\big[Dg(a)[h]\big]\big\| \\
&\qquad = \big\|f\big(b + Dg(a)[h] + \omega(h)\big) - f(b) - Df(b)\big[Dg(a)[h]\big]\big\| \\
&\qquad = \big\|Df(b)[\omega(h)] + \eta\big(Dg(a)[h] + \omega(h)\big)\big\| \\
&\qquad \le \|Df(b)\|\varepsilon\|h\| + \varepsilon\big(\|Dg(a)\|\,\|h\| + \varepsilon\|h\|\big) \\
&\qquad < \big(\|Df(b)\| + \|Dg(a)\| + 1\big)\varepsilon\|h\|,
\end{aligned}$$

which proves the theorem. □

For Gâteaux differentiability we have the following version:

Theorem 70. *Let X, Y, Z be normed linear spaces, $A \subset X$, and let $g\colon A \to Y$ be Gâteaux differentiable at $a \in A$. Suppose further that $g(A) \subset V$, where $V \subset Y$ is an open set, and $f\colon V \to Z$. If either f is Fréchet differentiable at $g(a)$, or f is Gâteaux differentiable at $g(a)$ and f is Lipschitz on V, then $f \circ g\colon A \to Z$ is Gâteaux differentiable at a and*

$$\delta(f \circ g)(a) = \delta f\big(g(a)\big) \circ \delta g(a).$$

Proof. The case when f is Fréchet differentiable at $g(a)$ follows easily from Theorem 69 used on f and $t \mapsto g(a+th)$ for any fixed $h \in X$. So suppose that f is Gâteaux differentiable at $g(a)$ and f is L-Lipschitz on V. Fix $h \in X$. Then for all $t \in \mathbb{K} \setminus \{0\}$ small enough such that $a+th \in A$ and $g(a)+t\delta g(a)[h] \in V$ we have

$$\frac{1}{t}\Big(f\big(g(a+th)\big)-f\big(g(a)\big)\Big)=\frac{1}{t}\Big(f\big(g(a+th)\big)-f\big(g(a)+t\delta g(a)[h]\big)\Big) \\ +\frac{1}{t}\Big(f\big(g(a)+t\delta g(a)[h]\big)-f\big(g(a)\big)\Big).$$

Now the second summand tends to $\delta f\big(g(a)\big)\big[\delta g(a)[h]\big]$ as $t \to 0$, while the first summand tends to 0. Indeed, from the Lipschitz property of f we get

$$\left\|\frac{1}{t}\Big(f\big(g(a+th)\big)-f\big(g(a)+t\delta g(a)[h]\big)\Big)\right\| \\ \leq L\left\|\frac{1}{t}\big(g(a+th)-g(a)-t\delta g(a)[h]\big)\right\|. \qquad \square$$

We remark that if the mapping g is Féchet differentiable at a and f is Gâteaux differentiable at $g(a)$, then the composition $f \circ g$ may not be Gâteaux differentiable at a, and even if it is, the chain rule formula may not hold. Consider the mapping $g\colon \mathbb{R} \to \mathbb{R}^2$, $g(x)=(x,x^2)$, and $f\colon \mathbb{R}^2 \to \mathbb{R}$ defined as $f(x,y)=0$ for $y \neq x^2$, $f(x,x^2)=\operatorname{sgn} x$ (or $f(x,x^2)=x$).

Proposition 71. *Let X, Y be normed linear spaces, $U \subset X$, $f\colon U \to Y$, and $C \subset U$ convex. Suppose that f is Gâteaux differentiable at each point of C and $\sup_{x\in C}\|\delta f(x)\| = K < +\infty$. Then f is K-Lipschitz on C.*

Proof. In case that X and Y are complex spaces, we may consider their real versions $X_{\mathbb{R}}$ and $Y_{\mathbb{R}}$, and the real Gâteaux derivative of the mapping f considered as a mapping from $X_{\mathbb{R}}$ to $Y_{\mathbb{R}}$. This change of structure has no impact on the differentiability: $\delta_{\mathbb{R}} f(x) \in \mathcal{L}(X_{\mathbb{R}}; Y_{\mathbb{R}})$ and $\delta_{\mathbb{R}} f(x) = \delta f(x)$. Thus we may assume without loss of generality that the spaces X, Y are real.

Fix any $a, b \in C$, put $h = b-a$, and define the function $g\colon [0,1] \to Y$ by $g(t) = f(a+th)$. Let $\phi \in S_{Y^*}$ be such that $\|f(b)-f(a)\| = \phi\big(f(b)-f(a)\big)$. We have $(\phi \circ g)'(t) = \phi\big(\delta f(a+th)[h]\big)$ for $t \in [0,1]$. Thus by the Mean value theorem there is $\xi \in (0,1)$ such that $\phi\circ g(1) - \phi\circ g(0) = (\phi\circ g)'(\xi)$. Hence $\|f(b)-f(a)\| = \phi\big(g(1)-g(0)\big) = (\phi\circ g)'(\xi) \leq \|\phi\|\|\delta f(a+\xi h)\|\|h\| \leq K\|b-a\|$. $\square$

The following variant of the previous proposition, where the convexity is slightly relaxed, can be useful in certain situations.

Proposition 72. *Let X, Y be normed linear spaces, $U \subset X$, $f\colon U \to Y$, $C \subset X$ convex, and let $D \subset C$ be such that $C \setminus D \subset U$ and for every $x, y \in C \setminus D$ there is $z \in C$ such that $Z = \operatorname{span}\{y-x, z-x\}$ is two-dimensional and $D \cap (x+Z)$ is countable. (This holds in particular if C is not a subset of a line and D is countable or discrete.) Suppose that f is Gâteaux differentiable at each point of $C \setminus D$ and $\sup_{x\in C\setminus D}\|\delta f(x)\| = K < +\infty$. Then f is K-Lipschitz on $C \setminus D$.*

Proof. Fix $x, y \in C \setminus D$ and let $z \in C$ be such that $Z = \operatorname{span}\{y - x, z - x\}$ is two-dimensional and $D \cap (x + Z)$ is countable. Set $u = \frac{1}{2}(x + y)$ and $v = \frac{z-u}{\|z-u\|}$. Let $\varepsilon > 0$ be arbitrary. Then the sets $B_1 = \{t \in [0, \varepsilon];\ [x, u + tv] \cap D \neq \emptyset\}$ and $B_2 = \{t \in [0, \varepsilon];\ [y, u + tv] \cap D \neq \emptyset\}$ are countable. Therefore there is $t_0 \in [0, \varepsilon] \setminus (B_1 \cup B_2)$. Set $w = u + t_0 v$. The line segments $[x, w]$ and $[y, w]$ lie in $C \setminus D$ and so by Proposition 71 we get $\|f(x) - f(y)\| \leq \|f(x) - f(w)\| + \|f(w) - f(y)\| \leq K(\|x - w\| + \|w - y\|) \leq K(\|x - u\| + t_0 + \|u - y\| + t_0) \leq K(\|x - y\| + 2\varepsilon)$. Since ε can be chosen arbitrarily small, the assertion follows. □

Fact 73. *Let X, Y be normed linear spaces, $U \subset X$ open, and let $f: U \to Y$ be a Gâteaux differentiable mapping such that its Gâteaux derivative $\delta f: U \to \mathcal{L}(X; Y)$ is continuous at $x \in U$. Then f is Fréchet differentiable at x.*

Proof. Define a mapping $g: U \to Y$ by $g(y) = f(y) - \delta f(x)[y - x]$. Notice that (using Fact 67) $\delta g(y) = \delta f(y) - \delta f(x)$. Choose $\varepsilon > 0$ and find $\eta > 0$ such that $\|\delta f(y) - \delta f(x)\| < \varepsilon$ whenever $y \in U$ satisfies $\|y - x\| < \eta$. Then, by Proposition 71, g is ε-Lipschitz on $U(x, \eta)$. Thus, for an arbitrary $h \in U(0, \eta)$,

$$\big\|f(x + h) - f(x) - \delta f(x)[h]\big\| = \|g(x + h) - g(x)\| \leq \varepsilon\|h\|. \qquad \square$$

Definition 74. Let X, Y be normed linear spaces over $\mathbb{K}$, $U \subset X$ an open set, let $f: U \to Y$, $a \in U$, and $k \in \mathbb{N}$. We say that f is twice Fréchet differentiable at a if $Df(x)$ exists for x in some neighbourhood of a and the mapping $x \mapsto Df(x)$ is Fréchet differentiable at a. We define the second Fréchet derivative of f at a by $D^2 f(a) = D(Df)(a)$. Then $D^2 f(a) \in \mathcal{L}\big(X; \mathcal{L}(X; Y)\big)$, but using the natural identification from Fact 9 we may consider $D^2 f(a)$ to be an element of $\mathcal{L}(^2X; Y)$.

Further, we define inductively the kth Fréchet derivative $D^k f(a)$ of f at a. We say that f is k-times Fréchet differentiable at a if $D^{k-1} f(x) \in \mathcal{L}(^{k-1}X; Y)$ exists for x in some neighbourhood of a and the mapping $x \mapsto D^{k-1} f(x)$ is Fréchet differentiable at a. We put $D^k f(a) = D(D^{k-1} f)(a)$. Using the identification from Fact 9 we may consider $D^k f(a)$ to be an element of $\mathcal{L}(^kX; Y)$. We denote by $D^k f(a)[x_1, \dots, x_k] \in Y$ the evaluation of $D^k f(a)$ at $(x_1, \dots, x_k) \in X^k$.

We say that f is C^k-smooth if $D^k f$ is continuous in the domain. We say that f is C^∞-smooth if $D^k f$ is continuous in the domain for every $k \in \mathbb{N}$. We denote by $C^k(U; Y)$ the set of all C^k-smooth mappings from U into Y, $k \in \mathbb{N} \cup \{\infty\}$. For convenience we put $C^0(U; Y) = C(U; Y)$. As usual, for functions, i.e. mappings into the scalars, we use the shortened notation $C^k(U) = C^k(U; \mathbb{K})$.

By abusing the notation slightly we write $f \in C^k(U; M)$, where M is a subset of a normed linear space Y, as a short for $f \in C^k(U; Y)$, $f(U) \subset M$.

The following fact follows immediately by induction.

Fact 75. *Let X, Y be normed linear spaces over $\mathbb{K}$, $U \subset X$ open, $k \in \mathbb{N}$, let $f, g: U \to Y$ be k-times Fréchet differentiable at $a \in U$, and $c \in \mathbb{K}$. Then $f + g$ is k-times Fréchet differentiable at a and $D^k(f + g)(a) = D^k f(a) + D^k g(a)$. Further, $c \cdot f$ is k-times Fréchet differentiable at a and $D^k(cf)(a) = cD^k f(a)$. In particular, $C^k(U; Y)$ is a linear subspace of $C(U; Y)$.*

Theorem 76. *Let X, Y be normed linear spaces, $U \subset X$ an open set, $f: U \to Y$, $k \in \mathbb{N}$, and $a \in U$. Suppose that $D^k f(a)$ exists. Then $D^k f(a) \in \mathcal{L}^s({}^k X; Y)$.*

Proof. Notice that for $g(x) = f(a+x)$ we have $D^k g(0) = D^k f(a)$ and so we may without loss of generality assume that $a = 0$.

Assume first that $k = 2$. Given $\varepsilon > 0$, choose $\delta > 0$ so that

$$\big\| Df(x) - Df(0) - D(Df)(0)[x] \big\| \le \varepsilon \|x\| \quad \text{for } x \in U(0, 2\delta).$$

For a fixed $v \in U(0,\delta)$ we define $g_v(u): U(0,\delta) \to Y$ by

$$g_v(u) = f(u+v) - f(u) - f(v) + f(0) - \big(D(Df)(0)[v]\big)[u].$$

Observe that (using Fact 67)

$$\begin{aligned} Dg_v(u) &= Df(u+v) - Df(u) - D(Df)(0)[v] \\ &= \big(Df(u+v) - Df(0) - D(Df)(0)[u+v]\big) \\ &\qquad - \big(Df(u) - Df(0) - D(Df)(0)[u]\big). \end{aligned}$$

Thus

$$\|Dg_v(u)\| \le \varepsilon \|u+v\| + \varepsilon \|u\| \le 2\varepsilon(\|u\| + \|v\|).$$

By Proposition 71, g_v is $2\varepsilon(\eta + \|v\|)$-Lipschitz on $B(0,\eta)$, $\eta < \delta$, and since $g_v(0) = 0$ we conclude that $\|g_v(u)\| \le 2\varepsilon(\|u\| + \|v\|)\|u\| \le 2\varepsilon(\|u\| + \|v\|)^2$. It follows that for any $w, z \in X$, $(w,z) \ne (0,0)$, we have

$$\lim_{t\to 0} \frac{\big\| f(tw+tz) - f(tw) - f(tz) + f(0) - D^2 f(0)[tw,tz] \big\|}{(\|tw\| + \|tz\|)^2} = 0$$

and so

$$\lim_{t\to 0} \frac{f(tw+tz) - f(tw) - f(tz) + f(0)}{t^2} = D^2 f(0)[w,z].$$

Interchanging the roles of w and z we see that $D^2 f(0)[z,w]$ is given by the same limit and hence $D^2 f(0)$ is a symmetric 2-linear mapping.

We proceed by induction. Assume that $k > 2$ and $D^{k-1} f(x) \in \mathcal{L}^s({}^{k-1}X; Y)$ for each x from some neighbourhood of a. Recall that each permutation is a composition of adjacent transpositions, therefore it suffices to show that given $u_1, \dots, u_k \in X$ and $p \in \{1, \dots, k-1\}$ we have

$$D^k f(0)[u_1, \dots, u_k] = D^k f(0)[u_1, \dots, u_{p-1}, u_{p+1}, u_p, u_{p+2}, \dots, u_k].$$

For $p = 1$ this follows from the case $k = 2$ used on the mapping $D^{k-2} f$: we obtain

$$\big(D^2(D^{k-2} f)(0)[u_1,u_2]\big)[u_3, \dots, u_k] = \big(D^2(D^{k-2} f)(0)[u_2,u_1]\big)[u_3, \dots, u_k].$$

For $p \ge 2$ we use the fact that

$$\big(D(D^{k-1} f)(0)[u_1]\big)[u_2, \dots, u_k] = D\big(D^{k-1} f(\cdot)[u_2, \dots, u_k]\big)(0)[u_1] \tag{17}$$

and the inductive hypothesis. The formula (17) can be proved for example by using the evaluation operator $\varepsilon_{u_2,\dots,u_k}: \mathcal{L}({}^{k-1}X; Y) \to Y$, $\varepsilon_{u_2,\dots,u_k}(M) = M(u_2, \dots, u_k)$, which is a bounded linear operator, together with the Chain rule (Theorem 69) and Fact 67. □

Suppose that X, Y are normed linear spaces and $f\colon U \to Y$, where U is an open subset of X. If $W = a + Z$ is an affine subspace of X, where $a \in X$ and $Z \subset X$ is a subspace of X, then we will usually identify $f\restriction_{W\cap U}$ with the mapping $g\colon Z\cap(U-a) \to Y$, $g(x) = f(x+a)$. Using this identification we say that $f\restriction_{W\cap U}$ is C^k-smooth meaning actually that g is C^k-smooth, and similarly we write (formally incorrectly) $D^k(f\restriction_{W\cap U})(x)$ meaning $D^k g(x-a)$. (Note that under this identification $D^k(f\restriction_{W\cap U})(x)$ does not depend on the choice of $a \in W$.) The next statement is easy, but important.

Fact 77. *Let X, Y be normed linear spaces, $U \subset X$ an open set, $f \in C^k(U;Y)$ for some $k \in \mathbb{N}$, let $Z \subset X$ be a subspace, and $a \in X$. Denote $W = a + Z$. Then*

$$D^k(f\restriction_{W\cap U})(x) = \big(D^k f(x)\big)\restriction_{Z^k} \in \mathcal{L}^s({}^k Z;Y)$$

for every $x \in W \cap U$.

Fact 78. *Let X, Y, Z be normed linear spaces, $L \in \mathcal{L}(Y;Z)$, $U \subset X$ an open set, and $k \in \mathbb{N}$. Let $f : U \to Y$ be k-times Fréchet differentiable at $a \in U$. Then $D^k(L\circ f)(a) = L \circ D^k f(a)$.*

Proof. We use induction on k. For $k = 1$ we use Theorem 69 and Fact 67. For the induction step from $k-1$ to k we compute

$$\begin{aligned}
&\big\|D^{k-1}(L\circ f)(a+h) - D^{k-1}(L\circ f)(a) - \big(L\circ D^k f(a)\big)[h,\cdot,\dots,\cdot]\big\|\\
&\quad= \big\|L\circ D^{k-1}f(a+h) - L\circ D^{k-1}f(a) - \big(L\circ D^k f(a)\big)[h,\cdot,\dots,\cdot]\big\|\\
&\quad\le \|L\|\big\|D^{k-1}f(a+h) - D^{k-1}f(a) - D^k f(a)[h,\cdot,\dots,\cdot]\big\| = o(\|h\|),\ h\to 0.
\end{aligned}$$

□

Proposition 79. *Let $X_1,\dots,X_n$, and Y be normed linear spaces and further let $M \in \mathcal{L}(X_1,\dots,X_n;Y)$. Put $X = X_1\oplus\cdots\oplus X_n$. Then $M \in C^\infty(X;Y)$. For $k \le n$ and all $x = (x_1,\dots,x_n) \in X$ we have*

$$D^k M(x)[(h_1^1,\dots,h_n^1),\dots,(h_1^k,\dots,h_n^k)] = \sum_\pi M(z_1^\pi,\dots,z_n^\pi),$$

where the summation is over all one-to-one mappings $\pi\colon \{1,\dots,k\}\to\{1,\dots,n\}$ and where

$$z_j^\pi = \begin{cases} h_j^l & \text{if } j = \pi(l) \text{ for some } l \in \{1,\dots,k\},\\ x_j & \text{otherwise.}\end{cases}$$

For $k > n$ we have $D^k M(x) = 0$ for all $x \in X$.

Proof. Without loss of generality we may consider X with the maximum norm. Fix $x = (x_1, \dots, x_n) \in X$. For $h = (h_1, \dots, h_n) \in X$ we have by (1)

$$\begin{aligned}\Big\| M(x+h) - M(x) - \sum_{j=1}^{n} M(x_1, \dots, x_{j-1}, h_j, x_{j+1}, \dots, x_n) \Big\| \\ \le \sum_{\substack{A \subset \{1,\dots,n\} \\ |A| \ge 2}} \Big\| M\Big(\big(\chi_A(j)h_j + (1-\chi_A(j))x_j\big)_{j=1}^{n}\Big)\Big\| \\ \le \sum_{\substack{A \subset \{1,\dots,n\} \\ |A| \ge 2}} \|M\| \|h\|^{|A|} \|x\|^{n-|A|} = o(\|h\|),\ h \to 0.\end{aligned}$$

Since clearly $h \mapsto \sum_{j=1}^{n} M(x_1, \dots, x_{j-1}, h_j, x_{j+1}, \dots, x_n) \in \mathcal{L}(X;Y)$, it follows that M is Fréchet differentiable at x and the formula for the first derivative holds.

We proceed by induction. Let $1 \le k < n$ and suppose that M is k-times Fréchet differentiable on X and the formula for $D^k M$ holds. Fix a one-to-one mapping $\pi : \{1, \dots, k\} \to \{1, \dots, n\}$. Let $p = n - k$ and $j_1 < \dots < j_p$ be the enumeration of $\{1, \dots, n\} \setminus \pi(\{1, \dots, k\})$. Set $M_\pi(x_{j_1}, \dots, x_{j_p})[h^1, \dots, h^k] = M(z_1^\pi, \dots, z_n^\pi)$. Then $M_\pi(x_{j_1}, \dots, x_{j_p}) \in \mathcal{L}({}^k X; Y)$ and $M_\pi \in \mathcal{L}\big(X_{j_1}, \dots, X_{j_p}; \mathcal{L}({}^k X; Y)\big)$. Thus by the first step of the proof M_π is Fréchet differentiable on $X_{j_1} \oplus \dots \oplus X_{j_p}$. Now if $T_\pi : X \to X_{j_1} \oplus \dots \oplus X_{j_p}$ is the canonical projection, then $M_\pi \circ T_\pi$ is Fréchet differentiable on X by Theorem 69. As $D^k M = \sum_\pi M_\pi \circ T_\pi$, it follows that M is $(k+1)$-times differentiable on X. The formula for $D^{k+1} M$ follows from the formula for DM_π from the first step. When $k = n$ we thus obtain

$$D^n M(x)[(h_1^1, \dots, h_n^1), \dots, (h_1^n, \dots, h_n^n)] = \sum_{\pi \in S_n} M(h_{\pi(1)}^1, \dots, h_{\pi(n)}^n),$$

where S_n is the set of all permutations of $\{1, \dots, n\}$. It follows that $x \mapsto D^n M(x)$ is a constant mapping, and hence $D^{n+1} M(x) = 0$ for every $x \in X$. □

Next, we are going to introduce the generalised partial derivatives.

Definition 80. Let $X_1, \dots, X_n, Y$ be normed linear spaces, $X = X_1 \oplus \dots \oplus X_n$, $A \subset X$, and let $f : A \to Y$. We say that the mapping f is Fréchet differentiable at a point $a = (a_1, \dots, a_n) \in A$ with respect to the kth variable if the mapping $g : X_k \cap (A - a) \to Y$ defined as

$$g(y) = f(a_1, \dots, a_{k-1}, a_k + y, a_{k+1}, \dots, a_n)$$

is Fréchet differentiable at $0 \in X_k$. The partial Fréchet derivative will be denoted by $D_k f(a_1, \dots, a_n) = Dg(0) \in \mathcal{L}(X_k; Y)$.

Proposition 81. *Let $X_1, \dots, X_n$, Y be normed linear spaces, $X = X_1 \oplus \dots \oplus X_n$, $A \subset X$, and let $f : A \to Y$ be Fréchet differentiable at $a \in A$. Then $D_k f(a)$, $1 \le k \le n$, exist and*

$$Df(a)[h] = \sum_{k=1}^{n} D_k f(a)[h_k] \quad \textit{for each } h = (h_1, \dots, h_n) \in X.$$

Proof. The fact that $D_k f(a)$ exist is immediate and the formula follows from the fact that

$$D_k f(a)[h_k] = Df(a)[(0,\dots,0,h_k,0,\dots,0)], \quad h_k \in X_k,$$

which follows immediately from the definition. □

Theorem 82. *Let $X_1,\dots,X_n$, Y be normed linear spaces, $X = X_1 \oplus\cdots\oplus X_n$, $a \in X$, and suppose that $D_k f$, $k = 1,\dots,n$ exist in a neighbourhood of a and are continuous at a. Then f is Fréchet differentiable at a.*

Proof. Consider X with the maximum norm. Choose an arbitrary $\varepsilon > 0$. Let $\delta > 0$ be such that $\|D_k f(a+h) - D_k f(a)\| < \frac{\varepsilon}{n}$, $k = 1,\dots,n$, whenever $h \in X$, $\|h\| < \delta$. Let $P_k\colon X \to X$, $1 \le k \le n$, be the projection onto the first k "coordinates", i.e. $P_k(h) = (h_1,\dots,h_k,0,\dots,0)$ for $h = (h_1,\dots,h_n)$, and put $P_0(h) = 0$. Then

$$\begin{aligned} f(a+h) - f(a) - \sum_{k=1}^{n} D_k f(a)[h_k] \\ = \sum_{k=1}^{n} f\big(a + P_k(h)\big) - f\big(a + P_{k-1}(h)\big) - D_k f(a)[h_k] \end{aligned} \tag{18}$$

for all $h \in X$, $\|h\| < \delta$. Given $h \in X$, $\|h\| < \delta$, by Proposition 71 used on a mapping $g_k(y) = f\big(a + P_{k-1}(h) + y\big) - D_k f(a)[y]$ for $y \in X_k$, $\|y\| < \delta$, we obtain

$$\begin{aligned} \big\|f\big(a + P_k(h)\big) - f\big(a + P_{k-1}(h)\big) - D_k f(a)[h_k]\big\| &= \|g_k(h_k) - g_k(0)\| \\ &\le \sup_{\|y\|\le\|h_k\|} \big\|D_k f\big(a + P_{k-1}(h) + y\big) - D_k f(a)\big\| \, \|h_k\| \le \frac{\varepsilon}{n}\|h_k\|. \end{aligned}$$

Combining this with (18) gives

$$\left\| f(a+h) - f(a) - \sum_{k=1}^{n} D_k f(a)[h_k] \right\| \le \frac{\varepsilon}{n}\big(\|h_1\| + \cdots + \|h_n\|\big) \le \varepsilon\|h\|,$$

which proves the claim. □

Theorem 83. *Let X, $Y_1,\dots,Y_n$, Z be normed linear spaces, $Y = Y_1 \oplus \cdots \oplus Y_n$, $U \subset X$ open, and let $g_j\colon U \to Y_j$ be Fréchet differentiable at $a \in U$ for each $j \in \{1,\dots,n\}$. Suppose further that $\big(g_1(x),\dots,g_n(x)\big) \in V$ for all $x \in U$, where $V \subset Y$ is an open set, and $f\colon V \to Z$ is Fréchet differentiable at $\big(g_1(a),\dots,g_n(a)\big)$. Then $h(x) = f\big(g_1(x),\dots,g_n(x)\big)$ is Fréchet differentiable at a and*

$$Dh(a) = \sum_{j=1}^{n} D_j f\big(g_1(a),\dots,g_n(a)\big) \circ Dg_j(a). \tag{19}$$

Moreover, if f, g_j, $j = 1,\dots,n$, are C^k-smooth, $k \in \mathbb{N} \cup \{\infty\}$, then h is C^k-smooth as well.

Note that $Dg_j(a) \in \mathcal{L}(X;Y_j)$, $D_j f\big(g_1(a),\dots,g_n(a)\big) \in \mathcal{L}(Y_j;Z)$, and $\circ$ represents the usual composition of linear operators.

Proof. Let $\phi\colon U \to Y$ be defined as $\phi(x) = \big(g_1(x), \dots, g_n(x)\big)$. Then it is easily seen that $D\phi(a) = \big(Dg_1(a), \dots, Dg_n(a)\big)$. Since $h = f \circ \phi$, by the Chain rule (Theorem 69) $Dh(a) = Df\big(\phi(a)\big) \circ D\phi(a)$ from which the formula (19) follows using Proposition 81.

To show the C^k-smoothness case we use induction on k. The case $k = 1$ is easily proved using the formula (19) so we move on to the induction step from k to $k+1$, assuming that f, g_j, $j = 1, \dots, n$, are C^{k+1}-smooth. By (19), Dh is a finite sum of expressions of the type $D_j f\big(g_1(\cdot), \dots, g_n(\cdot)\big) \circ Dg_j(\cdot)$, where $D_j f\colon V \to \mathcal{L}(Y_j; Z)$, $Dg_j\colon U \to \mathcal{L}(X; Y_j)$ are C^k-smooth. By the inductive assumption, $D_j f\big(g_1(\cdot), \dots, g_n(\cdot)\big)\colon U \to \mathcal{L}(Y_j; Z)$ is C^k-smooth as well. The composition operation $\circ\colon \mathcal{L}(Y_j; Z) \times \mathcal{L}(X; Y_j) \to \mathcal{L}(X; Z)$ is bilinear and bounded, and so C^∞-smooth by Proposition 79. By the inductive hypothesis (applied to $\circ$), $D_j f\big(g_1(\cdot), \dots, g_n(\cdot)\big) \circ Dg_j(\cdot)\colon U \to \mathcal{L}(X; Z)$ is C^k-smooth, so Dh is C^k-smooth and consequently h is C^{k+1}-smooth as stated. □

The closed formulae for higher derivatives of compositions can be deduced from Corollary 117 and the Polarisation formula.

The following is an easy application of Theorem 83.

Corollary 84. *Let X, Y be normed linear spaces over $\mathbb{K}$, $U \subset X$ an open set, and $k \in \mathbb{N} \cup \{\infty\}$. Let $f \in C^k(U)$ and $g \in C^k(U; Y)$. Then $f \cdot g \in C^k(U; Y)$. If moreover $f \neq 0$ on U, then $\frac{1}{f} \in C^k(U)$.*

Proof. Define $h\colon \mathbb{K} \times Y \to Y$ by $h(p, q) = p \cdot q$. Then $f \cdot g = h(f, g)$ and since h is a bounded bilinear mapping, the result follows using Theorem 83 and Proposition 79. Analogously for $\frac{1}{f}$. □

Theorem 85. *Let X be a normed linear space, Y a Banach space, $U \subset X$ an open convex bounded set, $k \in \mathbb{N}$, and let $f_n\colon U \to Y$ be a sequence of k-times Fréchet differentiable mappings. If the sequence $\{D^k f_n\}_{n=1}^\infty$ converges uniformly on U and there are $x_0, \dots, x_{k-1} \in U$ such that $\{D^j f_n(x_j)\}_{n=1}^\infty$ converges for each $j \in \{0, \dots, k-1\}$, then there is a k-times Fréchet differentiable mapping $f\colon U \to Y$ such that $D^j f_n \to D^j f$ uniformly on U for each $j \in \{0, \dots, k\}$.*

Proof. It suffices to prove the theorem for $k = 1$, the general case then follows by a straightforward induction. So assume that $Df_n \to g$ uniformly on U and let $R = \operatorname{diam} U$. For a given $\varepsilon > 0$ there is $n_0 \in \mathbb{N}$ such that $\|f_n(x_0) - f_m(x_0)\| \le \frac{\varepsilon}{2}$ and $\|Df_n(y) - Df_m(y)\| \le \frac{\varepsilon}{2R}$ for each $y \in U$, whenever $m, n \ge n_0$. It follows from Proposition 71 that the mappings $f_n - f_m$ are $\frac{\varepsilon}{2R}$-Lipschitz on U. Thus

$$\begin{aligned}\|f_n(x) - f_m(x)\| &\le \|f_n(x) - f_m(x) - f_n(x_0) + f_m(x_0)\| + \|f_n(x_0) - f_m(x_0)\| \\ &\le \frac{\varepsilon}{2R}\|x - x_0\| + \frac{\varepsilon}{2} \le \varepsilon\end{aligned}$$

for every $x \in U$ and $m, n \ge n_0$. Hence there is $f\colon U \to Y$ such that $f_n \to f$ uniformly on U.

Further, fix $x \in U$, and let V be a punctured neighbourhood of 0 such that $x+V \subset U$. Define $\eta_n(h) = \frac{1}{\|h\|}\big(f_n(x+h) - f_n(x) - Df_n(x)[h]\big)$ for $h \in V$. Similarly as above we get

$$\|\eta_m(h) - \eta_n(h)\| \leq \frac{1}{\|h\|}\big\|(f_n - f_m)(x+h) - (f_n - f_m)(x)\big\| \\ + \frac{1}{\|h\|}\big\|\big(Df_n(x) - Df_m(x)\big)[h]\big\| \leq \frac{\varepsilon}{R}$$

for every $h \in V$ and $m, n \geq n_0$. Hence $\eta_n \to \eta$ uniformly on $h \in V$, where $\eta(h) = \frac{1}{\|h\|}\big(f(x+h) - f(x) - g(x)[h]\big)$. By the Moore-Osgood theorem we have $\lim_{h\to 0}\eta(h) = \lim_{n\to\infty}\lim_{h\to 0}\eta_n(h) = 0$, which implies that f is Fréchet differentiable at x and $Df(x) = g(x)$. □

In the next lemma the completeness is crucial.

Lemma 86. *Let X, Y be isomorphic Banach spaces and let $\mathcal{U} \subset \mathcal{L}(X;Y)$ be the set consisting of all isomorphisms. Then $\mathcal{U}$ is open and $\Phi\colon \mathcal{U} \to \mathcal{L}(Y;X)$, $\Phi(T) = T^{-1}$, is a C^∞-smooth mapping.*

Proof. Let $T \in \mathcal{U}$, $S \in \mathcal{L}(X;Y)$, $\|S\| < \|T^{-1}\|^{-1}$. Then

$$(T+S)^{-1} = \sum_{n=0}^{\infty}(-1)^n(T^{-1}S)^nT^{-1}. \tag{20}$$

Indeed, since $\|T^{-1}S\| < 1$, the series is absolutely convergent. We have

$$(T+S)\sum_{n=0}^{\infty}(-1)^n(T^{-1}S)^nT^{-1} \\ = Id + ST^{-1} + \sum_{n=1}^{\infty}(-1)^n\big(S(T^{-1}S)^{n-1}T^{-1} + S(T^{-1}S)^nT^{-1}\big) = Id.$$

A similar calculation works when composing the sum with $T+S$ the other way round, which shows that (20) holds. It follows that $\mathcal{U}$ is open.

If $\|S\| < \|T^{-1}\|^{-1}$, then we have

$$\left\|\sum_{n=2}^{\infty}(-1)^n(T^{-1}S)^nT^{-1}\right\| \leq \frac{\|T^{-1}\|^3\|S\|^2}{1 - \|T^{-1}\|\|S\|}.$$

Thus

$$(T+S)^{-1} - T^{-1} = -T^{-1}ST^{-1} + \sum_{n=2}^{\infty}(-1)^n(T^{-1}S)^nT^{-1} \\ = -T^{-1}ST^{-1} + o(\|S\|),\ S \to 0,$$

which implies that Φ is Fréchet differentiable at T and

$$D\Phi(T)[S] = -T^{-1}ST^{-1}, \quad D\Phi(T) \in \mathcal{L}\big(\mathcal{L}(X;Y); \mathcal{L}(Y;X)\big).$$

To show that the mapping Φ is C^∞-smooth consider the bounded bilinear mapping $\Psi\colon \mathcal{L}(Y;X)\times\mathcal{L}(Y;X)\to\mathcal{L}\big(\mathcal{L}(X;Y);\mathcal{L}(Y;X)\big)$, $\Psi(P,Q)(S)=-P\circ S\circ Q$. By Proposition 79, Ψ is C^∞-smooth. We have

$$D\Phi(T)=\Psi(T^{-1},T^{-1})=\Psi\big(\Phi(T),\Phi(T)\big).$$

This formula shows that Φ is C^1-smooth. Further, by Theorem 83 it is clear that if Φ is C^k-smooth, then it is also C^{k+1}-smooth, which proves the claim via induction on k. □

Theorem 87 (Implicit function theorem)**.** *Let X be a normed linear space, Y and Z Banach spaces, $G\subset X\oplus Y$ an open set, and $f\in C^1(G;Z)$. Let $(x_0,y_0)\in G$ and let $D_2f(x_0,y_0)$ be an isomorphism of Y onto Z. Then there are open neighbourhoods U of x_0 in X and V of y_0 in Y such that there is a unique mapping $u\colon U\to V$ satisfying $f\big(x,u(x)\big)=f(x_0,y_0)$ for all $x\in U$, this mapping u is C^1-smooth on U, and*

$$Du(x)=-\Big(D_2f\big(x,u(x)\big)\Big)^{-1}\circ D_1f\big(x,u(x)\big) \tag{21}$$

for all $x\in U$. If f is moreover C^k-smooth for some $k\in\mathbb{N}\cup\{\infty\}$, then $u\in C^k(U;Y)$.

Proof. We may assume without loss of generality that $f(x_0,y_0)=0$. We denote $S=D_2f(x_0,y_0)\in\mathcal{L}(Y;Z)$ and define $g\colon G\to Y$ by

$$g(x,y)=y-S^{-1}\circ f(x,y).$$

Notice that $f(x,y)=0$ if and only if $g(x,y)=y$, and that $D_2g(x_0,y_0)=0$.

Since f and g are C^1-smooth, there exist an open neighbourhood U of x_0 and $r>0$ such that $U\times B(y_0,r)\subset G$ and for $x\in U$, $y\in B(y_0,r)$ the following holds: $D_2f(x,y)$ is an isomorphism, $\|D_2g(x,y)\|\le\frac12$, and $\|g(x,y_0)-y_0\|<\frac r2$. Put $V=U(y_0,r)$. By Proposition 71, $y\mapsto g(x,y)$ is $\frac12$-Lipschitz on $B(y_0,r)$ for every $x\in U$. It follows that $\|g(x,y)-y_0\|\le\|g(x,y)-g(x,y_0)\|+\|g(x,y_0)-y_0\|<\frac12\|y-y_0\|+\frac r2\le r$ whenever $x\in U$, $y\in B(y_0,r)$. So, if $x\in U$, then $y\mapsto g(x,y)$ is a contraction of $B(y_0,r)$ into $V\subset B(y_0,r)$ and thus by the Banach contraction principle there is a unique mapping $u\colon U\to V$ such that $g\big(x,u(x)\big)=u(x)$, or equivalently $f\big(x,u(x)\big)=0$.

Further, the mapping u is continuous. Indeed, choose any $a\in U$ and $\varepsilon>0$. Since by the choice of U the mapping $D_2f\big(a,u(a)\big)$ is an isomorphism, by the just proved part of the theorem applied to the mapping f restricted to $U\times\big(U(u(a),\varepsilon)\cap V\big)$ and a point $\big(a,u(a)\big)$, there is a neighbourhood W of a such that $\big(x,u(x)\big)\in U\times U(u(a),\varepsilon)$ whenever $x\in W$. Note that here we substantially used the uniqueness of the mapping u.

Next we prove that u is Fréchet differentiable on U. For a fixed $x\in U$ choose $\delta>0$ such that $B(x,\delta)\subset U$. Let $h\colon B(0,\delta)\to Y$ be the continuous mapping $h(z)=u(x+z)-u(x)$. We have $f\big(x+z,u(x)+h(z)\big)=0$, so by Proposition 81

$$\begin{aligned}&\big\|D_1f\big(x,u(x)\big)[z]+D_2f\big(x,u(x)\big)[h(z)]\big\|\\&\quad=\big\|f\big(x+z,u(x)+h(z)\big)-f\big(x,u(x)\big)-Df\big(x,u(x)\big)\big[\big(z,h(z)\big)\big]\big\|\\&\quad=o\Big(\big\|\big(z,h(z)\big)\big\|\Big),\ z\to0.\end{aligned}$$

Recall that $T = D_2 f\big(x, u(x)\big)$ is an isomorphism. We deduce

$$\begin{aligned}
&\big\|u(x+z) - u(x) + T^{-1} \circ D_1 f\big(x, u(x)\big)[z]\big\| \\
&\quad = \big\|h(z) + T^{-1} \circ D_1 f\big(x, u(x)\big)[z]\big\| \\
&\quad \le \|T^{-1}\| \, \big\|D_2 f\big(x, u(x)\big)[h(z)] + D_1 f\big(x, u(x)\big)[z]\big\| \\
&\quad = o\big(\|z\| + \|h(z)\|\big), \; z \to 0.
\end{aligned} \tag{22}$$

In particular, for z sufficiently small we obtain $\big\|h(z) + T^{-1} \circ D_1 f\big(x, u(x)\big)[z]\big\| \le \frac{1}{2}\big(\|z\| + \|h(z)\|\big)$, so by the triangle inequality

$$\|h(z)\| \le \|z\| + 2\big\|T^{-1} \circ D_1 f\big(x, u(x)\big)[z]\big\| \le \Big(1 + 2\big\|T^{-1} \circ D_1 f\big(x, u(x)\big)\big\|\Big)\|z\|.$$

Plugging this back into (22) we finally obtain

$$\big\|u(x+z) - u(x) + T^{-1} \circ D_1 f\big(x, u(x)\big)[z]\big\| = o(\|z\|), \; z \to 0,$$

which proves that u is Fréchet differentiable at x and (21) holds.

The C^k-smoothness of the mapping u follows by a similar bootstrapping argument as in the proof of Lemma 86. Denote by Φ the mapping from Lemma 86 and by $\Psi\colon \mathcal{L}(Z;Y) \times \mathcal{L}(X;Z) \to \mathcal{L}(X;Y)$ the composition mapping $\Psi(P,Q) = P \circ Q$, and note that both Φ and Ψ are C^∞-smooth. Then the formula (21) can be rewritten as

$$Du(x) = -\Psi\Big(\Phi\Big(D_2 f\big(x, u(x)\big)\Big), D_1 f\big(x, u(x)\big)\Big).$$

Since u is continuous on U, this formula immediately implies that $u \in C^1(U;Y)$. Now suppose that f is C^k-smooth. If u is C^j-smooth on U for some $j < k$, then from Theorem 83 and the above formula it follows that Du is C^j-smooth on U and hence u is C^{j+1}-smooth on U. Thus $u \in C^k(U;Y)$ by induction. □

Definition 88. Let Y be a normed linear space over $\mathbb{K}$. Then every $M \in \mathcal{L}({}^k\mathbb{K};Y)$ is of the form $(t_1,\dots,t_k) \mapsto t_1 \cdots t_k \cdot M(1,\dots,1)$ and so there is a canonical isometry $\mathcal{L}({}^k\mathbb{K};Y) \cong Y$ given by $M \mapsto M(1,\dots,1)$. Thus if $f : U \to Y$, where $U \subset \mathbb{K}$, is k-times Fréchet differentiable at $a \in U$, then each derivative $D^j f(a)$, $j = 1,\dots,k$, can be identified with an element of Y. We will use the traditional notation $f', f'', f^{(j)}$ for this identification, i.e. $D^j f(a)[t_1,\dots,t_j] = t_1 \cdots t_j f^{(j)}(a)$, which corresponds to the classical notion of derivative in one dimension.

Later, we will use the fundamental theorem of calculus in the Banach space setting. The integral below is the Bochner integral.

Theorem 89. *Let Y be a real Banach space, $[a,b] \subset \mathbb{R}$ an interval, and $f : [a,b] \to Y$ a continuous mapping which is C^1-smooth on (a,b). Then*

$$f(b) - f(a) = \int_a^b f'(t)\,\mathrm{d}t$$

whenever the right-hand side is defined (in particular whenever f' can be continuously extended to $[a,b]$).

Proof. For any $\phi \in Y^*$ we have

$$\phi\big(f(b) - f(a)\big) = \phi \circ f(b) - \phi \circ f(a) = \int_a^b (\phi \circ f)'(t)\,\mathrm{d}t$$
$$= \int_a^b \phi\big(f'(t)\big)\,\mathrm{d}t = \phi\Big(\int_a^b f'(t)\,\mathrm{d}t\Big),$$

where we used the real valued fundamental theorem of calculus (see e.g. [RudinW, Theorem 7.21]) and the fact that the Bochner integral commutes with continuous linear operators. Since Y^* separates the points of Y, the assertion follows. □

We close this section with a theorem on differentiation under the integral sign. By a slight abuse of notation in what follows we will denote by $D_1 f(x,t)$ the derivative of the mapping $z \mapsto f(z,t)$ at x, although f will not be defined on an open subset of a product of two normed linear spaces (as in Definition 80).

Theorem 90. *Let (Ω, μ) be a measure space, X a normed linear space, Y a Banach space, $U \subset X$ an open connected set, and $f : U \times \Omega \to Y$. Suppose that $t \mapsto f(x,t)$ is (strongly) μ-measurable for each $x \in U$ and $x \mapsto f(x,t)$ is Fréchet differentiable on U for μ-almost all $t \in \Omega$. Further, suppose that there is $g \in L_1(\mu)$ such that for μ-almost all $t \in \Omega$ we have $\|D_1 f(x,t)\| \le g(t)$ for all $x \in U$, and that $t \mapsto f(x_0,t)$ is Bochner integrable for some $x_0 \in U$. Then $t \mapsto f(x,t)$ is Bochner integrable for all $x \in U$, the mapping $F : U \to Y$, $F(x) = \int_\Omega f(x,t)\,\mathrm{d}\mu(t)$ is Fréchet differentiable on U, and $DF(x) = \int_\Omega D_1 f(x,t)\,\mathrm{d}\mu(t)$ for all $x \in U$. If $x \mapsto f(x,t)$ is moreover C^1-smooth on U for almost all $t \in \Omega$, then so is F.*

Proof. Fix any $x \in U$. Since U is connected and open, there is a polygonal path $[x_0, x_1, \dots, x_n]$ in U, where $x_n = x$. For almost all $t \in \Omega$ the mapping $y \mapsto f(y,t)$ is $g(t)$-Lipschitz on each of the line segments $[x_{j-1}, x_j]$ (Proposition 71). Hence $\|f(x,t)\| \le \|f(x_0,t)\| + g(t)\sum_{j=1}^n \|x_j - x_{j-1}\|$ for almost all $t \in \Omega$. Since by the assumptions $\|f(x_0,\cdot)\| \in L_1(\mu)$ and $g \in L_1(\mu)$, the mapping $t \mapsto f(x,t)$ is Bochner integrable.

Further, find $\delta > 0$ such that $U(x,\delta) \subset U$. Note that the mapping $y \mapsto f(y,t)$ is $g(t)$-Lipschitz on $U(x,\delta)$ for almost all $t \in \Omega$ again by Proposition 71. Define $r : U(0,\delta) \times \Omega \to Y$ by $r(h,t) = f(x+h,t) - f(x,t) - D_1 f(x,t)[h]$ whenever $D_1 f(x,t)$ is defined, $r(h,t) = 0$ otherwise (this occurs for t in a set of measure zero). Then for almost all $t \in \Omega$ we have $r(h,t) = o(\|h\|)$, $h \to 0$ and also $\|r(h,t)\| \le 2g(t)\|h\|$. Using the fact that the evaluation operator $\varepsilon_h : \mathcal{L}(X;Y) \to Y$ defined by $\varepsilon_h(L) = L(h)$ is a bounded linear operator and hence commutes with the Bochner integration, we obtain

$$\frac{\big\|F(x+h) - F(x) - \big(\int_\Omega D_1 f(x,t)\,\mathrm{d}\mu(t)\big)[h]\big\|}{\|h\|}$$
$$= \frac{\big\|\int_\Omega \big(f(x+h,t) - f(x,t) - D_1 f(x,t)[h]\big)\,\mathrm{d}\mu(t)\big\|}{\|h\|} \le \int_\Omega \frac{\|r(h,t)\|}{\|h\|}\,\mathrm{d}\mu(t).$$

By the Lebesgue dominated convergence theorem the right-hand side tends to zero as $h \to 0$, which shows the Fréchet differentiability. The C^1-smoothness follows by another application of the Lebesgue dominated convergence theorem. □

Corollary 91. *Let μ be a finite Borel measure on a compact space K, X a normed linear space, Y a Banach space, $U \subset X$ open, $f : U \times K \to Y$, and $k \in \mathbb{N} \cup \{\infty\}$. Suppose that $t \mapsto f(x,t)$ is Bochner integrable for all $x \in U$. Put $u_t(x) = f(x,t)$. Suppose that $u_t \in C^k(U;Y)$ for all $t \in K$ and $(x,t) \mapsto D^j u_t(x)$ is locally bounded on $U \times K$ for all $1 \le j \le k$ (this holds in particular if $(x,t) \mapsto D^j u_t(x)$ are continuous). Then the mapping $F(x) = \int_K f(x,t)\,\mathrm{d}\mu(t)$ is C^k-smooth on U and $D^j F(x) = \int_K D^j u_t(x)\,\mathrm{d}\mu(t)$ for $1 \le j \le k$.*

Proof. Fix $x \in U$. By the compactness there is $\delta > 0$ and $M > 0$ such that $\|Du_t(y)\| \le M$ for all $y \in U(x,\delta)$ and $t \in K$. Thus we may apply Theorem 90 on $U(x,\delta) \times K$ for $g(t) = M$. The higher order smoothness follows by an easy induction. □

4. Taylor polynomial

Since higher derivatives are symmetric multilinear mappings, they are in one-to-one correspondence with homogeneous polynomials. So using differentials, i.e. homogeneous polynomials, as objects representing derivatives leads to no loss of information. Moreover, we benefit from the fact that the domain of the differentials considered as polynomials will remain the same underlying space. In fact, the most important method for applying higher smoothness is through using the Taylor formula, which locally approximates the mapping by a polynomial.

Definition 92. Let X, Y be normed linear spaces, $U \subset X$ open, $f : U \to Y$, $k \in \mathbb{N}$, and $a \in U$. Suppose that $D^k f(a)$ exists. The kth differential of f at a is defined by $d^k f(a) = \widehat{D^k f(a)}$, i.e. it is the k-homogeneous polynomial which corresponds to the kth derivative of f at a. For the evaluation at $h \in X$ we use the notation $d^k f(a)[h] = D^k f(a)[{}^k h]$. For convenience we denote $d^0 f = f$, $df = d^1 f$.

Lemma 93. *Let X, Y be normed linear spaces, $U \subset X$ open, $f : U \to Y$, $k \in \mathbb{N}$, and $a \in U$. Then $d^k f(a)$ exists if and only if $D(d^{k-1} f)(a)$ exists. In this case $d^k f(a)[h] = \big(D(d^{k-1} f)(a)[h]\big)[h]$ for every $h \in X$.*

Proof. Let us denote by $I : \mathcal{P}({}^{k-1}X;Y) \to \mathcal{L}^s({}^{k-1}X;Y)$ the canonical isomorphism $I(P) = \check{P}$ from the Polarisation formula (Proposition 13). Then $D^{k-1} f = I \circ d^{k-1} f$ and $d^{k-1} f = I^{-1} \circ D^{k-1} f$ and so the equivalence follows from the Chain rule (Theorem 69). Also, $D(D^{k-1} f)(a) = I \circ D(d^{k-1} f)(a)$ (Fact 67), and so

$$\begin{aligned} d^k f(a)[h] &= D^k f(a)[{}^k h] = \big(D(D^{k-1} f)(a)[h]\big)[{}^{k-1}h] \\ &= I\big(D(d^{k-1} f)(a)[h]\big)[{}^{k-1}h] = \big(D(d^{k-1} f)(a)[h]\big)[h]. \end{aligned}$$ □

Notice that by the Polarisation formula (Proposition 13) $f \in C^k(U;Y)$ if and only if the mappings $x \mapsto d^j f(x)$, $j = 1, \dots, k$, are continuous on U.

Before going further and proving the main results of this section we have to make a slight digression and define the Gâteaux and directional derivatives of higher order. Although these notions are not as important for the general theory as the Fréchet derivative, they are indispensable for the computation of concrete derivatives (see Fact 98).

Definition 94. Let X, Y be normed linear spaces, $U \subset X$ open, $f: U \to Y$, and $x \in U$. We denote the (first) Gâteaux derivative of f by $\delta^1 f = \delta f$ and define inductively the kth Gâteaux derivative $\delta^k f(x) \in \mathcal{L}^s({}^kX;Y)$ at x provided that $\delta^{k-1} f$ exists in some neighbourhood of x and the limit

$$\delta^k f(x)[h_1, \dots, h_k] = \lim_{\substack{t \to 0 \\ t \in \mathbb{K}}} \frac{\delta^{k-1} f(x + th_1)[h_2, \dots, h_k] - \delta^{k-1} f(x)[h_2, \dots, h_k]}{t}$$

exists for all $h_1, \dots, h_k \in X$ and is a symmetric bounded k-linear mapping.

We say that a mapping is G^k-smooth in the domain if it has the kth Gâteaux derivative at each point therein. We say that a mapping is G^∞-smooth if it is G^k-smooth for each $k \in \mathbb{N}$.

By induction it is easy to see that if a mapping f is k-times Fréchet differentiable at a point x, then it is also k-times Gâteaux differentiable at x and $\delta^k f(x) = D^k f(x)$. The converse is false in general.

Definition 95. Let X, Y be normed linear spaces, $A \subset X$, $f: A \to Y$, and $x \in A$. We denote the directional derivative of f in the direction $h \in X$ by $\frac{\partial^1 f}{\partial h} = \frac{\partial f}{\partial h}$ and we define inductively the directional derivatives of higher order by

$$\frac{\partial^k f}{\partial h_k \partial h_{k-1} \dots \partial h_1}(x) = \frac{\partial}{\partial h_k} \left(\frac{\partial^{k-1} f}{\partial h_{k-1} \dots \partial h_1} \right)(x) \in Y,$$

where $h_1, \dots, h_k \in X$.

The order of h_js in taking the directional derivatives in general matters, but under various assumptions h_js can be reordered not affecting the final value. In particular, this is the case whenever the mappings $y \mapsto \frac{\partial^k f}{\partial h_{\pi(k)} \dots \partial h_{\pi(1)}}(y)$ are continuous on a neighbourhood of x for every permutation π of $\{1, \dots, k\}$, or whenever the mappings $y \mapsto \frac{\partial^k f}{\partial h_{\sigma(k)} \dots \partial h_{\sigma(1)}}(y)$ are continuous at x for all $\sigma \in \{1, \dots, k\}^k$. This follows from the classical theorems on differentiation in finite dimensions; the former fact follows by induction from the theorem on interchangeability of partial derivatives of the second order, the latter fact follows from the next theorem combined with Theorem 76 and Fact 98 below.

Theorem 96. *Let $U \subset \mathbb{R}^n$ be open, Y a normed linear space, $f: U \to Y$, and suppose that all the partial derivatives of f of order $k \in \mathbb{N}$ exist on U and are continuous at $x \in U$. Then f is k-times Fréchet differentiable at x.*

Sketch of proof. By shrinking the U if necessary we may assume that all the partial derivatives of order k are bounded on U. Without loss of generality we may assume that U is convex and hence by Proposition 71 the partial derivatives of order up to $k-1$ are coordinatewise Lipschitz on U and thus continuous on U. The statement now follows by using inductively Theorem 82 and Fact 98 below. □

The following fact follows from the definition using induction.

Fact 97. *Let X, Y be normed linear spaces over $\mathbb{K}$, $A \subset X$, $f\colon A \to Y$, $x \in A$, $h \in X$, and $k \in \mathbb{N}$. Then $\frac{\partial^k f}{\partial h \ldots \partial h}(x)$ exists if and only if the mapping $g(t) = f(x+th)$ (defined for $t \in \mathbb{K}$ satisfying $x + th \in A$) is k-times differentiable at 0. In this case $\frac{\partial^k f}{\partial h \ldots \partial h}(x) = g^{(k)}(0) = (g\restriction_{\mathbb{R}})^{(k)}(0)$.*

The following instrumental fact also follows by an easy induction.

Fact 98. *Let X, Y be normed linear spaces, $U \subset X$ open, $f\colon U \to Y$, $x \in U$, and suppose that $\delta^k f(x)$ exists. Then*

$$\delta^k f(x)[h_1, \ldots, h_k] = \frac{\partial^k f}{\partial h_1 \partial h_2 \ldots \partial h_k}(x).$$

This fact may be used in order to obtain the values of derivatives of k-times differentiable mappings. For example we can easily compute the derivatives of a polynomial.

Lemma 99. *Let X and Y be normed linear spaces and let $P \in \mathcal{P}(^nX;Y)$. Then $P \in C^\infty(X;Y)$,*

$$\frac{\partial P}{\partial h}(x) = n\check{P}(^{n-1}x, h), \quad \textit{and}$$
$$DP(x)[h] = n\check{P}(^{n-1}x, h)$$

for every $x, h \in X$. More generally, for $k \leq n$ we have

$$D^kP(x)[h_1, \ldots, h_k] = n(n-1)\cdots(n-k+1)\check{P}(^{n-k}x, h_1, \ldots, h_k),$$
$$d^kP(x)[h] = n(n-1)\cdots(n-k+1)\check{P}(^{n-k}x, {}^kh).$$

Thus $D^kP\colon X \to \mathcal{L}^s(^kX;Y)$ is a polynomial mapping from $\mathcal{P}\big(^{n-k}X; \mathcal{L}^s(^kX;Y)\big)$, while $d^kP\colon X \to \mathcal{P}(^kX;Y)$ is a polynomial mapping from $\mathcal{P}\big(^{n-k}X; \mathcal{P}(^kX;Y)\big)$. In particular $x \mapsto d^nP(x)$ is a constant mapping whose value is a polynomial $n!P$, and so $d^{n+1}P = 0$.

Proof. The polynomial P is C^∞-smooth by Proposition 79 and Theorem 83. By Lemma 21 we have

$$\frac{1}{t}\big(P(x+th) - P(x)\big) = \frac{1}{t}\left(\sum_{j=0}^{n-1}\binom{n}{j}t^{n-j}\check{P}(^jx, {}^{n-j}h)\right).$$

So limiting for $t \to 0$ we immediately obtain the first order formulae. Note that in the first formula for a fixed direction h the resulting mapping $x \mapsto \frac{\partial P}{\partial h}(x)$ belongs to $\mathcal{P}(^{n-1}X;Y)$, whereas in the second formula $DP \in \mathcal{P}\big(^{n-1}X; \mathcal{L}(X;Y)\big)$. The higher order formulae follow by a similar argument using induction and Fact 98. □

By an easy induction we obtain the following lemma.

Lemma 100. *Let X, Y be normed linear spaces, $U \subset X$, $f: U \to Y$, $x \in U$, and $h_1, \dots, h_k \in X$. Suppose that $\frac{\partial^k f}{\partial h_1 \partial h_2 \dots \partial h_k}(x)$ exists. Then*

$$\frac{\partial^k f}{\partial h_1 \partial h_2 \dots \partial h_k}(x) = \lim_{t_1 \to 0}\left(\lim_{t_2 \to 0}\left(\dots\left(\lim_{t_k \to 0} \frac{1}{t_1 \cdots t_k} \Delta^k f(x; t_1 h_1, \dots, t_k h_k)\right)\right)\right).$$

A version of the following theorem was proved by Jacek Bochnak in [Boc] under the additional assumption that the finite-dimensional restrictions are C^k-smooth.

Theorem 101. *Let X be a Banach space, Y a normed linear space, $U \subset X$ open, $f: U \to Y$, and let $k \in \mathbb{N}$ be such that $\dim X \geq k+1$. Suppose further that $\phi \circ f$ is Baire measurable for every $\phi \in Y^*$. Then f is G^k-smooth on U if and only if for every $(k+1)$-dimensional affine subspace $E \subset X$ the mapping $f \restriction_{E \cap U}$ is G^k-smooth. In this case*

$$\delta^k f(x) \restriction_{F^k} = \delta^k(f \restriction_{(x+F) \cap U})(x) \tag{23}$$

for every $x \in U$ and every subspace $F \subset X$ satisfying $\dim F \leq k+1$.

Proof. $\Rightarrow$ is clear.

$\Leftarrow$ We prove this implication by induction on k. Because the proof of the case $k = 1$ is almost identical to the induction step, we show only the induction step from $k-1$ to k. Fix $x \in U$ and let $M: X^k \to Y$ be defined as

$$M(h_1, \dots, h_k) = \delta^k(f \restriction_{(x+E) \cap U})(x)[h_1, \dots, h_k], \quad \text{where } E = \operatorname{span}\{h_j\}_{j=1}^k.$$

It is easy to check that $M \in L^s({}^k X; Y)$ using an analogue of Fact 77 for Gâteaux derivatives.

Next, we show that M is continuous. Let $\{t_n\}$ be a decreasing sequence of reals such that $t_n \to 0$ and $B(x, k t_1) \subset U$, and let

$$f_{n_1, \dots, n_k}(h_1, \dots, h_k) = \frac{1}{t_{n_1} \cdots t_{n_k}} \Delta^k f(x; t_{n_1} h_1, \dots, t_{n_k} h_k), \quad h_j \in B_X.$$

From Lemma 100 we get

$$M(h_1, \dots, h_k) = \lim_{n_1 \to \infty}\left(\lim_{n_2 \to \infty}\left(\dots\left(\lim_{n_k \to \infty} f_{n_1, \dots, n_k}(h_1, \dots, h_k)\right)\right)\right)$$

whenever $h_j \in B_X$. Choose any $\phi \in Y^*$. The functions $\phi \circ f_{n_1, \dots, n_k}: B_X^k \to \mathbb{K}$ are separately Baire measurable (i.e. Baire measurable in each variable) by Lemma 52. Thus $\phi \circ M \restriction_{B_X^k}$ is separately Baire measurable. Similarly we obtain that $\phi \circ M \restriction_{n B_X^k}$ is separately Baire measurable for any $n \in \mathbb{N}$ and consequently $\phi \circ M$ is separately Baire measurable. By Theorem 7 we can conclude that M is continuous.

Now choose arbitrary $h_1, \dots, h_k \in X$ and put $E = \operatorname{span}\{h_j\}_{j=1}^k$. Then by the inductive hypothesis $\delta^{k-1} f(y)[h_2, \dots, h_k] = \delta^{k-1}(f \restriction_{(x+E) \cap U})(y)[h_2, \dots, h_k]$ for every $y \in (x+E) \cap U$. Hence

$$\frac{\partial}{\partial h_1}(\delta^{k-1} f(\cdot)[h_2, \dots, h_k])(x) = \delta^k(f \restriction_{(x+E) \cap U})(x)[h_1, \dots, h_k] = M(h_1, \dots, h_k),$$

which shows that f is k-times Gâteaux differentiable at x and $\delta^k f(x) = M$. The formula (23) follows again from an analogue of Fact 77 for Gâteaux derivatives. □

We note that the assumption on the dimension of the affine subspaces in the previous theorem is in general necessary: the function $f(x,y) = \frac{x^2 y}{x^2+y^2}$, $f(0,0) = 0$, is C^∞-smooth on every line in $\mathbb{R}^2$, but it is not Gâteaux differentiable at the origin. However, the complex case is dramatically different, see Theorem 160. Similarly, the measurability assumption is not superfluous, as can be seen by considering any discontinuous linear functional.

Corollary 102. *Let X be a Banach space, Y a normed linear space, $U \subset X$ open, and $f\colon U \to Y$. Suppose further that $\phi \circ f$ is Baire measurable for every $\phi \in Y^*$. Then f is G^∞-smooth on U if and only if for every finite-dimensional subspace $Z \subset X$ the mapping $f\restriction_{Z\cap U}$ is G^∞-smooth.*

Proof. If X is finite-dimensional, then the statement is obvious. Otherwise just apply Theorem 101. □

Armed with the computational tools we can return to our main theme. The following result is essential for the theory.

Theorem 103 (Peano's form of Taylor's theorem). *Let X, Y be normed linear spaces, $U \subset X$ an open set, $f\colon U \to Y$, $a \in U$, $k \in \mathbb{N}$, and suppose that $D^k f(a)$ exists. Then*

$$\left\| f(x) - \sum_{j=0}^{k} \frac{1}{j!} d^j f(a)[x-a] \right\| = o(\|x-a\|^k),\ x \to a.$$

Proof. We use induction on k. For $k = 1$ the statement follows from the definition of the Fréchet derivative. Now suppose that $k > 1$ and the theorem holds for $k-1$. Put $R(x) = f(x) - \sum_{j=0}^{k} \frac{1}{j!} d^j f(a)[x-a]$. Then R is differentiable on a neighbourhood of a with DR continuous at a, so there is an open ball $V = U(a,\delta)$, $\delta > 0$, such that for every $x \in V$

$$\|R(x)\| = \|R(x) - R(a)\| \le \|x-a\| \sup_{y \in B(a,\|x-a\|)} \|DR(y)\|$$

by Proposition 71. To compute DR denote $P_j(y) = \frac{1}{j!} d^j f(a)[y]$. Then by Lemma 99

$$DP_j(y)[h] = \frac{1}{(j-1)!} D^j f(a)[{}^{j-1}y, h] = \frac{1}{(j-1)!} (D^{j-1}(Df)(a)[{}^{j-1}y])[h]$$

for every $y, h \in X$, and therefore $DP_j(y) = \frac{1}{(j-1)!} d^{j-1}(Df)(a)[y] \in \mathcal{L}(X;Y)$, $j = 1,\dots,k$. It follows that for every $y \in V$

$$DR(y) = Df(y) - \sum_{j=0}^{k-1} \frac{1}{j!} d^j (Df)(a)[y-a].$$

Using the inductive hypothesis on the mapping Df we conclude that $\|DR(y)\| = o(\|y-a\|^{k-1})$, $y \to a$. Therefore

$$\|R(x)\| \le \|x-a\| \sup_{y\in B(a,\|x-a\|)} \|DR(y)\| = o(\|x-a\|^k),\ x \to a. \qquad \square$$

The polynomial $\sum_{j=0}^{k} \frac{1}{j!} d^j f(a) \in \mathcal{P}^k(X;Y)$ is called the Taylor polynomial of order k of the mapping f at a.

Proposition 104. *Let X, Y be normed linear spaces, $U \subset X$ open, $f: U \to Y$, and $a \in U$. Suppose that there are $P_j \in P({}^jX;Y)$, $j = 0,\dots,k$, such that*

$$\left\| f(x) - \sum_{j=0}^{k} P_j(x-a) \right\| = o(\|x-a\|^k),\ x \to a.$$

Then the polynomials P_j are uniquely determined. In particular, if $D^k f(a)$ exists, then $P_j = \frac{1}{j!} d^j f(a)$, $j = 0,\dots,k$.

Proof. Suppose that there are polynomials $Q_j \in P({}^jX;Y)$, $j = 0,\dots,k$, such that $\|f(x) - \sum_{j=0}^{k} Q_j(x-a)\| = o(\|x-a\|^k)$, $x \to a$. Then

$$\left\| \sum_{j=0}^{k} (Q_j - P_j)(x-a) \right\| = o(\|x-a\|^k),\ x \to a.$$

Since the estimated mapping is a polynomial of degree at most k, it must be identically zero (Fact 48), and so by Corollary 32 we get $Q_j = P_j$. The final statement follows from Theorem 103. $\square$

The above proposition says that the Taylor polynomial of order k of a C^k-smooth mapping is uniquely determined among all polynomials of degree at most k by the property of giving the (best) estimate of order $o(\|x-a\|^k)$. This fact can be used conveniently for obtaining various formulae for higher derivatives, see e.g. Corollary 116, Corollary 117.

Next we show that if f is C^k-smooth, then the remainder in Taylor's theorem can be given by an integral formula.

Lemma 105. *Let $I \subset \mathbb{R}$ be an open interval, let X, Y, Z be real normed linear spaces, $k \in \mathbb{N}$, $f \in C^k(I;X)$, $g \in C^k(I;Y)$, and $\phi \in \mathcal{L}(X,Y;Z)$. Then*

$$\left(\sum_{j=0}^{k-1} (-1)^j \phi(f^{(j)}, g^{(k-j-1)}) \right)' = \phi(f, g^{(k)}) - (-1)^k \phi(f^{(k)}, g) \quad \text{on } I. \tag{24}$$

Proof. Using Theorem 83 and Fact 67 we get $\big(\phi(f_0, g_0)\big)' = \phi(f_0', g_0) + \phi(f_0, g_0')$ on I for any $f_0 \in C^1(I;X)$, $g_0 \in C^1(I;Y)$. Applying this formula to compute the derivative on the left-hand side of (24), the sum telescopes to the right-hand side of (24). $\square$

Proposition 106. *Let $I \subset \mathbb{R}$ be an open interval, Y a real Banach space, $k \in \mathbb{N}$, and $f \in C^k(I;Y)$. Then for any pair of points $a, x \in I$ we have*

$$f(x) = \sum_{j=0}^{k-1} \frac{(x-a)^j}{j!} f^{(j)}(a) + \int_a^x \frac{(x-t)^{k-1}}{(k-1)!} f^{(k)}(t)\,\mathrm{d}t.$$

Proof. By applying Lemma 105 to the bilinear mapping $\phi\colon Y \times \mathbb{R} \to Y$ defined by $\phi(p,q) = q \cdot p$, the function $g(t) = \frac{(x-t)^{k-1}}{(k-1)!}$, and the mapping f, using $g^{(k-j-1)}(t) = (-1)^{k-j-1}\frac{(x-t)^j}{j!}$ we obtain

$$\left(\sum_{j=0}^{k-1}(-1)^{k-1}\frac{(x-t)^j}{j!} f^{(j)}(t)\right)' = (-1)^{k+1}\frac{(x-t)^{k-1}}{(k-1)!} f^{(k)}(t), \quad t \in I.$$

Integrating both sides in t between a and x in the Bochner sense and using Theorem 89 we obtain the desired formula. □

Theorem 107 (Taylor formula)**.** *Let X be a normed linear space, Y a Banach space, $U \subset X$ an open convex set, $k \in \mathbb{N}$, and $f \in C^k(U;Y)$. Then for any $x \in U$ and $h \in X$ satisfying $x + h \in U$ we have*

$$f(x+h) = \sum_{j=0}^{k-1} \frac{1}{j!} d^j f(x)[h] + \left(\int_0^1 \frac{(1-t)^{k-1}}{(k-1)!} d^k f(x+th)\,\mathrm{d}t\right)[h].$$

Proof. We apply Proposition 106 to the mapping $g\colon (-\delta, 1+\delta) \to Y_{\mathbb{R}}$, where $Y_{\mathbb{R}}$ is the real version of the space Y in case that $\mathbb{K} = \mathbb{C}$, defined for a sufficiently small $\delta > 0$ by the formula $g(t) = f(x+th)$. Using Fact 98 and Fact 97 we obtain $g^{(j)}(t) = D^j f(x+th)[^j h] = d^j f(x+th)[h]$. Thus

$$f(x+h) = \sum_{j=0}^{k-1} \frac{1}{j!} d^j f(x)[h] + \int_0^1 \frac{(1-t)^{k-1}}{(k-1)!} d^k f(x+th)[h]\,\mathrm{d}t.$$

To finish, consider the evaluation operator $\varepsilon_h\colon \mathcal{P}(^kX;Y) \to Y$, $\varepsilon_h(P) = P(h)$. Then obviously $\varepsilon_h \in \mathcal{L}(\mathcal{P}(^kX;Y);Y)$ and hence the evaluation commutes with Bochner integration. □

Corollary 108. *Let X, Y be normed linear spaces, $U \subset X$ an open convex set, $k \in \mathbb{N}$, and $f \in C^k(U;Y)$. Then for any $x \in U$ and $h \in X$ satisfying $x + h \in U$ we have*

$$\left\| f(x+h) - \sum_{j=0}^{k} \frac{1}{j!} d^j f(x)[h] \right\| \le \frac{1}{k!}\left(\sup_{t\in[0,1]} \left\| d^k f(x+th) - d^k f(x)\right\|\right) \cdot \|h\|^k.$$

Proof. Working with the completion of Y if necessary we may use Theorem 107 to obtain

$$\left\| f(x+h) - \sum_{j=0}^{k} \frac{1}{j!} d^j f(x)[h] \right\| \leq \left\| \int_0^1 \frac{(1-t)^{k-1}}{(k-1)!} d^k f(x+th)\,\mathrm{d}t - \frac{1}{k!} d^k f(x) \right\| \cdot \|h\|^k \leq \left(\int_0^1 \frac{(1-t)^{k-1}}{(k-1)!} \left\| d^k f(x+th) - d^k f(x) \right\| \mathrm{d}t \right) \cdot \|h\|^k,$$

from which the estimate follows. □

Let us spell out the connections between our abstract formalism and the classical theory of partial derivatives. In case that the mapping f is defined on a subset of $\mathbb{K}^n$, i.e. $f(x) = f(x_1, \dots, x_n) = f\left(\sum_{j=1}^n x_j e_j\right)$, we will use the classical notation for the operator of partial derivatives (derivatives along the canonical basis vectors e_j)

$$\frac{\partial^{|\alpha|} f}{\partial x_1^{\alpha_1} \dots \partial x_n^{\alpha_n}}(x_0).$$

Let $U \subset \mathbb{K}^n$ be an open set and $f : U \to Y$, where Y is a normed linear space over $\mathbb{K}$. If f is m-times Fréchet differentiable at $a \in U$, then combining Proposition 23 and Fact 98 we obtain the classical equality

$$d^k f(a)[x] = \sum_{\alpha \in \mathcal{J}(n,k)} \binom{k}{\alpha} \frac{\partial^k f}{\partial x_1^{\alpha_1} \dots \partial x_n^{\alpha_n}}(a) x^\alpha$$

for $k \leq m$. Combining this with Theorem 103 we obtain the classical multivariate Taylor formula

$$f(a+x) = \sum_{k=0}^{m} \sum_{\alpha \in \mathcal{J}(n,k)} \frac{1}{\alpha_1! \cdots \alpha_n!} \frac{\partial^k f}{\partial x_1^{\alpha_1} \dots \partial x_n^{\alpha_n}}(a) x^\alpha + o(\|x\|^m),\ x \to 0.$$

Definition 109. Let X, Y be normed linear spaces, $U \subset X$ open, $f : U \to Y$, and $p \in [0, +\infty)$. We say that f is T^p-smooth at $x \in U$ if there exists a polynomial $P \in \mathcal{P}^{[p]}(X; Y)$ satisfying $P(0) = f(x)$ and

$$f(x+h) - P(h) = o(\|h\|^p),\ h \to 0. \tag{25}$$

We say that f is T^p-smooth on U if it is T^p-smooth at every point $x \in U$.

Notice a few things. First, the polynomial in (25) is uniquely determined by Proposition 104. Next, it is easy to see that T^p-smoothness implies T^q-smoothness for $q < p$. Finally, if f is T^1-smooth at x, then f is Fréchet differentiable at x with $Df(x) = P_1$, the 1-homogeneous term of P.

Theorem 103 implies that a C^k-smooth mapping is also T^k-smooth. The converse is not true in general: consider $f : \mathbb{R} \to \mathbb{R}$, $f(x) = x^{k+1} \sin \frac{1}{x^k}$, $f(0) = 0$. Then f is T^k-smooth but not even C^1-smooth. Nevertheless, under certain uniformity assumptions the converse does hold.

Theorem 110 (Converse Taylor theorem). *Let X, Y be normed linear spaces, $U \subset X$ an open set, $f : U \to Y$, and $k \in \mathbb{N}_0$. Then $f \in C^k(U; Y)$ if and only if f is a T^k-smooth mapping satisfying*

$$\lim_{\substack{(y,h)\to(x,0)\\ h\neq 0}} \frac{\|R(y,h)\|}{\|h\|^k} = 0 \tag{26}$$

for every $x \in U$, where $R(x,h) = f(x+h) - P^x(h)$ and where the polynomials $P^x \in \mathcal{P}^k(X; Y)$ come from the definition of T^k-smoothness of f at x. In this case $P^x = \sum_{j=0}^k \frac{1}{j!} d^j f(x)$.

Proof. $\Rightarrow$ If $f \in C^k(U; Y)$, then f is T^k-smooth, and $P^x = \sum_{j=0}^k \frac{1}{j!} d^j f(x)$ by Proposition 104. Fix $x \in U$ and choose any $\varepsilon > 0$. Let $\delta > 0$ be such that $B(x, 2\delta) \subset U$ and $\|d^k f(z) - d^k f(x)\| < \varepsilon$ for $z \in B(x, 2\delta)$. By Corollary 108

$$\|R(y,h)\| \le \frac{1}{k!}\left(\sup_{z\in B(y,\delta)} \left\|d^k f(z) - d^k f(y)\right\|\right) \cdot \|h\|^k \le \frac{2\varepsilon}{k!}\|h\|^k$$

whenever $y \in B(x, \delta)$ and $h \in B(0, \delta)$, from which (26) follows.

$\Leftarrow$ We use induction on k. For $k = 0$ the assertion is obvious, since both T^0-smoothness and C^0-smoothness mean just the continuity of f. So assume that $k \in \mathbb{N}$ and the theorem holds for $k - 1$.

Fix $x \in U$ and let $\delta > 0$ be such that $U(x, 2\delta) \subset U$. We have

$$\begin{aligned} f(x+h+y) &= P^x(h+y) + R(x, h+y), \\ f(x+h+y) &= P^{x+h}(y) + R(x+h, y) \end{aligned} \tag{27}$$

for all $h, y \in U(0, \delta)$. Set $q(h, y) = P^{x+h}(y) - P^x(h+y)$. Denote by P_j^z the j-homogeneous summands of P^z for $j = 0, \dots, k$. By Lemma 21 we can write $q(h, y) = \sum_{j=0}^k q_j(h, y)$, where

$$q_j(h, y) = P_j^{x+h}(y) - \sum_{l=j}^{k} \binom{l}{j} \widetilde{P_l^x}({}^{l-j}h, {}^j y).$$

Note that $q(h, \cdot) \in \mathcal{P}^k(X; Y)$ and $q_j(h, \cdot) \in \mathcal{P}({}^j X; Y)$, $j = 0, \dots, k$, for $h \in U(0, \delta)$. By subtracting the equalities (27) we obtain $q(h, y) = R(x, h+y) - R(x+h, y)$. Thus for any $h, y \in U(0, \delta)$ such that $\|y\| \le \|h\|$, $y \neq 0$, and $y \neq -h$ we obtain

$$\begin{aligned} \|q(h,y)\| &\le \|R(x, h+y)\| + \|R(x+h, y)\| \\ &\le \left(2^k \frac{\|R(x, h+y)\|}{\|h+y\|^k} + \frac{\|R(x+h, y)\|}{\|y\|^k}\right) \|h\|^k. \end{aligned}$$

It follows (using also simpler versions of the above estimate if $y = 0$ or $y = -h$) that

$$\|q(h, y)\| = o(\|h\|^k), \ (h, y) \to (0, 0), \|y\| \le \|h\|.$$

Recall that by Corollary 31 there are constants $K_{n,j}$ such that

$$\|Q_j(x)\| \le K_{n,j} \max_{0\le l\le n} \left\|Q\left(\tfrac{l}{n}x\right)\right\|$$

whenever $Q \in \mathcal{P}^n(X; Y)$, $x \in X$, and $j \in \{0, \dots, n\}$, where Q_j is the j-homogeneous summand of Q. Applying this we get $\|q_j(h, y)\| \le K_{k,j} \max_{0\le l\le k} \|q(h, \frac{l}{k}y)\|$ for all $h \in U(0, \delta)$, $y \in Y$, and $j \in \{0, \dots, k\}$. Therefore

$$\|q_j(h, y)\| = o(\|h\|^k), \ (h, y) \to (0, 0), \|y\| \le \|h\|.$$

So finally by taking the supremum over $y \in B(0, \|h\|)$ and using the j-homogeneity of $q_j(h, \cdot)$ we obtain

$$\begin{aligned} \|q_j(h, \cdot)\| &= \frac{1}{\|h\|^j} \sup_{\|y\|\le\|h\|} \|q_j(h, y)\| = \sup_{\|y\|\le\|h\|} \frac{\|q_j(h, y)\|}{\|h\|^j} \\ &= o(\|h\|^{k-j}), \ h \to 0 \end{aligned} \tag{28}$$

for each $j \in \{0, \dots, k\}$.

Since $q_k(h, \cdot) = P_k^{x+h} - P_k^x$, it follows that the mapping $x \mapsto P_k^x$ is continuous on U. Further, since for $h \ne 0$

$$\begin{aligned} \frac{\left\|f(y + h) - \sum_{j=0}^{k-1} P_j^y(h)\right\|}{\|h\|^{k-1}} &\le \frac{\|R(y, h)\| + \|P_k^y(h)\|}{\|h\|^{k-1}} \\ &\le \left(\frac{\|R(y, h)\|}{\|h\|^k} + \|P_k^y\|\right) \|h\|, \end{aligned}$$

the continuity of $x \mapsto P_k^x$ implies $\lim\limits_{\substack{(y,h)\to(x,0)\\ h\ne0}} \frac{\|f(y+h)-\sum_{j=0}^{k-1} P_j^y(h)\|}{\|h\|^{k-1}} = 0$ and so by the inductive hypothesis f is C^{k-1}-smooth and $P_j^x = \frac{1}{j!}d^j f(x)$, $j = 0, \dots, k-1$. Thus

$$q_{k-1}(h, y) = \frac{1}{(k-1)!}(d^{k-1}f(x + h)[y] - d^{k-1}f(x)[y]) - k\widetilde{P_k^x}(h, {}^{k-1}y)$$

and from (28) we get

$$\left\|d^{k-1}f(x + h) - d^{k-1}f(x) - k!\widetilde{P_k^x}(h, \cdot, \dots, \cdot)\right\| = o(\|h\|), \ h \to 0.$$

By Lemma 93 this means that $d^k f(x) = k! P_k^x$, which finishes the proof. □

Definition 111. Let X and Y be normed linear spaces, $A \subset X$, $f\colon A \to Y$, and $p \in [0, +\infty)$. We say that f is weakly T^p-smooth at $x \in A$ if there exists a polynomial $P \in \mathcal{P}^{[p]}(X; Y)$ satisfying $P(0) = f(x)$ and $f(x + th) - P(th) = o(|t|^p)$, $t \to 0$ for each $h \in X$. We say that f is weakly T^p-smooth on A if it is weakly T^p-smooth at every $x \in A$.

We say that f is directionally T^p-smooth at $x \in A$ if the mapping $t \mapsto f(x+th)$ is T^p-smooth at 0 for each $h \in X$, i.e. for each $h \in X$ there is $P^{x,h} \in \mathcal{P}^{[p]}(\mathbb{K};Y)$ satisfying $P^{x,h}(0) = f(x)$ and $f(x+th) - P^{x,h}(t) = o(|t|^p)$, $t \to 0$. Finally, we say that f is directionally T^p-smooth on A if it is directionally T^p-smooth at every $x \in A$.

Note that again all the polynomials in the above definitions are uniquely determined by Proposition 104. In particular, if the mapping f is directionally T^p-smooth at x, then $P^{x,h}(t) = P^{x,th}(1)$ for every $h \in X$ and $t \in \mathbb{R}$. Obviously, if f is T^p-smooth at x, then it is weakly T^p-smooth at x, and if f is weakly T^p-smooth at x, then it is directionally T^p-smooth at x. Also, f is weakly T^1-smooth at x if and only if f is Gâteaux differentiable at x.

Proposition 112. *Let X, Y be normed linear spaces, $U \subset X$ an open set, $f: U \to Y$, and $k \in \mathbb{N}$. If f is G^k-smooth at $x \in U$, then f is weakly T^k-smooth at x with $P = \sum_{j=0}^{k} \frac{1}{j!}\widehat{\delta^j f(x)}$.*

Proof. Follows from Fact 98 and Fact 97 combined with Theorem 103. □

The following fact is immediate.

Fact 113. *Let X, Y be normed linear spaces over $\mathbb{K}$, $A \subset X$, $p \in [0,+\infty)$, let $f, g: A \to Y$ be (weakly; directionally) T^p-smooth at $a \in A$, and $c \in \mathbb{K}$. Then $f+g$ and $c \cdot f$ are (weakly; directionally) T^p-smooth at a.*

Proposition 114. *Let X, Y be normed linear spaces over $\mathbb{K}$, $U \subset X$ an open set, $f: U \to \mathbb{K}$, $g: U \to Y$, $x \in U$, and $p \in [0,+\infty)$. Suppose that f and g are T^p-smooth at x with $f(x+h) - P(h) = o(\|h\|^p)$ and $g(x+h) - Q(h) = o(\|h\|^p)$, $h \to 0$, where $P = \sum_{n=0}^{k} P_n$, $P_n \in \mathcal{P}(^nX)$, $P(0) = f(x)$, $Q = \sum_{n=0}^{k} Q_n$, $Q_n \in \mathcal{P}(^nX;Y)$, $Q(0) = g(x)$, and $k = [p]$. Then the mapping $f \cdot g$ is T^p-smooth at x with $fg(x+h) - R(h) = o(\|h\|^p)$, $h \to 0$ for $R = \sum_{n=0}^{k} R_n$, $R_n \in \mathcal{P}(^nX;Y)$, where*

$$R_n = \sum_{j=0}^{n} P_{n-j} \cdot Q_j, \quad n = 0, \dots, k.$$

Proof. Using Fact 27 we collect the n-homogeneous terms up to order k and put the rest into the remainder to obtain

$$\begin{aligned} f(x+h)g(x+h) &= \left(\sum_{j=0}^{k} P_j(h) + o(\|h\|^p)\right)\left(\sum_{l=0}^{k} Q_l(h) + o(\|h\|^p)\right) \\ &= \sum_{n=0}^{k} \sum_{j+l=n} P_j(h)Q_l(h) + o(\|h\|^p), \ h \to 0. \end{aligned}$$ □

It is easy to see that a similar statement holds also in the weak T^p-smooth case.

Proposition 115. *Let X, Y, and Z be normed linear spaces, $U \subset X$ and $V \subset Y$ open sets, and $p \in [0, +\infty)$. Further, let $g\colon U \to V$ be T^p-smooth at $x \in U$ and let $f\colon V \to Z$ be T^p-smooth at $y = g(x)$ with $g(x+h) - Q(h) = o(\|h\|^p)$ and $f(y+h) - P(h) = o(\|h\|^p)$, $h \to 0$, where $P = f(y) + \sum_{n=1}^k P_n$ with $P_n \in \mathcal{P}(^nY; Z)$ and $Q = g(x) + \sum_{n=1}^k Q_n$ with $Q_n \in \mathcal{P}(^nX; Y)$ for $k = [p]$. Then the mapping $f \circ g$ is T^p-smooth at x and $f \circ g(x+h) - R(h) = o(\|h\|^p)$, $h \to 0$ for $R = \sum_{n=0}^k R_n$, $R_n \in \mathcal{P}(^nX; Z)$, where $R_0(h) = f \circ g(x)$ and*

$$R_n(h) = \sum_{j=1}^{n} \sum_{\substack{1 \le l_s \le n \\ \sum_{s=1}^{j} l_s = n}} \check{P}_j\big(Q_{l_1}(h), \dots, Q_{l_j}(h)\big), \quad n = 1, \dots, k.$$

Proof. Put $\omega(h) = f(y+h) - P(h)$ and $\eta(h) = g(x+h) - Q(h)$ on suitable neighbourhoods of the origins. Then

$$\begin{aligned} f\big(g(x+h)\big) &= f\left(g(x) + \sum_{l=1}^{k} Q_l(h) + \eta(h)\right) \\ &= f(y) + \sum_{j=1}^{k} P_j\left(\sum_{l=1}^{k} Q_l(h) + \eta(h)\right) + \omega\left(\sum_{l=1}^{k} Q_l(h) + \eta(h)\right). \end{aligned}$$

Using the fact that $\omega(0) = 0$ it is not difficult to check that the last summand is $o(\|h\|^p)$, $h \to 0$. Further, $\check{P}_j\big({}^{j-n}\big(\sum_{l=1}^k Q_l(h)\big), {}^n\eta(h)\big) = o(\|h\|^{np})$, $h \to 0$ for $j = 1, \dots, k$, $n = 1, \dots, j$. Hence by Lemma 21

$$f \circ g(x+h) = f(y) + \sum_{j=1}^{k} P_j\left(\sum_{l=1}^{k} Q_l(h)\right) + o(\|h\|^p), \ h \to 0.$$

Wc cxpand this using (1) into

$$f \circ g(x+h) = f(y) + \sum_{j=1}^{k} \sum_{\substack{1 \le l_s \le k \\ s = 1, \dots, j}} \check{P}_j\big(Q_{l_1}(h), \dots, Q_{l_j}(h)\big) + o(\|h\|^p).$$

The formulae for R_n now follow from Lemma 26. □

We note that similarly if g is assumed to be only weakly T^p-smooth at x, then $f \circ g$ is weakly T^p-smooth at x and the same formula for the composition holds.

Corollary 116 (Leibniz formula)**.** *Let X, Y be normed linear spaces, $U \subset X$ an open set, $f \in C^k(U)$, $g \in C^k(U; Y)$, and $a \in U$. Then*

$$d^k(f \cdot g)(a) = \sum_{j=0}^{k} \binom{k}{j} d^{k-j} f(a) \cdot d^j g(a).$$

Proof. By Corollary 84 $f \cdot g \in C^k(U;Y)$. By Theorem 103, Proposition 114, and Proposition 104 we have

$$\frac{1}{k!}d^k(f \cdot g)(a) = \sum_{j=0}^{k} \frac{1}{(k-j)!} d^{k-j} f(a) \cdot \frac{1}{j!} d^j g(a). \qquad \square$$

Corollary 117 (Chain rule). *Let X, Y, Z be normed linear spaces, $U \subset X$ and $V \subset Y$ open sets, $f \in C^k(V;Z)$, $g \in C^k(U;Y)$, $g(U) \subset V$, and $a \in U$. Then*

$$d^k(f \circ g)(a) = \sum_{j=1}^{k} \frac{1}{j!} \sum_{\substack{1 \le l_s \le k \\ \sum_{s=1}^{j} l_s = k}} \binom{k}{l_1, \dots, l_j} D^j f(g(a))[d^{l_1}g(a), \dots, d^{l_j}g(a)].$$

Proof. By Theorem 83, $f \circ g \in C^k(U;Z)$. By Theorem 103, Proposition 115, and Proposition 104 we have

$$\frac{1}{k!}d^k(f \circ g)(a) = \sum_{j=1}^{k} \sum_{\substack{1 \le l_s \le k \\ \sum_{s=1}^{j} l_s = k}} \frac{1}{j!} D^j f(g(a)) \left[\frac{1}{l_1!} d^{l_1}g(a), \dots, \frac{1}{l_j!} d^{l_j}g(a)\right],$$

from which the formula follows. $\square$

Expressing $d^k(f \circ g)$ using only $d^j f$ and $d^l g$ is in principle possible by using the Polarisation formula, however the resulting formula will not be a sum of homogeneous polynomials.

Theorem 118 ([JohaZa]). *Let X be a real Banach space, Y a real normed linear space, $U \subset X$ open, $f : U \to Y$, and $a \in U$. Suppose that $\phi \circ f$ is Baire measurable for each $\phi \in Y^*$, f is directionally T^k-smooth on U, and further*

$$R(a + ty, th) = o(t^k), \ t \to 0 \quad \textit{for each } y, h \in X, \tag{29}$$

where $R(x,h) = f(x+h) - P^{x,h}(1)$ and where the polynomials $P^{x,h} \in \mathcal{P}^k(\mathbb{R};Y)$ come from the definition of the directional T^k-smoothness of f at x. Then f is weakly T^k-smooth at a.

Proof. Without loss of generality we may assume that $a = 0$. Set $P(h) = P^{0,h}(1)$ for each $h \in X$. We show that $P \in \mathcal{P}^k(X;Y)$ using Corollary 55. First we show that $\phi \circ P$ is Baire measurable for each $\phi \in Y^*$. So fix $\phi \in Y^*$. For each $h \in X$ there are $g_0(h), \dots, g_k(h) \in \mathbb{R}$ such that $\phi \circ P^{0,h}(t) = \sum_{j=0}^{k} g_j(h)t^j$ for $t \in \mathbb{R}$. Note that $\phi \circ P(h) = \phi \circ P^{0,h}(1) = \sum_{j=0}^{k} g_j(h)$ and so it suffices to show that the functions $g_0, \dots, g_k : X \to \mathbb{R}$ are Baire measurable. Clearly $g_0(h) = \phi \circ f(0)$ for each $h \in X$. We proceed by induction, assuming that $g_0, \dots, g_{m-1}$ are Baire measurable for some $1 \le m \le k$. Since for a fixed $h \in X$ we have

$$\lim_{t \to 0} \frac{1}{t^m} \left(\phi \circ f(th) - \sum_{j=0}^{k} g_j(h)t^j\right) = \phi\left(\lim_{t \to 0} \frac{f(th) - P^{0,h}(t)}{t^m}\right) = 0,$$

it follows that $g_m(h) = \lim_{t\to 0} \frac{1}{t^m}\big(\phi \circ f(th) - \sum_{j=0}^{m-1} g_j(h)t^j\big)$. In particular,

$$g_m(h) = \lim_{n\to\infty} n^m \left(\phi \circ f\left(\frac{h}{n}\right) - \sum_{j=0}^{m-1} \frac{g_j(h)}{n^j}\right).$$

Hence on each bounded set g_m is a pointwise limit of a sequence of Baire measurable functions and so it is Baire measurable.

Now we show that P satisfies the formula (10). Fix $x, h \in X$ and define

$$q(t) = \sum_{j=0}^{k+1} (-1)^{k+1-j} \binom{k+1}{j} P(tx + jth).$$

Clearly $q \in \mathcal{P}^k(\mathbb{R}; Y)$. Let $r > 0$ be such that $t(x + jh) \in U$ for all $j \in \{0, \dots, k+1\}$ and $|t| \le r$. We have

$$\begin{aligned}\|q(t)\| &= \Bigg\| \sum_{j=0}^{k+1} (-1)^{k+1-j} \binom{k+1}{j} \Big(P(tx + jth) - f(tx + jth) \\ &\qquad\qquad + f(tx + jth) - P^{tx,th}(j) + P^{tx,th}(j) \Big) \Bigg\| \\ &\le \sum_{j=0}^{k+1} \binom{k+1}{j} \big(\|R(0, tx + jth)\| + \|R(tx, jth)\| \big)\end{aligned}$$

for all $|t| \le r$, since $\sum_{j=0}^{k+1} (-1)^{k+1-j} \binom{k+1}{j} P^{tx,th}(j) = 0$ by Corollary 55. This estimate combined with the assumption (29) implies that $q(t) = o(t^k)$, $t \to 0$. This means that q is a zero polynomial (Fact 48), and in particular $q(1) = 0$, which is what we wanted to prove. □

As a consequence of the above theorem we obtain that the Converse Taylor theorem holds even if we assume only directional T^k-smoothness:

Theorem 119 ([JohaZa]). *Let X be a real Banach space, Y a real normed linear space, $U \subset X$ an open set, $f : U \to Y$, and $k \in \mathbb{N}_0$. Then $f \in C^k(U; Y)$ if and only if f is a directionally T^k-smooth mapping such that $\phi \circ f$ is Baire measurable for each $\phi \in Y^*$ and*

$$\lim_{\substack{(y,h)\to(x,0) \\ h\neq 0}} \frac{\|R(y,h)\|}{\|h\|^k} = 0 \tag{30}$$

for every $x \in U$, where $R(x, h) = f(x + h) - P^{x,h}(1)$ and where the polynomials $P^{x,h} \in \mathcal{P}^k(X; Y)$ come from the definition of the directional T^k-smoothness of f at x. In this case $P^{x,h}(t) = \sum_{j=0}^{k} \frac{1}{j!} d^j f(x)[th]$.

Proof. It suffices to notice that the assumption (30) implies (29) in Theorem 118 and it also promotes the weak T^k-smoothness to T^k-smoothness. Hence we can apply Theorem 110. □

5. Smoothness classes

In this section we lay down definitions of various classes of smooth mappings which will be used throughout the book. We show some relations between them, including the uniform version of the Converse Taylor theorem, and prove some basic arithmetic and composition rules.

Let (P, ρ), (Q, σ) be metric spaces. The minimal modulus of continuity of a uniformly continuous mapping $f: P \to Q$ is defined as

$$\omega_f(\delta) = \sup\left\{\sigma(f(x), f(y));\ x, y \in P, \rho(x, y) \le \delta\right\} \quad \text{for } \delta \in \mathbb{R}_0^+.$$

Clearly, ω_f is continuous at 0.

A modulus is a non-decreasing function $\omega\colon [0, +\infty) \to [0, +\infty]$ continuous at 0 with $\omega(0) = 0$. The set of all moduli will be denoted by $\mathcal{M}$. We say that $f: P \to Q$ is uniformly continuous with modulus of continuity $\omega \in \mathcal{M}$ if $\omega_f \le \omega$.

An important subset of all moduli consists of the sub-additive moduli, which we shall denote by $\mathcal{M}_s$. A nice feature of $\omega \in \mathcal{M}_s$ is that it is real-valued and uniformly continuous with modulus of continuity ω. It is easy to check that the minimal modulus of continuity of a uniformly continuous mapping defined on a convex subset of a normed linear space is sub-additive. Note also that both the sets $\mathcal{M}$ and $\mathcal{M}_s$ are stable under composition.

Fact 120. *Let $\omega \in \mathcal{M}_s$. Then $\omega'_+(0) = \sup_{t>0} \frac{\omega(t)}{t}$. In particular, if ω is not identically zero, then for each $a > 0$ there is $L > 0$ such that $\omega(t) \ge Lt$ for $t \in [0, a]$.*

Note that it means that Lipschitz mappings on convex bounded sets have the best modulus of continuity.

Proof. If ω is identically zero, then $\omega'_+(0) = 0$. Otherwise $q = \sup_{t>0} \frac{\omega(t)}{t} > 0$. Assume that $q \in \mathbb{R}$ and choose any $\varepsilon > 0$. Let $s \in \mathbb{R}^+$ be such that $\frac{\omega(s)}{s} > q - \varepsilon$ and let $\delta > 0$ be such that $\omega(\delta) < s\varepsilon$. Choose any $t \in (0, \delta)$ and put $n = \left[\frac{s}{t}\right]$. Then $s - \delta < nt \le s$ and hence using the sub-additivity of ω we obtain

$$q - \varepsilon < \frac{\omega(s)}{s} \le \frac{\omega(nt)}{s} + \frac{\omega(s - nt)}{s} \le \frac{n\omega(t)}{s} + \frac{\omega(\delta)}{s} < \frac{\omega(t)}{t} + \varepsilon.$$

Since $\frac{\omega(t)}{t} \le q$, the assertion now follows. The case $q = +\infty$ can be dealt with similarly. □

Definition 121. Let X, Y be normed linear spaces, $V \subset X$ open, $f: V \to Y$, and $p \in [0, +\infty)$. We say that f is UT^p-smooth on V if there exists $\omega \in \mathcal{M}$ such that for each $x \in V$ there is a polynomial $P \in \mathcal{P}^{[p]}(X; Y)$ satisfying

$$\|f(x + h) - P(h)\| \le \omega(\|h\|)\|h\|^p \quad \text{for } x + h \in V.$$

We note that UT^p-smoothness implies UT^q-smoothness for $[p] \le q < p$, but in general it does not imply UT^q-smoothness for $q < p$ – the function $f(x) = x^3$ is UT^p-smooth on $\mathbb{R}$ for every $p \ge 2$ but it is not UT^1-smooth on $\mathbb{R}$.

Definition 122. Let X, Y be normed linear spaces, $U \subset X$ open, $f\colon U \to Y$, and $p \in [0, +\infty)$. We say that f is locally UT^p-smooth (UT^p_{loc}-smooth) on U if for each $x \in U$ there is a neighbourhood of x on which f is UT^p-smooth. We say that f is UT^p_{c}-smooth on U if for each $K \subset U$ compact there exists $\omega \in \mathcal{M}$ such that for each $x \in K$ there is a polynomial $P \in \mathcal{P}^{[p]}(X;Y)$ satisfying

$$\|f(x+h) - P(h)\| \le \omega(\|h\|)\|h\|^p \quad \text{for } x+h \in U.$$

Obviously, if f is UT^p_{loc}-smooth on U, then it is UT^p_{c}-smooth on U, and by Theorem 110 it is $C^{[p]}$-smooth on U. Further, if f is C^k-smooth on U, then it is UT^k_{c}-smooth on U. Indeed, for a given $K \subset U$ compact we set

$$\omega(\delta) = \frac{1}{k!} \sup_{\substack{x \in K \\ \|h\| \le \delta}} \|d^k f(x+h) - d^k f(x)\|$$

for all $0 \le \delta < \eta = \operatorname{dist}(K, X \setminus U)$ and $\omega(\delta) = +\infty$ for $\delta \ge \eta$. By the compactness it is easily seen that $\omega \in \mathcal{M}$. For a fixed $x \in K$ we may then apply Corollary 108 on $U(x, \eta)$.

Recall that a convex cone is a subset A of a real vector space satisfying $au + bv \in A$ for any $u, v \in A$, $a, b \in \mathbb{R}^+$. We will also use the term convex cone for a subset of $\mathcal{M}$ in an obvious sense, although it may not be a subset of a vector space.

Definition 123. Let X, Y be normed linear spaces, $U \subset X$ an open set, $k \in \mathbb{N}$, $f \in C^k(U;Y)$, and let $\Omega \subset \mathcal{M}$ be a convex cone. We say that f is $C^{k,\Omega}$-smooth on U if $d^k f$ is uniformly continuous on U with modulus ω for some $\omega \in \Omega$. We denote by $C^{k,\Omega}(U;Y)$ the vector space of all $C^{k,\Omega}$-smooth mappings from U into Y.

We say that f is locally $C^{k,\Omega}$-smooth ($C^{k,\Omega}_{\mathrm{loc}}$-smooth) on U if for each $x \in U$ there is a neighbourhood of x on which f is $C^{k,\Omega}$-smooth. We denote by $C^{k,\Omega}_{\mathrm{loc}}(U;Y)$ the vector space of all $C^{k,\Omega}_{\mathrm{loc}}$-smooth mappings from U into Y.

As usual, for functions, i.e. mappings into the scalars, we use the shortened notation $C^{k,\Omega}(U) = C^{k,\Omega}(U;\mathbb{K})$ and $C^{k,\Omega}_{\mathrm{loc}}(U) = C^{k,\Omega}_{\mathrm{loc}}(U;\mathbb{K})$.

Definition 124. Let X, Y be normed linear spaces, $U \subset X$ open, $f \in C^k(U;Y)$, $k \in \mathbb{N}$, and $\alpha \in [0,1]$. We say that f is $C^{k,\alpha}$-smooth on U if $f \in C^{k,\Omega}(U;Y)$ for $\Omega = \{Kt^\alpha;\ K \in \mathbb{R}^+\}$, i.e. $d^k f$ is α-Hölder on U. We say that f is $C^{k,+}$-smooth on U if $f \in C^k(U;Y)$ and $d^k f$ is uniformly continuous on U. The vector spaces of all respective mappings are denoted by $C^{k,\alpha}(U;Y)$ and $C^{k,+}(U;Y)$. In particular, $f \in C^{k,1}(U;Y)$ if $d^k f$ is Lipschitz on U.

Similarly we define the local versions $C^{k,\alpha}_{\mathrm{loc}}(U;Y)$, $C^{k,+}_{\mathrm{loc}}(U;Y)$ and the scalar valued spaces $C^{k,\alpha}(U)$, $C^{k,+}(U)$, $C^{k,\alpha}_{\mathrm{loc}}(U)$, and $C^{k,+}_{\mathrm{loc}}(U)$.

As an example, taking Ω to be the cone generated by $\{Kt^\alpha;\ K \in \mathbb{R}^+, \alpha \in (0,1]\}$, the space $C^{k,\Omega}(U;Y)$ for U bounded consists of mappings with Hölder kth derivative.

Clearly, $C^{k,\alpha}_{\mathrm{loc}}$-smoothness implies both $C^{k,\beta}_{\mathrm{loc}}$-smoothness and $C^{k,+}_{\mathrm{loc}}$-smoothness for $0 \leq \beta < \alpha \leq 1$, $C^{k,\alpha}$-smoothness implies $C^{k,+}$-smoothness, but $C^{k,\alpha}$-smoothness implies $C^{k,\beta}$-smoothness for $0 \leq \beta < \alpha \leq 1$ only on bounded sets. Further, C^{k+1}-smoothness implies $C^{k,1}_{\mathrm{loc}}$-smoothness (Proposition 71), and $C^{k,0}_{\mathrm{loc}}$-smoothness is just C^k-smoothness. $C^{k,0}$-smoothness means that $d^k f$ is bounded, so it implies $C^{k-1,1}$-smoothness in case that the domain is convex (Proposition 71). Similarly, $C^{k,+}$-smoothness implies $C^{k-1,1}$-smoothness on convex bounded sets. Finally, $C^{k,1}_{\mathrm{loc}}$-smoothness implies $C^{k,\Omega}_{\mathrm{loc}}$-smoothness for any Ω, and $C^{k,1}$-smoothness implies $C^{k,\Omega}$-smoothness on convex bounded sets for any Ω (Fact 120).

Corollary 108 implies the following relations: A $C^{k,\alpha}_{\mathrm{loc}}$-smooth mapping on U is UT^p_{loc}-smooth on U for every $p \in [0, k+\alpha)$; in general however it is not even $T^{k+\alpha}$-smooth (consider for example the function $t \mapsto \sqrt[3]{t^4}$, which is $C^{1,\frac{1}{3}}$-smooth on $\mathbb{R}$ but is not $T^{\frac{4}{3}}$-smooth at 0). Similarly, a $C^{k,\alpha}$-smooth mapping on a convex U is UT^p-smooth on U for every $p \in [0, k+\alpha)$, a $C^{k,+}_{\mathrm{loc}}$-smooth mapping on U is UT^k_{loc}-smooth on U, and a $C^{k,+}$-smooth mapping on a convex U is UT^k-smooth on U.

Moreover, if $d^k f$ is uniformly continuous on a convex U with modulus $\omega \in \mathcal{M}$, then f is UT^k-smooth on U with modulus $\frac{1}{k!}\omega$. The converse statement is contained in the next theorem.

Theorem 125. *Let X, Y be normed linear spaces, $U \subset X$ an open set, $f : U \to Y$, and $k \in \mathbb{N}$. Suppose that f is UT^k-smooth on U and the modulus ω from the definition of UT^k-smoothness is sub-additive. Let Ω be the convex cone generated by ω. Then $f \in C^{k,\Omega}(V;Y)$ for any bounded open $V \subset U$ satisfying* $\operatorname{dist}(V, X \setminus U) > 0$. *More precisely, $d^k f$ is uniformly continuous on V with modulus $m\omega$, where $m > 0$ is a constant depending on k,* $\operatorname{diam} V$, *and* $\operatorname{dist}(V, X \setminus U)$.

If moreover U is bounded and there exists $r > 0$ such that for each $x \in U$ there is $a \in U$ satisfying $\operatorname{conv}\big(\{x\} \cup B(a,r)\big) \subset U$, *then $f \in C^{k,\Omega}(U;Y)$. More precisely, $d^k f$ is uniformly continuous on U with modulus $c_k\big(\frac{\operatorname{diam} U}{r}\big)^k\omega$, where $c_k > 0$ is a constant depending only on k. The same holds if $U = X$ (in this case the modulus is just $c_k\omega$).*

Proof. First notice that $f \in C^k(U;Y)$ by Theorem 110. Let $V \subset U$ be a non-empty bounded open set that satisfies $\operatorname{dist}(V, X \setminus U) > 0$. Put $\rho = \operatorname{diam} V$ and $\varepsilon = \min\{\rho, \operatorname{dist}(V, X \setminus U)\}$. Fix any $x \in V$ and $h \in X$ such that $x + h \in V$. We use the notation from the proof of Theorem 110. For any $y \in X$ satisfying $\|y\| \leq \|h\|$ and $x + h + y \in U$ we have

$$\begin{aligned}
\|q(h,y)\| &\leq \|R(x, h+y)\| + \|R(x+h, y)\| \\
&\leq \omega(\|h+y\|)\|h+y\|^k + \omega(\|y\|)\|y\|^k \\
&\leq \omega(2\|h\|)2^k\|h\|^k + \omega(\|h\|)\|h\|^k \leq (2^{k+1}+1)\omega(\|h\|)\|h\|^k.
\end{aligned} \tag{31}$$

Therefore

$$\left\|d^k f(x+h)[y] - d^k f(x)[y]\right\| = k!\|q_k(h,y)\| \le k! K_{k,k} \max_{0\le l\le k} \left\|q\left(h, \tfrac{l}{k}y\right)\right\| \le k! K_{k,k}(2^{k+1}+1)\omega(\|h\|)\|h\|^k$$

for all $y \in X$ satisfying $\|y\| \le \frac{\varepsilon}{\rho}\|h\|$. Consequently, if $h \neq 0$,

$$\left\|d^k f(x+h) - d^k f(x)\right\| = \frac{1}{\left(\frac{\varepsilon}{\rho}\|h\|\right)^k} \sup_{\|y\|\le \frac{\varepsilon}{\rho}\|h\|} \left\|d^k f(x+h)[y] - d^k f(x)[y]\right\| \le \frac{\rho^k}{\varepsilon^k} k! K_{k,k}(2^{k+1}+1)\omega(\|h\|).$$

Note that in the case $U = X$ we can take the supremum over $\|y\| \le \|h\|$ (since $x + h + y$ always lies in the domain of f) and thus we obtain the estimate for any $x, h \in X$.

Finally, suppose that U is bounded and there exists $r > 0$ such that for each $x \in U$ there is $a \in U$ satisfying $\operatorname{conv}\left(\{x\} \cup B(a,r)\right) \subset U$. Fix any $x \in U$ and $h \in X \setminus \{0\}$ such that $x + h \in U$. Put $\alpha = \frac{r}{\operatorname{diam} U}$ and $s = \alpha\|h\|$, and note that $\alpha \le \frac{1}{2}$ and $s \le r$. Let $a \in U$ be such that $\operatorname{conv}\left(\{x+h\} \cup B(a,r)\right) \subset U$. Set $u = a$ if $\|a - x - h\| \le \|h\|$ and $u = x + h + \frac{a-x-h}{\|a-x-h\|}\|h\|$ otherwise. Note that $B(u,s) \subset U$. Indeed, if $u = a$, then we use the fact that $s \le r$. Otherwise, every $z \in B(u,s)$ can be expressed as a convex combination $z = \left(1 - \frac{\|h\|}{\|a-x-h\|}\right)(x+h) + \frac{\|h\|}{\|a-x-h\|}w$, where $w = a + (z-u)\frac{\|a-x-h\|}{\|h\|} \in B(a,r)$.

Now for any $y \in B(u-x-h, s)$ we have $x + h + y \in B(u,s) \subset U$ and $\|y\| \le \|u - x - h\| + s \le \|h\| + s = (1+\alpha)\|h\| < 2\|h\|$. Hence, similarly as in (31),

$$\begin{aligned}\|q(h,y)\| &\le \omega(\|h+y\|)\|h+y\|^k + \omega(\|y\|)\|y\|^k \\ &\le \omega(3\|h\|)(2+\alpha)^k\|h\|^k + \omega(2\|h\|)(1+\alpha)^k\|h\|^k \\ &\le \left(3(2+\alpha)^k + 2(1+\alpha)^k\right)\omega(\|h\|)\|h\|^k \le 5(2+\alpha)^k\omega(\|h\|)\|h\|^k.\end{aligned}$$

Thus using Lemma 47 we obtain

$$\begin{aligned}\left\|d^k f(x+h) - d^k f(x)\right\| &= \frac{1}{s^k} \sup_{y\in B(0,s)} \left\|d^k f(x+h)[y] - d^k f(x)[y]\right\| \\ &= \frac{k!}{s^k} \sup_{y\in B(0,s)} \|q_k(h,y)\| \le \frac{k^k}{s^k} \sup_{y\in B(u-x-h,s)} \|q(h,y)\| \\ &\le \frac{k^k}{s^k} 5(2+\alpha)^k\omega(\|h\|)\|h\|^k = 5k^k\left(1 + \frac{2}{\alpha}\right)^k \omega(\|h\|).\end{aligned}$$

□

In the convex case the assumption on the sub-additivity of the modulus ω can be dropped:

Corollary 126. *Let X, Y be normed linear spaces and let $U \subset X$ be an open convex set that is either bounded or has the property that there are $a \in X$, $r > 0$, and $\{u_n\}_{n\in\mathbb{N}} \subset X$, $\|u_n\| = n$ such that $B(a + u_n, nr) \subset U$ for each $n \in \mathbb{N}$ (this holds in particular if U contains an unbounded cone). If $f : U \to Y$ is UT^k-smooth, $k \in \mathbb{N}$, and ω is the modulus from the definition of UT^k-smoothness, then $f \in C^{k,\Omega}(U;Y)$, where Ω is the convex cone generated by ω. More precisely, in the bounded case $d^k f$ is uniformly continuous on U with modulus $c_k e_U^k \omega$, where $e_U = \frac{\operatorname{diam} U}{\sup\{r;\ B(a,r)\subset U\}}$ and $c_k > 0$ is a constant depending only on k; in the unbounded case $d^k f$ is uniformly continuous on U with modulus $c_k\big(1 + \frac{1}{r}\big)^k \omega$.*

Proof. First suppose that U is bounded and $B(a,r) \subset U$. By the convexity the assumption of Theorem 125 is satisfied and from the proof it is easily seen that we obtain $\|d^k f(x + h) - d^k f(x)\| \le 2k^k\Big(1 + \frac{2\operatorname{diam} U}{r}\Big)^k \omega(3\|h\|)$ for any $x, h \in U$. Thus $\omega_1(t) \le c_k e_U^k \omega(3t)$, where ω_1 is the minimal modulus of continuity of $d^k f$ on U. But since ω_1 is sub-additive, we get $\omega_1(t) \le 3\omega_1(\frac{1}{3}t) \le 3c_k e_U^k \omega(t)$.

In the unbounded case choose any $n \in \mathbb{N}$ and put $V = U \cap U(a, n + nr)$. Then $U(a+u_n, nr) \subset V$ and so by the previous part $f \in C^{k,\Omega}(V;Y)$ and $d^k f$ is uniformly continuous on V with modulus $m\omega$, where $m = c_k\big(\frac{2(n+nr)}{nr}\big)^k = 2^k c_k\big(1+\frac{1}{r}\big)^k$. Since this constant is independent of n, it follows that $f \in C^{k,\Omega}(U;Y)$ and $d^k f$ is uniformly continuous on the whole of U with modulus $m\omega$. □

The next theorem asserts that it actually suffices to test the $C^{k,+}$-smoothness only on every line. Compare with similar characterisations in Corollary 55 and Theorem 160; see also the example after Theorem 101.

Theorem 127. *Let X be a real Banach space, Y a real normed linear space, $U \subset X$ an open convex bounded set or $U = X$, $f : U \to Y$, and $k \in \mathbb{N}$. Suppose that $\phi \circ f$ is Baire measurable for each $\phi \in Y^*$ and there is a modulus ω such that the restriction of f to every line in U is $C^{k,+}$-smooth with its kth differential having modulus of continuity ω. Then $f \in C^{k,\Omega}(U;Y)$, where Ω is the convex cone generated by ω. More precisely, $d^k f$ is uniformly continuous with modulus $c_k e_U^k \omega$ (or just $c_k\omega$ if $U = X$), where $e_U = \frac{\operatorname{diam} U}{\sup\{r;\ B(a,r)\subset U\}}$ and $c_k > 0$ is a constant depending only on k.*

Proof. Taylor's theorem (Corollary 108) implies that f is directionally T^k-smooth with $\|R(x,h)\| \le \frac{1}{k!}\omega(\|h\|)\|h\|^k$, where $R(x,h) = f(x+h) - P^{x,h}(1)$ and where the polynomials $P^{x,h} \in \mathcal{P}^k(\mathbb{R};Y)$ come from the definition of the directional T^k-smoothness. Now it suffices to combine Theorem 118 and Corollary 126. □

Next we prove several statements that deal with operations with $C^{k,+}$-smooth mappings.

Proposition 128. *Let X, Y, Z be normed linear spaces, $U \subset X$ and $V \subset Y$ open sets, $k \in \mathbb{N}$, $f \in C^k(V;Z)$, $g \in C^k(U;Y)$, $g(U) \subset V$, and $A \subset U$. Assume further that one of the following conditions is satisfied:*

(i) *There is $\omega \in \mathcal{M}_s$ such that $d^j g$ is bounded and uniformly continuous on A with modulus ω and $d^j f$ is bounded and uniformly continuous on $g(A)$ with modulus ω for all $j \in \{1,\dots,k\}$, and g is Lipschitz on A.*

(ii) *There is $\omega \in \mathcal{M}$ such that $d^j g$ is bounded and uniformly continuous on A with modulus ω and $d^j f$ is bounded and Lipschitz on $g(A)$ for all $j \in \{1,\dots,k\}$, and g is uniformly continuous on A with modulus ω.*

Then there is $C > 0$ such that $d^j(f \circ g)$ is bounded and uniformly continuous on A with modulus $C\omega$ for all $j \in \{1,\dots,k\}$.

Note that the condition (i) above is automatically satisfied if the mappings f and g are $C^{k,\Omega}$-smooth and the sets A and V are convex and bounded (Proposition 71, Fact 120).

Proof. Let $M > 0$ be such that $d^j f$ is bounded by M on $g(A)$ and $d^j g$ is bounded by M on A for all $j \in \{1,\dots,k\}$. Fix $j \in \{1,\dots,k\}$ and $l_1,\dots,l_j \in \{1,\dots,k\}$. Let $L \in \mathbb{N}$ be such that g is L-Lipschitz on A in case (i), resp. $L > 0$ be such that $d^j f$ is L-Lipschitz on $g(A)$ in case (ii). Notice that in both of these cases we obtain $\|d^j f(g(x)) - d^j f(g(y))\| \le L\omega(\|x-y\|)$ for any $x, y \in A$. Using this together with Proposition 13 and (3) from the proof of Proposition 4 we can estimate

$$
\begin{aligned}
&\|D^j f(g(x))[d^{l_1}g(x),\dots,d^{l_j}g(x)] - D^j f(g(y))[d^{l_1}g(y),\dots,d^{l_j}g(y)]\| \\
&\quad\le \left\|\left(D^j f(g(x)) - D^j f(g(y))\right)[d^{l_1}g(x),\dots,d^{l_j}g(x)]\right\| \\
&\qquad + \|D^j f(g(y))[d^{l_1}g(x),\dots,d^{l_j}g(x)] - D^j f(g(y))[d^{l_1}g(y),\dots,d^{l_j}g(y)]\| \\
&\quad\le \frac{j^j}{j!}\|d^j f(g(x)) - d^j f(g(y))\| \cdot \|d^{l_1}g(x)\| \cdots \|d^{l_j}g(x)\| \\
&\qquad\qquad + 2^j(2M)^{j-1}\|D^j f(g(y))\| \max_{s=1,\dots,j} \|d^{l_s}g(x) - d^{l_s}g(y)\| \\
&\quad\le \frac{j^j}{j!}M^j L\omega(\|x-y\|) + \frac{j^j}{j!}2^{2j-1}M^j\omega(\|x-y\|) \\
&\quad\le \frac{j^j}{j!}M^j(L + 2^{2j-1})\omega(\|x-y\|)
\end{aligned}
$$

for any $x, y \in A$. Corollary 117 now finishes the proof. □

Proposition 129. *Let X, Y be normed linear spaces, $U \subset X$ an open set, $k \in \mathbb{N}$, $f \in C^k(U)$, and $g \in C^k(U;Y)$. Assume further that there are $A \subset U$ and $\omega \in \mathcal{M}$ such that $d^j f$, $d^j g$ are bounded and uniformly continuous on A with modulus ω for all $j \in \{0,\dots,k\}$. Then there is $C > 0$ such that $d^j(f \cdot g)$ is bounded and uniformly continuous on A with modulus $C\omega$ for all $j \in \{0,\dots,k\}$. If moreover* $\inf_A |f| > 0$, *then there is $C > 0$ such that $d^j \frac{1}{f}$ is bounded and uniformly continuous on A with modulus $C\omega$ for all $j \in \{0,\dots,k\}$.*

Proof. The first statement follows immediately from Corollary 116. The second statement follows from Proposition 128. □

Proposition 130. *Let X, Y be normed linear spaces, $U \subset X$ an open set, Γ an arbitrary set, and let $f: U \to \ell_\infty(\Gamma; Y)$ be locally bounded. Denote by R_α, $\alpha \in \Gamma$ the canonical projections $R_\alpha\colon \ell_\infty(\Gamma; Y) \to Y$ given by $R_\alpha((y_\gamma)) = y_\alpha$ and by $f_\gamma = R_\gamma \circ f$ the component mappings of f. Then f is T^p-smooth at $x \in U$ if and only if all f_γ, $\gamma \in \Gamma$ are equi-T^p-smooth at x, i.e. if and only if there are $\omega \in \mathcal{M}$, a neighbourhood V of the origin in X, and polynomials $P_\gamma \in \mathcal{P}^{[p]}(X; Y)$ satisfying $\|f_\gamma(x+h) - P_\gamma(h)\| \le \omega(\|h\|)\|h\|^p$ for all $h \in V$ and all $\gamma \in \Gamma$.*

Further, the mapping f is C^k-smooth with $d^k f$ uniformly continuous with modulus $\omega \in \mathcal{M}$ if and only if each f_γ, $\gamma \in \Gamma$, is C^k-smooth with $d^k f_\gamma$ uniformly continuous with modulus ω.

Proof. First we deal with the first part of the theorem. To see $\Rightarrow$ it suffices to notice that if P is the approximating polynomial from the definition of T^p-smoothness of f at x, then it suffices to put $\omega(\delta) = \sup_{h \in B(0,\delta)\setminus\{0\}} \frac{\|f(x+h)-P(h)\|}{\|h\|^p}$ and $P_\gamma = R_\gamma \circ P$. To prove $\Leftarrow$ let $K > 0$ and $\delta \in (0,1]$ be such that $\omega(\delta) \le K$ and $\|f(x+h)\| \le K$ for $h \in B(0,\delta)$. Then $\|P_\gamma(h)\| \le \|f_\gamma(x+h)\| + \omega(\|h\|)\|h\|^p \le 2K$ for $h \in B(0,\delta)$ and $\gamma \in \Gamma$. By Corollary 31 the set $\{P_\gamma\}_{\gamma\in\Gamma}$ is bounded in the space $\mathcal{P}^{[p]}(X; Y)$. If we set $P(h) = (P_\gamma(h))_{\gamma\in\Gamma}$, then $P \in \mathcal{P}^{[p]}(X; \ell_\infty(\Gamma; Y))$ by Fact 50. Clearly, $\|f(x+h) - P(h)\| \le \omega(\|h\|)\|h\|^p$ for $h \in V$.

As for the second part of the theorem, the implication $\Rightarrow$ follows from Fact 78.

$\Leftarrow$ Given $x \in U$, let $r > 0$ be such that $U(x,r) \subset U$. Corollary 108 implies that the collection $\{f_\gamma\}_{\gamma\in\Gamma}$ is equi-UT^k-smooth on $U(x,r)$. Thus by the first part of the proof f is UT^k-smooth on $U(x,r)$. By Theorem 110 the mapping f is C^k-smooth on $U(x,r)$. Finally, using Fact 78 we obtain

$$\begin{aligned}\big\|d^k f(x)[h] - d^k f(y)[h]\big\| &= \sup_{\gamma\in\Gamma} \big\|R_\gamma\big(d^k f(x)[h]\big) - R_\gamma\big(d^k f(y)[h]\big)\big\| \\ &= \sup_{\gamma\in\Gamma} \big\|d^k f_\gamma(x)[h] - d^k f_\gamma(y)[h]\big\| \le \omega(\|x-y\|)\end{aligned}$$

for any $x, y \in U$, $h \in B_X$. □

Corollary 131. *Let X, Y be normed linear spaces, $U \subset X$ an open set, $f: U \to Y$ a locally bounded mapping, $k \in \mathbb{N}$, $\omega \in \mathcal{M}$, and $A \subset B_{Y^*}$ a λ-norming set. If $\phi \circ f \in C^{k,+}(U)$ with $d^k(\phi \circ f)$ uniformly continuous with modulus ω for each $\phi \in A$, then $f \in C^{k,+}(U; Y)$ and $d^k f$ is uniformly continuous with modulus $\lambda\omega$.*

Proof. Let $T\colon Y \to \ell_\infty(A)$ be given by $T(y) = \big(\phi(y)\big)_{\phi\in A}$. Then T is clearly a linear isomorphism of Y onto $Z = T(Y)$ satisfying $\|T\| \le 1$ and $\|T^{-1}\| \le \lambda$. By Proposition 130 the mapping $T \circ f$ is C^k-smooth with $d^k(T \circ f)$ uniformly continuous with modulus ω. Since $f = T^{-1} \circ (T \circ f)$, Fact 78 finishes the proof. □

6. Power series and their convergence

In this section we will investigate the convergence of power series. We prove the Polynomial lemma of Leja and apply it to obtain that a power series convergent on some translate of an absorbing set has a positive radius of convergence. We begin with some auxiliary notions.

Definition 132. Let X be a vector space over $\mathbb{K}$. A subset $A \subset X$ is called balanced if $\lambda A \subset A$ for every $\lambda \in \mathbb{K}$, $|\lambda| \leq 1$.

Definition 133. Let X be a vector space over $\mathbb{K}$. A subset $A \subset X$ is called absorbing if for every $x \in X$ there is $r_x > 0$ such that $r_x A$ contains the line segment joining x and 0, i.e. the set $\{tx;\ t \in [0,1]\}$.

Equivalently, for every $x \in X$ there is $\lambda_x > 0$ such that $tx \in A$ for every $t \in [0, \lambda_x]$. Obviously, any superset of an absorbing set is also absorbing.

Lemma 134. *An F_σ absorbing set A in a Banach space X has a non-empty interior.*

Proof. Suppose that $A = \bigcup_{k \in \mathbb{N}} A_k$, where each A_k is closed. Obviously we have $X = \bigcup_{n \in \mathbb{N}} nA = \bigcup_{k,n \in \mathbb{N}} nA_k$, where each nA_k is closed. Hence by the Baire category theorem some nA_k (and thus also A) has a non-empty interior. □

Definition 135. Let X be a normed linear space over $\mathbb{K}$, Y a Banach space over $\mathbb{K}$, and $P_n \in \mathcal{P}(^nX; Y)$. Consider the power series

$$\sum_{n=0}^{\infty} P_n. \tag{32}$$

The set $\operatorname{Int}\{x \in X;\ \sum_{n=0}^{\infty} P_n(x) \text{ converges}\}$ is called the domain of convergence of the power series (32). The radius of convergence of the real power series

$$\sum_{n=0}^{\infty} \|P_n\| \lambda^n$$

is called the radius of norm convergence of the power series (32).

We start with some simple observations. If the power series (32) converges in $x \in X$ (or just if the sequence $\{P_n(x)\}$ is bounded), then it converges absolutely in λx (i.e. $\sum_{n=0}^{\infty} \|P(\lambda x)\|$ converges) for every $\lambda \in \mathbb{K}$, $|\lambda| < 1$. Denote by G the domain of convergence of (32). Since for every $x \in G$ there is $\mu > 1$ such that $\mu x \in G$, it follows that G is a balanced set and the power series (32) converges absolutely on G. Further, if $R > 0$ is the radius of norm convergence of (32), then the series is absolutely uniformly convergent on $B(0, r)$ for each $r < R$. On the other hand, if the power series (32) converges uniformly on $U(0, r)$, then its radius of norm convergence is at least r. Indeed, in this case there is $N \in \mathbb{N}$ such that $\left\|\sum_{n=m}^{\infty} P_n(x)\right\| \leq 1$ for $x \in U(0, r)$ and $m \geq N$. Hence if $m \geq N$ and $x \in U(0, r)$, then $\|P_m(x)\| \leq \left\|\sum_{n=m}^{\infty} P_n(x) - \sum_{n=m+1}^{\infty} P_n(x)\right\| \leq 2$ and so $\|P_m\| r^m \leq 2$, from which the statement follows.

We note that some authors use the term radius of uniform convergence rather than the radius of norm convergence, but we find it somewhat misleading, since there may be an abundance of open sets outside the ball $B(0, R)$, R being the radius of norm convergence, on which the series converges uniformly; see e.g. Example 137.

Recall the well-known Cauchy-Hadamard formula:

Proposition 136. *The radius R of norm convergence of the power series* (32) *satisfies the formula*

$$R = \frac{1}{\limsup_{n\to\infty} \| P_n \|^{\frac{1}{n}}}.$$

(Here we put $\frac{1}{0} = +\infty$.)

Example 137. Let $X = c_0$ (real or complex). Denote the coordinate functionals by e_n^* and put $P_n(x) = \left(e_n^*(x)\right)^n$, $n \in \mathbb{N}$, $P_0 = 0$. Then $P_n \in \mathcal{P}(^nX)$ (Fact 38). Since for each fixed $x \in X$ we have $|e_n^*(x)| \leq \frac{1}{2}$ for n sufficiently large, the power series (32) converges absolutely for every $x \in X$. Moreover, it is not difficult to check that for every $a \in X$ it converges absolutely uniformly on $B(a, r)$ for every $0 < r < 1$. However, since $\| P_n \| = 1$ for each $n \in \mathbb{N}$, by Proposition 136 the radius of norm convergence of (32) is equal to 1.

The following important lemma has powerful applications in the theory of power series and analytic mappings. It can be used for example to prove the Hartogs theorem or to give a constructive proof of the Riemann conformal mapping theorem.

Lemma 138 (Polynomial lemma; Franciszek Leja, [Lej1]). *For every $N \in \mathbb{N}$, $d > 0$, and every $q > 1$ there is $\delta = \delta(N, d, q) > 0$ such that if Y is a normed linear space over $\mathbb{K}$, $A_j \subset \mathbb{K}$ are connected compact sets with* $\operatorname{diam} A_j \geq d$, $j = 1, \dots, N$, *and $\Phi \subset \mathcal{P}(\mathbb{K}^N; Y) \setminus \{0\}$ is a family of polynomials pointwise bounded on the set $A = A_1 \times \cdots \times A_N \subset \mathbb{K}^N$, then there is $M > 0$ such that for every $P \in \Phi$ we have*

$$\sup_{x\in A^\delta} \| P(x) \| \leq M q^{\deg P},$$

where $A^\delta = \{x \in \mathbb{K}^N;\ \operatorname{dist}(x, A) < \delta\}$.

To prove this result we need first a technical lemma which asserts that given any subset of an interval with a positive Lebesgue measure, we can always find in this set any number of nodes for the Lagrange interpolation formula which are suitably spaced apart.

Lemma 139. *Let $0 < \varepsilon < r$ and $E \subset [0, r]$ be such that $\lambda(\overline{E}) \geq r - \frac{\varepsilon}{2}$. Further, let $\rho\colon [0, 1] \to [0, 1]$ be a non-decreasing function. Then for each $n \in \mathbb{N}$ there are points $x_0, \dots, x_n \in E$ satisfying $|x_j - x_k| \geq |\rho(\frac{j}{n}) - \rho(\frac{k}{n})|(r - \varepsilon)$ for all $j, k \in \{0, \dots, n\}$.*

Proof. Fix $n \in \mathbb{N}$. Define a function $\eta\colon [0, r] \to \mathbb{R}_0^+$ by $\eta(x) = \lambda(\overline{E} \cap [0, x])$. We find the points $x_0, \dots, x_n \in E$ by an inductive construction so that they satisfy the

Definition 3. Let $B \subset \mathbb{R}^n$ be a subset. We denote by $\mathcal{F}(B) \subset M(B)$ the vector space of all linear combinations of the Dirac measures δ_x, $x \in B$. So, for every $\phi \in \mathcal{F}(B)$ there exist unique finite sequences $\{a_j\}_{j=1}^k \subset \mathbb{R}$, $\{x_j\}_{j=1}^k \subset B$, x_j distinct, such that

$$\phi = \sum_{j=1}^{k} a_j \delta_{x_j}.$$

Because $\mathcal{F}(B)$ separates the points of $C(B)$ and vice versa, $\langle C(B), \mathcal{F}(B)\rangle$ forms a dual pair. If Y is a finite-dimensional subspace of $C(B)$, then every functional from Y^* has a representation as an element of $\mathcal{F}(B)$. In particular, every functional from $\mathcal{P}^d(\mathbb{R}^n)^*$, resp. $\mathcal{P}(^d\mathbb{R}^n)^*$, has a representation as an element of $\mathcal{F}(\mathbb{R}^n)$ using the formula

$$\langle f, \phi\rangle = \phi(f) = \sum_{j=1}^{k} a_j f(x_j), \quad \phi = \sum_{j=1}^{k} a_j \delta_{x_j}, \ f \in C(\mathbb{R}^n).$$

It follows that the linear operator $T\colon \mathcal{F}(\mathbb{R}^n) \to \mathcal{P}^d(\mathbb{R}^n)^*$, $T(\phi)(P) = \langle P, \phi\rangle$ is onto and hence $\mathcal{P}^d(\mathbb{R}^n)^*$ can be identified with $\mathcal{F}(\mathbb{R}^n)/\ker T = \mathcal{F}(\mathbb{R}^n)/\mathcal{K}^d$, where $\mathcal{K}^d \subset \mathcal{F}(\mathbb{R}^n)$ is the subspace consisting of all elements $\phi \in \mathcal{F}(\mathbb{R}^n)$ satisfying $\langle P, \phi\rangle = 0$ for all $P \in \mathcal{P}^d(\mathbb{R}^n)$. Similarly, $\mathcal{P}(^d\mathbb{R}^n)^* = \mathcal{F}(\mathbb{R}^n)/\mathcal{P}(^d\mathbb{R}^n)^\perp$. In order to simplify the language we are sometimes going to identify the elements $x \in \mathbb{R}^n$ with δ_x, calling them nodes for $\mathcal{P}^d(\mathbb{R}^n)$ (resp. $\mathcal{P}(^d\mathbb{R}^n)$).

Suppose that $y_1, \dots, y_r \in \mathbb{R}^n$. We say that the corresponding Dirac functionals $\delta_{y_1}, \dots, \delta_{y_r}$ are $\mathcal{P}^d(\mathbb{R}^n)$-independent (resp. $\mathcal{P}(^d\mathbb{R}^n)$-independent) if they are linearly independent as elements of $\mathcal{P}^d(\mathbb{R}^n)^*$ (resp. $\mathcal{P}(^d\mathbb{R}^n)^*$). In the alternative terminology, we say that the nodes $y_1, \dots, y_r \in \mathbb{R}^n$ are $\mathcal{P}^d(\mathbb{R}^n)$-independent (resp. $\mathcal{P}(^d\mathbb{R}^n)$-independent). It is clear from basic linear algebra that the nodes $y_1, \dots, y_r$ are $\mathcal{P}^d(\mathbb{R}^n)$-independent (resp. $\mathcal{P}(^d\mathbb{R}^n)$-independent) if and only if there exist polynomials $p_1, \dots, p_r \in \mathcal{P}^d(\mathbb{R}^n)$ (resp. $\mathcal{P}(^d\mathbb{R}^n)$) biorthogonal to $y_1, \dots, y_r$, i.e. polynomials satisfying $p_j(y_k) = \delta_{j,k}$ (the Kronecker delta).

Let $K \subset \mathbb{R}^n$ be a compact set and $y_1, \dots, y_N \in K$, and further suppose that $\{(p_k; \delta_{y_k})\}_{k=1}^N$ is a biorthogonal system in $\mathcal{P}^d(\mathbb{R}^n) \times \mathcal{P}^d(\mathbb{R}^n)^*$. We define a linear operator $L\colon C(K) \to \operatorname{span}\{p_1, \dots, p_N\}$ by the formula

$$L(f)(x) = \sum_{k=1}^{N} p_k(x) f(y_k), \quad f \in C(K),\ x \in K. \tag{1}$$

This operator is called the Lagrange interpolation operator. It is easily seen to be a bounded projection onto an N-dimensional subspace $\operatorname{span}\{p_1, \dots, p_N\}$ of $C(K)$. Hence for every $f \in C(K)$ there is a unique $g \in \ker L$ such that $f = L(f) + g$.

If the nodes $y_1, \dots, y_r$ are $\mathcal{P}^d(\mathbb{R}^n)$-independent (resp. $\mathcal{P}(^d\mathbb{R}^n)$-independent), then $r \le |\mathcal{J}(n,d)|$ (resp. $r \le |\mathcal{I}(n,d)|$). If $r = |\mathcal{J}(n,d)|$ (resp. $r = |\mathcal{I}(n,d)|$), the independent nodes form a basis of the dual space $\mathcal{P}^d(\mathbb{R}^n)^*$ (resp. $\mathcal{P}(^d\mathbb{R}^n)^*$) and we call $\{y_k\}_{k=1}^r$ the basic set of nodes for $\mathcal{P}^d(\mathbb{R}^n)$ (resp. $\mathcal{P}(^d\mathbb{R}^n)$). Basic linear algebra assures that any independent set of nodes can be enlarged into a basic set of nodes. It is also easy to see that if $\{y_k\}_{k=1}^r$ is $\mathcal{P}(^d\mathbb{R}^n)$-independent (resp. basic) set of nodes, then for any invertible $L \in \mathcal{L}(\mathbb{R}^n;\mathbb{R}^n)$ the set $\{L(y_k)\}_{k=1}^r$ has the same property. The same conclusion holds for $\mathcal{P}^d(\mathbb{R}^n)$-independent sets of nodes and any affine invertible operator $L\colon \mathbb{R}^n \to \mathbb{R}^n$.

The following proposition follows easily:

Proposition 4. *Let either $Z = \mathcal{P}(^d\mathbb{R}^n)$ and $r = |\mathcal{I}(n,d)|$, or $Z = \mathcal{P}^d(\mathbb{R}^n)$ and $r = |\mathcal{J}(n,d)|$. If $\{y_k\}_{k=1}^r$ is a basic set of nodes for Z and $\{p_k\}_{k=1}^r \subset Z$ are polynomials biorthogonal to $\{y_k\}_{k=1}^r$, then for each $P \in Z$*

$$P(x) = \sum_{k=1}^r P(y_k)p_k(x), \quad x \in \mathbb{R}^n.$$

A classical example of a basic set of nodes for $\mathcal{P}(^d\mathbb{R}^n)$ is the set of multi-indices $\mathcal{I}(n,d) \subset \mathbb{R}^n$.

Theorem 5 ([Bie], see also [Rez3, Proposition 2.11])**.** *The set $\mathcal{I}(n,d)$ is a basic set of nodes for $\mathcal{P}(^d\mathbb{R}^n)$.*

Proof. It suffices to find polynomials biorthogonal to the nodes $\{\alpha;\ \alpha \in \mathcal{I}(n,d)\}$. For $\alpha \in \mathcal{I}(n,d)$ put

$$p_\alpha(x) = \frac{1}{d^d \cdot \alpha_1! \cdots \alpha_n!} \prod_{j=1}^n \prod_{k=0}^{\alpha_j - 1} (d \cdot x_j - k(x_1 + \cdots + x_n)), \quad x \in \mathbb{R}^n.$$

Clearly, $p_\alpha \in \mathcal{P}(^d\mathbb{R}^n)$ and if $\beta \in \mathcal{I}(n,d)$, then

$$p_\alpha(\beta) = \frac{1}{d^d \cdot \alpha_1! \cdots \alpha_n!} \prod_{j=1}^n \prod_{k=0}^{\alpha_j - 1} (d \cdot \beta_j - kd).$$

Thus $p_\alpha(\alpha) = 1$. On the other hand, if $\beta \in \mathcal{I}(n,d)$, $\beta \neq \alpha$, then $\beta_j < \alpha_j$ for some $j \in \{0,\dots,n\}$ and so $p_\alpha(\beta) = 0$. □

The classical Lagrange interpolation theorem (Proposition 1.56) implies that any set of $d+1$ distinct points in $\mathbb{R}$ forms a basic set of nodes for $\mathcal{P}^d(\mathbb{R})$. Similarly to the homogeneous case above the following result holds.

Theorem 6. *The set $\mathcal{J}(n,d)$ is a basic set of nodes for $\mathcal{P}^d(\mathbb{R}^n)$.*

Proof. By Theorem 5, $\mathcal{I}(n+1,d)$ is a basic set of nodes for $\mathcal{P}(^d\mathbb{R}^{n+1})$. Since $\mathcal{I}(n+1,d)$ is a subset of the affine hyperplane $H = \big\{(x_1,\dots,x_{n+1});\ \sum_{j=1}^{n+1} x_j = d\big\}$, by Fact 2 it is a basic set of nodes for $\mathcal{P}^d(H)$ (slightly abusing the notation). Considering the affine isomorphism $T\colon H \to \mathbb{R}^n$, $T(x_1,\dots,x_{n+1}) = (x_1,\dots,x_n)$, we see immediately that $T(\mathcal{I}(n+1,d)) = \mathcal{J}(n,d)$. □

The following is a general characterisation of basic sets of nodes, see also [Lor], [Rez3].

Theorem 7 (Lagrange formula). *Let $d \in \mathbb{N}$ and either $Z = \mathcal{P}(^d\mathbb{R}^n)$ and $\mathcal{I} = \mathcal{I}(n,d)$, or $Z = \mathcal{P}^d(\mathbb{R}^n)$ and $\mathcal{I} = \mathcal{J}(n,d)$. Let $r = |\mathcal{I}|$ and $\mathcal{I} = \{\alpha_1, \dots, \alpha_r\}$. Then $\{y_k\}_{k=1}^r \subset \mathbb{R}^n$ is a basic set of nodes for Z if and only if*

$$\det \begin{pmatrix} y_1^{\alpha_1} & y_1^{\alpha_2} & \dots & y_1^{\alpha_r} \\ y_2^{\alpha_1} & y_2^{\alpha_2} & \dots & y_2^{\alpha_r} \\ \vdots & \vdots & & \vdots \\ y_r^{\alpha_1} & y_r^{\alpha_2} & \dots & y_r^{\alpha_r} \end{pmatrix} \neq 0.$$

In this case $\delta_x = \sum_{k=1}^r p_k(x)\delta_{y_k}$ in Z^* for every $x \in \mathbb{R}^n$, where $\{p_k(x)\}_{k=1}^r$ are the unique solutions of the system of linear equations

$$\sum_{k=1}^r p_k(x) y_k^\alpha = x^\alpha, \quad \alpha \in \mathcal{I}.$$

The functions $x \mapsto p_k(x)$ are polynomials from Z given by the formula

$$p_k(x) = \frac{\det \begin{pmatrix} y_1^{\alpha_1} & \dots & y_1^{\alpha_r} \\ \vdots & & \vdots \\ y_{k-1}^{\alpha_1} & \dots & y_{k-1}^{\alpha_r} \\ x^{\alpha_1} & \dots & x^{\alpha_r} \\ y_{k+1}^{\alpha_1} & \dots & y_{k+1}^{\alpha_r} \\ \vdots & & \vdots \\ y_r^{\alpha_1} & \dots & y_r^{\alpha_r} \end{pmatrix}}{\det \begin{pmatrix} y_1^{\alpha_1} & \dots & y_1^{\alpha_r} \\ \vdots & & \vdots \\ y_r^{\alpha_1} & \dots & y_r^{\alpha_r} \end{pmatrix}}.$$

Finally, $\{(p_k; \delta_{y_k})\}_{k=1}^r$ is a biorthogonal basis for $Z \times Z^*$ and the following Lagrange interpolation formula holds:

$$P(x) = \sum_{k=1}^r p_k(x) P(y_k), \quad P \in Z,\ x \in \mathbb{R}^n.$$

Proof. A set $\{y_k\}_{k=1}^r$ is a basic set of nodes for Z if and only if for every $x \in \mathbb{R}^n$ the equation $\delta_x = \sum_{k=1}^r p_k(x)\delta_{y_k}$ has a unique solution in the space Z^*. Since the set $\{x^\alpha;\ \alpha \in \mathcal{I}\}$ is a basis of Z, this holds if and only if the system

$$x^\alpha = \langle x^\alpha, \delta_x \rangle = \left\langle x^\alpha, \sum_{k=1}^r p_k(x)\delta_{y_k} \right\rangle = \sum_{k=1}^r p_k(x) y_k^\alpha, \quad \alpha \in \mathcal{I}$$

has a unique solution for every $x \in \mathbb{R}^n$. In the homogeneous case this means in particular that there is a unique solution for $x = 0$, which is equivalent to the fact that the matrix of the system is non-singular. To treat the non-homogeneous case assume that $\alpha_1 = (0, \dots, 0)$. Put $x = y_1$. Note that the first row of the augmented matrix equals $(1, \dots, 1)$. We eliminate the first column and note that the last column of the augmented matrix after the elimination equals $(1, 0, \dots, 0)^T$. It follows that this system has a unique solution if and only if its matrix is non-singular.

The rest of the theorem follows from elementary linear algebra and Proposition 4. □

2. Cubature formulae

In this section we give a simple but useful abstract formulation of representation and cubature formulae for linear functionals ϕ acting on $\mathcal{P}^d(\mathbb{R}^n)$. We then continue by studying the seemingly very special functionals ϕ and Φ defined on $\mathbb{R}$ and $\mathbb{R}^2$, which are closely related to Chebyshev polynomials. The formulae obtained in this way will be exploited in the next section in Harris's proof of a general Skalyga-Markov-type inequality. The following theorem will be instrumental for getting upper estimates on the numbers of nodes in various situations.

Theorem 8 (Constantin Carathéodory). *Let C be a compact convex subset of $\mathbb{R}^n$. Then every $x \in C$ is a convex combination of at most $n+1$ extreme points of C.*

Similarly, if $z = \sum_{j=1}^m x_j$, $x_j \in \mathbb{R}^n$, then $z = \sum_{k=1}^n y_k$, where each y_k is a non-negative multiple of some x_j.

Proof. We use induction on n. For $n = 1$ the set C is a closed interval and the statement is clear. Assume that the statement holds for $n-1$. Consider $C \subset \mathbb{R}^n$. Without loss of generality we may assume that $\operatorname{Int} C \neq \emptyset$, otherwise C lies in an affine hyperplane and the claim follows from the inductive hypothesis. If x is a boundary point of C, let H be a supporting hyperplane of C such that $x \in H \cap C$. By the inductive hypothesis, x is a convex combination of at most n extreme points of $H \cap C$, which are extreme points of C.

If $x \in \operatorname{Int} C$, choose an extreme point $z_1 \in C$ and let y be another boundary point of C that lies on the line going through x and z_1. By the same argument as above, y is a convex combination of at most n extreme points $z_2, \dots, z_k$ of C. Thus $y = \sum_{j=2}^k \alpha_j z_j$ with $\alpha_j \geq 0$ and $\sum_{j=2}^k \alpha_j = 1$. Now $x = \alpha y + (1-\alpha) z_0$ for some $\alpha \in (0,1)$, so $x = \sum_{j=2}^k \alpha\alpha_j z_j + (1-\alpha) z_1$, which is a convex combination of $k \leq n+1$ extreme points $z_1, \dots, z_k$.

In order to prove the last statement, it suffices to prove it for $m = n+1$ and then use a straightforward induction. Since x_j are linearly dependent, $\sum_{j=1}^{n+1} \rho_j x_j = 0$ for some $(\rho_1, \dots, \rho_{n+1}) \neq 0$. Without loss of generality $0 < \rho_{n+1} = \max_j \rho_j$. Then $1 \geq \frac{\rho_j}{\rho_{n+1}}$ for all $j \in \{1, \dots, n+1\}$ and

$$z = \sum_{j=1}^{n+1} \left(x_j - \frac{\rho_j}{\rho_{n+1}} x_j \right) = \sum_{j=1}^{n} \left(1 - \frac{\rho_j}{\rho_{n+1}} \right) x_j. \qquad \square$$

Proposition 9. *Let X be a real vector space, $x_1, \dots, x_n \in X$ lie in an affine hyperplane not containing the origin, and $x \in \operatorname{conv}\{x_1, \dots, x_n\}$. Then there is a linearly independent subset $A \subset \{x_1, \dots, x_n\}$ such that $x \in \operatorname{conv} A$.*

Proof. Without loss of generality assume $x \notin \{x_1, \dots, x_n\}$. Let $A \subset \{x_1, \dots, x_n\}$ be a subset with minimal cardinality such that $x \in \operatorname{conv} A$. Without loss of generality we may assume that $A = \{x_1, \dots, x_k\}$, $k \geq 2$. Then $x = \sum_{j=1}^{k} \alpha_j x_j$ where $\sum_{j=1}^{k} \alpha_j = 1$ and $\alpha_j > 0$, $j = 1, \dots, k$. Suppose that $\sum_{j=1}^{k} \lambda_j x_j = 0$ for some $\lambda_j \in \mathbb{R}$, $(\lambda_1, \dots, \lambda_k) \neq 0$. Without loss of generality we may assume that $\frac{|\lambda_k|}{\alpha_k} = \max\{\frac{|\lambda_j|}{\alpha_j};\ j = 1, \dots, k\}$. Then $\lambda_k \neq 0$ and hence $x_k = -\frac{1}{\lambda_k} \sum_{j=1}^{k-1} \lambda_j x_j$. Thus $x = \sum_{j=1}^{k-1} \alpha_j x_j - \frac{\alpha_k}{\lambda_k} \sum_{j=1}^{k-1} \lambda_j x_j = \sum_{j=1}^{k-1} (\alpha_j - \frac{\alpha_k}{\lambda_k} \lambda_j) x_j$, which is a contradiction with the minimality of A. Indeed, by the assumption there is a linear functional f on X such that $f(x_j) = 1$, $j = 1, \dots, k$. Therefore $1 = \sum_{j=1}^{k} \alpha_j f(x_j) = f(x) = \sum_{j=1}^{k-1} (\alpha_j - \frac{\alpha_k}{\lambda_k} \lambda_j)$ and $\alpha_j - \frac{\alpha_k}{\lambda_k} \lambda_j \geq 0$ by the property of $\frac{|\lambda_k|}{\alpha_k}$. □

Of course, different norms on $\mathbb{R}^n$ give rise to different (but equivalent) norms on $\mathcal{P}^d(\mathbb{R}^n)$. So we will consider n dimensional real Banach spaces X and the spaces of polynomials $\mathcal{P}^d(X)$. We obtain a useful representation of the dual elements as follows:

Proposition 10. *Let X be an n-dimensional real Banach space and $\phi \in \mathcal{P}^d(X)^*$ (resp. $\phi \in \mathcal{P}(^d X)^*$). Then ϕ admits a representation formula (as an element of $\mathcal{F}(B_X)$)*

$$\phi(P) = \sum_{k=1}^{N} \lambda_k P(x_k), \quad P \in \mathcal{P}^d(X) \text{ (resp. } P \in \mathcal{P}(^d X)) \tag{2}$$

with $\lambda_k \in \mathbb{R}$, $\sum_{k=1}^{N} |\lambda_k| = \|\phi\|$, and $x_k \in B_X$, $k = 1, \dots, N$.

Moreover, for every representation as above there exists a $\mathcal{P}^d(X)$-independent (resp. $\mathcal{P}(^d X)$-independent) subset of $\{x_k\}_{k=1}^{N}$ such that the representation above holds with this subset of nodes and with coefficients γ_k that satisfy $\operatorname{sgn} \gamma_k = \operatorname{sgn} \lambda_k$. In particular, we may assume without loss of generality that $N \leq |\mathcal{J}(n,d)|$ (resp. $N \leq |\mathcal{I}(n,d)|$).

Proof. We may assume $\|\phi\| = 1$. The space $\mathcal{P}^d(X)$ is linearly isometric to a subspace of $C(B_X)$. The linear operator $T\colon M(B_X) \to M(B_X)/\mathcal{P}^d(X)^\perp = \mathcal{P}^d(X)^*$ is w^*–w^* continuous and maps the unit ball onto the unit ball. Hence each extreme point of the unit ball in $\mathcal{P}^d(X)^*$ is an image under T of an extreme point of the unit ball in $M(B_X)$, [FHHMZ, Proposition 3.67]. The extreme points of the unit ball of $M(B_X)$ are precisely the ± 1 multiples of the Dirac functionals δ_x, $x \in B_X$, [FHHMZ, Lemma 3.116]. Thus each extreme point of the unit ball in $\mathcal{P}^d(X)^*$ can be represented as $\pm\delta_x$ for some $x \in B_X$. By Carathéodory's theorem (Theorem 8) there exist sets $\{x_k\}_{k=1}^{N} \subset B_X$ and $\{\lambda_k\}_{k=1}^{N} \subset \mathbb{R}$, $\sum_{k=1}^{N} |\lambda_k| = 1 = \|\phi\|$, where $N \leq |\mathcal{J}(n,d)|$, such that (2) holds. Analogously for the space $\mathcal{P}(^d X)$.

Now assume that $\phi = \sum_{k=1}^{N} \lambda_k \delta_{x_k}$ for some $\{x_k\}_{k=1}^{N} \subset B_X$, and $\sum_{j=1}^{k} |\lambda_k| = 1$. Then all the functionals $\operatorname{sgn} \lambda_k \delta_{x_k}$ with $\lambda_k \neq 0$ lie in a hyperplane supporting the unit ball of $\mathcal{P}^d(X)^*$ (resp. $\mathcal{P}(^d X)^*$) at ϕ and hence we may apply Proposition 9. □

Observe that if $\phi \in \mathcal{P}^d(X)^*$ with the representation as in Proposition 10 such that $\lambda_k \neq 0$, $k = 1, \dots, N$, then the polynomial $P \in \mathcal{P}^d(X)$, $\|P\| = 1$ for which $\phi(P) = \|\phi\|$ satisfies $P(x_k) = \operatorname{sgn} \lambda_k$, $k = 1, \dots, N$.

Definition 11. Let X be an n-dimensional real Banach space and let $\phi \in \mathcal{P}^d(X)^*$. The representation formula for ϕ from Proposition 10 is called a cubature formula provided that $\lambda_k > 0$, $k = 1, \dots, N$. The coefficients λ_k are called weights.

The next result follows immediately from Proposition 10 and the remark that follows it.

Proposition 12. *Let X be an n-dimensional Banach space and $\phi \in \mathcal{P}^d(X)^* \setminus \{0\}$. Then ϕ admits a representation by a cubature formula if and only if $\phi(P) = \|\phi\|$ for the constant polynomial P that equals* 1. *Moreover, the nodes $\{x_k\}_{k=1}^{N}$ in the cubature formula can be chosen $\mathcal{P}^d(X)$-independent and $N \leq |\mathcal{J}(n, d)|$.*

The next corollary is now immediate:

Corollary 13 (Vladimir L. Chakalov; [Cha]). *Let X be an n-dimensional Banach space and let μ be a finite Borel measure on B_X. Then the functional $\phi \in \mathcal{P}^d(X)^*$, $\phi(p) = \int p \, d\mu$ for $p \in \mathcal{P}^d(X)$ admits a cubature formula.*

Let $K \subset B_X$ be a compact set in an n-dimensional Banach space X. For the rest of this section we are going to investigate the properties of functionals $\phi \in C(K)^*$ given by a finite Borel measure μ on K. It follows from Chakalov's theorem that ϕ restricted to $\mathcal{P}^d(X)$ admits a cubature formula and by Proposition 10 the cardinality of the set of nodes in (2) can be assumed to satisfy $N \leq |\mathcal{J}(n, d)|$. We will be interested in finding lower bounds on N in (2).

Note that if $\operatorname{supp} \mu$ is not contained in a zero set of any non-trivial polynomial from $\mathcal{P}^d(\mathbb{R}^n)$, then the following lower estimate on the number of nodes holds:

$$N \geq \left|\mathcal{J}(n, [\tfrac{d}{2}])\right| = \binom{n + [\frac{d}{2}]}{n}. \tag{3}$$

Indeed, for any set of nodes $z_1, \dots, z_N$ in $\mathbb{R}^n$, where $N < \binom{n+[\frac{d}{2}]}{n}$, there exists a non-trivial polynomial $Q \in \{\delta_{z_1}, \dots, \delta_{z_n}\}^{\perp} \subset \mathcal{P}^{[\frac{d}{2}]}(\mathbb{R}^n)$, which means that Q has zeros at $z_1, \dots, z_N$ and $0 \neq Q^2 \in \mathcal{P}^d(\mathbb{R}^n)$. But $\phi(Q^2) > 0$, a contradiction with the existence of a cubature formula with nodes $z_1, \dots, z_N$. We remark that Hans Michael Möller (see [Mö]) has found a better estimate for centrally symmetric functionals from $\mathcal{P}^{2k-1}(\mathbb{R}^2)^*$ (i.e. functionals that vanish on polynomials that are odd functions), namely

$$N \geq \binom{k+1}{2} + \left[\frac{k}{2}\right]. \tag{4}$$

We now pass to the discussion of the classical case of $X = \mathbb{R}$ and the functional

$$\phi(f) = \int_{-1}^{1} f \, \mathrm{d}\mu, \quad f \in C([-1,1]),$$

where μ is a finite Borel measure on $[-1,1]$ such that $\operatorname{supp}\mu$ is an infinite set, see also [BE]. We show a method for finding an explicit cubature formula for this functional restricted to $\mathcal{P}^d(\mathbb{R})$. The first step is an easy construction of a (unique up to a sign) sequence of polynomials $\{P_n\}_{n=0}^{\infty}$, $\deg P_n = n$, satisfying the orthogonality relations

$$\int_{-1}^{1} P_m P_n \, \mathrm{d}\mu = \delta_{m,n}, \quad m, n \in \mathbb{N}_0. \tag{5}$$

This is done by the Gram-Schmidt orthogonalisation. Since P_0 is a constant and non-zero function, the orthogonality relations immediately imply that

$$\int_{-1}^{1} P_n \, \mathrm{d}\mu = 0 \quad \text{for } n \geq 1. \tag{6}$$

Further, each P_n has exactly n distinct roots in $[-1,1]$. Indeed, if this is not the case, then there is $Q \in \mathcal{P}^{n-1}(\mathbb{R})$ which has only simple roots and these are precisely at the points of $[-1,1]$ where P_n has roots of odd multiplicity. Then $P_n Q$ does not change sign on $[-1,1]$, a contradiction with (5).

Let $\zeta_0, \dots, \zeta_n$ be the roots of $P_{n+1}(t) = a_{n+1}(t-\zeta_0)\cdots(t-\zeta_n)$. Then the system $\left\{\frac{P_{n+1}(t)}{P'_{n+1}(\zeta_j)(t-\zeta_j)}\right\}_{j=0}^{n}$ consists of polynomials from $\mathcal{P}^n(\mathbb{R})$ which are biorthogonal to the nodes $\{\zeta_j\}_{j=0}^{n}$. Hence from Proposition 4 we obtain immediately the formula

$$\phi(P) = \sum_{k=0}^{n} \lambda_k P(\zeta_k), \quad P \in \mathcal{P}^n(\mathbb{R}), \tag{7}$$

where the weights satisfy

$$\lambda_k = \frac{1}{P'_{n+1}(\zeta_k)} \int_{-1}^{1} \frac{P_{n+1}(t)}{t-\zeta_k} \, \mathrm{d}\mu(t), \quad k = 0, \dots, n.$$

We will show that $\lambda_k > 0$. Since $\phi(P) = \|\phi\|$ for a constant polynomial P equal 1, the above is a cubature formula.

Moreover, the orthogonality relations can be easily applied to show that (7) actually holds for all polynomials $P \in \mathcal{P}^{2n+1}(\mathbb{R})$. Indeed, the Lagrange interpolation polynomial of $P \in \mathcal{P}^{2n+1}(\mathbb{R})$ at the nodes $\zeta_0, \dots, \zeta_n$ obtained by

$$L_n(P)(t) = \sum_{k=0}^{n} \frac{P(\zeta_k) P_{n+1}(t)}{P'_{n+1}(\zeta_k)(t-\zeta_k)}$$

is a polynomial of degree at most n which coincides with P at all points $\zeta_0, \dots, \zeta_n$. Hence $P - L_n(P) = P_{n+1} R$ for some polynomial R of degree at most n and the equality (7) for P now follows easily using (5). On the other hand, by (3) it is easy to see that there is no cubature formula for ϕ on $\mathcal{P}^{2n+2}(\mathbb{R})$ with at most $n+1$ nodes.

Now we can show that $\lambda_k > 0$. Indeed, the polynomial $t \mapsto \left(\frac{P_{n+1}(t)}{t-\xi_k}\right)^2$ belongs to $\mathcal{P}^{2n}(\mathbb{R})$ and hence by (7)

$$\frac{1}{(P'_{n+1}(\zeta_k))^2} \int_{-1}^{1} \left(\frac{P_{n+1}(t)}{t-\zeta_k}\right)^2 \mathrm{d}\mu(t) = \lambda_k, \quad k = 0, \ldots, n.$$

Let us describe in more detail the special case when $d\mu = w d\lambda$ for the weight function $w(x) = \frac{1}{\pi}\frac{1}{\sqrt{1-x^2}}$. This case plays an important role in the proof of the Markov-type inequalities below. We begin by describing the orthogonal system corresponding to this functional, which consists of suitable scalar multiples of the Chebyshev system of polynomials.

Definition 14. Let $n \in \mathbb{N}_0$. The uniquely determined polynomial T_n that satisfies $T_n(x) = \cos\bigl(n \arccos(x)\bigr)$ on $[-1, 1]$ is called the Chebyshev polynomial of degree n.

We note that indeed such a polynomial exists, is of degree n, and satisfies

$$T_n(x) = \sum_{j=0}^{n} t_j^n x^j, \quad \text{where} \quad \begin{cases} t_{n-(2k+1)}^n = 0, & k = 0, \ldots, [\frac{n-1}{2}], \\ t_{n-2k}^n = (-1)^k \sum_{j=k}^{[\frac{n}{2}]} \binom{n}{2j}\binom{j}{k}, & k = 0, \ldots, [\frac{n}{2}]. \end{cases}$$

In particular, $t_n^n = 2^{n-1}$ ([Lov, Problem 1.42a]). To get the above expansion it suffices to use the trigonometric formula

$$\cos(n\eta) = \cos^n \eta - \binom{n}{2} \cos^{n-2}\eta \sin^2\eta + \cdots + (-1)^{[\frac{n}{2}]} \binom{n}{2[\frac{n}{2}]} \cos^{n-2[\frac{n}{2}]}\eta \sin^{2[\frac{n}{2}]}\eta.$$

Combining with the identity $\sin^2\eta = 1 - \cos^2\eta$ and collecting the appropriate terms we obtain

$$\cos(n\eta) = \sum_{k=0}^{[\frac{n}{2}]} \left((-1)^k \sum_{j=k}^{[\frac{n}{2}]} \binom{n}{2j}\binom{j}{k} \right) \cos^{n-2k}\eta.$$

From the trigonometric formula $\cos(n\eta) = \frac{1}{2}(e^{in\eta} + e^{-in\eta})$ we obtain the identity

$$T_n(x) = \frac{1}{2}\left(\left(x + \sqrt{x^2-1}\right)^n + \left(x - \sqrt{x^2-1}\right)^n \right). \tag{8}$$

From the definition we immediately obtain the roots and the points of local extrema of T_n.

Fact 15. *The roots of T_n are at the points $\xi_k^n = \cos\bigl(\frac{(2k-1)\pi}{2n}\bigr)$, $k = 1, \ldots, n$. The local extrema of T_n on $[-1, 1]$ (which are also global there) are attained at the points $\eta_k^n = \cos\bigl(\frac{k\pi}{n}\bigr)$, $k = 0, \ldots, n$ and $T_n(\eta_k^n) = (-1)^k$. Note that $\eta_0^n = 1$, $\eta_n^n = -1$ are the endpoints of $[-1, 1]$. Moreover,*

$$\int_{-1}^{1} T_m(x) T_n(x) \frac{1}{\sqrt{1-x^2}} \mathrm{d}\lambda = \begin{cases} 0 & \textit{if } m \neq n, \\ \frac{\pi}{2} & \textit{if } m = n \neq 0, \\ \pi & \textit{if } m = n = 0. \end{cases}$$

Proof. The last relation follows easily from the well-known formulae

$$\int_0^\pi \cos m\theta \cos n\theta \,\mathrm{d}\theta = \begin{cases} 0 & \text{if } m \neq n, \\ \frac{\pi}{2} & \text{if } m = n \neq 0, \\ \pi & \text{if } m = n = 0, \end{cases}$$

via the change of variable $x = \cos\theta$. □

From now on we will keep the notation for the roots ξ_k^n and points of local extrema η_k^n of T_n from Fact 15.

Letting $P_0 = 1$, $P_n = \sqrt{2}T_n$ we get an orthonormal system of polynomials $\{P_n\}_{n=0}^\infty$ with respect to the weight function $w(x) = \frac{1}{\pi}\frac{1}{\sqrt{1-x^2}}$ on $[-1,1]$. Hence for the functional

$$\phi(p) = \frac{1}{\pi}\int_{-1}^{1} \frac{p(x)}{\sqrt{1-x^2}}\,\mathrm{d}\lambda, \quad p \in \mathcal{P}^{2n-1}(\mathbb{R})$$

the general formula (7) implies the following cubature formula with n nodes:

$$\phi(p) = \sum_{j=1}^{n} \frac{1}{n} p(\xi_j^n), \quad p \in \mathcal{P}^{2n-1}(\mathbb{R}). \tag{9}$$

To see that the weights are all equal to $\frac{1}{n}$ it suffices to check the validity of the formula for the polynomials in $\mathcal{P}^{n-1}(\mathbb{R})$ (see the discussion above), and in particular only for its basis $\{T_k\}_{k=0}^{n-1}$. Using (6) and the fact that $\sum_{j=1}^n T_k(\xi_j^n) = 0$, $k = 1,\dots,n-1$ (see [Ri, Exercise 1.5.25]) we get $\phi(T_k) = 0 = \sum_{j=1}^n \frac{1}{n}T_k(\xi_j^n)$, $k = 1,\dots,n-1$, and moreover $\phi(T_0) = 1 = \sum_{j=1}^n \frac{1}{n}$.

By using elementary manipulations of trigonometric functions (which were used to define the Chebyshev polynomials), one can prove another cubature formula for the same linear functional ϕ, using instead the points of extrema $\{\eta_k^n\}_{k=0}^n$ of T_n as nodes (see [Ri, Exercise 1.5.29]):

$$\phi(p) = \sum_{j=0}^{n} \lambda_j p(\eta_j^n) = \sum_{j=1}^{n-1} \frac{1}{n} p(\eta_j^n) + \frac{1}{2n}p(-1) + \frac{1}{2n}p(1), \; p \in \mathcal{P}^{2n-1}(\mathbb{R}). \tag{10}$$

An easy way to generate cubature formulae in higher dimensional spaces consists of taking the product measures.

Proposition 16. *Let X_k be an n_k-dimensional Banach space and let μ_k be a Borel measure on B_{X_k}, $k = 0,1$. Assume that $\{x_j^k\}_{j\in N_k}$, $k = 0,1$, are the sets of nodes for the cubature formulae*

$$\int_{B_{X_k}} P\,\mathrm{d}\mu_k = \sum_{j\in N_k} \lambda_j^k P(x_j^k), \quad P \in \mathcal{P}^d(X_k),\ k = 0,1.$$

Then $\{(x_j^0, x_l^1)\}_{(j,l)\in N_0\times N_1}$ *is a set of nodes such that the following cubature formula holds for the product measure functional:*

$$\int_{B_{X_0}\times B_{X_1}} P(x,y)\,\mathrm{d}\mu_0(x)\times d\mu_1(y) = \sum_{(j,l)\in N_0\times N_1} \lambda_j^0\lambda_l^1 P(x_j^0, x_l^1), \quad P\in \mathcal{P}^d(X_0\oplus_\infty X_1).$$

Moreover, if the respective sets of nodes are $\mathcal{P}^d(X_k)$*-independent, then the set of nodes* $\{(x_j^0, x_l^1)\}_{(j,l)\in N_0\times N_1}$ *is* $\mathcal{P}^{2d}(X_0\oplus_\infty X_1)$*-independent.*

Proof. By the Fubini theorem the formula holds for every $P(x,y) = P_0(x)P_1(y)$, where $\deg P_k \le d$, $k = 0,1$. Since the linear span of the polynomials of this form contains $\mathcal{P}^d(X_0\oplus_\infty X_1)$ as a subspace, the claim follows.

If the respective sets of nodes are $\mathcal{P}^d(X_k)$-independent, then they admit a biorthogonal system of polynomials $\{P_j^k\}_{j\in N_k}$ of degree at most d. It is easy to see that the set consisting of polynomials of degree at most $2d$ defined as $\{P_j^0(x)P_l^1(y)\}_{(j,l)\in N_0\times N_1}$ is biorthogonal to the set of nodes $\{(x_j, y_l)\}_{(j,l)\in N_0\times N_1}$, witnessing the fact that this set of nodes is $\mathcal{P}^{2d}(X_0\oplus_\infty X_1)$-independent. □

Suppose that $\{x_j^k\}_{j\in N_k}\subset B_{X_k}$ form basic sets of nodes for $\mathcal{P}^d(X_k)$, $k=0,1$. Given $u\in X_0$, $v\in X_1$ let $\delta_u = \sum_{j\in N_0}\alpha_j\delta_{x_j^0}$ in $\mathcal{P}^d(X_0)^*$ and $\delta_v = \sum_{j\in N_1}\beta_j\delta_{x_j^1}$ in $\mathcal{P}^d(X_1)^*$. Then

$$\begin{aligned}\delta_{(u,v)}(P) &= \delta_u\big(P(\cdot,v)\big) = \sum_{j\in N_0}\alpha_j\delta_{x_j^0}\big(P(\cdot,v)\big) = \sum_{j\in N_0}\alpha_j\delta_v\big(P(x_j^0,\cdot)\big)\\ &= \sum_{j\in N_0}\alpha_j\sum_{l\in N_1}\beta_l\delta_{x_l^1}\big(P(x_j^0,\cdot)\big) = \sum_{j\in N_0}\sum_{l\in N_1}\alpha_j\beta_l\delta_{(x_j^0,x_l^1)}(P)\end{aligned}$$

for any $P\in\mathcal{P}^d(X_0\oplus_\infty X_1)$. Therefore $\operatorname{span}\{(x_j^0,x_l^1)\}_{(j,l)\in N_0\times N_1} = \mathcal{P}^d(X_0\oplus X_1)^*$. Hence by Proposition 10 there is a subset $A\subset N_0\times N_1$ such that $\{(x_j^0,x_l^1)\}_{(j,l)\in A}$ is a basic set of nodes for $\mathcal{P}^d(X_0\oplus X_1)$ and moreover the product measure functional in Proposition 16 admits a cubature formula whose nodes are a subset thereof.

Let us apply these simple abstract observations to the concrete situation of Chebyshev nodes. One can combine any of the formulae (9) and (10) in order to find a cubature formula for the functional

$$\Phi(p) = \frac{1}{\pi^2}\int_{-1}^{1}\int_{-1}^{1} p(x,y)\frac{1}{\sqrt{1-x^2}}\frac{1}{\sqrt{1-y^2}}\,\mathrm{d}x\,\mathrm{d}y, \quad p\in\mathcal{P}^{2n-1}(\mathbb{R}^2).$$

Let $P_0 = 1$ and $P_n = \sqrt{2}T_n$ for $n\ge 1$ be the orthogonal system as above. We put $Q_{j,l}(x,y) = P_j(x)P_l(y)$. The system $\{Q_{j,l}\}_{0\le j+l\le 2n-1}$ is a basis of the vector space $\mathcal{P}^{2n-1}(\mathbb{R}^2)$, since by the Fubini theorem it is an orthonormal basis with respect to the inner product

$$\langle P,Q\rangle = \frac{1}{\pi^2}\int_{-1}^{1}\int_{-1}^{1} P(x,y)Q(x,y)\frac{1}{\sqrt{1-x^2}}\frac{1}{\sqrt{1-y^2}}\,\mathrm{d}x\,\mathrm{d}y.$$

Thus the functional Φ is determined by its values on these polynomials. Note that by the Fubini theorem $\Phi(Q_{j,l}) = \phi(P_j)\phi(P_l)$. Using Proposition 16 and (10) we get

$$\Phi(p) = \sum_{(r,q)\in\{0,\dots,n\}^2} \frac{1}{n^2} c_r c_q p(\eta_r^n, \eta_q^n), \quad p \in \mathcal{P}^{2n-1}(\mathbb{R}^2),$$

where $c_0 = c_n = \frac{1}{2}$ and $c_r = 1$ otherwise.

We proceed by finding optimal improvements to this cubature formula. Fix $n \in \mathbb{N}$. Let M_0 be the set of all even numbers from $\{0,\dots,n\}$ and $M_1 = \{0,\dots,n\} \setminus M_0$ be the set of all odd numbers from $\{0,\dots,n\}$. We also introduce a partition of $\{0,\dots,n\}^2$ into two disjoint sets, namely

$$Q_0^n = M_0 \times M_0 \cup M_1 \times M_1, \qquad Q_1^n = M_1 \times M_0 \cup M_0 \times M_1. \tag{11}$$

We use a method from [BojPe] to derive two optimal improvements of the above cubature formula (which were obtained originally in [X2] and [MoPa]). We introduce the following linear functionals on $\mathcal{P}^{2n-1}(\mathbb{R})$:

$$E(p) = \sum_{j\in M_0}^{n} \frac{1}{n} c_j p(\eta_j^n), \quad p \in \mathcal{P}^{2n-1}(\mathbb{R}),$$

$$O(p) = \sum_{j\in M_1}^{n} \frac{1}{n} c_j p(\eta_j^n), \quad p \in \mathcal{P}^{2n-1}(\mathbb{R}).$$

Then $\phi(p) = E(p)+O(p)$ for $p \in \mathcal{P}^{2n-1}(\mathbb{R})$. By (6), $0 = \phi(P_k) = O(P_k)+E(P_k)$ for $k = 1,\dots,n-1$. We claim that $E(P_k) = O(P_k) = 0$, $k = 1,\dots,n-1$.

We distinguish two cases. First assume $n = 2m$. Then $\eta_{2j-1}^{2m} = \cos\left(\frac{(2j-1)\pi}{2m}\right) = \xi_j^m$, $j = 1,\dots,m$. By (9),

$$0 = \phi(P_k) = \sum_{j=1}^{m} \frac{1}{n} P_k(\xi_j^m) = \sum_{j=1}^{m} \frac{1}{n} P_k(\eta_{2j-1}^n) = O(P_k), \quad k = 1,\dots,n-1.$$

Hence also $E(P_k) = 0$, $k = 1,\dots,n-1$. In the remaining case when $n = 2m-1$ we use

$$E(P_k) = \frac{1}{2n} P_k(\eta_0^n) + \frac{1}{n} \sum_{j=1}^{m-1} P_k(\eta_{2j}^n) = \frac{\sqrt{2}}{n}\left(\frac{1}{2} + \sum_{j=1}^{m-1} \cos\left(j\frac{2k\pi}{2m-1}\right)\right) = 0,$$

where the last equality is standard to check using $\cos t = \operatorname{Re} e^{it}$.

Next we define the following functionals:

$$A(p) = \sum_{(r,q)\in Q_0^n} \frac{1}{n^2} c_r c_q p(\eta_r^n, \eta_q^n), \quad p \in \mathcal{P}^{2n-1}(\mathbb{R}^2),$$

$$B(p) = \sum_{(r,q)\in Q_1^n} \frac{1}{n^2} c_r c_q p(\eta_r^n, \eta_q^n), \quad p \in \mathcal{P}^{2n-1}(\mathbb{R}^2).$$

It is clear that $\Phi = A + B$. We claim that $A(p) = B(p)$ for $p \in \mathcal{P}^{2n-1}(\mathbb{R}^2)$. It suffices to prove this claim for all polynomials from the basis $\{Q_{j,l}\}_{0\le j+l\le 2n-1}$ of $\mathcal{P}^{2n-1}(\mathbb{R}^2)$. But this is straightforward. Suppose e.g. that $l = \min\{j,l\}$. Clearly $l \le n-1$. If $l > 0$, then $E(P_l) = O(P_l) = 0$ and so $A(Q_{j,l}) = E(P_j)E(P_l) + O(P_j)O(P_l) = 0$ and $B(Q_{j,l}) = E(P_j)O(P_l) + O(P_j)E(P_l) = 0$. If $l = 0$, then $E(P_0) = \sum_{j\in M_0} \frac{1}{n}c_j = \frac{1}{2} = \sum_{j\in M_1} \frac{1}{n}c_j = O(P_0)$ and hence we obtain $A(Q_{j,0}) = \frac{1}{2}E(P_j) + \frac{1}{2}O(P_j) = B(Q_{j,0})$. Consequently $\Phi = 2A = 2B$, i.e.

$$\Phi(p) = \sum_{(r,q)\in Q_0^n} \frac{2}{n^2} c_r c_q\, p(\eta_r^n, \eta_q^n) = \sum_{(r,q)\in Q_1^n} \frac{2}{n^2} c_r c_q\, p(\eta_r^n, \eta_q^n), \quad p \in \mathcal{P}^{2n-1}(\mathbb{R}^2).$$

Denote by $N_k^n = \{(\eta_r^n, \eta_q^n);\ (r,q) \in Q_k^n\}$, $k = 0, 1$, the corresponding disjoint sets of nodes. Clearly $|N_1^n| = |N_0^n| = \frac{(n+1)^2}{2}$ for n odd and $|N_1^n| = |N_0^n| - 1 = \frac{n(n+2)}{2}$ for n even. Thus the cubature formulae above attain (for n even and $k = 1$) or exceed by at most one the lower bound for the number of nodes given by (4).

Yuan Xu and Lawrence Albert Harris managed to find explicit formulae for the biorthogonal polynomials to the above systems of nodes N_k^n. Namely, for $(r,q) \in Q_k^n$ we can choose

$$P_{r,q}(x,y) = \frac{8c_r c_q}{n^2} \sum_{l=0}^{n}{}'' \sum_{j=0}^{l}{}'' T_{l-j}(x)T_j(y)T_{l-j}(\eta_r^n)T_j(\eta_q^n) - \frac{c_r c_q}{n^2}\big(T_n(x)T_n(\eta_r^n) + T_n(y)T_n(\eta_q^n)\big).$$

Here the symbol $\sum''$ indicates that the first and the last terms of the sum should be divided by 2. Clearly, $P_{r,q} \in \mathcal{P}^n(\mathbb{R}^2)$. Define further $V_j \in \mathcal{P}^n(\mathbb{R}^2)$ as

$$V_j(x,y) = T_{n-j}(x)T_j(y) - (-1)^k T_j(x)T_{n-j}(y), \quad j = 0, \dots, n.$$

Denote $V = \operatorname{span}\{V_j\}_{j=0}^n$. We omit the proof of the next Lagrange interpolation result, which is based on not very difficult computations involving Chebyshev polynomials given in [X2] or [Harr4].

Theorem 17. *The set $\{P_{r,q};\ (r,q) \in Q_k^n\}$ is biorthogonal to the set of nodes N_k^n, while $V_j(\eta_r^n, \eta_q^n) = 0$ for all $(\eta_r^n, \eta_q^n) \in N_k^n$, $k = 0, 1$. Moreover,*

$$\mathcal{P}^n(\mathbb{R}^2) = \operatorname{span}\{P_{r,q};\ (r,q) \in Q_k^n\} \oplus V.$$

In particular, for every $p \in \mathcal{P}^n(\mathbb{R}^2)$ there is a unique polynomial $v \in V$ such that the following Lagrange interpolation formula holds:

$$p(x,y) = \sum_{(r,q)\in Q_k^n} p(\eta_r^n, \eta_q^n) P_{r,q}(x,y) + v(x,y), \quad p \in \mathcal{P}^n(\mathbb{R}^2),\ k = 0, 1.$$

3. Estimates related to Chebyshev polynomials

The main result generalising the classical Markov inequalities, due to Valentin Ivanovich Skalyga, is a general estimate on the norm of the kth derivative of a real polynomial in terms of the norm of the polynomial. We give an outline of the proof, due to Lawrence Albert Harris, leaving out the lengthy and non-trivial calculations involving certain properties of the Chebyshev polynomials. We begin the section by describing some easier one-dimensional results, where some of the ideas appear in a more transparent form.

The method of the proof of the following result of Werner Wolfgang Rogosinski leads naturally to the study of extremal polynomials, see [Ri]. The same basic approach will be used also in the proof of the theorem of Harris.

Lemma 18 (Rogosinski, [Rog]). *Let $x_0 < x_1 < \cdots < x_n$ be real numbers, $M_j \geq 0$, $j = 0, \ldots, n$, and let $P \in \mathcal{P}^n(\mathbb{R})$ be the unique polynomial determined by the conditions $P(x_j) = (-1)^{n-j} M_j$, $j = 0, \ldots, n$. Then for any polynomial p on $\mathbb{R}$ of degree at most n with complex coefficients satisfying $|p(x_j)| \leq M_j$, $j = 0, \ldots, n$, and for any $0 \leq k \leq n$ we have $|p^{(k)}(x)| \leq |P^{(k)}(x)|$ whenever $x \in \mathbb{R} \setminus (x_0, x_n)$. If there are $k \in \{0, \ldots, n\}$ and $x \in \mathbb{R} \setminus (x_0, x_n)$, $x \neq x_0$ and $x \neq x_n$ in case $k = 0$, such that $|p^{(k)}(x)| = |P^{(k)}(x)|$, then $p = \theta P$ for some $\theta \in \mathbb{C}$, $|\theta| = 1$.*

Proof. Suppose that the polynomial p satisfies the assumptions. Fix $k \in \{0, \ldots, n\}$. By the Lagrange interpolation formula (Proposition 1.56)

$$p^{(k)}(x) = \sum_{j=0}^{n} \frac{p(x_j)}{w_j(x_j)} w_j^{(k)}(x)$$

for all $x \in \mathbb{R}$, where $w_j(x) = \prod_{\substack{l=0 \\ l \neq j}}^{n} (x - x_l)$. Further, fix $x \in \mathbb{R} \setminus (x_0, x_n)$ (or $x \in \mathbb{R} \setminus [x_0, x_n]$ if $k = 0$; note that in this case the inequality trivially holds in x_0 and x_n). Notice that the leading coefficients of $w_j^{(k)}$ are positive and hence all the numbers $w_j^{(k)}(x)$, $j = 0, \ldots, n$, have the same sign and are non-zero. Since $\operatorname{sgn} w_j(x_j) = (-1)^{n-j}$, from the Lagrange interpolation formula we obtain

$$|P^{(k)}(x)| = \sum_{j=0}^{n} \frac{M_j}{|w_j(x_j)|} |w_j^{(k)}(x)| \geq |p^{(k)}(x)|.$$

Moreover, the equality clearly occurs only if there is $\theta \in \mathbb{C}$, $|\theta| = 1$ such that $p(x_j) = \theta M_j \operatorname{sgn} w_j(x_j) = \theta P(x_j)$, which implies that $p = \theta P$. □

Theorem 19 (Issai Schur, [Schu]). *Suppose that p is a polynomial on $\mathbb{R}$ of degree at most n with complex coefficients that satisfies $\max_{j=0,\ldots,n} |p(\eta_j^n)| \leq 1$. Then $|p^{(k)}(x)| \leq |T_n^{(k)}(x)|$ for all $x \in \mathbb{R} \setminus (-1, 1)$, $k = 0, \ldots, n$. If there are $k \in \{0, \ldots, n\}$ and $x \in \mathbb{R} \setminus (-1, 1)$, $x \neq \pm 1$ in case $k = 0$, such that $|p^{(k)}(x)| = |T_n^{(k)}(x)|$, then $p = \theta T_n$ for some $\theta \in \mathbb{C}$, $|\theta| = 1$.*

Proof. It suffices to apply Lemma 18 to the points $x_j = \eta^n_{n-j}$, $j = 0, \dots, n$ in the interval $[-1, 1]$, where T_n attains its extremal values by Fact 15, and the values $M_j = 1$. Clearly, T_n serves as the polynomial P in Lemma 18. □

The rest of the domain for $k > 0$ is dealt with in the much harder result of Richard James Duffin and Albert Charles Schaeffer.

Theorem 20 (Duffin and Schaeffer, [DuSc]). *Suppose that $p \in \mathcal{P}^n(\mathbb{R})$ satisfies $\max_{j=0,\dots,n} |p(\eta^n_j)| \le 1$. Then $\left|p^{(k)}(x)\right| \le T_n^{(k)}(1)$ for all $x \in [-1, 1]$, $k = 1, \dots, n$. Equality holds only if $p = \pm T_n$ and $x = \pm 1$.*

For the proof see [Ri, Theorem 2.24]. This result implies in particular the classical inequalities of Andrei Andreevich Markov and Vladimir Andreevich Markov (Theorem 80, Theorem 81), which are formulated using a stronger assumption, namely assuming that $\|p\| \le 1$.

Theorem 21 ([Er]). *Let p be a polynomial on $\mathbb{C}$ of degree at most n with real coefficients. Suppose that $\max_{k=0,\dots,n} |p(\eta^n_k)| \le 1$. Then $|p(z)| \le |T_n(z)|$ for all $z \in \mathbb{C}$, $|z| \ge 1$. If there is equality for some $z \in \mathbb{C}$, $|z| \ge 1$, $z \ne \pm 1$, then $p = T_n$ or $p = -T_n$.*

Proof. We follow the proof of Lemma 18 using the system of points $x_j = \eta^n_{n-j}$, $j = 0, \dots, n$. By the Lagrange interpolation formula (Proposition 1.56) for any $z \in \mathbb{C}$ we have

$$p(z) = \sum_{j=0}^{n} \frac{p(x_j)}{w_j(x_j)} w_j(z) = \left(\prod_{j=0}^{n} (z - x_j)\right) \sum_{j=0}^{n} \frac{p(x_j)}{w_j(x_j)} \frac{1}{z - x_j}. \tag{12}$$

Fix $z \in \mathbb{C}$, $|z| \ge 1$, $z \ne \pm 1$. Since $x_j \in [-1, 1]$, a classical result from elementary geometry implies that the vectors $z - x_0$ and $z - x_n$ form an acute (or right) angle and the remaining vectors $z - x_j$ lie inside this angle. The same fact remains true for the reciprocal values $\frac{1}{z - x_j}$, $j = 0, \dots, n$. In order to maximise the modulus of the last sum in (12), note that $w_j(x_j)$, $j = 0, \dots, n$ are real numbers, $\operatorname{sgn} w_j(x_j) = (-1)^{n-j}$, and $p(x_j)$ are by our assumption also real numbers. Thus the sum will have maximal modulus if and only if there is $\theta = \pm 1$ such that $p(x_j) = \theta(-1)^{n-j}$, $j = 0, \dots, n$. This holds only if $p = \theta T_n$. □

Theorem 22 (Pafnuty Lvovich Chebyshev; [Ri, Theorem 2.1]). *Let p be a polynomial on $\mathbb{R}$ such that $p(x) = x^n + a_{n-1}x^{n-1} + \dots + a_0$, $a_j \in \mathbb{C}$, $j = 0, \dots, n-1$. Then*

$$\|p\| \ge \max_{0 \le k \le n} |p(\eta^n_k)| \ge \frac{1}{2^{n-1}}.$$

In the last relation the equality holds if and only if $p = \frac{1}{2^{n-1}} T_n$.

Proof. Put $q = \frac{p}{\max_{0 \le k \le n} |p(\eta^n_k)|}$. By Theorem 19

$$\frac{n!}{\max_{0 \le k \le n} |p(\eta^n_k)|} = \left|q^{(n)}(1)\right| \le \left|T_n^{(n)}(1)\right| = n! 2^{n-1}$$

and the equality holds if and only if $q = \theta T_n$ for some $\theta \in \mathbb{C}$, $|\theta| = 1$. For the evaluation of $T_n^{(n)}(1)$ see the remark after Definition 14. □

Fact 23. *The Chebyshev polynomials satisfy*

$$\max_{x\in\mathbb{C},|x|\le 1} |T_n(x)| = |T_n(i)| = \frac{\left(1+\sqrt{2}\right)^n + \left(1-\sqrt{2}\right)^n}{2}.$$

Proof. It is clear from (8) that $T_n(x) = \frac{1}{2}x^n\Big(\Big(1+\sqrt{1-\frac{1}{x^2}}\Big)^n + \Big(1-\sqrt{1-\frac{1}{x^2}}\Big)^n\Big)$ for $x \in \mathbb{C}\setminus\{0\}$ (note that it does not matter which branch of the square root we use provided that both are the same), so by the Maximum modulus principle (Theorem 1.150)

$$\max_{x\in\mathbb{C},|x|\le 1} |T_n(x)| = \max_{x\in\mathbb{C},|x|=1} |T_n(x)| \le \max_{|\xi|=1} \frac{1}{2}\Big(\big(1+\sqrt{1-\xi}\,\big)^n + \big(1-\sqrt{1-\xi}\,\big)^n\Big).$$

Let $\sqrt{1-\xi} = \theta r$, where $r \ge 0$, $|\theta| = 1$. Then

$$\big(1+\sqrt{1-\xi}\,\big)^n + \big(1-\sqrt{1-\xi}\,\big)^n = \sum_{j=0}^{n} \binom{n}{j} r^j \left(\theta^j + (-1)^j\theta^j\right) = \sum_{j=0}^{[\frac{n}{2}]} \binom{n}{2j} r^{2j} 2\theta^{2j}.$$

On the other hand,

$$\left|\sum_{j=0}^{[\frac{n}{2}]} \binom{n}{2j} r^{2j} 2\theta^{2j}\right| \le \sum_{j=0}^{[\frac{n}{2}]} \binom{n}{2j} 2r^{2j}.$$

It is clear that for a given $r > 0$, the maximum of the modulus occurs if and only if $\theta = \pm 1$ and so it suffices to find $\max_{|\xi|=1}\left|\sqrt{1-\xi}\right|$. This occurs for $\xi = -1$, i.e. e.g. for $x = i$. This finishes the proof. □

Corollary 24 ([Er]). *Suppose that p is a polynomial on $\mathbb{C}$ of degree at most n with real coefficients satisfying* $\max_{j=0,\dots,n} |p(\eta_j^n)| \le 1$. *Then* $|p(z)| \le |T_n(i)|$ *for all* $z \in \mathbb{C}$, $|z| \le 1$.

Proof. Combine the Maximum modulus principle, Theorem 21, and Fact 23. □

The following result, due to Alain Durand in case of $X = Y = \mathbb{C}$, describes the growth of the maximum of the modulus of complex polynomial with the radius of the domain. Results of this type are treated e.g. in [Dur] or [Kl1]. The validity of the result for arbitrary complex normed linear spaces was observed in [La].

Proposition 25 ([Dur], [La]). *Let X, Y be complex normed linear spaces and let $P \in \mathcal{P}^n(X;Y)$. Then for every $r \in (0,1]$*

$$\|P\| \le \frac{1}{r^n} \sup_{x\in B(0,r)} \|P(x)\|.$$

Proof. Suppose $X = Y = \mathbb{C}$, $p(z) = \sum_{j=0}^{n} a_j z^j$, and let $q(z) = \sum_{j=0}^{n} \overline{a_j} z^{n-j}$. If $|z| = 1$, then $q(z) = z^n \overline{p(z)}$. Indeed, if $|z| = 1$, then $\overline{z^j} = z^{-j}$. In particular, by the Maximum modulus principle $\|p\| = \|q\|$. Let now $Q(z) = r^n q(\frac{z}{r})$. Observe that

$$\|Q\| = r^n \max_{|z|=1} \left| \frac{z^n}{r^n} \sum_{j=0}^{n} \overline{a_j} \left(\frac{r}{z}\right)^j \right| = \max_{|z|=1} \left| p\left(\frac{\overline{r}}{z}\right) \right| = \max_{|z|=r} |p(z)|.$$

On the other hand, $\|Q\| \geq \max_{|z|=r} |Q(z)| = r^n \|q\| = r^n \|p\|$. So, in combination with the above we get the desired estimate.

For the general case fix $x \in B_X$ and let $\phi \in S_{Y^*}$ be such that $\phi\big(P(x)\big) = \|P(x)\|$. Put $p(t) = \phi\big(P(tx)\big)$, $t \in \mathbb{C}$. Then $p \in \mathcal{P}^n(\mathbb{C})$ and so by the above we obtain $\|P(x)\| = p(1) \leq \frac{1}{r^n} \sup_{|t| \leq r} |p(t)| \leq \frac{1}{r^n} \|P\|$. □

Let us now prove a version of Erdős's result due to Muñoz, Sarantopoulos, and Tonge; compare with Theorem 1.63.

Theorem 26 ([MST]). *Let X, Y be real normed linear spaces and $P \in \mathcal{P}^n(X;Y)$. Then*

$$\|\tilde{P}\|_{\mathbb{C}} \leq 2^{\frac{n}{2}} |T_n(i)| \cdot \|P\|.$$

Proof. Fix $\varepsilon > 0$. By Proposition 25 there is $z = x + iy \in \tilde{X}$ with $\|z\|_{\mathbb{C}} \leq \frac{1}{\sqrt{2}}$ and

$$\|\tilde{P}(z)\| \geq (1-\varepsilon) \left(\frac{1}{\sqrt{2}}\right)^n \|\tilde{P}\|_{\mathbb{C}}. \tag{13}$$

By Fact 1.59 there exists a functional $\psi \in S_Y^*$ such that $\|\tilde{P}(z)\| = |\tilde{\psi}(\tilde{P}(z))|$. Put $q(t) = \tilde{\psi}\big(\tilde{P}(x+ty)\big)$, $t \in \mathbb{C}$. We have $q(t) = \psi\big(P(x+ty)\big)$ for $t \in \mathbb{R}$ and so q has real coefficients. Hence by Corollary 24

$$\max_{t \in \mathbb{C}, |t|=1} |q(t)| \leq |T_n(i)| \max_{t \in [-1,1]} |q(t)|. \tag{14}$$

But for $t \in [-1,1]$ using the Cauchy inequality we obtain

$$\begin{aligned} \|x + ty\| &= \sup_{\phi \in B_{X^*}} |\phi(x) + t\phi(y)| \\ &\leq \sqrt{1+t^2} \sup_{\phi \in B_{X^*}} \sqrt{\phi(x)^2 + \phi(y)^2} \leq \sqrt{2}\|x + iy\|_{\mathbb{C}} \leq 1. \end{aligned}$$

This shows that $\max_{t \in [-1,1]} |q(t)| \leq \|P\|$. Therefore combining (13) and (14) we obtain

$$\begin{aligned} (1-\varepsilon)\left(\frac{1}{\sqrt{2}}\right)^n \|\tilde{P}\|_{\mathbb{C}} &\leq \|\tilde{P}(z)\| = |q(i)| \leq \max_{t \in \mathbb{C}, |t|=1} |q(t)| \\ &\leq |T_n(i)| \max_{t \in [-1,1]} |q(t)| \leq |T_n(i)| \cdot \|P\|. \end{aligned}$$

This finishes the proof. □

In the rest of the section we outline the proof of a recent theorem of Lawrence Albert Harris ([Harr3] and [Harr4]) which implies the Skalyga-Markov inequalities (Theorem 28) for every real polynomial $P \in \mathcal{P}^n(X;Y)$.

We assume all polynomials over real spaces and with real coefficients. For $n, k \in \mathbb{N}$ denote $Q_k^n = Q_0^n$ for k even and $Q_k^n = Q_1^n$ for k odd, where Q_0^n and Q_1^n were introduced in (11), and let $N_k^n = \{(\eta_r^n, \eta_q^n);\ (r,q) \in Q_k^n\}$ (see also Section 2).

Theorem 27 ([Harr4]). *Let $n \in \mathbb{N}$, $k \in \{0,\dots,n\}$, and $p \in \mathcal{P}^n(\ell_\infty^2)$ satisfy $|p(x)| \le 1$ whenever $x \in N_k^n$. Then*

$$\left|d^k p(t,t)[1,-1]\right| \le T_n^{(k)}(t) \quad \textit{for } t \ge 1.$$

Sketch of proof. Fix $k \in \{0,\dots,n\}$ and $t \in \mathbb{R}$, $t \ge 1$ and consider the linear functional $L(p) = d^k p(t,t)[1,-1]$ on $\mathcal{P}^n(\ell_\infty^2)$. By Theorem 17 we have

$$p(x,y) = \sum_{(r,q)\in Q_k^n} p(\eta_r^n, \eta_q^n) P_{r,q}(x,y) + v(x,y),$$

for every $p \in \dim \mathcal{P}^n(\mathbb{R}^2)$, where $v \in V$ is a unique polynomial depending on p. The special form of $V_j \in V$ implies (use Corollary 1.116) that $L(V_j) = 0$, $j = 0,\dots,n$, and hence $L(v) = 0$ for $v \in V$. Thus

$$L(p) = \sum_{(r,q)\in Q_k^n} L(P_{r,q}) p(\eta_r^n, \eta_q^n) = \sum_{(r,q)\in Q_k^n} \lambda_{r,q}\, p(\eta_r^n, \eta_q^n) \tag{15}$$

for $p \in \mathcal{P}^n(\ell_\infty^2)$, where $\lambda_{r,q} = L(P_{r,q}) = d^k P_{r,q}(t,t)[1,-1]$.

The main difficulty now consists in proving the inequality

$$(-1)^r \lambda_{r,q} \ge 0 \quad \text{whenever } (r,q) \in Q_k^n.$$

The proof of this inequality is based on somewhat lengthy computations involving special properties of Chebyshev polynomials. We prefer to omit it, see [Harr4] for details. Let us see how to finish the proof once this inequality has been established. In fact, it follows immediately from the fact that the Chebyshev polynomials in one variable are extremal for the sets of nodes in the representation formula (15). Denote $S_n(x,y) = T_n(x)$. Since $S_n(\eta_r^n, \eta_q^n) = (-1)^r$ for $(r,q) \in Q_k^n$, we see that

$$T_n^{(k)}(t) = L(S_n) = \sum_{(r,q)\in Q_k^n} \lambda_{r,q} S_n(\eta_r^n, \eta_q^n) = \sum_{(r,q)\in Q_k^n} |\lambda_{r,q}| \ge |L(p)|$$

whenever p satisfies the assumptions of the theorem. □

The next optimal result of Valentin Ivanovich Skalyga is a generalisation of V. A. Markov's and Schur's theorems into the infinite-dimensional setting.

Theorem 28 ([Sk]). *Let X, Y be real normed linear spaces and $P \in \mathcal{P}^n(X;Y)$. Then for all $0 \le k \le n$*

$$\left\|d^k P\right\| \le T_n^{(k)}(1)\|P\|,$$

$$\left\|d^k P(x)\right\| \le T_n^{(k)}(\|x\|)\|P\| \quad \textit{for } x \in X,\ \|x\| \ge 1.$$

Proof. ([Harr3]) Consider the linear isometry $T\colon \ell_\infty^2 \to \ell_1^2$ defined by $T(x,y) = \left(\frac{x+y}{2}, \frac{x-y}{2}\right)$. Given any $p \in \mathcal{P}^n(\ell_1^2)$, $\|p\| \le 1$, we put $q = p \circ T \in \mathcal{P}^n(\ell_\infty^2)$. Then $d^k q(t,t)[1,-1] = d^k p(T(t,t))[T(1,-1)] = d^k p(t,0)[0,1]$. Hence by Theorem 27 we obtain $|d^k p(t,0)[0,1]| \le T_n^{(k)}(t)$ for $t \ge 1$.

To finish the proof we may without loss of generality assume that $\|P\| = 1$ and, by using the Hahn-Banach theorem and Fact 1.78, that $Y = \mathbb{R}$. Given any $z, y \in B_X$, let $S \in \mathcal{L}(\ell_1^2; X)$ be defined as $S(u,v) = uz + vy$. Then $\|S\| \le 1$. Put $p = P \circ S$. Then $p \in \mathcal{P}^n(\ell_1^2)$ and $\|p\| \le 1$. Also, $d^k p(t,0)[0,1] = d^k P(tz)[y]$. Hence the first inequality follows from the above using $t = 1$. The second inequality follows using $z = \frac{x}{\|x\|}$ and $t = \|x\|$. □

For completeness we remark that the value of the constant in the inequality is the following ([Ri, Exercise 1.5.6]):

$$T_n^{(k)}(1) = \frac{1}{(2k-1)!!}\prod_{j=0}^{k-1}(n^2 - j^2) = \frac{n^2(n^2-1)\cdots(n^2-(k-1)^2)}{1\cdot 3\cdot \cdots \cdot (2k-1)}.$$

From Lemma 1.99 and Corollary 1.17 we immediately obtain the following estimate on the norm of the differential of a *homogeneous* polynomial.

Theorem 29 ([Sa]). *Let X, Y be real normed linear spaces, $P \in \mathcal{P}(^nX; Y)$, $k \in \mathbb{N}$, and $k \le n$. Then*

$$\frac{1}{k!}\|d^k P\| \le \binom{n}{k}\sqrt{\frac{n^n}{(n-k)^{n-k}k^k}}\,\|P\|.$$

4. Polynomials and L_p-norms on $\mathbb{R}^n$

In this section we focus on real valued p-homogeneous polynomials on $\mathbb{R}^n$ and their close relationship with L_p-spaces. Among the corollaries of this study are some results on isometric embeddings of finite-dimensional spaces and in particular of Hilbert spaces into ℓ_p^n spaces, due to Yuri Illich Lyubich and Leonid N. Vaserstein, and due to Abraham Neyman. We formulate the recent isometric version of the Dvoretzky theorem due to Vladimir Leonidovich Dol'nikov and Roman Nikolaevich Karasev, and we prove the classical Birch theorem on the existence of null subspaces for odd degree homogeneous polynomials.

Definition 30. Let $(X, \|\cdot\|)$ be a real normed linear space. We say that the norm $\|\cdot\|$ is k-polynomial, $k \in \mathbb{N}$, if the function $x \mapsto \|x\|^k$ is a k-homogeneous polynomial. If the norm is k-polynomial for some $k \in \mathbb{N}$, then we say that it is polynomial.

Notice that if the space X has a k-polynomial norm, then k is necessarily even, as $\|tx\|^k = |t|^k\|x\|^k$. By Theorem 5.106 the canonical norm on $L_p(\mu)$ for p even is a p-polynomial norm. Hence every isometric subspace of an L_p space, p even, also has a polynomial norm.

We assume that the n-dimensional space $\mathbb{R}^n$ is equipped with the canonical inner product $\langle\cdot,\cdot\rangle$. We use the standard notation S^{n-1} for its unit sphere. Every $y \in \mathbb{R}^n$ canonically generates a polynomial $\langle y,\cdot\rangle^d \in \mathcal{P}({}^d\mathbb{R}^n)$. The following formula follows from Proposition 1.22:

$$\langle y,x\rangle^d = \sum_{\alpha\in\mathcal{J}(n,d)} \binom{d}{\alpha} y^\alpha x^\alpha$$

Definition 31. We introduce a natural inner product $[\cdot,\cdot]$ on the space $\mathcal{P}({}^d\mathbb{R}^n)$ as follows. For $P, Q \in \mathcal{P}({}^d\mathbb{R}^n)$, $P(x) = \sum_{\alpha\in\mathcal{J}(n,d)} a_\alpha x^\alpha$, $Q(x) = \sum_{\alpha\in\mathcal{J}(n,d)} b_\alpha x^\alpha$ we put

$$[P,Q] = \sum_{\alpha\in\mathcal{J}(n,d)} \binom{d}{\alpha}^{-1} a_\alpha b_\alpha.$$

It is clear that the space $(\mathcal{P}({}^d\mathbb{R}^n), [\cdot,\cdot])$ is a real $|\mathcal{J}(n,d)|$-dimensional Hilbert space. For $y \in \mathbb{R}^n$ and $P \in \mathcal{P}({}^d\mathbb{R}^n)$ we have $[P,\langle y,\cdot\rangle^d] = P(y)$. Note that $\operatorname{span}\{\langle y,\cdot\rangle^d;\ y \in \mathbb{R}^n\} = \mathcal{P}({}^d\mathbb{R}^n)$, otherwise there would be $P \neq 0$ satisfying $0 = [P,\langle y,\cdot\rangle^d] = P(y)$ for all $y \in \mathbb{R}^n$, a contradiction. It follows that the inner product $[\cdot,\cdot]$ is independent of the choice of the orthonormal coordinate basis in $\mathbb{R}^n$. The following is a simple, but important observation.

Fact 32. *There is a canonical isomorphism between polynomials $P \in \mathcal{P}({}^d\mathbb{R}^n)$ and functionals from $\mathcal{P}({}^d\mathbb{R}^n)^*$ given by the relation $P \mapsto [P,\cdot]$. The functional $\phi = \sum_{j=1}^k a_j\delta_{x_j} \in \mathcal{P}({}^d\mathbb{R}^n)^*$ corresponds to the polynomial $P = \sum_{j=1}^k a_j\langle x_j,\cdot\rangle^d$.*

Note that given $y_1,\dots,y_k \in \mathbb{R}^n$, it is easy to see that $\langle y_1,\cdot\rangle^d,\dots,\langle y_k,\cdot\rangle^d$ are linearly independent in the space $\mathcal{P}({}^d\mathbb{R}^n)$ if and only if the nodes $y_1,\dots,y_k$ are $\mathcal{P}({}^d\mathbb{R}^n)$-independent.

We introduce two important convex cones in $\mathcal{P}({}^d\mathbb{R}^n)$, but first we need some basic facts about the duality of cones. Let H be a real Hilbert space. Recall, that a dual cone to a set $A \subset H$ is the set

$$A^* = \{y \in H;\ \langle y,x\rangle \ge 0 \text{ for all } x \in A\}.$$

Fact 33. *Let H be a real Hilbert space and $A \subset H$. Then A^* is a closed convex cone. If A is a closed convex cone, then $(A^*)^* = A$.*

Proof. The first statement is obvious. Also, clearly $A \subset (A^*)^*$. Now suppose that A is a closed convex cone and there is $x \in (A^*)^* \setminus A$. By the separation theorem there is $y \in H$ such that $\langle y,z\rangle \ge 0$ for all $z \in A$ and $\langle y,x\rangle < 0$. It follows that $y \in A^*$, and consequently $x \notin (A^*)^*$, a contradiction. □

Note that for any $K \subset \mathbb{R}^n$ compact and any finite Borel measure μ on K the function $x \mapsto \int_K \langle y,x\rangle^d \,\mathrm{d}\mu(y)$ is a d-homogeneous polynomial on $\mathbb{R}^n$ which is convex and non-negative if d is even. (Indeed, $(x_1,\dots,x_d) \mapsto \int_K \langle y,x_1\rangle\cdots\langle y,x_d\rangle \,\mathrm{d}\mu(y)$ is a d-linear form on $\mathbb{R}^n$.)

following two conditions:

$$\begin{aligned} x_j &\geq x_{j-1} + \left(\rho\left(\frac{j}{n}\right) - \rho\left(\frac{j-1}{n}\right)\right)(r-\varepsilon), \quad j = 1, \dots, n, \\ \eta(x_j) &< \rho\left(\frac{j}{n}\right)(r-\varepsilon) + \frac{j+1}{n+1}\frac{\varepsilon}{4}, \quad j = 0, \dots, n. \end{aligned} \tag{33}$$

The assertion of the lemma will be satisfied thanks to the first condition.

To start, choose $x_0 \in E \cap \big[\inf E, \inf E + \frac{1}{n+1}\frac{\varepsilon}{4}\big)$. Note that since $\overline{E} \cap (0, \inf E) = \emptyset$, we have $\eta(\inf E) = 0$ and the condition (33) is satisfied for $j = 0$. Now assume that the points $x_0, \dots, x_{j-1}$ have been chosen. Put $a = x_{j-1} + \big(\rho(\frac{j}{n}) - \rho(\frac{j-1}{n})\big)(r-\varepsilon)$. As

$$\begin{aligned} \eta(a) &\leq \eta(x_{j-1}) + \left(\rho\left(\frac{j}{n}\right) - \rho\left(\frac{j-1}{n}\right)\right)(r-\varepsilon) \\ &< \rho\left(\frac{j}{n}\right)(r-\varepsilon) + \frac{j}{n+1}\frac{\varepsilon}{4} < r - \frac{\varepsilon}{2}, \end{aligned} \tag{34}$$

the set $E \cap [a, r]$ must be non-empty and so we may set $b = \inf(E \cap [a, r])$. Note that $\eta(b) = \eta(a)$. Finally, choose $x_j \in E \cap \big[b, b + \frac{1}{n+1}\frac{\varepsilon}{4}\big)$. Then $\eta(x_j) \leq \eta(b) + \frac{1}{n+1}\frac{\varepsilon}{4} = \eta(a) + \frac{1}{n+1}\frac{\varepsilon}{4}$, which combined with (34) gives (33). □

Proof of Lemma 138. The crucial part of the proof is the one-dimensional case, the general case is proved by an easy induction. So suppose $N = 1$ and let $d > 0$ and $q > 1$ be given. Put $r = \frac{d}{2}$ and

$$I(\xi) = \exp \int_0^1 \log\left(1 + \frac{3\xi}{rt^2}\right) \mathrm{d}t.$$

By the Lebesgue dominated convergence theorem we have $\lim\limits_{\xi \to 0+} I(\xi) = 1$ and hence we may find $\delta > 0$ such that $I(2\delta) < \sqrt{q}$.

Now suppose that Y is a normed linear space over $\mathbb{K}$, $A \subset \mathbb{K}$ is a connected compact set with $\operatorname{diam} A \geq d$, and $\Phi \subset \mathcal{P}(\mathbb{K}; Y)$ is a family of polynomials pointwise bounded on A. Fix an arbitrary $u \in A$. The connectedness of A and the fact that $\operatorname{diam} A \geq d$ imply the existence of a mapping $\varphi\colon [0, r) \to A$ which satisfies $|u - \varphi(t)| = t$ for each $t \in [0, r)$. We let $E_k = \{t \in [0, r);\ \|P(\varphi(t))\| \leq k \text{ for each } P \in \Phi\}$. Since Φ is pointwise bounded on A, $\bigcup_{k \in \mathbb{N}} E_k = [0, r)$. Put $\varepsilon = \min\{\delta, \frac{r}{2}\}$. As $E_k \subset E_{k+1}$, it follows that there is $K \in \mathbb{N}$ such that $\lambda(\overline{E_K}) \geq r - \frac{\varepsilon}{2}$. Denote $E = E_K$. Now we will use the Lagrange interpolation formula (Proposition 56) with nodes contained in $\varphi(E)$, exploiting the fact that the polynomials in Φ are uniformly bounded by the constant K on the set $\varphi(E)$. By choosing suitably distributed nodes we obtain estimates on the polynomials w_j appearing in the interpolation formula, and via them also on the polynomials in Φ.

Set $\rho(t) = t^2$. Fix $n \in \mathbb{N}$. By Lemma 139 there are points $t_0, \dots, t_n \in E$ such that $|t_j - t_k| \geq \frac{|j^2-k^2|}{n^2}(r-\varepsilon)$. Denote $x_j = \varphi(t_j)$, $j = 0, \dots, n$, and notice that

$$|x_j - x_k| \geq \big||\varphi(t_j) - u| - |u - \varphi(t_k)|\big| = |t_j - t_k| \geq \frac{|j^2 - k^2|}{n^2}(r-\varepsilon).$$

Hence

$$|w_j(x_j)| \geq (r-\varepsilon)^n \prod_{\substack{k=0\\k\neq j}}^{n} \frac{|j^2-k^2|}{n^2} \geq (r-\varepsilon)^n \frac{1}{2} \prod_{k=1}^{n} \frac{k^2}{n^2}.$$

To see the last inequality, observe that

$$\begin{aligned}\prod_{\substack{k=0\\k\neq j}}^{n} |j^2-k^2| = \prod_{\substack{k=0\\k\neq j}}^{n} |k-j||j+k| &= \big((n-j)(n-j-1)\cdots 1\cdot|-1|\cdots|-j|\big)\\ &\quad\cdot\big(j(j+1)\cdots(2j-1)(2j+1)\cdots(n+j)\big)\\ &= \frac{n!}{n(n-1)\cdots(n-j+1)} \cdot \frac{n!(n+1)\cdots(n+j)}{2}\\ &= \frac{1}{2}\left(\prod_{k=1}^{n} k^2\right)\frac{(n+1)\cdots(n+j)}{(n+1-j)\cdots n} \geq \frac{1}{2}\prod_{k=1}^{n} k^2.\end{aligned}$$

Further, for any $x \in U(u, 2\delta)$ we have

$$\begin{aligned}|x - x_k| &\leq |x-u| + |u - x_k| < 2\delta + t_k\\ &\leq 2\delta + t_n + t_k - t_n < 2\delta + r - \frac{n^2-k^2}{n^2}(r-\varepsilon)\\ &= 2\delta + \varepsilon + \frac{k^2}{n^2}(r-\varepsilon) = \left(\frac{k^2}{n^2} + \frac{2\delta+\varepsilon}{r-\varepsilon}\right)(r-\varepsilon) \leq \left(\frac{k^2}{n^2} + \frac{6\delta}{r}\right)(r-\varepsilon)\end{aligned}$$

and consequently

$$|w_j(x)| \leq (r-\varepsilon)^n \prod_{k=1}^{n}\left(\frac{k^2}{n^2} + \frac{6\delta}{r}\right) \quad \text{for } j \in \{0, \dots, n\}.$$

Thus

$$\left|\frac{w_j(x)}{w_j(x_j)}\right| \leq 2R_n^n, \quad \text{where } R_n = \left(\prod_{k=1}^{n} \frac{\frac{k^2}{n^2} + \frac{6\delta}{r}}{\frac{k^2}{n^2}}\right)^{\frac{1}{n}}.$$

Now Proposition 56 implies that for any $P \in \Phi \cap \mathcal{P}^n(\mathbb{K}; Y)$ we have

$$P(x) = \sum_{j=0}^{n} \frac{w_j(x)}{w_j(x_j)} P(x_j), \quad x \in \mathbb{K},$$

and hence $\sup_{x\in U(u,2\delta)} \|P(x)\| \leq 2(n+1)KR_n^n$.

Since $\log R_n = \frac{1}{n}\sum_{k=1}^{n} \log\left(1+\frac{6\delta}{r\frac{k^2}{n^2}}\right)$, we have $\lim_{n\to\infty} R_n = I(2\delta) < \sqrt{q}$. Thus there is $n_0 \in \mathbb{N}$ such that $R_n < \sqrt{q}$ and $2(n+1) \le (\sqrt{q})^n$ for $n \ge n_0$. So, setting $L = \max\{\max_{n<n_0} 2(n+1)KR_n^n, K\}$ we obtain $\sup_{x\in U(u,\delta)} \|P(x)\| < Lq^n$ whenever $P \in \Phi \cap \mathcal{P}^n(\mathbb{K};Y)$.

We have shown that for every $u \in A$ there exists a constant $L > 0$ such that $\sup_{x\in U(u,2\delta)} \|P(x)\| < Lq^n$ whenever $P \in \Phi \cap \mathcal{P}^n(\mathbb{K};Y)$. We finish the proof of the case $N = 1$ by covering A by open balls of radius δ and using the compactness of the set A.

Next we use induction to prove the lemma for an arbitrary dimension $N \in \mathbb{N}$. Without loss of generality we will use the maximum norm on $\mathbb{K}^N$. Suppose that the statement holds for all dimensions up to $N-1$. Let $d > 0$ and $q > 1$ be given and put $\delta = \min\{\delta(1,d,\sqrt{q}), \delta(N-1,d,\sqrt{q})\}$. Further, suppose that Y is a normed linear space over $\mathbb{K}$, $A_j \subset \mathbb{K}$ are connected compact sets with $\operatorname{diam} A_j \ge d$ for $j = 1,\dots,N$, and $\Phi \subset \mathcal{P}(\mathbb{K}^N;Y)$ is a family of polynomials pointwise bounded on $A = A_1\times\cdots\times A_N$. Denoting $B = A_1\times\cdots\times A_{N-1} \subset \mathbb{K}^{N-1}$ we have $A = B\times A_N$. For convenience we use the notation for polynomials on $\mathbb{K}^N$ using coordinates. For each $x \in A_N$ we consider the family $\Phi_x = \{P_x;\ P \in \Phi\} \subset \mathcal{P}(\mathbb{K}^{N-1};Y)$, where $P_x(z) = P(z_1,\dots,z_{N-1},x)$ for $z = (z_1,\dots,z_{N-1}) \in \mathbb{K}^{N-1}$. Notice that by Fact 38 we have $\deg P_x \le \deg P$. Therefore by the inductive hypothesis there is $M_x > 0$ such that $\sup_{z\in B^\delta} \|P(z,x)\| \le M_x(\sqrt{q})^{\deg P}$ for all $P \in \Phi$.

Now we set $\Psi = \{x \mapsto (\sqrt{q})^{-\deg P}\cdot P(z,x);\ P \in \Phi, z \in B^\delta\} \subset \mathcal{P}(\mathbb{K};Y)$. We have just shown that this family is pointwise bounded on A_N. So by the first step of the induction there is $M > 0$ such that $\sup_{x\in A_N^\delta} \|(\sqrt{q})^{-\deg P}\cdot P(z,x)\| \le M(\sqrt{q})^{\deg P}$ holds for all $P \in \Phi$ and $z \in B^\delta$, from which the assertion of the lemma easily follows. □

We remark that the assumption in Lemma 138 on Φ being pointwise bounded on the whole of A can by somewhat relaxed to require only pointwise boundedness "almost everywhere" on A in a sense which can be easily deduced from the proof.

Theorem 140 ([Hill], [AO]). *Let X, Y be Banach spaces and $P_n \in \mathcal{P}(^nX;Y)$ for $n \in \mathbb{N}_0$. If there is an absorbing set $A \subset X$ such that the power series* (32) *converges on $a + A$ for some $a \in X$, then the radius of norm convergence of* (32) *is positive. In particular, the domain of convergence of the power series* (32) *is either empty, or a balanced neighbourhood of zero.*

Proof. By the assumption for each $x \in S_X$ there exists $\delta_x > 0$ such that the series $\sum_{n=0}^{\infty} P_n(a+tx)$ converges for $t \in [-\delta_x,\delta_x]$. It follows that the sequence of polynomials $\{t \mapsto P_n(a+tx)\}_{n=0}^{\infty}$ is pointwise bounded on $[-\delta_x,\delta_x]$. By the Polynomial lemma (Lemma 138) there is $M_x \ge 1$ such that for each $n \in \mathbb{N}$ we have $\sup_{t\in[-\delta_x,\delta_x]} \|P_n(a+tx)\| < M_x 2^n$. Thus $\|P_n(\delta_x x)\| < e^n M_x 2^n$ by Lemma 47 and setting $\lambda_x = \frac{\delta_x}{2eM_x}$ we obtain

$$\|P_n(tx)\| < 1 \quad \text{whenever } 0 \leq t \leq \lambda_x, n \in \mathbb{N}.$$

We put $C = \{tx;\ x \in S_X, 0 \leq t \leq \lambda_x\}$, which is an absorbing set in X. Because the polynomials P_n are continuous,

$$\|P_n(y)\| \leq 1 \quad \text{whenever } y \in \overline{C}, n \in \mathbb{N}.$$

By Lemma 134 the absorbing set $\overline{C}$ contains a ball $B(z, r)$ for some $r > 0$. Now by Lemma 47

$$\|P_n\| = \frac{1}{r^n} \sup_{x \in B(0,r)} \|P_n(x)\| \leq \frac{1}{r^n} \frac{n^n}{n!} \sup_{x \in B(z,r)} \|P_n(x)\| < \frac{e^n}{r^n}$$

for every $n \in \mathbb{N}$. Therefore by Proposition 136 the radius of norm convergence of (32) is at least $\frac{r}{e}$. □

We note that the fact that if the power series converges on some absorbing set, then it has a positive radius of norm convergence is substantially easier to prove, as we do not need to use the Polynomial lemma to shift the estimates to the origin.

Further, we remark that the assumption of completeness of X in the previous theorem is crucial, as we show in the next example.

Example 141. Let $X = c_{00}$, i.e. the linear subspace of c_0 consisting of finitely supported vectors. Denote the coordinate functionals by e_n^* and put $P_n(x) = \left(ne_n^*(x)\right)^n$, $n \in \mathbb{N}$, $P_0 = 0$. Then $P_n \in \mathcal{P}(^nX)$ (Fact 38) and $\|P_n\| = n^n$. Clearly, the power series (32) converges absolutely in every $x \in X$, but its radius of norm convergence $R = 0$. Moreover, if we denote $f = \sum_{n=0}^{\infty} P_n$, then $\delta f(0) = e_1^*$ but f is not Fréchet differentiable at zero (consider the vectors $h_k = \sum_{j=1}^{k} \frac{1}{k} e_j$).

Lemma 142. *Let Y be a Banach space over $\mathbb{K}$, $X = \mathbb{K}^N$, $N \in \mathbb{N}$, $P_n \in \mathcal{P}(^nX; Y)$ for $n \in \mathbb{N}_0$, and let G be the domain of convergence of the power series* (32). *Then for every $a \in G$ there are a neighbourhood V of a, $0 < \lambda < 1$, and $M > 0$ such that $\sup_{x \in V} \|P_n(x)\| \leq M\lambda^n$ for each $n \in \mathbb{N}$.*

Proof. Given $a \in G$ let $D_1, \dots, D_N \subset \mathbb{K}$ be closed discs with positive radii such that $V = D_1 \times \cdots \times D_N \subset G$ is a neighbourhood of a. By the compactness of V there is $\mu > 1$ such that $\mu^2 V \subset G$. Put $\lambda = \frac{1}{\mu}$. Since $\{P_n\}$ is pointwise bounded on $\mu^2 V$, by the Polynomial lemma (Lemma 138) there is $M > 0$ such that $\sup_{x \in \mu^2 V} \|P_n(x)\| \leq M\mu^n$ for each $n \in \mathbb{N}$. So $\|P_n(x)\| = \mu^{-2n} \|P_n(\mu^2 x)\| \leq M\lambda^n$ for any $x \in V$, $n \in \mathbb{N}$. □

Corollary 143. *Let X, Y be Banach spaces,* $\dim X < \infty$, *and $P_n \in \mathcal{P}(^nX; Y)$ for $n \in \mathbb{N}_0$. Then the power series* (32) *converges in its domain of convergence absolutely locally uniformly.*

We will show in Section 8 that the last two results hold without the restriction to finite-dimensional spaces.

Next we take a look at complexifications. We start with rather easy observation.

Lemma 144. *Let X be a real normed linear space and Y a real Banach space, let $P_n \in \mathcal{P}(^nX;Y)$ for $n \in \mathbb{N}_0$, and let $\tilde{P}_n \in \mathcal{P}(^n\tilde{X};\tilde{Y})$ be their complexifications. If $R > 0$ is the radius of norm convergence of the power series* (32), *then the radius of norm convergence of the power series $\sum_{n=0}^{\infty} \tilde{P}_n$ is at least $\frac{R}{2}$.*

Proof. Combine Proposition 136 with Theorem 63. □

To obtain more subtle information on the domain of convergence of the complexified series we have to employ the Polynomial lemma again.

Lemma 145. *Let X be a real normed linear space and Y a real Banach space, let $P_n \in \mathcal{P}(^nX;Y)$ for $n \in \mathbb{N}_0$, and let $\tilde{P}_n \in \mathcal{P}(^n\tilde{X};\tilde{Y})$ be their complexifications. If $G \subset X$ is the domain of convergence of the power series* (32), *then the power series $\sum_{n=0}^{\infty} \tilde{P}_n$ converges on an open set $\tilde{G} \subset \tilde{X}$ containing G.*

Proof. Choose any $a \in G$. Let $r > 0$ and $\mu > 1$ be such that $\mu^2 U(a, 2r) \subset G$ and let $\delta = \delta(1, 2, \mu)$ be the constant from the Polynomial lemma (Lemma 138). If $x \in U_X(a, r)$, $y \in B_X(0, r)$, then $\mu^2(x + ty) \in G$ for every $t \in [-1, 1]$. It follows that the sequence of polynomials $\{t \mapsto \tilde{P}_n(\mu^2(x + ty))\}_{n=1}^{\infty}$ is pointwise bounded on $[-1, 1]$. By the Polynomial lemma (Lemma 138) there is $M > 0$ such that $\|\tilde{P}_n(\mu^2(x + ty))\| \le M\mu^n$ whenever $t \in \mathbb{C}$, $\operatorname{dist}(t, [-1, 1]) < \delta$. In particular, $\|\tilde{P}_n(x + ity)\| \le M\mu^{-n}$ for $t \in (-\delta, \delta)$. Thus the power series $\sum_{n=0}^{\infty} \tilde{P}_n$ converges on $V_a = U_X(a, r) + iU_X(0, \delta r)$. Now it suffices to put $\tilde{G} = \bigcup_{a \in G} V_a$. □

We end this section with a preliminary result concerning the differentiability properties of the power series. Much more will be obtained in Section 8 (Theorem 168).

Theorem 146. *Let X be a normed linear space and Y a Banach space. Suppose that the power series* (32) *has a positive radius of norm convergence R. Let the mapping $f : U(0, R) \to Y$ be defined as $f(x) = \sum_{n=0}^{\infty} P_n(x)$. Then $f \in C^{\infty}(U(0, R); Y)$ and*

$$d^k f(x) = \sum_{n=k}^{\infty} d^k P_n(x) \tag{35}$$

for every $x \in U(0, R)$ and $k \in \mathbb{N}$. Further, the power series in (35) *have radius of norm convergence R and $P_n = \frac{1}{n!} d^n f(0)$.*

Proof. First we show the statement about the radii of norm convergence. Using Lemma 99 and Corollary 17 we obtain

$$\|DP_n\| = \sup_{x,h \in B_X} \left\| n\check{P}_n(^{n-1}x, h) \right\| \le n \sqrt{\frac{n^n}{(n-1)^{n-1}}} \|P_n\|.$$

On the other hand, $\|DP_n\| = \sup_{x,h \in B_X} \left\| n\check{P}_n(^{n-1}x, h) \right\| \ge n \sup_{x \in B_X} \|P_n(x)\| = n\|P_n\|$. Thus by the Cauchy-Hadamard formula the power series in (35) for $k = 1$ has a radius of norm convergence R. For higher derivatives it follows immediately by induction. An application of Theorem 85 now finishes the proof except for the fact that $P_n = \frac{1}{n!} d^n f(0)$, which follows easily from Lemma 99. □

We note that the assumption that $R > 0$ is necessary for the smoothness of f, as can be seen from Example 141.

7. Complex mappings

In this section we collect some standard results of the classical theory of complex mappings in one or several variables with values in a general complex Banach space which will be needed in the sequel.

Definition 147. Let X be a complex normed linear space, Y a complex Banach space, $U \subset X$ an open set, and $f : U \to Y$. We say that f is holomorphic if f is Fréchet differentiable on U. We denote by $H(U;Y)$ the vector space of all holomorphic mappings from U into Y. As usual, for functions, i.e. mappings into $\mathbb{C}$, we use the shortened notation $H(U) = H(U;\mathbb{C})$.

Theorem 148 ([Dun]). *Let Y be a complex Banach space, $U \subset \mathbb{C}$, and $f : U \to Y$. Then f is holomorphic if and only if $\phi \circ f$ is holomorphic for every $\phi \in Y^*$.*

Proof. $\Rightarrow$ is clear.

$\Leftarrow$ Fix an arbitrary $x \in U$. We prove that the derivative $f'(x)$ exists by showing that the mapping $z \mapsto \frac{1}{z}\big(f(x+z) - f(x)\big)$ satisfies the Bolzano-Cauchy criterion. To this end fix some $r > 0$ such that $B(x,r) \subset U$ and let $\gamma(t) = x + re^{it}$, $t \in [0, 2\pi]$. For any $y, z \in U(0,r) \setminus \{0\}$, $y \neq z$, and any $\phi \in Y^*$ we have

$$\begin{aligned}
&\phi\Big(\frac{1}{z-y}\Big(\frac{f(x+z)-f(x)}{z} - \frac{f(x+y)-f(x)}{y}\Big)\Big)\\
&\quad = \frac{1}{z-y}\Big(\frac{\phi\circ f(x+z) - \phi\circ f(x)}{z} - \frac{\phi\circ f(x+y) - \phi\circ f(x)}{y}\Big)\\
&\quad = \frac{1}{2\pi i}\int_\gamma \frac{\phi\circ f(w)}{(w-x)(w-x-z)(w-x-y)}\,dw,
\end{aligned}$$

where we used the Cauchy formula on the holomorphic function $\phi \circ f$. Denote $M_\phi = \max_\gamma |\phi \circ f|$. Then

$$\sup_{\substack{y,z\in B(0,\frac{r}{2})\setminus\{0\}\\ y\neq z}} \Big|\phi\Big(\frac{1}{z-y}\Big(\frac{f(x+z)-f(x)}{z} - \frac{f(x+y)-f(x)}{y}\Big)\Big)\Big| \leq \frac{4M_\phi}{r^2}.$$

Hence by the Banach-Steinhaus uniform boundedness principle there is $M \in \mathbb{R}$ such that

$$\Big\|\frac{f(x+z)-f(x)}{z} - \frac{f(x+y)-f(x)}{y}\Big\| \leq M|z-y|$$

whenever $y, z \in B\big(0, \frac{r}{2}\big) \setminus \{0\}$, which proves our claim. □

The proof of the following theorem can be obtained by formal modifications of the complex valued case using the Bochner path integral, which is defined for a piecewise continuously differentiable curve (path) $\gamma \colon [\alpha, \beta] \to \mathbb{C}$ by

$$\int_\gamma f(z)\,\mathrm{d}z = \int_\alpha^\beta \gamma'(t) f\big(\gamma(t)\big)\,\mathrm{d}t.$$

Theorem 149 ([Hill, Theorem 3.11.3, ff.]). *Let Y be a complex Banach space and $f \in H(U(a,r);Y)$ for some $a \in \mathbb{C}$, $r > 0$. Then $f \in C^\infty(U(a,r);Y)$ and*

$$f(x) = \sum_{n=0}^{\infty} \frac{1}{n!} d^n f(a)[x-a]$$

for every $x \in U(a,r)$.

The next statement, well-known for complex valued functions, is obtained easily by using compositions of f with linear functionals from Y^*.

Theorem 150 (Maximum principle, [Hill, Section 3.13]). *Let Y be a complex Banach space, $U \subset \mathbb{C}$ a non-empty connected bounded open subset, $f \in C(\overline{U};Y)$, and $f \in H(U;Y)$. Then the function $z \mapsto \|f(z)\|$ attains its maximum at a boundary point of U.*

We will need further some basic results on the mappings of several complex variables.

Theorem 151 ([Dieu, Theorem 9.10.1, Theorem 9.3.6]). *Let Y be a complex Banach space, $U \subset \mathbb{C}^n$ an open set, and $f \in H(U;Y)$. Then $f \in C^\infty(U;Y)$.*

Theorem 152 ([Dieu, Theorem 9.9.4]). *Let Y be a complex Banach space, $U \subset \mathbb{C}^n$ an open set, and $f \colon U \to Y$. If f is continuous and f is holomorphic in every variable, then f is holomorphic.*

Theorem 152 holds even without the continuity assumption, but the proof is substantially harder. This is a celebrated result of Fritz Hartogs. A nice proof based on the Polynomial lemma (Lemma 138) can be found in [Lej2].

Theorem 153 (Fritz Hartogs, [Hart]). *Let Y be a complex Banach space, $U \subset \mathbb{C}^n$ an open set, and $f \colon U \to Y$. If f is holomorphic in every variable, then f is holomorphic.*

8. Analytic mappings

In this section we switch our framework to Banach spaces. One of the reasons is that we have seen in Example 141 that the sum of the power series on a non-complete space may not be Fréchet differentiable. We proceed with the theory of holomorphic mappings by giving a list of equivalent conditions in Theorem 160. This result is used to complete also the real analytic case by giving several characterisations and proving the basic properties including the Implicit function theorem.

Definition 154. Let X, Y be Banach spaces over $\mathbb{K}$, $U \subset X$ open, and $f : U \to Y$. We say that f is analytic, if for every $a \in U$ there exist polynomials $P_n \in \mathcal{P}(^nX; Y)$ and $\delta > 0$ such that

$$f(x) = \sum_{n=0}^{\infty} P_n(x - a)$$

for all $x \in U(a, \delta)$. We denote by $C^\omega(U; Y)$ the vector space of all analytic mappings from U into Y. If $\mathbb{K} = \mathbb{R}$, then we will sometimes stress this fact by saying that f is real analytic. As usual, for functions, i.e. mappings into the scalars, we use the shortened notation $C^\omega(U) = C^\omega(U; \mathbb{K})$.

Note that by Theorem 140 the power series in the definition has a positive radius of norm convergence, and hence it converges uniformly to f on a neighbourhood of a. As a corollary of Theorem 146 we obtain the following theorem.

Theorem 155. *Let X, Y be Banach spaces over $\mathbb{K}$, $U \subset X$ open, and $f \in C^\omega(U; Y)$. Then $f \in C^\infty(U; Y)$ and for each $a \in U$ the power series expansion of f at a is unique and it is given by the Taylor series:*

$$f(x) = \sum_{n=0}^{\infty} \frac{1}{n!} d^n f(a)[x - a] \tag{36}$$

for x in some neighbourhood of a. Further, $d^k f \in C^\omega\big(U; \mathcal{P}(^kX; Y)\big)$ for each $k \in \mathbb{N}$.

The following fact follows easily from the definition and Fact 18, and moreover from Theorem 155 and Fact 77 for the real part.

Fact 156. *Let X, Y be Banach spaces over $\mathbb{K}$, $U \subset X$ open, and $f \in C^\omega(U; Y)$. Then $f\restriction_{U \cap W}$ is analytic for every closed affine subspace $W \subset X$.*

If X, Y are real Banach spaces, $U \subset \tilde{X}$ an open set, and $f \in C^\omega(U; \tilde{Y})$ such that $f(U \cap X) \subset Y$, then $f\restriction_{U \cap X} \in C^\omega(U \cap X; Y)$.

The analyticity has strong implications concerning the global behaviour.

Theorem 157. *Let X, Y be Banach spaces over $\mathbb{K}$, $U \subset X$ an open connected set, and $f, g \in C^\omega(U; Y)$. If $f = g$ on some non-empty open subset of U, then $f = g$ on U.*

Proof. Let $G = \operatorname{Int}\{x \in U;\ f(x) = g(x)\}$. Then G is a non-empty open subset of U. We show that G is relatively closed in U. The connectedness of U will then imply $G = U$. So let $z \in U$ be an accumulation point of G. Since $f - g = 0$ on the open set G, we conclude that $d^n f = d^n g$ on G for all $n \in \mathbb{N}_0$. Using the continuity of the differentials we obtain $d^n f(z) = d^n g(z)$, $n \in \mathbb{N}_0$, and so $f = g$ on a neighbourhood of z by Theorem 155. Therefore $z \in G$, which is what we wanted. □

Corollary 158. *Let X be a real Banach space, Y a complex Banach space, $U \subset \tilde{X}$ an open connected set, and $f, g \in C^\omega(U; Y)$. If $f = g$ on some non-empty subset of $U \cap X$ which is relatively open in X, then $f = g$ on U.*

Proof. Put $h = f - g$ and let $a \in U \cap X$ be such that $h = 0$ on a neighbourhood of a in X. Further, put $P_n = \frac{1}{n!}d^n h(a)$ and let V be a neighbourhood of a in $\tilde{X}$ such that $h(x) = \sum_{n=0}^{\infty} P_n(x-a)$ for $x \in V$. Obviously, $d^n(h\restriction_{U\cap X})(a) = 0$, and since $d^n h(a)\restriction_X = d^n(h\restriction_{U\cap X})(a)$ by Fact 77, we conclude using Proposition 61 that $P_n = 0$ for all $n \in N_0$. Thus $h = 0$ on V, and by Theorem 157 $f = g$ on U. □

Corollary 159. *Let X, Y be Banach spaces over $\mathbb{K}$, $U \subset X$ an open connected set, and let $f \in C^\omega(U;Y)$ be such that $\overline{\operatorname{span}}\, f(U) = Y$. Then $\overline{\operatorname{span}}\, f(V) = Y$ for every non-empty open $V \subset U$.*

Proof. Assume that there is a non-empty open $V \subset U$ such that $\overline{\operatorname{span}}\, f(V) \neq Y$. Then there is $\phi \in S_{Y^*}$ such that $\phi(\overline{\operatorname{span}}\, f(V)) = \{0\}$. Since clearly $\phi \circ f \in C^\omega(U)$ and $\phi \circ f(x) = 0$ for all $x \in V$, it follows that $\phi \circ f = 0$ on U (Theorem 157). Hence $\phi = 0$, a contradiction. □

The following important theorem says among other things that for complex spaces the notion of an analytic mapping coincides with the notion of a holomorphic mapping.

Theorem 160. *Let X, Y be complex Banach spaces, $U \subset X$ open, and $f: U \to Y$. The following statements are equivalent:*

(i) *f is analytic.*
(ii) *$f \in C^\infty(U;Y)$.*
(iii) *f is holomorphic.*
(iv) *$\phi \circ f$ is holomorphic for every $\phi \in Y^*$.*
(v) *f is Gâteaux differentiable and $\phi \circ f$ is Baire measurable for every $\phi \in Y^*$.*
(vi) *$f\restriction_{E\cap U}$ is holomorphic for every one-dimensional affine subspace $E \subset X$ and $\phi \circ f$ is Baire measurable for every $\phi \in Y^*$.*
(vii) *f is G^∞-smooth on U.*

Proof. (i)⇒(ii) follows from Theorem 155. (ii)⇒(iii)⇒(iv) and (iii)⇒(v) are obvious. (iv)⇒(vi) follows from Theorem 148. (v)⇒(vi) follows from the fact that Gâteaux and Fréchet differentiability are equivalent on one-dimensional spaces.

(vi)⇒(vii) By the Hartogs theorem (Theorem 153) $f\restriction_{Z\cap U}$ is holomorphic for every finite-dimensional subspace $Z \subset X$. Thus f is G^∞-smooth by Theorem 151 and Corollary 102.

It remains to prove (vii)⇒(i). Fix $a \in U$, let $\eta > 0$ be such that $U(a,\eta) \subset U$, and put $P_n = \frac{1}{n!}\widehat{\delta^n f(a)}$. Let $x \in U(a,\eta)$, $x \neq a$, and set $g(\zeta) = f(a + \zeta(x-a))$ for $\zeta \in \mathbb{C}$, $|\zeta| < \eta/\|x-a\|$. From the definition of the Gâteaux derivative we obtain $g'(\zeta) = \delta f(a+\zeta(x-a))[x-a]$. Thus g is holomorphic and using Theorem 149, Fact 98, and Fact 97 we obtain

$$f(x) = g(1) = \sum_{n=0}^{\infty} \frac{1}{n!} d^n g(0)[1] = \sum_{n=0}^{\infty} P_n(x-a),$$

which means that f is analytic. □

We note that if we assume continuity rather than Baire measurability in the previous theorem, we can prove it using Theorem 152 instead of the much deeper Hartogs theorem.

The proof above yields immediately the next statement.

Theorem 161. *Let X, Y be complex Banach spaces, $U \subset X$ open, $f \in H(U;Y)$, and $a \in U$. Then the formula* (36) *holds for every $x \in U$ satisfying $a + \lambda(x-a) \in U$ for every $|\lambda| \le 1$. In particular, the formula* (36) *holds whenever $B(a, \|x-a\|) \subset U$.*

The following theorem is of fundamental importance in complex analysis.

Theorem 162. *Let X, Y be complex Banach spaces, $U \subset X$ an open set, and let $f_n \in H(U;Y)$, $n \in \mathbb{N}$, be a sequence pointwise weakly convergent to a mapping $f: U \to Y$. If the restrictions of f_n to each one-dimensional affine subspace of X converge weakly locally uniformly, then f is holomorphic.*

Proof. Choose an arbitrary $\phi \in Y^*$. Then $\phi \circ f$ is Borel, since $\phi \circ f_n$ converges pointwise to $\phi \circ f$. Further, let $E \subset X$ be a one-dimensional affine subspace. Then $\phi \circ (f_n \restriction_{E \cap U}) \to \phi \circ (f \restriction_{E \cap U})$ locally uniformly on $E \cap U$. Thus the function $\phi \circ (f \restriction_{E \cap U})$ is holomorphic by the classical version of the theorem for holomorphic functions of one complex variable (e.g. [RudinW, Theorem 10.28]). Now it suffices to apply Theorem 160. □

Theorem 163 (Cauchy formula)**.** *Let X, Y be complex Banach spaces, $U \subset X$ an open set, and $f \in H(U;Y)$. Let $a \in U$, $h \in X$, and $r > 0$ be such that $a + \zeta h \in U$ for each $|\zeta| \le r$. Put $\gamma(t) = re^{it}$, $t \in [0, 2\pi]$. Then*

$$d^k f(a+\tau h)[h] = \frac{k!}{2\pi i} \int_\gamma \frac{f(a+\zeta h)}{(\zeta-\tau)^{k+1}}\, \mathrm{d}\zeta$$

for every $|\tau| < r$ and $k \in \mathbb{N}_0$.

Proof. Let $\phi \in Y^*$ and put $g(\zeta) = \phi \circ f(a+\zeta h)$. Then g is a holomorphic function on a neighbhourhood of the disc $\{\zeta \in \mathbb{C};\ |\zeta| \le r\}$ and so by the Cauchy formula for holomorphic functions of one complex variable

$$g^{(k)}(\tau) = \frac{k!}{2\pi i} \int_\gamma \frac{g(\zeta)}{(\zeta-\tau)^{k+1}}\, \mathrm{d}\zeta.$$

Since the functional ϕ commutes both with differentiation (Fact 78) and integration, we obtain

$$\phi\big(d^k f(a+\tau h)[h]\big) = d^k(\phi \circ f)(a+\tau h)[h] = g^{(k)}(\tau) = \phi\bigg(\frac{k!}{2\pi i} \int_\gamma \frac{f(a+\zeta h)}{(\zeta-\tau)^{k+1}}\, \mathrm{d}\zeta\bigg).$$

The desired formula now follows from the fact that Y^* separates the points of Y. □

Corollary 164 (Cauchy estimates)**.** *Let X, Y be complex Banach spaces, $U \subset X$ an open set, $f \in H(U;Y)$, and $a \in U$. If f is bounded by M on $B(a,R) \subset U$, $R > 0$, then for each $k \in \mathbb{N}$*

$$\|d^k f(a)\| \le \frac{Mk!}{R^k}.$$

Corollary 165. *Let X, Y be complex Banach spaces, $U \subset X$ open, $f \in H(U;Y)$, and $U(a,R) \subset U$. The radius of norm convergence of the Taylor series of f at a is at least R if and only if f is bounded on $B(a,r)$ for each $r < R$.*

Proof. $\Leftarrow$ follows from the Cauchy estimates and the Cauchy-Hadamard formula.
$\Rightarrow$ follows from Theorem 161. □

Corollary 166. *Let X, Y be complex Banach spaces, $U \subset X$ an open set, and $f_n \in H(U;Y)$, $n \in \mathbb{N}$. If the sequence $\{f_n\}$ converges locally uniformly on U to a mapping $f : U \to Y$, then f is holomorphic on U and $d^k f_n \to d^k f$ locally uniformly on U for every $k \in \mathbb{N}$.*

Proof. The fact that f is holomorphic follows from Theorem 162. Fix $a \in U$ and let $r > 0$ be such that $f_n \to f$ uniformly on $B(a,2r)$. For every $k \in \mathbb{N}$ and $x \in B(a,r)$ we obtain using the Cauchy estimates

$$\left\|d^k f_n(x) - d^k f(x)\right\| = \left\|d^k(f_n - f)(x)\right\| \le \frac{k!}{r^k} \sup_{y \in B(a,2r)} \left\|(f_n - f)(y)\right\|,$$

which implies that $d^k f_n \to d^k f$ uniformly on $B(a,r)$. □

The following is an immediate consequence of Corollary 165.

Corollary 167. *Let X, Y be complex Banach spaces, $U \subset X$ an open set, and $f \in H(U;Y)$. Suppose that the radius of norm convergence of the Taylor series of f at $a \in U$ is at least R and $U(a,R) \subset U$. Let $x \in U(a,R)$. Then the radius of norm convergence of the Taylor series of f at x is at least $R - \|x - a\|$.*

Although it is natural to expect that the sum of the power series is an analytic mapping, this fact is rather difficult to prove.

Theorem 168. *Let X, Y be Banach spaces over $\mathbb{K}$ and $P_n \in \mathcal{P}(^nX;Y)$ for $n \in \mathbb{N}_0$. Let $U \subset X$ be the domain of convergence of the power series*

$$\sum_{n=0}^{\infty} P_n. \tag{37}$$

Put $f(x) = \sum_{n=0}^{\infty} P_n(x)$ for $x \in U$. Then f is analytic on U. Moreover, for every $a \in U$ there is an open balanced neighbourhood of zero V containing a such that

$$\sum_{n=0}^{\infty} \sup_{x \in V} \|P_n(x)\| < +\infty.$$

In particular, the series (37) converges absolutely locally uniformly on U. Further,

$$d^k f(x) = \sum_{n=k}^{\infty} d^k P_n(x) \quad \textit{for every } x \in U,\ k \in \mathbb{N}. \tag{38}$$

Proof. We deal with the real and complex case separately. First suppose that $\mathbb{K} = \mathbb{C}$. By Corollary 143 the power series (37) converges locally uniformly on $U \cap Z$ for every two-dimensional subspace Z of X. Thus f is analytic by Theorem 162.

Put $\varphi(x) = \max_{t\in\mathbb{C},|t|\le 1}\|f(tx)\|$ for $x \in U$. Recall that U is balanced and so φ is well-defined. Using the compactness of the unit disc in $\mathbb{C}$ it is not difficult to check that φ is continuous on U. Suppose $a \in U$. Then $W = \{x \in U;\ \varphi(x) < \varphi(a) + 1\}$ is an open balanced neighbourhood of zero containing a. Let $0 < \lambda < 1$ be such that $a \in \lambda W$ and put $V = \lambda W$. By Theorem 155 we have $P_n = \frac{1}{n!}d^n f(0)$. Using the Cauchy formula (Theorem 163) we obtain $\|P_n(y)\| \le \varphi(y) < \varphi(a) + 1$ for $y \in W$, $n \in \mathbb{N}_0$. Hence $\|P_n(x)\| = \lambda^n\|P_n(x/\lambda)\| < (\varphi(a) + 1)\lambda^n$ for $x \in V$, $n \in \mathbb{N}_0$.

Finally, the formula (38) follows from Corollary 166.

In case $\mathbb{K} = \mathbb{R}$, by Lemma 145 there is an open set $\tilde{U} \subset \tilde{X}$ containing U such that the complexified series $\sum_{n=0}^{\infty}\tilde{P}_n$ converges on $\tilde{U}$. By the first part of the proof this series converges absolutely locally uniformly to an analytic mapping on $\tilde{U}$. Thus f is analytic by Fact 156 and the formula (38) follows using Fact 77 and the fact that if $Q_n \to Q$ for $Q_n, Q \in \mathcal{P}(^k\tilde{X};\tilde{Y})$, then $Q_n\restriction_X \to Q\restriction_X$ in $\mathcal{P}(^kX;\tilde{Y}_{\mathbb{R}})$. □

Note that the series in Example 137 converges to an analytic function on the whole c_0, although its radius of norm convergence is equal to 1.

The next few results deal with the real analytic mappings.

Theorem 169. *Let X, Y be real Banach spaces, $U \subset X$ open, and let $f: U \to Y$ be such that $\phi\circ f$ is Baire measurable for every $\phi \in Y^*$. Suppose that each restriction of f to a finite-dimensional subspace of X is G^∞-smooth, and moreover each restriction of f to a one-dimensional affine subspace of X is analytic. Then f is analytic.*

Proof. By Corollary 102 f is G^∞-smooth. Fix $a \in U$ and put

$$g(x) = \sum_{n=0}^{\infty}\frac{1}{n!}\widehat{\delta^n f(a)}[x-a]. \tag{39}$$

As the one-dimensional restrictions of f are analytic, for every $h \in S_X$ there is $\lambda_h > 0$ such that $f(a + th) = g(a + th)$ whenever $|t| < \lambda_h$. This means that the series in (39) converges for $x - a$ in an absorbing set, and hence it converges for $x - a$ in a neighbourhood of zero by Theorem 140. Thus there is $\delta > 0$ such that $U(a,\delta) \subset U$ and g is an analytic mapping on $U(a,\delta)$ (Theorem 168). Now for every $x \in U(a,\delta)$ we have $f\restriction_{(a+\mathrm{span}\{x-a\})\cap U(a,\delta)} = g\restriction_{(a+\mathrm{span}\{x-a\})\cap U(a,\delta)}$ by Theorem 157, in particular $f(x) = g(x)$. This finishes the proof. □

Theorem 170. *Let X, Y be real Banach spaces, $U \subset X$ open, and $f \in C^\omega(U;Y)$. Suppose that the radius of norm convergence of the Taylor series of f at $a \in U$ is at least R and $U\big(a,\frac{R}{\sqrt{2}}\big) \subset U$. Let $x \in U\big(a,\frac{R}{\sqrt{2}}\big)$. Then the radius of norm convergence of the Taylor series of f at x is at least $\frac{R}{\sqrt{2}} - \|x - a\|$.*

Proof. Assume without loss of generality that $a = 0$. Let $\sum_{n=0}^{\infty}P_n$, $P_n \in \mathcal{P}(^nX;Y)$ be the Taylor series of f at 0. Fix any $x \in U\big(0,\frac{R}{\sqrt{2}}\big)$ and $y \in U\big(x,\frac{R}{\sqrt{2}} - \|x\|\big)$. By

Lemma 21

$$f(y) = \sum_{n=0}^{\infty} P_n(y) = \sum_{n=0}^{\infty} P_n(y-x+x) = \sum_{n=0}^{\infty} \sum_{k=0}^{n} \binom{n}{k} \check{P}_n\big({}^k(y-x), {}^{n-k}x\big). \quad (40)$$

Notice that $\frac{n^n}{k^k(n-k)^{n-k}} \le 2^n$ for each $k \in \{0, \dots, n\}$. Thus by Corollary 17

$$\begin{aligned}
&\sum_{n=0}^{\infty} \sum_{k=0}^{n} \binom{n}{k} \|\check{P}_n\big({}^k(y-x), {}^{n-k}x\big)\| \\
&\qquad \le \sum_{n=0}^{\infty} \sum_{k=0}^{n} \binom{n}{k} \sqrt{\frac{n^n}{k^k(n-k)^{n-k}}} \|P_n\| \|y-x\|^k \|x\|^{n-k} \qquad (41) \\
&\qquad \le \sum_{n=0}^{\infty} \sqrt{2^n} \|P_n\| \big(\|y-x\| + \|x\|\big)^n < +\infty,
\end{aligned}$$

as $\limsup_{n\to\infty} \big(\sqrt{2^n}\|P_n\|\big)^{\frac{1}{n}} \le \frac{\sqrt{2}}{R}$. It follows that the generalised series in (40) is absolutely convergent, so we may rearrange its members to get $f(y) = \sum_{n=0}^{\infty} Q_n(y-x)$, where Q_n are given by the absolutely convergent series

$$Q_n(z) = \sum_{k=n}^{\infty} \binom{k}{n} \check{P}_k({}^n z, {}^{k-n}x).$$

Since this series converges for every $z \in U\big(0, \frac{R}{\sqrt{2}} - \|x\|\big)$, it converges for every $z \in X$ by the homogeneity. Thus $Q_n \in \mathcal{P}({}^n X; Y)$ by Theorem 29. Finally, choose any $0 < \lambda < \frac{R}{\sqrt{2}} - \|x\|$. Then

$$\sum_{n=0}^{\infty} \|Q_n\| \lambda^n \le \sum_{n=0}^{\infty} \sum_{k=n}^{\infty} \binom{k}{n} \sqrt{2^k} \|P_k\| \|x\|^{k-n} \lambda^n < +\infty,$$

as it is a rearrangement of a generalised series that converges absolutely by (41). This proves the claim. □

Theorem 171. *Let X and Y be real Banach spaces, $U \subset X$ an open set, and let $f \in C^\omega(U;Y)$. Then there exists an open set $\tilde{U} \subset \tilde{X}$ satisfying $\tilde{U} \cap X = U$ and a unique mapping $\tilde{f} \in H(\tilde{U}; \tilde{Y})$ such that $\tilde{f} \restriction_U = f$. More precisely, if we denote the radius of norm convergence of the Taylor series of f at a by R_a, then we can take $\tilde{U} = \bigcup_{a\in U} U_{\tilde{X}}(a, r_a)$, where $r_a = \min\{\frac{R_a}{2}, \operatorname{dist}(a, X \setminus U)\}$. The extension is then given by*

$$\tilde{f}(x) = \sum_{n=0}^{\infty} \tilde{P}_n(x-a) \quad (42)$$

for $x \in U_{\tilde{X}}(a, r_a)$, where $P_n = \frac{1}{n!} d^n f(a)$.

Proof. Define the mapping $\tilde{f}_a \colon U_{\tilde{X}}(a, r_a) \to \tilde{Y}$ by the formula (42) and note that $\tilde{f}_a \in H(U_{\tilde{X}}(a, r_a); \tilde{Y})$ by Lemma 144 and Theorem 168, and $\tilde{f}_a = f$ on $U_X(a, r_a)$ by Theorem 161. Finally, we define $\tilde{f}$ by $\tilde{f}(x) = \tilde{f}_a(x)$ whenever $x \in U_{\tilde{X}}(a, r_a)$ for some $a \in U$. The mapping $\tilde{f}$ is well-defined. Indeed, if $U_{\tilde{X}}(a, r_a) \cap U_{\tilde{X}}(b, r_b) \neq \emptyset$ for some points $a, b \in U$, then we have $\|a - b\| < r_a + r_b$. It follows that the set $U_{\tilde{X}}(a, r_a) \cap U_{\tilde{X}}(b, r_b) \cap X = U_X(a, r_a) \cap U_X(b, r_b)$ is a non-empty subset of U relatively open in X. Thus $\tilde{f}_a = \tilde{f}_b$ on $U_{\tilde{X}}(a, r_a) \cap U_{\tilde{X}}(b, r_b)$ by Corollary 158. Obviously $\tilde{f}$ is a holomorphic extension of f. Corollary 158 implies that the extension is uniquely determined on each $U_{\tilde{X}}(a, r_a)$ and thus also on $\tilde{U}$. □

We note that although the radius of norm convergence of (42) is at least $\frac{R_a}{2}$, it may be impossible to define the extension on all of $U_{\tilde{X}}(a, \frac{R_a}{2})$ – consider the function $\operatorname{sgn} x \in C^\omega(\mathbb{R} \setminus \{0\})$.

It is clear that the theorems that consider C^1-smoothness (notably the Implicit function theorem) give immediately results on holomorphic mappings. Considering the real analyticity, we can transfer many statements by extending the real analytic mapping using Theorem 171, applying the holomorphic version, and then appealing to Fact 156. Thus we obtain the following results.

Theorem 172. *Let $X_1, \dots, X_n$, Y, Z be real Banach spaces, $X = X_1 \oplus \dots \oplus X_n$, $U \subset X$ an open set, and $f \in C^\omega(U; Y)$. Let $V \subset Z$ be open and let $g_j \in C^\omega(V; X_j)$, $j = 1, \dots, n$, be such that $(g_1(z), \dots, g_n(z)) \in U$ for all $z \in V$. Then the mapping $h(z) = f(g_1(z), \dots, g_n(z))$ is analytic on V.*

Proof. It suffices to extend the mappings $f, g_j, j = 1, \dots, n$, to holomorphic mappings $\tilde{f}, \tilde{g}_j$ in the complex neighbourhoods $\tilde{U}$ of U, resp. $\tilde{V}$ of V by Theorem 171. Using continuity we may also assume $(\tilde{g}_1(z), \dots, \tilde{g}_n(z)) \in \tilde{U}$ for all $z \in \tilde{V}$. By Theorem 83 the mapping $h(z) = \tilde{f}(\tilde{g}_1(z), \dots, \tilde{g}_n(z))$ is holomorphic on $\tilde{V}$, so its restriction to V is analytic by Fact 156. □

In the same way using Corollary 84 we prove the following proposition.

Proposition 173. *Let X, Y be real Banach spaces, $U \subset X$ open, $f \in C^\omega(U)$, and $g \in C^\omega(U; Y)$. Then $f \cdot g \in C^\omega(U; Y)$. If moreover $f \neq 0$ on U, then $\frac{1}{f} \in C^\omega(U)$.*

Theorem 174 (Implicit function theorem – analytic case)**.** *Let X, Y, Z be Banach spaces, $G \subset X \oplus Y$ an open set, and $f \in C^\omega(G; Z)$. Let $(x_0, y_0) \in G$ and $D_2 f(x_0, y_0)$ be an isomorphism of Y onto Z. Then there are open neighbourhoods U of x_0 in X and V of y_0 in Y such that there is a unique mapping $u \colon U \to V$ satisfying $f(x, u(x)) = f(x_0, y_0)$ for all $x \in U$ and such that this mapping u is analytic on U.*

Proof. Theorem 87 covers the complex case, so suppose that the spaces X, Y, Z are real. By Theorem 87 there are open neighbourhoods W of x_0 in X and V of y_0 in Y, and a unique mapping $u \colon W \to V$ satisfying $f(x, u(x)) = f(x_0, y_0)$ for all $x \in W$. Now set $G_1 = W \times V$ and consider the holomorphic extension $\tilde{f} \colon \tilde{G}_1 \to \tilde{Z}$. Since

$\widetilde{X \oplus Y}$ is isomorphic to $\tilde{X} \oplus \tilde{Y}$, we may view $\tilde{f}$ as a mapping from $\tilde{G}_1 \subset \tilde{X} \oplus \tilde{Y}$. So using Proposition 81 and (42) we get

$$\begin{aligned} D_2\tilde{f}(x_0, y_0)[h] &= D\tilde{f}(x_0, y_0)[(0, h)] = \widetilde{Df(x_0, y_0)}[(0, h)] \\ &= Df(x_0, y_0)[(0, \operatorname{Re} h)] + iDf(x_0, y_0)[(0, \operatorname{Im} h)] \\ &= D_2 f(x_0, y_0)[\operatorname{Re} h] + iD_2 f(x_0, y_0)[\operatorname{Im} h] = \widetilde{D_2 f(x_0, y_0)}[h] \end{aligned}$$

for any $h \in \tilde{Y}$. Thus $D_2\tilde{f}(x_0, y_0)$ is an isomorphism of $\tilde{Y}$ onto $\tilde{Z}$ and we can apply the holomorphic version of the Implicit function theorem. We obtain open neighbourhoods $\tilde{U}$ of x_0 in $\tilde{X}$ and $\tilde{V}$ of y_0 in $\tilde{Y}$, and a unique mapping $\tilde{u}\colon \tilde{U} \to \tilde{V}$ satisfying $\tilde{f}\big(x, \tilde{u}(x)\big) = f(x_0, y_0)$ whenever $x \in \tilde{U}$, and such that $\tilde{u}$ is holomorphic on $\tilde{U}$. Due to the uniqueness we infer that $u(x) = \tilde{u}(x)$ for $x \in U = \tilde{U} \cap W$, so u is analytic on U by Fact 156. □

Exploiting the fact that a holomorphic mapping cannot oscillate too fast we can obtain explicit estimates on the size of the region where the Implicit function theorem gives the solution to $f(x, y) = f(x_0, y_0)$.

Theorem 175. *Let X, Y, Z be complex Banach spaces, $G \subset X \oplus Y$ an open set, and $f \in H(G; Z)$. Let $(x_0, y_0) \in G$ and $S = D_2 f(x_0, y_0)$ be an isomorphism of Y onto Z. Further, let $R > 0$ and $M > 0$ be such that $B(x_0, R) \times B(y_0, R) \subset G$ and $\|f(x, y)\| \le M$ for every $x \in B(x_0, R)$, $y \in B(y_0, R)$. Assume that $\|S^{-1}\| < a$ and $0 < r < R + 2aM - 2\sqrt{aM(R + aM)}$. Put $c = \frac{4aMr}{(R-r)^2}$, $s = \min\{r, R\frac{(1-c)r}{aM+(1-c)r}\}$. Then there exists a unique mapping $u\colon B(x_0, s) \to B(y_0, r)$ such that it satisfies $f\big(x, u(x)\big) = f(x_0, y_0)$ for $x \in B(x_0, s)$. Moreover, u is holomorphic on $U(x_0, s)$.*

Proof. The proof is the same as that of Theorem 87, but with explicit estimates on the size of the neighbourhood U. Assume for convenience $f(x_0, y_0) = 0$. Define $g\colon G \to Y$ by

$$g(x, y) = y - S^{-1} \circ f(x, y).$$

Note that r is chosen so that $c < 1$ and moreover $r < R$. Consider the space $X \oplus Y$ with the maximum norm and put $B = B(x_0, r) \times B(y_0, r)$. Using Proposition 71, Proposition 13, and finally the Cauchy estimates (Corollary 164), we can estimate for any $(x, y) \in B$

$$\begin{aligned} \|D_2 f(x, y) - S\| &\le \|Df(x, y) - Df(x_0, y_0)\| \\ &\le \sup_{w \in B} \|D^2 f(w)\| \max\big\{\|x - x_0\|, \|y - y_0\|\big\} \\ &\le \sup_{w \in B} 2\|d^2 f(w)\| r \le \frac{4Mr}{(R - r)^2} = \frac{c}{a} < \frac{1}{a} < \frac{1}{\|S^{-1}\|}. \end{aligned} \tag{43}$$

It follows that $D_2 f(x, y)$ is an isomorphism for every $(x, y) \in B$ (see the proof of Lemma 86).

Now fix any $x \in B(x_0, s)$. For arbitrary $y, z \in B(y_0, r)$, $y \neq z$, using Proposition 71, Fact 78, and the estimate (43) we get

$$\begin{aligned}\|g(x,z) - g(x,y)\| &\le \sup_{v \in B(y_0,r)} \|D_2 g(x,v)\| \|z - y\| \\ &= \sup_{v \in B(y_0,r)} \|Id - S^{-1} \circ D_2 f(x,v)\| \|z - y\| \\ &\le \sup_{v \in B(y_0,r)} \|S^{-1}\| \|S - D_2 f(x,v)\| \|z - y\| < c\|z - y\|.\end{aligned}$$

Thus the mapping $y \mapsto g(x,y)$ is a contraction. Furthermore, for $y \in B(y_0, r)$ with $(x, y) \neq (x_0, y_0)$ we can similarly estimate

$$\begin{aligned}\|g(x,y) - y_0\| &\le \|g(x,y) - g(x,y_0)\| + \|g(x,y_0) - y_0\| \\ &\le \|g(x,y) - g(x,y_0)\| + \|S^{-1}\| \|f(x,y_0) - f(x_0,y_0)\| \\ &< c\|y - y_0\| + a \sup_{w \in B(x_0,s)} \|D_1 f(w,y_0)\| \|x - x_0\| \\ &\le cr + \frac{aM}{R - s} s \le cr + (1 - c)r = r.\end{aligned}$$

Hence the mapping $y \mapsto g(x,y)$ is a contraction of $B(y_0, r)$ into $U(y_0, r)$ and we may proceed as in the proof of Theorem 87. □

In case $Y = Z = \mathbb{C}$ we can get slightly better estimates by a different approach to the proof, namely using the Rouché theorem instead of the Banach contraction principle. It is also worth noticing that in this case we can get an explicit formula for the local solution of $f(x, y) = 0$.

Theorem 176. *Let X be a complex Banach space, $U \subset X$ and $V \subset \mathbb{C}$ open sets, and $f \in H(U \times V)$. Further, let $x_0 \in U$, $y_0 \in V$, and let $R > 0$, $S > 0$, and $M > 0$ be such that $B(x_0, S) \subset U$, $B(y_0, R) \subset V$, and $|f(x,y)| \le M$ for every $x \in B(x_0, S)$, $y \in B(y_0, R)$. Assume that $f(x_0, y_0) = 0$, $|D_2 f(x_0, y_0)| \ge a > 0$, and $0 < r < R\frac{aR}{aR+M}$. Put $c = ar - \frac{Mr^2}{R(R-r)}$ and $s = S\frac{c}{c+M}$. Then there is a unique mapping $u\colon B(x_0, s) \to B(y_0, r)$ satisfying $f\big(x, u(x)\big) = 0$ for $x \in B(x_0, s)$. Moreover, $u \in H\big(U(x_0, s); U(y_0, r)\big)$ and*

$$u(x) = \frac{1}{2\pi i} \int_\gamma y \frac{D_2 f(x,y)}{f(x,y)} \, \mathrm{d}y$$

for $x \in U(x_0, s)$, where $\gamma(t) = y_0 + re^{it}$, $t \in [0, 2\pi]$.

Proof. Denote $g(y) = f(x_0, y)$. Then $g \in H(V)$, $g(y_0) = 0$, $|g'(y_0)| \ge a$, and $|g(y)| \le M$ for $y \in B(y_0, R)$. Thus, using Theorem 149 and the Cauchy estimates

(Corollary 164),

$$\begin{aligned}|g(y)| &= \left|\sum_{n=0}^{\infty}\frac{1}{n!}d^n g(y_0)[y-y_0]\right| \geq |g'(y_0)||y-y_0| - \sum_{n=2}^{\infty}\frac{1}{n!}|g^{(n)}(y_0)||y-y_0|^n \\ &\geq a|y-y_0| - \sum_{n=2}^{\infty}\frac{M}{R^n}|y-y_0|^n = a|y-y_0| - M\frac{|y-y_0|^2}{R(R-|y-y_0|)} \\ &= |y-y_0|\left(a - \frac{M|y-y_0|}{R(R-|y-y_0|)}\right)\end{aligned}$$

for every $y \in U(y_0, R)$. Therefore $|g(y)| > 0$ for every $y \in B(y_0, r) \setminus \{y_0\}$, which means that the only root of g in $B(y_0, r)$ is at y_0 and this root is simple. Moreover,

$$|g(y)| \geq r\left(a - \frac{Mr}{R(R-r)}\right) = c > 0 \quad \text{for } y \in \mathbb{C}, |y-y_0| = r.$$

Fix any $x \in B(x_0, s)$. Then for every $y \in \mathbb{C}$, $|y - y_0| = r$, we obtain (using the Cauchy estimates again)

$$\begin{aligned}|f(x,y) - g(y)| &= |f(x,y) - f(x_0,y)| \\ &\leq \sup_{w \in B(x_0,s)} \|D_1 f(w,y)\| \|x - x_0\| < \frac{Ms}{S-s} = c \leq |g(y)|.\end{aligned}$$

By the Rouché theorem (e.g. [RudinW, Theorem 10.43]) it follows that the function $y \mapsto f(x,y)$ has the same number of roots (counting with multiplicity) in $B(y_0, r)$ as the function g and these roots lie in $U(y_0, r)$. Hence there is a unique $u(x) \in B(y_0, r)$ for which $f\big(x, u(x)\big) = 0$, this $u(x)$ lies in $U(y_0, r)$, and it is a simple root of $f(x, \cdot)$.

Further, with x still fixed, let $W \subset V$ be an open neighbourhood of $B(y_0, r)$ such that $f(x, y) \neq 0$ for $y \in W \setminus \{u(x)\}$. It is easy to see that the function $y \mapsto y\frac{D_2 f(x,y)}{f(x,y)}$ is holomorphic in $W \setminus \{u(x)\}$ and its residue at $u(x)$ is $u(x)$. Thus the formula for $u(x)$ follows from the Residue theorem. Finally, using Corollary 91 we can conclude that u is differentiable on $U(x_0, s)$ and so it is holomorphic. □

9. Notes and remarks

Section 1. The material here is classical and can be found in many books. We refer to [Din], where the unexpectedly complicated history of the concept of polynomial is described in more detail. The definition of polynomials using multilinear forms came about in the early 1930's and was known to Stefan Banach and the Polish school [Bana], [MO1], [MO2]. There exist several concise survey papers on polynomials and geometry of the underlying space, e.g. [GuJL], [Aro3]. Corollary 55 spells out the equivalence of several definitions of the notion of polynomial on infinite-dimensional spaces, which were proposed by the early researchers in this area (e.g. [Fré1], [Fré2], [MO1], [MO2]). We refer to [Din] for detailed historical comments. The Lagrange interpolation formula goes back to Edward Waring around 1779.

Section 2. This section is based on the paper of Gustavo A. Muñoz, Yannis Sarantopoulos, and Andrew Morton Tonge [MST]. Our definition of the complexification norm is originally due to Angus Ellis Taylor and appeared in [MW]. We refer to the comprehensive paper [MST] for the references regarding various complexification methods used in Banach spaces. While the complex operations are obtained naturally, the complex norm can be introduced in infinitely many non-isometric (but of course real-isomorphic) ways. As it turns out, no definition fulfills all the naturally expected properties such as passing to subspaces, quotients, and following the natural lattice extensions. This subject is thoroughly treated in the paper [MST]. Extensions of polynomials into complexifications were used by A. E. Taylor in [Tay1]. The main Theorem 63 generalises the classical inequalities of Pafnuty Lvovich Chebyshev and Vladimir Andreevich Markov. The problem of finding the norm of a complex extension of a polynomial is a special case of a more general problem of estimating the value $P(x)$ (or the value of the derivatives of P at x) based on some information about $P \restriction_B$, where $x \in X$, and $B \subset X$. This forms a vast body of research in the finite-dimensional setting. We refer to books [BE], [Kl1], [Kra], [MMR]. Corollary 64 is valid in the classical case when $P(t) = \sum_{k=0}^{n} a_k t^k$ with optimal constants $|a_n| \leq 2^{n-1}\|P\|$, and $|a_{n-1}| \leq 2^{n-2}\|P\|$. This is due to P. L. Chebyshev and V. A. Markov.

Section 3. Here we collected classical results and formulae related to higher derivatives of mappings. The various notions of derivative of a mapping between topological vector spaces have been thoroughly investigated in [AvSm1], [AvSm2]. Our presentation mostly follows the line of Jean Dieudonné's book [Dieu] together with the survey papers [AvSm1], [AvSm2], and Herbert Federer's book [Fed]. The principal tools consist of Proposition 79 and the formula in Theorem 83, which can be efficiently combined to bootstrap higher smoothness into various results.

Section 4. Our presentation of higher order Gâteaux derivatives was inspired by [Boc]. In contrast with this work we assume the boundedness of the derivatives as a part of the definition. For the important role the Gâteaux derivative plays in the theory of non-separable real Banach spaces see [Fab3]. The Converse Taylor theorem (Theorem 110) seems to appear in [LjSo], see also [AlDia].

Section 6. The Polynomial lemma of Franciszek Leja (Lemma 138) has many important applications, including the proof of the Hartogs theorem. The main Theorem 140 was established by Einar Hille [Hill] in the complex case, and in the real case in [AO]. Our approach is partially based on the presentation in [Sic].

Section 7. Most of the results and their proofs can be found in [Dieu]. An in-depth presentation of the theory of several complex variables is in [Kra]. Another classical source is [Hill], where Banach space valued holomorphic mappings of one complex variable are developed. Holomorphic vector valued mappings of one complex variable have been studied by A. E. Taylor in [Tay2], [Tay3], and also in [Hill]. However, as pointed out in [Tay2], the possibility of such theory was known earlier to Norbert Wiener and others. Theorem 152 is due to A. E. Taylor.

Section 8. Many prominent mathematicians from the turn of the nineteenth century have contributed to the development of the general theory of holomorphic mappings in infinite-dimensional spaces. For the historical remarks we refer to [Tay4], [Tay5]. In a systematic way it was presented by E. Hille in [Hill], see also [Na]. The case of real analytic mappings was first treated by Andrzej Alexiewicz and Władysław Orlicz in [AO]. Our approach is partially inspired by [Boc] and [Sic].

The case of the infinite-dimensional domain is treated e.g. in [Din], [Mu], [Na]. A variant of Theorem 171 was used in [AO]. Further analysis of the convergence of the power series expansion of a holomorphic function in $\mathbb{C}^n$ is given in Section 3.6.

Problem 177. It seems to be unknown if Theorem 170 holds with R in place of $\frac{R}{\sqrt{2}}$ similarly as in the complex case.

Chapter 2

Basic properties of polynomials on $\mathbb{R}^n$

In the present chapter we focus on some basic properties of polynomials on $\mathbb{R}^n$. The first section is devoted to the duality theory for spaces of polynomials on finite-dimensional Banach spaces. The space $\mathcal{P}^d(\mathbb{R}^n)$ consisting of polynomials of degree at most d on $\mathbb{R}^n$ has finite dimension. It is clear that the Dirac functionals form a separating set in the dual of this space, and so by the basic principles of linear algebra there exists a basis of the dual $\mathcal{P}^d(\mathbb{R}^n)^*$ consisting of Dirac functionals. Such sets, called basic sets of nodes, are characterised in the classical Lagrange formula (Theorem 7).

Section 2 focuses on the abstract as well as concrete representation formulae for linear functionals on the polynomial spaces, in particular the cubature formulae representing positive functionals. We present a simple proof of the well-known Chakalov theorem. The main part of the section is devoted to cubature formulae on $\mathbb{R}$ and $\mathbb{R}^2$ related to Chebyshev polynomials. This topic seems to be rather special, but the choice is justified in Section 3.

In Section 3 we continue the study of Chebyshev polynomials resulting in the outline of the proof of the Skalyga-Markov-type inequality, an optimal upper estimate for the higher derivatives of real polynomials on Banach spaces spelled out in Theorem 28. Amazingly, it turns out that the nth Chebyshev polynomial T_n in one real variable is the extremal case in this situation. More precisely, the norm of the kth derivative of any real polynomial of norm one and degree at most n on any real Banach space is bounded above by the kth derivative $T_n^{(k)}(1)$.

In Section 4 we show the connection between homogeneous polynomials and the isometric theory of finite-dimensional Banach spaces. In particular, every homogeneous polynomial is a finite sum of powers of functionals, which leads to the theory of isometric embeddings of finite-dimensional spaces into ℓ_p-spaces where p is even. We present, without proof, a recent isometric version of the Dvoretzky theorem, and its close (but easier) relative, the Birch theorem. These results assure that an N-dimensional real Banach space whose unit sphere is a sub-level set of some p-homogeneous polynomial contains an isometric copy of a k-dimensional Hilbert space, where k depends only on the values N and p. Moreover, for a fixed p the value of k can be arbitrarily large, provided N is chosen large enough.

In Section 5 we describe an alternative approach to real polynomials, using restrictions to finite-dimensional subspaces of the domain, and the connection with differentiation. We cover the basic theory of identities for polynomials. The main result, Theorem 49, implies in particular that a continuous mapping on the unit ball of a Banach space is a restriction of a polynomial of degree at most n provided each affine one-dimensional restriction is a polynomial of degree at most n. The theorem also provides a simple test, a formula that has to be satisfied for all pairs of vectors from X, for the validity of these conditions.

The last section is devoted to the study of a relatively simple but very useful complex averaging method, which leads to many interesting and sometimes optimal estimates of coefficients of complex polynomials. Besides that, it will provide us with important ingredients for several other infinite-dimensional results later on, including the connection between separating polynomials and cotype of the space.

1. Spaces of polynomials on $\mathbb{R}^n$

In this section we investigate some basic properties of the spaces of polynomials on $\mathbb{R}^n$. To begin with, observe that by Proposition 1.23 the set of polynomials $\{x^\alpha;\ \alpha \in \mathcal{I}(n,d)\}$ is a basis of the vector space $\mathcal{P}({}^d\mathbb{R}^n)$. Hence

$$\dim \mathcal{P}({}^d\mathbb{R}^n) = |\mathcal{I}(n,d)| = \binom{n+d-1}{n-1}.$$

Definition 1. By $\mathcal{J}(n,d) = \bigcup_{j=0}^{d} \mathcal{I}(n,j)$ we denote the set of all multi-indices from $\mathbb{N}_0^n$ of order at most d.

By Propositions 1.43 and 1.23, $\mathcal{P}^d(\mathbb{R}^n) = \bigoplus_{j=0}^{d} \mathcal{P}({}^j\mathbb{R}^n)$ and the set of polynomials $\{x^\alpha;\ \alpha \in \mathcal{J}(n,d)\}$ is a basis of $\mathcal{P}^d(\mathbb{R}^n)$. It follows from the well-known combinatorial identity (the hockey-stick identity, see e.g. [Lov, Problem 1.42b]) that

$$\dim \mathcal{P}^d(\mathbb{R}^n) = |\mathcal{J}(n,d)| = \sum_{j=0}^{d} \binom{n+j-1}{n-1} = \binom{n+d}{n} = |\mathcal{I}(n+1,d)|.$$

The following fact is immediate.

Fact 2. *There is a natural linear isomorphism $T\colon \mathcal{P}({}^d\mathbb{R}^{n+1}) \to \mathcal{P}^d(\mathbb{R}^n)$ given by the restriction $T(P) = P\restriction_E$, where $E = \{x \in \mathbb{R}^{n+1};\ x_{n+1} = 1\}$ is an affine hyperplane which is isomorphic to $\mathbb{R}^n$. In other words, applying T to a d-homogeneous polynomial means replacing the $(n+1)$-st coordinate by the constant 1. The inverse T^{-1} acts on the polynomials $x^\alpha \in \mathcal{P}^d(\mathbb{R}^n)$, $|\alpha| \le d$ as follows: $T^{-1}(x^\alpha) = x^\alpha x_{n+1}^{d-|\alpha|}$.*

In our study we are going to exploit the fact that for a given norm on $\mathbb{R}^n$ the spaces of polynomials $\mathcal{P}^d(\mathbb{R}^n)$, $d \in \mathbb{N}_0$ canonically form an increasing chain of subspaces of the Banach space $C(B_{\mathbb{R}^n})$. Since $B_{\mathbb{R}^n}$ is a compact space, $C(B_{\mathbb{R}^n})^* = M(B_{\mathbb{R}^n})$ is a Banach space of Radon measures on $B_{\mathbb{R}^n}$.

Definition 34. Let $\mathcal{P}_{n,d} = \{P \in \mathcal{P}(^d\mathbb{R}^n);\ P(x) \geq 0 \text{ for all } x \in \mathbb{R}^n\}$. Further, let $\mathcal{Q}_{n,d} \subset \mathcal{P}(^d\mathbb{R}^n)$ be the set of all $P \in \mathcal{P}(^d\mathbb{R}^n)$ for which there exists a finite Borel measure μ on S^{n-1} such that

$$P(x) = \int_{S^{n-1}} \langle y, x\rangle^d \,\mathrm{d}\mu(y). \tag{16}$$

It is clear from the definition that $\mathcal{P}_{n,d}$ is a closed convex cone and $\mathcal{Q}_{n,d}$ is a convex cone. Further, since $\{\langle\alpha,\cdot\rangle^d;\ \alpha \in \mathcal{I}(n,d)\} \subset \mathcal{Q}_{n,d}$, Theorem 5 implies that $\mathcal{P}(^d\mathbb{R}^n) = \mathcal{Q}_{n,d} - \mathcal{Q}_{n,d}$. Hence $\mathcal{Q}_{n,d}$ does not lie in an affine hyperplane and so by the convexity $\operatorname{Int}\mathcal{Q}_{n,d} \neq \emptyset$.

For d odd these cones are trivial: $\mathcal{P}_{n,d} = \emptyset$ and $\mathcal{Q}_{n,d} = \mathcal{P}(^d\mathbb{R}^n)$ (the latter follows from the fact that in this case the cone $\mathcal{Q}_{n,d}$ is symmetric). For d even $\mathcal{Q}_{n,d} \subset \mathcal{P}_{n,d} \cap \{P \in \mathcal{P}(^d\mathbb{R}^n);\ P \text{ is convex}\} \subset \mathcal{P}_{n,d}$ and these cones are not equal (for $p \neq 2, n > 2$; see Theorem 47). Consequently if d is even, then also $\operatorname{Int}\mathcal{P}_{n,d} \neq \emptyset$. Finally, if $P \in \mathcal{Q}_{n,d}$, d even, then $P^{\frac{1}{d}}$ is quasi-convex and homogeneous and hence it is a semi-norm on $\mathbb{R}^n$.

Theorem 35. *$P \in \mathcal{Q}_{n,d}$ if and only if there exist $y_1, \dots, y_N \in \mathbb{R}^n$ such that*

$$P(x) = \sum_{k=1}^{N} \langle y_k, x\rangle^d, \ x \in \mathbb{R}^n.$$

Moreover we may assume that $N \leq |\mathcal{I}(n,d)|$. Further, $\mathcal{Q}_{n,d}$ is a closed convex cone.

Proof. $\Leftarrow$ It suffices to put $\mu = \sum_{k=1}^N \|y_k\|^d \delta_{\frac{y_k}{\|y_k\|}}$.

$\Rightarrow$ This follows immediately from Chakalov's theorem (Corollary 13) used on $\phi(p) = \int_{S^{n-1}} p\,\mathrm{d}\mu$. The estimate $N \leq |\mathcal{I}(n,d)|$ follows from Proposition 10.

Finally, it suffices to prove the last statement for d even. Suppose that $P_j \in \mathcal{Q}_{n,d}$, $P_j \to P \in \mathcal{P}(^d\mathbb{R}^n)$, and $P_j = \sum_{k=1}^N \langle y_{j,k}, \cdot\rangle^d$, where $N \leq |\mathcal{I}(n,d)|$ (add zeros if necessary). Then $\|y_{j,l}\|^{2d} \leq \sum_{k=1}^N \langle y_{j,k}, y_{j,l}\rangle^d = P_j(y_{j,l}) \leq \|P_j\|\|y_{j,l}\|^d$ for each $l \in \{1, \dots, N\}$. Therefore the set $\{y_{j,k};\ j \in \mathbb{N}, 1 \leq k \leq N\}$ is bounded and passing to a suitable subsequence we may assume that $\lim_{j\to\infty} y_{j,k} = y_k \in \mathbb{R}^n$, $k = 1, \dots, N$. It follows that $P = \sum_{k=1}^N \langle y_k, \cdot\rangle^d$ and so $\mathcal{Q}_{n,d}$ is closed. □

Proposition 36. *The cones $\mathcal{P}_{n,d}$ and $\mathcal{Q}_{n,d}$ are dual to each other, i.e.*

$$\mathcal{P}_{n,d} = \{P \in \mathcal{P}(^d\mathbb{R}^n);\ [P,Q] \geq 0 \text{ for all } Q \in \mathcal{Q}_{n,d}\},$$
$$\mathcal{Q}_{n,d} = \{Q \in \mathcal{P}(^d\mathbb{R}^n);\ [P,Q] \geq 0 \text{ for all } P \in \mathcal{P}_{n,d}\}.$$

Proof. The first relation follows from Theorem 35 and the second from Fact 33. □

We now pass to the connection between homogeneous polynomials of even degree and isometric subspaces of ℓ_p, p even.

Proposition 37. *Let $p \in \mathbb{N}$ be even and $P \in \mathcal{Q}_{n,p}$. If $P(x) = \sum_{k=1}^{N} \langle y_k, x\rangle^p$, $y_k \in \mathbb{R}^n$, then there exists a linear mapping $T\colon \mathbb{R}^n \to \ell_p^N$ such that $P(x) = \|T(x)\|^p$. In particular, if $P^{\frac{1}{p}}$ is a norm, then the Banach space with the polynomial norm $(\mathbb{R}^n, P^{\frac{1}{p}})$ is isometric to a subspace of ℓ_p^N.*

Proof. It suffices to let $T(x) = \big(\langle y_1, x\rangle, \dots, \langle y_N, x\rangle\big)$. □

In fact, the above simple result has a converse, so it is a characterisation of finite-dimensional isometric subspaces of L_p-spaces.

Theorem 38 ([El1], [DJP]). *Let $p \in \mathbb{N}$ be even and $X \subset L_p([0,1])$ be an n-dimensional subspace. Then there exist $y_1, \dots, y_N \in X^*$, $N \le |\mathcal{I}(n,p)|$, such that*

$$\|f\|^p = \sum_{j=1}^{N} \langle y_j, f\rangle^p, \quad f \in X$$

and so the polynomial $\|\cdot\|^p \restriction_X$ belongs to $\mathcal{Q}_{n,p}$ on $X \cong \mathbb{R}^n$. In particular, X is isometric to a subspace of ℓ_p^N.

Proof. We introduce an averaging linear operator $T_k\colon L_p([0,1]) \to L_p([0,1])$ for $k \in \mathbb{N}$ by the formula

$$T_k(f)(t) = k \int_{[\frac{j}{k}, \frac{j+1}{k})} f(s)\,\mathrm{d}s \quad \text{for } t \in I_{k,j} = \big[\tfrac{j}{k}, \tfrac{j+1}{k}\big),\ j = 0, \dots, k-1.$$

We have $\|T_k(f)\|^p = \sum_{j=0}^{k-1} \frac{k^p}{k} \big|\int_{I_{k,j}} f\big|^p \le \sum_{j=0}^{k-1} k^{p-1} \frac{1}{k^{p-1}} \int_{I_{k,j}} |f|^p = \|f\|^p$ (we used the Hölder inequality) and so $\|T_k\| \le 1$. Further, $T_k(f) \to f$ for each $f \in L_p([0,1])$ (check it for characteristic functions of intervals and use the density of simple functions). Thus $T_k(f) \to f$ uniformly on compact subsets of $L_p([0,1])$. Indeed, if this is not the case, then by passing to a subsequence we may assume that there are $\{f_k\} \subset L_p([0,1])$, $f_k \to f$, and $\varepsilon > 0$ such that $\|T_k(f_k) - f_k\| \ge \varepsilon$. But then $\|T_k(f_k) - f_k\| \le \|T_k\|\|f_k - f\| + \|T_k(f) - f\| + \|f - f_k\| \to 0$, a contradiction.

Let $\langle\cdot,\cdot\rangle$ be an inner product on X and let $y_{k,j} \in X$ satisfy $\langle y_{k,j}, f\rangle = k^{1-\frac{1}{p}} \int_{I_{k,j}} f$ for each $f \in X$, $j = 0, \dots, k-1$, $k \in \mathbb{N}$. Then $\|T_k(f)\|^p = \sum_{j=0}^{k-1} \langle y_{k,j}, f\rangle^p$ for $f \in X$ and hence $f \mapsto \|T_k(f)\|^p \in \mathcal{Q}_{n,p}$ on X. Since $\mathcal{Q}_{n,p}$ is closed by Theorem 35, it follows that $f \mapsto \|f\|^p \in \mathcal{Q}_{n,p}$ on X. This together with Proposition 37 finishes the proof. □

Proposition 39. *Let $p \in \mathbb{N}$ be even. The following statements are equivalent:*

(i) *$P \in \mathcal{Q}_{n,p}$ and $P^{\frac{1}{p}}$ is a norm on $\mathbb{R}^n$.*

(ii) *$P \in \mathcal{Q}_{n,p}$ and every (equivalently some) finite Borel measure μ on S^{n-1} satisfying* (16) *is not supported by a proper subspace of $\mathbb{R}^n$.*

(iii) *$P^{\frac{1}{p}}$ is a norm on $\mathbb{R}^n$ and $(\mathbb{R}^n, P^{\frac{1}{p}})$ is isometric to a subspace of $L_p([0,1])$.*

Proof. (i)⇒(ii). Suppose that a finite Borel measure μ on S^{n-1} satisfies the formula (16) and $\operatorname{supp}\mu \subset X$ for some proper subspace $X \subset \mathbb{R}^n$. If $0 \neq x \in X^{\perp}$, then $P(x) = \int_{X\cap S^{n-1}} \langle y, x\rangle^p \, d\mu(y) = 0$, a contradiction.

(ii)⇒(i). Let μ be a finite Borel measure satisfying (16) whose support is not contained in any proper subspace $X \subset \mathbb{R}^n$. Then $\mu(\{y;\ \langle y, x\rangle \neq 0\}) > 0$ for any $x \in \mathbb{R}^n \setminus \{0\}$. Thus $P(x) > 0$ and (i) follows.

The equivalence of (iii) with the previous statements follows from Proposition 37 and Theorem 38. □

Lemma 40 ([LV]). *Let $p \in \mathbb{N}$ be even and let σ be the spherical probability measure on S^{n-1} (i.e. the probability measure that is invariant with respect to the action of the orthogonal group on S^{n-1}). Then*

$$\int_{S^{n-1}} \langle u, x\rangle^p \, d\sigma(u) = \gamma_{n,p}\langle x, x\rangle^{\frac{p}{2}} = \gamma_{n,p}\|x\|_{\ell_2}^p,$$

where

$$\gamma_{n,p} = \frac{1\cdot 3\cdot \dots \cdot (p-1)}{n(n+2)\cdots(n+p-2)}.$$

Proof. First, note that the value of the integral is independent of $x \in S^{n-1}$. Indeed, this is immediate from the invariance of σ with respect to the orthogonal group. Moreover, the integral as a function of x is a p-homogeneous polynomial. These facts combined imply immediately that the statement holds with some constant $\gamma_{n,p} > 0$. For the precise value of $\gamma_{n,p}$ we refer to [LV]. □

Theorem 41 ([GoSe], [LV], [Rez3]). *Let $n, N, p \in \mathbb{N}$, p even, and let σ be the spherical probability measure on S^{n-1}. The following statements are equivalent:*

(i) *There are $y_1, \dots, y_N \in \mathbb{R}^n$ such that $\langle x, x\rangle^{\frac{p}{2}} = \sum_{j=1}^N \langle y_j, x\rangle^p$ for all $x \in \mathbb{R}^n$,*

(ii) *There are $y_1, \dots, y_N \in \mathbb{R}^n$ such that $\int_{S^{n-1}} P(u)\, d\sigma(u) = \sum_{j=1}^N P(y_j)$ for all $P \in \mathcal{P}(^p\mathbb{R}^n)$.*

(iii) *ℓ_2^n is isometric to a subspace of ℓ_p^N.*

Proof. Notice that the mapping $y \mapsto \langle y, \cdot\rangle^p$ is continuous on $\mathbb{R}^n$ and for a fixed $P \in \mathcal{P}(^p\mathbb{R}^n)$ the mapping $Q \mapsto [P, Q]$ is linear. Using Lemma 40 and the fact that Bochner integration commutes with bounded linear operators we obtain for any $P \in \mathcal{P}(^p\mathbb{R}^n)$ the following identities:

$$\int_{S^{n-1}} P(u)\, d\sigma(u) = \int_{S^{n-1}} [P, \langle u, \cdot\rangle^p]\, d\sigma(u) = \left[P, \int \langle u, \cdot\rangle^p \, d\sigma(u)\right] = \gamma_{n,p}[P, \|\cdot\|_{\ell_2}^p],$$

$$\sum_{j=1}^N P(y_j) = \sum_{j=1}^N [P, \langle y_j, \cdot\rangle^p] = \left[P, \sum_{j=1}^N \langle y_j, \cdot\rangle^p\right].$$

Hence (i) is equivalent to the (homogeneous) cubature formula (ii). By Proposition 37, (i) implies (iii). To see the converse let $T : \ell_2^n \to \ell_p^N$ be a linear isometry. Then we find $y_j \in \mathbb{R}^n$, $j = 1, \dots, N$, such that $\langle y_j, x\rangle = T(x)_j$ (the jth coordinate of $T(x)$) for each $x \in \mathbb{R}^n$. □

For given $n, p \in \mathbb{N}$, where p is even, denote by $N(n,p)$ the smallest N such that any of the equivalent conditions in Theorem 41 holds.

Proposition 42 ([LV], [DJP]). *If $n, p \in \mathbb{N}$ and p is even, then*

$$\left|\mathcal{I}(n,\tfrac{p}{2})\right| = \binom{n+\frac{p}{2}-1}{n-1} \le N(n,p) \le \binom{n+p-1}{n-1} = |\mathcal{I}(n,p)|.$$

Proof. Consider the functional $\phi(P) = \int_{S^{n-1}} P(u)\,\mathrm{d}\sigma(u) \in \mathcal{P}(^p\mathbb{R}^n)^*$, where σ is the spherical probability measure on S^{n-1}. There is $Q \in \mathcal{P}(^p\mathbb{R}^n)$ such that $\phi(P) = [Q,P]$ for each $P \in \mathcal{P}(^p\mathbb{R}^n)$. Clearly, the functional ϕ is positive for all $0 \ne P \in \mathcal{P}_{n,p}$, so by Proposition 36 we have $Q \in \mathcal{Q}_{n,p}$. By Theorem 35 there exist $\{y_j\}_{j=1}^N$, $N \le |\mathcal{I}(n,p)|$, such that $Q = \sum_{j=1}^n \langle y_j,\cdot\rangle^p$, and so $\phi(P) = [Q,P] = \sum_{j=1}^n P(y_j)$ and (ii) in Theorem 41 holds.

On the other hand, suppose that $\dim \mathcal{P}(^{\frac{p}{2}}\mathbb{R}^n) = \left|\mathcal{I}(n,\frac{p}{2})\right| > N(n,p)$. Then for any given set $\{y_j\}_{j=1}^{N(n,p)} \subset S^{n-1}$ there exists a non-zero polynomial $P \in \mathcal{P}(^{\frac{p}{2}}\mathbb{R}^n)$ with $P(y_j) = 0$. But $P^2 \in \mathcal{P}(^p\mathbb{R}^n)$ is a non-zero polynomial such that the right-hand side in (ii) in Theorem 41 is zero while the left-hand side is positive, a contradiction. □

Theorem 43 ([LV]). *Suppose that $p,q \in [1,\infty)$, $p \ne q$, $m,n \in \mathbb{N}$, $n \ge m \ge 2$, and there is an isometric embedding of ℓ_p^m into ℓ_q^n. Then $p = 2$ and q is an even integer.*

Proof. It suffices to prove the result for $m = 2$. Suppose that $T\colon \ell_p^2 \to \ell_q^n$ is an isometry and $T(1,0) = (x_1,\dots,x_n)$, $T(0,1) = (y_1,\dots,y_n)$. Let s be the number of the non-zero coordinates of the vector $(x_1,\dots,x_n)$. Without loss of generality we may assume that $x_k \ne 0$ for $k \le s$ and $x_k = 0$ for $s+1 \le k \le n$. Note that there is $1 \le j \le s$ such that $y_j \ne 0$. Indeed, otherwise $2^{\frac{q}{p}} = \|(1,1)\|_p^q = \|T(1,1)\|_q^q = \sum_{k=1}^s |x_k|^q + \sum_{k=s+1}^n |y_k|^q = \|T(1,0)\|^q + \|T(0,1)\|^q = 2$, a contradiction with the fact that $p \ne q$.

Set $z_k = \frac{y_k}{x_k}$ for $k \le s$. Then for any $t \in \mathbb{R}$, $|t| < a = \min_{1\le k\le s} \frac{1}{|z_k|}$ we have

$$\begin{aligned}\xi_0|t|^q + \sum_{k=1}^s \xi_k(1+tz_k)^q &= \sum_{k=1}^n |x_k + ty_k|^q = \|T(1,t)\|_q^q \\ &= \|(1,t)\|_p^q = (1+|t|^p)^{\frac{q}{p}},\end{aligned} \tag{17}$$

where $\xi_k = |x_k|^q > 0$ for $1 \le k \le s$ and $\xi_0 = \sum_{k=s+1}^n |y_k|^q \ge 0$. If $p = 1$, then $q > 1$ and in this case the left-hand side of the above identity is differentiable at $t = 0$, while the right-hand side is not, which is a contradiction. So we can assume that $p > 1$.

Put $\varphi(t) = \xi_0 t^q + \sum_{k=1}^s \xi_k(1+tz_k)^q$ and $\psi(t) = (1+t^p)^{\frac{q}{p}}$. Then φ is holomorphic on $\Omega_1 = \mathbb{C} \setminus \{(-\infty,0] \cup [a,+\infty)\}$ and ψ is holomorphic on a connected $\Omega_2 \subset \Omega_1$ containing $(0,a)$ and some neighbourhood of the half-line $re^{\frac{i\pi}{p}}$, $r \ge 1$, without this half-line. By the uniqueness theorem we have $\varphi = \psi$ on Ω_2. However, φ is continuous in $2e^{\frac{i\pi}{p}}$, while $\lim_{t\to 2e^{\frac{i\pi}{p}}} \psi(t)$ exists only if $\frac{p}{q} \in \mathbb{N}$. This shows that $q = lp$ for some $l \in \mathbb{N}$, $l > 1$.

Now using the Taylor expansion on the left-hand side of (17), for $t \in \mathbb{R}^+$ small we obtain

$$\sum_{j=1}^{\infty} d_0 t^j = 1 + lt^p + \binom{l}{2} t^{2p} + \cdots + \binom{l}{l-1} t^{(l-1)p} + (1 - \xi_0) t^{lp},$$

where $d_0 = \sum_{k=1}^{s} \xi_k > 0$, $d_1 = \sum_{k=1}^{s} q \xi_k z_k$, and $d_2 = \sum_{k=1}^{s} \frac{1}{2} q(q-1) \xi_k z_k^2 > 0$. Passing to a limit as $t \to 0+$ in the above formula we obtain $d_0 = 1$, so we have $1 + d_1 t + o(t) = 1 + o(t)$, $t \to 0+$, since $p > 1$. Thus $d_1 = 0$. Finally, this means that $d_2 t^2 + o(t^2) = lt^p + o(t^p)$, $t \to 0+$. Since $d_2 \neq 0$, we conclude that $p = 2$. □

By combining Theorem 41 and Proposition 42 we see that given $k \in \mathbb{N}$ and p even, for large enough N the space ℓ_p^N contains a subspace isometric to ℓ_2^k. This fundamental observation can be traced back to Hilbert [Hilb]. Hilbert proved this result in the course of his positive solution to Waring's problem in number theory. The isometric embedding result has the following recent remarkable generalisation due to Vladimir Leonidovich Dol'nikov and Roman Nikolaevich Karasev.

Theorem 44 ([DoKar]). *Suppose that $k, p \in \mathbb{N}$ and p is even. Then there exists $n(p,k) \in \mathbb{N}$ such that for any $P \in \mathcal{P}(^p\mathbb{R}^N)$ with $N \geq n(p,k)$ there exists a k-dimensional subspace $Y \subset \mathbb{R}^N$ such that $P \restriction_Y$ is a multiple of the polynomial $x \mapsto \langle x, x \rangle^{\frac{p}{2}}$. In particular, if $P^{\frac{1}{p}}$ is a polynomial norm on an N-dimensional Banach space X, then X contains isometrically a k-dimensional Euclidean subspace.*

The proof relies on deep results from algebraic topology and it is beyond the scope of this book. Analogous result also holds for p odd. This is the celebrated theorem of Birch [Bir]. In the odd case of course $P \restriction_Y = 0$ identically. For p odd it is shown in [DoKar] that

$$n(p,k) \leq k + \dim \mathcal{P}^p(\mathbb{R}^k) = k + \binom{k+p}{k}.$$

The proof of this result is again beyond the scope of this book. We are going to present a somewhat weaker earlier estimate which is accessible via the well-known Borsuk-Ulam antipodal theorem.

Theorem 45 ([ArHál]). *Let $k, N, p \in \mathbb{N}$, p odd, $N > k!(\log_2 N)^k \binom{k+p-1}{k-1}$ and $N > 1$, and let $Q \in \mathcal{P}(^p\mathbb{R}^n)$. Then there exists a k-dimensional subspace $Y \subset \mathbb{R}^N$ such that $Q \restriction_Y = 0$.*

One of the ingredients of the proof is a combinatorial lemma. Let $k \in \mathbb{N}$. A mapping $\sigma \colon A \to \{1, \ldots, k\}$ is called a k-colouring of a set A.

Lemma 46. *Let A be a finite set, $|A| = l > 1$, and $k \in \mathbb{N}$, $k \leq l$. Then there exist $\eta(k,l) \in \mathbb{N}$, $\eta(k,l) \leq k!(\log_2 l)^k$, and a system $\{\sigma_j\}_{j=1}^{\eta(k,l)}$ of k-colourings of A such that for every $B \subset A$, $|B| = k$, there exists a colouring σ_j satisfying $|\sigma_j(B)| = k$, i.e. each member of B has a different colour.*

Proof. For formal reasons we will work with colourings by $\{0,\dots,k-1\}$ rather than $\{1,\dots,k\}$. Clearly, $\eta(1,l) = \eta(l,l) = 1$ for any $l \in \mathbb{N}$, so from now on we may assume $l > 2$. We will use induction on k and for the proof we will also require additionally that if $2^{n-1} < l \le 2^n$, then $|\sigma_j^{-1}(s)| \le 2^{n-1}$ for each $s \in \{0,\dots,k-1\}$, $j = 1,\dots,\eta(k,l)$.

We start with $k = 2$. Suppose that $A = \{1,\dots,l\}$ and $2^{n-1} < l \le 2^n$. We define the system $\{\sigma_j\}_{j=1}^n$ of Rademacher-like functions by $\sigma_j(m) = 1$ if and only if $[\frac{m-1}{2^{n-j}}]$ is even. It is easy to see that this system satisfies the required conditions and so we may set $\eta(2,l) = n = \lceil \log_2 l \rceil \le 2(\log_2 l)^2$.

Now suppose that the statement holds for some $k \in \mathbb{N}$. Let $\{\sigma_j\}_{j=1}^{\eta(k,l)}$ be a system of k-colourings of A satisfying the required conditions. For each $s = 0,\dots,k-1$ let $\{\beta_r^{j,s}\}$ be the system of 2-colourings of the set $\sigma_j^{-1}(s)$ from the first step. We extend these colourings by the value 0 to the whole A. Finally we set $\sigma_r^{j,s} = \sigma_j + (k-s)\beta_r^{j,s}$ for each $j = 1,\dots,\eta(k,l)$, $s = 0,\dots,k-1$, and $r = 1,\dots,\eta(2,|\sigma_j^{-1}(s)|)$. This is a system of $(k+1)$-colourings of A which clearly satisfies the cardinality condition on $(\sigma_r^{j,s})^{-1}(m)$. Now if $B \subset A$, $|B| = k$, and $a \in A \setminus B$, then choose j such that $|\sigma_j(B)| = k$. Let $p \in B$ be such that $\sigma_j(a) = \sigma_j(p)$ and denote this value by q. Let r be such that $\beta_r^{j,q}(a) \ne \beta_r^{j,q}(p)$. Then $\left|\sigma_r^{j,q}(B \cup \{a\})\right| = k+1$ and so the system $\{\sigma_r^{j,s}\}$ satisfies the requirements.

Finally, if $2^{n-1} < l \le 2^n$, then from the above and using the inductive hypothesis on $\eta(k,l)$ we obtain $\eta(k+1,l) \le k\eta(k,l)\eta(2,2^{n-1}) \le kk!(\log_2 l)^k(n-1) \le (k+1)!(\log_2 l)^{k+1}$. □

Proof of Theorem 45. For a given vector $v = (v_1,\dots,v_k) \in \mathbb{R}^k$ and a k-colouring σ of the set $\{1,\dots,l\}$ we define a linear operator $v^\sigma : \mathbb{R}^l \to \mathbb{R}^l$ by the formula $v^\sigma(x_1,\dots,x_l) = (v_{\sigma(1)}x_1,\dots,v_{\sigma(l)}x_l)$. We may assume that $Q \ne 0$ and using the continuity we can choose the basis $\{e_1,\dots,e_N\}$ of $\mathbb{R}^N$ such that $Q(e_j) \ne 0$, $j = 1,\dots,N$. Consider the system of k-colourings $\{\sigma_j\}_{j=1}^{\eta(k,N)}$ of the set $\{1,\dots,N\}$ from Lemma 46 and the basic set of nodes $\mathcal{J}(k,p)$ for $\mathcal{P}({}^p\mathbb{R}^k)$ (Theorem 5). Let $M = \{(j,\alpha);\ 1 \le j \le \eta(k,N), \alpha \in \mathcal{J}(k,p)\}$. We fix a basis $\{e_{(j,\alpha)}\}_{(j,\alpha)\in M}$ in $\mathbb{R}^{|M|}$ and form a p-homogeneous polynomial $S : \mathbb{R}^N \to \mathbb{R}^{|M|}$ by

$$S(x) = \sum_{j=1}^{\eta(k,N)} \sum_{\alpha \in \mathcal{J}(k,p)} Q\big(\alpha^{\sigma_j}(x)\big)e_{j,\alpha}.$$

The mapping S is clearly odd (meaning $S(-x) = -S(x)$) and by the assumption $N - 1 \ge k!(\log_2 N)^k\binom{k+p-1}{k-1} \ge |M|$. By the Borsuk-Ulam antipodal theorem [Mat, Theorem 2.1.1] there exists a non-zero element $z = (z_1,\dots,z_N) \in \mathbb{R}^N$ such that $S(z) = 0$. Denote $B = \operatorname{supp} z$. We first claim that $|B| > k$.

Indeed, otherwise there exists some σ_j such that $|\sigma_j(B)| = |B|$. Choose any $b \in B$ and let $\beta = (0,\dots,0,p,0,\dots,0) = pe_{\sigma_j(b)} \in \mathcal{J}(k,p)$. Clearly, $\beta^{\sigma_j}(z) = pz_be_b$ and so $Q\big(\beta^{\sigma_j}(z)\big) \ne 0$, a contradiction with $S(z) = 0$.

Thus $|B| > k$ and we can find σ_j such that $|\sigma_j(B)| = k$, which means that setting $y_i = P_{\sigma_j^{-1}(i)}(z)$ (the projections that selects only coordinates in $\sigma_j^{-1}(i)$) we have $y_i \neq 0$. Next, define a polynomial R on $\mathbb{R}^k$ by $R(t_1, \ldots, t_k) = Q\big(\sum_{i=1}^k t_i y_i\big)$. Since $S(z) = 0$, it is immediate that $R(\alpha) = Q\big(\alpha^{\sigma_j}(z)\big) = 0$ for all $\alpha \in \mathcal{I}(k, p)$. In other words, R vanishes at all points of a basic set of nodes and so $R = 0$ on $\mathbb{R}^k$. To finish the proof, it suffices to choose $Y = \operatorname{span}\{y_i\}_{i=1}^k$. □

Note that $\lim_{N\to\infty} \frac{N}{(\log_2 N)^k} = +\infty$ and so there always is $N \in \mathbb{N}$ satisfying the requirements of Theorem 45. In order to give an explicit estimate for the asymptotic dependence of N on the values of k and p let us note that $N(p, k) = (k + p)^{3k}$ satisfies the requirements of Theorem 45 provided that $k + p \geq 10$.

The results given earlier underline the close connection between polynomial norms and L_p-spaces for p even. In particular, every p-homogeneous polynomial on $\mathbb{R}^n$ can be represented as a difference of two p-homogeneous polynomials which are themselves a restriction of the canonical ℓ_p-norm to a suitable subspace. It would be tempting to conjecture that every p-polynomial norm is a restriction of an L_p-norm to a suitable subspace, but this is false. In fact, passing to infinite-dimensional spaces it is false even in the isomorphic sense (see Notes and remarks in Chapter 5).

Theorem 47 ([Ne]). *Let $p \in \mathbb{N}$, $p \neq 2$ be even, $k, n \in \mathbb{N}$, $2 < k \leq n$. Then there exists an n-dimensional Banach space $X_{n,k}$ such that every $(k-1)$-dimensional subspace of $X_{n,k}$ is isometrically embeddable in ℓ_p^N, $N \leq |\mathcal{I}(k-1, p)|$, but there is a k-dimensional subspace of $X_{n,k}$ which is not isometrically embeddable in $L_p([0, 1])$.*

Let $n \in \mathbb{N}$ and let $p \in \mathbb{N}$ be even. Given a subspace $E \subset \mathbb{R}^n$ we denote $\mathcal{P}_{n,p}(E) = \mathcal{P}_{n,p} \cap \operatorname{span}\{\langle y, \cdot\rangle^p;\ y \in E\}$. For $1 \leq k \leq n$ let $\mathcal{P}_{n,p}(k)$ be the convex cone generated by the set

$$\{P;\ P \in \mathcal{P}_{n,p}(E) \text{ for some subspace } E \subset \mathbb{R}^n, \dim E \leq k\}.$$

Note that $\mathcal{Q}_{n,p} = \mathcal{P}_{n,p}(1) \subset \mathcal{P}_{n,p}(2) \subset \cdots \subset \mathcal{P}_{n,p}(n) = \mathcal{P}_{n,p}$ and similarly as in the proof of Theorem 35 it can be shown that all these cones are closed.

Lemma 48 ([Ne]). *Let $n, p \in \mathbb{N}$ and $p \neq 2$ be even. Then $\mathcal{P}_{n,p}(k-1) \subsetneq \mathcal{P}_{n,p}(k)$ for every $1 < k \leq n$.*

Proof. Observe that to prove that $h \notin \mathcal{P}_{n,p}(k)$ for some $0 \neq h \in \mathcal{P}_{n,p}$ it suffices to show that if $E \subset \mathbb{R}^n$, $\dim E \leq k$, and $g \in \mathcal{P}_{n,p}(E)$ satisfies $g \leq h$ on $\mathbb{R}^n$, then $g = 0$.

First assume that $k = 2$. Let $h(x) = h(x_1, \ldots, x_n) = x_1^2 x_2^{p-2}$. Then clearly $h \in \mathcal{P}_{n,p}(2)$ and $h(e_j) = 0$ for each $j \in \{1, \ldots, n\}$. Now if $g \in \mathcal{P}_{2,p}(E)$ for some one-dimensional E, then $g(x) = a\langle y, x\rangle^p$ for some $y \in \mathbb{R}^n$, $a \in \mathbb{R}$. Therefore if $0 \leq g \leq h$, then either $a = 0$ or $e_1, \ldots, e_n \in \ker\langle y, \cdot\rangle$, and so $g = 0$.

Now suppose that $k \geq 3$ and denote $p = 2l$. Let $h \in \mathcal{P}({}^l\mathbb{R}^n)$ be defined by

$$h(x) = h(x_1, \ldots, x_n) = \sum_{j=1}^k (-1)^{j+1} x_j^l.$$

Then $h^2 \in \mathcal{P}_{n,p}(k)$. We will show that $h^2 \notin \mathcal{P}_{n,p}(k-1)$. Let $E \subset \mathbb{R}^n$ be a $(k-1)$-dimensional subspace and $g \in \mathcal{P}_{n,p}(E)$, $g \leq h^2$. We will show that $g = 0$, which will finish the proof. As g is a polynomial, in order to prove that $g = 0$ it suffices to prove that g vanishes on some open subset of $\mathbb{R}^n$.

Note that $\dim E^{\perp} = n - k + 1$ and hence there is $z \in E^{\perp}$ such that $z_j \neq 0$ for some $j \in \{1, \dots, k\}$. We have $g(x + tz) = g(x)$ for every $x \in \mathbb{R}^n$ and $t \in \mathbb{R}$. We distinguish the cases $h(z) \neq 0$ and $h(z) = 0$.

In the first case let $M = \{x \in \mathbb{R}^n;\ h(z)h(x) < 0\}$. Clearly, M is open and non-empty, as $(1, 0, \dots, 0) \in M$ or $(0, 1, 0, \dots, 0) \in M$. Choose an arbitrary $x \in M$ and observe that $\varphi\colon t \mapsto h(x + tz)$ is a polynomial of degree l, $\varphi(0) = h(x)$, and $\lim_{t\to\infty} \frac{\varphi(t)}{t^l} = h(z)$. Thus there exists some $t \in [0, +\infty)$ for which $h(x + tz) = 0$. We have $0 \leq g(x + tz) \leq h^2(x + tz) = 0$, thus $g(x) = g(x + tz) = 0$ for any $x \in M$, which finishes the proof in this case.

Now let $h(z) = 0$. Then for each $x \in \mathbb{R}^n$ we have

$$\begin{aligned} h(x + tz) &= \sum_{j=1}^{k} (-1)^{j+1} \left(\sum_{m=0}^{l} \binom{l}{m} x_j^m t^{l-m} z_j^{l-m} \right) \\ &= t^{l-1} \sum_{j=1}^{k} (-1)^{j+1} z_j^{l-1} x_j + o(t^{l-1}), \ |t| \to +\infty. \end{aligned}$$

First assume that l is even. The set $M = \{x \in \mathbb{R}^n;\ \sum_{j=1}^{k} (-1)^{j+1} z_j^{l-1} x_j \neq 0\}$ is open and non-empty. Thus for each $x \in M$ the polynomial $t \mapsto h(x + tz)$ changes sign on $\mathbb{R}$ and so there is some $t \in \mathbb{R}$ satisfying $h(x + tz) = 0$ and we can finish the proof as above.

Finally, we claim that if $l > 1$ is odd, then the set

$$M = \left\{ x \in \mathbb{R}^n;\ h(x) \sum_{j=1}^{k} (-1)^{j+1} z_j^{l-1} x_j < 0 \right\}$$

is open and non-empty. We distinguish two cases. First, suppose that there exist $i, j \in \{1, \dots, k\}$ such that $(-1)^i \neq (-1)^j$ and $|z_i| > |z_j|$. Then there exist suitable $0 < x_i < x_j$ such that $x_i^l - x_j^l < 0$ while $z_i^{l-1} x_i - z_j^{l-1} x_j > 0$, and so $M \neq \emptyset$. In the opposite case $|z_1| = |z_2| = \dots = |z_k| \neq 0$. But then $(1, \frac{3}{2}, 1, 0, \dots, 0) \in M$, since $l > 1$. As for all $x \in M$ again $t \mapsto h(x + tz)$ changes sign on $[0, \infty)$, we can finish the proof as above. □

Proof of Theorem 47. By Lemma 48 and the separation theorem there is $P \in \mathcal{P}(^p\mathbb{R}^n)$ such that $[P, Q] > 0$ for all $Q \in \mathcal{P}_{n,p}(k-1) \setminus \{0\}$ and $[P, R] < 0$ for some $R \in \mathcal{P}_{n,p}(k)$. In particular, $P(x) = [P, \langle x, \cdot\rangle^p] > 0$ for each $x \in \mathbb{R}^n \setminus \{0\}$. Let $E \subset \mathbb{R}^n$ be a subspace of dimension $k - 1$. For any $Q \in \mathcal{P}_{n,p}(E)$ we have $[P \restriction_E, Q \restriction_E] = [P, Q] \geq 0$ and hence $P \restriction_E \in \mathcal{Q}_{k-1,p}$ on E (Proposition 36). Thus $(P \restriction_E)^{\frac{1}{p}}$ is a p-polynomial norm on E and E is isometric to a subspace of ℓ_p^N for some $N \leq |\mathcal{J}(k-1,p)|$ (Proposition 37). Since this holds for any subspace $E \subset \mathbb{R}^n$ of dimension $k-1 \geq 2$, the function $P^{\frac{1}{p}}$ is a norm on $\mathbb{R}^n$ and we put $X_{n,k} = (\mathbb{R}^n, P^{\frac{1}{p}})$.

On the other hand, let $R = \sum_{j=1}^{m} R_j$, where $0 \neq R_j \in \mathcal{P}_{n,p}(E_j)$, $\dim E_j \leq k$. Since $0 > [P,R] = \sum_{j=1}^{m}[P,R_j]$ and $[P,R_j] > 0$ whenever $\dim E_j < k$, it follows that there is $l \in \{1,\ldots,m\}$ such that $[P,R_l] < 0$ and $\dim E_l = k$. Thus $P\restriction_{E_l} \notin \mathcal{Q}_{k,p}$ on E_l (Proposition 36) and hence $(E_l, P^{\frac{1}{p}})$ is not isometric to a subspace of $L_p([0,1])$ (Proposition 39). □

5. Polynomial identities

The section is devoted to the study of polynomial identities, a subject inspired by Corollary 1.55 and the characterisation of the Hilbert space by the parallelogram identity. We are interested in general polynomials in connection with linear dependence of the sets of nodes and the Lagrange interpolation formula. In this section all spaces are real unless stated otherwise.

We begin by proving a local version of Corollary 1.55 (and also a version for complex spaces). We remark that the proof below is analytic rather than algebraic and completely different from that of Corollary 1.55.

Theorem 49. *Let X, Y be real normed linear spaces, $U \subset X$ a convex set with non-empty interior, and let $f\colon U \to Y$ be continuous. The following statements are equivalent:*

(i) *There is $P \in \mathcal{P}^n(X;Y)$ such that $f = P\restriction_U$.*

(ii) *$\Delta^{n+1} f(x;h_1,\ldots,h_{n+1}) = 0$ for all $x, h_j \in X$ such that the difference is defined.*

(iii) *$\Delta^{n+1} f(x;h,\ldots,h) = 0$ for all $x, h \in X$ such that the difference is defined.*

(iv) *$f\restriction_{E\cap U}$ is a restriction of a polynomial of degree not exceeding n for every affine one-dimensional subspace E of X.*

(v) *f is G^{n+1}-smooth and $\delta^{n+1} f = 0$ on* $\operatorname{Int} U$.

If X and Y are complex spaces, then all the statements are equivalent if we additionally assume in (ii) and (iii) that f is Gâteaux differentiable on $\operatorname{Int} U$.

The additional assumptions in the complex case cannot be dropped, as witnessed by the complex function $f(z) = \operatorname{Re} z$. The main step of the proof of the above theorem is contained in the following result.

Theorem 50. *Let X, Y be real normed linear spaces, $U \subset X$ a convex set with non-empty interior, let $f\colon U \to Y$ be continuous, and $n \in \mathbb{N}_0$. There is $P \in \mathcal{P}^n(X;Y)$ such that $f = P\restriction_U$ if and only if*

$$\sum_{k=0}^{n+1} (-1)^{n+1-k} \binom{n+1}{k} f(x+kh) = 0 \tag{18}$$

for all $x, h \in X$ such that $x + kh \in U$, $k = 0,\ldots,n+1$.

Proof. The case $n = 0$ is trivial, so we may assume that $n \geq 1$.

$\Rightarrow$ This implication follows from the Polarisation formula (Proposition 1.13) by setting $a = x + \frac{n+1}{2}h$ and $x_j = \frac{h}{2}$.

$\Leftarrow$ Without loss of generality we may assume that U is open and $0 \in U$. Further, it suffices to prove this implication only for $X = \mathbb{R}^N$, $N \leq n + 1$. The general case then follows: For any subspace $E \subset X$ with $\dim E = n + 1$ let $P_E \in \mathcal{P}^n(E; Y)$ be such that $P_E \restriction_{E \cap U} = f \restriction_{E \cap U}$. If $E, F \subset X$ are any two subspaces of dimension $n + 1$, then $P_E \restriction_{E \cap F \cap U} = f \restriction_{E \cap F \cap U} = P_F \restriction_{E \cap F \cap U}$ and hence $P_E \restriction_{E \cap F} = P_F \restriction_{E \cap F}$ by Fact 1.39. So we may define $P(x) = P_E(x)$, where $E \subset X$ is any subspace of dimension $n + 1$ containing x. Then $P \in P^n(X; Y)$ by Fact 1.36, and since f is continuous, so is P (Proposition 1.40).

First consider the case $N = 1$. Let $M \in \mathbb{N}$ be large enough so that $\{\frac{j}{2^M}\}_{j=0}^n \subset U$. For each $m \geq M$ let P_m be the unique polynomial of degree at most n satisfying $P_m(\frac{j}{2^m}) = f(\frac{j}{2^m})$ for $j = 0, \dots, n$ (Proposition 1.56). Assume that m is large enough so that $\frac{n+1}{2^m} \in U$. Since both P_m and f satisfy (18), using $x = 0$ and $h = \frac{1}{2^m}$ we conclude that $P_m(\frac{n+1}{2^m}) = f(\frac{n+1}{2^m})$. Repeating this argument inductively we deduce that $f(\frac{j}{2^m}) = P_m(\frac{j}{2^m})$ for all $j \in \mathbb{Z}$ such that $\frac{j}{2^m} \in U$. For any $m, r \geq M$ the set $\{\frac{j}{2^m}\}_{j \in \mathbb{Z}} \cap \{\frac{j}{2^r}\}_{j \in \mathbb{Z}} \cap U$ has at least $n + 1$ elements. Thus P_m and P_r agree on a set of cardinality at least $n+1$ and hence $P_m = P_r$ on the whole $\mathbb{R}$. Since f is assumed to be continuous and $\{\frac{j}{2^m}\}_{m \in \mathbb{N}, j \in \mathbb{Z}}$ contains a dense subset of U, the mapping $f = P_M \restriction_U$ is a restriction of a continuous polynomial of degree at most n.

Let us pass to the case $N > 1$. Denote by W the vector space of all continuous mappings from U into Y satisfying (18). Fix $g \in W$ and $h \in \mathbb{R}^N$. By the one-dimensional part of the proof, for any $x \in U$ the mapping $t \mapsto g(x + th)$ defined on a neighbourhood of zero is a restriction of a polynomial and so $\frac{\partial g}{\partial h}(x)$ exists. We show that $\frac{\partial g}{\partial h} \in W$. Clearly, $\frac{\partial g}{\partial h}$ satisfies (18). It remains to show that it is a continuous mapping.

Fix $x \in U$ and $\varepsilon > 0$. Without loss of generality we may assume that $\|h\| \leq 1$. There is $\eta > 0$ such that $\|g(x+th) - g(x) - t\frac{\partial g}{\partial h}(x)\| < \frac{\varepsilon}{2n^2}|t|$ whenever $|t| \leq \eta$. Since g is uniformly continuous on $B(x, 2\eta)$, there is $0 < \delta \leq \eta$ such that $\|g(y) - g(z)\| < \frac{\varepsilon}{2n^2}\eta$ whenever $y, z \in B(x, 2\eta)$, $\|y - z\| < \delta$. Choose an arbitrary $y \in U(x, \delta)$ and define $\varphi(t) = g(y + th) - g(x) - t\frac{\partial g}{\partial h}(x)$ for $t \in [-\eta, \eta]$. Since by the first part of the proof φ is a polynomial of degree at most n, by the Markov inequality (Theorem 80; note that it holds also for vector-valued polynomials, which can be easily shown by composing with functionals from Y^*; see also more involved Theorem 28)

$$\left\| \frac{\partial g}{\partial h}(y) - \frac{\partial g}{\partial h}(x) \right\| = \|\varphi'(0)\| \leq \frac{n^2}{\eta} \max_{t \in [-\eta, \eta]} \|\varphi(t)\|$$

$$\leq \frac{n^2}{\eta} \max_{t \in [-\eta, \eta]} \left(\left\| g(y + th) - g(x + th) \right\| + \left\| g(x + th) - g(x) - t\frac{\partial g}{\partial h}(x) \right\| \right) < \varepsilon.$$

This proves that $\frac{\partial g}{\partial h}$ is continuous and so $\frac{\partial g}{\partial h} \in W$.

It follows that any mapping in W (and in particular f) has continuous directional derivatives of all orders in all directions. Thus by Theorem 1.96 the mapping f is C^∞-smooth on U. Since $t \mapsto f(th)$ is a polynomial of degree at most n, using the Taylor formula (Theorem 1.107) and the uniqueness of the Taylor polynomial (Proposition 1.104) we obtain

$$f(th) = \sum_{k=0}^{n} \frac{t^k}{k!} d^k P(0)[h]$$

for each $h \in U$ and $t \in [0,1]$. Therefore $f = \sum_{k=0}^{n} \frac{1}{k!} d^k f(0)$ on U and so f is a restriction of a polynomial of degree at most n. □

Proof of Theorem 49. (i)⇒(ii) follows from Lemma 1.52 and the Polarisation formula (Proposition 1.13) by setting $a = x + \sum_{j=1}^{n+1} \frac{h_j}{2}$ and $x_j = \frac{h_j}{2}$.

(ii)⇒(iii) is trivial.

(iii)⇒(i) and (iv)⇒(i) in the real case follow from Lemma 1.52 and Theorem 50.

(i)⇒(iv) is clear.

(i)⇒(v) follows from Lemma 1.99.

(v)⇒(i) Fix any $x \in \operatorname{Int} U$. Given $h \in X$ such that $x + h \in \operatorname{Int} U$ we define $g(t) = f(x + th)$. By Fact 1.97, Fact 1.98 and the Taylor formula (Theorem 1.107) used on g we obtain $f(x+h) = \sum_{j=0}^{n} \frac{1}{j!} \widehat{\delta^j f(x)}[h]$. We finish by using Lemma 1.34.

Finally, assume that the spaces X and Y are complex and either (iii) or (iv) holds. Then f is C^∞-smooth on $\operatorname{Int} U$ by Theorem 1.160. From the real case we already know that f is a restriction of a polynomial from $X_{\mathbb{R}}$ into $Y_{\mathbb{R}}$ (the real versions of the spaces) of degree at most n. Thus for each $x \in \operatorname{Int} U$ and $h \in X$ we have $d^{n+1} f(x)[h] = 0$ by Fact 1.97 and so (v) holds. □

In the rest of this section we indicate a general approach to linear identities which characterise polynomials of prescribed degree. It turns out that these identities are independent of the underlying normed linear spaces and are closely related to the Lagrange formulae.

Definition 51. Let X, Y be normed linear spaces and $f \in C(X;Y)$. We say that f is compatible with $\phi \in \mathcal{F}(\mathbb{R}^m)$, $\phi = \sum_{k=1}^{n} a_k \delta_{x_k}$, if

$$\langle f \circ L, \phi \rangle = \sum_{k=1}^{n} a_k f(L(x_k)) = 0 \quad \text{for all } L \in \mathcal{L}(\mathbb{R}^m; X).$$

Remark 52. Clearly, $f \in C(X;Y)$ is compatible with $\phi = \sum_{k=1}^{n} a_k \delta_{x_k} \in \mathcal{F}(\mathbb{R}^m)$, where $x_k = (x_k^1, \dots, x_k^m)$, if and only if

$$\sum_{k=1}^{n} a_k f\left(\sum_{j=1}^{m} x_k^j z_j \right) = 0 \quad \text{for every } z_1, \dots, z_m \in X.$$

In particular, Fréchet's theorem (Theorem 50) is equivalent to saying that f is a polynomial of degree at most n if and only if f is compatible with the element

$\phi_n^F \in \mathcal{F}(\mathbb{R}^2)$ given by

$$\phi_n^F = \sum_{k=0}^{n+1} (-1)^{n+1-k} \binom{n+1}{k} \delta_{(1,k)}. \tag{19}$$

The following fact is a simple consequence of the definition.

Fact 53. *Let $\phi \in \mathcal{F}(\mathbb{R}^m)$, $\phi = \sum_{j=1}^k a_j \delta_{x_j}$. Further, let $L \in \mathcal{L}(\mathbb{R}^m; \mathbb{R}^n)$ and $\psi \in \mathcal{F}(\mathbb{R}^n)$, $\psi = \sum_{j=1}^k a_j \delta_{L(x_j)}$. Finally, let X, Y be normed linear spaces and $f \in C(X; Y)$. If f is compatible with ϕ, then f is compatible with ψ as well.*

Easy examples show that the implication in the previous fact cannot be reversed, see [HK].

Lemma 54. *Let $\phi \in \mathcal{F}(\mathbb{R}^m)$. The following statements are equivalent:*

(i) *For all normed linear spaces X, Y each $P \in \mathcal{P}(^d X; Y)$ (resp. $P \in \mathcal{P}^d(X; Y)$) is compatible with ϕ.*
(ii) *Every $P \in \mathcal{P}(^d \mathbb{R}^m)$ (resp. every $P \in \mathcal{P}^d(\mathbb{R}^m)$) is compatible with ϕ.*
(iii) *$\langle P, \phi \rangle = 0$ for every $P \in \mathcal{P}(^d \mathbb{R}^m)$ (resp. for every $P \in \mathcal{P}^d(\mathbb{R}^m)$).*

Proof. (i)$\Rightarrow$(ii)$\Rightarrow$(iii) are clear.

(iii)$\Rightarrow$(ii) Let $P \in \mathcal{P}(^d \mathbb{R}^m)$. If $L \in \mathcal{L}(\mathbb{R}^m; \mathbb{R}^m)$, then $P \circ L \in \mathcal{P}(^d \mathbb{R}^m)$, hence $\langle P \circ L, \phi \rangle = 0$, and therefore P is compatible with ϕ.

(ii)$\Rightarrow$(i) Let X, Y be normed linear spaces and $P \in \mathcal{P}(^d X; Y)$. Let $L \in \mathcal{L}(\mathbb{R}^m; X)$ and choose an arbitrary $f \in Y^*$. Then $f \circ P \circ L \in \mathcal{P}(^d \mathbb{R}^m)$, and therefore we have $0 = \langle f \circ P \circ L, \phi \rangle = f(\langle P \circ L, \phi \rangle)$. Since f was arbitrary, we conclude that $\langle P \circ L, \phi \rangle = 0$.

The proof of the non-homogeneous case is identical. □

Lemma 55. *Let X, Y be normed linear spaces and let $P = \sum_{k=0}^d P_k \in \mathcal{P}^d(X; Y)$, where $P_k \in \mathcal{P}(^k X; Y)$. If P is compatible with $\phi \in \mathcal{F}(\mathbb{R}^m)$, then each P_k is compatible with ϕ.*

Proof. Fix $L \in \mathcal{L}(\mathbb{R}^m; X)$. For each $t \in \mathbb{R}$ the mapping $g_t(z) = tL(z)$ belongs to $\mathcal{L}(\mathbb{R}^m; X)$ and hence

$$0 = \langle P \circ g_t, \phi \rangle = \sum_{k=0}^d t^k \langle P_k \circ L, \phi \rangle.$$

The right-hand side is a Y-valued polynomial in t. Thus each $\langle P_k \circ L, \phi \rangle = 0$. □

Lemma 56 ([Rez1]). *Let X and Y be normed linear spaces, $\phi \in \mathcal{F}(\mathbb{R}^m)$, and let $0 \neq P \in \mathcal{P}(^d X; Y)$. Then P is compatible with ϕ if and only if the polynomial $t \mapsto t^d$ from $\mathcal{P}(^d \mathbb{R})$ is compatible with ϕ.*

Proof. $\Rightarrow$ There exists a one-dimensional subspace $E \subset X$ such that $P \restriction_E = at^d$, $a \neq 0$. For every $L\colon \mathbb{R}^m \to E$ we have $\langle P \circ L, \phi \rangle = 0$. Consequently, $t \mapsto t^d$ is compatible with ϕ.

$\Leftarrow$ If $t \mapsto t^d$ is compatible with ϕ, then so is every $\langle y, \cdot\rangle^d$, where $y \in \mathbb{R}^m$. Indeed, $\langle y, \cdot\rangle^d$ is a composition of a linear projection of $\mathbb{R}^m$ onto $\mathbb{R}$ with $t \mapsto t^d$. Since $\{\langle y, \cdot\rangle^d;\ y \in \mathbb{R}^m\}$ forms a basis of $\mathcal{P}(^d\mathbb{R}^m)$, each $Q \in \mathcal{P}(^d\mathbb{R}^m)$ is compatible with ϕ. Lemma 54 then finishes the proof. □

Corollary 57. *An element $\phi = \sum_{k=1}^{n} a_k \delta_{x_k} \in \mathcal{F}(\mathbb{R}^n)$ is compatible with $t \mapsto t^d$ (or any other non-zero d-homogeneous polynomial) if and only if $\langle P, \phi\rangle = 0$ for all $P \in \mathcal{P}(^d\mathbb{R}^n)$.*

The following result can be viewed as a further abstraction of the cubature formulae treated in Section 2, where the functional is represented by the polynomial Q, and the evaluation at the nodes have been replaced by evaluation at a given element P and its linear translates.

Corollary 58. *Let $0 \neq P \in \mathcal{P}(^d\mathbb{R}^n)$. Then for any $Q \in \mathcal{P}(^d\mathbb{R}^n)$ there exist a finite collection of $L_k \in \mathcal{L}(\mathbb{R}^n;\mathbb{R}^n)$ and $\varepsilon_k \in \{-1, 1\}$, $k = 1, \dots, r$, $r \le |\mathcal{J}(n,d)|$, such that $Q = \sum_{k=1}^{r} \varepsilon_k P \circ L_k$.*

Proof. By contradiction, suppose that $H = \operatorname{span}\{P \circ L;\ L \in \mathcal{L}(\mathbb{R}^n;\mathbb{R}^n)\}$ is a proper subspace of $\mathcal{P}(^d\mathbb{R}^n)$, i.e. there exist some $Q \in \mathcal{P}(^d\mathbb{R}^n) \setminus H$ and a linear functional $\phi \in \mathcal{F}(\mathbb{R}^n)$ which is zero on H and $\langle Q, \phi\rangle \neq 0$. Then P is compatible with ϕ but Q is not. This contradicts Lemma 56. To obtain the estimate $r \le |\mathcal{J}(n,d)|$ use Theorem 8. □

The above corollary is analogous to Theorem 41. Indeed, choose $0 \neq y \in \mathbb{R}^n$, put $P = \langle y, \cdot\rangle^p$ and $Q(x) = \|x\|_{\ell_2}^p$, and apply Corollary 58. The difference is that here we do not have any additional assumptions because we allow linear combinations with both signs.

Next, we investigate the properties of $\phi \in \mathcal{F}(\mathbb{R}^n)$ which lead to compatibility (Theorem 60). From this, the theorem of Pascual Jordan and John von Neumann (Corollary 61) follows immediately. The result was proved by Bruce Reznick under the assumption that the continuous mapping f is homogeneous. In the proof of Theorem 60 we will use the following result of William Harold Wilson.

Theorem 59 ([Wi]). *Let X, Y be normed linear spaces, $f \in C(X;Y)$, $n \in \mathbb{N}_0$, and let $\phi = \sum_{k=0}^{n+1} a_k \delta_{x_k} \in \mathcal{F}(\mathbb{R}^2)$. Suppose that $a_0 \neq 0$ and for every $0 < k \le n+1$ the vectors x_0 and x_k are linearly independent. Let $p+2$ be the number of distinct (unoriented) directions determined by the vectors $x_0, \dots, x_{n+1}$. If f is compatible with ϕ, then f is compatible with ϕ_p^F from* (19).

Proof. Suppose that $x_j = (r_j, s_j)$, $j = 0, \dots, n+1$. By Fact 53 we may assume without loss of generality that $r_0 = 0$ and $s_0 = 1$. Then $r_j \neq 0$ for $1 \le j \le n+1$. Moreover, we may also assume that the vectors $x_0, \dots, x_{p+1}$ are pairwise linearly independent. Put $I_k = \{1 \le j \le n+1;\ x_j \in \operatorname{span}\{x_k\}\}$ for $k = 1, \dots, n+1$ and $J_k = \{1, \dots, n+1\} \setminus (I_1 \cup \dots \cup I_k)$ for $k = 0, \dots, p+1$. Finally, set $\Delta_{k,j} = s_j - r_j \frac{s_k}{r_k}$ for $k, j \in \{1, \dots, n+1\}$ and note that $\Delta_{k,j} = 0$ if and only if x_k and x_j are linearly dependent which is if and only if $j \in I_k$.

Fix an arbitrary $z \in X$ and define inductively

$$\psi_0(x, y) = a_0 f(y) + \sum_{j=1}^{n+1} a_j f(r_j x + s_j y)$$

and $\psi_{m+1}(x, y) = \psi_m\big(x - \frac{s_{m+1}}{r_{m+1}} z, y + z\big) - \psi_m(x, y)$ for $x, y \in X$, $m = 0, \dots, p$. By Remark 52 we have $\psi_0(x, y) = 0$ for every $x, y \in X$ and thus also $\psi_m(x, y) = 0$ for every $x, y \in X$, $m = 1, \dots, p+1$. Using the fact that $\Delta_{m+1,j} = 0$ if and only if $j \in I_{m+1}$ it is easy to check by induction that

$$\psi_m(x, y) = a_0 \sum_{k=0}^{m} (-1)^{m-k} \binom{m}{k} f(y + kz) + \sum_{j \in J_m} a_j \sum_{A \subset \{1,\dots,m\}} (-1)^{m+|A|} f\Big(r_j x + s_j y + \sum_{k \in A} \Delta_{k,j} z\Big).$$

Since $J_{p+1} = \emptyset$, we obtain

$$a_0 \sum_{k=0}^{p+1} (-1)^{p+1-k} \binom{p+1}{k} f(y + kz) = 0$$

for every $y, z \in X$, and therefore f is compatible with ϕ_p^F. □

Theorem 60 ([Rez1], [HK]). *Let X, Y be normed linear spaces, $f \in C(X; Y)$, and $\phi = \sum_{k=0}^{n+1} a_k \delta_{x_k} \in \mathcal{F}(\mathbb{R}^m)$, $m \geq 2$, $n \in \mathbb{N}_0$. Suppose that $a_{n+1} \neq 0$ and for every $k \neq n+1$ the vectors x_k and x_{n+1} are linearly independent. Let $p+2$ be the number of distinct directions determined by the vectors $x_0, \dots, x_{n+1}$. If f is compatible with ϕ, then f is a polynomial of degree at most p.*

Proof. First suppose that $m = 2$. Since f is compatible with ϕ, it is compatible with ϕ_p^F by Theorem 59. By Remark 52 the mapping f is a polynomial of degree at most p.

Now we consider the case $m > 2$. By Fact 53 it is enough to find a $T \in \mathcal{L}(\mathbb{R}^m; \mathbb{R}^2)$ such that $\psi = \sum_{k=0}^{n+1} a_k \delta_{T(x_k)}$ satisfies the assumptions of the previous case, i.e. such that the vectors $T(x_k)$ and $T(x_{n+1})$ are linearly independent for each $k \neq n+1$. (The number of distinct directions determined by $T(x_0), \dots, T(x_{n+1})$ is clearly less than or equal to $p+2$.) This is easily done as follows: Let $E_k = \operatorname{span}\{x_k, x_{n+1}\} \subset \mathbb{R}^m$, $k \in \{0, \dots, n\}$, be a system of two-dimensional subspaces of $\mathbb{R}^m$. There exists an $(m-2)$-dimensional subspace $F \subset \mathbb{R}^m$ such that $F \cap E_k = \{0\}$, $k \in \{0, \dots, n\}$. (Equivalently, $F + E_k = \mathbb{R}^m$). Then the orthogonal projection T in $\mathbb{R}^m$ with kernel F and the two-dimensional range $F^\perp \subset \mathbb{R}^m$ clearly satisfies the requirements. □

Corollary 61 (Pascual Jordan, John von Neumann; [JN]). *Let X be a real Banach space such that*

$$\|x + y\|^2 + \|x - y\|^2 = 2\|x\|^2 + \|y\|^2 \quad \textit{for all } x, y \in X.$$

Then X is isometric to a Hilbert space.

Proof. The function $f(x) = \|x\|^2$ satisfies $f(x+h)+f(x-h)-2f(x)-2f(h) = 0$, i.e. f is compatible with $\phi = \delta_{(1,1)} + \delta_{(1,-1)} - 2\delta_{(1,0)} - 2\delta_{(0,1)} \in \mathcal{F}(\mathbb{R}^2)$. By Theorem 60, f is a polynomial of degree 2. Thus $\|x\| = \sqrt{M(x,x)}$, where M is a bounded bilinear form on X. □

If $\phi \in \mathcal{F}(\mathbb{R}^m)$ and X, Y are normed linear spaces, then the set of all mappings in $C(X;Y)$ which are compatible with ϕ is a subspace of the space of polynomials (Theorem 60). We are now ready to describe this space more precisely.

Theorem 62 ([HK]). *Suppose $\phi = \sum_{k=0}^{n+1} a_k \delta_{x_k} \in \mathcal{F}(\mathbb{R}^m)$, where $x_0, \dots, x_{n+1}$ are pairwise linearly independent vectors. Then there exists $A \subset \{0, \dots, n\}$ such that if X, Y are normed linear spaces and $f \in C(X;Y)$, then f is compatible with ϕ if and only if $f = \sum_{k \in A} P_k$ for some $P_k \in \mathcal{P}({}^kX;Y)$.*

Proof. Let A be the set of all $k \in \{0, \dots, n\}$ for which there exist normed linear spaces X, Y and a non-zero polynomial from $\mathcal{P}({}^kX;Y)$ which is compatible with ϕ. By Lemma 56, if $k \in A$, then for all normed linear spaces X and Y each polynomial from $\mathcal{P}({}^kX;Y)$ is compatible with ϕ and the same holds also for their linear combinations. Let now X, Y be normed linear spaces and $P: X \to Y$ be a continuous mapping compatible with ϕ. By Theorem 60 the mapping P is a polynomial of degree at most n, say $P = \sum_{k=0}^{n} P_k$, $P_k \in \mathcal{P}({}^kX;Y)$. If $P_k \neq 0$ for some $k \in \{0, \dots, n\}$, then it follows from Lemma 55 that $k \in A$. Hence $P = \sum_{k \in A} P_k$. □

More can be said if the points $x_0, \dots, x_{n+1}$ lie in an affine hyperplane not containing 0.

Lemma 63. *Suppose $\phi = \sum_{k=0}^{n+1} a_k \delta_{x_k} \in \mathcal{F}(\mathbb{R}^m)$, where $x_0, \dots, x_{n+1}$ lie in an affine hyperplane not containing 0. If every polynomial from $\mathcal{P}({}^d\mathbb{R}^m)$ is compatible with ϕ, then the same holds for every polynomial from $\mathcal{P}^d(\mathbb{R}^m)$.*

Proof. Let H be an affine hyperplane in $\mathbb{R}^m$ which contains $x_0, \dots, x_{n+1}$ and does not contain 0. Suppose that every polynomial from $\mathcal{P}({}^d\mathbb{R}^m)$ is compatible with ϕ. If $P \in \mathcal{P}^d(\mathbb{R}^m)$, then there exists $Q \in \mathcal{P}({}^d\mathbb{R}^m)$ such that $Q\restriction_H = P\restriction_H$ (see Fact 2). Since Q is compatible with ϕ, we see that $\langle P, \phi \rangle = \langle Q, \phi \rangle = 0$. By Lemma 54 every polynomial from $\mathcal{P}^d(\mathbb{R}^m)$ is compatible with ϕ. □

Theorem 64 ([HK]). *Suppose $\phi = \sum_{k=0}^{n+1} a_k \delta_{x_k} \in \mathcal{F}(\mathbb{R}^m)$, where $x_0, \dots, x_{n+1}$ lie in an affine hyperplane not containing 0. If $\sum_{k=0}^{n+1} a_k = 0$, then there exists $p \in \mathbb{N}_0$, $p \leq n$, such that if X, Y are normed linear spaces and $f \in C(X;Y)$, then f is compatible with ϕ if and only if f is a polynomial of degree at most p. If $\sum_{k=0}^{n+1} a_k \neq 0$, then there is no non-zero mapping compatible with ϕ.*

Proof. Since $x_0, \dots, x_{n+1}$ are pairwise linearly independent, the assumptions of Theorem 62 are satisfied. Let $A \subset \{0, \dots, n\}$ be the set from Theorem 62. If $\sum_{k=0}^{n+1} a_k = 0$, then $t \mapsto 1$, $t \in \mathbb{R}$, is compatible with ϕ and therefore A is non-empty. Let $p = \max A$. Since every polynomial from $\mathcal{P}({}^p\mathbb{R}^m)$ is compatible with ϕ, by Lemma 63 so is every polynomial from $\mathcal{P}^p(\mathbb{R}^m)$. Hence $A = \{0, \dots, p\}$. This argument also shows that if

A is non-empty, then $t \mapsto 1$ is compatible with ϕ and consequently $\sum_{k=0}^{n+1} a_k = 0$. Hence if $\sum_{k=0}^{n+1} a_k \neq 0$, then there is no non-zero mapping compatible with ϕ. □

In order to generate linear identities we rely on Theorem 7. In fact, the Lagrange formula is an expression of linear dependence of functionals in the dual of $\mathcal{P}^d(\mathbb{R}^m)$. Let $\{x_k\}_{k=1}^r \subset \mathbb{R}^m$ be a basic set of nodes for $\mathcal{P}^d(\mathbb{R}^m)$ and let $\{p_k\}_{k=1}^r \subset \mathcal{P}^d(\mathbb{R}^m)$ be biorthogonal to $\{x_k\}_{k=1}^r$. For any fixed $z \in \mathbb{R}^m \setminus \{x_k\}_{k=1}^r$ Theorem 7 implies $P(z) = \sum_{k=1}^r p_k(z)P(x_k)$ for every $P \in \mathcal{P}^d(\mathbb{R}^m)$. It follows that every polynomial $P \in \mathcal{P}^d(\mathbb{R}^m)$ is compatible with $\phi = -\delta_z + \sum_{k=1}^r p_k(z)\delta_{x_k}$ (Lemma 54).

Lemma 65. *Suppose that $\{x_k\}_{k=1}^r \subset \mathbb{R}^m$ is a basic set of nodes for $\mathcal{P}^d(\mathbb{R}^m)$ and $\{p_k\}_{k=1}^r \subset \mathcal{P}^d(\mathbb{R}^m)$ is biorthogonal to $\{x_k\}_{k=1}^r$. Further, let $z \in \mathbb{R}^m \setminus \{x_k\}_{k=1}^r$ and $\phi = -\delta_z + \sum_{k=1}^r p_k(z)\delta_{x_k}$. If every $P \in \mathcal{P}^l(\mathbb{R}^m)$ is compatible with ϕ, then $l \leq d$.*

Proof. Assume without loss of generality that $p_1(z) \neq 0$. Clearly, there exists $R \in \mathcal{P}^1(\mathbb{R}^m)$ (an affine function on $\mathbb{R}^m$) such that $R(x_1) \neq 0$ and $R(z) = 0$. Set $P = p_1 R \in \mathcal{P}^{d+1}(\mathbb{R}^m)$. Then $P(x_1) \neq 0$, $P(x_k) = 0$ for $k = 2, \dots, r$, $P(z) = 0$, and so $\langle P, \phi \rangle = p_1(z)P(x_1) \neq 0$. Hence P is not compatible with ϕ, and thus $l \leq d$. □

The following theorem describes a method of generating linear identities which characterise polynomials of degree at most d.

Theorem 66 ([HK]). *Let $\{x_k\}_{k=1}^r \subset \mathbb{R}^m$ be a basic set of nodes for $\mathcal{P}^d(\mathbb{R}^m)$, and let $\{p_k\}_{k=1}^r \subset \mathcal{P}^d(\mathbb{R}^m)$ be biorthogonal to $\{x_k\}_{k=1}^r$. Further, let $z \in \mathbb{R}^m \setminus \{x_k\}_{k=1}^r$ and $\phi = -\delta_z + \sum_{k=1}^r p_k(z)\delta_{x_k}$. Let $T: \mathbb{R}^m \to \mathbb{R}^n$, $n > m$, be an affine one-to-one mapping such that $0 \notin T(\mathbb{R}^m)$. Then $a_1 = p_1(z), \dots, a_r = p_r(z)$ are the unique coefficients with the following property: Let $\psi = -\delta_{T(z)} + \sum_{k=1}^r a_k \delta_{T(x_k)}$. If X, Y are normed linear spaces and $f \in C(X; Y)$, then f is compatible with ψ if and only if f is a polynomial of degree at most d.*

Proof. Since $T(x_1), \dots, T(x_r), T(z)$ lie in an affine hyperplane not containing 0, the assumptions of Theorem 64 are satisfied. Thus it suffices to prove the theorem for $X = \mathbb{R}^n$ and $Y = \mathbb{R}$, and it also follows that the space of those continuous $f: \mathbb{R}^n \to \mathbb{R}$ which are compatible with ψ is either $\mathcal{P}^l(\mathbb{R}^n)$ for some $l \in \mathbb{N}_0$, or a trivial space.

If $P \in \mathcal{P}^d(\mathbb{R}^n)$, then $P \circ T \in \mathcal{P}^d(\mathbb{R}^m)$, so $P \circ T$ is compatible with ϕ, and therefore $\langle P, \psi \rangle = 0$. By Lemma 54 every member of $\mathcal{P}^d(\mathbb{R}^n)$ is compatible with ψ. Hence the space of compatible functions is non-trivial and $l \geq d$. On the other hand, if $P \in \mathcal{P}^l(\mathbb{R}^m)$, then $P \circ T^{-1}: T(\mathbb{R}^m) \to \mathbb{R}$ can be extended to a member of $\mathcal{P}^l(\mathbb{R}^n)$, which is compatible with ψ by the definition of l. It follows from Lemma 54 that every polynomial from $\mathcal{P}^l(\mathbb{R}^m)$ is compatible with ϕ. By Lemma 65 we conclude that $l \leq d$. Theorem 7 then yields the uniqueness part. □

6. Estimates of coefficients of polynomials

In this section we show some applications of a simple complex averaging technique which leads to many useful and sharp coefficient estimates. Our goal is to find estimates which are independent of the number of variables, i.e. of the dimension of the domain, as these estimates can be easily applied to polynomials defined on the classical Banach spaces c_0 or ℓ_p.

We begin with a few result for complex polynomials analogous to the Polarisation formula. Let us define the generalised Rademacher system. For given $n, k \in \mathbb{N}$ let $r_1^{n,k}(l) = e^{i2\pi\frac{m}{n}}$ for $l \in \mathbb{N}, mn^{k-1} \leq (l-1) \bmod n^k < (m+1)n^{k-1}, 0 \leq m \leq n-1$, and put $r_j^{n,k}(l) = r_1^{n,k}(n^{j-1}l)$, $l \in \{1,\dots,n^k\}$, $j = 2,\dots,k$. We note that the system can be defined also on [0, 1] or more generally on a suitable probability space, but we chose to work with the discrete domain to underline the discrete nature of the methods used here.

Lemma 67 ([ALRT], [Din]). *Let $n, k \in \mathbb{N}$. The generalised Rademacher system has the following property: If $m_1, \dots, m_k \in \mathbb{N}_0$, then the average*

$$\frac{1}{n^k}\sum_{l=1}^{n^k} r_1^{n,k}(l)^{m_1}\cdots r_k^{n,k}(l)^{m_k} = \begin{cases} 1 & \text{if } m_j \equiv 0 \pmod{n},\ j = 1,\dots,k, \\ 0 & \text{otherwise.} \end{cases}$$

Proof. The first case is clear, since $r_j^{n,k}(l)^n = 1$ for every $l \in \{1,\dots,n^k\}$ and $j = 1,\dots,k$. To prove the other case let s be the biggest number in $\{1,\dots,k\}$ such that $m_s \not\equiv 0 \pmod{n}$. Then

$$\sum_{l=1}^{n^k} r_1^{n,k}(l)^{m_1}\cdots r_k^{n,k}(l)^{m_k}$$
$$= \sum_{q=1}^{n^{s-1}}\left(r_1^{n,k}(qn^{k-s+1})^{m_1}\cdots r_{s-1}^{n,k}(qn^{k-s+1})^{m_{s-1}}\sum_{l=1}^{n^{k-s+1}} r_s^{n,k}(l)^{m_s}\right) = 0$$

since $\sum_{j=0}^{n-1} e^{i2\pi\frac{j}{n}m_s} = 0$. □

The following two results should be compared with the Polarisation formula (Proposition 1.13, Lemma 1.33), where $k = n$ and $\varepsilon_j = r_j^{2,n}$.

Proposition 68 ([ALRT]). *Let X, Y be complex vector spaces, $n \in \mathbb{N}$, and let $P \in P(^nX; Y)$. Then*

$$\check{P}(^{n_1}x_1,\dots,{}^{n_k}x_k)$$
$$= \frac{n_1!\cdots n_k!}{n!}\frac{1}{n^k}\sum_{l=1}^{n^k} r_1^{n,k}(l)^{n-n_1}\cdots r_k^{n,k}(l)^{n-n_k} P\left(\sum_{j=1}^{k} r_j^{n,k}(l)x_j\right)$$

for all $x_1,\dots,x_k \in X$, $n_1,\dots,n_k \in \mathbb{N}$ with $n_1 + \cdots + n_k = n$, $1 \leq k \leq n$.

Proof. Without loss of generality we may assume that $k \geq 2$. Using the Multinomial formula (Proposition 1.22) and Lemma 67 we obtain

$$\frac{1}{n^k}\sum_{l=1}^{n^k} r_1^{n,k}(l)^{n-n_1}\cdots r_k^{n,k}(l)^{n-n_k} P\bigg(\sum_{j=1}^{k} r_j^{n,k}(l)x_j\bigg)$$

$$= \frac{1}{n^k}\sum_{l=1}^{n^k}\sum_{\alpha\in\mathcal{J}(k,n)}\binom{n}{\alpha} r_1^{n,k}(l)^{n-n_1+\alpha_1}\cdots r_k^{n,k}(l)^{n-n_k+\alpha_k}\check{P}({}^{\alpha_1}x_1,\ldots,{}^{\alpha_k}x_k)$$

$$= \sum_{\alpha\in\mathcal{J}(k,n)}\binom{n}{\alpha}\check{P}({}^{\alpha_1}x_1,\ldots,{}^{\alpha_k}x_k)\bigg(\frac{1}{n^k}\sum_{l=1}^{n^k} r_1^{n,k}(l)^{n-n_1+\alpha_1}\cdots r_k^{n,k}(l)^{n-n_k+\alpha_k}\bigg)$$

$$= \binom{n}{n_1,\ldots,n_k}\check{P}({}^{n_1}x_1,\ldots,{}^{n_k}x_k)$$

from which the formula follows. □

Lemma 69. *Let X, Y be complex vector spaces, $n \in \mathbb{N}$, and let $P \in P^n(X;Y)$ be such that $P = \sum_{m=0}^{n} P_m$, $P_m \in P({}^mX;Y)$. Then for every $x_1,\ldots,x_k \in X$*

$$P_0(0) + \sum_{j=1}^{k} P_n(x_j) = \frac{1}{n^k}\sum_{l=1}^{n^k} P\bigg(\sum_{j=1}^{k} r_j^{n,k}(l)x_j\bigg).$$

Proof. By Lemma 67

$$\frac{1}{n^k}\sum_{l=1}^{n^k} P\bigg(\sum_{j=1}^{k} r_j^{n,k}(l)x_j\bigg)$$

$$= P_0(0) + \sum_{m=1}^{n}\frac{1}{n^k}\sum_{l=1}^{n^k}\sum_{\substack{1\le j_s\le k\\ s=1,\ldots,m}} r_{j_1}^{n,k}(l)\cdots r_{j_m}^{n,k}(l)\check{P}_m(x_{j_1},\ldots,x_{j_m})$$

$$= P_0(0) + \sum_{j=1}^{k} P_n(x_j).$$ □

The next results improves the estimate from Corollary 1.17 for complex multilinear mappings.

Corollary 70 ([Harr1]). *Let X and Y be complex normed linear spaces and let $M \in \mathcal{L}^s({}^nX;Y)$. Then*

$$\big\|M({}^{n_1}x_1,\ldots,{}^{n_k}x_k)\big\| \le \frac{n_1!\cdots n_k!}{n_1^{n_1}\cdots n_k^{n_k}}\frac{n^n}{n!}\|\widehat{M}\|$$

for all $x_1,\ldots,x_k \in B_X$, $n_1,\ldots,n_k \in \mathbb{N}$ with $n_1+\cdots+n_k = n$, $1 \le k \le n$.

Proof. Let $x_1, \ldots, x_k \in B_X$ and $n_1, \ldots, n_k \in \mathbb{N}$ with $n_1 + \cdots + n_k = n$. Put $z_j = \frac{n_j}{n} x_j$. Then $\left\| \sum_{j=1}^k r_j^{n,k}(l) z_j \right\| \le \sum_{j=1}^k \|z_j\| = \sum_{j=1}^k \frac{n_j}{n} \|x_j\| \le 1$. Hence by Proposition 68

$$\begin{aligned} \left\| M({}^{n_1}x_1, \ldots, {}^{n_k}x_k) \right\| &= \frac{n^n}{n_1^{n_1} \cdots n_k^{n_k}} \left\| M({}^{n_1}z_1, \ldots, {}^{n_k}z_k) \right\| \\ &\le \frac{n^n}{n_1^{n_1} \cdots n_k^{n_k}} \frac{n_1! \cdots n_k!}{n!} \frac{1}{n^k} \sum_{l=1}^{n^k} \left\| \widehat{M}\left(\sum_{j=1}^k r_j^{n,k}(l) z_j \right) \right\| \\ &\le \frac{n_1! \cdots n_k!}{n_1^{n_1} \cdots n_k^{n_k}} \frac{n^n}{n!} \| \widehat{M} \|. \end{aligned}$$

□

Theorem 71 ([AG], [ABE]). *Let $p \in \mathcal{P}^n(\mathbb{C}^N)$ and*

$$p(x_1, \ldots, x_N) = \sum_{j=1}^N a_j x_j^n + \sum_{\substack{\beta \in \mathcal{J}(N,n) \\ \beta \neq (0,\ldots,0,n,0,\ldots,0)}} a_\beta x^\beta.$$

Then

$$\sum_{j=1}^N |a_j| \le \max_{z_j \in \mathbb{C}, |z_j| = 1} |p(z_1, \ldots, z_N)| \le \|p\|_\infty.$$

In particular, in the complex scalar case, if $P \in \mathcal{P}({}^n c_0)$, then $\sum_{j=1}^\infty |P(e_j)| \le \|P\|_\infty$.

Proof. Assume without loss of generality that the constant term C of p is real and non-negative. Let $\lambda_j \in \mathbb{C}$ be such that $|\lambda_j| = 1$ and $a_j \lambda_j^n = |a_j|$, $j = 1, \ldots, N$. Using Lemma 69 we obtain

$$\sum_{j=1}^N |a_j| \le C + \sum_{j=1}^N a_j \lambda_j^n = C + \sum_{j=1}^N p_n(\lambda_j e_j) = \frac{1}{n^N} \sum_{l=1}^{n^N} p\left(\sum_{j=1}^N r_j^{n,N}(l) \lambda_j e_j \right),$$

where p_n denotes the n homogeneous summand of p and $\{e_j\}_{j=1}^N$ is the canonical basis of $\mathbb{C}^N$. This finishes the proof of the finite-dimensional case. The infinite-dimensional case follows immediately. □

The following result is an optimal generalisation of the Chebyshev theorem (Theorem 22), which corresponds to $N = 1$.

Theorem 72 ([MST]). *Let $p \in \mathcal{P}^n(\mathbb{R}^N)$ and*

$$p(x_1, \ldots, x_N) = \sum_{j=1}^N a_j x_j^n + \sum_{\substack{\beta \in \mathcal{J}(N,n) \\ \beta \neq (0,\ldots,0,n,0,\ldots,0)}} a_\beta x^\beta.$$

Then

$$\sum_{j=1}^N |a_j| \le 2^{n-1} \max_{x_j \in [-1,1]} |p(x_1, \ldots, x_N)|$$

and the constant 2^{n-1} is the best possible.

Proof. Denote by p_n the n-homogeneous term of p and apply Theorem 71 on $\widetilde{p_n}$. Note that Taylor's norm on $\mathbb{C}^N$ which arises from the supremum norm on $\mathbb{R}^N$ is the canonical supremum norm on $\mathbb{C}^N$. Thus

$$\sum_{j=1}^{N} |a_j| \le \|\widetilde{p_n}\|_{\infty,\mathbb{C}} \le 2^{n-1}\|p\|_\infty$$

by Theorem 1.63. Finally, the polynomial $q(x_1,\dots,x_N) = T_n(x_1) + \dots + T_n(x_N)$ shows that the estimate is sharp. □

Theorem 73 ([ABE]). *Let $p \in \mathcal{P}(^n\mathbb{R}^N)$ and*

$$p(x_1,\dots,x_N) = \sum_{j=1}^{N} a_j x_j^n + \sum_{\substack{\beta\in\mathcal{J}(N,n)\\ \beta\neq(0,\dots,0,n,0,\dots,0)}} a_\beta x^\beta.$$

Then

$$\sum_{j=1}^{N} |a_j| \le 4n^2 \max_{x_j\in[0,1]} |p(x_1,\dots,x_N)|.$$

In particular, in the real scalar case, if $P \in \mathcal{P}(^n c_0)$, then $\sum_{j=1}^{\infty}|P(e_j)| \le 4n^2\|P\|_\infty$.

Proof. First, choose a subset $L \subset \{1,\dots,N\}$ such that a_j, $j \in L$ have the same sign and $\left|\sum_{j\in L} a_j\right| \ge \frac{1}{2}\sum_{j=1}^{N}|a_j|$. Without loss of generality we may assume that $L = \{1,\dots,m\}$. From now on consider p restricted to the first m variables only. Fix $t \in (0,1)$ and let $\{v_j\}_{j=1}^m$ be independent random variables on $[0,1]$ taking values 1 with probability t and 0 with probability $1-t$. (We can take for example non-symmetric Rademacher-like functions, where $v_1 = \chi_{[0,t)}$, $v_2 = \chi_{[0,t^2)\cup[t,t+t(1-t))}$ and so on.) Note that $v_j^k = v_j$ for each $k \in \mathbb{N}$, $j \in \{1,\dots,m\}$. Let us denote by l_β the number of non-zero components in $\beta \in \mathcal{J}(m,n)$ and put $A_j = \sum_{\beta:\, l_\beta = j} a_\beta$, $j = 2,\dots,n$. Note that in our notation $A_1 = a_1 + \ldots + a_m$. Then

$$\int_0^1 p\big(v_1(s),\dots,v_m(s)\big)\,ds = \sum_{\beta\in\mathcal{J}(m,n)} a_\beta \int_0^1 v_1^{\beta_1}\cdots v_m^{\beta_m} = \sum_{\beta\in\mathcal{J}(m,n)} a_\beta \int_0^1 v_1^{\beta_1}\cdots\int_0^1 v_m^{\beta_m}$$
$$= A_1 t + A_2 t^2 + \dots + A_n t^n.$$

Let us denote $q(t) = A_1 t + A_2 t^2 + \dots + A_n t^n$ and put $r(t) = q(\frac{t+1}{2})$. Using Markov's inequality (Theorem 80) we get

$$\max_{t\in[0,1]}|q(t)| = \max_{t\in[-1,1]}|r(t)| \ge \frac{1}{n^2}\max_{t\in[-1,1]}|r'(t)|$$
$$= \frac{1}{2n^2}\max_{t\in[0,1]}|q'(t)| \ge \frac{1}{2n^2}|q'(0)| = \frac{1}{2n^2}|A_1|.$$

It follows that for every $\varepsilon > 0$ there is $t \in (0,1)$ such that $|q(t)| > \frac{1}{2n^2}|A_1| - \varepsilon$ and considering $\{v_j\}$ corresponding to this t there must be $s \in [0,1]$ such that

$p\big(v_1(s), \dots, v_m(s)\big) > \frac{1}{2n^2}|A_1| - \varepsilon$. Hence

$$\sum_{j=1}^{N} |a_j| \le 2|A_1| < 4n^2 \max_{x_j \in [0,1]} |p(x_1, \dots, x_m)| + 4n^2\varepsilon,$$

which proves the claim. □

The best multiplicative constant in Theorem 73 appears to be unknown, even asymptotically. A lower estimate follows from the next example.

Example 74. Define by induction a sequence of polynomials by

$$p_0(x_1) = x_1,$$
$$p_{n+1}(x_1, \dots, x_{2^{n+1}}) = p_n(x_1, \dots, x_{2^n})^2 - p_n(x_{2^n+1}, \dots, x_{2^{n+1}})^2.$$

Then for every $n \in \mathbb{N}_0$, p_n is a 2^n-homogeneous polynomial in 2^n variables satisfying $|p_n(x_1, \dots, x_{2^n})| \le 1$ for $x_1, \dots, x_{2^n} \in [-1, 1]$. The coefficients of $x_j^{2^n}$ are equal to ± 1 and hence the sum of their absolute values is 2^n. Thus in Theorem 73 the constant grows at least linearly.

Proposition 75 ([Za2]). *Consider the complex scalar field. Let* $1 < p < \infty$, $n \in \mathbb{N}$, $n < p$, *and* $P \in \mathcal{P}(^n\ell_p)$. *Then*

$$\|(P(e_j))\|_{\frac{p}{p-n}} = \left(\sum_{j=1}^{\infty} |P(e_j)|^{\frac{p}{p-n}}\right)^{\frac{p-n}{p}} \le \|P\|.$$

Proof. ([ALRT], [Din, Proposition 1.58]) For each $j \in \mathbb{N}$ let $\lambda_j \in \mathbb{C}$ be such that $\lambda_j^n P(e_j) = |P(e_j)|^{\frac{p}{p-n}}$ and $\lambda_j = 0$ if $P(e_j) = 0$. Note that $|\lambda_j|^p = |P(e_j)|^{\frac{p}{p-n}}$. For a fixed $k \in \mathbb{N}$ we have by Lemma 69

$$\sum_{j=1}^{k} |P(e_j)|^{\frac{p}{p-n}} = \sum_{j=1}^{k} \lambda_j^n P(e_j) = \sum_{j=1}^{k} P(\lambda_j e_j) = \frac{1}{n^k} \sum_{l=1}^{n^k} P\left(\sum_{j=1}^{k} r_j^{n,k}(l) \lambda_j e_j\right)$$
$$\le \|P\| \left(\sum_{j=1}^{k} |\lambda_j|^p\right)^{\frac{n}{p}} = \|P\| \left(\sum_{j=1}^{k} |P(e_j)|^{\frac{p}{p-n}}\right)^{\frac{n}{p}}$$

from which the inequality follows. □

For the real scalar field, passing to the complexification using Theorem 1.63 we obtain an extra multiplicative constant $2^{n-1}C$ in Proposition 75, where C is the constant of equivalence of the canonical norm of the complex ℓ_p and the Taylor norm on $\tilde{\ell}_p$. The optimal result again seems to be unknown.

Another application of the averaging procedure will be useful later on when dealing with the cotype of Banach spaces.

Theorem 76 (Kahane's inequality, [DJT, Theorem 11.1]). *Let $0 < p, q < \infty$. There is a constant $K_{p,q} > 0$ such that for any normed linear space X and any $x_1, \dots, x_n \in X$*

$$\left(\frac{1}{2^n} \sum_{\varepsilon_j = \pm 1} \left\| \sum_{j=1}^{n} \varepsilon_j x_j \right\|^q \right)^{\frac{1}{q}} \le K_{p,q} \left(\frac{1}{2^n} \sum_{\varepsilon_j = \pm 1} \left\| \sum_{j=1}^{n} \varepsilon_j x_j \right\|^p \right)^{\frac{1}{p}}.$$

Notice that for $p \ge q$ this is a consequence of the Hölder inequality and in this case may take $K_{p,q} = 1$. We also remark for later use that $K_{1,2} = \sqrt{2}$, [DJT, p. 227].

Lemma 77. *Let X, Y be vector spaces over $\mathbb{K}$ and $P \in P^n(X; Y)$, $n \ge 2$. Then for every $x_1, \dots, x_k \in X$*

$$\sum_{j=1}^{k} P(x_j) = \frac{1}{2^{(n-1)k}} \sum_{\substack{\varepsilon_j^m = \pm 1 \\ 1 \le j \le k \\ 1 \le m \le n-1}} \check{P}\Bigg(\sum_{j=1}^{k} \varepsilon_j^1 x_j, \sum_{j=1}^{k} \varepsilon_j^1 \varepsilon_j^2 x_j, \sum_{j=1}^{k} \varepsilon_j^2 \varepsilon_j^3 x_j, \dots$$

$$\dots, \sum_{j=1}^{k} \varepsilon_j^{n-2} \varepsilon_j^{n-1} x_j, \sum_{j=1}^{k} \varepsilon_j^{n-1} x_j \Bigg).$$

Proof. The right-hand side can be expanded into

$$\frac{1}{2^{(n-1)k}} \sum_{\substack{\varepsilon_j^m = \pm 1 \\ 1 \le j \le k \\ 1 \le m \le n-1}} \sum_{\substack{1 \le j_s \le k \\ s = 1, \dots, n}} \varepsilon_{j_1}^1 \varepsilon_{j_2}^1 \varepsilon_{j_2}^2 \varepsilon_{j_3}^2 \varepsilon_{j_3}^3 \cdots \varepsilon_{j_{n-1}}^{n-2} \varepsilon_{j_{n-1}}^{n-1} \varepsilon_{j_n}^{n-1} \check{P}(x_{j_1}, \dots, x_{j_n}).$$

By interchanging the sums and applying Lemma 67 (where n of the lemma is 2 and k of the lemma is $(n-1)k$) we can see that the only non-zero summands of the outer sum are those, for which $j_1 = j_2 = \dots = j_n$. This implies the desired formula. □

Proposition 78. *Let $n \in \mathbb{N}$ and $p \in (0, +\infty)$. There is a constant $C_{p,n} > 0$ such that if X, Y are normed linear spaces, $P \in \mathcal{P}(^nX; Y)$, and $x_1, \dots, x_k \in X$, then*

$$\left\| \sum_{j=1}^{k} P(x_j) \right\| \le C_{p,n} \|P\| \left(\frac{1}{2^k} \sum_{\varepsilon_j = \pm 1} \left\| \sum_{j=1}^{k} \varepsilon_j x_j \right\|^p \right)^{\frac{n}{p}}.$$

Proof. It is easy to see that

$$\frac{1}{2^{(n-1)k}} \sum_{\substack{\varepsilon_j^m = \pm 1 \\ 1 \le j \le k \\ 1 \le m \le n-1}} \left\| \sum_{j=1}^{k} \varepsilon_j^l \varepsilon_j^{l+1} x_j \right\|^n = \frac{1}{2^k} \sum_{\substack{\varepsilon_j = \pm 1 \\ 1 \le j \le k}} \left\| \sum_{j=1}^{k} \varepsilon_j x_j \right\|^n$$

Thus using Lemma 77, the (generalised) Hölder inequality [HLP, Theorem 11], and the Polarisation formula (Proposition 1.13) we obtain

$$\begin{aligned}\left\|\sum_{j=1}^{k} P(x_j)\right\| &\le \|\check{P}\|\frac{1}{2^{(n-1)k}}\sum_{\varepsilon_j^m=\pm 1}\left\|\sum_{j=1}^{k}\varepsilon_j^1 x_j\right\|\left\|\sum_{j=1}^{k}\varepsilon_j^1\varepsilon_j^2 x_j\right\|\cdots\\ &\qquad\cdots\left\|\sum_{j=1}^{k}\varepsilon_j^{n-2}\varepsilon_j^{n-1}x_j\right\|\left\|\sum_{j=1}^{k}\varepsilon_j^{n-1}x_j\right\|\\ &\le \|\check{P}\|\frac{1}{2^{(n-1)k}}\left(\sum_{\varepsilon_j^m=\pm 1}\left\|\sum_{j=1}^{k}\varepsilon_j^1 x_j\right\|^n\right)^{\frac{1}{n}}\left(\sum_{\varepsilon_j^m=\pm 1}\left\|\sum_{j=1}^{k}\varepsilon_j^1\varepsilon_j^2 x_j\right\|^n\right)^{\frac{1}{n}}\cdots\\ &\qquad\cdots\left(\sum_{\varepsilon_j^m=\pm 1}\left\|\sum_{j=1}^{k}\varepsilon_j^{n-1}x_j\right\|^n\right)^{\frac{1}{n}}\\ &= \frac{n^n}{n!}\|P\|\frac{1}{2^k}\sum_{\varepsilon_j=\pm 1}\left\|\sum_{j=1}^{k}\varepsilon_j x_j\right\|^n.\end{aligned}$$

Now it suffices to apply Kahane's inequality (Theorem 76). □

We finish this section with another important application of the averaging technique, this time applied to the structural properties of spaces of polynomials.

Proposition 79 ([DiD], [DiGo]). *Let X be a Banach space over $\mathbb{K}$ with an unconditional Schauder basis and let $P \in \mathcal{P}(^nX)$ be a polynomial such that $|P(x_j)| \ge a > 0$, $j \in \mathbb{N}$, for some normalised block basic sequence $\{x_j\}$. Then $\mathcal{P}(^nX)$ contains ℓ_∞.*

Proof. Let $\{e_j\}$ be the normalised unconditional Schauder basis of X. There exist sequences of integers $n_1 < m_1 < n_2 < m_2 < \cdots$ such that $\operatorname{supp} x_j \subset [n_j, m_j]$. Denote by $\pi_j : X \to X$ the projection $\pi_j\left(\sum_{l=1}^{\infty} a_l e_l\right) = \sum_{l=n_j}^{m_j} a_l e_l$ and define $P_j \in \mathcal{P}(^nX)$ by $P_j = P \circ \pi_j$.

Assume first that $\mathbb{K} = \mathbb{C}$. Fix $(a_j) \in B_{\ell_\infty}$ and $x \in X \setminus \{0\}$ and find $\zeta_j \in \mathbb{C}$ such that $\zeta_j^n = \frac{\overline{P_j(x)}}{|P_j(x)|}$, $j \in \mathbb{N}$. For any $k \in \mathbb{N}$ we obtain using Lemma 69

$$\begin{aligned}\sum_{j=1}^{k}|a_j P_j(x)| &\le \sum_{j=1}^{k}|P_j(x)| = \sum_{j=1}^{k}P_j(\zeta_j x) = \frac{1}{n^k}\sum_{l=1}^{n^k}P\left(\sum_{j=1}^{k}r_j^{n,k}(l)\zeta_j\pi_j(x)\right)\\ &\le \|P\|\max_{1\le l\le n^k}\left\|\sum_{j=1}^{k}r_j^{n,k}(l)\zeta_j\pi_j(x)\right\|^n \le (2K)^n\|P\|\|x\|^n,\end{aligned}$$

where K is the unconditional basis constant of $\{e_j\}$. It follows that the sum $\sum_{j=1}^{\infty} a_j P_j$ converges pointwise to a polynomial $Q \in \mathcal{P}(^nX)$ with $\|Q\| \le (2K)^n\|P\|$ (Theorem 1.29).

On the other hand, choosing $m \in \mathbb{N}$ so that $|a_m| \geq \frac{1}{2}\|(a_j)\|_\infty$ and $\zeta \in \mathbb{K}$ with $\zeta^n = \frac{\overline{a_m}}{|a_m|}$, we have

$$\|Q\| = \sup_{x \in B_X} \left|\sum_{j=1}^{\infty} a_j P_j(x)\right| \geq \left|\sum_{j=1}^{\infty} a_j P_j(\zeta x_m)\right| = |a_m||P(x_m)| \geq \frac{a}{2}\|(a_j)\|_\infty.$$

The real case follows immediately by passing to the complexification. □

7. Notes and remarks

Section 1. A thorough introduction into the spaces of polynomials on $\mathbb{R}^n$ can be found in the paper [Rez3] of Bruce Reznick. If $n = 1$ the basic sets of nodes for $\mathcal{P}^d(\mathbb{R})$ are characterised by the classical one-dimensional Lagrange theorem (Proposition 1.56) as any sets of distinct points of cardinality $d + 1$. The problem of finding simple and general geometrical conditions forcing the sets of nodes to be independent (or basic) for $\mathcal{P}({}^d\mathbb{R}^n)$ and $\mathcal{P}^d(\mathbb{R}^n)$ when $n \geq 2$ appears to be open. On one hand, the Lebesgue measure of the point configurations of a given cardinality (which does not exceed the dimension of the respective space of polynomials) which are linearly dependent as nodes equals to zero. Thus "randomly chosen" sets of nodes will be linearly independent. However, finding concrete examples of independent nodes and their biorthogonal polynomials (and associated Lagrange interpolation) seems to be a difficult task, see [GaSa] for a recent survey of available techniques.

One of the main goals is finding good explicit examples of sets of nodes and their biorthogonal polynomials so that the norm of the projection in (1) will be as small as possible, leading to efficient polynomial approximations for given classes of continuous functions. In the one-dimensional case, by a classical result of Viktor Fedorovich Nikolaev ([Kor, pp. 150–153]) there is a lower estimate: there is $c > 0$ such that $\|L_n\| > c \log n$, where L_n is any bounded linear projection from $C([0, 1])$ onto $\mathcal{P}^n(\mathbb{R})$. It turns out that the projections based on the Lagrange interpolation using Chebyshev polynomials satisfy an upper estimate of the same order of growth $\log n$. For more on the projection constants in several variables see e.g. John Charles Mason's paper [Mas1] or the survey [Mas2] which contains many further references. Pavel Petrovich Korovkin's book [Kor] serves as a nice introduction to approximation theory respecting the functional analytic point of view.

Multivariate approximation theory, in particular the Lagrange interpolation, is studied in [Lor] or [BDVX], where some estimates of the norms of the projections are given. For example, for the square $[-1, 1]^2$ the best order of the norm of the projection is $O((\log n)^2)$, and it is attained by the so called Padua basic set of nodes.

Section 2. The topic of cubature formulae was historically motivated in part by the need for efficient numerical formulae for evaluation of integrals, see [Boj] for a nice historical account. Corollary 13 is a celebrated theorem of Vladimir L. Chakalov [Cha]. Many proofs have been published, see [CF] and references therein. Our proof is original and perhaps one of the shortest in the literature. There is a large body of literature on the subject of the location of the nodes in the representation formula, as well as on the minimal N in (2) in the multivariate case. We point out that determining whether $\phi \in \mathcal{P}^d(X)^*$ admits a cubature formula, i.e. corresponds to an integral with respect to positive measure, based on the knowledge of the values $\phi(x^\alpha)$, $\alpha \in \mathcal{J}(n,d)$ is the well-known moment (or more precisely truncated moment) problem. In [Rez3] several topics closely related to the moment problems in several variables, and the problems of representation of positive semi-definite forms as sums of powers of forms of lower degree are studied (see also [El1]).

We refer to [Ri], [MH], or [BE] for information on Chebyshev polynomials. We refer to [X1] and references therein for the bivariate (or multivariate) situation, which is much more complicated, but which still admits a theory using the same basic approach of orthogonal polynomials. See [GaSa] for a survey on Lagrange interpolation.

Section 3. This section is devoted to various estimates concerning polynomials and their derivatives on $\mathbb{R}$. There are numerous books devoted to this broad subject, e.g. [BE], [MMR], [Kl1], [Dur]. There is a vast body of literature on the subject of Chebyshev polynomials and Markov inequalities, see e.g. [RS], [Ri], [BE]. It is interesting to point out that the inequality of A. A. Markov was first conjectured by the famous Russian chemist Dmitri Ivanovich Mendeleev. We refer to [Sk] for a thorough historical account.

Theorem 80 (Andrei Andreevich Markov, [MarkoA]). *Let $p \in \mathcal{P}^n(\mathbb{R})$. Then*

$$\max_{x\in[-1,1]} |p'(x)| \le n^2 \max_{x\in[-1,1]} |p(x)|.$$

The equality holds if and only if $p = \pm T_n$.

Theorem 81 (Vladimir Andreevich Markov, [MarkoV]). *Let $p \in \mathcal{P}^n(\mathbb{R})$. Then*

$$\begin{aligned}\max_{x\in[-1,1]} \left|p^{(k)}(x)\right| &\le T_n^{(k)}(1) \max_{x\in[-1,1]} |p(x)| \\ &= \frac{n^2(n^2-1)\cdots(n^2-(k-1)^2)}{1\cdot 3\cdot\cdots\cdot(2k-1)} \max_{x\in[-1,1]} |p(x)|.\end{aligned}$$

The equality holds if and only if $p = \pm T_n$.

At the beginning of the section we followed an elementary approach of Werner Wolfgang Rogosinski in [Rog] and Pál Erdős in [Er], in order to expose the basic idea of finding extremal polynomials for the given functional on the space of polynomials. The same basic idea is also used later in the proof of Harris's theorem. The main Theorem 28 was first proved by Valentin Ivanovich Skalyga in [Sk]. In the complex setting there is the following optimal result of Maciej Klimek [Kl2]: If $P \in \mathcal{P}^n(\mathbb{C}^N)$, then

$$\sup_{\substack{|z_j|=1\\ j=1,\dots,N}} |P(z_1,\dots,z_N)| \le \left(1+\sqrt{2}\right)^n \sup_{t_j\in[-1,1]} |P(t_1,\dots,t_N)|.$$

The next problem is posed in [MST].

Problem 82. Is it possible to improve the estimate in Theorem 26, for arbitrary real Banach spaces, by leaving out the factor $2^{\frac{n}{2}}$?

Section 4. The homogeneous cubature formulae have been studied e.g. in [GoSe], [LV], [Rez3] and the references therein. The problem consists of finding concrete representing nodes and weights for a positive functional defined on homogeneous polynomials of a fixed degree. In the case when the nodes are placed on the unit sphere S^{n-1} and the weights are all equal we speak of the spherical design. These are rather rare objects, see e.g. [Bann] and the references therein.

The presentation of the closely related theory of isometric embeddings of ℓ_2^n into ℓ_p^N comes from [LV], where some of these results were proven for the first time, see also [JL, Chapter 21] and references therein. Theorem 38 was proved many times independently, e.g. [El1], [Rez3], and [DJP]. The existence of an isometric imbedding of ℓ_2^n into ℓ_p^N for large enough $N = N(n, p)$ goes back to David Hilbert [Hilb], in the course of his positive solution of the famous Waring's problem. D. Hilbert's result in number theory claims that for every given $k \in \mathbb{N}$ even there exists an integer $N(k)$ such that every integer $n \in \mathbb{N}$ can be expressed as a sum of at most $N(k)$ of kth powers of integers. For more on this and the proof of Hilbert's theorem see [El2].

There is a large amount of literature on the closely inter-related subjects of Lagrange interpolation and its use in numerical mathematics, isometric embeddings of Banach spaces, error-correcting codes, and optimal designs; for references see e.g. [JL, Chapter 21]. The remarkable Theorem 44 was proved in [DoKar] by Vladimir Leonidovich Dol'nikov and Roman Nikolaevich Karasev, solving a problem of Mikhail Leonidovich Gromov and Vitali Davidovich Milman from [Mi]. It was conjectured in [Mi] that $n(p,k) \sim k^d$. This result is an isometric analogue of the celebrated Dvoretzky theorem. The suggested approach to the problem was via the so called Knaster conjecture.

We recall that the Knaster conjecture claimed (in particular) that for any configuration of $n + 1$ points $\{x_j\}_{j=0}^n$ on S^n and any continuous real valued function f on S^n there exists an orthogonal transformation T such that $f(T(x_0)) = \cdots = f(T(x_n))$. This claim seems to be quite reasonable, by the heuristic argument of "counting the dimension". Unfortunately, counterexamples to the Knaster conjecture have been found in [KS] by Boris Sergeevich Kashin and Stanisław J. Szarek. It is still possible though that the Knaster conjecture holds true for sets containing less than $n + 1$ points, or sets with particular geometrical properties. This would lead to an alternative proof of Theorem 44, and perhaps also to some explicit estimates.

Problem 83. Find a reasonable explicit estimate for $n(p,k)$ in Theorem 44 in the case of even p.

We point out that the related problem of characterising isometric subspaces of L_p-spaces for non-integer p has been treated e.g. in [Kri], [DJP], and references therein. The following problem arises naturally as a finite-dimensional reformulation of a certain renorming question that we will encounter in later chapters.

Problem 84. Let $r, R \in \mathbb{R}$, $k \in \mathbb{N}$ be given. Are there some $s, S \in \mathbb{R}$, $K \in \mathbb{N}$ with the following property: Let X be an n-dimensional Banach space, $P \in \mathcal{P}(^k X)$ be a polynomial such that $r < \inf_{B_X} P \leq \sup_{B_X} P < R$. Then there is a convex polynomial $Q \in \mathcal{P}(^K X)$ such that $s < \inf_{B_X} Q \leq \sup_{B_X} Q < S$.

There seem to be no analogous results for good polynomial behaviour of quotient spaces.

Problem 85. Let X be a quotient of ℓ_p^n, $p \in \mathbb{N}$ even. Is the quotient norm on X again a polynomial norm?

Section 5. The theory of functional equations and identities in the complex plane was initiated by Maurice Fréchet [Fré1], [Fré2], William Harold Wilson [Wi], and Stanisław Mazur and Władysław Orlicz [MO1], [MO2]. The proof of Theorem 50 is new.

A nice application of the theory of linear identities to the structure of Banach spaces, Corollary 61, was made by Pascual Jordan and John von Neumann. This theorem has served as a model for results characterising classes of Banach spaces by means of polynomial norms, e.g. [Gi], [Koe], [Rez1], [Rez2], and the references therein. This theory is closely related to the isometric Banach space theory, see e.g. [JL, Chapter 21] and references therein. Theorem 60 was first formulated by B. Reznick [Rez1] for polynomial norms on Banach spaces. A related approach to characterising subspaces of L_p-spaces using systems of inequalities can be found e.g. in [Kri] and references therein.

Section 6. Theorem 71 was shown in [AG] for homogeneous polynomials and in general in [ABE]. We have mostly followed the presentation in [ALRT] in this section. Proposition 78 is in [DGZ, Lemma V.4.5].

Chapter 3

Weak continuity of polynomials and estimates of coefficients

The present chapter is devoted to the study of polynomials between Banach spaces with the emphasis laid on their various weak continuity properties and coefficient estimates. The quantitative results, especially in Section 5, are formulated in a completely elementary language, but their study relies on several advanced tools; in particular tensor products and (p, q)-summing operators, which are outlined in the respective sections in order to keep the flow of the theory uninterrupted. These topics are covered with full proofs by the well-known books [DefFl], [DJT] devoted to these areas.

The first section deals with some rather basic results concerning the theory of projective and injective tensor products of Banach spaces. We discuss, mostly without detailed proofs, the duality theory for these spaces and the related universality properties for these spaces as well as the symmetric tensor products and their duality with polynomial spaces.

In the second section we first give a brief summary of the theory of uniform spaces and their topologies. We use the language of uniformities in order to study the weakly uniformly (or weakly sequentially) continuous polynomials. The class of Banach spaces with the Dunford-Pettis property plays an important role here. We apply the tensor product theory to prove the result of Raymond A. Ryan, showing that every weakly compact polynomial from a space with the DPP maps weakly Cauchy sequences to norm convergent ones.

In Section 3 we study in detail the weak continuity properties in Banach spaces not containing ℓ_1. Using the classical Rosenthal's ℓ_1-theorem and its relatives we show that weak sequential continuity on bounded sets coincides with full (resp. uniform) continuity for continuous mappings. We also give some results on quotients of Banach spaces which contain ℓ_1 in the spirit of the well-known fact that such spaces have a quotient isomorphic to ℓ_2.

In Section 4 we introduce the language of the theory of (p, q)-summing operators and mention some connections with tensor products. This abstract theory provides precise quantitative estimates concerning the properties of sequences of vectors in some classical Banach spaces (such as L_p, $C(K)$) and their images under linear operators. We formulate, without proof, some of the fundamental and deep results of the theory, introducing also the notions of type and cotype.

In Section 5 we introduce the theory of multiple $(p;q)$-summing operators, which is a generalisation of the (p,q)-summing operators into the multilinear and polynomial setting. Drawing on the results from Sections 1, 4, and some other analytic tools such as the complex interpolation we outline the proof of a recent theorem of Andreas Defant and Pablo Sevilla-Peris, which gives optimal estimates on the coefficients of polynomials in $\mathcal{P}({}^n c_0; \ell_p)$ and contains several classical inequalities as special cases.

In the last section the so-called Bohr radius of absolute convergence of power series in finite-dimensional Banach spaces is studied.

In this chapter except for Sections 3 and 6 all the spaces are over $\mathbb{K}$, i.e. we consider both real and complex scalars.

1. Tensor products and spaces of multilinear mappings

The goal of this section is to collect some definitions and elementary facts concerning tensor products and symmetric tensor products. Tensor products offer an important point of view of polynomials and multilinear mappings. Moreover, they are an indispensable tool for handling polynomial duality and (p,q)-summing operators.

Given vector spaces $X_1, \ldots, X_n$ over $\mathbb{K}$ we let Λ be the vector space of all formal linear combinations $\sum_{k=1}^N a_k(x_1^k \otimes \cdots \otimes x_n^k)$, $a_k \in \mathbb{K}$, $x_j^k \in X_j$, where we identify $\sum_{k=1}^N a_k(x_1^k \otimes \cdots \otimes x_n^k)$ and $\sum_{k=1}^N a_{\pi(k)}(x_1^{\pi(k)} \otimes \cdots \otimes x_n^{\pi(k)})$ for any permutation π of $\{1,\ldots,N\}$, and similarly $\sum_{k=1}^{N+1} a_k(x_1^k \otimes \cdots \otimes x_n^k)$ and $\sum_{k=1}^N a_k(x_1^k \otimes \cdots \otimes x_n^k)$ for $a_{N+1} = 0$. By Λ_0 we denote the linear subspace of Λ spanned by the vectors

$$\begin{gathered}(x_1 \otimes \cdots \otimes (ax_k) \otimes \cdots \otimes x_n) - a(x_1 \otimes \cdots \otimes x_n),\\ (x_1 \otimes \cdots \otimes (x_k + y_k) \otimes \cdots \otimes x_n)\\ - (x_1 \otimes \cdots \otimes x_k \otimes \cdots \otimes x_n) - (x_1 \otimes \cdots \otimes y_k \otimes \cdots \otimes x_n),\end{gathered}$$

where $k \in \{1,\ldots,n\}$, $x_j, y_j \in X_j$, $a \in \mathbb{K}$. Then the quotient space Λ/Λ_0 is called the tensor product $X_1 \otimes \cdots \otimes X_n = \bigotimes_{j=1}^n X_j$. It is easily verified that tensor products are associative, so they can be built up inductively starting from the tensor product of a pair of vector spaces. Note also that by the definition of Λ_0 each $z \in X_1 \otimes \cdots \otimes X_n$ has a representation $z = \sum_{j=1}^k x_1^j \otimes \cdots \otimes x_n^j$. An element of $X_1 \otimes \cdots \otimes X_n$ that admits a representation $x_1 \otimes \cdots \otimes x_n$ is called an elementary tensor. The following is a fundamental observation: Given $\phi_j \in X_j^\#$, $j = 1,\ldots,n$, the function

$$\sum_{j=1}^k a_j(x_1^j \otimes \cdots \otimes x_n^j) \mapsto \sum_{j=1}^k a_j \phi_1(x_1^j) \cdots \phi_n(x_n^j) \tag{1}$$

is a linear form on the vector space Λ. Thus we obtain a useful criterion for distinguishing the vectors in a tensor product:

Proposition 1. *Let $X_1, \dots, X_n$ be vector spaces and let $A_j \subset X_j^\#$ be subsets that separate the points of X_j, $j = 1, \dots, n$. Then $\sum_{j=1}^k a_j x_1^j \otimes \cdots \otimes x_n^j = 0$ in the space $X_1 \otimes \cdots \otimes X_n$ if and only if*

$$\sum_{j=1}^{k} a_j \phi_1(x_1^j) \cdots \phi_n(x_n^j) = 0$$

for every choice of $\phi_j \in A_j$.

Proof. $\Rightarrow$ The form (1) clearly vanishes on the generating vectors of Λ_0 and thus also on Λ_0.

$\Leftarrow$ By the definition of the space Λ_0 it is easily seen that we may assume that the vectors $\{x_l^j\}_{j=1}^k$ are linearly independent for each $l = 1, \dots, n$. Using an elementary linear algebra we obtain functionals $\{\phi_l^j\}_{j=1}^k \subset \operatorname{span} A_l$ biorthogonal to $\{x_l^j\}_{j=1}^k$. Hence $a_l = \sum_{j=1}^k a_j \phi_1^l(x_1^j) \cdots \phi_n^l(x_n^j) = 0$ for each $l = 1, \dots, k$. □

Further, we define an n-linear mapping $\otimes\colon X_1 \times \cdots \times X_n \to \bigotimes_{j=1}^n X_j$ by the formula $\otimes(x_1, \dots, x_n) = x_1 \otimes \cdots \otimes x_n$.

Theorem 2 (Universality of the tensor product – algebraic case)**.** *Let $X_1, \dots, X_n, Y$ be vector spaces. For every n-linear mapping $M \in L(X_1, \dots, X_n; Y)$ there exists a unique linear operator $L_M \in L(X_1 \otimes \cdots \otimes X_n; Y)$ such that $M = L_M \circ \otimes$:*

$$\begin{array}{ccc} X_1 \times \cdots \times X_n & \xrightarrow{\;M\;} & Y \\ {\scriptstyle\otimes}\big\downarrow & \nearrow_{L_M} & \\ X_1 \otimes \cdots \otimes X_n & & \end{array}$$

The operator L_M satisfies

$$L_M(x_1 \otimes \cdots \otimes x_n) = M(x_1, \dots, x_n). \tag{2}$$

Proof. Using Proposition 1 it is easily seen that $\otimes$ maps products of linearly independent sets to linearly independent sets. Let A_j be the basis of X_j. We define L_M by the formula (2) on $\otimes(A_1 \times \cdots \times A_n)$ and extend it linearly onto the whole tensor product. Using the multilinearity it can be checked that the formula (2) still holds and it clearly uniquely determines L_M. □

The operator L_M corresponding to M in the above theorem is called the linearisation of M. The following result is immediate.

Theorem 3. *Let $X_1, \dots, X_n$ be vector spaces. Given $M \in L(X_1, \dots, X_n; \mathbb{K})$ and $z = \sum_{j=1}^k x_1^j \otimes \cdots \otimes x_n^j \in X_1 \otimes \cdots \otimes X_n$ we put*

$$\langle M, z \rangle = \sum_{j=1}^{k} M(x_1^j, \dots, x_n^j) = \sum_{j=1}^{k} L_M(x_1^j \otimes \cdots \otimes x_n^j) = L_M(z).$$

Then $\langle L(X_1, \dots, X_n; \mathbb{K}), X_1 \otimes \cdots \otimes X_n \rangle$ forms a dual pair.

Tensor products of Banach spaces admit many non-equivalent natural norms. We will describe two important examples, namely the projective and the injective tensor norm. These norms can be shown to be in some sense extreme cases, as every "reasonable" tensor norm is bounded below by the injective norm and bounded above by the projective norm, [Ry3, Proposition 6.1].

Definition 4. Let $X_1, \dots, X_n$ be normed linear spaces. The projective tensor norm π on $X_1 \otimes \cdots \otimes X_n$ is defined by the formula

$$\pi(z) = \sup\{\langle M, z\rangle;\ M \in \mathcal{L}(X_1, \dots, X_n; \mathbb{K}), \|M\| \le 1\}, \quad z \in X_1 \otimes \cdots \otimes X_n.$$

The projective tensor product, denoted by $X_1 \otimes_\pi \cdots \otimes_\pi X_n$, is the completion of the normed linear space $(X_1 \otimes \cdots \otimes X_n, \pi)$.

A very useful alternative description of the projective tensor product and its norm is the following, see [Ry3, Proposition 2.8].

Proposition 5. *Let $X_1, \dots, X_n$ be normed linear spaces. Then for every vector $z \in X_1 \otimes_\pi \cdots \otimes_\pi X_n$ there exist bounded sequences $\{x_l^j\}_{j=1}^\infty \subset X_l$, $l = 1, \dots, n$, such that $z = \sum_{j=1}^\infty x_1^j \otimes \cdots \otimes x_n^j$ is an absolutely convergent series and*

$$\pi(z) = \inf\left\{\sum_{j=1}^\infty \|x_1^j\| \cdots \|x_n^j\|;\ z = \sum_{j=1}^\infty x_1^j \otimes \cdots \otimes x_n^j\right\}.$$

Furthermore, $\pi(x_1 \otimes \cdots \otimes x_n) = \|x_1\| \cdots \|x_n\|$ for every $x_j \in X_j$, $j = 1, \dots, n$.

This can be translated into a simple geometrical description of the unit ball of the projective tensor product:

$$B_{X_1 \otimes_\pi \cdots \otimes_\pi X_n} = \overline{\operatorname{conv}}\, \otimes(B_{X_1} \times \cdots \times B_{X_n}). \tag{3}$$

In particular, $\otimes\colon X_1 \times \cdots \times X_n \to X_1 \otimes_\pi \cdots \otimes_\pi X_n$ is a bounded n-linear mapping of norm 1. From the above it follows that the projective norm is defined so that the universality property of the tensor product remains valid also in the topological sense:

Theorem 6 (Universality of the tensor product – continuous case)**.** *Let $X_1, \dots, X_n$, and Y be normed linear spaces. For every $M \in \mathcal{L}(X_1, \dots, X_n; Y)$ there exists a unique $L_M \in \mathcal{L}(X_1 \otimes_\pi \cdots \otimes_\pi X_n; Y)$ such that $M = L_M \circ \otimes$. The operator L_M satisfies* (2) *and the mapping $M \mapsto L_M$ is an isometry of the spaces $\mathcal{L}(X_1, \dots, X_n; Y)$ and $\mathcal{L}(X_1 \otimes_\pi \cdots \otimes_\pi X_n; Y)$.*

In particular, if $Y = \mathbb{K}$, then we get a simple but important duality relation.

Theorem 7. *Let $X_1, \dots, X_n$ be normed linear spaces. Then*

$$(X_1 \otimes_\pi \cdots \otimes_\pi X_n)^* = \mathcal{L}(X_1, \dots, X_n; \mathbb{K}),$$

where the evaluation is given by $\langle M, x_1 \otimes \cdots \otimes x_n\rangle = M(x_1, \dots, x_n)$.

If $n = 2$, then the canonical identification $\mathcal{L}(X_1, X_2; \mathbb{K}) = \mathcal{L}(X_1; X_2^*)$ of Fact 1.9 leads to an equivalent dual representation:

Fact 8. *Let X, Y be normed linear spaces. Then $(X \otimes_\pi Y)^* = \mathcal{L}(X; Y^*)$, where the evaluation is given by $\langle L, x \otimes y\rangle = L(x)(y)$.*

Let us pass to the description of another, non-equivalent, natural norm on the tensor product space.

Definition 9. Let $X_1, \dots, X_n$ be normed linear spaces. The injective tensor norm ε on $X_1 \otimes \cdots \otimes X_n$ is defined as follows:

$$\varepsilon\left(\sum_{j=1}^{k} x_1^j \otimes \cdots \otimes x_n^j\right) = \sup_{\substack{x_j^* \in B_{X_j^*}\\ j=1,\dots,n}} \left|\sum_{j=1}^{k} x_1^*(x_1^j) \cdots x_n^*(x_n^j)\right|.$$

By $X_1 \otimes_\varepsilon \cdots \otimes_\varepsilon X_n$ we denote the injective tensor product, that is the completion of the normed linear space $(X_1 \otimes \cdots \otimes X_n, \varepsilon)$.

It is easy to show that both the injective and the projective tensor products are associative, i.e. $(X \otimes_\varepsilon Y) \otimes_\varepsilon Z = X \otimes_\varepsilon (Y \otimes_\varepsilon Z)$, resp. $(X \otimes_\pi Y) \otimes_\pi Z = X \otimes_\pi (Y \otimes_\pi Z)$, and this holds by induction for any number of factors.

In the finite-dimensional case we have the following identification.

Proposition 10 ([FHHMZ, Proposition 16.8])**.** *Let X be a finite-dimensional Banach space and Y a Banach space. Then $\underbrace{X^* \otimes_\varepsilon \cdots \otimes_\varepsilon X^*}_{n\text{-times}} \otimes_\varepsilon Y$ is canonically isometric to $\mathcal{L}(^nX; Y)$ via the identification which maps every $(x_1^* \otimes \cdots \otimes x_n^*) \otimes y$ to an n-linear mapping $\left(\prod_{j=1}^n x_j^*(\cdot)\right) y$.*

Next, we look at the tensor products of linear operators.

Proposition 11. *Let X_j, Y_j, $j = 1, \dots, n$, be vector spaces and $T_j \in L(X_j; Y_j)$. Then there is a unique linear operator $S \in L(X_1 \otimes \cdots \otimes X_n; Y_1 \otimes \cdots \otimes Y_n)$ with the property*

$$S(x_1 \otimes \cdots \otimes x_n) = T_1(x_1) \otimes \cdots \otimes T_n(x_n).$$

The operator is denoted by $S = T_1 \otimes \cdots \otimes T_n$.

Proof. This follows from Theorem 2 used on an n-linear mapping $M(x_1, \dots, x_n) = T_1(x_1) \otimes \cdots \otimes T_n(x_n)$. □

Proposition 12. *Let X_j, Y_j, $j = 1, \dots, n$, be normed linear spaces and $T_j \in \mathcal{L}(X_j; Y_j)$. Then*

$$\left\|T_1 \otimes \cdots \otimes T_n \colon (X_1 \otimes_\pi \cdots \otimes_\pi X_n) \to (Y_1 \otimes_\pi \cdots \otimes_\pi Y_n)\right\| = \|T_1\| \cdots \|T_n\|,$$
$$\left\|T_1 \otimes \cdots \otimes T_n \colon (X_1 \otimes_\varepsilon \cdots \otimes_\varepsilon X_n) \to (Y_1 \otimes_\varepsilon \cdots \otimes_\varepsilon Y_n)\right\| = \|T_1\| \cdots \|T_n\|.$$

Proof. The first equality follows from Proposition 5. The second equality follows from the fact that $x_1^*(T_1(x_1)) \cdots x_n^*(T_n(x_n)) = T_1^*(x_1^*)(x_1) \cdots T_n^*(x_n^*)(x_n)$. □

We continue by collecting some facts on symmetric tensor products and their close relationship with polynomials. Recall that the symmetrisation of the tensor product mapping, $\otimes_s\colon X\times\cdots\times X\to X\otimes\cdots\otimes X$, is a symmetric n-linear mapping given by

$$\otimes_s(x_1,\dots,x_n)=\frac{1}{n!}\sum_{\eta\in S_n}\otimes(x_{\eta(1)},\dots,x_{\eta(n)})=\frac{1}{n!}\sum_{\eta\in S_n}x_{\eta(1)}\otimes\cdots\otimes x_{\eta(n)},$$

where S_n is the set of all permutations of $\{1,\dots,n\}$. We will also use the notation $\otimes_s(x_1,\dots,x_n)=x_1\otimes_s\cdots\otimes_s x_n$ and $\otimes^n x=\otimes(^n x)=x\otimes\cdots\otimes x$. The Polarisation formula (Proposition 1.13) yields that $\otimes^n_s X=\operatorname{span}\{\otimes^n x;\ x\in X\}$. The space $\otimes^n_s X$ is called a symmetric tensor product and the elements of $\otimes^n_s X$ are called symmetric tensors. When $\otimes^n_s X$ is equipped with the projective norm inherited from its superspace $\otimes^n_\pi X$, its completion becomes a closed subspace $\otimes^n_{\pi,s}X$ of $\otimes^n_\pi X$. Then the linearisation $\sigma^n_X\colon\ \otimes^n_\pi X\to\otimes^n_{\pi,s}X$ of $\otimes_s$ from Theorem 6 is a projection of norm 1. Thus we obtain the following:

Theorem 13 (Universality of the symmetric tensor product). *Let X, Y be normed linear spaces. For every symmetric $M\in\mathcal{L}^s(^nX;Y)$ there exists a unique linear operator $L_M\in\mathcal{L}(\otimes^n_{\pi,s}X;Y)$ such that $M=L_M\circ\otimes_s=L_M\circ\sigma^n_X\circ\otimes$. The mapping $M\mapsto L_M$ is an isometry of the spaces $\mathcal{L}^s(^nX;Y)$ and $\mathcal{L}(\otimes^n_{\pi,s}X;Y)$. These facts are expressed by the following commutative diagram:*

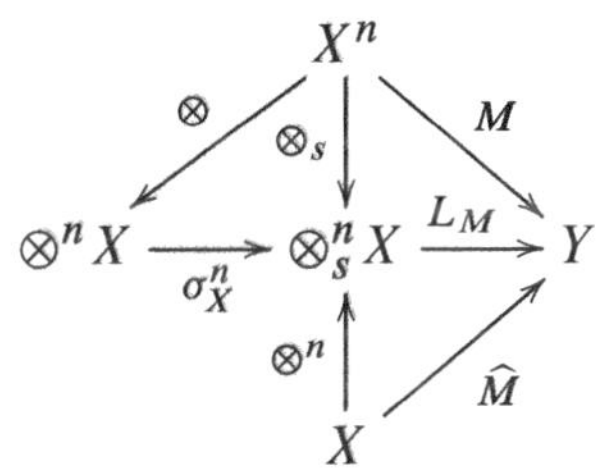

The following is immediate, using also Fact 8.

Corollary 14. *Let X, Y be normed linear spaces. Then the spaces $\mathcal{P}(^nX;Y)$ and $\mathcal{L}(\otimes^n_{\pi,s}X;Y)$ are canonically isomorphic. In particular,*

$$(\otimes^n_{\pi,s}X)^*=\mathcal{P}(^nX)$$

in the isomorphic sense, where the evaluation is given by $\langle P,\otimes^n x\rangle=P(x)$. More generally,

$$\big((\otimes^n_{\pi,s}X)\otimes_\pi Y\big)^*=\mathcal{L}(\otimes^n_{\pi,s}X;Y^*)=\mathcal{P}(^nX;Y^*)$$

in the isomorphic sense, where the evaluation is given by $\langle P,\otimes^n x\otimes y\rangle=P(x)(y)$.

2. Weak continuity and spaces of polynomials

In this section we introduce the weak continuity properties for continuous mappings. For the most part of this section we will be interested in the special case of polynomials.

We apply tensor products to the study of spaces of polynomials and their properties which are reflected by the weak continuity properties of individual polynomials.

We remind the reader that all topological spaces in this book are assumed to be Hausdorff. Parallel to the concept of topology there is the notion of uniformity and of uniform spaces, see [Eng, Chapter 8] or [Cho]. Since this concept is perhaps slightly less known we recall some of its basic properties. All the general results below can be found in [Eng, Chapter 8]. Let S be a non-empty set. Given a subset U of $S \times S$, denote $-U = \{(y, x);\ (x, y) \in U\}$, and if V is another subset of $S \times S$, let

$$U + V = \big\{(x, z);\ \text{there exists } y \in S \text{ such that } (x, y) \in U, (y, z) \in V\big\}.$$

A *uniformity* $\mathcal{U}$ in S is a filter consisting of subsets of $S \times S$ that satisfy the following:

(U1) $\Delta \subset U$ for all $U \in \mathcal{U}$, where $\Delta = \{(x, x);\ x \in S\}$;
(U2) if $U \in \mathcal{U}$, then $-U \in \mathcal{U}$;
(U3) for each $U \in \mathcal{U}$ there exists $V \in \mathcal{U}$ such that $V + V \subset U$;
(U4) $\bigcap_{U \in \mathcal{U}} U = \Delta$.

The pair $(S, \mathcal{U})$ is called a *uniform space*. Every uniform space $(S, \mathcal{U})$ becomes canonically a topological space with the basis of the topology consisting of all sets $U(x) = \{y \in S;\ (y, x) \in U\}$ for $x \in S, U \in \mathcal{U}$. The topology so defined is called the topology induced by the uniformity.

Every metric space (P, ρ) is a uniform space. Indeed, it suffices to consider the family $\mathcal{U}$ of all supersets of sets of the form $\{(x, y) \in P \times P;\ \rho(x, y) < 1/n\}, n \in \mathbb{N}$. Then $\mathcal{U}$ is a uniformity on P such that the induced topology is the topology defined by the metric. Every topological vector space (X, τ) has a unique translation-invariant uniformity defined on X that induces the topology τ. The uniformity $\mathcal{U}$ is defined as follows: $U \subset X \times X$ belongs to $\mathcal{U}$ if and only if $U = \{(x, y) \in X \times X;\ x - y \in V\}$, where V is a neighbourhood of 0. The case we are going to deal with mostly is the weak topology on a Banach space (X, w) or its restriction to some subset of X. We will call the canonical uniformity on (X, w) the weak uniformity on X (or on its subsets).

Topological spaces whose topology arises from a suitably chosen uniformity are called uniformisable. Uniformisable topological spaces are characterised as being Tychonov (i.e. completely regular) spaces. Equivalently, they are homeomorphic to subsets of $[0, 1]^{\Gamma}$. In general, the uniformity on a uniformisable space is not uniquely determined by the topology, but every compact topological space admits a unique uniformity.

A mapping f from a uniform space $(S, \mathcal{U})$ into another uniform space $(T, \mathcal{V})$ is called *uniformly continuous* if for every $V \in \mathcal{V}$ there exists $U \in \mathcal{U}$ such that $(f(x), f(y)) \in V$ whenever $(x, y) \in U$. Obviously, every uniformly continuous mapping is continuous when S and T are endowed with their induced topologies. If $S \subset T$ is a subset of a uniform space $(T, \mathcal{U})$, then S is naturally a uniform space by using the restricted uniformity $\mathcal{V} = \{V;\ V = U \cap S \times S, U \in \mathcal{U}\}$. The identity mapping $Id : S \to T$ is then uniformly continuous. Note that any continuous linear mapping between topological vector spaces is w–w uniformly continuous.

A net $\{x_\gamma\}_{\gamma\in\Gamma}$ in a uniform space $(S, \mathcal{U})$ is said to be *Cauchy* if given $U \in \mathcal{U}$ there exists $\gamma_0 \in \Gamma$ such that $(x_\alpha, x_\beta) \in U$ for all $\alpha, \beta \in \Gamma$, $\alpha \succeq \gamma_0$, $\beta \succeq \gamma_0$. It is easy to see that a subnet of a Cauchy net is again a Cauchy net. Note that a net $\{x_\gamma\}_{\gamma\in\Gamma}$ in a topological vector space X is weakly Cauchy if and only if for each $f \in X^*$ the net $\{f(x_\gamma)\}_{\gamma\in\Gamma}$ is Cauchy (or equivalently convergent).

A uniform space $(S, \mathcal{U})$ is said to be *complete* if every Cauchy net in S converges. For every uniform space $(S, \mathcal{U})$ there exists a unique (up to a uniform isomorphism) completion $(\tilde{S}, \tilde{\mathcal{U}})$, i.e. a complete uniform space containing S as a dense subset. A uniformly continuous mapping has a unique extension to a uniformly continuous mapping between the respective completions.

Let $(S, \mathcal{U})$ be a uniform space and $U \in \mathcal{U}$. A subset A of S is said to be U-dense in S if for every $x \in S$ there is $y \in A$ such that $x \in U(y)$, i.e. $(x, y) \in U$. The space $(S, \mathcal{U})$ is called *totally bounded* if for every $U \in \mathcal{U}$ there exists a finite U-dense subset of S. The space $(S, \mathcal{U})$ is totally bounded if and only if every net in S has a Cauchy subnet ([Cho, Proposition 5.23]). A uniform space is compact if and only if it is totally bounded and complete. Consequently, if $B \subset X$ is a bounded subset of a Banach space equipped with the uniformity inherited from (X, w), then its completion coincides with the w^*-closure of B in X^{**} equipped with the w^*-uniformity $(\overline{B}^{w^*}, w^*)$. The completion of a totally bounded space is totally bounded and hence compact. A continuous mapping from a compact topological space (X, τ) is automatically uniformly continuous in the unique uniformity on X and the unique uniformity on its compact range.

A mapping between uniform spaces is called *Cauchy-continuous* (resp. *sequentially Cauchy-continuous*) if it maps Cauchy nets to Cauchy nets (resp. Cauchy sequences to Cauchy sequences). In particular, a function (or more generally a mapping into a complete space) is (sequentially) Cauchy-continuous if and only if it maps Cauchy nets (sequences) to convergent ones. It is easy to see that a uniformly continuous mapping is Cauchy-continuous (and hence also sequentially Cauchy-continuous). Further, a (sequentially) Cauchy-continuous mapping $f: (X, \mathcal{U}) \to (Y, \mathcal{V})$ is (sequentially) continuous. Indeed, given a net $\{x_\gamma\}_{\gamma\in\Gamma} \subset X$ converging to x, consider the directed set $\Lambda = \Gamma \times \{0, 1\}$ with the lexicographic ordering and set $x_{\gamma,0} = x_\gamma$ and $x_{\gamma,1} = x$. Then the net $\{x_\alpha\}_{\alpha\in\Lambda}$ is Cauchy and hence also $\{f(x_\alpha)\}_{\alpha\in\Lambda}$ is Cauchy. Now if $V \in \mathcal{V}$, then there is $\gamma_0 \in \Gamma$ such that $(f(x_{\gamma,0}), f(x_{\gamma,1})) \in V$ whenever $\gamma \succeq \gamma_0$, i.e. $f(x_\gamma) \in V(f(x))$. A Cauchy-continuous mapping from a uniform space X to a complete uniform space can be extended to a continuous mapping on the completion of X ([Schec, Theorem 19.27]). The following observation is of great importance.

Proposition 15. *Let X be a totally bounded uniform space, Y a uniform space, and $f: X \to Y$ a Cauchy-continuous mapping. Then f is uniformly continuous.*

Proof. The mapping f can be extended to a continuous mapping $F: \tilde{X} \to \tilde{Y}$, where $\tilde{X}, \tilde{Y}$ are the completions of X, Y. Since $\tilde{X}$ is compact, F is uniformly continuous, and thus so is f. □

A closed, convex, and bounded subset of a normed linear space X will be abbreviated as a CCB set. Of great importance is the fact that a bounded set $B \subset X$ equipped with the uniformity inherited from (X, w) is totally bounded, which follows from the fact that $\overline{B}^{w^*} \subset X^{**}$ is w^*-compact. In particular, a mapping from B into a Banach space Y is w–$\|\cdot\|$ uniformly continuous if and only if it maps weakly Cauchy nets to convergent nets (Proposition 15).

We continue by giving a list of various notions of weak continuity that will be used in our investigations. Some of these classes have been introduced and studied by Richard Martin Aron and his co-authors, e.g. in [AHV], [AP], [Aro1]; see also [Din].

Definition 16. Let X be a normed linear space, Y a Banach space, and $U \subset X$ a convex set. By $\mathcal{C}(U;Y)$ we denote the space $C(U;Y)$ endowed with the locally convex topology τ_b of uniform convergence on CCB subsets of U.

By $\mathcal{C}_{\mathrm{w}}(U;Y)$ we denote the linear subspace of $\mathcal{C}(U;Y)$ consisting of all mappings that are w–$\|\cdot\|$ continuous on CCB subsets of U.

By $\mathcal{C}_{\mathrm{wu}}(U;Y)$ we denote the linear subspace of $\mathcal{C}(U;Y)$ consisting of all mappings that are w–$\|\cdot\|$ uniformly continuous on CCB subsets of U. (In other words, $f \in \mathcal{C}_{\mathrm{wu}}(U;Y)$ if and only if for any CCB set V and any $\varepsilon > 0$ there are $\delta > 0$ and $\phi_1, \dots, \phi_k \in B_{X^*}$ such that $\|f(x) - f(y)\| < \varepsilon$ whenever $x, y \in V$ are such that $|\phi_j(x-y)| < \delta$ for $j = 1, \dots, k$.)

By $\mathcal{C}_{\mathrm{wsc}}(U;Y)$ we denote the linear subspace of $\mathcal{C}(U;Y)$ consisting of all mappings that are w–$\|\cdot\|$ sequentially continuous on CCB subsets of U, i.e. that map weakly convergent sequences in CCB subsets of U to convergent sequences in Y.

By $\mathcal{C}_{\mathrm{wsC}}(U;Y)$ we denote the linear subspace of $\mathcal{C}(U;Y)$ consisting of all mappings that are w–$\|\cdot\|$ sequentially Cauchy-continuous on CCB subsets of U, i.e. that map weakly Cauchy sequences in CCB subsets of U to convergent sequences in Y.

By $\mathcal{C}_{\mathrm{K}}(U;Y)$ we denote the linear subspace of $\mathcal{C}(U;Y)$ consisting of all mappings that map CCB subsets of U to relatively compact sets in Y.

By $\mathcal{C}_{\mathrm{wK}}(U;Y)$ we denote the linear subspace of $\mathcal{C}(U;Y)$ consisting of all mappings that map CCB subsets of U to relatively weakly compact sets in Y.

We use the usual convention for the notation when the range space is scalars, e.g. $\mathcal{C}_{\mathrm{wsc}}(U) = \mathcal{C}_{\mathrm{wsc}}(U;\mathbb{K})$.

We remark that if U is closed (in particular if $U = X$), then the topology on $\mathcal{C}(U;Y)$ is the topology of uniform convergence on bounded subsets of U; moreover, we may replace the CCB sets in the definitions above by bounded sets and $\mathcal{C}_{\mathrm{wsc}}(U;Y)$ (resp. $\mathcal{C}_{\mathrm{wsC}}(U;Y)$) are just mappings w–$\|\cdot\|$ sequentially continuous on U (resp. w–$\|\cdot\|$ sequentially Cauchy-continuous). Also, if X^* is separable, then it is easy to see that the weak uniformity on B_X is induced by the metric $\rho(x,y) = \sum_{n=1}^{\infty} \frac{1}{2^n} |f_n(x-y)|$ for some sequence $\{f_n\}$ dense in B_{X^*}. Thus in this case $\mathcal{C}_{\mathrm{wsc}}(U;Y) = \mathcal{C}_{\mathrm{w}}(U;Y)$ and $\mathcal{C}_{\mathrm{wsC}}(U;Y) = \mathcal{C}_{\mathrm{wu}}(U;Y)$ (Proposition 15).

It is easy to prove that $\mathcal{C}_{\Psi}(U;Y)$, where Ψ is one of the properties w, wu, wsc, wsC, K, is a closed subspace of $\mathcal{C}(U;Y)$. In the case of $\mathcal{C}_{\mathrm{wK}}(U;Y)$ we additionally have to employ the following lemma.

Lemma 17 (Alexander Grothendieck; [FHHMZ, Lemma 13.32]). *Let X be a Banach space and $A \subset X$. If for every $\varepsilon > 0$ there is a weakly compact set $A_\varepsilon \subset X$ such that $A \subset A_\varepsilon + \varepsilon B_X$, then A is relatively weakly compact.*

The following inclusions hold for any Banach space Y and any convex subset U of a normed linear space:

$$\begin{array}{ccccc} & \nearrow & \mathcal{C}_{\mathrm{K}}(U;Y) & \subset & \mathcal{C}_{\mathrm{wK}}(U;Y) \\ \mathcal{C}_{\mathrm{wu}}(U;Y) & \subset & \mathcal{C}_{\mathrm{w}}(U;Y) & \subset & \mathcal{C}_{\mathrm{wsc}}(U;Y) \\ & \searrow & \mathcal{C}_{\mathrm{wsC}}(U;Y) & \nearrow & \end{array}$$

All the inclusions are obvious except for the one from the following lemma.

Lemma 18 ([AP]). *Let X be a normed linear space, Y a Banach space, and $U \subset X$ a convex subset. Then $\mathcal{C}_{\mathrm{wu}}(U;Y) \subset \mathcal{C}_{\mathrm{K}}(U;Y)$.*

Proof. Let $f \in \mathcal{C}_{\mathrm{wu}}(U;Y)$, $B \subset U$ a CCB subset, and $\varepsilon > 0$. There exist $\delta > 0$ and $\phi_1, \dots, \phi_n \in X^*$ such that $\|f(x) - f(y)\| < \varepsilon$ if $x, y \in B$ and $\max_j |\phi_j(x-y)| < \delta$. Choose a finite set $\{x_j\}_{j=1}^k \subset B$ such that $\{(\phi_1(x_j), \dots, \phi_n(x_j));\ j \le k\}$ forms a δ-dense set in the totally bounded set $\{(\phi_1(x), \dots, \phi_n(x));\ x \in B\}$. Then clearly $\{f(x_j);\ j \le k\}$ is a finite ε-dense set in $f(B)$, so the latter set is relatively compact. □

Proposition 19. *Let X be a normed linear space, Y a Banach space, $U \subset X$ a convex set, and $f \in \mathcal{C}_{\mathrm{K}}(U;Y)$. Let Ψ be one of the properties w, wsc, wsC, wu. Then $f \in \mathcal{C}_{\Psi}(U;Y)$ if and only if $\phi \circ f \in \mathcal{C}_{\Psi}(U)$ for every $\phi \in Y^*$, i.e. if and only if f is w–w (sequentially/Cauchy/uniformly) continuous on CCB subsets of U.*

Proof. $\Rightarrow$ is clear.

$\Leftarrow$ We give a proof in the case of $\mathcal{C}_{\mathrm{wu}}(U;Y)$, as the other cases are similar. Assuming the contrary, there are a CCB set $B \subset U$ and $\varepsilon > 0$ such that for each finite subset $F \subset B_{X^*}$ and for each $\delta > 0$ there exist $x_{(F,\delta)}, y_{(F,\delta)} \in B$ such that $\max_{g \in F} |g(x_{(F,\delta)} - y_{(F,\delta)})| < \delta$ and $\|f(x_{(F,\delta)}) - f(y_{(F,\delta)})\| \ge \varepsilon$. Consider a directed set $\Gamma = \{(F,\eta);\ F \subset B_{X^*}, |F| < \infty, \eta > 0\}$ endowed with the partial ordering $(F,\eta) \succeq (E,\delta)$ provided that $F \supset E$ and $\eta \le \delta$. Because $f(B)$ is relatively compact, there exists a cofinal subset $\Lambda \subset \Gamma$ such that the nets $\{f(x_\gamma)\}_{\gamma \in \Lambda}$ and $\{f(y_\gamma)\}_{\gamma \in \Lambda}$ are convergent. Let $v = \lim_{\gamma \in \Lambda} f(x_\gamma)$ and $w = \lim_{\gamma \in \Lambda} f(y_\gamma)$. Then $\|v - w\| \ge \varepsilon$.

Choose a functional $\phi \in B_{Y^*}$ with $\phi(v - w) \ge \varepsilon$. By the assumption there are a finite $F \subset B_{X^*}$ and $\delta > 0$ such that $|\phi \circ f(x) - \phi \circ f(y)| < \frac{\varepsilon}{2}$ whenever $x, y \in B$ and $\max_{g \in F} |g(x - y)| < \delta$. Since Λ is cofinal in Γ, there is $\alpha \in \Lambda$ such that $\alpha \succeq (F,\delta)$, $\|f(x_\alpha) - v\| < \frac{\varepsilon}{4}$, and $\|f(y_\alpha) - w\| < \frac{\varepsilon}{4}$. It follows that $|\phi \circ f(x_\alpha) - \phi \circ f(y_\alpha)| > \frac{\varepsilon}{2}$, which is a contradiction. □

Similarly we have the following.

Proposition 20. *Let X be a normed linear space, Y a Banach space, $U \subset X$, and $f\colon U \to Y$. Then f is w–$\|\cdot\|$ sequentially continuous if and only if f is w–w sequentially continuous and maps weakly compact subsets of U to compact sets. In particular, if U is a convex subset of a Banach space, then $f \in \mathcal{C}_{\mathrm{wsc}}(U;Y)$ if and only if f is w–w sequentially continuous on CCB subsets of U and maps weakly compact CCB subsets of U to compact sets.*

Proof. $\Rightarrow$ The mapping f is clearly w–w sequentially continuous. Let $K \subset U$ be weakly compact and $\{x_n\}$ a sequence in K. There is a subsequence $\{x_{n_k}\}$ weakly convergent to some $x \in K$. Then $f(x_{n_k}) \to f(x)$. It follows that $f(K)$ is compact.

$\Leftarrow$ By contradiction, suppose that there exist $\varepsilon > 0$ and $x_n \overset{w}{\to} x$ such that $\|f(x_n) - f(x)\| \geq \varepsilon$ for each $n \in \mathbb{N}$. The set $\{x_n;\ n \in \mathbb{N}\} \cup \{x\}$ is weakly compact, hence $\{f(x_n);\ n \in \mathbb{N}\} \cup \{f(x)\}$ is compact. So there is a subsequence $\{x_{n_k}\}$ such that $f(x_{n_k}) \to z \in Y$. By the assumption $f(x_{n_k}) \overset{w}{\to} f(x)$, and so $z = f(x)$, a contradiction. □

Having defined the general concepts, we introduce the corresponding classes of polynomials in a natural way. Let X be a normed linear space, Y a Banach space, and let Ψ be one of the properties w, wu, wsc, wsC, K, wK. We define

$$\begin{aligned} \mathcal{P}_\Psi({}^nX;Y) &= \mathcal{P}({}^nX;Y) \cap \mathcal{C}_\Psi(X;Y), \\ \mathcal{P}^n_\Psi(X;Y) &= \mathcal{P}^n(X;Y) \cap \mathcal{C}_\Psi(X;Y), \text{ and} \\ \mathcal{P}_\Psi(X;Y) &= \mathcal{P}(X;Y) \cap \mathcal{C}_\Psi(X;Y). \end{aligned}$$

Finally, we introduce the multilinear versions, using the product topology on the Cartesian power X^n, which can be naturally considered as a normed linear space $X \oplus \cdots \oplus X$:

$$\mathcal{L}_\Psi({}^nX;Y) = \mathcal{L}({}^nX;Y) \cap \mathcal{C}_\Psi(X \oplus \cdots \oplus X;Y).$$

The spaces $\mathcal{P}({}^nX;Y)$ and $\mathcal{P}^n(X;Y)$ are closed in $\mathcal{C}(X;Y)$ (i.e. in the topology of uniform convergence on bounded sets; see the proofs of Theorems 1.29 and 1.46 and notice that because of the uniform convergence we do not need to use the Uniform boundedness principle) and similarly $\mathcal{L}({}^nX;Y)$ is closed in $\mathcal{C}(X \oplus \cdots \oplus X;Y)$. Therefore $\mathcal{P}_\Psi({}^nX;Y)$ is a closed subspace of $\mathcal{P}({}^nX;Y)$, $\mathcal{P}^n_\Psi(X;Y)$ is a closed subspace of $\mathcal{P}^n(X;Y)$, and finally $\mathcal{L}_\Psi({}^nX;Y)$ is a closed subspace of $\mathcal{L}({}^nX;Y)$, where Ψ is one of the properties w, wu, wsc, wsC, K, wK. The Polarisation formula (Proposition 1.13) gives readily the following useful fact.

Fact 21. *Let X be a normed linear space, Y a Banach space, and let Ψ be one of the properties w, wu, wsc, wsC, K, wK. If $M \in \mathcal{L}_\Psi({}^nX;Y)$, then $\widehat{M} \in \mathcal{P}_\Psi({}^nX;Y)$. If $P \in \mathcal{P}({}^nX;Y)$, then $P \in \mathcal{P}_\Psi({}^nX;Y)$ if and only if $\check{P} \in \mathcal{L}_\Psi({}^nX;Y)$.*

Proposition 22. *Let X be a normed linear space, Y a Banach space, and let Ψ be one of the properties w, wu, wsc, wsC, K, wK. Let $P \in \mathcal{P}^n(X;Y)$, $P = \sum_{j=0}^n P_j$, $P_j \in \mathcal{P}({}^jX;Y)$. The following statements are equivalent:*

(i) *$P\restriction_B \in \mathcal{C}_\Psi(B;Y)$ for some ball $B \subset X$.*
(ii) *$P_j \in \mathcal{P}_\Psi({}^jX;Y)$ for each $j = 0,\dots,n$.*
(iii) *$P \in \mathcal{P}_\Psi(X;Y)$.*

Proof. (ii)⇒(iii)⇒(i) is obvious. To prove (i)⇒(ii), suppose that $B = B(a,r)$ and define $Q(x) = P(a+x)$. Then $Q \restriction_{B(0,r)} \in \mathscr{C}_\Psi(B(0,r);Y)$. Let Q_j be the j-homogeneous summand of Q. Then $Q_j \restriction_{B(0,r)} \in \mathscr{C}_\Psi(B(0,r);Y)$, $j = 0,\dots,n$, by Corollary 1.31. Hence $Q_j \in \mathscr{P}_\Psi({}^jX;Y)$ by the homogeneity. It follows that $Q \in \mathscr{P}_\Psi(X;Y)$ and thus also $P \in \mathscr{P}_\Psi(X;Y)$. □

We note that the operators in $\mathscr{L}_{\mathrm{wsc}}(X;Y)$ are also called completely continuous operators. From Proposition 20 we obtain the following corollary.

Corollary 23. *Let X be a normed linear space and Y be a Banach space. Then $T \in \mathscr{L}_{\mathrm{wsc}}(X;Y)$ if and only if T maps weakly compact sets to compact sets.*

Combining Proposition 19 and Lemma 18 with the obvious fact that $X^* \subset \mathscr{C}_{\mathrm{wu}}(X)$ we obtain the following corollary.

Corollary 24. *Let X be a normed linear space and let Y be a Banach space. Then $\mathscr{L}_{\mathrm{K}}(X;Y) = \mathscr{L}_{\mathrm{wu}}(X;Y)$.*

Fact 25. *Let X be a normed linear space and Y a Banach space. Then*

$$\begin{aligned}
M \in \mathscr{L}_{\mathrm{K}}({}^nX;Y) &\quad \text{if and only if} \quad L_M \in \mathscr{L}_{\mathrm{K}}(\otimes_\pi^n X;Y),\\
M \in \mathscr{L}_{\mathrm{wK}}({}^nX;Y) &\quad \text{if and only if} \quad L_M \in \mathscr{L}_{\mathrm{wK}}(\otimes_\pi^n X;Y),\\
P \in \mathscr{P}_{\mathrm{K}}({}^nX;Y) &\quad \text{if and only if} \quad L_{\check{P}} \in \mathscr{L}_{\mathrm{K}}(\otimes_{\pi,s}^n X;Y),\\
P \in \mathscr{P}_{\mathrm{wK}}({}^nX;Y) &\quad \text{if and only if} \quad L_{\check{P}} \in \mathscr{L}_{\mathrm{wK}}(\otimes_{\pi,s}^n X;Y).
\end{aligned}$$

Proof. Clearly, the statements on the right-hand side imply the statements on the left-hand side. The reverse implications follow by (3), Theorems 6 and 13, the fact that in every locally convex space the convex hull of a totally bounded set is a totally bounded set, and Fact 21. □

Theorem 26 ([AHV]). *Let $X_1,\dots,X_n$, and Y be normed linear spaces and let $M \in \mathscr{L}(X_1,\dots,X_n;Y)$. The following statements are equivalent:*

(i) *The mapping M is w–$\|\cdot\|$ continuous on bounded sets, resp. w–$\|\cdot\|$ sequentially continuous.*
(ii) $\lim_{\gamma\in\Gamma} M(x_\gamma^1,\dots,x_\gamma^n) = 0$ *whenever $\{x_\gamma^k\}_{\gamma\in\Gamma} \subset X_k$, $k = 1,\dots,n$ are weakly convergent bounded nets and at least one of them is weakly null, and Γ is any directed set, resp. $\Gamma = (\mathbb{N},\le)$.*
(iii) $\lim_{\gamma\in\Gamma} M(x_\gamma^1,\dots,x_\gamma^n) = 0$ *whenever $\{x_\gamma^k\}_{\gamma\in\Gamma} \subset X_k$, $k = 1,\dots,n$ are weakly Cauchy bounded nets and at least one of them is weakly null, and Γ is any directed set, resp. $\Gamma = (\mathbb{N},\le)$.*
(iv) *M maps weakly Cauchy bounded nets to Cauchy nets, resp. weakly Cauchy sequences to Cauchy sequences.*

Proof. (i)⇒(ii), (iii)⇒(ii), and (iv)⇒(i) are clear.

(ii)⇒(i) Let $\{x^k_\gamma\}_{\gamma\in\Gamma} \subset X_k$ be bounded nets such that $x^k_\gamma \xrightarrow{w} x^k$. Then

$$\|M(x^1_\gamma,\dots,x^n_\gamma) - M(x^1,\dots,x^n)\| \\ \le \sum_{j=1}^{n} \big\| M(x^1,\dots,x^{j-1},x^j_\gamma - x^j, x^{j+1}_\gamma,\dots,x^n_\gamma)\big\|.$$

Each of the summands on the right-hand side has a bounded weakly null net in one of the coordinates, so by (ii) the right-hand side tends to zero in limit with respect to $\gamma \in \Gamma$.

(iii)⇒(iv) By contradiction, assume that there are weakly Cauchy bounded nets $\{x^k_\gamma\}_{\gamma\in\Gamma} \subset X_k$ and an $\varepsilon > 0$ such that for every $\gamma \in \Gamma$ there are $\alpha(\gamma) \succeq \gamma$ and $\beta(\gamma) \succeq \gamma$ with $\|M(x^1_{\alpha(\gamma)},\dots,x^n_{\alpha(\gamma)}) - M(x^1_{\beta(\gamma)},\dots,x^n_{\beta(\gamma)})\| \ge \varepsilon$. But

$$\|M(x^1_{\alpha(\gamma)},\dots,x^n_{\alpha(\gamma)}) - M(x^1_{\beta(\gamma)},\dots,x^n_{\beta(\gamma)})\| \\ \le \sum_{j=1}^{n} \big\| M(x^1_{\beta(\gamma)},\dots,x^{j-1}_{\beta(\gamma)},x^j_{\alpha(\gamma)} - x^j_{\beta(\gamma)}, x^{j+1}_{\alpha(\gamma)},\dots,x^n_{\alpha(\gamma)})\big\|.$$

Each of the summands on the right-hand side has a bounded weakly null net in one of the coordinates, so by (iii) the right-hand side tends to zero in limit with respect to $\gamma \in \Gamma$, which is a contradiction.

(i)⇒(iii) We use induction on n. If $n = 1$, then the statement is trivially satisfied. The inductive step from $n-1$ to n is proved by contradiction. Assume that $\{x^k_\gamma\}_{\gamma\in\Gamma} \subset X_k$ are bounded weakly Cauchy nets for all $k = 1,\dots,n$, $\{x^1_\gamma\}_{\gamma\in\Gamma}$ is weakly null, and $\|M(x^1_\gamma,\dots,x^n_\gamma)\| > \varepsilon > 0$ for all $\gamma \in \Gamma_1$, where $\Gamma_1 \subset \Gamma$ is some cofinal subset. For any $x \in X_n$ the mapping $(z_1,\dots,z_{n-1}) \mapsto M(z_1,\dots,z_{n-1},x) \in \mathcal{L}(X_1,\dots,X_{n-1};Y)$ satisfies (i). Therefore for every $\alpha \in \Gamma_1$ there is $\beta \in \Gamma_1$, $\beta = \beta(\alpha) \succeq \alpha$ such that $\|M(x^1_\beta,\dots,x^{n-1}_\beta,x^n_\alpha)\| < \frac{\varepsilon}{2}$. Whence

$$\frac{\varepsilon}{2} \le \|M(x^1_\beta,\dots,x^n_\beta) - M(x^1_\beta,\dots,x^{n-1}_\beta,x^n_\alpha)\| = \|M(x^1_\beta,\dots,x^{n-1}_\beta,x^n_\beta - x^n_\alpha)\|.$$

We replace the original collection of weakly Cauchy nets by a new collection of weakly Cauchy nets $\{y^1_\alpha\}_{\alpha\in\Gamma_1},\dots,\{y^n_\alpha\}_{\alpha\in\Gamma_1}$, where $y^k_\alpha = x^k_{\beta(\alpha)}$, $k < n$, $\alpha \in \Gamma_1$, and $y^n_\alpha = x^n_{\beta(\alpha)} - x^n_\alpha$. For $k < n$ these new nets are subnets of the original ones (Γ_1 is cofinal in Γ). The net $\{y^n_\alpha\}$ is a bounded weakly null net. This reduces the problem to the case when at least two of the nets $\{y^k_\alpha\}_{\alpha\in\Gamma_1}$, $k = 1,\dots,n$, are weakly null. By repeating this argument $(n-2)$-times (passing to cofinal subsets of the index set $\Gamma_{n-1} \subset \cdots \subset \Gamma_1 \subset \Gamma$) we arrive at a collection of bounded weakly null nets $\{z^k_\alpha\}_{\alpha\in\Gamma_{n-1}}$, $k = 1,\dots,n$, such that $\|M(z^1_\alpha,\dots,z^n_\alpha)\| \ge \frac{\varepsilon}{2^{n-1}}$ for all $\alpha \in \Gamma_{n-1}$. This is a contradiction, because $\Gamma_{n-1} \subset \Gamma$ is a cofinal subset. □

Combining this with Proposition 15 and Fact 21 we obtain the following corollary.

Corollary 27 ([AHV]). *Let X be a normed linear space, Y a Banach space, and $n \in \mathbb{N}$. Then*

$$\begin{aligned}\mathcal{L}_{\mathrm{w}}({}^nX;Y) &= \mathcal{L}_{\mathrm{wu}}({}^nX;Y),\\ \mathcal{L}_{\mathrm{wsc}}({}^nX;Y) &= \mathcal{L}_{\mathrm{wsC}}({}^nX;Y),\\ \mathcal{P}_{\mathrm{w}}({}^nX;Y) &= \mathcal{P}_{\mathrm{wu}}({}^nX;Y),\\ \mathcal{P}_{\mathrm{wsc}}({}^nX;Y) &= \mathcal{P}_{\mathrm{wsC}}({}^nX;Y).\end{aligned}$$

From this and the relations shown earlier we obtain the following inclusions:

$$\begin{array}{l} \qquad\qquad\qquad\qquad\quad {}_{\subset}\ \mathcal{P}_{\mathrm{K}}(X;Y) \subset \mathcal{P}_{\mathrm{wK}}(X;Y)\\ \mathcal{P}_{\mathrm{wu}}(X;Y) = \mathcal{P}_{\mathrm{w}}(X;Y) \subset \mathcal{P}_{\mathrm{wsc}}(X;Y) = \mathcal{P}_{\mathrm{wsC}}(X;Y)\end{array}$$

$$\begin{array}{l} \qquad\qquad\qquad\qquad\qquad\qquad\qquad\qquad\quad {}_{\subset}\ \mathcal{L}_{\mathrm{wK}}(X;Y)\\ \mathcal{L}_{\mathrm{wu}}(X;Y) = \mathcal{L}_{\mathrm{K}}(X;Y) = \mathcal{L}_{\mathrm{w}}(X;Y) \subset \mathcal{L}_{\mathrm{wsc}}(X;Y) = \mathcal{L}_{\mathrm{wsC}}(X;Y)\end{array}$$

We remark that unlike the $\|\cdot\|$–$\|\cdot\|$ continuity, generally it is not sufficient to check the w–$\|\cdot\|$ continuity of polynomials only at the origin.

Example 28 ([Aro2]). Let $P \in \mathcal{P}({}^3\ell_2)$ be defined as $P(x) = x_1 \sum_{n=2}^{\infty} x_n^2$. Then the restriction of P to any bounded set is weakly continuous at the origin, but P is not weakly sequentially continuous. Indeed, $e_1 + e_n \xrightarrow{w} e_1$, but $P(e_1 + e_n) = 1$ and $P(e_1) = 0$.

Proposition 29. *Let X be a normed linear space, Y a Banach space, and let Ψ be one of the properties w, wu, wsc, wsC, K, wK. If $n \in \mathbb{N}$ is such that $\mathcal{P}({}^nX;Y) = \mathcal{P}_{\Psi}({}^nX;Y)$, then $\mathcal{P}^n(X;Y) = \mathcal{P}^n_{\Psi}(X;Y)$.*

Proof. First we prove the statement for wsc polynomials. The proof will be in two steps. First we show that any polynomial in $\mathcal{P}^n(X;Y)$ is weakly sequentially continuous at every point except at zero. Let $k \in \mathbb{N}$, $k < n$, and $P \in \mathcal{P}({}^kX;Y)$. Further, let $\{x_j\} \subset X$ be such that $x_j \xrightarrow{w} z \neq 0$. Let $f \in X^*$ be such that $f(z) = 1$. Put $Q(x) = f(x)^{n-k} P(x)$ for $x \in X$. Then by our assumption $Q \in \mathcal{P}_{\mathrm{wsc}}({}^nX;Y)$. Since $P(x_j) = Q(x_j)/f(x_j)^{n-k}$ for j large enough, we conclude that $P(x_j) \to P(z)$.

Now let $P \in \mathcal{P}^n(X;Y)$ and assume that $x_j \xrightarrow{w} 0$. Let $y \in X \setminus \{0\}$ and define $Q(x) = P(x - y)$ for $x \in X$. Then $Q \in \mathcal{P}^n(X;Y)$ (Lemma 1.34). Thus by the first part of the proof $P(x_j) = Q(y + x_j) \to Q(y) = P(0)$.

The case of $\mathcal{P}^n_{\mathrm{w}}(X;Y)$ is identical using bounded nets, the cases of wu and wsC are clear from Corollary 27. The proof for compact (resp. weakly compact) polynomials is similar: First we show that $P(V)$ is relatively compact for any $P \in \mathcal{P}^n(X;Y)$ and any CCB set $V \subset X$ not containing the origin. Fix such a set V and let $k \in \mathbb{N}$, $k < n$, and $P \in \mathcal{P}({}^kX;Y)$. Let $f \in X^*$ be such that $\inf f(V) = \delta > 0$. Put $Q(x) = f(x)^{n-k} P(x)$ for $x \in X$. Then by our assumption $Q \in \mathcal{P}_{\mathrm{K}}({}^nX;Y)$. Since $P(x) = Q(x)/f(x)^{n-k}$ for $x \in V$ and $0 \le \frac{1}{f^{n-k}}(V) \le \frac{1}{\delta^{n-k}}$, it is easy to see that $P(V)$ is relatively compact in Y.

Finally, if $P \in \mathcal{P}^n(X;Y)$ and V is a CCB set with $0 \in V$, then there is $y \in X$ satisfying $-y \notin V$. Define $Q \in \mathcal{P}^n(X;Y)$ by $Q(x) = P(x-y)$ for $x \in X$. Then by the first part of the proof $P(V) = Q(y+V)$ is relatively compact in Y. □

The next two lemmata have numerous applications as they allow one to conclude, under suitable assumptions, that the origin is the point of weak discontinuity.

Lemma 30 ([CHL]). *Let X be a normed linear space and Y a Banach space such that $\mathcal{P}({}^{n-1}X;Y) = \mathcal{P}_{\mathrm{wsc}}({}^{n-1}X;Y)$. If $P \in \mathcal{P}({}^nX;Y) \setminus \mathcal{P}_{\mathrm{wsc}}({}^nX;Y)$, then there is a weakly null sequence $\{y_k\}_{k=1}^\infty \subset S_X$ such that* $\inf_k \|P(y_k)\| > 0$.

Proof. Suppose that $\{x_k\}_{k=1}^\infty \subset X$ is weakly convergent to $x \in X$ and satisfies $\|P(x_k) - P(x)\| \geq \varepsilon > 0$ for all $k \in \mathbb{N}$. Put $y_k = x_k - x$. By Lemma 1.21

$$P(x_k) - P(x) = P(y_k) + \sum_{j=1}^{n-1} \binom{n}{j} \check{P}({}^jx, {}^{n-j}y_k)$$

for each $k \in \mathbb{N}$. The assumption and Proposition 29 imply $\lim_{k\to\infty} \check{P}({}^jx, {}^{n-j}y_k) = 0$ for each $j \in \{1, \dots, n-1\}$. It follows that there is $k_0 \in \mathbb{N}$ such that $\|P(y_k)\| \geq \frac{\varepsilon}{2}$ for $k \geq k_0$. Finally, since $\{y_k\}_{k=k_0}^\infty$ is bounded and non-zero, we can pass to the normalised sequence. □

Lemma 31 ([D'H]). *Let X, Y be Banach spaces such that X does not contain ℓ_1 and $\mathcal{P}({}^{n-1}X;Y) = \mathcal{P}_{\mathrm{K}}({}^{n-1}X;Y)$. If $P \in \mathcal{P}({}^nX;Y) \setminus \mathcal{P}_{\mathrm{K}}({}^nX;Y)$, then there is a weakly null sequence $\{y_k\}_{k=1}^\infty \subset S_X$ such that $\{P(y_k)\}_{k=1}^\infty$ is not relatively compact.*

Proof. By Rosenthal's ℓ_1-theorem (Theorem 71) there are an $\varepsilon > 0$ and a weakly Cauchy sequence $\{x_k\}_{k=1}^\infty \subset B_X$ such that $\|P(x_k) - P(x_l)\| \geq \varepsilon$ for all $k, l \in \mathbb{N}$, $k \neq l$. By Lemma 1.21

$$P(x_k - x_l) = P(x_k) + \sum_{j=1}^{n-1} \binom{n}{j} (-1)^j \check{P}({}^jx_l, {}^{n-j}x_k) + (-1)^n P(x_l).$$

The polynomials $x \mapsto \sum_{j=1}^{n-1} \binom{n}{j} (-1)^j \check{P}({}^jx, {}^{n-j}x_k)$ are compact by the assumption and Proposition 29, and so by passing to subsequences and then diagonalising there is an infinite $M \subset \mathbb{N}$ such that the limits

$$y_k = \lim_{\substack{l\to\infty \\ l\in M}} \sum_{j=1}^{n-1} \binom{n}{j} (-1)^j \check{P}({}^jx_l, {}^{n-j}x_k)$$

exist for each $k \in \mathbb{N}$. Thus for each $k \in \mathbb{N}$ we can choose $m_k \geq k$ such that $\left\| y_k - \sum_{j=1}^{n-1} \binom{n}{j} (-1)^j \check{P}({}^jx_l, {}^{n-j}x_k) \right\| < \frac{\varepsilon}{8}$ whenever $l \in M$, $l \geq m_k$. Then

$$\begin{aligned} &\|P(x_k - x_l) - P(x_p - x_r)\| \\ &\qquad > \left\| P(x_k) + y_k + (-1)^n P(x_l) - P(x_p) - y_p - (-1)^n P(x_r) \right\| - \frac{\varepsilon}{4} \end{aligned}$$

whenever $k, p \in \mathbb{N}$, $l, r \in M$, $l \geq m_k$, and $r \geq m_p$. Suppose that $k, l, p \in \mathbb{N}$ are given and let $z = P(x_l) + (-1)^n(P(x_k) + y_k - P(x_p) - y_p)$. Then there is $r_{k,l,p} \in \mathbb{N}$ such that $\|P(x_r) - z\| \geq \frac{\varepsilon}{2}$ for all $r \geq r_{k,l,p}$. Whence $\|P(x_k - x_l) - P(x_p - x_r)\| > \frac{\varepsilon}{4}$ whenever $k, p \in \mathbb{N}$, $l, r \in M$, $l \geq m_k$, and $r \geq \max\{m_p, r_{k,l,p}\}$. Now it suffices to find $l_k \in M$ such that $l_k \geq \max\{m_k, r_{1,l_1,k}, \dots, r_{k-1,l_{k-1},k}\}$ and put $z_k = x_k - x_{l_k}$. Then $\{z_k\}$ is weakly null and $\{P(z_k)\}$ is an $\frac{\varepsilon}{4}$-separated sequence.

Finally, by passing to a subsequence we may assume that $\|z_k\| \to c > 0$. We set $y_k = \frac{z_k}{\|z_k\|}$. Then $\{y_k\}$ is weakly null and

$$\|P(y_k) - P(y_l)\| = \left\| \frac{1}{\|z_k\|^n} P(z_k) - \frac{1}{\|z_l\|^n} P(z_l) \right\|$$

$$\geq \frac{1}{c^n}\|P(z_k) - P(z_l)\| - \|P(z_k)\| \left| \frac{1}{c^n} - \frac{1}{\|z_k\|^n} \right| - \|P(z_l)\| \left| \frac{1}{c^n} - \frac{1}{\|z_l\|^n} \right| > \frac{\varepsilon}{8c^n}$$

for all k, l large enough, $k \neq l$. □

We recall a fundamental result from the theory of Schauder bases. For the proof see [LiTz2, Proposition 1.a.11] and the remark after.

Proposition 32 (Bessaga-Pełczyński selection principle). *Suppose that X is a Banach space with a Schauder basis $\{(e_n; f_n)\}$. If $\{x_n\} \subset X$ is a sequence that satisfies $\lim_{n\to\infty} f_k(x_n) = 0$ for each $k \in \mathbb{N}$ (in particular if $\{x_n\}$ is weakly null) and $\limsup \|x_n\| > 0$, and if $\{\varepsilon_k\} \subset \mathbb{R}^+$, then there are a block basic sequence $\{u_k\} \subset X$ and a subsequence $\{x_{n_k}\}$ such that $\|x_{n_k} - u_k\| \leq \varepsilon_k$ for each $k \in \mathbb{N}$. In particular, for a suitably chosen $\{\varepsilon_k\}$ the sequence $\{x_{n_k}\}$ is a basic sequence equivalent to $\{u_k\}$.*

The following folklore result will be useful several times.

Proposition 33. *Let X be a Banach space, $Y = \ell_p$, $1 \leq p < \infty$, or $Y = c_0$, and suppose there is a non-compact operator $T \in \mathcal{L}(X; Y)$. Then there are $S \in \mathcal{L}(X; Y)$ and a normalised basic sequence $\{x_n\} \subset X$ such that $S(x_n) = e_n$, $n \in \mathbb{N}$, where $\{e_n\}$ is the canonical basis of Y. If X does not contain ℓ_1, then $\{x_n\}$ may be chosen to be weakly null. If $Y = \ell_1$, then S is in fact onto.*

Proof. Denote by $\{f_n\}$ the functionals biorthogonal to $\{e_n\}$. Let $\{y_n\}_{n=1}^{\infty} \subset B_X$ be such that $\{T(y_n)\}$ is ε-separated for some $\varepsilon > 0$. By passing to subsequences and then diagonalising we may assume that for each $k \in \mathbb{N}$ the sequence $\{f_k(T(y_n))\}_{n=1}^{\infty}$ is convergent. Further, we may assume that $\overline{\operatorname{span}}\{y_n\}$ is a subspace of a Banach space with a Schauder basis (e.g. $C([0,1])$) with coordinate functionals $\{\phi_n\}_{n=1}^{\infty}$. Hence similarly by passing to further subsequence we may assume that for each $k \in \mathbb{N}$ the sequence $\{\phi_k(y_n)\}_{n=1}^{\infty}$ is convergent. Put $z_n = y_{n+1} - y_n$, $n \in \mathbb{N}$. Then $\varepsilon \leq \|T(z_n)\| \leq \|T\|\|z_n\|$ and so $\frac{\varepsilon}{\|T\|} \leq \|z_n\| \leq 2$. Thus we may set $x_n = \frac{z_n}{\|z_n\|}$, $n \in \mathbb{N}$.

We have $\lim_{n\to\infty} f_k(T(x_n)) = 0$ and $\lim_{n\to\infty} \phi_k(x_n) = 0$ for each $k \in \mathbb{N}$. Thus from Proposition 32 it follows that by passing to a subsequence we may assume that $\{x_n\}$ is a basic sequence. Further, from Proposition 32 combined with [FHHMZ, Theorem 4.23] and [FHHMZ, Proposition 4.45] it follows that by passing to another subsequence we may assume that $\{T(x_n)\}$ is a basic sequence equivalent to a block basis of $\{e_n\}$ (and consequently equivalent to $\{e_n\}$) and moreover $Z = \overline{\operatorname{span}}\{T(x_n)\}$ is complemented in Y. Let $P: Y \to Z$ be a projection and $R: Z \to Y$ be an isomorphism satisfying $R(T(x_n)) = e_n$. Then $S = R \circ P \circ T: X \to Y$ satisfies $S(x_n) = e_n$.

If X does not contain ℓ_1, then by Rosenthal's ℓ_1-theorem (Theorem 71) $\{y_n\}$ can be chosen to be weakly Cauchy and consequently $\{x_n\}$ is weakly null. If $Y = \ell_1$, then $\sum_{n=1}^{\infty} a_n x_n$ clearly converges absolutely in X whenever $(a_n) \in \ell_1$ and hence $S\bigl(\sum_{n=1}^{\infty} a_n x_n\bigr) = \sum_{n=1}^{\infty} a_n e_n$. □

We will also use a non-linear version of the previous proposition (typically it is used by putting $g = T \circ f$ for some non-linear non-compact mapping f).

Proposition 34. *Let $1 < p < \infty$ and let $\{y_n\} \subset \ell_p$ be a bounded sequence that is not relatively compact. Then there are an affine transformation $T \in \mathcal{P}^1(\ell_p; \ell_p)$ and a subsequence $\{y_{n_k}\}$ such that $T(y_{n_k}) = e_k$, $k \in \mathbb{N}$, where $\{e_k\}$ is the canonical basis of ℓ_p.*

Proof. We may assume that $\{y_n\}$ is ε-separated for some $\varepsilon > 0$. Using the weak compactness and separability we may assume that $y_n \overset{w}{\to} y \in \ell_p$. Thus from Proposition 32 combined with [FHHMZ, Theorem 4.23] and [FHHMZ, Proposition 4.45] it follows that there is a subsequence $\{y_{n_k}\}$ such that $\{y_{n_k} - y\}$ is a basic sequence equivalent to a block basis of $\{e_k\}$ (and consequently equivalent to the basis $\{e_k\}$) and moreover $Z = \overline{\operatorname{span}}\{y_{n_k} - y\}$ is complemented in ℓ_p. Let $P: \ell_p \to Z$ be a projection and $R: Z \to \ell_p$ be an isomorphism satisfying $R(y_{n_k} - y) = e_k$. Then $T(x) = R \circ P(x - y)$ is the desired affine transformation. □

For ℓ_1 we have a stronger statement: the transformation T can be even linear. For the proof we need the following folklore result (of course the interesting case is $p = 1$).

Proposition 35. *Let $1 \le p < \infty$ and let $\{x_n\} \subset \ell_p$ be a semi-normalised basic sequence. Then there is a subsequence $\{x_{n_k}\}$ equivalent to the canonical basis of ℓ_p such that $\overline{\operatorname{span}}\{x_{n_k}\}$ is complemented in ℓ_p.*

Proof. By passing to a subsequence we may assume that there is $y \in \ell_p$ such that $x_n \overset{w^*}{\to} y$. Put $y_n = x_n - y$. Then $y_n \overset{w^*}{\to} 0$ and by Proposition 32 combined with [FHHMZ, Theorem 4.23] and [FHHMZ, Proposition 4.45] it follows that by passing to a subsequence we may assume that $\{y_n\}$ is a basic sequence equivalent to the canonical basis of ℓ_p and $Y = \overline{\operatorname{span}}\{y_n\}$ is complemented in ℓ_p by the projection $Q: \ell_p \to Y$. If $y = 0$ (as is the case when $p > 1$, since by the reflexivity $x_n \overset{w}{\to} y$), then the proof is finished. So assume that $p = 1$ and $y \neq 0$.

Since $\{y_n\}$ is a basic sequence, by removing one of the vectors from $\{y_n\}$ if necessary we may assume that $y \notin Y$. Similarly, since $\{x_n\}$ is also a basic sequence, we may assume that $y \notin Z = \overline{\operatorname{span}}\{x_n\}$. Notice that $Z \subset Y \oplus \operatorname{span}\{y\}$. Let $f, g \in \ell_1^*$ be such that $f(y) = g(y) = 1$, $Y \subset \ker f$, and $Z \subset \ker g$. Then $R: \ell_1 \to Y \oplus \operatorname{span}\{y\}$, $R(x) = f(x)\bigl(y - Q(y)\bigr) + Q(x)$ and $S: Y \oplus \operatorname{span}\{y\} \to Z$, $S(x) = x - g(x)y$ are projections, and hence we may define a projection $P: \ell_1 \to Z$ by $P = S \circ R$.

Finally, assuming $\sum_{n=1}^{\infty} a_n x_n$ converges we get $f\left(\sum_{n=1}^{\infty} a_n x_n\right) = \sum_{n=1}^{\infty} a_n \in \mathbb{K}$. Consequently $\sum_{n=1}^{\infty} a_n y_n = \sum_{n=1}^{\infty} a_n x_n - y \sum_{n=1}^{\infty} a_n$ converges and hence $\{x_n\}$ is equivalent to the canonical basis of ℓ_1. □

Proposition 36. *Let $\{y_n\} \subset \ell_1$ be a bounded sequence that is not relatively compact. Then there are $T \in \mathcal{L}(\ell_1; \ell_1)$ and a subsequence $\{y_{n_k}\}$ such that $T(y_{n_k}) = e_k$, $k \in \mathbb{N}$, where $\{e_k\}$ is the canonical basis of ℓ_1.*

Proof. By passing to a subsequence we may assume that $\{y_n\}$ is δ-separated for some $\delta > 0$. By Schur's theorem ([FHHMZ, Theorem 5.36]) $\{y_n\}$ does not contain weakly Cauchy subsequences and so by Rosenthal's ℓ_1-theorem we may assume that $\{y_n\}$ is a basic sequence equivalent to $\{e_n\}$. By Proposition 35 (passing to a further subsequence) there is a projection $P : \ell_1 \to \overline{\operatorname{span}}\{y_n\}$. Let $S : \overline{\operatorname{span}}\{y_n\} \to \ell_1$ be the isomorphism satisfying $S(y_n) = e_n$. It suffices to put $T = S \circ P$. □

Recall a result of Czesław Bessaga and Aleksander Pełczyński, [FHHMZ, Theorem 4.44]: Let X be a Banach space such that X^* contains c_0. Then X contains a complemented subspace isomorphic to ℓ_1 (and hence X^* actually contains a complemented subspace isomorphic to ℓ_∞). Applying this result to the duality relation in Corollary 14 we get the next result.

Theorem 37 ([D'H]). *Suppose that X is a normed linear space and $n \in \mathbb{N}$. Then $\mathcal{P}(^n X; \ell_1) = \mathcal{P}_{\mathrm{K}}(^n X; \ell_1)$ if and only if the space $\mathcal{P}(^n X)$ does not contain c_0.*

Proof. $\Rightarrow$ Assume that $\mathcal{P}(^n X)$ contains c_0. By Corollary 14 the space $\mathcal{P}(^n X)$ is isomorphic to $(\otimes^n_{\pi,s} X)^*$, and so $\otimes^n_{\pi,s} X$ contains a complemented subspace isomorphic to ℓ_1 by the Bessaga-Pełczyński theorem. Hence there is a $T \in \mathcal{L}(\otimes^n_{\pi,s} X; \ell_1)$ which is non-compact. Consequently $\widehat{T \circ \otimes_s} \notin \mathcal{P}_{\mathrm{K}}(^n X; \ell_1)$ by Fact 25.

$\Leftarrow$ If $\mathcal{P}(^n X; \ell_1) \neq \mathcal{P}_{\mathrm{K}}(^n X; \ell_1)$, then by Fact 25 there is a non-compact operator $T \in \mathcal{L}(\otimes^n_{\pi,s} X; \ell_1)$. Thus ℓ_1 is a quotient of $\otimes^n_{\pi,s} X$ by Proposition 33. By the duality and Corollary 14 the space $\mathcal{P}(^n X)$ then contains ℓ_∞, which is a contradiction. □

We recall some classical concepts and results concerning the weak sequential continuity properties of linear operators.

Definition 38. We say that a normed linear space X has the Schur property (or that X is a Schur space) if every weakly convergent sequence in X is norm convergent.

The space ℓ_1 is a Schur space [FHHMZ, Theorem 5.36]. It is clear that the Schur property passes to subspaces. Further, in a Schur space X every weakly Cauchy sequence is norm Cauchy. Indeed, assuming the contrary, there are a weakly Cauchy sequence $\{x_n\} \subset X$, $\varepsilon > 0$, and increasing sequences $\{n_k\} \subset \mathbb{N}$, $\{m_k\} \subset \mathbb{N}$ such that $\|x_{n_k} - x_{m_k}\| \geq \varepsilon$ for all $k \in \mathbb{N}$. Then $\{x_{n_k} - x_{m_k}\}_{k=1}^{\infty}$ is weakly null and hence also norm null, a contradiction. Consequently, $\mathcal{P}(X; Y) = \mathcal{P}_{\mathrm{wsC}}(X; Y)$ for every Schur space X and every Banach space Y, and more generally if U is a convex subset of a Banach space with the Schur property and Y is a Banach space, then $\mathcal{C}(U; Y) = \mathcal{C}_{\mathrm{wsC}}(U; Y)$. We also remark that every infinite-dimensional Banach space X with the Schur property is ℓ_1-saturated: Assume that X does not contain ℓ_1.

Then by Rosenthal's ℓ_1-theorem any sequence in B_X has a subsequence that is weakly Cauchy and hence also norm Cauchy. This means that X is finite-dimensional.

Definition 39. We say that a normed linear space X has the Dunford-Pettis property (DPP for short) if for every Banach space Y every $T \in \mathcal{L}_{\mathrm{wK}}(X;Y)$ maps weakly compact sets to compact sets.

Every infinite-dimensional reflexive Banach space clearly fails the DPP. It follows from Corollary 23 that a normed linear space X has the DPP if and only if $\mathcal{L}_{\mathrm{wK}}(X;Y) \subset \mathcal{L}_{\mathrm{wsc}}(X;Y)$ for every Banach space Y (which is by Corollary 27 equivalent to $\mathcal{L}_{\mathrm{wK}}(X;Y) \subset \mathcal{L}_{\mathrm{wsC}}(X;Y)$). Thus every Schur space has the DPP.

There are many characterisations of the DPP. We are going to use the following one for Banach spaces.

Proposition 40 (Alexander Grothendieck). *Let X be a Banach space. The following statements are equivalent:*

(i) *X has the Dunford-Pettis property.*
(ii) $\lim_{n\to\infty} f_n(x_n) = 0$ *whenever* $\{x_n\}_{n=1}^{\infty} \subset X$, $\{f_n\}_{n=1}^{\infty} \subset X^*$ *are both weakly null.*
(iii) $\lim_{n\to\infty} f_n(x_n) = 0$ *whenever* $\{x_n\}_{n=1}^{\infty} \subset X$ *is weakly Cauchy and* $\{f_n\}_{n=1}^{\infty} \subset X^*$ *is weakly null.*
(iv) $\lim_{n\to\infty} f_n(x_n) = 0$ *whenever* $\{x_n\}_{n=1}^{\infty} \subset X$ *is weakly null and* $\{f_n\}_{n=1}^{\infty} \subset X^*$ *is weakly Cauchy.*

Proof. For (i) $\Leftrightarrow$ (ii) see [FHHMZ, Proposition 13.42].

(ii)$\Rightarrow$(iii) Assume the contrary. Then there are $\varepsilon > 0$, a weakly Cauchy $\{x_n\} \subset X$, and a weakly null $\{f_n\} \subset X^*$ such that $f_n(x_n) \geq \varepsilon$ for every $n \in \mathbb{N}$. Since $\{f_n\}$ is weakly null, there is a subsequence $\{f_{n_k}\}$ that satisfies $f_{n_k}(x_k) < \frac{\varepsilon}{2}$ for each $k \in \mathbb{N}$. We put $y_k = x_{n_k} - x_k$. Then $\{y_k\}$ is weakly null and so $f_{n_k}(y_k) \to 0$. But $f_{n_k}(y_k) = f_{n_k}(x_{n_k}) - f_{n_k}(x_k) > \frac{\varepsilon}{2}$ for each $k \in \mathbb{N}$, a contradiction.

The rest of the implications is obvious or similar. □

From this it is clear that a Banach space X has the DPP if X^* has the DPP, in particular c_0 has the DPP. The converse is false in general. The DPP also does not in general pass to subspaces or quotients. Polynomials on spaces with the DPP will be investigated later. The following theorem collects some classical results on the DPP, see [AK], [Dies1], [BourgJ2, Corollary 1.30].

Theorem 41.

(i) *$\mathcal{L}_1$-spaces and $\mathcal{L}_\infty$-spaces have the DPP. In particular $L_1(\mu)$ and $C(K)$, K compact, have the DPP.*
(ii) *Subspaces of $c_0(\Gamma)$ have the DPP.*
(iii) *Schur spaces have the DPP.*
(iv) *Let X be a Banach space. X^* is a Schur space if and only if X has the DPP and does not contain ℓ_1.*
(v) *If a Banach space X has the DPP and $Y \subset X$ is a subspace that does not contain ℓ_1, then X/Y has the DPP.*

Recall that a series $\sum_{n=1}^{\infty} x_n$ in a Banach space is called unconditionally convergent provided that the series $\sum_{n=1}^{\infty} x_{\sigma(n)}$ converges regardless of the permutation σ of the set $\mathbb{N}$. It is not hard to show that a series $\sum_{n=1}^{\infty} x_n$ is unconditionally convergent if and only if $\sum_{n=1}^{\infty} a_n x_n$ converges regardless of the scalar values $|a_n| \le 1$ (see e.g. [FHHMZ, Exercises 1.37–39]). In connection with this let us also recall the following simple fact.

Fact 42 ([FHHMZ, Exercise 1.37]). *Suppose that X is a normed linear space and $x_1, \dots, x_n \in X$. Then*

$$\max_{\varepsilon_j = \pm 1} \left\| \sum_{j=1}^{n} \varepsilon_j x_j \right\| \le \sup_{|a_j| \le 1} \left\| \sum_{j=1}^{n} a_j x_j \right\| = \sup_{\phi \in B_{X^*}} \sum_{j=1}^{n} |\phi(x_j)|.$$

In the case of real scalars all quantities are equal.

The terminology for the following notion is traditional, albeit somewhat misleading.

Definition 43. A sequence $\{x_n\}_{n=1}^{\infty}$ in a Banach space is called weakly unconditionally Cauchy (meaning weakly unconditionally Cauchy *summable*; wuC for short) provided that the series $\sum_{n=1}^{\infty} x_{\sigma(n)}$ is weakly Cauchy regardless of the permutation σ of the set $\mathbb{N}$.

In view of the remarks preceding the definition it is easy to see that a sequence $\{x_n\}$ in a Banach space X is weakly unconditionally Cauchy if and only if the series $\sum_{n=1}^{\infty} \phi(x_n)$ converges absolutely for every $\phi \in X^*$. It follows that $\{x_n\}$ is weakly unconditionally Cauchy if and only if $\sum_{n=1}^{\infty} a_n x_n$ is weakly Cauchy regardless of the scalar values $|a_n| \le 1$. Thus a subsequence of a weakly unconditionally Cauchy sequence is again a weakly unconditionally Cauchy sequence. Also note that a weakly unconditionally Cauchy sequence is weakly null and thus bounded. Furthermore, the convergence of the series $\sum_{n=1}^{\infty} \phi(x_n)$ is uniform in the following sense:

Fact 44. *A sequence $\{x_n\}_{n=1}^{\infty}$ in a Banach space X is weakly unconditionally Cauchy if and only if there is $K > 0$ such that for every $\phi \in X^*$*

$$\sum_{n=1}^{\infty} |\phi(x_n)| \le K \|\phi\|.$$

Proof. Define an operator $T : X^* \to \ell_1$ by $T(\phi) = \sum_{n=1}^{\infty} \phi(x_n) e_n$. Then T is clearly linear an has a closed graph, so it is bounded, from which the inequality follows. □

Bounded linear operators obviously map wuC sequences to wuC sequences. The canonical basis $\{e_j\}$ of the space c_0 is intimately connected with wuC sequences. Clearly, $\{e_j\}$ is wuC, and $\{x_j\}$ is wuC in a Banach space X if and only if there is a bounded linear operator $T : c_0 \to X$, $T(e_j) = x_j$ (Facts 42 and 44). The fundamental role of the canonical basis of c_0 for wuC sequences is apparent from the following two results.

Theorem 45 (Władysław Orlicz and Billy James Pettis, [DJT, Theorem 1.8]). *Suppose that $\sum_{n=1}^{\infty} x_n$ is a series in a Banach space. Then the following statements are equivalent:*

(i) *$\sum_{n=1}^{\infty} x_n$ converges unconditionally.*
(ii) *$\sum_{k=1}^{\infty} x_{n_k}$ converges for every increasing sequence $\{n_k\} \subset \mathbb{N}$.*
(iii) *$\sum_{k=1}^{\infty} x_{n_k}$ converges weakly for every increasing sequence $\{n_k\} \subset \mathbb{N}$.*

Note however, that the conditions above are not equivalent to the statement that the series $\sum_{n=1}^{\infty} x_n$ converges weakly unconditionally, as the example $x_1 = e_1$, $x_n = e_n - e_{n-1}$ in c_0 shows. In view of the next theorem, this is essentially the only counterexample.

Theorem 46 (Czesław Bessaga and Aleksander Pełczyński, [FHHMZ, Corollary 4.52]). *Let X be a Banach space and let $\{x_n\}_{n=1}^{\infty} \subset X$ be a weakly unconditionally Cauchy sequence. Then either $\sum_{n=1}^{\infty} x_n$ is unconditionally convergent, or $\{x_n\}_{n=1}^{\infty}$ has a block sequence equivalent to the canonical basis of c_0. In particular, the following statements are equivalent:*

(i) *X does not contain c_0.*
(ii) *If $\{x_n\} \subset X$ is a sequence such that the set $\left\{\sum_{j=1}^{n} \varepsilon_j x_j;\ \varepsilon_j = \pm 1, n \in \mathbb{N}\right\}$ is bounded, then $\sum_{n=1}^{\infty} x_n$ is unconditionally convergent.*
(iii) *If $\{x_n\} \subset X$ is a sequence such that the set $S = \left\{\sum_{j=1}^{n} \varepsilon_j x_j;\ \varepsilon_j = \pm 1, n \in \mathbb{N}\right\}$ is bounded, then S is relatively compact.*

These concepts will be studied in more detail in Section 4, now we restrict ourselves to stating a classical result which motivates much of the theory of smooth mappings and which will be generalised in the course of the present chapter and the book. For the proof see [DU, Theorem VI.2.15, Corollary VI.2.17] and [HMVZ, Theorem 7.6]. Note that the implication (i)⇒(ii) characterises the DPP.

Theorem 47. *Let Y be a Banach space, K a compact space, and $T \in \mathcal{L}(C(K); Y)$. The following statements are equivalent:*

(i) *$T \in \mathcal{L}_{\mathrm{wK}}(C(K); Y)$.*
(ii) *$T \in \mathcal{L}_{\mathrm{wsC}}(C(K); Y)$.*
(iii) *If $\{x_n\} \subset C(K)$ is weakly unconditionally Cauchy, then $\sum_{n=1}^{\infty} T(x_n)$ converges unconditionally.*
(iv) *If a subspace $Z \subset C(K)$ is isomorphic to c_0, then $T\restriction_Z$ is not an isomorphism.*

Moreover, if K is a scattered compact space, then the above statements are equivalent to $T \in \mathcal{L}_{\mathrm{K}}(C(K); Y)$.

We continue by introducing the notions of upper and lower estimate for sequences, which lead to rather precise results on the weak sequential continuity of polynomials. These notions will also play an important role in the asymptotic approach to polynomials in Chapter 4.

Definition 48. Let $1 \le p, q \le \infty$. We say that a sequence $\{x_j\}_{j=1}^{\infty}$ in a Banach space over $\mathbb{K}$ has an upper p-estimate (resp. lower q-estimate) if there exists $C > 0$ such that for every $n \in \mathbb{N}$ and every $a_1, \dots, a_n \in \mathbb{K}$

$$\left\| \sum_{j=1}^{n} a_j x_j \right\| \le C \left(\sum_{j=1}^{n} |a_j|^p \right)^{\frac{1}{p}}, \tag{4}$$

respectively

$$\left\| \sum_{j=1}^{n} a_j x_j \right\| \ge C \left(\sum_{j=1}^{n} |a_j|^q \right)^{\frac{1}{q}},$$

where the right-hand side is replaced by $\max\limits_{j=1,\dots,n} |a_j|$ if $p = \infty$, resp. $q = \infty$.

Clearly a sequence has an upper 1-estimate if and only if it is bounded and an upper p-estimate implies an upper r-estimate for every $r < p$. Similarly, a lower q-estimate implies a lower r-estimate for every $r > q$. A sequence $\{x_j\}$ with a lower ∞-estimate satisfies $\inf_j \|x_j\| > 0$ and if $\{x_j\}$ is a basic sequence with $\inf_j \|x_j\| > 0$, then it has a lower ∞-estimate. Obviously a sequence has both upper and lower p-estimates if and only if it is a basic sequence equivalent to the canonical basis of ℓ_p. Thus if a sequence has both an upper p-estimate and a lower q-estimate, then necessarily $q \ge p$.

Suppose that the sequence $\{x_j\}$ has an upper p-estimate with the constant C. It is easy to see that if $(a_j) \in \ell_p$ (or c_0 if $p = \infty$), then the series $\sum_{j=1}^{\infty} a_j x_j$ converges and $\left\| \sum_{j=1}^{\infty} a_j x_j \right\| \le C \|(a_j)\|_{\ell_p}$. The following fact is clear.

Fact 49. *Let X be a Banach space and $1 \le p, q < \infty$. A sequence $\{x_j\}_{j=1}^{\infty} \subset X$ has an upper p-estimate if and only if the linear operator $T\colon \ell_p \to X$, $T(e_j) = x_j$, is bounded. A sequence $\{x_j\}_{j=1}^{\infty} \subset X$ has a lower q-estimate if and only if the linear operator $T\colon \overline{\operatorname{span}}\{x_j\} \to \ell_q$, $T(x_j) = e_j$, is bounded. In case $p = \infty$ we replace ℓ_p by c_0 and analogously for $q = \infty$.*

The next fact easily follows from the above using duality.

Fact 50. *Let X be a Banach space and $\{(x_j; x_j^*)\}_{j=1}^{\infty} \subset X \times X^*$ be a biorthogonal sequence. Let $1 \le p, q \le \infty$, $\frac{1}{p} + \frac{1}{q} = 1$. If $\{x_j\}$ has an upper p-estimate, then $\{x_j^*\}$ has a lower q-estimate. If $\{x_j\}$ has a lower q-estimate, then $\{x_j^*\}$ has an upper p-estimate in $\left(\overline{\operatorname{span}}\{x_j\}\right)^*$, i.e. in a quotient of X^*.*

We will use the following observation.

Fact 51. *A sequence $\{x_j\}$ in Banach space X has an upper ∞-estimate if and only if it is weakly unconditionally Cauchy.*

Proof. If $\{x_j\}$ is wuC, then it has an upper ∞-estimate by Fact 44. On the other hand, given $\phi \in X^*$ choose $\alpha_j \in \mathbb{K}$, $|\alpha_j| = 1$ such that $\phi(\alpha_j x_j) = |\phi(x_j)|$. Then $\sum_{j=1}^{n} |\phi(x_j)| = \phi\left(\sum_{j=1}^{n} \alpha_j x_j\right) \le \|\phi\| \left\| \sum_{j=1}^{n} \alpha_j x_j \right\| \le C \|\phi\|$ for any $n \in \mathbb{N}$. □

Theorem 52 ([Pełl], [ALRT], [GJ1]). *Let X, Y be Banach spaces and $P \in \mathcal{P}(^nX;Y)$. If $n < p < \infty$, then P maps sequences with an upper p-estimate to sequences with an upper $\frac{p}{n}$-estimate.*

Proof. Let $\tilde{P} \in \mathcal{P}(^n\tilde{X};\tilde{Y})$ be the complex extension of P in the case of real scalars. Suppose that $\{x_j\}_{j=1}^{\infty} \subset X$ has the upper p-estimate with the constant $C > 0$ and let $a_1, \dots, a_k \in \mathbb{K}$. Combining Lemma 2.69 and Theorem 1.63 we obtain

$$\left\|\sum_{j=1}^{k} a_j P(x_j)\right\| = \left\|\sum_{j=1}^{k} \tilde{P}\left(a_j^{\frac{1}{n}} x_j\right)\right\| = \frac{1}{n^k}\left\|\sum_{l=1}^{n^k} \tilde{P}\left(\sum_{j=1}^{k} r_j^{n,k}(l) a_j^{\frac{1}{n}} x_j\right)\right\|$$

$$\le \frac{1}{n^k}\sum_{l=1}^{n^k}\|\tilde{P}\|\left\|\sum_{j=1}^{k} r_j^{n,k}(l) a_j^{\frac{1}{n}} x_j\right\|^n \le C^n 2^{n-1}\|P\|\left(\sum_{j=1}^{k}|a_j|^{\frac{p}{n}}\right)^{\frac{n}{p}}.$$

□

As we have seen above, wuC sequences coincide with sequences having an upper ∞-estimate, so the next theorem is a natural extension of the previous theorem and it is proved by an identical argument.

Theorem 53 ([GGu3], [GJ1]). *Let X and Y be Banach spaces. Then any polynomial $P \in \mathcal{P}(X;Y)$ satisfying $P(0) = 0$ maps wuC sequences to wuC sequences.*

Definition 54 ([KO2], [GJ1]). Let $1 \le p \le \infty$. We say that a Banach space has property S_p (resp. T_q) if every normalised weakly null sequence has a subsequence with an upper p-estimate (resp. lower q-estimate).

Obviously every Banach space has property S_1 and T_∞. If a Banach space has property S_p, then it has also property S_r for every $r < p$, and if it has property T_q, then it has also property T_r for every $r > q$. Both properties pass to subspaces. The space $\ell_p(\Gamma)$ has both property S_p and T_p, the space $c_0(\Gamma)$ has both property S_∞ and T_∞, Proposition 32 and [FHHMZ, Proposition 4.45]. The property S_∞ is equivalent to saying that every normalised weakly null sequence contains a subsequence equivalent to the canonical basis of c_0. If a Banach space has both property S_p and T_q, then necessarily $q \ge p$.

A normed linear space X is said to have the hereditary Dunford-Pettis property (hereditary DPP) if every closed subspace of X has the Dunford-Pettis property. We mention that Schur spaces, the subspaces of $c_0(\Gamma)$ (Theorem 41), as well as of the Hagler tree space [Hag2], have the hereditary DPP.

It was noted in [Cem] that property S_∞ for a Banach space coincides with the hereditary DPP. The difficult implication depends on some deep work of John Hancock Elton (see [O] and for more recent exposition [LóTo]) and it was observed in [Dies1]. We refer to [KO2] (see also [KO1] for the case $p = \infty$) for the proof of the next closely related theorem.

Theorem 55 ([KO2]). *Let $1 \le p \le \infty$. If a Banach space X has property S_p, then the constant C in the estimate (4) for the subsequence may be chosen independently of the normalised weakly null sequence $\{x_j\}_{j=1}^{\infty}$.*

We remark that an analogous theorem for property T_q fails to be true. It was observed in [Gon1] how to use spaces X_λ constructed in [JO] to obtain a counterexample.

Theorem 56. *Let X be a Banach space and let $\{f_n\} \subset X^*$ be a bounded sequence. The following statements are equivalent:*

(i) *$\{f_n\}$ is not a relatively compact set.*
(ii) *There are a subsequence $\{g_n\}$ of $\{f_n\}$ and an (infinite-dimensional) subspace $Y \subset X$ such that $\{g_n \restriction_Y\} \subset Y^*$ is a semi-normalised w^*-null sequence.*
(iii) *There is a semi-normalised basic sequence $\{x_n\} \subset X$ which is biorthogonal to a subsequence of $\{f_n\}$.*

Moreover, we may assume in addition that $\{x_n\}$ is either weakly null or equivalent to the canonical basis of ℓ_1.

Proof. (iii)⇒(i) Let $\{g_n\}$ be the subsequence of $\{f_n\}$ biorthogonal to $\{x_n\}$. Then $\|g_m - g_n\| \geq (g_m - g_n)\big(\frac{x_m}{\|x_m\|}\big) = \frac{1}{\|x_m\|}$ for $m, n \in \mathbb{N}$, $m \neq n$.

(i)⇒(ii) Since $\{f_n\}$ is not totally bounded, there is an $\varepsilon > 0$ such that $\{f_n\}$ is not contained in a union of finitely many balls of radius 2ε. Hence by passing to a subsequence we may assume that $\operatorname{dist}(\operatorname{span}\{f_1, \dots, f_n\}, f_{n+1}) \geq \varepsilon$. Using [FHHMZ, Proposition 2.7] together with Helly's theorem we can find a semi-normalised sequence $\{y_n\}_{n=1}^\infty \subset X$ such that $f_k(y_n) = 0$ for $1 \leq k < n$ and $f_n(y_n) = 1$, $n \in \mathbb{N}$.

By passing to further subsequences and then to a diagonal sequence we may assume that $\lim_{k\to\infty} f_k(y_n) = t_n \in \mathbb{K}$ for all $n \in \mathbb{N}$. If $\{t_n\}$ contains infinitely many zeros, then we are done, so from now on we assume that $t_n \neq 0$ for all $n \in \mathbb{N}$. Let $A, B > 0$ be such that $\|y_n\| \leq A$ and $\|f_n\| \leq B$ for all $n \in \mathbb{N}$. We distinguish two cases. If there is $\delta > 0$ such that $|t_n| \geq \delta$ for all $n \in \mathbb{N}$, then we set $z_n = y_n - \frac{t_n}{t_{n+1}} y_{n+1}$ and $g_n = f_n$. Then $\|z_n\| \leq A + \frac{AB}{\delta} A$ and $g_n(z_n) = 1$. In the opposite case by passing to a subsequence we may assume that $\left|\frac{t_{n+1}}{t_n}\right| \leq \frac{1}{2AB}$. We set $z_n = y_{n+1} - \frac{t_{n+1}}{t_n} y_n$ and $g_n = f_{n+1}$. Then $\|z_n\| \leq A + \frac{1}{2B}$ and $|g_n(z_n)| \geq \frac{1}{2}$.

Finally, we put $Y = \overline{\operatorname{span}}\{z_n\}$. Clearly, $\lim_{k\to\infty} g_k(z_n) = 0$ for all $n \in \mathbb{N}$ and $\|g_n \restriction_Y\| \geq \frac{1}{\|z_n\|} |g_n(z_n)|$. Hence $\{g_n \restriction_Y\}$ is semi-normalised and w^*-null.

(ii)⇒(iii) The set $\{g_n \restriction_Y\}$ is not relatively compact. Indeed, otherwise there would be a subsequence converging to some $g \in Y^*$. This contradicts the fact that this subsequence is w^*-null (and so $g = 0$) and semi-normalised. As in the previous part of the proof we can then find a semi-normalised sequence $\{y_j\}_{j=1}^\infty \subset Y$ such that $g_k(y_j) = 0$ for $1 \leq k < j$ and $g_j(y_j) = 1$, $j \in \mathbb{N}$. Let $a, A > 0$ be such that $2a \leq \|y_j\| \leq A$ for all $j \in \mathbb{N}$.

We construct a sequence $\{z_n\}_{n=1}^\infty \subset Y$ by induction, along with increasing sequences $\{k_j^n\}_{j=1}^\infty \subset \mathbb{N}$, $n \in \mathbb{N}_0$, such that $\{k_j^n\}_{j=1}^\infty$ is a subsequence of $\{k_j^{n-1}\}_{j=1}^\infty$ and $k_1^n > k_1^{n-1}$ for each $n \in \mathbb{N}$. Put $k_j^0 = j$, $j \in \mathbb{N}$. Suppose that z_l and $\{k_j^l\}_{j=1}^\infty$, $1 \leq l < n$ are already defined for some $n \in \mathbb{N}$. Since $\{g_{k_j^{n-1}} \restriction_Y\}_{j=1}^\infty$ is w^*-null, there is a subsequence $\{k_j^n\}_{j=1}^\infty$ of $\{k_j^{n-1}\}_{j=1}^\infty$ such that $k_1^n > k_1^{n-1}$, $|g_{k_j^n}(y_{k_1^{n-1}})| \leq \frac{a}{A} \frac{1}{2^{j+1}}$ and $|g_{k_j^n}(y_{k_l^n})| \leq \frac{1}{2^{j+1}}$ for all $1 \leq l < j$, $j \in \mathbb{N}$. Put $a_1 = g_{k_1^n}(y_{k_1^{n-1}})$ and

$a_j = g_{k_j^n}(y_{k_1^{n-1}}) - \sum_{l=1}^{j-1} a_l g_{k_j^n}(y_{k_l^n})$, and note that $|a_j| \le \frac{a}{A}\frac{1}{2^j}$. Finally, set $z_n = y_{k_1^{n-1}} - \sum_{j=1}^{\infty} a_j y_{k_j^n}$. Then $a \le \|z_n\| \le A + a$, $g_{k_1^{n-1}}(z_n) = 1$, $g_{k_1^j}(z_n) = 0$ for $0 \le j < n-1$, and $g_{k_j^n}(z_n) = 0$ for all $j \in \mathbb{N}$.

Put $h_n = g_{k_1^{n-1}}$. By the construction, $\{(z_n; h_n)\}_{n=1}^{\infty}$ is a biorthogonal system. To finish the proof, by Rosenthal's ℓ_1-theorem we may assume by passing to a subsequence that either $\{z_n\}$ is equivalent to the canonical basis of ℓ_1 (in this case we set $x_n = z_n$) or $\{z_n\}$ is weakly Cauchy. In the latter case we set $x_n = z_{2n} - z_{2n-1}$ noting that $\{(x_n; h_{2n})\}_{n=1}^{\infty}$ is a biorthogonal system. Then $\{x_n\}$ is weakly null and we can pass to a final subsequence of $\{x_n\}$ which will be basic (Proposition 32). □

Corollary 57 ([KO1], [GJ1]). *Let X be a Banach space which does not contain ℓ_1 and $1 \le p \le \infty$. If X has property S_p, then X^* has property T_q, where $\frac{1}{p} + \frac{1}{q} = 1$.*

Proof. Let $\{x_n^*\}_{n=1}^{\infty}$ be a normalised weakly null sequence in X^*. By Theorem 56 there are a semi-normalised weakly null sequence $\{y_k\}_{k=1}^{\infty}$ and a subsequence $\{y_k^*\}_{k=1}^{\infty}$ of $\{x_n^*\}_{n=1}^{\infty}$ which is biorthogonal to $\{y_k\}_{k=1}^{\infty}$. By the assumption, $\{y_k\}_{k=1}^{\infty}$ has a further subsequence $\{y_{k_j}\}_{j=1}^{\infty}$ with an upper p-estimate. By Fact 50, $\{y_{k_j}^*\}_{j=1}^{\infty}$ has a lower q-estimate as required. □

We remark that the dual statement that property T_q of X implies property S_p of X^* does not hold: It is observed in [GJ1] that a Banach space constructed in [JO] is a counterexample.

Corollary 58 ([GJ1]). *Let X be a Banach space with property S_p, $1 < p \le \infty$. If $n < p$, then $\mathcal{P}^n(X) = \mathcal{P}^n_{\mathrm{wsC}}(X)$.*

Proof. By Corollary 27 and Lemma 30 it suffices to show that $\lim_{k\to\infty}|P(x_k)| = 0$ for any weakly null sequence $\{x_k\} \subset X$ and any $P \in \mathcal{P}(^nX)$, $1 \le n < p$. By contradiction, suppose that $|P(x_k)| > \delta > 0$, $k \in \mathbb{N}$, for some weakly null sequence $\{x_k\}$. By passing to a subsequence we may assume that $\{x_k\}$ has an upper p-estimate. By Theorem 52 (resp. Theorem 53) we conclude that $\{P(x_k)\}$ has an upper $\frac{p}{n}$-estimate. By considering the sums $\sum_{j=1}^{m} \frac{1}{t_{k_j}} t_{k_j}$ it is easy to see that any sequence $\{t_k\}_{k=1}^{\infty}$ in $\mathbb{C}$ with an upper r-estimate, $r > 1$, is convergent to 0, which is a contradiction. □

Since $\ell_p(\Gamma)$ has property S_p and $c_0(\Gamma)$ has property S_∞, we obtain the next result. (Recall also that in case of the separable dual the weak uniformity on bounded sets is metrisable.)

Corollary 59 ([Bog], [Pełl]). *Let Γ be any set, $1 < p < \infty$, and $n \in \mathbb{N}$, $n < p$. Then*

$$\mathcal{P}^n(\ell_p(\Gamma)) = \mathcal{P}^n_{\mathrm{wsC}}(\ell_p(\Gamma)),$$
$$\mathcal{P}(c_0(\Gamma)) = \mathcal{P}_{\mathrm{wsC}}(c_0(\Gamma)).$$

In particular, in the separable case

$$\mathcal{P}^n(\ell_p) = \mathcal{P}^n_{\mathrm{wu}}(\ell_p),$$
$$\mathcal{P}(c_0) = \mathcal{P}_{\mathrm{wu}}(c_0).$$

On the other hand, if $n \ge p$, then $x \mapsto \sum_{j=1}^{\infty} x_j^n \in \mathcal{P}(^n\ell_p) \setminus \mathcal{P}_{\mathrm{wsc}}(^n\ell_p)$.

Corollary 60 ([AAD]). *Let T^* denote the (original) Tsirelson space. Then we have $\mathcal{P}(T^*) = \mathcal{P}_{\rm wu}(T^*)$.*

Proof. By [CasaSh, Proposition V.10] each bounded block basis of the canonical basis of T^* has an upper p-estimate for every $p < \infty$. Thus T^* has property S_p for every $p < \infty$ by Proposition 32 and we may apply Corollary 58. □

Corollary 61. *Let $Q \in \mathcal{P}^n(\ell_p)$, $n < p < \infty$, and $\varepsilon > 0$. Then there exists $N \in \mathbb{N}$ such that $\|Q - Q \circ P_N\| \le \varepsilon$, where $P_N : \ell_p \to \ell_p$ is the projection onto the first N coordinates.*

Proof. By contradiction, suppose that there exists a sequence $\{x_k\} \subset B_{\ell_p}$ such that $|Q(x_k) - Q(P_k(x_k))| > \varepsilon$. By the weak compactness there is a subsequence $\{x_{k_j}\}$ that converges weakly to x. Thus also $P_{k_j}(x_{k_j}) \xrightarrow{w} x$ (check the convergence on the coordinate functionals), which contradicts the fact that Q is weakly sequentially continuous (Corollary 59). □

Corollary 62. *Let Γ be an uncountable set and $P \in \mathcal{P}({}^n c_0(\Gamma))$, $n \in \mathbb{N}$, resp. $P \in \mathcal{P}({}^n\ell_p(\Gamma))$, $1 \le n < p < \infty$. Then P is countably supported in the sense that the set $\{\gamma \in \Gamma;\ P(e_\gamma) \neq 0\}$ is countable.*

Proof. Assume the contrary. Then there are $\varepsilon > 0$ and an infinite set $\Gamma_0 \subset \Gamma$ such that $|P(e_\gamma)| > \varepsilon$ for all $\gamma \in \Gamma_0$. But then there is a sequence $\{\gamma_n\} \subset \Gamma_0$ such that $e_{\gamma_n} \xrightarrow{w} 0$, a contradiction with Corollary 59. □

Theorem 63 ([Pełl], [ALRT],[GGu3], [GJ1]). *Let X be a Banach space with property S_p, $1 < p < \infty$, and Y a Banach space with property T_q, $1 \le q < \infty$. Then $\mathcal{P}^n(X;Y) = \mathcal{P}^n_{\rm wsC}(X;Y)$ for $n < \frac{p}{q}$. If X has property S_∞ and Y has no subspace isomorphic to c_0, then $\mathcal{P}(X;Y) = \mathcal{P}_{\rm wsC}(X;Y)$.*

Proof. By Lemma 30 it suffices to show that $\lim_{k\to\infty}\|P(x_k)\| = 0$ for any weakly null sequence $\{x_k\}_{k=1}^\infty$ in X and any $P \in \mathcal{P}({}^nX;Y)$, $1 \le n < \frac{p}{q}$ (resp. any $n \in \mathbb{N}$). Assume the contrary. Then there are $n < \frac{p}{q}$ (resp. $n \in \mathbb{N}$), $P \in \mathcal{P}({}^nX;Y)$, $\varepsilon > 0$, and a normalised weakly null sequence $\{x_k\}$ satisfying $\|P(x_k)\| \ge \varepsilon$ for all $k \in \mathbb{N}$.

Consider the case $p < \infty$ first. By the assumptions we may assume without loss of generality that $\{x_k\}$ has an upper p-estimate. By Theorem 52, $\{P(x_k)\}$ has an upper $\frac{p}{n}$-estimate. Note that by Corollary 58 the sequence $\{P(x_k)\}$ is also weakly null. Hence some subsequence of it has a lower q-estimate, which is not possible as $q < \frac{p}{n}$.

Suppose now that $p = \infty$. By the assumptions and Fact 51 we may assume without loss of generality that $\{x_k\}$ is wuC. By Theorem 53, $\{P(x_k)\}$ is also wuC. Thus $\sum_{k=1}^\infty P(x_k)$ is convergent by the Bessaga-Pełczyński theorem (Theorem 46), which is a contradiction. □

Corollary 64 ([AAD], [GJ1]). *Let T^* be the (original) Tsirelson space and Y a Banach space with property T_q, $1 \le q < \infty$. Then $\mathcal{P}(T^*;Y) = \mathcal{P}_{\rm wu}(T^*;Y)$.*

Proof. Combine Theorem 63 with the fact that T^* has property S_p for every $p < \infty$, [CasaSh, Proposition V.10], and use the separability of $(T^*)^*$. □

Note that by Proposition 4.36 the previous result applies in particular to any Y of a non-trivial cotype. If Y is of a non-trivial type, then even much stronger results hold, see Theorem 6.54 and Theorem 6.32.

We will use the following classical theorem. For the proof see e.g. [Con, Theorem VI.5.5].

Theorem 65 (Vera Ruvimovna Gantmacher). *Let X, Y be Banach spaces and let $T \in \mathcal{L}(X;Y)$. The following statements are equivalent:*

(i) *T is weakly compact.*
(ii) *$T^{**}(X^{**}) \subset Y$.*
(iii) *T^* is weakly compact.*

Lemma 66. *Let X, Y be Banach spaces and let $T = (T_k)_{k=1}^{\infty} \in \mathcal{L}(X; c_0(\mathbb{N}; Y))$. Then T is weakly compact if and only if the following two conditions hold:*

(i) *$T_k \in \mathcal{L}_{\mathrm{wK}}(X;Y)$ for each $k \in \mathbb{N}$ and*
(ii) *$\lim\limits_{k\to\infty} \|T_k^{**}(F)\| = 0$ for any $F \in X^{**}$.*

Proof. In our case $T^{**}\colon X^{**} \to \ell_\infty(\mathbb{N}; Y^{**})$, $T^{**} = (T_k^{**})_{k=1}^{\infty}$, so the equivalence is clear from Theorem 65. □

Theorem 67 (Raymond A. Ryan, [Ry1]). *Let X be a normed linear space. The following statements are equivalent:*

(i) *X has the Dunford-Pettis property.*
(ii) *$\mathcal{L}_{\mathrm{wK}}(X;Y) \subset \mathcal{L}_{\mathrm{wsC}}(X;Y)$ for every Banach space Y.*
(iii) *$\mathcal{L}_{\mathrm{wK}}({}^nX;Y) \subset \mathcal{L}_{\mathrm{wsC}}({}^nX;Y)$ for every Banach space Y and every $n \in \mathbb{N}$.*
(iv) *$\mathcal{P}_{\mathrm{wK}}(X;Y) \subset \mathcal{P}_{\mathrm{wsC}}(X;Y)$ for every Banach space Y.*

Proof. (iii)⇒(iv)⇒(ii)⇔(i) are clear.

We prove (i)⇒(iii) by induction. For $n = 1$ it is just the implication (i)⇒(ii). Assume that (iii) holds for $n \in \mathbb{N}$ and let $M \in \mathcal{L}_{\mathrm{wK}}({}^{n+1}X;Y)$. By Theorem 26 it suffices to show that given weakly convergent sequences $\{x_k^j\}_{k=1}^{\infty} \subset X$, $j = 1,\dots,n$, and a weakly null sequence $\{y_k\}_{k=1}^{\infty} \subset X$ we have $\lim_{k\to\infty} M(x_k^1,\dots,x_k^n,y_k) = 0$.

We will use the following notation: For any $S \in \mathcal{L}({}^{n+1}X;Z)$, where Z is a normed linear space, and for $x \in X$ we define $S[x] \in \mathcal{L}({}^nX;Z)$ by $S[x](z_1,\dots,z_n) = S(z_1,\dots,z_n,x)$. Clearly $\|S[x]\| \le \|S\|\|x\|$. We put $A_k = M[y_k]$ for $k \in \mathbb{N}$. For any $z_1,\dots,z_n \in X$ the linear mapping $x \mapsto M(z_1,\dots,z_n,x)$ is weakly compact, so $\lim_{k\to\infty} M(z_1,\dots,z_n,y_k) = 0$ by the Dunford-Pettis property. Hence $A = (A_k)_{k=1}^{\infty} \in \mathcal{L}({}^nX; c_0(\mathbb{N};Y))$ with $\|A\| \le \|M\| \sup_{k\in\mathbb{N}} \|y_k\|$. We will show that $A \in \mathcal{L}_{\mathrm{wK}}({}^nX; c_0(\mathbb{N};Y))$, which enables us to finish the proof by induction. Indeed, by the inductive hypothesis we obtain $A \in \mathcal{L}_{\mathrm{wsC}}({}^nX; c_0(\mathbb{N};Y))$ and so $\lim_{k\to\infty} A(x_k^1,\dots,x_k^n) = (w_j)_{j=1}^{\infty} \in c_0(\mathbb{N};Y)$. Since

$$\begin{aligned}\|M(x_k^1,\dots,x_k^n,y_k)\| &= \|A_k(x_k^1,\dots,x_k^n)\| \le \|A_k(x_k^1,\dots,x_k^n) - w_k\| + \|w_k\| \\ &\le \sup_{j\in\mathbb{N}} \|A_j(x_k^1,\dots,x_k^n) - w_j\| + \|w_k\|\end{aligned}$$

and this upper estimate tends to zero with $k \to \infty$, we are done.

Fact 25 implies that $A \in \mathcal{L}_{\mathrm{wK}}({}^nX; c_0(\mathbb{N}; Y))$ is equivalent to the linearisation $L_A \in \mathcal{L}(\otimes_\pi^n X; c_0(\mathbb{N}; Y))$ being weakly compact. Clearly $L_A = (L_{A_k})_{k=1}^\infty$, where $L_{A_k} \in \mathcal{L}(\otimes_\pi^n X; Y)$ are the linearisations of A_k. By Fact 25, $L_{A_k} \in \mathcal{L}_{\mathrm{wK}}(\otimes_\pi^n X; Y)$, and hence by Lemma 66 it remains to show that $\lim_{k\to\infty} \|L_{A_k}^{**}(\phi)\| = 0$ for every $\phi \in (\otimes_\pi^n X)^{**} = \mathcal{L}({}^nX)^*$ (Theorem 7).

To do this, fix $\phi \in \mathcal{L}({}^nX)^*$ and consider the operator $T_\phi\colon X \to \mathcal{L}({}^{n+1}X)^*$ defined by

$$\langle T_\phi(x), S\rangle = \langle \phi, S[x]\rangle, \quad S \in \mathcal{L}({}^{n+1}X).$$

It is easy to see that T_ϕ is well-defined, linear, and bounded. So for every $\psi \in Y^*$

$$\begin{aligned}\langle L_{A_k}^{**}(\phi), \psi\rangle &= \langle \phi, L_{A_k}^{*}(\psi)\rangle = \langle \phi, \psi \circ L_{A_k}\rangle = \langle \phi, \psi \circ A_k\rangle \\ &= \langle T_\phi(y_k), \psi \circ M\rangle = \langle T_\phi(y_k), \psi \circ L_M\rangle = \langle L_M^{**}(T_\phi(y_k)), \psi\rangle.\end{aligned}$$

Therefore $L_{A_k}^{**}(\phi) = (L_M^{**} \circ T_\phi)(y_k)$ for all $k \in \mathbb{N}$. Fact 25 implies that the operator L_M is weakly compact and so $L_M^{**} \in \mathcal{L}_{\mathrm{wK}}((\otimes_\pi^{n+1} X)^{**}; Y)$ by Theorem 65. Thus $L_M^{**} \circ T_\phi \in \mathcal{L}_{\mathrm{wK}}(X; Y) \subset \mathcal{L}_{\mathrm{wsc}}(X; Y)$ by the DPP. Since $\{y_k\}$ is weakly null, it follows that $\lim_{k\to\infty} L_{A_k}^{**}(\phi) = \lim_{k\to\infty} (L_M^{**} \circ T_\phi)(y_k) = 0$. □

So if X is a normed linear space with the DPP and Y is any Banach space, then we can update our chains of inclusions as follows:

$$\mathcal{P}_{\mathrm{wu}}(X;Y) = \mathcal{P}_{\mathrm{w}}(X;Y) \subset \mathcal{P}_{\mathrm{K}}(X;Y) \subset \mathcal{P}_{\mathrm{wK}}(X;Y) \subset \mathcal{P}_{\mathrm{wsc}}(X;Y) = \mathcal{P}_{\mathrm{wsC}}(X;Y)$$
$$\mathcal{L}_{\mathrm{wu}}(X;Y) = \mathcal{L}_{\mathrm{w}}(X;Y) = \mathcal{L}_{\mathrm{K}}(X;Y) \subset \mathcal{L}_{\mathrm{wK}}(X;Y) \subset \mathcal{L}_{\mathrm{wsc}}(X;Y) = \mathcal{L}_{\mathrm{wsC}}(X;Y)$$

It follows that if X is a normed linear space with the Dunford-Pettis property, then $\mathcal{P}(X) = \mathcal{P}_{\mathrm{K}}(X) = \mathcal{P}_{\mathrm{wsC}}(X)$. Thus in this case any $P \in \mathcal{P}(X;Y)$, where Y is any Banach space, maps weakly Cauchy sequences to weakly Cauchy sequences (consider the compositions with linear functionals). This is a result of Aleksander Pełczyński, [Peł4]. By a similar argument to that of Theorem 67 we get a nice and useful variant below.

Theorem 68 ([GGu1]). *Let X be a normed linear space with the DPP and Y a Banach space such that $\mathcal{L}(X;Y) = \mathcal{L}_{\mathrm{wK}}(X;Y)$. Then $\mathcal{P}(X;Y) = \mathcal{P}_{\mathrm{wsC}}(X;Y)$.*

Proof. We prove that $\mathcal{L}({}^nX;Y) = \mathcal{L}_{\mathrm{wsC}}({}^nX;Y)$ by induction on n. For $n = 1$ this follows from the assumption and the DPP. Assume that the statement holds for $n \in \mathbb{N}$ and let $M \in \mathcal{L}({}^{n+1}X;Y)$. By Theorem 26 it suffices to show that given weakly convergent sequences $\{x_k^j\}_{k=1}^\infty \subset X$, $j = 1, \dots, n+1$, where $\{x_k^1\}$ is weakly null, we have $\lim_{k\to\infty} M(x_k^1, \dots, x_k^{n+1}) = 0$.

Define $T_k \in \mathcal{L}(X;Y)$ by $T_k(x) = M(x_k^1, \dots, x_k^n, x)$ and $T\colon X \to c_0(\mathbb{N};Y)$ by $T(x) = (T_k(x))_{k=1}^\infty$. By Theorem 26 the operator T maps indeed into $c_0(\mathbb{N};Y)$. Clearly T is a linear operator and $\|T\| \le \|M\| \sup_{k\in\mathbb{N}} \|x_k^1\| \cdots \|x_k^n\|$. We claim that T is weakly compact, and consequently $T \in \mathcal{L}_{\mathrm{wsC}}(X; c_0(\mathbb{N};Y))$ by the DPP. Then

$\lim_{k\to\infty} T(x_k^{n+1}) = (w_j)_{j=1}^{\infty} \in c_0(\mathbb{N}; Y)$. Since

$$\|M(x_k^1,\dots,x_k^{n+1})\| = \|T_k(x_k^{n+1})\| \le \|T_k(x_k^{n+1}) - w_k\| + \|w_k\| \\ \le \sup_{j\in\mathbb{N}} \|T_j(x_k^{n+1}) - w_j\| + \|w_k\|$$

and this upper estimate tends to zero with $k \to \infty$, we are done.

To prove the claim note that by the assumption $T_k \in \mathcal{L}_{\mathrm{wK}}(X;Y)$ for each $k \in \mathbb{N}$, so by Lemma 66 it remains to show that $\lim_{k\to\infty}\|T_k^{**}(F)\| = 0$ for every $F \in X^{**}$. To do this, fix $F \in X^{**}$. By the assumption the linear operator $z \mapsto M(x_1,\dots,x_n,z)$ is weakly compact and by Theorem 65 so is its second adjoint, which then maps into Y. Thus we may define a mapping $Q_F : X^n \to Y$ by

$$Q_F(x_1,\dots,x_n) = \big(z \mapsto M(x_1,\dots,x_n,z)\big)^{**}(F).$$

It is easy to see that $Q_F \in \mathcal{L}(^nX;Y)$ and $T_k^{**}(F) = Q_F(x_k^1,\dots,x_k^n)$. Indeed, $\langle T_k^{**}(F),\psi\rangle = \langle F, T_k^*(\psi)\rangle = \langle F, M(x_k^1,\dots,x_k^n,\cdot)^*(\psi)\rangle = \langle Q_F(x_k^1,\dots,x_k^n),\psi\rangle$ for any $\psi \in Y^*$. As $Q_F \in \mathcal{L}_{\mathrm{wsC}}(^nX;Y)$ by the inductive hypothesis, we obtain $\lim_{k\to\infty} T_k^{**}(F) = \lim_{k\to\infty} Q_F(x_k^1,\dots,x_k^n) = 0$, since $\{x_k^1\}$ is weakly null. □

Corollary 69 ([Peł4]). *Let K be a compact space and Y a Banach space with no subspace isomorphic to c_0. Then $\mathcal{P}((C(K);Y) = \mathcal{P}_{\mathrm{wsC}}(C(K);Y)$.*

Proof. By Theorem 41, $C(K)$ spaces have the DPP. By the classical Theorem 47 we have $\mathcal{L}(C(K);Y) = \mathcal{L}_{\mathrm{wK}}(C(K);Y)$, so the result follows from Theorem 68. □

The next proposition shows that the compactness passes from the mapping to its derivatives.

Proposition 70. *Let X be a normed linear space, Y a Banach space, $U \subset X$ open, and let $f : U \to Y$ be T^k-smooth at $a \in U$ with the approximating polynomial $P \in \mathcal{P}^k(X;Y)$. If $P \notin \mathcal{P}_{\mathrm{K}}(X;Y)$, resp. $P \notin \mathcal{P}_{\mathrm{wK}}(X;Y)$, then $f(U)$ is not relatively compact, resp. relatively weakly compact.*

Proof. Without loss of generality we may assume that $a = 0$. First assume that $P \notin \mathcal{P}_{\mathrm{K}}(X;Y)$. Since T^k-smoothness implies T^m-smoothness for $m < k$, we may without loss of generality assume that $P = Q + P_k$, where $Q \in \mathcal{P}_{\mathrm{K}}^{k-1}(X;Y)$ and $P_k \in \mathcal{P}(^kX;Y) \setminus \mathcal{P}_{\mathrm{K}}(^kX;Y)$. There are $\varepsilon > 0$ and $\{x_n\} \subset B_X$ such that $\{P_k(x_n)\}$ is an ε-separated set. By the definition of T^k-smoothness there is $\delta > 0$ such that $\|f(h) - P(h)\| \le \frac{\varepsilon}{4}\|h\|^k$ whenever $h \in X$, $\|h\| \le \delta$. Since Q is compact, by passing to a subsequence we may assume that there is $y \in Y$ such that $\|Q(\delta x_n) - y\| < \frac{\varepsilon}{8}\delta^k$ for all $n \in \mathbb{N}$. Therefore

$$\|f(\delta x_m) - f(\delta x_n)\| \ge \|P_k(\delta x_m) - P_k(\delta x_n)\| - \|Q(\delta x_m) - Q(\delta x_n)\| \\ - \|f(\delta x_m) - P(\delta x_m)\| - \|f(\delta x_n) - P(\delta x_n)\| \ge \frac{\varepsilon}{4}\delta^k$$

whenever $m \ne n$, which means that $f(U)$ is not relatively compact.

In case that $P \notin \mathcal{P}_{\mathrm{wK}}(X;Y)$ we may similarly assume that $P = Q + P_k$, where $Q \in \mathcal{P}^{k-1}_{\mathrm{wK}}(X;Y)$ and $P_k \in \mathcal{P}({}^kX;Y) \setminus \mathcal{P}_{\mathrm{wK}}({}^kX;Y)$. There is a net $\{x_\alpha\} \subset B_X$ such that $\{P_k(x_\alpha)\}$ has no weak cluster point. Let $y \in Y^{**}$ be a w^*-cluster point of $\{P_k(x_\alpha)\}$ in Y^{**}. Clearly $y \notin Y$. Let $\varepsilon = \operatorname{dist}(y, Y) > 0$. There is $\delta > 0$ such that $\|f(h) - P(h)\| \leq \frac{\varepsilon}{2}\|h\|^k$ whenever $h \in X$, $\|h\| \leq \delta$. Note that in particular it means that $f(\delta B_X)$ is bounded. Let $\{x_\beta\}$ be a subnet of $\{x_\alpha\}$ such that $P_k(x_\beta) \xrightarrow{w^*} y$. By the compactness there is a subnet $\{x_\gamma\}$ of $\{x_\beta\}$ such that $Q(\delta x_\gamma) \xrightarrow{w} u$ for some $u \in Y$ and $f(\delta x_\gamma) \xrightarrow{w^*} z$ for some $z \in Y^{**}$. We have $\|f(\delta x_\gamma) - P(\delta x_\gamma)\| \leq \frac{\varepsilon}{2}\delta^k$ and $f(\delta x_\gamma) - P(\delta x_\gamma) \xrightarrow{w^*} z - u - \delta^k y$, which implies $\|z - u - \delta^k y\| \leq \frac{\varepsilon}{2}\delta^k$. Since $\operatorname{dist}(\delta^k y, Y) = \varepsilon\delta^k$, it follows that $z - u \notin Y$ and hence $z \notin Y$. This means that $\{f(\delta x_\gamma)\}$ has no weak cluster point. □

3. Weak continuity and ℓ_1

In this section we continue our study of weakly continuous mappings initiated in Section 2, focusing on the case when the domain does not contain ℓ_1. All the spaces in this section are real. We start by recalling some important classical results concerning Banach spaces not containing ℓ_1.

Theorem 71 (Rosenthal's ℓ_1-theorem [Ros2], [FHHMZ, Theorem 5.37]). *Every bounded sequence in a Banach space has a subsequence which is either weakly Cauchy or equivalent to the canonical basis of ℓ_1.*

Recall that a function f on a topological space T is of Borel class 1 if $f^{-1}(G)$ is an F_σ set for every open set $G \subset \mathbb{R}$. It is easy to see that a pointwise limit of a sequence of continuous functions on T is of Borel class 1. If T is metrisable, then also the converse statement holds, namely f is of Borel class 1 if and only if f is a pointwise limit of a sequence of continuous functions on T ([Kec, Theorem 24.10]).

Theorem 72 (Edward Wilfred Odell and Haskell Paul Rosenthal, [OR]; [FHHMZ, Theorem 5.40]). *Let X be a separable Banach space. The following statements are equivalent:*

(i) *X does not contain ℓ_1.*
(ii) *Every element of $B_{X^{**}}$ is a w^*-limit of a sequence in B_X.*
(iii) $\operatorname{card} X^{**} = \operatorname{card} X = \mathfrak{c}$.

Definition 73. A subset A of a topological space T is called relatively countably compact in T if every sequence in A has a cluster point (in T).

A topological space T is said to have *countable tightness* if for every $A \subset T$ and $x \in \overline{A}$ there is a countable set $S \subset A$ such that $x \in \overline{S}$.

A topological space T is called *angelic* if for every relatively countably compact $A \subset T$ the following two conditions hold:

(i) A is relatively compact,
(ii) given $x \in \overline{A}$ there is a sequence $\{x_n\} \subset A$ such that $x_n \to x$.

It is not difficult to show that the above properties are hereditary (i.e. they pass to arbitrary subsets). Note that if X is a Banach space and (B_{X^*}, w^*) is angelic, then the points in the weak* closure of a bounded set $A \subset X^*$ are w^*-limits of sequences from A. The following observation notes that the same holds for weak closures in X provided that the bidual is angelic.

Fact 74. *Let X be a normed linear space such that $(B_{X^{**}}, w^*)$ is angelic. If $A \subset X$ is bounded and $x \in \overline{A}^w$, then there is a sequence $\{x_n\} \subset A$ such that $x_n \xrightarrow{w} x$.*

Proof. Without loss of generality we may assume that $A \subset B_X$. Then A is relatively w^*-compact in $B_{X^{**}}$. We have $x \in \overline{A}^{(X,w)} \subset \overline{A}^{(X^{**},w^*)} \subset B_{X^{**}}$ and so there is a sequence $\{x_n\} \subset A$ such that $x_n \xrightarrow{w^*} x$ in X^{**}. But this means that $x_n \xrightarrow{w} x$ in X. □

We will rely on the following fundamental result.

Theorem 75 (Irving Kaplansky [FHHMZ, Theorem 3.54]). *If X is a normed linear space, then the topological space (X, w) has countable tightness.*

If a compact space K does not have countable tightness (e.g. $K = [0, \omega_1]$), then for $X = C(K)$ we have $K \subset (S_{X^*}, w^*)$, and so (X^*, w^*) does not have countable tightness either. Thus Kaplansky's theorem in general does not hold for the weak* topology. Nevertheless, for some Banach spaces $(B_{X^{**}}, w^*)$ is angelic, which is stronger (Fact 74) than the countable tightness of (B_X, w); see e.g. [EW]. An important example is provided by separable Banach spaces not containing ℓ_1. Let us describe the situation in more detail.

Recall that a Polish space is a separable and completely metrisable topological space, and a Souslin space is a topological space that is a continuous image of a Polish space.

Theorem 76 (Jean Bourgain, David H. Fremlin, and Michel Talagrand, [BFT]; see [FHHMZ, Theorem 5.49]). *If S is a Souslin space, then the space of functions on S of Borel class 1 with the topology of pointwise convergence is angelic.*

As a corollary we obtain the following result.

Theorem 77. *If X is a separable Banach space that does not contain ℓ_1, then (X^{**}, w^*) is angelic.*

Proof. As X is separable, (B_{X^*}, w^*) is a Polish space, [FHHMZ, Proposition 3.101]. Since the topological sum of a sequence of Polish spaces is a Polish space, it follows that (X^*, w^*) is a Souslin space (as a continuous image of $\bigoplus_{n=1}^{\infty}(n B_{X^*}, w^*)$). Consider the canonical embedding $(X^{**}, w^*) \subset \mathbb{R}^{X^*}$. Since $X \subset C\big((X^*, w^*)\big)$, in view of Theorem 72 the elements of X^{**} are functions on (X^*, w^*) of Borel class 1. Hence we are in a position to apply Theorem 76. □

Corollary 78. *Let X be a Banach space that does not contain ℓ_1. If $A \subset X$ is bounded and $x \in \overline{A}^w$, then there is a sequence $\{x_n\} \subset A$ such that $x_n \xrightarrow{w} x$.*

Proof. By Kaplansky's theorem (Theorem 75) there exists a countable $S \subset A$ such that $x \in \overline{S}^w$. Let $Y = \overline{\text{span}}(S \cup \{x\})$. Then Y is separable and hence (Y^{**}, w^*) is angelic by Theorem 77. Fact 74 finishes the proof. □

We now continue in developing our theory of weakly continuous mappings. Theorems 79 and 81 below are strengthenings of the observation mentioned after Definition 16: if X is a normed linear space with X^* separable, then the weak uniformity on bounded subsets of X is metrisable and hence $\mathcal{C}_{\text{wsc}}(U; Y) = \mathcal{C}_{\text{w}}(U; Y)$ and $\mathcal{C}_{\text{wsC}}(U; Y) = \mathcal{C}_{\text{wu}}(U; Y)$ for $U \subset X$ convex and Y any Banach space.

Theorem 79 ([CHL], [FGL], [Gut]). *Let X, Y be Banach spaces such that X does not contain ℓ_1 and $U \subset X$ a convex set. Then $\mathcal{C}_{\text{wsc}}(U; Y) = \mathcal{C}_{\text{w}}(U; Y)$.*

Proof. We only need to prove that $\mathcal{C}_{\text{wsc}}(U; Y) \subset \mathcal{C}_{\text{w}}(U; Y)$. By contradiction, let $f \in \mathcal{C}_{\text{wsc}}(U; Y) \setminus \mathcal{C}_{\text{w}}(U; Y)$. Then there are a CCB set $V \subset U$, $\varepsilon > 0$, $A \subset V$, and $x \in \overline{A}^w$ such that $\operatorname{dist}(f(x), f(A)) > \varepsilon$. By Corollary 78 there is a sequence $\{x_n\} \subset A$ such that $x_n \xrightarrow{w} x$. As $f \in \mathcal{C}_{\text{wsc}}(U; Y)$, we have $\lim f(x_n) = f(x)$, a contradiction. □

Combining Theorem 79 and Corollary 27 we obtain the next result.

Corollary 80 ([AHV]). *Let X, Y be Banach spaces such that X does not contain ℓ_1. Then $\mathcal{P}_{\text{wsc}}(X; Y) = \mathcal{P}_{\text{wu}}(X; Y)$.*

Theorem 81 ([CHL]). *Let X, Y be Banach spaces such that X does not contain ℓ_1 and let $U \subset X$ be a convex set. Then $\mathcal{C}_{\text{wsC}}(U; Y) = \mathcal{C}_{\text{wu}}(U; Y)$.*

Proof. We only need to prove that $\mathcal{C}_{\text{wsC}}(U; Y) \subset \mathcal{C}_{\text{wu}}(U; Y)$. By contradiction, let $f \in \mathcal{C}_{\text{wsC}}(U; Y)$ and let $V \subset U$ be a CCB set such that f is not w–$\|\cdot\|$ uniformly continuous on V. Then there is $\varepsilon > 0$ such that for each finite $\Phi \subset B_{X^*}$ and each $n \in \mathbb{N}$ there are $x_{\Phi,n}, y_{\Phi,n} \in V$ satisfying $\max_{\phi \in \Phi} |\phi(x_{\Phi,n} - y_{\Phi,n})| < \frac{1}{n}$ and $\|f(x_{\Phi,n}) - f(y_{\Phi,n})\| \geq \varepsilon$. Put $A = \{x_{\Phi,n} - y_{\Phi,n};\ \Phi \subset B_{X^*} \text{ finite}, n \in \mathbb{N}\}$, which is a bounded set. Clearly $0 \in \overline{A}^w$ and so by Corollary 78 there is a weakly null sequence $\{z_k\} \subset A$ satisfying $z_k = x_k - y_k$, $\|f(x_k) - f(y_k)\| \geq \varepsilon$ for all $k \in \mathbb{N}$. Using Rosenthal's ℓ_1-theorem (Theorem 71), by passing to subsequences we may assume that $\{x_k\}$ and $\{y_k\}$ are weakly Cauchy. Since the sequence $x_1, y_1, x_2, y_2, \ldots$ is then also weakly Cauchy, we have reached a contradiction. □

The previous results do not hold in general: Since ℓ_1 has the Schur property, $\mathcal{L}(\ell_1; \ell_1) = \mathcal{L}_{\text{wsc}}(\ell_1; \ell_1) = \mathcal{L}_{\text{wsC}}(\ell_1; \ell_1)$, while $Id \notin \mathcal{L}_{\text{K}}(\ell_1; \ell_1) = \mathcal{L}_{\text{wu}}(\ell_1; \ell_1)$.

Let X be a Banach space that does not contain ℓ_1, Y a Banach space, and $U \subset X$ a convex set. By combining the above results with the inclusions obtained earlier in this chapter we conclude the following relations:

$$\begin{array}{c} \qquad\qquad\qquad\qquad\qquad\quad \subset \mathcal{C}_{\text{K}}(U; Y) \subset \mathcal{C}_{\text{wK}}(U; Y) \\ \mathcal{C}_{\text{wu}}(U; Y) = \mathcal{C}_{\text{wsC}}(U; Y) \subset \mathcal{C}_{\text{w}}(U; Y) = \mathcal{C}_{\text{wsc}}(U; Y) \end{array}$$

$$\mathcal{P}_{\text{wu}}(X; Y) = \mathcal{P}_{\text{w}}(X; Y) = \mathcal{P}_{\text{wsC}}(X; Y) = \mathcal{P}_{\text{wsc}}(X; Y) \subset \mathcal{P}_{\text{K}}(X; Y) \subset \mathcal{P}_{\text{wK}}(X; Y)$$

$$\mathcal{L}_{\text{wu}}(X; Y) = \mathcal{L}_{\text{w}}(X; Y) = \mathcal{L}_{\text{wsC}}(X; Y) = \mathcal{L}_{\text{wsc}}(X; Y) = \mathcal{L}_{\text{K}}(X; Y) \subset \mathcal{L}_{\text{wK}}(X; Y)$$

Note also that if X does not contain ℓ_1 and has the DPP (which is equivalent to X^* being a Schur space, Theorem 41), then all the spaces of polynomials above are equal.

By combining Theorem 81 with Corollary 59 and Theorem 63 we obtain the following improvement of the non-separable case:

Corollary 82 ([Peł1], [ALRT],[GGu3], [GJ1]). *Let Γ be any set, $1 < p < \infty$, and $n \in \mathbb{N}$, $n < p$. Then $\mathcal{P}^n(\ell_p(\Gamma)) = \mathcal{P}^n_{\mathrm{wu}}(\ell_p(\Gamma))$ and $\mathcal{P}(c_0(\Gamma)) = \mathcal{P}_{\mathrm{wu}}(c_0(\Gamma))$.*

Further, if Λ is any set, $1 \le q < \infty$, and $n < \frac{p}{q}$, then $\mathcal{P}^n\big(\ell_p(\Gamma);\ell_q(\Lambda)\big) = \mathcal{P}^n_{\mathrm{wu}}\big(\ell_p(\Gamma);\ell_q(\Lambda)\big)$ and $\mathcal{P}\big(c_0(\Gamma);\ell_q(\Lambda)\big) = \mathcal{P}_{\mathrm{wu}}\big(c_0(\Gamma);\ell_q(\Lambda)\big)$.

For the definition of type and some elementary properties see Section 4.

Theorem 83 ([CHL]). *Let X, Y be Banach spaces such that X does not contain ℓ_1 and Y^* is of type $p > 1$. Let $\frac{1}{p} + \frac{1}{q} = 1$ and suppose that $\mathcal{P}({}^{nd}X) = \mathcal{P}_{\mathrm{wsc}}({}^{nd}X)$ for some $n, d \in \mathbb{N}$, $d > q$. Then $\mathcal{P}^n(X;Y) = \mathcal{P}^n_{\mathrm{wu}}(X;Y)$.*

Proof. By Corollary 80 it suffices to prove that $\mathcal{P}({}^mX;Y) = \mathcal{P}_{\mathrm{wsc}}({}^mX;Y)$ for all $1 \le m \le n$. We proceed by contradiction. Let $1 \le m \le n$ be minimal such that $\mathcal{P}({}^mX;Y) \ne \mathcal{P}_{\mathrm{wsc}}({}^mX;Y)$. By Lemma 30 there exist $P \in \mathcal{P}({}^mX;Y)$, a normalised weakly null sequence $\{x_j\} \subset X$, and $\delta > 0$ such that $\|P(x_j)\| > \delta$, $j \in \mathbb{N}$. By the assumption and Proposition 29, $\phi \circ P \in \mathcal{P}_{\mathrm{wsc}}({}^mX)$ for any $\phi \in Y^*$, so we conclude that $\{P(x_j)\}$ is weakly null in Y. By Corollary 4.43 there exist a subsequence $\{y_k\}$ of $\{P(x_j)\}$ and a polynomial $Q \in \mathcal{P}({}^dY)$ such that $Q(y_k) = 1$, $k \in \mathbb{N}$. Hence $Q \circ P \in \mathcal{P}({}^{md}X) \setminus \mathcal{P}_{\mathrm{wsc}}({}^{md}X)$, a contradiction with Proposition 29. □

To be able to lift weakly Cauchy sequences from quotients we need the following lemma.

Lemma 84 (Robert Henry Lohman, [Loh]). *Let X be a Banach space, $Y \subset X$ a subspace that does not contain ℓ_1, and $Q\colon X \to X/Y$ the canonical quotient mapping. If $\{y_n\} \subset B_{X/Y}$ is weakly Cauchy, then there is a weakly Cauchy sequence $\{x_n\} \subset X$ such that $\lim\|x_n\| \le 1$ and $\{Q(x_n)\}$ is a subsequence of $\{y_n\}$.*

Proof. By contrary, assume that $\{y_n\} \subset B_{X/Y}$ is weakly Cauchy and there is no weakly Cauchy sequence $\{x_n\} \subset X$ such that $\lim\|x_n\| \le 1$ and $\{Q(x_n)\}$ is a subsequence of $\{y_n\}$. For each $n \in \mathbb{N}$ let $x_n \in X$ be such that $Q(x_n) = y_n$ and $\|x_n\| \le 1 + \frac{1}{n}$. By passing to a subsequence we may assume that $\lim\|x_n\| \le 1$. By the assumption $\{x_n\}$ has no weakly Cauchy subsequence and so by Rosenthal's ℓ_1-theorem we may assume that $\{x_n\}$ is a basic sequence equivalent to the canonical basis $\{e_n\}$ of ℓ_1.

Put $z_n = y_{2n} - y_{2n-1}$. Since $z_n \xrightarrow{w} 0$, by Mazur's theorem ([FHHMZ, Corollary 3.46]) applied to the sets $\{z_n;\ n \ge p\}$, $p \in \mathbb{N}$ there are an increasing sequence $\{p_n\} \subset \mathbb{N}$ and a sequence $\{v_n\} \subset X/Y$ such that $v_n \in \operatorname{conv}\{z_j;\ p_n \le j < p_{n+1}\}$ and $v_n \to 0$. Let $\alpha_j \ge 0$ be such that $v_n = \sum_{j=p_n}^{p_{n+1}-1} \alpha_j z_j$ and $\sum_{j=p_n}^{p_{n+1}-1} \alpha_j = 1$ for all $n \in \mathbb{N}$, and put $u_n = \sum_{j=p_n}^{p_{n+1}-1} \alpha_j (x_{2j} - x_{2j-1})$. Then $\{u_n\}$ is a block basis of $\{x_n\}$

and so it is a basic sequence equivalent to a (semi-normalised) block basis of $\{e_n\}$; thus it is equivalent to $\{e_n\}$ ([FHHMZ, Proposition 4.45]). Since $Q(u_n) = v_n$, it follows that $\operatorname{dist}(u_n, Y) = \|v_n\| \to 0$. Thus by passing to a subsequence we may find vectors $w_n \in Y$ sufficiently close to u_n so that $\{w_n\}$ is a basic sequence equivalent to $\{u_n\}$ ([FHHMZ, Theorem 4.23]). In particular, Y contains ℓ_1, which is a contradiction. □

Lemma 85. *Let X be a Banach space, $Y \subset X$ a separable subspace that does not contain ℓ_1, and $Q\colon X \to X/Y$ the canonical quotient mapping. If $\{y_n\} \subset B_{X/Y}$ is weakly Cauchy, then there is a weakly Cauchy sequence $\{x_n\} \subset X$ such that $\lim\|x_{2k-1}\| \le 1$, $\lim\|x_{2k}\| \le 3$, $\{Q(x_{2k-1})\}$ is a subsequence of $\{y_{2k-1}\}$, and $\{Q(x_{2k})\}$ is a subsequence of $\{y_{2k}\}$.*

Proof. By Lemma 84 there are weakly Cauchy sequences $\{u_k\} \subset X$, $\{v_k\} \subset X$ such that $\lim\|u_k\| \le 1$, $\lim\|v_k\| \le 1$, $\{Q(u_k)\}$ is a subsequence of $\{y_{2k-1}\}$, and $\{Q(v_k)\}$ is a subsequence of $\{y_{2k}\}$. By the w^*-completeness of $2B_{X^{**}}$ (which follows from the w^*-compactness) there exist limits $u = w^*\text{-}\lim u_k \in X^{**}$ and $v = w^*\text{-}\lim v_k \in X^{**}$. Since Q^{**} is w^*–w^* continuous, we have $Q^{**}(u - v) = w^*\text{-}\lim Q^{**}(u_k - v_k) = w\text{-}\lim Q(u_k - v_k) = 0$. So $u - v \in \ker Q^{**} = Y^{**}$ (consider the canonical embedding $Y^{**} \subset X^{**}$ and note that $F \in X^{**}$ belongs to Y^{**} if and only if $F(f) = 0$ for every $f \in Y^{\perp}$, and that Q^* is onto $Y^{\perp}$). Clearly $\|u - v\| \le 2$ and so by the Goldstine theorem $u - v \in 2B_{Y^{**}} = \overline{2B_Y}^{(Y^{**}, w^*)}$. By Theorem 77 the compact $(2B_{Y^{**}}, w^*)$ is angelic and so there is a sequence $\{z_k\} \subset 2B_Y$ such that $u - v = w^*\text{-}\lim z_k$. Then $w^*\text{-}\lim v_k + z_k = u$, $Q(v_k + z_k) = Q(v_k)$, and by passing to a subsequence we may assume that $\lim\|v_k + z_k\| \le 3$. So the sequence $\{x_n\}$ defined by $x_{2k-1} = u_k$, $x_{2k} = v_k + z_k$ satisfies the requirements. □

We will use the following criterion several times throughout the book.

Fact 86. *Let $\{x_n\}$ be a sequence in a metric space P. Then $\{x_n\}$ is not Cauchy if and only if it has a subsequence $\{y_n\}$ such that* $\operatorname{dist}\big(\{y_{2k-1};\ k \in \mathbb{N}\}, \{y_{2k};\ k \in \mathbb{N}\}\big) > 0$.

Proof. $\Leftarrow$ is clear.

$\Rightarrow$ Without loss of generality we may assume that P is complete. If $\{x_n;\ n \in \mathbb{N}\}$ is totally bounded, then the sequence $\{x_n\}$ has two distinct cluster points $x, y \in P$ and we choose the subsequence $\{y_n\}$ so that $y_{2k} \to x$ and $y_{2k+1} \to y$. Otherwise there are $\varepsilon > 0$ and an infinite ε-separated subset of $\{x_n;\ n \in \mathbb{N}\}$, so the statement follows. □

The following result goes in the same direction as Theorem 41(v).

Proposition 87 ([CHL]). *Let X be a Banach space and let $Y \subset X$ be a separable subspace that does not contain ℓ_1. Let Z be a Banach space and suppose that $\mathcal{P}(^nX; Z) = \mathcal{P}_{\mathrm{wsc}}(^nX; Z)$ for some $n \in \mathbb{N}$. Then $\mathcal{P}^n(X/Y; Z) = \mathcal{P}^n_{\mathrm{wsc}}(X/Y; Z)$.*

Proof. Assuming the contrary, there is $P \in \mathcal{P}(^mX/Y; Z) \setminus \mathcal{P}_{\mathrm{wsC}}(^mX/Y; Z)$ for some $1 \le m \le n$ (Corollary 27). Therefore by Fact 86 there is a weakly Cauchy sequence $\{y_j\} \subset X/Y$ such that $\operatorname{dist}\big(\{P(y_{2k-1});\ k \in \mathbb{N}\}, \{P(y_{2k});\ k \in \mathbb{N}\}\big) > 0$. Denote by $Q\colon X \to X/Y$ the canonical quotient mapping. By Lemma 85 there is a weakly Cauchy sequence $\{x_j\} \subset X$ such that $\{Q(x_{2k-1})\}$ is a subsequence of $\{y_{2k-1}\}$

and $\{Q(x_{2k})\}$ is a subsequence of $\{y_{2k}\}$. However, since by Proposition 29 and the assumption $P \circ Q \in \mathcal{P}_{\mathrm{wsC}}({}^m X; Z)$, the sequence $\{P(Q(x_j))\}$ is Cauchy, which is a contradiction. □

We close this section by showing that the assumption that X does not contain ℓ_1 in the above results cannot be avoided. This fact is closely connected with the classical result of Aleksander Pełczyński. For the proof see [HMVZ, Theorem 7.29].

Theorem 88 (Aleksander Pełczyński [Peł5], James Hagler [Hag1]). *A Banach space X has a subspace isomorphic to ℓ_1 if and only if X^* has a subspace isomorphic to $L_1([0,1])$.*

In particular, it follows that if X has a subspace isomorphic to ℓ_1, then X^* has a subspace isomorphic to ℓ_2 ([FHHMZ, Theorem 4.53]). Since on ℓ_2 the weak and weak* topologies coincide, we conclude by duality the following fact.

Corollary 89. *Let X be a Banach space that has a subspace isomorphic to ℓ_1. Then X has a quotient isomorphic to ℓ_2.*

We are going to prove a useful variant of this corollary.

Lemma 90 ([D'H]). *Let X be a Banach space that has a subspace isomorphic to ℓ_1 and $p \geq 2$. Then there exist $T \in \mathcal{L}(X^{**}; \ell_p)$ and a normalised basic sequence $\{x_j\}_{j=1}^{\infty} \subset X$ equivalent to the canonical basis of ℓ_1 such that $T(x_j) = e_j$, where $\{e_j\}_{j=1}^{\infty}$ is the canonical basis of ℓ_p.*

Proof. It suffices to prove the case $p = 2$, as then we can compose T with the formal identity $Id\colon \ell_2 \to \ell_p$, which is a bounded linear operator. Let $R\colon \ell_2 \to L_1([0,1])$ be an isomorphic embedding, [FHHMZ, Theorem 4.53]. By Theorem 88 there is an isomorphic embedding $S\colon L_1([0,1]) \to X^*$. Put $f_j = S \circ R(e_j)$, $j \in \mathbb{N}$, where $\{e_j\}$ is the canonical basis of ℓ_2. Then $\{f_j\}$ is a basic sequence in X^* equivalent to $\{e_j\}$ and so it is weakly null. By Theorem 56 we may assume (passing to a subsequence) that there is a sequence $\{x_j\} \subset X$ biorthogonal to $\{f_j\}$. It is easy to see that $\{(R(e_j); S^*(x_j))\}$ is a biorthogonal system in $L_1([0,1]) \times L_\infty([0,1])$. Since $L_1([0,1])$ has the DPP and $\{R(e_j)\}$ is weakly null, by Proposition 40 the sequence $\{S^*(x_j)\}$ has no weakly Cauchy subsequence and hence the same holds for $\{x_j\}$. By Rosenthal's ℓ_1-theorem there is a subsequence of $\{x_j\}$ equivalent to the canonical basis of ℓ_1, which we still denote by $\{x_j\}$. We finish the proof by setting $T = (S \circ R)^*$. Since $\{T(x_j)\}$ is biorthogonal to $\{e_j\}$, it follows that $T(x_j) = e_j$. Finally we can normalise the sequence $\{x_j\}$ and compose T with the isomorphism $U\colon \ell_2 \to \ell_2$, $U\left(\sum a_j e_j\right) = \sum a_j \|x_j\| e_j$. □

Proposition 91 ([D'H]). *Let X be a Banach space that has a subspace isomorphic to ℓ_1 and let $k \in \mathbb{N}$, $k \geq 2$. Then there exist a polynomial $P \in \mathcal{P}({}^k X)$ and a normalised basic sequence $\{x_j\}_{j=1}^{\infty} \subset X$ equivalent to the canonical basis of ℓ_1 such that*

$$P\left(\sum_{j=1}^{\infty} a_j x_j\right) = \sum_{j=1}^{\infty} a_j^k$$

for every $(a_j)_{j=1}^{\infty} \in \ell_1$. In particular, the restriction of P to bounded sets is not weakly continuous at the origin, so $\mathcal{P}({}^k X) \neq \mathcal{P}_{\mathrm{w}}({}^k X)$.

Proof. Let $T\colon X \to \ell_k$ be (the restriction of) the linear operator and $\{x_j\} \subset S_X$ the basic sequence from Lemma 90. Define $Q \in \mathcal{P}(^k\ell_k)$ by $Q(y) = \sum_{j=1}^{\infty} y_j^k$ and put $P = Q \circ T$. To prove the last assertion let $W = \{x \in X;\ |\phi_j(x)| < \delta, j = 1, \dots, n\}$ be a weak neighbourhood of the origin given by $\delta > 0$ and $\phi_1, \dots, \phi_n \in X^*$. By the boundedness there is a subsequence $\{x_{j_s}\}_{s=1}^{\infty}$ such that $\{\phi_l(x_{j_s})\}_{s=1}^{\infty}$ converges for each $l = 1, \dots, n$. Thus there are distinct $k, l, m \in \mathbb{N}$ such that $|\phi_j(x_k) - \phi_j(x_l)| < \delta$ and $|\phi_j(x_k) - \phi_j(x_m)| < \delta$, $j = 1, \dots, n$. So for $x = x_k - \frac{1}{2}x_l - \frac{1}{2}x_m$ we have $x \in B(0,2) \cap W$ but $P(x) \ge 1 - \frac{2}{2^k} \ge \frac{1}{2}$. □

4. (p,q)-summing operators

In this section we survey some results concerning the (p,q)-summing operators, some of which will be needed in the proof of the main results in this chapter, i.e. the estimates of the coefficients of polynomials in $\mathcal{P}(^nc_0;\ell_p)$. Most of these results with full proofs can be found in [DJT], or in some cases in [FHHMZ] or [Dies2]. The theory has classical roots, based on the study of unconditional convergence of series in Banach spaces. The next classical theorem has played an important role in the theory of summing operators.

Theorem 92 (Władysław Orlicz, [Dies2, p. 108])**.** *Let $\sum_{n=1}^{\infty} x_n$ be an unconditionally convergent series in $L_p([0,1])$, $1 \le p \le 2$. Then $\sum_{n=1}^{\infty} \|x_n\|^2 < +\infty$.*

We introduce the basic concepts of the theory of summing operators.

Definition 93. Let X be a Banach space and let $1 \le p < \infty$. A sequence $\{x_j\}_{j=1}^{\infty} \subset X$ is called p-summable if $\sum_{j=1}^{\infty} \|x_j\|^p < +\infty$. We define its p-summable norm by

$$\|(x_j)_{j=1}^{\infty}\|_p = \left(\sum_{j=1}^{\infty} \|x_j\|^p\right)^{\frac{1}{p}}.$$

A sequence $\{x_j\}_{j=1}^{\infty} \subset X$ is called weakly p-summable if $\sum_{j=1}^{\infty} |\phi(x_j)|^p < +\infty$ for every $\phi \in X^*$. Analogously as in Fact 44 we can show that $\{x_j\}$ is weakly p-summable if and only if $\sup_{\phi \in B_{X^*}} \sum_{j=1}^{\infty} |\phi(x_j)|^p < +\infty$. Thus we can define its weakly p-summable norm by

$$\|(x_j)_{j=1}^{\infty}\|_p^w = \sup_{\phi \in B_{X^*}} \left(\sum_{j=1}^{\infty} |\phi(x_j)|^p\right)^{\frac{1}{p}}.$$

We denote by $\ell_p(\mathbb{N};X)$ the space of all p-summable sequences in X equipped with the p-summable norm. Similarly, we denote by $\ell_p^w(\mathbb{N};X)$ the space of all weakly p-summable sequences in X equipped with the weakly p-summable norm.

To shorten the notation we will use $\ell_p(X) = \ell_p(\mathbb{N}; X)$ and $\ell_p^w(X) = \ell_p(\mathbb{N}; X)$ in the rest of the chapter. This will hopefully not lead to a confusion with the usual notation $\ell_p(\Gamma) = \ell_p(\Gamma; \mathbb{K})$.

The spaces $\ell_p(X)$ and $\ell_p^w(X)$ are Banach spaces ([DJT, Chapter 2]). By Fact 44 we see that weakly 1-summable sequences coincide with wuC sequences. Hence the Orlicz theorem in combination with the Bessaga-Pełczyński theorem (Theorem 46) states that every weakly 1-summable sequence in $L_p([0,1])$, $1 \le p \le 2$, is 2-summable.

Proposition 94. *Let X be a Banach space, $1 < p \le \infty$, and $\frac{1}{p} + \frac{1}{q} = 1$. A sequence $\{x_j\}_{j=1}^\infty \subset X$ has an upper p-estimate with constant $C > 0$ if and only if it is weakly q-summable with $\|(x_j)\|_q^w \le C$.*

Proof. Suppose that $\sup_{\phi \in B_{X^*}} \left(\sum_{j=1}^\infty |\phi(x_j)|^q\right)^{\frac{1}{q}} = K < +\infty$. By the Hölder inequality

$$\left\|\sum_{j=1}^n a_j x_j\right\| = \sup_{\phi \in B_{X^*}} \phi\left(\sum_{j=1}^n a_j x_j\right) = \sup_{\phi \in B_{X^*}} \sum_{j=1}^n a_j \phi(x_j) \le K \|(a_j)_{j=1}^n\|_{\ell_p}.$$

Hence each weakly q-summable sequence has an upper p-estimate. To see the converse, let $C > 0$ be the constant from the upper estimate and $\phi \in B_{X^*}$. We use the classical duality of ℓ_p and ℓ_q to obtain

$$\left(\sum_{j=1}^n |\phi(x_j)|^q\right)^{\frac{1}{q}} = \sup_{\|(a_j)\|_{\ell_p} \le 1} \sum_{j=1}^n a_j \phi(x_j) = \sup_{\|(a_j)\|_{\ell_p} \le 1} \phi\left(\sum_{j=1}^n a_j x_j\right) \le C. \qquad \square$$

Corollary 95. *Let X be a Banach space, $1 < p \le \infty$, and $\frac{1}{p} + \frac{1}{q} = 1$. Then the space $\mathcal{L}(\ell_p; X)$ (resp. $\mathcal{L}(c_0; X)$ if $p = \infty$) is isometrically isomorphic to $\ell_q^w(X)$ via the canonical isomorphism $T \mapsto (T(e_j))_{j=1}^\infty$.*

Corollary 96. *Let $1 < p \le \infty$, let $\{e_j\}$ be the canonical basis of ℓ_p (resp. c_0 if $p = \infty$), and $\frac{1}{p} + \frac{1}{q} = 1$. Then $\|(e_j)\|_q^w = 1$.*

We denote by $\ell_p^{w0}(X) \subset \ell_p^w(X)$ the closed subspace consisting of the elements $(x_k)_{k=1}^\infty$ such that $\lim_{n \to \infty} \|(0, \dots, 0, x_n, x_{n+1}, \dots)\|_p^w = 0$.

Proposition 97 ([DefFl, Section 8.2], [DFS, Corollary 1.1.12])**.** *Let X be a Banach space and $1 \le p < \infty$. Then $\ell_p \otimes_\varepsilon X$ is isometric to the space $\ell_p^{w0}(X)$ via the natural mapping $I : \ell_p \otimes_\varepsilon X \to \ell_p^{w0}(X)$ defined on a dense subspace by*

$$I\left(\sum_{j=1}^n \left(\sum_{l=1}^m a_l^j e_l\right) \otimes x_j\right) = \left(\sum_{j=1}^n a_1^j x_j, \dots, \sum_{j=1}^n a_m^j x_j, 0, 0, \dots\right).$$

The vector valued sequence space $\ell_p(X)$ can be identified, in the isometric and canonical fashion, with the tensor product space $\ell_p \otimes_{\Delta_p} X$, where the tensor norm Δ_p is defined so that the canonical embedding $\ell_p \otimes_{\Delta_p} X \to \ell_p(X)$ which is given by $(a_j)_{j=1}^\infty \otimes x \mapsto (a_j x)_{j=1}^\infty$ becomes an isometry, [DefFl, Chapter 7].

Definition 98 ([MiPe]). Let X, Y be Banach spaces. Given $1 \leq p,q < \infty$ and an operator $T \in \mathcal{L}(X;Y)$, suppose that there is $C > 0$ such that

$$\left(\sum_{j=1}^{n} \|T(x_j)\|^p\right)^{\frac{1}{p}} \leq C \sup_{\phi \in B_{X^*}} \left(\sum_{j=1}^{n} |\phi(x_j)|^q\right)^{\frac{1}{q}}$$

for every $x_1, \dots, x_n \in X$, $n \in \mathbb{N}$. Then we say that T is (p,q)-summing. The infimum of C for which the inequality holds is denoted by $\pi_{p,q}(T)$, and it is called the (p,q)-summing norm of T.

By considering the sequences $\{\lambda_j x\}$ it is easy to see that if T is a non-zero operator that is (p,q)-summing, then $q \leq p$. Clearly a (p,q)-summing operator maps weakly q-summable sequences to p-summable sequences and by using the closed graph theorem similarly as in Fact 44 we can see that this is actually a characterisation. It is easy to see that composing any (p,q)-summing operator T with a bounded operator S from the left or from the right preserves the summing property and yields $\pi_{p,q}(S \circ T) \leq \|S\| \pi_{p,q}(T)$, resp. $\pi_{p,q}(T \circ S) \leq \|S\| \pi_{p,q}(T)$.

If $T \in \mathcal{L}(X;Y)$ and $q \leq p$, then the formal identity $Id : \ell_q \to \ell_p$ is bounded and by Proposition 12 we conclude that $Id \otimes T : \ell_q \otimes_\varepsilon X \to \ell_p \otimes_\varepsilon Y$ is a bounded operator. If T is (p,q)-summing, then the range space $\ell_p \otimes_\varepsilon Y$ may be equipped with the stronger Δ_p-norm still preserving the boundedness of $Id \otimes T$. In this case we are going to use the slightly inaccurate notation $Id \otimes T : \ell_q \otimes_\varepsilon X \to \ell_p(Y)$, which is very convenient for inductive arguments involving multiple tensor products, since $\ell_p^n \otimes_{\Delta_p} \ell_p^m \otimes_{\Delta_p} X = \ell_p^{nm} \otimes_{\Delta_p} X = \ell_p^{nm}(X)$. The following result gives a number of equivalent conditions for an operator to be (p,q)-summing.

Theorem 99. *Let X, Y be Banach spaces. Given $1 \leq p,q < \infty$, $C > 0$, and an operator $T \in \mathcal{L}(X;Y)$, the following statements are equivalent:*

(i) *T is (p,q)-summing and $\pi_{p,q}(T) \leq C$.*

(ii) *For every $\{x_j\}_{j=1}^{\infty} \subset X$*

$$\left(\sum_{j=1}^{\infty} \|T(x_j)\|^p\right)^{\frac{1}{p}} \leq C \sup_{\phi \in B_{X^*}} \left(\sum_{j=1}^{\infty} |\phi(x_j)|^q\right)^{\frac{1}{q}}.$$

(iii) $\|Id \otimes T : \ell_q^n \otimes_\varepsilon X \to \ell_p^n(Y)\| \leq C$ *for every $n \in \mathbb{N}$.*

(iv) $\|Id \otimes T : \ell_q \otimes_\varepsilon X \to \ell_p(Y)\| \leq C$.

(v) *For every $L \in \mathcal{L}(\ell_{q^*}; X)$, $\frac{1}{q} + \frac{1}{q^*} = 1$ (or $L \in \mathcal{L}(c_0; X)$ if $q = 1$) the operator $T \circ L$ is (p,q)-summing and $\pi_{p,q}(T \circ L) \leq C\|L\|$.*

Sketch of proof. Statements (i) and (ii) are clearly equivalent. Statement (iv) follows from (ii) using the identification from Proposition 97 and the remarks above. On the other hand, (iv) implies (iii), which readily implies (i) by using Proposition 97. (i)⇒(v) is clear and for the reverse implication fix $\{x_j\} \subset \ell_q^w(X)$ and let $L \in \mathcal{L}(\ell_{q^*}; X)$ (resp. $\mathcal{L}(c_0; X)$) be such that $L(e_j) = x_j$ and $\|L\| = \|(x_j)\|_q^w$ (Corollary 95).

Then $\left(\sum_{j=1}^n \|T(x_j)\|^p\right)^{\frac{1}{p}} = \left(\sum_{j=1}^n \|T \circ L(e_j)\|^p\right)^{\frac{1}{p}} \le C\|L\| \|(e_j)\|_q^w = C\|L\| = C\|(x_j)\|_q^w$, using also Corollary 96. □

The next result is a generalisation of Orlicz's theorem. Given $1 \le p \le q \le \infty$, $p < \infty$ we define the number

$$r(p,q) = \frac{2}{1 + 2\left(\frac{1}{p} - \max\left\{\frac{1}{q}, \frac{1}{2}\right\}\right)}.$$

Theorem 100 ([Ben], [Carl]). *If $1 \le p \le q \le \infty$, $p < \infty$, and $r = r(p,q)$, then $Id : \ell_p \to \ell_q$ is $(r, 1)$-summing.*

This fundamental result can be reformulated as follows.

Theorem 101. *Let $1 \le p \le q \le \infty$, $p < \infty$, and $r = r(p,q)$. There is $C > 0$ such that for every $L \in \mathcal{L}(c_0; \ell_p)$*

$$\left(\sum_{j=1}^{\infty} \|L(e_j)\|_q^r\right)^{\frac{1}{r}} \le C\|L\|.$$

Proof. Let $X = \ell_p$, $Y = \ell_q$, and $T = Id : X \to Y$. Then T is $(r, 1)$-summing by Theorem 100, so $T \circ L$ is also $(r, 1)$-summing and $\pi_{r,1}(T \circ L) \le \pi_{r,1}(T)\|L\|$. The statement then follows using Corollary 96. □

In Section 5 we will show an extension of this result for multilinear mappings.

The summing operators are intimately related to the notions of type and cotype, introduced by Jørgen Hoffmann-Jørgensen in [Ho] (type), and by Bernard Maurey and Gilles Pisier in [MaPi2] (cotype).

Definition 102. We say that a normed linear space X is of type p, $1 \le p < \infty$, if there exists a $C > 0$ such that for every $x_1, \dots, x_n \in X$

$$\left(\frac{1}{2^n} \sum_{\varepsilon_j = \pm 1} \left\| \sum_{j=1}^n \varepsilon_j x_j \right\|^p\right)^{\frac{1}{p}} \le C\left(\sum_{j=1}^n \|x_j\|^p\right)^{\frac{1}{p}}. \tag{5}$$

The smallest C allowed here is called the type-p constant of X and is denoted by $T_p(X)$.

We say that a normed linear space X is of cotype q, $1 \le q < \infty$, if there exists a $C > 0$ such that for every $x_1, \dots, x_n \in X$

$$\left(\sum_{j=1}^n \|x_j\|^q\right)^{\frac{1}{q}} \le C\left(\frac{1}{2^n} \sum_{\varepsilon_j = \pm 1} \left\| \sum_{j=1}^n \varepsilon_j x_j \right\|^q\right)^{\frac{1}{q}}. \tag{6}$$

The smallest C allowed here is called the cotype-q constant of X and is denoted by $C_q(X)$.

Clearly, every normed linear space X is of type 1 with $T_1(X) = 1$ and it is of cotype ∞ (in this case the inequality in the definition naturally transforms to $\max_j \|x_j\| \leq C \sum_{j=1}^n \|x_j\|$) with $C_\infty(X) = 1$. Further, using the Hölder inequality it is easily seen that if X is of type p, then it is also of type r for any $1 \leq r \leq p$ and $T_r(X) \leq T_p(X)$, and if X is of cotype q, then it is also of cotype s for any $s \geq q$ and $C_s(X) \leq C_q(X)$. By considering $x_j = x$ for all j, Khintchine's inequality ([FHHMZ, Lemma 4.54]) implies than no non-trivial normed linear space can be of type $p > 2$ or cotype $q < 2$. Thus we say that a normed linear space is of a non-trivial type if it is of type $p \in (1, 2]$ and it is of a non-trivial cotype if it is of cotype $q \in [2, +\infty)$.

The generalised parallelogram law in a Hilbert space H implies that H is of type 2 and cotype 2 with $T_2(H) = C_2(H) = 1$. More generally, the infinite-dimensional space $L_p(\mu)$, $1 \leq p < \infty$ is of type $\min\{p, 2\}$ and cotype $\max\{p, 2\}$ and both are the best possible ([AK, Theorem 6.2.14]). Moreover, $T_p(L_p(\mu)) = C_q(L_q(\mu)) = 1$ for $1 \leq p \leq 2$ and $q \geq 2$.

By Kahane's inequality (Theorem 2.76) we see that we can replace the exponent p on the left-hand side of (5) by any other exponent $1 \leq r < \infty$ (of course the constant C changes to $K_{p,r}C$) and similarly we can replace the exponent q on the right-hand side of (6) by any other exponent $1 \leq r < \infty$. Of particular interest is $r = 1$ or $r = 2$.

A simple consequence of Fact 42 is the following observation, which can be viewed as another generalisation of Orlicz's theorem.

Proposition 103. *If a Banach space X is of cotype q, then Id_X is $(q, 1)$-summing and $\pi_{q,1}(Id_X) \leq C_q(X)$.*

Obviously both the type and cotype are isomorphic invariants, are inherited by subspaces, and pass to finite direct sums. If X is of type p, then X^* is of cotype q, where $\frac{1}{p} + \frac{1}{q} = 1$, and $C_q(X^*) \leq T_p(X)$, see e.g. [AK, Proposition 6.2.12]. (We also remark that the converse does not hold: the space ℓ_1 is of cotype 2, while c_0 is not of a non-trivial type, see the remarks after Theorem 4.2.) If X has an equivalent norm with modulus of smoothness of power type p, then it is of type p ([DGZ, Proposition IV.5.10]), and if X has an equivalent norm with modulus of convexity of power type q, then it is of cotype q ([DGZ, Proposition IV.5.12]). These results can be reversed provided that X is a super-reflexive Banach lattice ([LiTz3, Chapter 1]).

Theorem 104 ([MaPi2], [DJT, Lemma 14.10])**.** *Let X be a Banach space of cotype q. Then for any measure μ the space $L_q(\mu; X)$ is of cotype q with the constant $C_q(L_q(\mu; X)) = C_q(X)$.*

For the convenience of the reader we also recall a version of the classical Minkowski inequality, see e.g. [HLP, Section 2.11]: Let $a_{j,k} \in \mathbb{R}$, $j = 1, \ldots, M$, $k = 1, \ldots, N$, and $0 < r \leq s < \infty$. Then

$$\left(\sum_{j=1}^{N}\left(\sum_{k=1}^{M}|a_{jk}|^r\right)^{\frac{s}{r}}\right)^{\frac{1}{s}} \leq \left(\sum_{k=1}^{M}\left(\sum_{j=1}^{N}|a_{jk}|^s\right)^{\frac{r}{s}}\right)^{\frac{1}{r}}. \tag{7}$$

5. Estimates of coefficients of multilinear mappings

In the present section we obtain precise estimates on the values of coefficients of multilinear mappings and homogeneous polynomials in $\mathcal{P}({}^n c_0; \ell_p)$, leading to a quantitative (and polynomial) version of condition (iii) in Theorem 47. These results are based on a systematic use of the properties of $(p,1)$-summing operators, whose relevance to the problem of estimating the coefficients can be seen from Proposition 1.24.

Definition 105. Let $X_1, \dots, X_n, Y$ be Banach spaces and $1 \le p, q_1, \dots, q_n < \infty$. We say that $M \in \mathcal{L}(X_1, \dots, X_n; Y)$ is multiple $(p; q_1, \dots, q_n)$-summing if there is a constant $K > 0$ such that for every choice of $m_1, \dots, m_n \in \mathbb{N}$ and $x_j^k \in X_k$, $1 \le j \le m_k$, $1 \le k \le n$,

$$\left(\sum_{\substack{1 \le j_k \le m_k \\ 1 \le k \le n}} \|M(x_{j_1}^1, \dots, x_{j_n}^n)\|^p \right)^{\frac{1}{p}} \le K \prod_{k=1}^n \|(x_j^k)_{j=1}^{m_k}\|_{q_k}^w. \tag{8}$$

In this case we define the multiple $(p; q_1, \dots, q_n)$-summing norm of M by

$$\pi_{p;q_1,\dots,q_n}(M) = \inf\{K;\ K \text{ satisfies } (8)\}.$$

A multiple $(p; q, \dots, q)$-summing multilinear mapping will be called for short multiple $(p;q)$-summing and the norm will be denoted by $\pi_{p;q}$. The next fact is straightforward.

Fact 106. *Let $X_1, \dots, X_n$, Y be Banach spaces and $1 \le p, q_1, \dots, q_n < \infty$. Then $M \in \mathcal{L}(X_1, \dots, X_n; Y)$ is multiple $(p; q_1, \dots, q_n)$-summing if and only if there is a constant $K > 0$ such that for every choice of $m_2, \dots, m_n \in \mathbb{N}$ and $x_j^k \in X_k$, $1 \le j \le m_k$, $2 \le k \le n$, with $\|(x_j^k)_{j=1}^{m_k}\|_{q_k}^w \le 1$ the associated linear operator $S\colon X_1 \to \ell_p^{m_2 \cdots m_n}(Y)$ given by*

$$S(x) = \big(M(x, x_{j_2}^2, \dots, x_{j_n}^n)\big)_{\substack{1 \le j_k \le m_k \\ 2 \le k \le n}} \tag{9}$$

is (p, q_1)-summing and satisfies

$$\pi_{p,q_1}(S) \le K. \tag{10}$$

In that case $\pi_{p;q_1,\dots,q_n}(M) = \inf\{K;\ K \text{ satisfies } (10)\}$.

Theorem 107 ([BPV]). *Let $X_1, \dots, X_n$ be Banach spaces and let Y be a Banach space of cotype q. Then every $M \in \mathcal{L}(X_1, \dots, X_n; Y)$ is multiple $(q;1)$-summing and $\pi_{q;1}(M) \le C_q(Y)^n \|M\|$.*

Proof. For $n = 1$ it suffices to notice that $M = Id_Y \circ M$ and apply Proposition 103. We proceed by induction. Suppose the statement has been proved for $n-1$. Fix $m_2, \dots, m_n \in \mathbb{N}$ and elements $x_j^k \in X_k$, $1 \le j \le m_k$, with $\|(x_j^k)_{j=1}^{m_k}\|_1^w \le 1$, $k = 2, \dots, n$. Using Fact 106 it suffices to show that $\pi_{q,1}(S) \le C_q(Y)^n \|M\|$, where S is defined by (9). By the inductive hypothesis for every $x \in X_1$ the mapping

$M_x \in \mathcal{L}(X_2, \ldots, X_n; Y)$ given by $M_x(x_2, \ldots, x_n) = M(x, x_2, \ldots, x_n)$ is multiple $(q; 1)$-summing and

$$\pi_{q;1}(M_x) \leq C_q(Y)^{n-1} \|M_x\| \leq C_q(Y)^{n-1} \|M\| \|x\|.$$

So for every $x \in X_1$

$$\|S(x)\| \leq \pi_{q;1}(M_x) \leq C_q(Y)^{n-1} \|M\| \|x\|,$$

whence $\|S\| \leq C_q(Y)^{n-1} \|M\|$. By Theorem 104, $\ell_q^{m_2 \cdots m_n}(Y)$ is of cotype q with the constant $C_q(\ell_q^{m_2 \cdots m_n}(Y)) = C_q(Y)$. Then from the linear case (i.e. $n = 1$) we obtain that S is $(q, 1)$-summing and

$$\pi_{q,1}(S) \leq C_q(Y) C_q(Y)^{n-1} \|M\| = C_q(Y)^n \|M\|. \qquad \square$$

Using $x_j^k = e_j$ in (8), where $\{e_j\}$ is the canonical basis of c_0, and combining the above theorem with Corollary 96 we conclude the following result.

Corollary 108. *Let Y be a Banach space of cotype q. Then every $M \in \mathcal{L}({}^n c_0; Y)$ satisfies*

$$\left(\sum_{j_1, \ldots, j_n \in \mathbb{N}} \|M(e_{j_1}, \ldots, e_{j_n})\|^q \right)^{\frac{1}{q}} \leq C_q(Y)^n \|M\|,$$

where $\{e_j\}_{j=1}^{\infty}$ is the canonical basis of c_0.

Since $C_q(\ell_q) = 1$ for $q \geq 2$, we obtain the following corollary.

Corollary 109 ([BPV]). *If $2 \leq q < \infty$, then every $M \in \mathcal{L}({}^n c_0; \ell_q)$ satisfies*

$$\left(\sum_{j_1, \ldots, j_n \in \mathbb{N}} \|M(e_{j_1}, \ldots, e_{j_n})\|^q \right)^{\frac{1}{q}} \leq \|M\|,$$

where $\{e_j\}_{j=1}^{\infty}$ is the canonical basis of c_0.

Corollary 110. *Let Y be a Banach space of cotype q and $P \in \mathcal{P}({}^n c_0; Y)$. If P is given by the formal expression*

$$P(x) = \sum_{\alpha \in \mathcal{J}(\infty, n)} x^\alpha y_\alpha$$

(see Proposition 1.24), then

$$\left(\sum_{\alpha \in \mathcal{J}(\infty, n)} \|y_\alpha\|^q \right)^{\frac{1}{q}} \leq C C_q(Y)^n \|\check{P}\|,$$

where $C = \max_{\beta \in \mathcal{J}(n,n)} \binom{n}{\beta}$.

Proof. By Proposition 1.24 each y_α, $\alpha = (\alpha_1, \alpha_2, \dots)$ is uniquely determined by

$$y_\alpha = \binom{n}{\alpha} \check{P}({}^{\alpha_1}e_1, {}^{\alpha_2}e_2, \dots),$$

where $\{e_j\}_{j=1}^\infty$ is the canonical basis of c_0. Hence

$$\Big(\sum_{\alpha \in \mathcal{J}(\infty,n)} \|y_\alpha\|^q\Big)^{\frac{1}{q}} \le C\Big(\sum_{j_1,\dots,j_n \in \mathbb{N}} \|\check{P}(e_{j_1},\dots,e_{j_n})\|^q\Big)^{\frac{1}{q}} \le C C_q(Y)^n \|\check{P}\|.$$

□

In [DeSe] the above estimates were further improved for $Y = \ell_q$, $q < 2$ (spaces of cotype 2). The method of proof relies on complex interpolation. Of course, the real case then follows by passing to the complexification, paying the price by getting worse constants in the estimates.

Theorem 111 ([DeSe]). *Let $n \in \mathbb{N}$ and $1 \le p \le q \le \infty$, $p < \infty$. There is $C > 0$ such that for every $M \in \mathcal{L}({}^n c_0; \ell_p)$*

$$\Big(\sum_{j_1,\dots,j_n \in \mathbb{N}} \|M(e_{j_1},\dots,e_{j_n})\|_q^\rho\Big)^{\frac{1}{\rho}} \le C\|M\|,$$

where $\{e_j\}_{j=1}^\infty$ is the canonical basis of c_0 and

$$\rho = \begin{cases} \frac{2n}{n+2\left(\frac{1}{p}-\max\left\{\frac{1}{q},\frac{1}{2}\right\}\right)} & \text{if } p < 2, \\ p & \text{if } p \ge 2. \end{cases}$$

The proof of the theorem is based on two lemmata together with Theorem 100. Note that while Theorem 99 gives an estimate on the operator norm of $Id \otimes T$, the next lemma gives an estimate on the summing norm of this operator.

Lemma 112 ([DeSe]). *Let X, Y be Banach spaces and suppose that Y is of cotype 2. If $T \in \mathcal{L}(X;Y)$ is $(r,1)$-summing for some $1 \le r < \infty$, then for each $m, n \in \mathbb{N}$ the operator $Id \otimes T \colon \otimes_\varepsilon^n \ell_1^m \otimes_\varepsilon X \to \ell_2^{m^n}(Y)$ is also $(r,1)$-summing and*

$$\pi_{r,1}(Id \otimes T) \le \big(\sqrt{2}C_2(Y)\big)^n \pi_{r,1}(T).$$

Proof. We use induction on n. Let us consider first the case $n = 1$. Take any $x_1,\dots,x_N \in \ell_1^m \otimes_\varepsilon X$ and let $x_k = \sum_{j=1}^m e_j \otimes x_k(j)$ where $x_k(j) \in X$ for $k = 1,\dots,N$. We want to show that

$$\Big(\sum_{k=1}^N \|(Id \otimes T)(x_k)\|_{\ell_2^m(Y)}^r\Big)^{\frac{1}{r}} \le \sqrt{2}C_2(Y)\pi_{r,1}(T) \sup_{\phi \in B_{(\ell_1^m \otimes_\varepsilon X)^*}} \sum_{k=1}^N |\phi(x_k)|.$$

Using successively the cotype 2 of the space Y and Kahane's inequality (with $K_{1,2} \le \sqrt{2}$), the Minkowski inequality (7), the fact that T is $(r,1)$-summing, and

Fact 42 we get the estimate

$$\begin{aligned}
\Big(\sum_{k=1}^{N}\|(Id\otimes T)(x_k)\|^r_{\ell_2^m(Y)}\Big)^{\frac{1}{r}} &= \Big(\sum_{k=1}^{N}\Big(\Big(\sum_{j=1}^{m}\|T(x_k(j))\|^2\Big)^{\frac{1}{2}}\Big)^r\Big)^{\frac{1}{r}}\\
&\le \sqrt{2}C_2(Y)\Big(\sum_{k=1}^{N}\Big(\frac{1}{2^m}\sum_{\varepsilon_j=\pm 1}\Big\|\sum_{j=1}^{m}\varepsilon_j T(x_k(j))\Big\|\Big)^r\Big)^{\frac{1}{r}}\\
&\le \sqrt{2}C_2(Y)\frac{1}{2^m}\sum_{\varepsilon_j=\pm 1}\Big(\sum_{k=1}^{N}\Big\|\sum_{j=1}^{m}\varepsilon_j T(x_k(j))\Big\|^r\Big)^{\frac{1}{r}}\\
&\le \sqrt{2}C_2(Y)\pi_{r,1}(T)\frac{1}{2^m}\sum_{\varepsilon_j=\pm 1}\sup_{\phi\in B_{X^*}}\sum_{k=1}^{N}\Big|\phi\Big(\sum_{j=1}^{m}\varepsilon_j x_k(j)\Big)\Big|\\
&\le \sqrt{2}C_2(Y)\pi_{r,1}(T)\sup_{|a_k|\le 1}\sup_{|\eta_j|\le 1}\Big\|\sum_{j=1}^{m}\eta_j\sum_{k=1}^{N}a_k x_k(j)\Big\|.
\end{aligned}$$

On the other hand, by Fact 42 and the definition of the injective tensor norm

$$\begin{aligned}
\sup_{\phi\in B_{(\ell_1^m\otimes_\varepsilon X)^*}}\sum_{k=1}^{N}|\phi(x_k)| &= \sup_{|a_k|\le 1}\Big\|\sum_{k=1}^{N}a_k x_k\Big\|_{\ell_1^m\otimes_\varepsilon X}\\
&= \sup_{|a_k|\le 1}\sup_{\substack{\eta\in B_{\ell_\infty^m}\\ \phi\in B_{X^*}}}\Big|\sum_{k=1}^{N}\sum_{j=1}^{m}a_k\eta_j\phi(x_k(j))\Big|\\
&= \sup_{\substack{|a_k|\le 1\\ \eta\in B_{\ell_\infty^m}}}\sup_{\phi\in B_{X^*}}\Big|\phi\Big(\sum_{k=1}^{N}\sum_{j=1}^{m}a_k\eta_j x_k(j)\Big)\Big|\\
&= \sup_{\substack{|a_k|\le 1\\ \eta\in B_{\ell_\infty^m}}}\Big\|\sum_{k=1}^{N}\sum_{j=1}^{m}a_k\eta_j x_k(j)\Big\|.
\end{aligned}$$

This finishes the proof for $n = 1$.

Now let $n > 1$ and assume that the statement of the lemma holds for $n-1$. Put $U = \otimes_\varepsilon^{n-1}\ell_1^m\otimes_\varepsilon X$, $V = \ell_2^{m^{n-1}}(Y)$, and $S = Id\otimes T\colon U\to V$. By Theorem 104, V is of cotype 2 with $C_2(V) = C_2(Y)$. Further, by the inductive hypothesis S is $(r,1)$-summing. Thus by the case $n = 1$ and the inductive hypothesis

$$\begin{aligned}
\pi_{r,1}\big(Id\otimes S\colon \ell_1^m\otimes_\varepsilon U\to \ell_2^m(V)\big) &\le \sqrt{2}C_2(V)\pi_{r,1}(S)\\
&\le \sqrt{2}C_2(Y)\big(\sqrt{2}C_2(Y)\big)^{n-1}\pi_{r,1}(T),
\end{aligned}$$

which completes the proof. □

In the proof of the next lemma we will make use of the following observation. Let $T \in \mathcal{L}(X;Y)$ for Banach spaces X, Y, and $1 \leq p,q,\rho \leq \infty$, $\frac{1}{p}+\frac{1}{q}=1$. Consider $S_T \in \mathcal{L}\big(\mathcal{L}({}^n\ell_p^m;X);\ell_\rho^{m^n}(Y)\big)$ defined by $S_T(M)=\big(T\circ M(e_{j_1},\dots,e_{j_n})\big)_{j_1,\dots,j_n=1}^m$. Then using the identification in Proposition 10 it is not difficult to see that S_T can be identified with an operator $Id\otimes T \in \mathcal{L}\big((\otimes_\varepsilon^n \ell_q^m)\otimes_\varepsilon X;\ell_\rho^{m^n}(Y)\big)$ and both operators have the same norm.

Lemma 113 ([DeSe]). *Let X, Y be complex Banach spaces where Y is of cotype 2, $M \in \mathcal{L}({}^n c_0;X)$, and let $T\in\mathcal{L}(X;Y)$ be $(r,1)$-summing for some $1\leq r\leq 2$. Then*

$$\Bigg(\sum_{j_1,\dots,j_n\in\mathbb{N}} \|T\circ M(e_{j_1},\dots,e_{j_n})\|^\rho\Bigg)^{\frac{1}{\rho}} \leq \big(\sqrt{2}C_2(Y)\big)^{n-1}\pi_{r,1}(T)\|M\|,$$

where $\rho=\frac{2n}{n+2(\frac{1}{r}-\frac{1}{2})}$ and $\{e_j\}_{j=1}^\infty$ is the canonical basis of c_0.

Proof. By the remark preceding the lemma it suffices to show that for each $m\in\mathbb{N}$

$$\big\|Id\otimes T\colon (\otimes_\varepsilon^n\ell_1^m)\otimes_\varepsilon X\to \ell_\rho^{m^n}(Y)\big\| \leq \big(\sqrt{2}C_2(Y)\big)^{n-1}\pi_{r,1}(T).$$

We use induction on n. For $n=1$ the statement is clear from the fact that T is $(r,1)$-summing and Theorem 99.

For the induction we use the notation $\rho=\rho_n$ (as r is fixed). Let us assume that the estimate holds for $n-1$. Denote $L=Id\otimes T\colon(\otimes_\varepsilon^n\ell_1^m)\otimes_\varepsilon X\to(\otimes_\varepsilon^n\ell_1^m)\otimes_\varepsilon Y$. In what follows we will obtain estimates on the norm of L in various norms on the target space, which will allow us to use another estimate on the norm using interpolation. First, Lemma 112 implies

$$\pi_{r,1}\big(Id\otimes T\colon(\otimes_\varepsilon^{n-1}\ell_1^m)\otimes_\varepsilon X\to\ell_2^{m^{n-1}}(Y)\big)\leq\big(\sqrt{2}C_2(Y)\big)^{n-1}\pi_{r,1}(T),$$

hence by Theorem 99

$$\big\|L\colon(\otimes_\varepsilon^n\ell_1^m)\otimes_\varepsilon X\to\ell_r^m(\ell_2^{m^{n-1}}(Y))\big\|\leq\big(\sqrt{2}C_2(Y)\big)^{n-1}\pi_{r,1}(T). \tag{11}$$

On the other hand, as we shall see, from the induction we are able to obtain estimates for the target space $\ell_{\rho_{n-1}}^{m^{n-1}}(\ell_2^m(Y))$. Our goal is to get an estimate of the norm of L in the target space $\ell_2^m(\ell_{\rho_{n-1}}^{m^{n-1}}(Y))$, which will be amenable to an easy interpolation with (11). To this end we use the Minkowski inequality (7), which however transposes the order of summation. To counter this effect, we will first compose L with the transposition operator.

So, denoting $S=Id\otimes T\colon\ell_1^m\otimes_\varepsilon X\to\ell_2^m(Y)$, Lemma 112 implies that S is $(r,1)$-summing and $\pi_{r,1}(S)\leq\sqrt{2}C_2(Y)\pi_{r,1}(T)$. Recall that $\ell_2^m(Y)$ is of cotype 2 with $C_2(\ell_2^m(Y))=C_2(Y)$ by Theorem 104. Thus by the inductive assumption used on S in place of T we obtain

$$\begin{aligned}\big\|Id\otimes S\colon(\otimes_\varepsilon^{n-1}\ell_1^m)\otimes_\varepsilon(\ell_1^m\otimes_\varepsilon X)\to\ell_{\rho_{n-1}}^{m^{n-1}}(\ell_2^m(Y))\big\|&\\ \leq\big(\sqrt{2}C_2(Y)\big)^{n-1}\pi_{r,1}(T).&\end{aligned}\tag{12}$$

Next, let $Q\colon \ell_{\rho_{n-1}}^{m^{n-1}}(\ell_2^m(Y)) \to \ell_2^m(\ell_{\rho_{n-1}}^{m^{n-1}}(Y))$ be defined by $Q\big(((y_{jk})_{j=1}^m)_{k=1}^{m^{n-1}}\big) = ((y_{jk})_{k=1}^{m^{n-1}})_{j=1}^m$. Then $\|Q\| \le 1$ by the Minkowski inequality (7), using also the fact that $\rho_{n-1} \le 2$. Further, let $t\colon \ell_1^m \otimes_\varepsilon (\otimes_\varepsilon^{n-1}\ell_1^m) \to (\otimes_\varepsilon^{n-1}\ell_1^m) \otimes_\varepsilon \ell_1^m$ be defined by $t(e_{j_1} \otimes (e_{j_2} \otimes \cdots \otimes e_{j_n})) = (e_{j_2} \otimes \cdots \otimes e_{j_n}) \otimes e_{j_1}$. Clearly, $\|t\| = 1$, and so also $\|t \otimes Id_X\| = 1$ (using the injective tensor product). Moreover, observe that $L = Q \circ L \circ (t \otimes Id_X)$ in the algebraic sense. Thus in combination with (12) we obtain

$$\big\|L\colon (\otimes_\varepsilon^n \ell_1^m) \otimes_\varepsilon X \to \ell_2^m(\ell_{\rho_{n-1}}^{m^{n-1}}(Y))\big\| \le \big(\sqrt{2}C_2(Y)\big)^{n-1}\pi_{r,1}(T).$$

Finally, we use the complex interpolation on the above estimate and (11) with $\theta = \frac{1}{n}$. By [BerLö, Theorem 4.1.2] and [BerLö, Theorem 4.2.1] we get the estimate

$$\big\|L\colon (\otimes_\varepsilon^n \ell_1^m)\otimes_\varepsilon X \to \big[\ell_2^m(\ell_{\rho_{n-1}}^{m^{n-1}}(Y)), \ell_r^m(\ell_2^{m^{n-1}}(Y))\big]_{\frac{1}{n}}\big\| \le \big(\sqrt{2}C_2(Y)\big)^{n-1}\pi_{r,1}(T).$$

But by [BerLö, Theorem 5.1.2],

$$\big[\ell_2^m(\ell_{\rho_{n-1}}^{m^{n-1}}(Y)), \ell_r^m(\ell_2^{m^{n-1}}(Y))\big]_{\frac{1}{n}} = \ell_\mu^m\big([\ell_{\rho_{n-1}}^{m^{n-1}}(Y), \ell_2^{m^{n-1}}(Y)]_{\frac{1}{n}}\big) = \ell_\mu^m(\ell_\nu^{m^{n-1}}(Y)),$$

where

$$\frac{1}{\mu} = \frac{1-\frac{1}{n}}{2} + \frac{\frac{1}{n}}{r} \quad\text{and}\quad \frac{1}{\nu} = \frac{1-\frac{1}{n}}{\rho_{n-1}} + \frac{\frac{1}{n}}{2}.$$

Since $\mu = \nu = \rho_n$, this completes the proof. □

Proof of Theorem 111. The case $2 \le p \le q \le \infty$ follows from Corollary 109 together with the fact that $\|y\|_q \le \|y\|_p$ for any $y \in \ell_p$.

Case $1 \le p \le q \le 2$: By Theorem 100 the identity $Id\colon \ell_p \to \ell_q$ is $(r,1)$-summing, where $\frac{1}{r} = \frac{1}{2} + \frac{1}{p} - \frac{1}{q}$. Using Lemma 113 together with the fact that ℓ_q is of cotype 2 we get the desired estimate.

Case $1 \le p < 2 < q \le \infty$: By Theorem 100 we know that $Id\colon \ell_p \to \ell_2$ is $(p,1)$-summing. We then apply Lemma 113 combined with the fact that $\|y\|_q \le \|y\|_2$ for any $y \in \ell_2$. □

By taking $p = 1$ and $q = 2$ in Theorem 111 we immediately obtain the following special case.

Corollary 114 (Henri Frédéric Bohnenblust and Einar Hille, [BH]). *If $M \in \mathcal{L}({}^n c_0;\mathbb{C})$, then*

$$\Bigg(\sum_{j_1,\dots,j_n\in\mathbb{N}} |M(e_{j_1},\dots,e_{j_n})|^{\frac{2n}{n+1}}\Bigg)^{\frac{n+1}{2n}} \le \big(\sqrt{2}\big)^{n-1}\|M\|.$$

The constant in the above corollary is due to Alexander Munro Davie [Da] and Sten Kaijser [Kai]. For $n = 2$ this result is known as Littlewood's $\frac{4}{3}$-formula.

The polynomial case of Theorem 111 follows readily. The proof is the same as that of Corollary 110.

Corollary 115 ([DeSe]). *Let $n \in \mathbb{N}$ and $1 \leq p \leq q \leq \infty$, $p < \infty$. Then there is $C > 0$ such that if $P \in \mathcal{P}({}^n c_0; \ell_p)$ is given by the formal expression*

$$P(x) = \sum_{\alpha \in \mathcal{J}(\infty,n)} x^\alpha y_\alpha,$$

then

$$\left(\sum_{\alpha \in \mathcal{J}(\infty,n)} \|y_\alpha\|_q^\rho \right)^{\frac{1}{\rho}} \leq C \|P\|,$$

where ρ is defined in Theorem 111.

6. Bohr radius

Let Y be a complex Banach space and $U \subset \mathbb{C}^n$. According to [Dieu, Section 9.3] a mapping $f: U \to Y$ is said to be analytic if for every $a = (a_1, \dots, a_n) \in U$ there are $r_1, \dots, r_n > 0$ and $y_\alpha \in Y$, $\alpha = (\alpha_1, \dots, \alpha_n) \in \mathbb{N}_0^n$, such that

$$f(z) = \sum_{\alpha \in \mathbb{N}_0^n} (z_1 - a_1)^{\alpha_1} \cdots (z_n - a_n)^{\alpha_n} y_\alpha$$

for every $z = (z_1, \dots, z_n)$ in some open polydisc $U(a_1, r_1) \times \dots \times U(a_n, r_n) \subset U$ and the power series is absolutely convergent there. Formally this definition is much stronger than our definition in Section 1.8, which only requires the convergence of the abstract power series, in which all the n-homogeneous terms are "pre-summed". However, from the Cauchy formula it easily follows that the Taylor expansion of a holomorphic mapping on $U \subset \mathbb{C}^n$ converges also in this stronger sense on any polydisc contained in U. Indeed, from Theorem 1.163 we obtain for any $\alpha \in \mathbb{N}_0^n$

$$\begin{aligned} y_\alpha &= \frac{1}{\alpha_1! \cdots \alpha_n!} \frac{\partial^{|\alpha|} f}{\partial z_1^{\alpha_1} \dots \partial z_n^{\alpha_n}}(a) \\ &= \frac{1}{(2\pi i)^n} \int_{\gamma_1} \cdots \int_{\gamma_n} \frac{f(a_1 + z_1, \dots, a_n + z_n)}{z_1^{\alpha_1+1} \cdots z_n^{\alpha_n+1}} \, \mathrm{d}z_n \cdots \mathrm{d}z_1, \end{aligned} \tag{13}$$

where $\gamma_j \colon [0, 2\pi] \to \mathbb{C}$, $\gamma_j(t) = r_j e^{it}$, $j = 1, \dots, n$. Thus there is $M > 0$ such that $\|y_\alpha\| \leq \frac{M}{r_1^{\alpha_1} \cdots r_n^{\alpha_n}}$ for every $\alpha \in \mathbb{N}_0^n$, from which the absolute convergence in $U(a_1, r_1) \times \dots \times U(a_n, r_n)$ follows. See also [Dieu, Theorem 9.10.1].

In this section we are going to be concerned with the absolute convergence of the classical ("expanded") power series in finite-dimensional complex Banach spaces. Since such power series are in fact generalised series, their convergence is always understood as an absolute convergence. Let X be a complex n-dimensional Banach space with a basis $\{e_j\}_{j=1}^n$ satisfying

$$\left\| \sum_{j=1}^n \lambda_j a_j e_j \right\| \leq \left\| \sum_{j=1}^n a_j e_j \right\| \quad \text{for all } a_j, \lambda_j \in \mathbb{C}, |\lambda_j| \leq 1, \, j = 1, \dots, n, \tag{14}$$

and denote by $D = U_X$ its open unit ball. All coordinates below are understood with respect to this basis. The Bohr radius of D, denoted by $K(D)$, is defined as a supremum of $r \in [0, 1]$ satisfying the following: for any power series $\sum_{\alpha \in \mathbb{N}_0^n} a_\alpha z^\alpha$, $a_\alpha \in \mathbb{C}$, absolutely convergent in D and satisfying $\sup_{z \in D} \left|\sum_{\alpha \in \mathbb{N}_0^n} a_\alpha z^\alpha\right| \le 1$ we have $\sup_{z \in rD} \sum_{\alpha \in \mathbb{N}_0^n} |a_\alpha z^\alpha| \le 1$.

Let $\sum_{\alpha \in \mathbb{N}_0^n} a_\alpha z^\alpha$, $a_\alpha \in \mathbb{C}$, be absolutely convergent in D. In what follows we need some estimates on the coefficients a_α. Put $f(z) = \sum_{\alpha \in \mathbb{N}_0^n} a_\alpha z^\alpha$ and assume that f is bounded on D. Fix $\alpha \in \mathbb{N}_0^n$ and $0 < r < 1$, and let $x \in r\overline{D}$ be such that $|x^\alpha| = \max_{r\overline{D}} |z^\alpha|$. Since $B(0, |x_1|) \times \cdots \times B(0, |x_n|) \subset D$ by (14), using (13) with $\gamma_j(t) = |x_j| e^{it}$ we obtain $|a_\alpha| \le \frac{\sup_{z \in D} |f(z)|}{|x_1|^{\alpha_1} \cdots |x_n|^{\alpha_n}}$ and consequently

$$|a_\alpha| \le \frac{\sup_{z \in D} |f(z)|}{\sup_{z \in D} |z^\alpha|}. \tag{15}$$

We will improve this estimate in the next lemma.

Lemma 116 (Norbert Wiener, [AM], [BoKh]). *Let D be an open unit ball of an n-dimensional Banach space with a basis $\{e_j\}_{j=1}^n$ satisfying the estimate* (14) *and let $f(z) = \sum_{\alpha \in \mathbb{N}_0^n} a_\alpha z^\alpha$ be a holomorphic function on D with $|f(z)| < 1$ for all $z \in D$. Then*

$$|a_\alpha| \le \frac{1 - |a_0|^2}{\sup_{z \in D} |z^\alpha|}$$

for all $\alpha \in \mathbb{N}_0^n$, $\alpha \ne 0$.

Proof. Fix $\beta \in \mathbb{N}_0^n$, $\beta \ne 0$. Let $k = |\beta| > 0$ and let ω be a primitive kth root of unity. Then

$$\frac{1}{k} \sum_{j=1}^{k} \omega^{jm} = \begin{cases} 1 & \text{if } m \equiv 0 \pmod{k}, \\ 0 & \text{otherwise.} \end{cases}$$

Let $g(z) = \frac{1}{k} \sum_{j=1}^k f(\omega^j z)$. Then $|g(z)| < 1$ for all $z \in D$ and by the property of ω above we have

$$g(z) = \sum_{|\alpha| \equiv 0 \bmod k} a_\alpha z^\alpha = a_0 + \sum_{|\alpha| = k} a_\alpha z^\alpha + \cdots.$$

Next, we put $\varphi(z) = \frac{z - a_0}{1 - \overline{a_0} z}$. Because $|a_0| = |f(0)| < 1$, the function φ is a holomorphic isomorphism of the open unit disc in $\mathbb{C}$, [RudinW, Theorem 12.4]. Further, $\varphi(z) = \frac{1}{1 - |a_0|^2}(z - a_0) + o(z - a_0)$, $z \to a_0$, and hence

$$h(z) = \varphi \circ g(z) = \sum_{|\alpha| = k} \frac{a_\alpha}{1 - |a_0|^2} z^\alpha + \sum_{|\alpha| > k} b_\alpha z^\alpha. \tag{16}$$

It is still $|h(z)| < 1$ for all $z \in D$. Applying the estimate (15) to the function h on D we get $\frac{|a_\beta|}{1 - |a_0|^2} \le \frac{1}{\sup_{z \in D} |z^\beta|}$, which implies the desired estimate. □

Theorem 117 ([Ai], [Boa], [BoKh]). *Let $1 \le p \le \infty$ and $D = U_{\ell_p^n}$. If $1 \le p < 2$, then*

$$\frac{1}{3e^{\frac{1}{3}}}\left(\frac{1}{n}\right)^{1-\frac{1}{p}} \le K(D) < 3\left(\frac{\log n}{n}\right)^{1-\frac{1}{p}}.$$

If $2 \le p \le \infty$, then

$$\frac{1}{3\sqrt{n}} \le K(D) < 2\sqrt{\frac{\log n}{n}}.$$

Partial proof. Let $f(z) = \sum_{\alpha\in\mathbb{N}_0^n} a_\alpha z^\alpha$ be a holomorphic function on D satisfying $|f(z)| < 1$ for all $z \in D$. Suppose first that $1 \le p < 2$. A straightforward calculation shows that $\sup_{z\in D}|z^\alpha| = \left(\frac{\alpha^\alpha}{|\alpha|^{|\alpha|}}\right)^{\frac{1}{p}} \le 1$. By Lemma 116

$$|a_\alpha| \le (1-|a_0|^2)\left(\frac{|\alpha|^{|\alpha|}}{\alpha^\alpha}\right)^{\frac{1}{p}} \le (1-|a_0|^2)\frac{|\alpha|^{|\alpha|}}{\alpha^\alpha} \le (1-|a_0|^2)\frac{|\alpha|^{|\alpha|}}{\alpha!}$$

for any $\alpha \in \mathbb{N}_0^n$, $\alpha \ne 0$. Observe that $\|z\|_{\ell_1^n}^k = \sum_{|\alpha|=k}\binom{k}{\alpha}|z^\alpha|$ for any $z \in X, k \in \mathbb{N}$ (Proposition 1.22). Thus it follows from the above estimate that for any $z \in X, k \in \mathbb{N}$,

$$\begin{aligned}\sum_{|\alpha|=k}|a_\alpha z^\alpha| &\le (1-|a_0|^2)\sum_{|\alpha|=k}\frac{k^k}{\alpha!}|z^\alpha| \\ &= (1-|a_0|^2)\frac{k^k}{k!}\sum_{|\alpha|=k}\binom{k}{\alpha}|z^\alpha| = (1-|a_0|^2)\frac{k^k}{k!}\|z\|_{\ell_1^n}^k.\end{aligned}$$

By the Hölder inequality $\|z\|_{\ell_1^n} \le n^{1-\frac{1}{p}}\|z\|_{\ell_p^n}$. Hence if $\|z\|_{\ell_p^n} \le r = (\frac{1}{n})^{1-\frac{1}{p}}q$, then

$$\sum_{\alpha\in\mathbb{N}_0^n}|a_\alpha z^\alpha| \le |a_0| + (1-|a_0|^2)\sum_{k=1}^{\infty}\frac{k^k}{k!}q^k.$$

It is easy to verify that

$$\max_{t\in\mathbb{R}}\left(t + \tfrac{1}{2}(1-t^2)\right) \le 1. \tag{17}$$

Thus $\sum_{\alpha\in\mathbb{N}_0^n}|a_\alpha z^\alpha| \le 1$ for all $z \in rB_{\ell_p^n}$ if q is chosen so that $\sum_{k=1}^{\infty}\frac{k^k}{k!}q^k = \frac{1}{2}$. (We refer to [Boa] for the precise computation $q = \frac{1}{3}e^{-\frac{1}{3}}$.)

Now we consider the case $p \ge 2$. Assume first that $D = rU_{\ell_\infty^n}$, $r > 1$. Let μ be the normalised Lebesgue measure on the torus $T = \{z \in \mathbb{C}^n;\ |z_j| = 1\} \subset D$. The monomials z^α are orthonormal in $L_2(\mu)$, so using the Bessel inequality on (16) we get $\sum_{|\alpha|=k}\left|\frac{a_\alpha}{1-|a_0|^2}\right|^2 \le \int_T |h|^2\,d\mu \le 1$ and hence $(\sum_{|\alpha|=k}|a_\alpha|^2)^{\frac{1}{2}} \le 1-|a_0|^2$ for each $k \in \mathbb{N}$. The Cauchy-Schwartz inequality now implies that

$$\sum_{\alpha\in\mathbb{N}_0^n}|a_\alpha z^\alpha| = |a_0| + \sum_{k=1}^{\infty}\sum_{|\alpha|=k}|a_\alpha z^\alpha| \le |a_0| + (1-|a_0|^2)\sum_{k=1}^{\infty}\left(\sum_{|\alpha|=k}|z^\alpha|^2\right)^{\frac{1}{2}}.$$

But $\sum_{|\alpha|=k}|z^\alpha|^2 \le \left(\sum_{j=1}^n |z_j|^2\right)^k$ (Proposition 1.22), so if $z \in \frac{1}{3}B_{\ell_2^n}$, then

$$\sum_{\alpha\in\mathbb{N}_0^n} |a_\alpha z^\alpha| \le |a_0| + (1-|a_0|^2)\sum_{k=1}^{\infty}\frac{1}{3^k} = |a_0| + \frac{1}{2}(1-|a_0|^2) \le 1,$$

using (17) again. The lower estimate for $K(U_{\ell_p^n})$ then follows from the fact that $n^{-\frac{1}{p}}B_{\ell_\infty^n} \subset B_{\ell_p^n}$ and $n^{\frac{1}{p}-\frac{1}{2}}B_{\ell_p^n} \subset B_{\ell_2^n}$.

The proof of the upper estimates consists of constructions of particular forms and polynomials using probabilistic techniques, see [Boa] for details. □

7. Notes and remarks

Section 1. For the theory of tensor products we refer to the books [DefFl], [Ry3], or [DFS], and for the basics see also [FHHMZ, Chapter 16]. Symmetric tensor products were introduced by Raymond A. Ryan [Ry2], for more see e.g. [AnFl], [Fl]. Applications of this theory to complex analysis on infinite-dimensional spaces can be found in [Din], where various topological spaces of holomorphic functions and polynomials are studied.

Section 2. The various weak continuity notions and results which occupy the beginning of the section were studied by Richard Martin Aron and his co-authors, e.g. in [AHV], [AP], [Aro1]; see [Din, Notes to Chapter 1] for references to a large amount of related literature. Our approach however is based on the systematic use of the theory of uniform spaces and the related results on uniformly continuous mappings. This simplifies some proofs and sometimes leads to stronger results. A good source of information on uniform spaces is e.g. [Eng].

The pioneering paper in this area seems to be Aleksander Pełczyński's [Peł1], which was inspired by Stanisław Mazur, and complemented Jaroslav Kurzweil's work in [Kur1]. It contains in particular the well-known Corollary 59 claiming that $\mathcal{P}(c_0) = \mathcal{P}_{\mathrm{wsc}}(c_0)$, due independently to Witold M. Bogdanowicz [Bog]. It seems to have passed unnoticed that this result follows from much earlier and more precise coefficient estimates of Henri Frédéric Bohnenblust and Einar Hille, Corollary 114.

Let us mention briefly some additional results on this property. In the papers [CCG] and [Far] several characterisations of the wsc property were obtained:

Let X be a Banach space and $n \in \mathbb{N}$. The following statements are equivalent:

(i) $\mathcal{P}(^nX) = \mathcal{P}_{\mathrm{wsc}}(^nX)$.
(ii) Each $P \in \mathcal{P}(^nX)$ is weakly continuous on every weakly compact set $A \subset X$.
(iii) $\theta\colon X \to \otimes_{\pi,s}^n X$, $\theta(x) = \otimes^n x$ is w–w sequentially continuous.
(iv) If $\{x_k\}$ is weakly null in X, then $\{\otimes^n x_k\}_{k=1}^\infty$ is weakly null in $\otimes_{\pi,s}^n X$.

Sets of weak continuity and wsc continuity of polynomials have been studied e.g. in [BR], [ArDim], [CGG]. In particular, if $P \in \mathcal{P}^2(X)$ is weakly continuous at the origin, then it is wsc ([ArDim]). More generally, in [FGL] it is shown that $\mathcal{C}_{\mathrm{wsc}}(X;Y)$

consists precisely of the functions that are weakly continuous when restricted to every weakly compact set in X.

Theorem 37 was motivated by a result due to [DL] which claims the following: Let X be a Banach space. Then $\mathcal{P}_K({}^nX;\ell_p) = \mathcal{P}({}^nX;\ell_p)$ for every $n \in \mathbb{N}$, $1 < p < \infty$, if and only if $\mathcal{P}({}^nX)$ does not contain c_0 for every $n \in \mathbb{N}$.

Similar results were obtained in the linear case by Nigel J. Kalton in [Kal] and Kamil John [John], [EJ], motivated by some early work in [Th], [Ton], [TW] which concerns the complementation property of $\mathcal{L}_K(X;Y)$ in $\mathcal{L}(X;Y)$.

The Dunford-Pettis property was introduced and studied by Alexander Grothendieck in [Gr] and Nelson Dunford and Billy James Pettis in [DuPe], who proved that $C(K)$ spaces and $L_1(\mu)$ enjoy the DPP, see e.g. [AK], [Dies1].

The classical Theorem 47 combines several structural results concerning linear operators from $C(K)$ spaces due to A. Grothendieck [Gr] and A. Pełczyński [Peł2]. We refer to [DU, p. 180] for more historical remarks.

The formulation of Theorem 56 appears to be new, although the ideas used for the proof have been around for a long time.

Versions of Theorem 63 have been reproved by many authors. The oldest version seems to be that of A. Pełczyński ([Peł1], [ALRT], [GGu3], [GJ1]). For more results on polynomial images of various types of sequences see e.g. [GGu3], [GGu2], [GGu5], [GGu4].

A. Pełczyński [Peł4], [Peł3] initiated the study of the wsc property for polynomials on spaces with the DPP, proving in particular that $\mathcal{P}(X) = \mathcal{P}_{\mathrm{wsc}}(X)$ whenever X has the DPP. The main result in this direction is Theorem 67, due to R. Ryan. On the other hand, the DPP is far from being a necessary condition to conclude the wsc property for all scalar valued polynomials. In [CGG] examples of Banach spaces failing the DPP for which $\mathcal{L}({}^nX) = \mathcal{L}_{\mathrm{wsc}}({}^nX)$ have been studied in detail.

Problem 118 ([CGG]). Let X be a Banach space such that $\mathcal{P}({}^nX) = \mathcal{P}_{\mathrm{wsc}}({}^nX)$. Is then $\mathcal{L}({}^nX) = \mathcal{L}_{\mathrm{wsc}}({}^nX)$?

By Theorem 67 and Theorem 7 we see that if X has the DPP and $\{x_j^k\}_{j=1}^{\infty}$ are weakly Cauchy sequences in X, $1 \le k \le n$, then $\{x_j^1 \otimes \cdots \otimes x_j^n\}_{j=1}^{\infty}$ is weakly Cauchy in $X \otimes_\pi \cdots \otimes_\pi X$ (the converse is not true in general, [Dies1]). In [BomVi] it is proved that the space $C(K_1) \otimes_\pi \cdots \otimes_\pi C(K_n)$ has the DPP if and only if all K_j, $1 \le j \le n$ are scattered. This is closely related with Fact 25: Consider $X = C(K)$ for non-scattered K. Since $X \otimes_\pi X$ lacks the DPP, there exist a Banach space Y and $T \in \mathcal{L}_{\mathrm{wK}}(\otimes_\pi^2 X;Y) \setminus \mathcal{L}_{\mathrm{wsc}}(\otimes_\pi^2 X;Y)$. However, as X has the DPP (Theorem 41), by Theorem 67, $T \circ \otimes \in \mathcal{L}_{\mathrm{wK}}({}^2X;Y) \subset \mathcal{L}_{\mathrm{wsc}}({}^2X;Y)$. This provides an example such that $M \in \mathcal{L}_{\mathrm{wsc}}({}^2X;Y)$ but $L_M \notin \mathcal{L}_{\mathrm{wsc}}(\otimes_\pi^2 X;Y)$.

On the other hand, observe that

$$M \in \mathcal{L}_{\mathrm{wu}}({}^nX;Y) \quad \text{implies} \quad L_M \in \mathcal{L}_{\mathrm{wu}}(\otimes_\pi^n X;Y).$$

Indeed, by Lemma 18 and Fact 25 we have $L_M \in \mathcal{L}_K(\otimes_\pi^n X;Y) = \mathcal{L}_{\mathrm{wu}}(\otimes_\pi^n X;Y)$. The implication cannot be reversed in general, as L_M is always weakly uniformly continuous for $Y = \mathbb{R}$.

According to [FaJ], a sequence $\{x_k\} \subset X$ is said to be $\mathcal{P}$-null provided that $P(x_k) \to 0$ for every $P \in \mathcal{P}(X)$, $P(0) = 0$. A Banach space X is said to have the polynomial DPP if $\phi_k(x_k) \to 0$ holds for every pair of sequences $\{x_k\}$ that is $\mathcal{P}$-null in X and $\{\phi_k\}$ weakly null in X^*. It is shown there, and independently in [JP], that in super-reflexive spaces $\mathcal{P}$-null sequences coincide with norm null ones. It is shown in [GGu5] that Schreier's pace S has the polynomial DPP (even hereditary), although it does not have the DPP. In this paper $\mathcal{P}$-convergent sequences in S are characterised. These sequences do not coincide with either weakly or norm convergent ones.

A Banach space X is said to be polynomially Schur if every $\mathcal{P}$-null sequence is norm null. (Polynomially Schur spaces has been introduced in [CCG] under a different name – the Λ-property.) Note that by Corollary 4.43 if X is of a non-trivial type (which is equivalent to X^* being of a non-trivial type), then X is polynomially Schur. It is shown in [CCG], [FaJ], that if X has the DPP, then X is polynomially Schur if and only if X is Schur.

The reflexivity of spaces of polynomials has been also extensively studied, see [Far] and the monograph [Din] for more results. R. Ryan obtained in [Ry2] the following characterisation: If X is a reflexive Banach space with the approximation property (Definition 7.18), then the following statements are equivalent:

(i) $\mathcal{P}(^nX)$ is reflexive.
(ii) $\mathcal{P}(^nX) = \mathcal{P}_{\mathrm{wsc}}(^nX)$.
(iii) Each $P \in \mathcal{P}(^nX)$ attains its norm on B_X.

More generally, [Far], [GJ1], if X and Y are reflexive Banach spaces with the approximation property, then $\mathcal{P}(^nX;Y) = \mathcal{P}_{\mathrm{wsc}}(^nX;Y)$ if and only if $\mathcal{P}(^nX;Y)$ is reflexive. The next problem seems rather subtle.

Problem 119 ([DiZa]). Is there a reflexive Banach space X such that $\mathcal{P}(X) = \mathcal{P}_{\mathrm{wsc}}(X)$ and simultaneously $\mathcal{P}(X^*) = \mathcal{P}_{\mathrm{wsc}}(X^*)$?

In [GGu4] the actions of polynomials on wuC sequences are studied. Recall that a Banach space X has the Grothendieck property if $\mathcal{L}(X;c_0) = \mathcal{L}_{\mathrm{wK}}(X;c_0)$. In [GGu2] it is shown that for every $n \in \mathbb{N}$, $n > 1$ the following statements are equivalent:

(i) $\mathcal{P}(^nX)$ is reflexive.
(ii) $\mathcal{P}(^nX;c_0) = \mathcal{P}_{\mathrm{wK}}(^nX;c_0)$.
(iii) $\otimes^n_{\pi,s}X$ is a Grothendieck space.

Similarly, for every $n \in \mathbb{N}$, $n > 1$ we have $\mathcal{P}(^nX;c_0) = \mathcal{P}_{\mathrm{wsc}}(^nX;c_0)$ if and only if $\mathcal{L}(X;c_0) = \mathcal{L}_{\mathrm{wsc}}(X;c_0)$.

Concerning other structural results for the spaces of polynomials, it is shown in [Bl] that $\mathcal{P}(^kX;Y)$ is isomorphic to a complemented subspace of $\mathcal{P}(^nX;Y)$ for every X, Y, and $k < n$. By [DíD], $\mathcal{P}(^nX) \cong \mathcal{L}(^nX)$, $n \in \mathbb{N}$ holds for Banach spaces X isomorphic to their cartesian square, see also [AnFl]. The structure of $\mathcal{L}(^n\ell_p)$ has been studied in detail in [AriFa].

Section 3. The deep theory concerning Banach spaces not containing ℓ_1, due to Haskell Paul Rosenthal, Edward Wilfred Odell, and Jean Bourgain, David H. Fremlin, Michel Talagrand, is presented in detail in [FHHMZ]. A long list of equivalent conditions to X not containing ℓ_1 is given in [Ros4]. In particular, this condition is equivalent to $\mathcal{L}_{\mathrm{wsc}}(X;Y) = \mathcal{L}_{\mathrm{K}}(X;Y)$ for every Banach space Y. The importance of this condition in our theory cannot be exaggerated. As a first example, consider Theorem 81 due to [CHL], which is one of the key tools in the theory of $\mathcal{W}$-spaces, among other things. Theorem 83 streamlines results in [AAD], [AAF], [GJ1]. The opposite condition, i.e. containing ℓ_1, is also very useful especially in dealing with stabilisation problems for polynomials as we will see in Chapter 4. Indeed, the well-known Corollary 89 to Theorem 88 of A. Pełczyński, [Peł5], states that X has a quotient isomorphic to ℓ_2 whenever X is a Banach space containing ℓ_1. A variant of this result, Proposition 91, is instrumental for building diagonal polynomials on X, which are indispensable for proving the main results on infinite-dimensional polynomial algebras in Chapter 4.

Section 4. For the theory of summing operators we refer to any of the following books: [DJT], [Pis4], [Tom2], [Pie], [JL]. Proposition 103 has a converse proved by M. Talagrand [Tal3], with some earlier results in [MaPi2]: If $q > 2$, then X is of cotype q if and only if Id_X is $(q,1)$-summing. For $q = 2$ the result is false, [Tal2]. Theorem 100 was proved independently by Grahame Bennett [Ben] and Bernd Carl [Carl]. An important special case when $p = 1$ was proved earlier by Stanisław Kwapień [Kw1].

Section 5. The whole section is based on [BPV] and [DPS]. The main result, Theorem 111, is optimal in the sense that the value of $\rho(n,p,q)$, for given $n \in \mathbb{N}$, $1 \le p \le q \le \infty$, $p < \infty$ cannot be improved (i.e. made smaller). The proof of the optimality requires a construction of a certain multilinear form by using random coefficients. The theory has many deep connections with other directions of research. We suggest the reader to check the recent works of Andreas Defant and his co-authors in this area, e.g. [DGM], [DeKal], [DFOOS], [DMP], [DefFr].

Section 6. The study of the radius of absolute convergence of complex power series started with a theorem of Harald Bohr stating that $K(U_{\mathbb{C}}) = \frac{1}{3}$. In the multidimensional case there are several alternative definitions of the notion of Bohr radius (cf. [DMP], [DePr], [BoKh], [Boa]). We adopted the one in [DGM], and we refer to [DT1] for historical comments and more references. Theorem 117 is due to Lev Aizenberg [Ai] for $p = 1$, and to [Boa] for $1 \le p \le 2$, by using essentially the same argument. It is due to [BoKh] for $p \ge 2$. The optimality of the asymptotic growth of the upper estimates can again be established using random techniques. For example, [Boa], if $1 \le p \le \infty$ and if n, d are integers larger than 1, then there exists a polynomial $P \in \mathcal{P}(^d\mathbb{C}^n)$ of the form (the signs are chosen using random techniques)

$$P(z) = \sum_{|\alpha|=d} \pm \binom{d}{\alpha} z^\alpha$$

such that the supremum of $|P(z)|$ on $B_{\ell_p^n}$ is at most

$$\sqrt{32\log(6d)}\begin{cases} \sqrt{n}(d!)^{1-\frac{1}{p}} & \text{if } 1 \le p \le 2, \\ n^{\frac{1}{2}+(\frac{1}{2}-\frac{1}{p})d}(d!)^{\frac{1}{2}} & \text{if } 2 \le p \le \infty. \end{cases}$$

Let X be a complex n-dimensional space with a basis $\{e_j\}$ satisfying (14). In [DefFr] the following inequality has been obtained:

$$\frac{1}{4e}\frac{1}{d(X,\ell_1^n)} \le K(B_X),$$

where $d(X,Y)$ denotes the Banach-Mazur distance of X and Y. It is known that $d(\ell_1^n,\ell_p^n) \asymp n^{1-\frac{1}{\min(p,2)}}$, [Tom2], so this result contains, in an asymptotic sense, Theorem 117 as a special case.

The problem of Bohr radius is closely connected with the problem of the existence of an unconditional basis in the space $\mathcal{P}(^nX)$, where X is an infinite-dimensional Banach space and $n \ge 2$. This direction was researched e.g. in [DT1] and [DGM]. The problem was finally settled in [DeKal] showing that $\mathcal{P}(^nX)$, $n \ge 2$ has an unconditional basis if and only if X is finite-dimensional. For bases in the multilinear case see e.g. [DiZa].

Chapter 4

Asymptotic properties of polynomials

It this chapter we employ the concept of finite representability to the study of polynomials, their asymptotic behaviour, and the linear structure of the underlying spaces. We describe the ultrapower construction for a Banach space X, which leads to a much larger Banach space $(X)_{\mathcal{U}}$ (possibly containing X^{**}) that is finitely representable in X and that is well-suited for constructions of uniformly continuous mappings. We give, without proof, several deep results from the local theory of Banach spaces, that will be used later in our investigation of the structure of C^k-smooth Banach spaces.

Another important tool is the spreading model construction for X, which leads to spaces with a sub-symmetric basis that capture the asymptotic behaviour of infinite sequences in X. It is suited particularly well for the study of the upper and lower estimates of sequences, the Banach-Saks, and the p-Banach-Saks properties. These results are then applied to the weak sequential continuity properties of polynomials, relying also on the concept of $\mathcal{P}^n$-null sequences in Banach spaces.

Some of the main results on Banach spaces with separating polynomials are given in Section 4. We show that if a Banach space has a separating d-homogeneous polynomial and a sub-symmetric basis, then it is isomorphic to some ℓ_p for p even and $d = kp$, $k \in \mathbb{N}$. Moreover, by a fundamental result of Robert Deville, every Banach space with a separating polynomial contains such ℓ_p.

While spreading model techniques give us a very good idea of the limit behaviour of any uniformly continuous function at infinity when restricted to suitable subspaces, the results remain finite-dimensional in nature. In Section 5 we show that for polynomials on ℓ_p we are able to obtain a sub-symmetric behaviour for full tail vectors. These results can be applied to the case of polynomials on general Banach spaces. We show that for an arbitrary polynomial P on a Banach space X there is an infinite-dimensional subspace Y of X such that the restriction of P to Y is either separating or asymptotically (and this cannot be strengthened) zero in a strong sense.

The last Sections 6 and 7 are devoted to the study of algebras $\mathcal{A}_n(X)$ of polynomials generated by polynomials of degree at most n on a Banach space X. Section 6 is devoted to the proof of our main technical tool, namely a finite-dimensional lemma which claims that the symmetric polynomial $s_n^N(x) = \sum_{j=1}^N x_j^n$ on $\mathbb{R}^N$ is not in the uniform closure of suitably defined sub-symmetric sub-algebra of $\mathcal{A}_{n-1}(\mathbb{R}^N)$ provided that N is large enough. We proceed by applying this result in Section 7 to infinite-dimensional spaces using spreading model techniques. The main result implies in particular that $\overline{\mathcal{A}_1(\ell_p)} = \cdots = \overline{\mathcal{A}_{n-1}(\ell_p)} \subsetneqq \overline{\mathcal{A}_n(\ell_p)} \subsetneqq \overline{\mathcal{A}_{n+1}(\ell_p)} \subsetneqq \cdots$, where $n = \lceil p \rceil$.

1. Finite representability and ultraproducts

The important concept of finite representability has become one of the key notions in modern Banach space theory (see [FHHMZ, Section 6.1] for some basic information). The principle of local reflexivity gives a very precise description of the fact that X^{**} is always finitely representable in X. We describe a very general technique of building Banach spaces which are finitely representable in X and have additional properties. The ultrapower construction leads to a large Banach space $(X)_{\mathcal{U}}$ (possibly containing also X^{**}) that is well-suited for constructions of uniformly continuous mappings. The section ends with a list of several deep results from the local theory of Banach spaces that will be used later in our investigation of the structure of C^k-smooth Banach spaces.

Definition 1. Let X, Y be normed linear spaces. We say that Y is crudely finitely representable in X if there is $K > 0$ such that for every finite-dimensional subspace F of Y there is a linear isomorphism $T\colon F \to T(F) \subset X$ satisfying $\|T\|\|T^{-1}\| \leq K$. We say that Y is finitely representable in X if for every $\varepsilon > 0$ the space Y is crudely finitely representable in X with the constant $K = 1 + \varepsilon$.

Observe that finite representability (and also crude) is a transitive relation. (Crude) finite representability clearly preserves properties that depend only on finite-dimensional subspaces (even in a uniform way independent of dimension). For example it preserves type or cotype, and moreover if Y is finitely representable in X, then $T_p(Y) \leq T_p(X)$ and $C_q(Y) \leq C_q(X)$. One of the key results in this area is the following principle of local reflexivity ([LindeRo], [JRZ]):

Theorem 2 (principle of local reflexivity, [FHHMZ, Theorem 6.3]). *Let X be a Banach space, $E \subset X^{**}$ and $F \subset X^*$ finite-dimensional subspaces, and $\varepsilon > 0$. Then there is a linear isomorphism T of E onto $T(E) \subset X$ such that $\|T\|\|T^{-1}\| \leq 1 + \varepsilon$, $f(T(\phi)) = \phi(f)$ for any $f \in F$ and $\phi \in E$, and T is the identity on $E \cap X$. In particular, X^{**} is finitely representable in X.*

According to this principle many properties of finite-dimensional subspaces of X pass to X^{**} unchanged. Thus the type or cotype of X passes to X^{**}, including the best constants. It should be noted that a Banach space X is of a non-trivial type if and only if X^* is of a non-trivial type, [DJT, Corollary 13.7, Theorem 13.10]. The same is not true for cotype: the space ℓ_1 is of cotype 2 but ℓ_∞ is not of a non-trivial cotype. Also, type is stronger than cotype in the following sense: If a Banach space X is of a non-trivial type, then it is also of a non-trivial cotype. Indeed, by the above, X^* is then also of a non-trivial type and so X^{**} (and thus also X as its subspace) is of a non-trivial cotype. The converse does not hold (ℓ_1 is of cotype 2 but it is not of a non-trivial type).

It is well-known that every normed linear space is finitely representable in c_0, [FHHMZ, Theorem 6.2]. The celebrated result of Aryeh Dvoretzky, [Dv], states that ℓ_2 is finitely representable in every infinite-dimensional normed linear space. For the proof see e.g. [AK, Theorem 11.3.13]. For a normed linear space X we define $p(X) = \sup\{p;\ X \text{ is of type } p\}$ and $q(X) = \inf\{q;\ X \text{ is of cotype } q\}$.

Theorem 3. *Let X be an infinite-dimensional Banach space. Then $\ell_{p(X)}$ and $\ell_{q(X)}$ are finitely representable in X.*

This theorem is due to [MaPi2] in case that X is of a non-trivial type, resp. cotype; for the proof see [MS, Theorem 13.2]. The case $p(X) = 1$, i.e. when X is not of a non-trivial type, is due to [Pis1]. The case $q(X) = \infty$, i.e. when X is not of a non-trivial cotype, comes from [MaPi1]. For the proof see [AK, Theorem 11.1.14].

Next we review some basic facts on ultraproducts of Banach spaces. Let I be an arbitrary index set. Recall that an ultrafilter on I is called principal if it is of the form $\{A \subset I;\ \alpha \in A\}$ for some fixed element $\alpha \in I$. A non-principal ultrafilter is also called a free ultrafilter. Further, an ultrafilter is called countably complete if it is closed under countable intersections. It is immediate that an ultrafilter $\mathcal{U}$ is countably incomplete if and only if there are $A_n \in \mathcal{U}$, $n \in \mathbb{N}$ such that $A_1 \supset A_2 \supset A_3 \supset \cdots$ and $\bigcap_{n=1}^{\infty} A_n = \emptyset$. Every free ultrafilter on $\mathbb{N}$ is countably incomplete, as it contains the Fréchet filter (all subsets of $\mathbb{N}$ with a finite complement).

Recall that if X is a topological space, I an index set, $\{x_\alpha\}_{\alpha\in I} \subset X$, and $\mathcal{U}$ an ultrafilter on I, then $\lim_{\mathcal{U}} x_\alpha = x \in X$ (the limit with respect to the ultrafilter $\mathcal{U}$) if $\{\alpha \in I;\ x_\alpha \in V\} \in \mathcal{U}$ for every neighbourhood V of x; this limit always exists if X is compact.

Let X_α, $\alpha \in I$, be normed linear spaces for some index set I. Consider the normed linear space $\ell_\infty(I;\{X_\alpha\}) = \{(x_\alpha)_{\alpha\in I};\ x_\alpha \in X_\alpha, \sup_{\alpha\in I}\|x_\alpha\| < +\infty\}$ with the supremum norm. If $\mathcal{U}$ is an ultrafilter on I, then $(x_\alpha) \mapsto \lim_{\mathcal{U}}\|x_\alpha\|$ is a continuous seminorm on $\ell_\infty(I;\{X_\alpha\})$ and so $N_{\mathcal{U}}(\{X_\alpha\}) = \big\{(x_\alpha) \in \ell_\infty(I;\{X_\alpha\});\ \lim_{\mathcal{U}}\|x_\alpha\| = 0\big\}$ is a closed subspace.

Definition 4 ([DaKr]). The normed linear space $\ell_\infty(I;\{X_\alpha\})/N_{\mathcal{U}}(\{X_\alpha\})$ is called the ultraproduct of $\{X_\alpha\}_{\alpha\in I}$ and it is denoted by $(X_\alpha)_{\mathcal{U}}$. In the important special case when $X_\alpha = X$ for all $\alpha \in I$ we call this space the ultrapower of X and denote it by $(X)_{\mathcal{U}}$.

The element of $(X_\alpha)_{\mathcal{U}}$ represented by $(x_\alpha) \in \ell_\infty(I;\{X_\alpha\})$ will be denoted by $(x_\alpha)_{\mathcal{U}}$. It is easy to see that the canonical quotient norm is given by $\|(x_\alpha)_{\mathcal{U}}\| = \lim_{\mathcal{U}}\|x_\alpha\|$. The mapping $x \mapsto (x_\alpha)_{\mathcal{U}}$, where $x_\alpha = x$ for all $\alpha \in I$, clearly defines a canonical isometric embedding of X into $(X)_{\mathcal{U}}$. If X_α are all Banach spaces, then $(X_\alpha)_{\mathcal{U}}$ is obviously also a Banach space. The following proposition is not difficult to prove.

Proposition 5 ([Ster], [DJT, Theorem 8.13]). *Let X be a normed linear space and $\mathcal{U}$ an ultrafilter. Then $(X)_{\mathcal{U}}$ is finitely representable in X.*

Theorem 6 ([Ster]). *Suppose that a normed linear space Y is finitely representable in a normed linear space X. Then there exists an ultrafilter $\mathcal{U}$ such that Y is isometric to a subspace of $(X)_{\mathcal{U}}$.*

Proof. Let $I = \{(E,n);\ E \subset Y$ is a finite-dimensional subspace, $n \in \mathbb{N}\}$ and pick $T_\alpha \in \mathcal{L}(E;X)$ with $\|T_\alpha\| = 1$ and $\|T_\alpha^{-1}\| < 1 + \frac{1}{n}$ for $\alpha = (E,n) \in I$. We put a partial ordering on I according to the relation $(E,m) \preceq (F,n)$ if $E \subset F$ and $m \le n$. The order filter $\mathcal{F}$ on the directed set $(I,\preceq)$ is a filter generated by all the sets of the form $\{\beta \in I;\ \beta \succeq \alpha\}$, $\alpha \in I$. Finally, let $\mathcal{U}$ be an ultrafilter on I extending $\mathcal{F}$. The embedding $T\colon Y \to (X)_{\mathcal{U}}$ is now defined by $T(y) = (x_\alpha)_{\mathcal{U}}$, where for each $\alpha = (E, n) \in I$

$$x_\alpha = \begin{cases} T_\alpha(y) & \text{if } y \in E, \\ 0 & \text{otherwise.} \end{cases}$$

It is easy to check that T is linear and an isometry. □

By combining the previous theorem with the fact that X^{**} is finitely representable in X we see that there is an ultrafilter $\mathcal{U}$ such that X^{**} is isometric to a subspace of $(X)_{\mathcal{U}}$. By utilising the full power of the principle of local reflexivity we obtain an isometric embedding with additional properties:

Corollary 7. *Let X be a Banach space. There are an ultrafilter $\mathcal{U}$ and an isometric embedding $T\colon X^{**} \to (X)_{\mathcal{U}}$ such that $T(x) = (x)_{\mathcal{U}}$ for each $x \in X$ and $w^*\text{-}\lim_{\mathcal{U}} x_\alpha = \psi$ for every $\psi \in X^{**}$ and every $\{x_\alpha\} \subset X$ such that $T(\psi) = (x_\alpha)_{\mathcal{U}}$.*

Proof. Let I be the collection of all triples (E, F, n) where $n \in \mathbb{N}$ and $E \subset X^{**}$, $F \subset X^*$ are finite-dimensional subspaces with $E \cap X \neq \{0\}$. By the principle of local reflexivity for each $\alpha = (E, F, n) \in I$ there is an isomorphism $T_\alpha\colon E \to T_\alpha(E) \subset X$ such that $1 \le \|T_\alpha\| \le 1 + \frac{1}{n}$, $1 \le \|T_\alpha^{-1}\| \le 1 + \frac{1}{n}$, $f(T_\alpha(\phi)) = \phi(f)$ for any $f \in F$ and $\phi \in E$, and T_α is the identity on $E \cap X$. We put a partial ordering on I according to the relation $(E_1, F_1, n_1) \preceq (E_2, F_2, n_2)$ if $E_1 \subset E_2$, $F_1 \subset F_2$, and $n_1 \le n_2$. The order filter $\mathcal{F}$ on the directed set $(I, \preceq)$ is a filter generated by all the sets of the form $\{\beta \in I;\ \beta \succeq \alpha\}$, $\alpha \in I$. Finally, let $\mathcal{U}$ be an ultrafilter on I extending $\mathcal{F}$. The embedding $T\colon X^{**} \to (X)_{\mathcal{U}}$ is now defined by $T(y) = (x_\alpha)_{\mathcal{U}}$, where for each $\alpha = (E, F, n) \in I$

$$x_\alpha = \begin{cases} T_\alpha(y) & \text{if } y \in E, \\ 0 & \text{otherwise.} \end{cases}$$

It is easy to check that T has the required properties. □

To be able to extend mappings from X to the ultrapowers, we first need to look at their domains. Let A be a subset of a normed linear space X and $\mathcal{U}$ an ultrafilter on I. By $(A)_{\mathcal{U}}$ we denote the set $\{(x_\alpha)_{\mathcal{U}} \in (X)_{\mathcal{U}};\ x_\alpha \in A \text{ for all } \alpha \in I\}$.

Proposition 8. *Let X be a normed linear space, $A \subset X$, and $\mathcal{U}$ an ultrafilter on I.*

(i) *If A is bounded, then so is $(A)_{\mathcal{U}}$.*
(ii) *If A is convex, then so is $(A)_{\mathcal{U}}$.*
(iii) *If $\mathcal{U}$ is countably incomplete or if A is closed, then $(A)_{\mathcal{U}}$ is closed.*
(iv) *If $z \in (X)_{\mathcal{U}}$, then there is a representation $z = (x_\alpha)_{\mathcal{U}}$ such that $\|x_\alpha\| = \|z\|$ for all $\alpha \in I$. In particular, $(S_X)_{\mathcal{U}} = S_{(X)_{\mathcal{U}}}$.*

Proof. (i) and (ii) are trivial.

To prove (iii), let $B_n \in \mathcal{U}$ be such that $B_1 \supset B_2 \supset B_3 \supset \cdots$ and $\bigcap_{n=1}^{\infty} B_n = \emptyset$ in case that $\mathcal{U}$ is countably incomplete; put $B_n = I$ for all $n \in \mathbb{N}$ otherwise. Let $(x_\alpha)_{\mathcal{U}} \in \overline{(A)_{\mathcal{U}}}$. There are $(x_\alpha^n)_{\mathcal{U}} \in (A)_{\mathcal{U}}$ such that $x_\alpha^n \in A$ for all $n \in \mathbb{N}$, $\alpha \in I$ and

$\|(x_\alpha^n)_{\mathcal{U}}-(x_\alpha)_{\mathcal{U}}\| < \frac{1}{n}$. Put $C_n = \{\alpha \in I;\ \|x_\alpha^n - x_\alpha\| < \frac{2}{n}\}$ and $D_n = B_n \cap \bigcap_{j=1}^n C_j$. Then $D_n \in \mathcal{U}$ for each $n \in \mathbb{N}$ and $D_1 \supset D_2 \supset D_3 \supset \cdots$. Put $D = \bigcap_{n=1}^\infty D_n$ and note that $D = \emptyset$ in case that $\mathcal{U}$ is countably incomplete. Thus by the assumption $x_\alpha \in A$ for each $\alpha \in D$.

If $\alpha \in D_1 \setminus D$, then there is a unique $n \in \mathbb{N}$ such that $\alpha \in D_n \setminus D_{n+1}$ and we put $y_\alpha = x_\alpha^n$. Set $y_\alpha = x_\alpha$ for $\alpha \in D$ and $y_\alpha = x_\alpha^1$ for $\alpha \notin D_1$, and note that $y_\alpha \in A$ for all $\alpha \in I$. Fix $n \in \mathbb{N}$. For each $\alpha \in D_n \setminus D$ there is a unique $m \in \mathbb{N}$, $m \geq n$ such that $\alpha \in D_m \setminus D_{m+1}$. Thus $\|y_\alpha - x_\alpha\| = \|x_\alpha^m - x_\alpha\| < \frac{2}{m} \leq \frac{2}{n}$. Hence $D_n \subset \{\alpha \in I;\ \|y_\alpha - x_\alpha\| < \frac{2}{n}\}$. It follows that $\lim_{\mathcal{U}} \|y_\alpha - x_\alpha\| = 0$ and so $(x_\alpha)_{\mathcal{U}} = (y_\alpha)_{\mathcal{U}} \in (A)_{\mathcal{U}}$.

(iv) Let $z = (z_\alpha)_{\mathcal{U}}$ and fix $e \in X$ with $\|e\| = \|z\|$. Put $x_\alpha = \|z\| \frac{z_\alpha}{\|z_\alpha\|}$ if $z_\alpha \neq 0$, $x_\alpha = e$ otherwise. Then $\lim_{\mathcal{U}} \|x_\alpha - z_\alpha\| = \lim_{\mathcal{U}} \big|\|z\| - \|z_\alpha\|\big| = 0$ and hence $(x_\alpha)_{\mathcal{U}} = (z_\alpha)_{\mathcal{U}}$. □

Now if X is a normed linear space, Y a uniform space, $A \subset X$, $f\colon A \to Y$ a mapping that is uniformly continuous on bounded sets and maps bounded sets to relatively compact sets, and $\mathcal{U}$ an ultrafilter, then we can define $(f)_{\mathcal{U}}\colon (A)_{\mathcal{U}} \to Y$ by

$$(f)_{\mathcal{U}}\big((x_\alpha)_{\mathcal{U}}\big) = \lim_{\mathcal{U}} f(x_\alpha).$$

It is easy to check that $(f)_{\mathcal{U}}$ is well-defined by the uniform continuity. Moreover, $(f)_{\mathcal{U}}\big((x)_{\mathcal{U}}\big) = f(x)$ for any $x \in X$, so we may view $(f)_{\mathcal{U}}$ as an extension of f.

In view of Corollary 7 this extension can be in particular used to extend mappings with domain in X to mappings with domain in X^{**}. Such an extension depends of course on the ultrafilter $\mathcal{U}$ as well as on the particular embedding of X^{**} into $(X)_{\mathcal{U}}$ (via the principle of local reflexivity). However, if X, Y are normed linear spaces and $f\colon X \to Y$ is w–w uniformly continuous (in addition to being $\|\cdot\|$–$\|\cdot\|$ uniformly continuous on bounded sets), then f has a unique w^*–w^* uniformly continuous extension $g\colon X^{**} \to Y^{**}$ (see the remarks preceding Proposition 3.15). Let T be the embedding from Corollary 7 and consider the extension $(f)_{\mathcal{U}}\colon (X)_{\mathcal{U}} \to (Y^{**}, w^*)$. Then given $\psi \in X^{**}$ we have $g(\psi) = w^*\text{-}\lim_{\mathcal{U}} f(T(\psi)_\alpha) = (f)_{\mathcal{U}}\big(T(\psi)\big)$ and so g and $(f)_{\mathcal{U}}$ coincide on X^{**}. Therefore the extension (on X^{**}) is independent of $\mathcal{U}$ and T. In particular, if $L \in \mathcal{L}(X;Y)$, then $(L)_{\mathcal{U}} \restriction_{X^{**}} = L^{**}$.

The next result is also useful for preserving uniform properties with respect to finite representability.

Theorem 9. *Let X, Y be normed linear spaces and let τ be either the norm, weak, or weak star (in case that $(Y, \|\cdot\|)$ is a dual space) topology on Y. Further, let $A \subset X$, and let $f\colon A \to Y$ be a mapping that is uniformly continuous on bounded sets and maps bounded sets to relatively τ-compact sets. If $\mathcal{U}$ is an ultrafilter on I, then $(f)_{\mathcal{U}}\colon (A)_{\mathcal{U}} \to (Y, \tau)$ has the following properties:*

(i) *If f is uniformly continuous with a modulus $\omega \in \mathcal{M}$ that is continuous, then so is $(f)_{\mathcal{U}}$.*

(ii) *If $f \in \mathcal{P}({}^nX;Y)$, then $(f)_{\mathcal{U}} \in \mathcal{P}({}^n(X)_{\mathcal{U}};Y)$, and if $f \in \mathcal{P}^n(X;Y)$, then $(f)_{\mathcal{U}} \in \mathcal{P}^n((X)_{\mathcal{U}};Y)$. Moreover, in both cases $\|(f)_{\mathcal{U}}\| = \|f\|$.*

(iii) *If f is T^k-smooth at x with the approximating polynomial P, then $(f)_{\mathcal{U}}$ is T^k-smooth at $(x)_{\mathcal{U}}$ with the approximating polynomial $(P)_{\mathcal{U}}$. In particular, if f is Fréchet differentiable at x, then $(f)_{\mathcal{U}}$ is Fréchet differentiable at $(x)_{\mathcal{U}}$ with $D(f)_{\mathcal{U}}((x)_{\mathcal{U}}) = (Df(x))_{\mathcal{U}}$.*

(iv) *Suppose that $A = X$ or that A is an open convex bounded set. If $\Omega \subset \mathcal{M}_s$ is a convex cone and $f \in C^{k,\Omega}(A;Y)$, then $(f)_{\mathcal{U}} \in C^{k,\Omega}(\mathrm{Int}(A)_{\mathcal{U}};Y)$. Moreover, $d^j(f)_{\mathcal{U}}(x)[h] = w^*\text{-}\lim_{\mathcal{U}} d^j f(x_\alpha)[h_\alpha]$ (in Y^{**} if $\tau \neq w^*$) for any $j = 1,\dots,k$, $x = (x_\alpha)_{\mathcal{U}} \in \mathrm{Int}(A)_{\mathcal{U}}$ with $x_\alpha \in A$, and $h = (h_\alpha)_{\mathcal{U}} \in (X)_{\mathcal{U}}$.*

Proof. (i) Notice that $\|\cdot\|$ is τ-lower semi-continuous. Let $x = (x_\alpha)_{\mathcal{U}} \in (A)_{\mathcal{U}}$, $y = (y_\alpha)_{\mathcal{U}} \in (A)_{\mathcal{U}}$ with $x_\alpha, y_\alpha \in A$. Denote $\delta = \|x - y\|$ and let $\varepsilon > 0$ be arbitrary. Then $V = \{\alpha \in I;\ \|x_\alpha - y_\alpha\| < \delta + \varepsilon\} \in \mathcal{U}$. Since $\|f(x_\alpha) - f(y_\alpha)\| \le \omega(\delta+\varepsilon)$ for all $\alpha \in V$, it follows that $\|(f)_{\mathcal{U}}(x) - (f)_{\mathcal{U}}(y)\| \le \lim_{\mathcal{U}} \|f(x_\alpha) - f(y_\alpha)\| \le \omega(\delta+\varepsilon)$ for all $\varepsilon > 0$. The continuity of ω now implies that $\|(f)_{\mathcal{U}}(x) - (f)_{\mathcal{U}}(y)\| \le \omega(\delta)$.

(ii) First we show the homogeneous case. Let us define $M\colon (X)^n_{\mathcal{U}} \to (Y,\tau)$ by $M\big((x^1_\alpha)_{\mathcal{U}},\dots,(x^n_\alpha)_{\mathcal{U}}\big) = \lim_{\mathcal{U}} \check{f}(x^1_\alpha,\dots,x^n_\alpha)$. It is clear that M is an n-linear symmetric mapping and $(f)_{\mathcal{U}} = \widehat{M}$, so $(f)_{\mathcal{U}} \in P(^n(X)_{\mathcal{U}};Y)$. Obviously this also implies that $(f)_{\mathcal{U}} \in P^n((X)_{\mathcal{U}};Y)$ in the non-homogeneous case. To show the boundedness (and the equality of the norms) let $x \in B_{(X)_{\mathcal{U}}}$. There are $x_\alpha \in B_X$ such that $x = (x_\alpha)_{\mathcal{U}}$. Then $\|(f)_{\mathcal{U}}(x)\| \le \lim_{\mathcal{U}} \|f(x_\alpha)\| \le \|f\|$.

(iii) Proposition 3.70 implies that P maps bounded sets to relatively τ-compact sets. Thus $(P)_{\mathcal{U}}$ is well-defined by the τ-limit and it is a polynomial of degree at most k by (ii). Let $\varepsilon > 0$. There is $\delta > 0$ such that $\|f(x+h) - P(h)\| \le \varepsilon\|h\|^k$ for all $h \in X$, $\|h\| \le \delta$. If $(h_\alpha)_{\mathcal{U}} \in (X)_{\mathcal{U}}$ is such that $\|(h_\alpha)_{\mathcal{U}}\| \le \delta$, then by Proposition 8 we may assume that $\|h_\alpha\| \le \delta$ for all $\alpha \in I$. Then $\|f(x+h_\alpha) - P(h_\alpha)\| \le \varepsilon\|h_\alpha\|^k$ and by passing to a limit we obtain

$$\begin{aligned}\big\|(f)_{\mathcal{U}}\big((x)_{\mathcal{U}} + (h_\alpha)_{\mathcal{U}}\big) - (P)_{\mathcal{U}}\big((h_\alpha)_{\mathcal{U}}\big)\big\| &\le \lim_{\mathcal{U}} \|f(x+h_\alpha) - P(h_\alpha)\| \\ &\le \varepsilon \lim_{\mathcal{U}} \|h_\alpha\|^k = \varepsilon\|(h_\alpha)_{\mathcal{U}}\|^k.\end{aligned}$$

(iv) For $x \in A$ denote by P_x the Taylor polynomial of the mapping f of order k at x. Let $\omega \in \Omega$ be the modulus of continuity of $d^k f$. Corollary 1.108 implies that $\|f(x+h) - P_x(h)\| \le \frac{1}{k!}\omega(\|h\|)\|h\|^k$ for all $x \in A$, $h \in X$ with $x + h \in A$. Put $Z = Y$ if $\tau = w^*$, $Z = Y^{**}$ otherwise. Fix $(x_\alpha)_{\mathcal{U}} \in \mathrm{Int}(A)_{\mathcal{U}}$ with $x_\alpha \in A$ and define $P_{(x_\alpha)_{\mathcal{U}}}\colon (X)_{\mathcal{U}} \to Z$ by the formula

$$P_{(x_\alpha)_{\mathcal{U}}}\big((h_\alpha)_{\mathcal{U}}\big) = w^*\text{-}\lim_{\mathcal{U}} P_{x_\alpha}(h_\alpha).$$

Using the uniform continuity of $x \mapsto P_x$ and the convexity of A it is not difficult to check that $P_{(x_\alpha)_{\mathcal{U}}}$ is well-defined and continuous. Assume that the spaces X and Y are real. Let $(y_\alpha)_{\mathcal{U}}, (h_\alpha)_{\mathcal{U}} \in (X)_{\mathcal{U}}$. By Theorem 2.50

$$\sum_{j=0}^{k+1} (-1)^{k+1-j} \binom{k+1}{j} P_{x_\alpha}(y_\alpha + jh_\alpha) = 0$$

for all $\alpha \in I$. Consequently, by passing to a w^*-limit with respect to the ultrafilter $\mathcal{U}$,

$$\sum_{j=0}^{k+1} (-1)^{k+1-j} \binom{k+1}{j} P_{(x_\alpha)_\mathcal{U}}\big((y_\alpha)_\mathcal{U} + j(h_\alpha)_\mathcal{U}\big) = 0$$

and so by Theorem 2.50 again we may conclude that $P_{(x_\alpha)_\mathcal{U}} \in \mathcal{P}^k\big((X)_\mathcal{U}; Z\big)$. Given $(h_\alpha)_\mathcal{U} \in (X)_\mathcal{U}$ such that $(x_\alpha)_\mathcal{U} + (h_\alpha)_\mathcal{U} \in \operatorname{Int}(A)_\mathcal{U}$, we may without loss of generality assume that $x_\alpha + h_\alpha \in A$ for all $\alpha \in I$. Then

$$\|f(x_\alpha + h_\alpha) - P_{x_\alpha}(h_\alpha)\| \le \frac{1}{k!}\omega(\|h_\alpha\|)\|h_\alpha\|^k$$

for all $\alpha \in I$. Since the topology on Y considered as a subspace of (Z, w^*) is weaker than or equal to τ, we have $(f)_\mathcal{U}\big((x_\alpha)_\mathcal{U} + (h_\alpha)_\mathcal{U}\big) = w^*\text{-}\lim_\mathcal{U} f(x_\alpha + h_\alpha)$. Thus by passing to a limit in the inequality above, using also the w^*-lower semi-continuity of the norm on Z and the continuity of ω, we obtain

$$\big\|(f)_\mathcal{U}\big((x_\alpha)_\mathcal{U} + (h_\alpha)_\mathcal{U}\big) - P_{(x_\alpha)_\mathcal{U}}\big((h_\alpha)_\mathcal{U}\big)\big\| \le \frac{1}{k!}\omega(\|(h_\alpha)_\mathcal{U}\|)\|(h_\alpha)_\mathcal{U}\|^k.$$

So we may apply Corollary 1.126 to obtain $(f)_\mathcal{U} \in C^{k,\Omega}(\operatorname{Int}(A)_\mathcal{U}; Y)$.

The complex case can be shown similarly as in (ii). (Of course, the method of (ii) works in the real case too. The use of Theorem 2.50 is nevertheless formally somewhat shorter.) □

Often in various constructions we rely on a reasonably rich supply of functions of a given class that have certain separating properties.

Definition 10. Let X be a set, $A, B \subset X$, Y a metric space, and $f : X \to Y$. We say that f separates A from B if $\operatorname{dist}\big(f(A), f(B)\big) > 0$. To simplify the notation, if A consists of a singleton $\{x\}$ we say that f separates x from B. Let X be a normed linear space and $f : X \to \mathbb{R}$. We say that f is a separating function if there is $x \in X$ and $r > 0$ such that f separates $\{x\}$ from $X \setminus B(x, r)$.

Note that if a set S of functions on a normed linear space X is stable under affine transformations of the domain (in particular if S is any of the smoothness spaces from Section 1.5), then there is a separating function in S if and only if there is a function in S that separates 0 from $X \setminus B_X$. An example of a separating function is a bump function (or shortly a bump) – a real function on X with a bounded non-empty support. From Proposition 8 it immediately follows that if $f : X \to \mathbb{R}$ is uniformly continuous on bounded sets such that it separates 0 from $X \setminus B_X$, then $(f)_\mathcal{U}$ separates 0 from $(X)_\mathcal{U} \setminus B_{(X)_\mathcal{U}}$. Combining this with Theorems 6 and 9 we obtain the next corollary.

Corollary 11. *Let $\Omega \subset \mathcal{M}_s$ be a convex cone and let X be a normed linear space that admits a $C^{k,\Omega}$-smooth bump, resp. separating polynomial in $\mathcal{P}^n(X)$, resp. separating polynomial in $\mathcal{P}(^nX)$. If a normed linear space Y is finitely representable in X, then Y admits a $C^{k,\Omega}$-smooth bump, resp. separating polynomial in $\mathcal{P}^n(Y)$, resp. separating polynomial in $\mathcal{P}(^nY)$.*

We proceed by introducing the important classes of $\mathcal{L}_p$-spaces, which will play an important role in the sequel (in particular in the case $p = \infty$). We recall that the Banach-Mazur distance between the normed linear spaces X and Y is defined as $d(X, Y) = \inf\{\|T\|\|T^{-1}\|;\ T \in \mathcal{L}(X; Y) \text{ is an isomorphism}\}$.

Definition 12 ([LP]). Let $1 \leq p \leq \infty$. A Banach space X is said to be a $\mathcal{L}_{p,\lambda}$-space, $\lambda \geq 1$, if for every finite-dimensional subspace E of X there is a finite-dimensional subspace F of X, $E \subset F$, such that $d(F, \ell_p^n) \leq \lambda$, $n = \dim F$. We say that X is a $\mathcal{L}_p$-space if X is a $\mathcal{L}_{p,\lambda}$-space for some $\lambda \geq 1$.

$\mathcal{L}_p$-spaces have been extensively studied in the literature. Although their definition seems to be purely finite-dimensional, these spaces have very strong structural properties. We recall some of the properties of this class of spaces for the future use. For more information and references we refer to [JL, Chapters 3 and 36].

Theorem 13 ([LP], [LindeRo]). *Let $1 \leq p \leq \infty$. Then $L_p(\mu)$ is a $\mathcal{L}_{p,\lambda}$-space for every $\lambda > 1$. Moreover, every $\mathcal{L}_p$-space is isomorphic to a subspace of $L_p(\mu)$ for some measure μ; if $1 < p < \infty$, then it is isomorphic to a complemented subspace of $L_p(\mu)$. On the other hand, a complemented subspace of a $\mathcal{L}_p$-space is either a $\mathcal{L}_p$-space or it is isomorphic to a Hilbert space.*

If X is a $\mathcal{L}_p$-space, $1 \leq p \leq \infty$, then $(X)_{\mathcal{U}}$ is again a $\mathcal{L}_p$-space for any ultrafilter $\mathcal{U}$, [BenLi, Proposition F.3] or [BourgJ2, Proposition 1.22]. Moreover, if $X = L_p(\mu)$, $1 \leq p < \infty$, then $(X)_{\mathcal{U}}$ is isometric to $L_p(\nu)$ for some measure ν, [He, Theorem 3.3]. Since any separable subspace of $L_p(\nu)$, $1 \leq p < \infty$ is isometric to a subspace of some separable $L_p(\nu_1)$ ([Woj, Proposition III.A.2]), which is isometric to a subspace of $L_p([0, 1])$ ([JL, pp. 14–15]), it follows that if a separable Banach space is finitely representable in $L_p([0, 1])$, then it is isometric to a subspace of $L_p([0, 1])$.

We recall that a Banach space is a $\mathcal{P}_\lambda$-space if it is λ-complemented in every superspace. The following should be compared with Proposition 7.78.

Proposition 14 ([FHHMZ, Proposition 5.13]). *Let X be a Banach space and $\lambda \geq 1$. The following statements are equivalent:*

(i) *X is a $\mathcal{P}_\lambda$-space.*
(ii) *For every Banach space Y, every $T \in \mathcal{L}(Y; X)$, and every Banach space $Z \supset Y$ there is an extension $\overline{T} \in \mathcal{L}(Z; X)$ of T with $\|\overline{T}\| \leq \lambda\|T\|$.*
(iii) *For every Banach spaces Z and $Y \supset X$, and for every $T \in \mathcal{L}(X; Z)$ there is an extension $\overline{T} \in \mathcal{L}(Y; Z)$ of T with $\|\overline{T}\| \leq \lambda\|T\|$.*

Theorem 15 ([LP], [LindeRo], [Li1]). *Let K be a compact space. Then $C(K)$ is a $\mathcal{L}_{\infty,\lambda}$-space for every $\lambda > 1$. Every complemented subspace of $C(K)$ is a $\mathcal{L}_\infty$-space. If X is a $\mathcal{L}_{\infty,\lambda}$-space, then X^{**} is a $\mathcal{P}_\lambda$-space.*

Theorem 16 ([LindeRo], [Steg1]). *Let X be a Banach space. Then X^* is isomorphic to ℓ_1 if and only if X is a separable $\mathcal{L}_\infty$-space that does not contain ℓ_1.*

It turns out that $\mathcal{L}_\infty$-spaces, and in particular isomorphic preduals of ℓ_1, need not contain c_0, [BD], [BourgJ2]. In fact, $\mathcal{L}_\infty$-spaces may even be saturated with spaces isomorphic to ℓ_2, [Hay4]. Every Banach space with a separable dual can be embedded into some isomorphic predual of ℓ_1, [FOS].

2. Spreading models

In this section we develop some basic facts concerning the spreading model construction for a Banach space X, which leads to a Banach space with a sub-symmetric basis which captures the asymptotic behaviour of infinite sequences in X. We then pass to the closely related concepts of the Banach-Saks and the p-Banach-Saks properties. These results will be later applied to the weak sequential continuity properties of polynomials.

Definition 17. Let $K \geq 1$. We say that a sequence $\{x_n\}_{n=1}^{\infty}$ in a normed linear space is K-spreading if

$$\left\|\sum_{j=1}^{k} a_j x_{m_j}\right\| \leq K \left\|\sum_{j=1}^{k} a_j x_{n_j}\right\|$$

whenever $k \in \mathbb{N}$, $a_1, \ldots, a_k$ are arbitrary scalars, and $m_j, n_j \in \mathbb{N}$ are such that $m_1 < m_2 < \cdots < m_k$, $n_1 < n_2 < \cdots < n_k$.

In particular, a sequence $\{x_n\}$ is 1-spreading if and only if

$$\left\|\sum_{j=1}^{k} a_j x_{n_j}\right\| = \left\|\sum_{j=1}^{k} a_j x_j\right\|$$

whenever $k \in \mathbb{N}$, $a_1, \ldots, a_k$ are any scalars, and $n_j \in \mathbb{N}$ satisfy $n_1 < n_2 < \cdots < n_k$. Note that from Rosenthal's ℓ_1-theorem (Theorem 3.71) it follows that any K-spreading sequence in a Banach space is either equivalent to the canonical basis of ℓ_1, or it is weakly Cauchy (use the fact that the linear operator $T\colon \operatorname{span}\{x_{n_j}\} \to \operatorname{span}\{x_j\}$, $T(x_{n_j}) = x_j$ is bounded and hence w–w uniformly continuous).

Proposition 18 ([BeaLa, Proposition I.4.2]). *Let $\{e_n\}$ be a K-spreading sequence in a Banach space X. Then $\{e_n\}$ is a basic sequence if and only if it is not weakly convergent to a non-zero element of X. If moreover $\{e_n\}$ is weakly null, then $\{e_n\}$ is an unconditional basic sequence.*

Definition 19. A Schauder basis $\{e_n\}_{n=1}^{\infty}$ of a Banach space is called symmetric if $\{e_{\pi(n)}\}_{n=1}^{\infty}$ is equivalent to $\{e_n\}_{n=1}^{\infty}$ for any permutation π of $\mathbb{N}$. A Schauder basis $\{e_n\}_{n=1}^{\infty}$ of a Banach space is called sub-symmetric if it is unconditional and $\{e_{n_k}\}_{k=1}^{\infty}$ is equivalent to $\{e_n\}_{n=1}^{\infty}$ for every increasing sequence $\{n_k\}_{k=1}^{\infty} \subset \mathbb{N}$.

We remark that a symmetric basis is automatically unconditional, and in fact sub-symmetric, [Sin, Proposition II.22.2]. It can be shown using the Uniform boundedness principle that a sub-symmetric basis is K-spreading for some $K \geq 1$ (the unconditionality here is substantial), and similarly if $\{e_n\}$ is a symmetric basis of a Banach space X, then there is $K \geq 1$ such that

$$\frac{1}{K}\left\|\sum_{n=1}^{\infty} a_n e_n\right\| \leq \left\|\sum_{n=1}^{\infty} a_n e_{\pi(n)}\right\| \leq K\left\|\sum_{n=1}^{\infty} a_n e_n\right\|$$

for every $\sum_{n=1}^{\infty} a_n e_n \in X$ and every permutation π of $\mathbb{N}$, [Sin, Theorems II.21.2 and II.22.1]. Further, it is easy to check that if $\{e_n\} \subset X$ is a sub-symmetric basis that is K-spreading, then the sequence $\{f_n\} \subset X^*$ biorthogonal to $\{e_n\}$ is a sub-symmetric basic sequence that is CK-spreading, where C is the unconditional basis constant of $\{e_n\}$.

Definition 20. Let X be a Banach space with a Schauder basis $\{e_n\}_{n=1}^{\infty}$. For any mapping $\sigma\colon \mathbb{N} \to \mathbb{N}$ denote $T_\sigma\big(\sum_{n=1}^{\infty} a_n e_n\big) = \sum_{n=1}^{\infty} a_n e_{\sigma(n)}$ for those vectors $\sum_{n=1}^{\infty} a_n e_n \in X$, for which the sum on the right converges.
A subset $U \subset X$ is called symmetric (resp. sub-symmetric) if $T_\sigma(U) \subset U$ for any permutation $\sigma\colon \mathbb{N} \to \mathbb{N}$ (resp. for any increasing mapping $\sigma\colon \mathbb{N} \to \mathbb{N}$). Let $U \subset X$ be a symmetric (resp. sub-symmetric) set and A any set. A mapping $f : U \to A$ is called symmetric (resp. sub-symmetric) if $f\big(T_\sigma(x)\big) = f(x)$ for all $x \in U$ and for any permutation $\sigma\colon \mathbb{N} \to \mathbb{N}$ (resp. for any increasing mapping $\sigma\colon \mathbb{N} \to \mathbb{N}$).

These notions will typically be applied to functions whose domain is a Banach space with a symmetric (resp. sub-symmetric) basis, or a subspace of a space with a Schauder basis consisting of finitely supported vectors.

Definition 21. Let $\{x_n\}$ be a sequence in a Banach space X. We say that a sequence $\{e_n\}$ in a Banach space Y is a spreading model of the sequence $\{x_n\}$ if for every $\varepsilon > 0$ and $k \in \mathbb{N}$ there is an $N \in \mathbb{N}$ such that

$$(1-\varepsilon)\left\|\sum_{j=1}^{k} a_j e_j\right\| \leq \left\|\sum_{j=1}^{k} a_j x_{n_j}\right\| \leq (1+\varepsilon)\left\|\sum_{j=1}^{k} a_j e_j\right\|$$

for all $N \leq n_1 < n_2 < \cdots < n_k$ and all scalars $a_1, \dots, a_k$.

Note that a spreading model is a 1-spreading sequence. Moreover, if $\{x_n\}$ is a basic sequence with a spreading model $\{e_n\}$, then $\{e_n\}$ is also a basic sequence with a basis constant bounded above by that of $\{x_n\}$. It is easy to see that if a spreading model $\{e_n\}$ of $\{x_n\}$ is a basic sequence, then it is uniquely determined in the sense that if $\{u_n\}$ is a spreading model of $\{x_n\}$, then $\{u_n\}$ is a basic sequence and $T \in \mathcal{L}(\overline{\operatorname{span}}\{e_n\}; \overline{\operatorname{span}}\{u_n\})$, $T(e_n) = u_n$ is an isometry. Clearly if a sequence has a spreading model, then its subsequences have the same spreading model. For a given sequence $\{x_n\}$ a spreading model may not exist. However, we have the following fundamental result.

Theorem 22 (Antoine Brunel and Louis Sucheston, [BrSu1]). *Let X be a Banach space and suppose that $\{x_n\} \subset X$ is a bounded sequence such that $\{x_n;\ n \in \mathbb{N}\}$ is not relatively compact. Then $\{x_n\}$ has a subsequence with a spreading model.*

The proof is based on repeated use of Ramsey's theorem and can be found e.g. in [FHHMZ, Theorem 6.6] (the main idea of the crucial step is exposed in the proof of a simpler Theorem 79).

The following fact follows almost immediately from the definition.

Fact 23. *Let X, Y be Banach spaces and $\{x_n\} \subset X$ a sequence with a spreading model $\{e_n\} \subset Y$. Then $\overline{\operatorname{span}}\{e_n\}$ is finitely representable in X.*

The proof of the next proposition can be found in [FHHMZ, Theorem 6.7].

Proposition 24. *Let X be a Banach space and $\{x_n\} \subset X$ a weakly null sequence with a spreading model $\{e_n\}$. Then $\{e_n\}$ is a sub-symmetric basic sequence with the unconditional basis constant at most 2.*

By passing to subsequences and diagonalising we obtain the following useful observation.

Proposition 25. *Let X, Y be a Banach spaces and $\{x_n\} \subset X$ a sequence with a spreading model $\{e_n\} \subset Y$. Let $\{\varepsilon_k\}_{k=1}^{\infty} \subset \mathbb{R}^+$ and $\{N_k\}_{k=1}^{\infty} \subset \mathbb{N}$. There is a subsequence $\{y_n\}$ of $\{x_n\}$ such that*

$$(1-\varepsilon_k)\left\|\sum_{j=1}^{N_k} a_j e_j\right\| \le \left\|\sum_{j=1}^{N_k} a_j y_{n_j}\right\| \le (1+\varepsilon_k)\left\|\sum_{j=1}^{N_k} a_j e_j\right\| \tag{1}$$

for all $k \le n_1 < n_2 < \cdots < n_{N_k}$ and all scalars $a_1, \ldots, a_{N_k}$.

Later on we will make use of the following additional result. We prefer to omit the proof, as it can be obtained by modifying the proof of Theorem 22, working simultaneously with the norm $\|\cdot\|$ on X and P and keeping in mind that homogeneous polynomials form a closed set in the topology of uniform convergence on bounded sets (see also [GJ2, Theorem 1.8]).

Theorem 26. *Let X be a Banach space, $\{x_n\} \subset X$ a semi-normalised basic sequence, $P \in \mathcal{P}(^d X)$, $\{\varepsilon_k\}_{k=1}^{\infty} \subset \mathbb{R}^+$, and $\{N_k\}_{k=1}^{\infty} \subset \mathbb{N}$. There are a subsequence $\{y_n\}$ of $\{x_n\}$ with a spreading model $\{e_n\} \subset Y = \overline{\operatorname{span}}\{e_n\}$ and a sub-symmetric polynomial $Q \in \mathcal{P}(^d Y)$ such that (1) holds and*

$$\left| Q\left(\sum_{j=1}^{N_k} a_j e_j\right) - P\left(\sum_{j=1}^{N_k} a_j y_{n_j}\right)\right| \le \varepsilon_k$$

for all $k \le n_1 < n_2 < \cdots < n_{N_k}$ and all scalars $a_1, \ldots, a_{N_k}$ with $\sum_{j=1}^{N_k} a_j y_{n_j} \in B_X$.

The behaviour of spreading models with respect to duality is described in the following theorem.

Theorem 27 ([BeaLa, Corollaries III.1.5 and III.1.6]). *Let X, Y be Banach spaces and $\{x_n\} \subset X$ a weakly null basic sequence with a spreading model $\{e_n\} \subset Y$. Further, let $\{x_n^*\} \subset X^*$ be a basic sequence biorthogonal to $\{x_n\}$ and $\{e_n^*\} \subset Y^*$ a basic sequence biorthogonal to $\{e_n\}$. Then there exists an infinite set $M \subset \mathbb{N}$ such that the sequence $\{x_n^*\}_{n\in M} \subset \left(\overline{\operatorname{span}}\{x_n\}_{n\in M}\right)^*$ has a spreading model $\{e_n^*\}_{n=1}^\infty$. In particular, the sequence $\{e_n^*\}_{n=1}^\infty$ is a spreading model of a sequence in a quotient of X^*.*

Definition 28. A sequence $\{x_n\}_{n=1}^\infty$ in a normed linear space is called a Banach-Saks sequence if the sequence $\left\{\frac{1}{N}\sum_{n=1}^N x_n\right\}_{N=1}^\infty$ is convergent. We say that $\{x_n\}_{n=1}^\infty$ is a hereditarily Banach-Saks sequence provided that every subsequence of $\{x_n\}_{n=1}^\infty$ is a Banach-Saks sequence.

Every convergent sequence is easily seen to be hereditarily Banach-Saks. The canonical basis of c_0 is an example of a non-convergent hereditarily Banach-Saks sequence. Haskell Rosenthal proved that every weakly null sequence in a Banach space has a subsequence which is either hereditarily Banach-Saks, or its spreading model is equivalent to the canonical basis of ℓ_1 ([BeaLa, Proposition II.6.2]). Hence if X is a Banach space, then either X is a Schur space, or X contains a normalised weakly null hereditarily Banach-Saks sequence, or X contains a normalised weakly null sequence and every spreading model of every normalised weakly null sequence is equivalent to the canonical basis of ℓ_1.

Definition 29. Let X be a Banach space. We say that X has the Banach-Saks property if every bounded sequence in X has a Banach-Saks subsequence. We say that X has the weak Banach-Saks property if every weakly null sequence in X has a Banach-Saks subsequence.

It follows readily from the James theorem [FHHMZ, Corollary 3.131] that every Banach space with the Banach-Saks property is reflexive. On the other hand, there exist reflexive spaces without the Banach-Saks property (the Baernstein space or the (Figiel-Johnson) Tsirelson space T, [CasaSh, p. 18]). Every Schur space (and in particular ℓ_1) enjoys the weak Banach-Saks property, but ℓ_1 clearly does not have the Banach-Saks property. Consequently, no Schur space has the Banach-Saks property. On the other hand, for reflexive spaces the weak Banach-Saks property clearly coincides with the Banach-Saks property. The spaces with the weak Banach-Saks property can be characterised, thanks to Rosenthal's results mentioned above, as those spaces whose spreading models of weakly null sequences are not equivalent to the canonical basis of ℓ_1 ([BeaLa, Theorem II.3.8]). If X has the Banach-Saks property, then every bounded sequence in X has even a hereditarily Banach-Saks subsequence, and similarly if X has the weak Banach-Saks property, then every weakly null sequence in X has even a hereditarily Banach-Saks subsequence, [Rak], see also [Dies2, p. 253, pp. 192–197]. The weak Banach-Saks property can be conveniently refined using the following concept.

Definition 30. Let $1 < p < \infty$. A sequence $\{x_n\}_{n=1}^{\infty}$ in a normed linear space is called a p-Banach-Saks sequence if there is $C > 0$ such that

$$\left\|\sum_{n=1}^{N} x_n\right\| \leq CN^{\frac{1}{p}}$$

for all $N \in \mathbb{N}$. We say that $\{x_n\}_{n=1}^{\infty}$ is a hereditarily p-Banach-Saks sequence if for some $C > 0$ every subsequence of $\{x_n\}_{n=1}^{\infty}$ is a p-Banach-Saks sequence with the constant C.

We say that a Banach space X has the weak p-Banach-Saks property if there exists a $C > 0$ such that every normalised weakly null sequence in X has a p-Banach-Saks subsequence with the constant C.

It is immediate that a p-Banach-Saks sequence is an r-Banach-Saks sequence for $r < p$ and it is also a Banach-Saks sequence. Thus the weak p-Banach-Saks property implies the weak Banach-Saks property. It follows readily from the Gurarii-James theorem [FHHMZ, Theorem 9.25] that every super-reflexive Banach space has the weak p-Banach-Saks property for some $p \in (1, \infty)$, hence it has the Banach-Saks property.

Fact 31. *Let $1 < p < \infty$ and let $\{x_n\}$ be a hereditarily p-Banach-Saks sequence in a normed linear space X. Then there is a $C > 0$ such that for any subsequence $\{y_n\}_{n=1}^{\infty}$ of $\{x_n\}_{n=1}^{\infty}$, any sequence $\{a_n\}_{n=1}^{\infty}$ of scalars with $|a_n| \leq 1$, $n \in \mathbb{N}$, and any $N \in \mathbb{N}$*

$$\left\|\sum_{n=1}^{N} a_n y_n\right\| \leq CN^{\frac{1}{p}}.$$

Proof. Assume that $\{x_n\}_{n=1}^{\infty}$ is a hereditarily p-Banach-Saks sequence with a constant C. Let $\{y_n\}$ be a subsequence of $\{x_n\}$ and fix $N \in \mathbb{N}$. Assume first that X is real. By the Bauer maximum principle

$$M = \max_{|a_n| \leq 1} \left\|\sum_{n=1}^{N} a_n y_n\right\|$$

is attained at an extremal point of $B_{\ell_\infty^N}$, i.e. for some $\{a_n\}_{n=1}^{N} \subset \{-1, 1\}$. Thus

$$M \leq \max_{A \subset \{1,\ldots,N\}} \left(\left\|\sum_{n \in A} y_n\right\| + \left\|\sum_{n \notin A} -y_n\right\|\right) \leq 2CN^{\frac{1}{p}}.$$

For the complex case use the above estimate for the real and imaginary part. □

Clearly, if a sequence has an upper p-estimate, then it is hereditarily p-Banach-Saks. Conversely, we have the following:

Lemma 32. *Let $1 < p < \infty$ and let $\{x_n\}_{n=1}^{\infty}$ be a hereditarily p-Banach-Saks sequence in a Banach space. Then $\{x_n\}_{n=1}^{\infty}$ has an upper r-estimate for any $1 < r < p$.*

Proof. Let q be such that $\frac{1}{p}+\frac{1}{q}=1$, let C be the constant from Fact 31, and denote $D=\sum_{n=0}^{\infty}2^{nq(\frac{r}{p}-1)}\in\mathbb{R}$. Choose $N\in\mathbb{N}$ and any scalars $a_1,\dots,a_N$ with $\sum_{j=1}^{N}|a_j|^r\le 1$. Define $B_n=\{j\in\mathbb{N};\ 2^{-(n+1)}<|a_j|\le 2^{-n}\}$, $n\in\mathbb{N}_0$. Then, using the Hölder inequality,

$$\begin{aligned}\left\|\sum_{j=1}^{N}a_jx_j\right\| &\le \sum_{n=0}^{\infty}\left\|\sum_{j\in B_n}a_jx_j\right\| \le C\sum_{n=0}^{\infty}2^{-n}|B_n|^{\frac{1}{p}} = C\sum_{n=0}^{\infty}2^{n(\frac{r}{p}-1)}2^{-\frac{nr}{p}}|B_n|^{\frac{1}{p}}\\ &\le CD^{\frac{1}{q}}\left(\sum_{n=0}^{\infty}2^{-nr}|B_n|\right)^{\frac{1}{p}} = 2^{\frac{r}{p}}CD^{\frac{1}{q}}\left(\sum_{n=0}^{\infty}2^{-(n+1)r}|B_n|\right)^{\frac{1}{p}}\\ &\le 2^{\frac{r}{p}}CD^{\frac{1}{q}}\left(\sum_{j=1}^{N}|a_j|^r\right)^{\frac{1}{p}} = 2^{\frac{r}{p}}CD^{\frac{1}{q}}.\end{aligned}$$

□

Lemma 33. *Let X be a Banach space, $\{x_n\}\subset X$ a sequence with a spreading model $\{e_n\}$, and $1<p<\infty$. Then $\{x_n\}$ has a hereditarily p-Banach-Saks subsequence if and only if $\{e_n\}$ is p-Banach-Saks.*

Proof. $\Rightarrow$ is easy to see.

$\Leftarrow$ Assume that $\{e_n\}$ is a p-Banach-Saks sequence with a constant K. We may assume without loss of generality that $\{e_n\}$ is normalised. Let $\{y_n\}$ be the subsequence of $\{x_n\}$ from Proposition 25 with $\varepsilon_k=\frac{1}{2}$ and $N_k=2^k$. Note that then $\|y_n\|\le 2$ for all $n\in\mathbb{N}$. For $N\in\mathbb{N}$ set

$$C_N=\frac{4\lceil\log_2 N\rceil}{N^{\frac{1}{p}}}+\frac{2}{N^{\frac{1}{p}}}\left\|\sum_{j=1}^{N}e_j\right\|.$$

Note that $\limsup_{N\to\infty}C_N\le 2K$ and hence $C=\sup_{N\in\mathbb{N}}C_N\in\mathbb{R}$. Let $\{n_j\}\subset\mathbb{N}$ be an increasing sequence and $N\in\mathbb{N}$. Putting $m=\lceil\log_2 N\rceil$ we have using (1)

$$\begin{aligned}\frac{1}{N^{\frac{1}{p}}}\left\|\sum_{j=1}^{N}y_{n_j}\right\| &\le \frac{1}{N^{\frac{1}{p}}}\left\|\sum_{j=1}^{m}y_{n_j}\right\|+\frac{1}{N^{\frac{1}{p}}}\left\|\sum_{j=m+1}^{N}y_{n_j}\right\| \le \frac{2m}{N^{\frac{1}{p}}}+\frac{2}{N^{\frac{1}{p}}}\left\|\sum_{j=m+1}^{N}e_j\right\|\\ &\le \frac{2m}{N^{\frac{1}{p}}}+\frac{2}{N^{\frac{1}{p}}}\left\|\sum_{j=1}^{N}e_j\right\|+\frac{2}{N^{\frac{1}{p}}}\left\|\sum_{j=1}^{m}e_j\right\| \le \frac{4m}{N^{\frac{1}{p}}}+\frac{2}{N^{\frac{1}{p}}}\left\|\sum_{j=1}^{N}e_j\right\| \le C.\end{aligned}$$

□

Clearly, an upper p-estimate or a lower q-estimate passes from a sequence to its spreading model. Conversely, the following holds.

Corollary 34 ([GJ1], [FaJ]). *Let X be a Banach space, let $\{x_n\}\subset X$ be a sequence with a spreading model $\{e_n\}$, and $1<p,q<\infty$. If $\{e_n\}$ is a p-Banach-Saks sequence, then $\{x_n\}$ has a subsequence with an upper r-estimate for all $1<r<p$. If $\{e_n\}$ has a lower q-estimate and $\{x_n\}$ is weakly null, then $\{x_n\}$ has a subsequence with a lower r-estimate for all $r>q$.*

Proof. The first statement follows by combining Lemmata 33 and 32. The second statement follows by combining Theorem 27 and Fact 3.50 with the first statement. □

We say that a sequence $\{x_n\}$ has an upper p-estimate for non-negative coefficients if there exists $C > 0$ such that $\|\sum_{n=1}^N a_n x_n\| \le C(\sum_{n=1}^N a_n^p)^{\frac{1}{p}}$ for every $N \in \mathbb{N}$ and every $a_1, \dots, a_N \in \mathbb{R}_0^+$; similarly for a lower q-estimate. Since a spreading model of a weakly null sequence is unconditional (Proposition 24), we conclude the following corollary (compare this result also with Theorem 6.71).

Corollary 35. *Let $1 < p, q < \infty$ and let $\{x_n\} \subset X$ be a semi-normalised weakly null sequence in a Banach space with an upper p-estimate (resp. lower q-estimate) for non-negative coefficients. Then $\{x_n\}$ has a subsequence with an upper r-estimate for all $1 < r < p$ (resp. lower r-estimate for all $r > q$).*

It is easy to see that if a Banach space X with an unconditional basis is of type p (resp. of cotype q), then it has property S_p (resp. T_q). The converse is false even for spaces with a symmetric basis, as shown by Gilles Pisier ([LiTz3, Example 1.f.19]). This observation leads to the next result.

Proposition 36. *If a Banach space X is of type $p > 1$, then X has the weak p-Banach-Saks property and it has property S_r for all $1 < r < p$. If a Banach space X is of cotype $q < \infty$, then X has property T_r for all $r > q$.*

Proof. Let $\{x_n\} \subset X$ be a normalised weakly null sequence. By passing to a subsequence we may without loss of generality assume that $\{x_n\}$ has a spreading model $\{e_n\} \subset Y = \overline{\operatorname{span}}\{e_n\}$ such that $\{e_n\}$ is an unconditional basis of Y with the unconditional basis constant at most 2 (Proposition 24). Since Y is finitely representable in X (Fact 23), Y is of type p (resp. cotype q). Hence $\{e_n\}$ has an upper p-estimate with the constant $2T_p(X)$ (resp. lower q-estimate with the constant $2C_q(X)$). The result for type now follows from Lemma 33 and Lemma 32 (notice that in the proof Lemma 33 the constant C in this case depends only on $T_p(X)$). The case of cotype follows from Corollary 34. □

3. Polynomials and p-estimates

In our study of polynomials and their relationship with the Banach space structure we are lead to make a fundamental distinction between weakly sequentially continuous ones (which are in some sense considered the easy ones) and the rest. Let X be a Banach space that does not contain ℓ_1 and $1 < p < \infty$. By Fact 3.49 if a semi-normalised basic sequence $\{f_j\}_{j=1}^\infty$ in X^* has an upper p-estimate, then the linear operator $S\colon \ell_p \to X^*$, $S(e_j) = f_j$ is bounded. By the reflexivity of ℓ_p there is $T \in \mathcal{L}(X; \ell_q)$, $\frac{1}{p} + \frac{1}{q} = 1$, such that $T^* = S$. Since S is non-compact, so is T. By Proposition 3.33 there are $L \in \mathcal{L}(X; \ell_q)$ and a weakly null sequence $\{x_n\} \subset X$ such that $L(x_n) = e_n$. We define $Q \in \mathcal{P}({}^{\lceil q \rceil}\ell_q)$ by $Q(u) = \sum_{j=1}^\infty u_j^{\lceil q \rceil}$ and put

$P = Q \circ L \in \mathcal{P}(^{\lceil q \rceil}X)$. Then P is a prototype of a non-wsc polynomial. Refinements of this basic approach are the subject of the present section.

Unlike linear operators, continuous polynomials are not necessarily w–w continuous. For example, $P \in \mathcal{P}(^2\ell_2; \ell_1)$, $P\big((x_k)_{k=1}^{\infty}\big) = (x_k^2)_{k=1}^{\infty}$ does not map weakly null sequences to weakly null sequences.

Fact 37. *Let $\{e_n\}$ be a semi-normalised unconditional basic sequence in a Banach space X. If $\{e_n\}$ is not weakly null, then it has a subsequence with a lower 1-estimate. Consequently, the subsequence is equivalent to the canonical basis of ℓ_1.*

Proof. We may assume without loss of generality that there is $f \in S_{X^*}$ such that $\operatorname{Re} f(e_n) \geq \alpha > 0$ for all $n \in \mathbb{N}$. Denote by K the unconditional basis constant of $\{e_n\}$. Let $\{a_n\}_{n=1}^{N}$ be a finite sequence of scalars. Then

$$\left\| \sum_{n=1}^{N} a_n e_n \right\| \geq \frac{1}{2K} \left\| \sum_{n=1}^{N} |a_n| e_n \right\| \geq \frac{1}{2K} \left| f\left(\sum_{n=1}^{N} |a_n| e_n \right) \right|$$
$$= \frac{1}{2K} \left| \sum_{n=1}^{N} |a_n| f(e_n) \right| \geq \frac{\alpha}{2K} \sum_{n=1}^{N} |a_n|. \qquad \square$$

We will generalise this fact to the polynomial setting.

Definition 38. A sequence $\{x_k\}_{k=1}^{\infty}$ in a normed linear space is called $\mathcal{P}^n$-null if $\lim_{k\to\infty} P(x_k) = 0$ for every $P \in \mathcal{P}^n(X)$, $P(0) = 0$.

Lemma 39. *Let X be a real Banach space, $n \in \mathbb{N}$, $n \geq 2$, and let $\{x_k\}_{k=1}^{\infty}$ be a $\mathcal{P}^{n-1}$-null sequence in X. If $P(x_k) \not\to 0$ for some $P \in \mathcal{P}(^nX)$, then $\{x_k\}$ has a subsequence with*

- *a lower n-estimate for non-negative coefficients,*
- *a lower r-estimate for every $r > n$,*
- *a lower n-estimate if n is even.*

Proof. ([Dev1]) Without loss of generality we may suppose that $P(x_k) \geq 1$ for all $k \in \mathbb{N}$. Put $\delta_k = \frac{1}{2k^n n! 2^k}$, $k \in \mathbb{N}$ and observe that for a fixed $m \in \mathbb{N}$ we have $\sum_{1 \leq k_1 \leq \cdots \leq k_n = m} \delta_m \leq \frac{1}{2n! 2^m}$. Hence $\sum_{1 \leq k_1 \leq \cdots \leq k_n} \delta_{k_n} \leq \frac{1}{2n!}$. We will construct an increasing sequence of integers $\{n_k\}_{k=1}^{\infty}$ such that $|\check{P}(x_{k_1}, \ldots, x_{k_n})| < \delta_l$ whenever $\{k_j\}_{j=1}^{n} \subset \{n_k\}$, $k_1 \leq \cdots \leq k_n$, $k_1 < k_n$, and $k_n = n_l$. To this end, put $n_1 = 1$ and continue by induction. Suppose that $n_1, \ldots, n_m$ are already defined for some $m \in \mathbb{N}$. For each $k_1, \ldots, k_r \in \mathbb{N}$ the mapping $x \mapsto \check{P}(x_{k_1}, \ldots, x_{k_r}, {}^{n-r}x)$ is a homogeneous polynomial of degree less than n. Thus there exists $n_{m+1} \in \mathbb{N}$, $n_{m+1} > n_m$ such that $|\check{P}(x_{k_1}, \ldots, x_{k_r}, {}^{n-r}x_{n_{m+1}})| < \delta_{m+1}$ whenever $r < n$, $\{k_1, \ldots, k_r\} \subset \{n_1, \ldots, n_m\}$, and $k_1 \leq \cdots \leq k_r$.

Denote $y_j = x_{n_j}$, $j \in \mathbb{N}$. Let $a_1, \dots, a_N \in \mathbb{R}$ and moreover $a_j \geq 0$, $j = 1, \dots, N$ in case that n is odd. Then

$$\begin{aligned}
P\left(\sum_{j=1}^{N} a_j y_j\right) &\geq \sum_{j=1}^{N} a_j^n P(y_j) - \sum_{\substack{1\leq k_1\leq\cdots\leq k_n\leq N\\ k_1<k_n}} n!\left|a_{k_1}\cdots a_{k_n}\check{P}(y_{k_1},\dots,y_{k_n})\right| \\
&\geq \sum_{j=1}^{N} a_j^n - n! \sum_{\substack{1\leq k_1\leq\cdots\leq k_n\leq N\\ k_1<k_n}} |a_{k_1}\cdots a_{k_n}|\delta_{k_n} \\
&\geq \sum_{j=1}^{N} a_j^n - n! \sum_{\substack{1\leq k_1\leq\cdots\leq k_n\leq N\\ k_1<k_n}} \frac{a_{k_1}^n + \cdots + a_{k_n}^n}{n}\delta_{k_n} \\
&\geq \sum_{j=1}^{N} a_j^n \left(1 - n! \sum_{\substack{1\leq k_1\leq\cdots\leq k_n\\ k_1<k_n}} \delta_{k_n}\right) \\
&\geq \frac{1}{2}\sum_{j=1}^{N} a_j^n.
\end{aligned}$$

Thus

$$\left\|\sum_{j=1}^{N} a_j y_j\right\|^n \geq \frac{1}{\|P\|} P\left(\sum_{j=1}^{N} a_j y_j\right) \geq \frac{1}{2\|P\|}\sum_{j=1}^{N} a_j^n.$$

This proves the first and the third statement. The second statement follows from the first one and Corollary 35. □

Note that the assumptions in the above lemma preclude $\{x_k\}_{k=1}^{\infty}$ from being equivalent to the canonical basis of ℓ_1 and the statement implies the existence of a non-compact bounded linear operator from a suitable subspace of X into ℓ_r.

In the case of an unconditional basis we obtain somewhat stronger result (we do not need to pass to a subsequence and we allow also complex scalars).

Lemma 40 ([GJ2]). *Let X be a Banach space and $\{x_k\}_{k=1}^{\infty} \subset X$ an unconditional basic sequence. If there are $P \in \mathcal{P}(^nX)$ and $c > 0$ such that $|P(x_k)| \geq c$ for all $k \in \mathbb{N}$, then $\{x_k\}$ has a lower n-estimate.*

Proof. It is easy to check that if X is real, then $\{x_k\}$ is also an unconditional basic sequence in $\tilde{X}$. Thus we may without loss of generality assume that the space X is complex. Fix any $a_1, \dots, a_N \in \mathbb{C}$ and choose $\alpha_1, \dots, \alpha_N \in \mathbb{C}$, $|\alpha_j| = 1$, such that

$\alpha_j^n a_j^n P(x_j) = |a_j|^n |P(x_j)|$. Then using Lemma 2.69 we obtain

$$\begin{aligned} c\sum_{j=1}^{N}|a_j|^n &\le \sum_{j=1}^{N}|a_j|^n|P(x_j)| = \sum_{j=1}^{N}\alpha_j^n a_j^n P(x_j) = \sum_{j=1}^{N}P(\alpha_j a_j x_j) \\ &= \frac{1}{n^N}\sum_{l=1}^{n^N}P\left(\sum_{j=1}^{N}r_j^{n,N}(l)\alpha_j a_j x_j\right) \\ &\le \frac{1}{n^N}\sum_{l=1}^{n^N}\|P\|\left\|\sum_{j=1}^{N}r_j^{n,N}(l)\alpha_j a_j x_j\right\|^n \\ &\le (2C)^n\|P\|\left\|\sum_{j=1}^{N}a_j x_j\right\|^n, \end{aligned}$$

where C is the unconditional basis constant of $\{x_k\}$. □

Corollary 41. *Let X be a Banach space with a sub-symmetric basis $\{e_n\}_{n=1}^{\infty}$ which fails to have a lower k-estimate for some $k \in \mathbb{N}$. Let $S \in \mathcal{L}(X; X)$ be the right shift operator and let $x \in X$ be finitely supported. Then no subsequence of $\{S^n(x)\}_{n=1}^{\infty}$ has a lower k-estimate and $\{S^n(x)\}_{n=1}^{\infty}$ is $\mathcal{P}^k$-null.*

Proof. Denote by K the sub-symmetric basis constant of $\{e_n\}$. Let $x = \sum_{j=1}^{m} c_j e_j$ and denote $C = \sum_{j=1}^{m}|c_j|$. Suppose that there is an increasing sequence $\{k_n\}_{n=1}^{\infty} \subset \mathbb{N}$ such that $\{S^{k_n}(x)\}_{n=1}^{\infty}$ has a lower k-estimate with a constant L. Let $a_1, \dots, a_N$ be arbitrary scalars. Then

$$\begin{aligned} L\left(\sum_{n=1}^{N}|a_n|^k\right)^{\frac{1}{k}} &\le \left\|\sum_{n=1}^{N}a_n S^{k_n}(x)\right\| = \left\|\sum_{n=1}^{N}a_n\sum_{j=1}^{m}c_j e_{k_n+j}\right\| \\ &\le \sum_{j=1}^{m}|c_j|\left\|\sum_{n=1}^{N}a_n e_{k_n+j}\right\| \le CK\left\|\sum_{n=1}^{N}a_n e_n\right\|, \end{aligned}$$

which is a contradiction.

To prove the second statement we use induction on k. For $k = 1$ it follows from Fact 37 and the first statement (notice that there is a subsequence of $\{S^n(x)\}_{n=1}^{\infty}$ which is a block basic sequence of $\{e_n\}$ and so it is unconditional). Now suppose that $k > 1$ and the statement holds for $k-1$. If $\{e_n\}_{n=1}^{\infty}$ does not have a lower k-estimate, then it does not have a lower $(k-1)$-estimate and so by the inductive hypothesis $\{S^n(x)\}_{n=1}^{\infty}$ is $\mathcal{P}^{k-1}$-null. Hence it is $\mathcal{P}^k$-null by Lemma 40. □

Proposition 42 ([FaJ]). *Let X be a Banach space, $p > 1$, and $\frac{1}{p} + \frac{1}{q} = 1$. Assume that X^* does not contain ℓ_1 and has property S_p. Then each semi-normalised basic sequence $\{x_k\}_{k=1}^{\infty} \subset X$ has a subsequence $\{x_{n_k}\}_{k=1}^{\infty}$ such that there exists a bounded linear operator $T\colon X \to \ell_q$ satisfying $T(x_{n_k}) = e_k$, where $\{e_k\}$ is the canonical basis of ℓ_q. Hence for each $n \ge q$ there is $P \in \mathcal{P}(^nX)$ such that $P(x_{n_k}) = 1$ for all $k \in \mathbb{N}$.*

Proof. Let $\{x_k^*\}_{k=1}^\infty \subset X^*$ be a sequence biorthogonal to $\{x_k\}_{k=1}^\infty$. By passing to a subsequence we may assume by Rosenthal's ℓ_1-theorem (Theorem 3.71) that $\{x_k^*\}$ is weakly Cauchy. Put $z_k^* = x_{2k}^* - x_{2k+1}^*$. Then $\{z_k^*\}$ is weakly null. Since X^* has property S_p, we may assume by passing to a subsequence that $\{z_k^*\}$ has an upper p-estimate, so by Fact 3.49 the linear operator $S\colon \ell_p \to X^*$, $S(f_k) = z_k^*$ is bounded, where $\{f_j\}$ is the canonical basis of ℓ_p. Since ℓ_p is reflexive, S is w^*–w^* continuous and so there is $T \in \mathcal{L}(X;\ell_q)$ such that $T^* = S$. Then $T(x_{2k}) = e_k$. Indeed, $\langle f_j, T(x_{2k})\rangle = \langle S(f_j), x_{2k}\rangle = \langle z_j^*, x_{2k}\rangle = \delta_{j,k}$. Let $n \geq q$ and $Q\colon \ell_q \to \mathbb{K}$ be defined by $Q(u) = \sum_{j=1}^\infty u_j^n$. Then $Q \in \mathcal{P}(^n\ell_q)$ by Theorem 1.29 and hence $P = Q \circ T \in \mathcal{P}(^nX)$. Clearly $P(x_{2k}) = 1$ for all $k \in \mathbb{N}$. □

Corollary 43 ([FaJ]). *Let X be a Banach space such that X^* is of type $p > 1$, $\frac{1}{p} + \frac{1}{q} = 1$, and let $\{x_k\}_{k=1}^\infty \subset X$ be a semi-normalised basic sequence. Then for each $s > q$ there is a subsequence $\{x_{n_k}\}_{k=1}^\infty$ such that there exists a bounded linear operator $T\colon X \to \ell_s$ satisfying $T(x_{n_k}) = e_k$, where $\{e_k\}$ is the canonical basis of ℓ_s. Further, there is a subsequence $\{x_{n_k}\}_{k=1}^\infty$ such that for each $n \in \mathbb{N}$, $n > q$ there is $P \in \mathcal{P}(^nX)$ such that $P(x_{n_k}) = 1$ for all $k \in \mathbb{N}$.*

Proof. If X^* is of type $p > 1$, then X^* does not contain ℓ_1, and by Proposition 36, X^* has property S_r for every $1 < r < p$. Thus we may apply Proposition 42. For the second part we may use a fixed $s \in (q, [q] + 1)$. □

Proposition 44 ([GJ2]). *Let X be a real Banach space that admits a $C^{k,\alpha}$-smooth bump, $k \in \mathbb{N}$, $\alpha \in (0, 1]$. Then any normalised $\mathcal{P}^k$-null sequence in X has a subsequence with an upper $(k + \alpha)$-estimate.*

Proof. Let $f \in C^{k,\alpha}(X)$ be such that $f(0) = 0$ and $f(x) \geq 2$ for $\|x\| \geq 1$. For each $x \in X$ denote by $P(x) \in \mathcal{P}^k(X)$ the polynomial $h \mapsto T(x)[h] - f(x)$, where $T(x)$ is the Taylor polynomial of order k of f at x. Put $q = k + \alpha$. By Corollary 1.108 there is $K \geq 1$ such that $\left|f(x + h) - f(x) - P(x)[h]\right| \leq K\|h\|^q$ for all $h \in X$. Let $\{x_n\}_{n=1}^\infty \subset X$ be a normalised $\mathcal{P}^k$-null sequence in X. We will construct a subsequence $\{x_{n_j}\}_{j=1}^\infty$ by induction.

Suppose that $n_1 < n_2 < \cdots < n_{m-1} \in \mathbb{N}$ are already defined for some $m \in \mathbb{N}$. The mapping $x \mapsto P(x)$ is continuous and so the set

$$\mathcal{Q}_m = \left\{Q \in \mathcal{P}^k(X);\ Q(h) = P\left(\sum_{j=1}^{m-1} a_j x_{n_j}\right)[a_m h], |a_j| \leq 1, j = 1, \ldots, m\right\}$$

is a compact subset of $\mathcal{P}^k(X)$. Since $\{x_n\}$ is $\mathcal{P}^k$-null, it follows that there is $n_m \in \mathbb{N}$, $n_m > n_{m-1}$ such that $|Q(x_{n_m})| \leq \frac{1}{2^m}$ for all $Q \in \mathcal{Q}_m$.

Now if $a_1, \ldots, a_N \in \mathbb{R}$ are not all zero, we set $a = \left(\sum_{j=1}^N |a_j|^q\right)^{\frac{1}{q}}$ and $b_j = \frac{a_j}{aK^{1/q}}$. Then $\sum_{j=1}^N |b_j|^q = \frac{1}{K}$ and $|b_j| \leq 1$, $j = 1, \ldots, N$. Hence using the definition of $\mathcal{Q}_l$

we can estimate

$$f\left(\sum_{j=1}^{N} b_j x_{n_j}\right) \le \sum_{l=1}^{N}\left| f\left(\sum_{j=1}^{l} b_j x_{n_j}\right) - f\left(\sum_{j=1}^{l-1} b_j x_{n_j}\right)\right|$$
$$\le \sum_{l=1}^{N} K|b_l|^q + \left| P\left(\sum_{j=1}^{l-1} b_j x_{n_j}\right)[b_l x_{n_l}]\right| \le \sum_{l=1}^{N} K|b_l|^q + \frac{1}{2^l} < 2,$$

which implies that $\left\|\sum_{j=1}^{N} b_j x_{n_j}\right\| < 1$ and so $\left\|\sum_{j=1}^{N} a_j x_{n_j}\right\| < K^{\frac{1}{q}}\left(\sum_{j=1}^{N} |a_j|^q\right)^{\frac{1}{q}}$.

□

4. Separating polynomials. Symmetric and sub-symmetric polynomials

In the present section we combine the spreading model techniques developed in the previous sections with the averaging technique for polynomials to obtain some of the main results on Banach spaces with separating polynomials. We show that if X has a separating d-homogeneous polynomial and a sub-symmetric basis, then it is isomorphic to some ℓ_p for p even integer and $d = kp, k \in \mathbb{N}$. Moreover, by a fundamental result of R. Deville every Banach space with a separating polynomial contains such ℓ_p. We show that for an arbitrary polynomial P on X there is an infinite-dimensional subspace Y of X such that the restriction of P to Y is either separating or asymptotically (and this cannot be strengthened) zero in a strong sense.

We start with a few facts concerning separating polynomials.

Fact 45 ([Kur1], [FPWZ]). *Let X be a real normed linear space and let $P \in \mathcal{P}^n(X)$, $z \in X$, and $r > 0$ be such that P separates z from $\{x \in X;\ \|x - z\| = r\}$. Then X admits*

- *a polynomial $Q \in \mathcal{P}^n(X)$ such that $Q(0) = 0$ and $\inf_{S_X} Q > 0$,*
- *a separating polynomial of degree at most $2n$ that separates 0 from $X \setminus B(0, s)$ for every $s > 0$,*
- *a homogeneous separating polynomial that separates 0 from $X \setminus B(0, s)$ for every $s > 0$.*

Proof. If $\dim X < \infty$, then the assertions are trivial, so assume that $\dim X = \infty$. Put $q(x) = P(rx + z) - P(z)$. Then $q(0) = 0$ and since S_X is connected we have either $\inf_{S_X} q > 0$, in which case we put $Q = q$, or $\sup_{S_X} q < 0$, in which case we put $Q = -q$. Let $\eta = \inf_{S_X} Q$ and let $Q_j \in \mathcal{P}(^jX)$, $j = 1, \dots, n$, be such that $Q = \sum_{j=1}^{n} Q_j$. Put $R = \sum_{j=1}^{n} Q_j^2$ and $S = \sum_{j=1}^{n} Q_j^{2n!/j}$. From Fact 1.28 it follows that $R \in \mathcal{P}^{2n}(X)$ and $S \in \mathcal{P}(^{2n!}X)$.

Let $s > 0$ and fix $x \in X$, $\|x\| > s$. There is $j \in \{1, \dots, n\}$ such that $Q_j\left(\frac{x}{\|x\|}\right) \ge \frac{\eta}{n}$. Then $R(x) \ge Q_j(x)^2 \ge \|x\|^{2j}\left(\frac{\eta}{n}\right)^2 \ge \min\{s^2, s^{2n}\}\left(\frac{\eta}{n}\right)^2$. Similarly, as $Q_l(x)^{2n!/l} \ge 0$

for each $l \in \{1, \dots, n\}$, it follows that $S(x) \geq Q_j(x)^{2n!/j} \geq \|x\|^{2n!}\big(\frac{\eta}{n}\big)^{2n!/j} \geq s^{2n!}\big(\min\{1, \frac{\eta}{n}\}\big)^{2n!}$. $\square$

We note that the polynomial $(x, y) \mapsto 2x^2 - x^4 + y^4$ on $X = \ell_2^2$ separates 0 from S_X but does not separate 0 from $X \setminus B(0, r)$ for any $r > 0$. Further, the polynomial $(x, y) \mapsto \sum_{j=1}^{\infty} x_j^4 + \sum_{j=1}^{\infty} y_j^6$ is a separating polynomial on $\ell_4 \oplus \ell_6$ of degree 6, but as it follows from Corollary 61 below, homogeneous separating polynomials on this space are of degree $12k$, $k \in \mathbb{N}$.

Notice that a separating polynomial is of degree at least 2 and that if a real Banach space X admits a separating polynomial of degree 2, then X is isomorphic to a Hilbert space. Indeed, by Fact 45 there is $P \in \mathcal{P}^2(X)$ such that $P(0) = 0$ and $\inf_{S_X} P > 0$. Letting $Q(x) = P(x) + P(-x)$ we have $Q \in \mathcal{P}(^2X)$ and $\eta = \inf_{S_X} Q > 0$. Hence $\eta\|x\|^2 \leq Q(x) = \check{Q}(x, x) \leq \|Q\|\|x\|^2$ for every $x \in X$. It follows that $\check{Q}$ is an inner product on X which gives rise to an equivalent norm on X. In connection with this notice also that every non-negative 2-homogeneous polynomial is convex (use Lemma 1.21).

Suppose that a real normed linear space X admits a separating k-homogeneous polynomial P that is convex. Then $\sqrt[k]{P}$ is an equivalent norm on X. Indeed, it is easily seen to be a quasi-convex positively homogeneous even function and hence it is a norm on X. Such a norm is of course a k-polynomial norm and it is obviously an analytic norm.

Notice also that if a real normed linear space admits a separating polynomial, then by composition with a suitable smooth real function we can produce a C^{∞}-smooth bump with all derivatives bounded (in particular it is $C^{k,1}$-smooth for all $k \in \mathbb{N}$).

The results in Chapter 3 immediately imply the next statement. We note though that later we show much stronger results.

Corollary 46. *Suppose that $p \in [1, +\infty)$ is not an even integer. Then there is no separating polynomial on ℓ_p of degree at most p.*

Proof. Suppose that $P \in \mathcal{P}^n(\ell_p)$ is a separating polynomial, where $n \leq p$. By Fact 45 we may assume that $P(0) = 0$ and $\inf_{S_X} P > 0$. Put $Q(x) = P(x) + P(-x)$. Then Q is of even degree and hence $\deg Q < p$. Further, $Q(0) = 0$ and $\inf_{S_X} Q > 0$. If $p = 1$, then $Q = 0$, which is impossible. For $p > 1$ we have $e_n \to 0$ weakly but $\inf_{n \in \mathbb{N}} Q(e_n) > 0$, a contradiction with Corollary 3.59. $\square$

Fact 47. *Let X be a real normed linear space and let $\{P_j\}_{j=1}^k \subset \mathcal{P}^n(X)$, $z \in X$, and $r > 0$ be such that $\inf\{\max_j |P_j(x) - P_j(z)|;\ \|x - z\| = r\} > 0$. Then X admits*

- *a family of non-constant homogeneous polynomials $\{Q_j;\ 1 \leq j \leq 2kn\}$ of degree at most n such that $\inf_{X \setminus B(0,s)} \max_j Q_j > 0$ for every $s > 0$,*
- *a separating polynomial of degree at most $2n$ that separates 0 from $X \setminus B(0, s)$ for every $s > 0$.*

Proof. We put $q_j(x) = P_j(rx + z) - P_j(z)$, $j = 1, \dots, k$. Then $q_j(0) = 0$ and $\eta = \inf_{S_X} \max_j |q_j| > 0$. Let $q_j = \sum_{l=1}^{n} Q_{j,l}$, where $Q_{j,l} \in \mathcal{P}(^lX)$, $j = 1, \dots, k$.

Then the family $\{\pm Q_{j,l};\ 1 \le j \le k, 1 \le l \le n\}$ satisfies the requirements. Indeed, given $s > 0$ and $x \in X \setminus B(0,s)$ there is $j \in \{1,\dots,k\}$ such that $\left|q_j\left(\frac{x}{\|x\|}\right)\right| \ge \eta$. It follows that there is $l \in \{1,\dots,n\}$ such that $\left|Q_{j,l}\left(\frac{x}{\|x\|}\right)\right| \ge \frac{\eta}{n}$ and consequently $|Q_{j,l}(x)| \ge \|x\|^l \frac{\eta}{n} \ge \min\{s, s^n\}\frac{\eta}{n}$. To see the second statement it suffices to take the polynomial $\sum_{j,l} Q_{j,l}^2$. □

A finite family of polynomials that satisfies the assumptions of Fact 47 is called a separating family of polynomials.

Fact 48. *Let X be a normed linear space and Y a finite-codimensional subspace of X. If Y admits a separating family of polynomials of degree at most n (resp. separating n-homogeneous polynomial), then so does X.*

Proof. Clearly we may assume that $\dim X = \infty$ and so $n > 1$. Using induction on the codimension of Y it suffices to prove the statement for $\operatorname{codim} Y = 1$. Let $f \in S_{X^*}$ be such that $Y = \ker f$ and $v \in X$ such that $f(v) = 1$. Suppose that $\{P_1,\dots,P_k\} \subset \mathcal{P}^n(Y)$ are polynomials such that $P_j(0) = 0$, $j = 1,\dots,k$, and $\inf_{Y\setminus B_Y} \max_j P_j \ge 1$. Put $Q_0(x) = \|v\|^2 f(x)^2$ and $Q_j(x) = P_j\big(x - f(x)v\big)$ for $x \in X$, $j = 1,\dots,k$. Then $Q_j(0) = 0$, $j = 0,\dots,k$. Let $x \in X$, $\|x\| > 2$. If $\|x - f(x)v\| > 1$, then there is $j \in \{1,\dots,k\}$ such that $P_j\big(x - f(x)v\big) \ge 1$ and so $Q_j(x) \ge 1$. Otherwise we have $|f(x)|\|v\| \ge 1$ and so $Q_0(x) \ge 1$.

In the case of a separating polynomial $P \in \mathcal{P}(^nY)$, $\inf_{S_Y} P \ge 1$, we can take $Q(x) = P\big(x - f(x)v\big) + \|v\|^n f(x)^n$. □

As we have seen above, in general the existence of a separating polynomial of degree n does not imply the existence of a separating n-homogeneous polynomial. However, in some special cases this is true.

Proposition 49. *Let X be an infinite-dimensional Banach space with a separating polynomial of degree at most $k \in \mathbb{N}$. If $\mathcal{P}^{k-1}(X) = \mathcal{P}^{k-1}_{\mathrm{wsc}}(X)$ (in particular, if X has property S_k), then X admits a k-homogeneous separating polynomial.*

Proof. By Theorem 5.44 we may assume that X is separable. Since X is an Asplund space (Theorem 5.2), X^* is separable. Let $\{f_n\}$ be a dense sequence in S_{X^*}. We may assume that $P = \sum_{j=1}^k P_j$, where $P_j \in \mathcal{P}(^jX)$, and that $\inf_{S_X} P > 0$. Assume that there is no k-homogeneous separating polynomial on X. By Fact 48 finite-codimensional subspaces of X do not admit a k-homogeneous separating polynomial either. Therefore for each $n \in \mathbb{N}$ there is $x_n \in S_X \cap \bigcap_{j=1}^n \ker f_j$ such that $|P_k(x_n)| < \frac{1}{n}$. Then $\{x_n\}$ is weakly null and so $P_j(x_n) \to 0$ for each $j = 1,\dots,k$, which is a contradiction. If X has property S_k, then we appeal to Corollary 3.58. □

Fact 50. *Let X be a vector space, $\varphi\colon X \to \mathbb{R}$ a convex function, $x_1,\dots,x_n \in X$, and $m \in \mathbb{N}$, $m \le n$. Then*

$$\frac{1}{2^m}\sum_{\varepsilon_j=\pm1}\varphi\left(\sum_{j=1}^m \varepsilon_j x_j\right) \le \frac{1}{2^n}\sum_{\varepsilon_j=\pm1}\varphi\left(\sum_{j=1}^n \varepsilon_j x_j\right).$$

Proof. We have $\varphi(y) = \varphi\big(\frac{1}{2}(y+x) + \frac{1}{2}(y-x)\big) \le \frac{1}{2}\big(\varphi(y+x) + \varphi(y-x)\big)$ for any $x, y \in X$. The statement now follows by a trivial induction on n. □

Theorem 51 ([Dev1]). *Let X be a Banach space with a separating family of polynomials $\{P_l\}_{l=1}^m$. Then X is of cotype $q = \max_l \deg P_l$.*

Proof. By Fact 47 we may assume that P_l are homogeneous and $\max_l P_l(x) \ge 1$ for $x \in S_X$. Let $K = \max_{1\le k\le q} C_{q,k}$, where $C_{q,k}$ are the constants from Proposition 2.78. To show that X is of cotype q choose arbitrary $x_1, \dots, x_n \in X \setminus \{0\}$. Since the inequality in the definition of cotype is homogeneous, we may assume without loss of generality that $x_1, \dots, x_n \in B_X$ and $x_1 \in S_X$. From Fact 50 it follows that $\frac{1}{2^n}\sum_{\varepsilon_j=\pm 1}\big\|\sum_{j=1}^n \varepsilon_j x_j\big\|^q \ge \|x_1\|^q = 1$.

For each $j = 1, \dots, n$ there is $l(j) \in \{1, \dots, m\}$ such that $P_{l(j)}\big(\frac{x_j}{\|x_j\|}\big) \ge 1$. Since $\sum_{j=1}^n \|x_j\|^q = \sum_{k=1}^m \sum_{l(j)=k} \|x_j\|^q$, it follows that there exists $k \in \{1, \dots, m\}$ such that $\sum_{l(j)=k} \|x_j\|^q \ge \frac{1}{m}\sum_{j=1}^n \|x_j\|^q$. Denote $A = l^{-1}(k) \subset \{1, \dots, n\}$ and let $r = \deg P_k \le q$. Then, using Proposition 2.78 and Fact 50,

$$
\begin{aligned}
\sum_{j=1}^n \|x_j\|^q &\le m\sum_{j\in A}\|x_j\|^q \le m\sum_{j\in A}\|x_j\|^r \le m\sum_{j\in A} P_k(x_j) \\
&\le mK\|P_k\|\left(\frac{1}{2^{|A|}}\sum_{\varepsilon_j=\pm 1}\Big\|\sum_{j\in A}\varepsilon_j x_j\Big\|^q\right)^{\frac{r}{q}} \\
&\le mK\|P_k\|\left(\frac{1}{2^n}\sum_{\varepsilon_j=\pm 1}\Big\|\sum_{j=1}^n\varepsilon_j x_j\Big\|^q\right)^{\frac{r}{q}} \le mK\|P_k\|\frac{1}{2^n}\sum_{\varepsilon_j=\pm 1}\Big\|\sum_{j=1}^n\varepsilon_j x_j\Big\|^q.
\end{aligned}
$$

□

It is of course easier to work with polynomials that have some regular structure. In what follows we will be dealing with symmetric and sub-symmetric polynomials. These are defined on (sub-)symmetric linear subspaces of spaces with Schauder basis. In particular, if X is a normed linear space with a Schauder basis $\{e_n\}_{n=1}^\infty$, then the linear subspace $\operatorname{span}\{e_n\}$ is both symmetric and sub-symmetric. Of course, if $\{e_n\}$ is K-spreading (resp. symmetric), then X itself is sub-symmetric (resp. symmetric). From Corollary 1.31 we immediately obtain the following observation.

Fact 52. *Let X be a sub-symmetric (resp. symmetric) linear subspace of a normed linear space with a Schauder basis, Y a vector space, and $P \in P(X;Y)$. Then P is sub-symmetric (resp. symmetric) if and only if all of its homogeneous summands are sub-symmetric (resp. symmetric).*

For $n, d \in \mathbb{N}$ we denote $\mathcal{I}^+(n,d) = \{\alpha \in \mathcal{I}(n,d);\ \alpha_j > 0, j = 1, \dots, n\}$, $\mathcal{I}^+(d) = \bigcup_{n=1}^d \mathcal{I}^+(n,d)$, and $\mathcal{N}(n) = \{\rho = (\rho_1, \dots, \rho_n) \in \mathbb{N}^n;\ \rho_1 < \rho_2 < \dots < \rho_n\}$. Clearly there is a natural bijection between $\mathcal{I}(\infty, d)$ and $\bigcup_{n=1}^d \mathcal{I}^+(n,d) \times \mathcal{N}(n)$ such that $(\alpha, \rho) \in \mathcal{I}^+(n,d) \times \mathcal{N}(n)$ corresponds to a multi-index in $\mathcal{I}(\infty, d)$ whose ρ_jth coordinate equals α_j, $j = 1, \dots, n$. Thus if X is a normed linear space with a Schauder basis $\{e_j\}_{j=1}^{\infty}$, Y a vector space, and $P \in P({}^dX; Y)$, then by Proposition 1.24 there is a unique collection of vectors $\{y_{\alpha,\rho};\ \alpha \in \mathcal{J}^+(n,d), \rho \in \mathcal{N}(n), n = 1, \dots, d\} \subset Y$ such that the formula

$$P(x) = \sum_{\substack{\alpha \in \mathcal{J}^+(n,d) \\ \rho \in \mathcal{N}(n) \\ n=1,\dots,d}} x_{\rho_1}^{\alpha_1} \cdots x_{\rho_n}^{\alpha_n} y_{\alpha,\rho} \tag{2}$$

holds for every finitely supported vector $x = \sum x_j e_j \in X$, where

$$y_{\alpha,\rho} = \binom{d}{\alpha} \check{P}({}^{\alpha_1}e_{\rho_1}, \dots, {}^{\alpha_n}e_{\rho_n}). \tag{3}$$

Definition 53. Let X be a normed linear space with a Schauder basis $\{e_j\}_{j=1}^{\infty}$ and denote $X_0 = \operatorname{span}\{e_j\}_{j=1}^{\infty}$. For $\alpha \in \mathcal{J}^+(n,d)$ we define the polynomial $P_\alpha \in P({}^dX_0)$ by

$$P_\alpha(x) = \sum_{\rho \in \mathcal{N}(n)} x_{\rho_1}^{\alpha_1} \cdots x_{\rho_n}^{\alpha_n}$$

for all $x = \sum x_j e_j \in X_0$. Clearly, each P_α is a sub-symmetric polynomial and it is called an elementary sub-symmetric polynomial. Further, we denote $s_d = P_{(d)}$, i.e.

$$s_d(x) = \sum_{j=1}^{\infty} x_j^d$$

for all $x = \sum x_j e_j \in X_0$. Clearly, each s_d is a symmetric polynomial and it is called a power sum symmetric polynomial.

In general the elementary sub-symmetric (or power sum symmetric) polynomials may not be bounded. They are, however, in the presence of a lower estimate.

Fact 54. *Let X be a Banach space with a Schauder basis $\{e_j\}_{j=1}^{\infty}$ that has a lower q-estimate, $1 \le q < \infty$, with a constant C. Let $\alpha \in \mathcal{J}^+(n,d)$ be such that $\alpha_l \ge q$, $l = 1, \dots, n$. Then*

$$|P_\alpha(x)| \le \prod_{l=1}^{n} \left(\sum_{j=1}^{k} |x_j|^{\alpha_l} \right) \le \left(\sum_{j=1}^{k} |x_j|^q \right)^{\frac{d}{q}} \le \frac{1}{C^d} \|x\|^d$$

for every finitely supported $x \in X$, $x = \sum_{j=1}^{k} x_j e_j$. In particular, P_α can be uniquely extended to a continuous d-homogeneous polynomial on the whole of X, which is then sub-symmetric in case that $\{e_j\}$ is K-spreading.

Similarly, each s_d, $d \ge q$, can be uniquely extended to a continuous d-homogeneous polynomial on the whole of X, which is then symmetric in case that $\{e_j\}$ is symmetric. Conversely, if $\{e_j\}$ is unconditional and s_d is bounded on $\operatorname{span}\{e_j\}$, *then $\{e_j\}$ has a lower d-estimate.*

Proof. The estimates are clear using $\sum_{j=1}^{k}|x_j|^{\alpha_l} \le \big(\sum_{j=1}^{k}|x_j|^q\big)^{\alpha_l/q}$. The extension follows from Proposition 1.25. Considering the assumption of the unconditional basis in the last statement, check s_1 on c_0 with the summing basis. □

For a sub-symmetric homogeneous polynomial the coefficients $y_{\alpha,\rho}$ in (2) do not depend on ρ by (3) and the Polarisation formula. Thus we obtain the following.

Fact 55. *Let X be a normed linear space with a Schauder basis $\{e_j\}_{j=1}^{\infty}$, Y a vector space, and $X_0 = \operatorname{span}\{e_j\}$. If $P \in P(^dX_0; Y)$ is sub-symmetric, then there is a unique collection of vectors $\{y_\alpha;\ \alpha \in \mathcal{I}^+(d)\} \subset Y$ such that for every $x \in X_0$*

$$P(x) = \sum_{\alpha\in\mathcal{I}^+(d)} P_\alpha(x) y_\alpha. \tag{4}$$

The coefficients y_α are given by $y_\alpha = \binom{d}{\alpha}\check{P}(^{\alpha_1}e_1,\dots,{}^{\alpha_n}e_n)$ when $\alpha \in \mathcal{I}^+(n,d)$.

The usefulness of sub-symmetric polynomials stems from the fact that the space of sub-symmetric polynomials is finite-dimensional (the previous fact), while a general polynomial can be asymptotically approximated by a sub-symmetric one (Theorem 26).

The following is a classical theorem, see e.g. [Sta, Corollary 7.7.2].

Theorem 56. *Let X be a normed linear space with a Schauder basis $\{e_j\}_{j=1}^{\infty}$, Y a vector space, and $X_0 = \operatorname{span}\{e_j\}$. If $P \in P(^dX_0; Y)$ is symmetric, then there is a unique polynomial $r \in \mathcal{P}(\mathbb{K}^d; Y)$ such that for every $x \in X_0$*

$$P(x) = r\big(s_1(x),\dots,s_d(x)\big). \tag{5}$$

Proof. Since P is symmetric, $\check{P}(^{\alpha_1}e_{l_1},\dots,{}^{\alpha_n}e_{l_n}) = \check{P}(^{\alpha_1}e_1,\dots,{}^{\alpha_n}e_n)$ for every $\alpha \in \mathcal{I}^+(n,d)$ and every distinct $l_1,\dots,l_n \in \mathbb{N}$ (use the Polarisation formula). For any $x = \sum_{j=1}^{m} a_j e_j \in X_0$ we have

$$\begin{aligned}
P(x) &= \sum_{\substack{1\le k_j\le m\\ j=1,\dots,d}} a_{k_1}\cdots a_{k_d}\check{P}(e_{k_1},\dots,e_{k_d})\\
&= \sum_{n=1}^{d}\sum_{\substack{\alpha\in\mathcal{I}^+(n,d)\\ 1\le l_1,\dots,l_n\le m\\ \text{distinct}}} c_\alpha a_{l_1}^{\alpha_1}\cdots a_{l_n}^{\alpha_n}\check{P}(^{\alpha_1}e_{l_1},\dots,{}^{\alpha_n}e_{l_n})\\
&= \sum_{n=1}^{d}\sum_{\alpha\in\mathcal{I}^+(n,d)} c_\alpha\check{P}(^{\alpha_1}e_1,\dots,{}^{\alpha_n}e_n)\sum_{\substack{1\le l_1,\dots,l_n\le m\\ \text{distinct}}} a_{l_1}^{\alpha_1}\cdots a_{l_n}^{\alpha_n},
\end{aligned}$$

where

$$c_\alpha = \binom{d-1}{\alpha_1-1}\binom{d-\alpha_1-1}{\alpha_2-1}\binom{d-\alpha_1-\alpha_2-1}{\alpha_3-1}\cdots\binom{\alpha_n-1}{\alpha_n-1}.$$

This can be seen as follows: for each $\gamma = (k_1,\dots,k_d) \in \{1,\dots,m\}^d$ we enumerate the distinct values in γ in the order as they appear, denoting them $l_1,\dots,l_n$, and we set

α_j to the number of occurrences of l_j in γ, $j = 1, \dots, n$. The number of different γs that lead to the same $(l_1, \dots, l_n)$ and $\alpha \in \mathcal{J}^+(n,d)$ is given by c_α, as $k_1 = l_1$ and the other $\alpha_1 - 1$ occurrences of l_1 can be distributed into $d-1$ places. Disregarding the already distributed values l_1, we now have $d - \alpha_1$ places available for the rest and we proceed inductively.

So, it is clearly sufficient to prove that each of the polynomials Q_α, $\alpha \in \mathcal{J}^+(n,k)$, $n \le k$, given by

$$Q_\alpha(x) = \sum_{\substack{1 \le l_1, \dots, l_n \le m \\ \text{distinct}}} a_{l_1}^{\alpha_1} \cdots a_{l_n}^{\alpha_n}$$

for any $x = \sum_{j=1}^m a_j e_j \in X_0$ can be expressed in the form (5). This can be done by induction on n. For $n = 1$ there is nothing to prove. Let $n \in \mathbb{N}$, $\alpha \in \mathcal{J}^+(n,k)$, and assume that the statement holds for each Q_β, $\beta \in \mathcal{J}^+(n-1,l)$, for every $l \ge n-1$. Then

$$\begin{aligned} Q_\alpha\Big(\sum_{j=1}^m a_j e_j\Big) &= \Big(\sum_{l_1=1}^m a_{l_1}^{\alpha_1}\Big) \sum_{\substack{1 \le l_2, \dots, l_n \le m \\ \text{distinct}}} a_{l_2}^{\alpha_2} \cdots a_{l_n}^{\alpha_n} \\ &\quad - \sum_{\substack{1 \le l_2, \dots, l_n \le m \\ \text{distinct}}} a_{l_2}^{\alpha_1+\alpha_2} \cdots a_{l_n}^{\alpha_n} - \dots - \sum_{\substack{1 \le l_2, \dots, l_n \le m \\ \text{distinct}}} a_{l_2}^{\alpha_2} \cdots a_{l_n}^{\alpha_1+\alpha_n} \end{aligned}$$

and we may use the inductive hypothesis on each of the summands. This finishes the proof of the existence of the polynomial r satisfying (5).

Considering the uniqueness it clearly suffices to show that $r = 0$ if $P = 0$. Let $Z = \operatorname{span}\{e_1, \dots, e_d\}$ and define $F: Z \to \mathbb{K}^d$ by $F(x) = (s_1(x)\restriction_Z, \dots, s_d(x)\restriction_Z)$. Then the Jacobian J_F of F is a constant multiple of the Vandermonde determinant. Thus J_F is non-zero on an open subset of Z. By the Inverse mapping theorem ([Dieu, Theorem 10.2.5]) there is an open $U \subset Z$ such that $F(U)$ is open. It follows that $r = 0$ on $F(U)$ and so $r = 0$ on the whole of $\mathbb{K}^d$ by Fact 1.39. □

The next theorem was proved in an unpublished version of [HaHá] for ℓ_p spaces. The averaging proof of the slightly more general formulation here is due to [Gon2].

Theorem 57. *Let X be a Banach space with a sub-symmetric Schauder basis $\{e_j\}_{j=1}^\infty$ and let $P \in \mathcal{P}(^dX)$ be a sub-symmetric polynomial. If P satisfies the formula* (4) *and $q = \min\{\alpha_j;\ \alpha \in \mathcal{J}^+(d), y_\alpha \ne 0\}$, then $\{e_j\}$ has a lower q-estimate.*

Proof. The proof is similar to the proof of Lemma 40. Since the basis is sub-symmetric, it is automatically semi-normalised and in particular it is bounded by $M \ge 1$. It is easy to check that if X is real, then $\{e_j\}$ is also a sub-symmetric basis of the complexification $\tilde{X}$ and the extension $\tilde{P}$ is sub-symmetric. Thus we may without loss of generality assume that X is complex. Further, we assume for simplicity that there is $\alpha = (\alpha_1, \dots, \alpha_k) \in \mathcal{J}^+(d)$ such that $y_\alpha \ne 0$ and $\alpha_1 = q$. Put $b = \check{P}(^{\alpha_1}e_1, \dots, {}^{\alpha_k}e_k)$. By Fact 55, $y_\alpha = \binom{d}{\alpha} b$ and hence $b \ne 0$. Since P is sub-symmetric, $\check{P}(^{\alpha_1}e_{n_1}, \dots, {}^{\alpha_k}e_{n_k}) = b$ for any $n_1, \dots, n_k \in \mathbb{N}$, $n_1 < \dots < n_k$.

Now fix any $x_1, \dots, x_N \in \mathbb{C}$ and choose $\xi_1, \dots, \xi_N \in \mathbb{C}$, $|\xi_j| = 1$, such that $\xi_j^q x_j^q b = |x_j^q b|$. Then using Lemma 2.69 we obtain

$$
\begin{aligned}
|b| \sum_{j=1}^{N} |x_j|^q &= \sum_{j=1}^{N} \xi_j^q x_j^q \check{P}({}^{\alpha_1}e_j, {}^{\alpha_2}e_{N+2}, \dots, {}^{\alpha_k}e_{N+k}) \\
&= \sum_{j=1}^{N} \check{P}\big({}^{q}(\xi_j x_j e_j), {}^{\alpha_2}e_{N+2}, \dots, {}^{\alpha_k}e_{N+k}\big) \\
&= \frac{1}{q^N} \sum_{l=1}^{q^N} \check{P}\left({}^{q}\left(\sum_{j=1}^{N} r_j^{q,N}(l) \xi_j x_j e_j\right), {}^{\alpha_2}e_{N+2}, \dots, {}^{\alpha_k}e_{N+k}\right) \\
&\le (2C)^q M^d \|\check{P}\| \left\| \sum_{j=1}^{N} x_j e_j \right\|^q,
\end{aligned}
$$

where C is the unconditional basis constant of $\{e_j\}$ □

Corollary 58. *Let $P \in \mathcal{P}({}^d\ell_p)$, $1 \le p < \infty$, be sub-symmetric. If P satisfies* (4), *then $\alpha_j \ge p$, $j = 1, \dots, n$ for every $\alpha = (\alpha_1, \dots, \alpha_n) \in \mathcal{J}^+(d)$ with $y_\alpha \neq 0$.*

For symmetric polynomials we have the following version.

Theorem 59 ([GGJ]). *Let X be a Banach space with a symmetric Schauder basis $\{e_j\}_{j=1}^{\infty}$ and $q = \min\{k \in \mathbb{N};\ \{e_j\}$ has a lower k-estimate$\}$. Suppose that $P \in \mathcal{P}({}^dX)$ is symmetric. If $d < q$, then $P = 0$. If $d \ge q$, then there is a unique polynomial $r \in \mathcal{P}(\mathbb{K}^{d-q+1})$ such that $P(x) = r\big(s_q(x), \dots, s_d(x)\big)$ for every $x \in X$.*

Proof. Let $\alpha \in \mathcal{J}^+(n,d)$ for some $1 \le n \le d$. Suppose that $\alpha_1 < q$ and put $Q(x) = \check{P}({}^{\alpha_1}x, {}^{\alpha_2}e_2, \dots, {}^{\alpha_n}e_n)$. Then $Q \in \mathcal{P}({}^{\alpha_1}X)$ and $Q(e_j) = Q(e_1)$ for all $j > n$ by the symmetry of P. Now if $Q(e_1) \neq 0$, then $\{e_j\}_{j=n+1}^{\infty}$ has a lower α_1-estimate by Lemma 40. It is not difficult to check that then also $\{e_j\}_{j=1}^{\infty}$ has a lower α_1-estimate, which contradicts the definition of q. Similarly we can see that $\check{P}({}^{\alpha_1}e_1, \dots, {}^{\alpha_n}e_n) = 0$ whenever $\alpha_j < q$ for some $j \in \{1, \dots, n\}$. The result now immediately follows from the proof of Theorem 56. □

Proposition 60 ([GGJ]). *Let X be a Banach space with a K-spreading (resp. symmetric) basis $\{e_j\}_{j=1}^{\infty}$. If X admits a k-homogeneous separating polynomial, then there exists a k-homogeneous sub-symmetric (resp. symmetric) separating polynomial on X.*

Proof. If the basis is K-spreading, then the spreading model from Theorem 26 is equivalent to $\{e_j\}$ and so we can transfer the sub-symmetric polynomial Q onto X. In the symmetric case we may assume that the norm on X satisfies $\left\|\sum_{j=1}^{\infty} a_j e_{\pi(j)}\right\| = \left\|\sum_{j=1}^{\infty} a_j e_j\right\|$ for every permutation π of $\mathbb{N}$ ([LiTz2, p. 114]). Further, we may assume that $P \in \mathcal{P}({}^kX)$ satisfies $P(x) \ge 1$ for $x \in S_X$. We define $Q_n \in \mathcal{P}({}^kX)$ by

$$
Q_n\left(\sum_{j=1}^{\infty} a_j e_j\right) = \frac{1}{n!} \sum_{\pi \in S_n} P\left(\sum_{j=1}^{n} a_{\pi(j)} e_j\right),
$$

where S_n is the set of all permutations of $\{1, \dots, n\}$. It is easily seen that Q_n is symmetric on $\operatorname{span}\{e_j;\ 1 \le j \le n\}$ and $Q_n(x) \ge 1$ for $x \in S_X$, $\max\operatorname{supp} x \le n$. Since $\{Q_n;\ n \in \mathbb{N}\}$ is equi-Lipschitz on bounded sets, by passing to subsequences and diagonalising we can find a subsequence $\{Q_{n_j}\}$ such that it converges pointwise on a countable set dense in X. Consequently, it converges for every $x \in X$. Thus $Q(x) = \lim_{j\to\infty} Q_{n_j}(x)$ is a k-homogeneous symmetric separating polynomial (Theorem 1.29). □

Corollary 61 ([HaHá]). *Suppose that ℓ_p admits a d-homogeneous separating polynomial. Then p is an even integer and d is an integer multiple of p.*

Proof. By Proposition 60 we may assume that there is a symmetric d-homogeneous polynomial P on ℓ_p with $P(x) \ge 1$ whenever $\|x\| = 1$. Let $n = \lceil p \rceil$. By Theorem 59 there is $r \in \mathcal{P}(\mathbb{K}^{d-n+1})$ such that $P(x) = r\big(s_n(x), \dots, s_d(x)\big)$. By Proposition 1.23, $r(x_1, \dots, x_{d-n+1}) = \sum_{\alpha\in\mathcal{I}(d-n+1,D)} y_\alpha x^\alpha$ for some $D \in \mathbb{N}$. Put $u_m = \sum_{j=1}^m \frac{(-1)^j}{m^{1/p}} e_j \in \ell_p$, $m \in \mathbb{N}$. Then $\|u_m\| = 1$ and $\lim_{m\to\infty} s_l(u_m) = 0$ for all $l > p$. Since $P(u_m) \ge 1$, it follows that $p = n$ and $y_\alpha \ne 0$ for some $\alpha = (\alpha_1, 0, \dots, 0)$. Since P is d-homogeneous, we necessarily have $p\alpha_1 = d$. Also, if p is odd, then $|s_p(u_m)| \le \frac{1}{m} \to 0$, again a contradiction with $P(u_m) \ge 1$. □

Corollary 62. *If a Banach space X admits a separating polynomial, then $p(X) = 2$ and $q(X)$ is an even integer.*

Proof. By Fact 45 the space X has a homogenous separating polynomial. By Theorem 3, $\ell_{p(X)}$ and $\ell_{q(X)}$ are finitely representable in X. Hence these spaces have a homogeneous separating polynomial thanks to Corollary 11. Corollary 61 finishes the proof. □

We are going to show in Chapter 5 that in fact X is of type 2 and cotype $q(X)$.

Corollary 63 ([Kur1], [GJ2]). *Let X be a Banach space with a sub-symmetric basis $\{e_i\}_{j=1}^\infty$ and a separating polynomial. Then $\{e_j\}$ is equivalent to the canonical basis of some ℓ_p, p an even integer.*

Proof. By Fact 45 and Proposition 60 we may assume that there is a sub-symmetric $P \in \mathcal{P}(^dX)$ with $P(x) \ge 1$ whenever $\|x\| = 1$. Let P be given by (4). By Theorem 57 there is $p \in \mathbb{N}$ such that $\{e_j\}$ has a lower p-estimate with a constant C. By Fact 54 we have

$$\|x\|^d \le P(x) = \sum_{\alpha\in\mathcal{J}^+(d)} P_\alpha(x) y_\alpha \le \sum_{\alpha\in\mathcal{J}^+(d)} |y_\alpha| \bigg(\sum_{j=1}^k |x_j|^p\bigg)^{\frac{d}{p}} \le \frac{1}{C^d}\|x\|^d \sum_{\alpha\in\mathcal{J}^+(d)} |y_\alpha|$$

for any $x = \sum_{j=1}^k x_j e_j \in X$. Finally, p must be even by Corollary 61. □

Theorem 64 ([GGo]). *Let X be a Banach space with a separating d-homogeneous polynomial. Then every normalised weakly null sequence $\{x_j\}_{j=1}^\infty$ in X has a subsequence equivalent to the canonical basis of ℓ_p, p even integer, such that $p \le q(X)$ and d is an integer multiple of p.*

Proof. Let n be the greatest integer such that $\{x_j\}_{j=1}^{\infty}$ is $\mathcal{P}^{n-1}$-null. Clearly $n \le d$. By Proposition 44 we can assume that $\{x_j\}_{j=1}^{\infty}$ has an upper n-estimate. Passing to further subsequence using Theorem 22 we may assume that $\{x_j\}_{j=1}^{\infty}$ has a spreading model $\{e_j\}_{j=1}^{\infty}$ that is sub-symmetric by Proposition 24. Let $Y = \overline{\operatorname{span}}\{e_j\}$. By Corollary 11 and Fact 23, Y admits a d-homogeneous separating polynomial. By Corollaries 63 and 61, $\{e_j\}_{j=1}^{\infty}$ is equivalent to the canonical basis of ℓ_p, where p is an even integer and d is an integer multiple of p. The upper n-estimate clearly passes to $\{e_j\}_{j=1}^{\infty}$, so we have $n \le p$.

By Lemma 39, $\{x_j\}_{j=1}^{\infty}$ has a subsequence with a lower r-estimate for every $r > n$. Hence $\{e_j\}_{j=1}^{\infty}$ has a lower r-estimate for every $r > n$, so $n \ge p$. It follows that $n = p$, which is an even integer. Thus by Lemma 39 a subsequence of $\{x_j\}_{j=1}^{\infty}$ has a lower p-estimate. Consequently, this subsequence is equivalent to the canonical basis of ℓ_p.

The fact that $q(\ell_p) = p \le q(X)$ follows automatically, as the cotype passes to subspaces. □

Corollary 65. *Let X be a Banach space with a separating polynomial. Let r be the smallest number such that ℓ_r embeds into X. Then X has properties S_r and $T_{q(X)}$, and for every p such that ℓ_p embeds into X we have $r \le p \le q(X)$.*

Our results obtained so far can be used to get an important structural step needed for the proof of Deville's fundamental theorem (Theorem 5.67). But first we recall the classical Mazur's technique for constructing basic sequences.

Lemma 66 (Stanisław Mazur). *Let X be a normed linear space, Y a finite-dimensional subspace of X, and $\varepsilon > 0$. Then there is a finite-codimensional subspace Z of X such that $\|y\| \le (1+\varepsilon)\|y+z\|$ for every $y \in Y$ and $z \in Z$.*

Proof. Assume additionally that $\varepsilon \le 1$. Let $\{y_j\}_{j=1}^{m}$ be an $\frac{\varepsilon}{2}$-dense set in S_Y and let $f_j \in S_{X^*}$ be such that $f_j(y_j) = 1$ for $j = 1,\dots,m$. Put $Z = \bigcap_{j=1}^{m} \ker f_j$. Given $y \in Y$, $y \ne 0$ there is $j \in \{1,\dots,m\}$ such that $\left\|\frac{y}{\|y\|} - y_j\right\| < \frac{\varepsilon}{2}$. Therefore $\frac{1}{\|y\|}\|y+z\| > \left\|y_j + \frac{z}{\|y\|}\right\| - \frac{\varepsilon}{2} \ge f_j\big(y_j + \frac{z}{\|y\|}\big) - \frac{\varepsilon}{2} = 1 - \frac{\varepsilon}{2} \ge \frac{1}{1+\varepsilon}$ for any $z \in Z$. □

One of the main tools is the following diagonalisation lemma. Its proof is similar to that of Lemma 39.

Lemma 67. *Let X be an infinite-dimensional Banach space and let $q \in \mathbb{N}$ be such that X does not have a separating family of polynomials of degree less than q. Let $\{P_l\}_{l=1}^{m} \subset \mathcal{P}^q(X)$ be a finite family of non-constant homogeneous polynomials. Then there is a normalised basic sequence $\{x_n\} \subset X$ such that if $x = \sum_{j=M}^{N} a_j x_j \in B_X$, then $|P_l(x)| \le \frac{1}{2M}$ whenever $\deg P_l < q$ and*

$$\left|P_l(x) - \sum_{j=M}^{N} a_j^q P_l(x_j)\right| \le \frac{1}{2M}$$

whenever $\deg P_l = q$. Moreover, if X^* is separable, then $\{x_n\}$ can be chosen to be weakly null, and if X has a Schauder basis, then $\{x_n\}$ can be chosen as its block basis.

Proof. First note that by Fact 48 no finite-codimensional subspace of X has a separating family of polynomials of degree less than q. If X^* is separable, then we choose a sequence $\{f_j\}_{j=1}^{\infty}$ dense in S_{X^*}; otherwise we just set $f_j = 0$, $j \in \mathbb{N}$. Let $N_l = \deg P_l$ and $\delta_k = \frac{1}{2^{k+1}q!4^q k^q}$. We construct a normalised basic sequence $\{x_n\}_{n=1}^{\infty} \subset X$ with the basis constant at most 2 such that $|\check{P}_l(x_{k_1}, \dots, x_{k_{N_l}})| < \delta_{k_{N_l}}$ whenever $1 \le l \le m$, $1 \le k_1 \le \dots \le k_{N_l}$, and $N_l < q$ or $k_1 < k_{N_l}$. In other words, for every $n \in \mathbb{N}$

$$|R_{(n,l;k_1,\dots,k_r)}(x_n)| < \delta_n$$
$$\text{whenever } 1 \le l \le m,\ 0 \le r < N_l,\ N_l - r < q,\ k_1 \le \dots \le k_r < n, \quad (6)$$

where $R_{(n,l;k_1,\dots,k_r)}(x) = \check{P}_l(x_{k_1}, \dots, x_{k_r}, {}^{N_l - r}x)$, which is a homogeneous polynomial of degree less than q. To this end let $\{\varepsilon_n\} \subset \mathbb{R}^+$ be such that $\prod_{n=1}^{\infty}(1 + \varepsilon_n) \le 2$. We construct the sequence $\{x_n\}$ by induction so that for every $n \in \mathbb{N}$ it moreover satisfies

$$\|y\| \le (1 + \varepsilon_n)\|y + \lambda x_{n+1}\| \quad \text{for all } y \in \operatorname{span}\{x_1, \dots, x_n\} \text{ and } \lambda \in \mathbb{K},$$
$$f_j(x_n) = 0 \quad \text{for every } j \le n.$$

The last condition clearly ensures that $\{x_n\}$ is weakly null in case that X^* is separable.

The first step is identical to the inductive step except for setting $Z = X$, so we describe just the inductive step from $n - 1$ to n. Let Z be the finite-codimensional subspace of X from Lemma 66 used on $Y = \operatorname{span}\{x_1, \dots, x_{n-1}\}$ and $\varepsilon = \varepsilon_{n-1}$. Put $W = Z \cap \bigcap_{j=1}^{n} \ker f_j$. Then W is finite-codimensional and so by the assumption there is $x_n \in S_W$ such that (6) holds, as the family of polynomials $R_{(n,l;k_1,\dots,k_r)}$ considered in (6) is finite. Finally, $\left\|\sum_{j=1}^{n} a_j x_j\right\| \le (1 + \varepsilon_n) \cdots (1 + \varepsilon_{k-1}) \left\|\sum_{j=1}^{k} a_j x_j\right\| \le 2\left\|\sum_{j=1}^{k} a_j x_j\right\|$ for all $n < k$ and all $a_1, \dots, a_k \in \mathbb{K}$. Thus $\{x_n\}$ is a basic sequence with the basis constant at most 2.

Choose any $x = \sum_{j=M}^{N} a_j x_j \in B_X$. Since the basis constant of $\{x_n\}$ is at most 2, it follows that $|a_j| \le 4$, $j = M, \dots, N$. Thus if $\deg P_l < q$, then

$$\begin{aligned}|P_l(x)| &\le \sum_{M \le k_1 \le \dots \le k_{N_l} \le N} N_l! \left|a_{k_1} \cdots a_{k_{N_l}} \check{P}_l(x_{k_1}, \dots, x_{k_{N_l}})\right| \\ &\le \sum_{M \le k_1 \le \dots \le k_{N_l} \le N} q! 4^q \delta_{k_{N_l}} \le \sum_{k=M}^{N} q! 4^q k^{N_l} \delta_k \le \frac{1}{2^M},\end{aligned}$$

while if $\deg P_l = q$, then

$$\begin{aligned}\left|P_l(x) - \sum_{j=M}^{N} a_j^q P_l(x_j)\right| &\le \sum_{\substack{M \le k_1 \le \dots \le k_q \le N \\ k_1 < k_q}} q! \left|a_{k_1} \cdots a_{k_q} \check{P}_l(x_{k_1}, \dots, x_{k_q})\right| \\ &\le \sum_{M \le k_1 \le \dots \le k_q \le N} q! 4^q \delta_{k_q} \le \frac{1}{2^M}.\end{aligned}$$

In case that X has a Schauder basis $\{(e_j; e_j^*)\}_{j=1}^{\infty}$, then in the inductive step we can take $Z = \bigcap_{j=1}^{k} \ker e_j^*$, where $k = \max \operatorname{supp} x_{n-1}$, and notice that since the family of polynomials considered in (6) is equi-Lipschitz on B_X, we can perturb x_n to have a finite support. □

We finish with a lemma that provides one of the steps in the proof of the important structural Theorem 5.67.

Lemma 68. *Let X be an infinite-dimensional Banach space with a separating family of polynomials $\{P_l\}_{l=1}^m$ of degree at most $q = q(X)$. Then X has a subspace isomorphic to ℓ_q.*

Proof. We may clearly assume that X is separable and that the polynomials P_l are homogeneous, $P_l(0) = 0$, and $\max_l P_l(x) \geq m$ for $x \in S_X$ (Fact 47). Since X is an Asplund space (Fact 47 and Theorem 5.2), X^* is separable. By Theorem 51 there is no separating family of polynomials of degree less than q on X. Let $\{x_n\}$ be the normalised weakly null basic sequence from Lemma 67. Put $M = \max_l \|P_l\|$ and $A = \{1 \leq l \leq m;\ \deg P_l = q\}$. Choose any $x = \sum_{j=1}^N a_j x_j \in S_X$. Then

$$m \leq \max_{1\leq l\leq m} P_l(x) \leq \sum_{l=1}^{m} |P_l(x)| \leq \sum_{l\in A}\sum_{j=1}^{N} |a_j|^q |P_l(x_j)| + \frac{m}{2} \leq mM \sum_{j=1}^{N} |a_j|^q + \frac{m}{2}.$$

Thus $\sum_{j=1}^N |a_j|^q \geq \frac{1}{2M}$ whenever $\sum_{j=1}^N a_j x_j \in S_X$. We conclude that $\{x_n\}$ has an upper q-estimate. By Theorem 64, $\{x_n\}$ has a subsequence equivalent to the canonical basis of ℓ_p for some $p \leq q$. So necessarily $p = q$ and the proof is completed. □

5. Stabilisation of polynomials

Our goal in this section is to improve the results concerning spreading properties of polynomials on ℓ_p, $1 \leq p < \infty$ and its canonical basis $\{e_j\}$. These subtler results are then applied to polynomials on general Banach spaces, leading to infinite-dimensional stabilisation results. In this section all the spaces are real.

Note that every spreading model of $\{e_j\}$ is again equal to $\{e_j\}$ (in the isometric sense). Using Theorem 26 and passing finitely many times to subsequences, for every $P \in \mathcal{P}^d(\ell_p)$ there exist an increasing sequence $M = \{m_j\} \subset \mathbb{N}$ and a sub-symmetric polynomial $Q \in \mathcal{P}^d(\ell_p)$ such that for every $\varepsilon > 0$ and $k \in \mathbb{N}$ there is $N \in \mathbb{N}$ such that

$$\left| Q\left(\sum_{j=1}^{k} x_j e_j\right) - P\left(\sum_{j=1}^{k} x_j e_{n_j}\right) \right| \leq \varepsilon \tag{7}$$

for all $N \leq n_1 < n_2 < \cdots < n_k$, $n_j \in M$ and all $\sum_{j=1}^k x_j e_{n_j} \in B_{\ell_p}$. We show that in fact a substantially stronger result holds:

Theorem 69 ([HaHá]). *Let $P \in \mathcal{P}^d(\ell_p)$, $1 \le p < \infty$. There are a subsequence $\{e_{n_j}\}$ of the canonical basis $\{e_j\}$ and a sub-symmetric polynomial $Q \in \mathcal{P}^d(\ell_p)$ such that for every $\varepsilon > 0$ there is $N \in \mathbb{N}$ such that*

$$\left| Q\left(\sum_{j=N}^{\infty} x_j e_j \right) - P\left(\sum_{j=N}^{\infty} x_j e_{n_j} \right) \right| \le \varepsilon$$

for every $\sum_{j=N}^{\infty} x_j e_{n_j} \in B_{\ell_p}$. In other words,

$$\lim_{N\to\infty} \left\| (Q-P)\restriction_{\overline{\operatorname{span}}\{e_{n_j}\}_{j=N}^{\infty}} \right\| = 0.$$

Let $M = \{m_j\} \subset \mathbb{N}$ be an increasing sequence. Note that every polynomial $P \in P^d(\ell_p)$ also belongs to $P^d(\ell_p(M))$ by restriction. We will work with the polynomials using their coordinate representation. By $\ell_p^0(M)$ we denote the subspace $\operatorname{span}\{e_j\}_{j\in M} \subset \ell_p(M)$ of finitely supported vectors. Further, let us denote (using the notation from (2))

$$\mathcal{K}(M,d) = \big\{(\alpha,\rho);\ \alpha \in \mathcal{I}^+(n,k), \rho \in \mathcal{N}(n), \rho \subset M, n = 1,\dots,k,\ k = 1,\dots,d\big\} \\ \cup\big\{((0),(m_1))\big\}.$$

Recall that, similarly as in (2), by Proposition 1.24 for every $P \in P^d(\ell_p^0(M))$ there is a unique collection of coefficients $\{a_{\alpha,\rho}\}_{(\alpha,\rho)\in\mathcal{K}(M,d)} \subset \mathbb{R}$ such that

$$P(x) = \sum_{(\alpha,\rho)\in\mathcal{K}(M,d)} a_{\alpha,\rho} x_{\rho_1}^{\alpha_1} \cdots x_{\rho_n}^{\alpha_n} \tag{8}$$

for every $x = \sum x_j e_j \in \ell_p^0(M)$, and conversely any collection $\{a_{\alpha,\rho}\}_{(\alpha,\rho)\in\mathcal{K}(M,d)} \subset \mathbb{R}$ gives rise to a polynomial $P \in P^d(\ell_p^0(M))$ defined by (8). If P is continuous, then it is easy to see from (3) that the set of its coefficients is bounded and moreover the function that to each $P \in \mathcal{P}^d(\ell_p^0(M))$ assigns $\sup_{(\alpha,\rho)\in\mathcal{K}(M,d)} |a_{\alpha,\rho}|$ for the coefficients corresponding to P is a norm on the space $\mathcal{P}^d(\ell_p^0(M))$. Note that if $P \in P^d(\ell_p^0(M))$ is given by (8) and M_1 is a subsequence of M, then the restriction of P to $\ell_p^0(M_1)$ is given by the formula (8) restricted to $\mathcal{K}(M_1,d)$.

The formula for the canonical norm on ℓ_p hints that the exponent p plays an important role here. We say that a power $x_{\rho_1}^{\alpha_1} \cdots x_{\rho_n}^{\alpha_n}$ is a large power if $\alpha_j \ge p$ for all $j \in \{1,\dots,n\}$, otherwise we say that it is a small power. We say that $P \in P^d(\ell_p^0(M))$ consists only of large powers if in the formula (8) only the large powers have non-zero coefficients. Similarly we say that P consists only of small powers if in the formula (8) only the small powers have non-zero coefficients.

The main idea is that for polynomials on $\ell_p^0(M)$ consisting only of large powers there is a good estimate using the ℓ_p-norm (Lemma 70), and in particular these polynomials are continuous and thus by Proposition 1.25 extendable to $\ell_p(M)$. This leads to the proof of Theorem 69 for these polynomials (Lemma 72). On the other hand we show that if we restrict a continuous polynomial to a suitable subspace, then the small powers are essentially negligible (Proposition 74).

The main part of the proof relies on an inductive argument. For this we also need a slight weakening of the previous definitions. We say that $P \in P^d(\ell_p^0(M))$, $M = \{m_j\}$ consists almost only of large powers provided that in the formula (8) the non-zero coefficients correspond to the powers $x_{\rho_1}^{\alpha_1} \cdots x_{\rho_n}^{\alpha_n}$ satisfying $\alpha_j \geq p$ for $j = 2, \dots, n$ and either $\alpha_1 \geq p$ or $\rho_1 = m_1$.

Lemma 70. *Let $M \subset \mathbb{N}$ be an increasing sequence and $P \in P^d(\ell_p^0(M))$, $1 \leq p < \infty$. If P consists almost only of large powers and the set of its coefficients in* (8) *is bounded, then P is continuous and*

$$\|P\| \leq 2^{d+1} \sup_{(\alpha,\rho)\in\mathcal{K}(M,d)} |a_{\alpha,\rho}|.$$

Proof. We may assume without loss of generality that $M = \mathbb{N}$. Let us denote $C = \sup_{(\alpha,\rho)\in\mathcal{K}(\mathbb{N},d)} |a_{\alpha,\rho}|$ and assume that $x \in \ell_p^0$, $x = \sum_{j=1}^N x_j e_j$, $\|x\| \leq 1$. Note that

$$1 \geq \left(\sum_{j=2}^N |x_j|^p\right)^n \geq \sum_{\substack{2\leq\rho_1,\dots,\rho_n\leq N \\ \rho_1,\dots,\rho_n \text{ distinct}}} |x_{\rho_1}|^p \cdots |x_{\rho_n}|^p = n! \sum_{\substack{\rho\in\mathcal{N}(n) \\ \rho_1>1,\ \rho_n\leq N}} |x_{\rho_1}|^p \cdots |x_{\rho_n}|^p.$$

Therefore

$$\begin{aligned}
|P(x)| &\leq \sum_{(\alpha,\rho)\in\mathcal{K}(\mathbb{N},d)} |a_{\alpha,\rho}||x_{\rho_1}|^{\alpha_1} \cdots |x_{\rho_n}|^{\alpha_n} \\
&\leq |a_{(0),(1)}| + \sum_{k=1}^{d}\sum_{n=1}^{k} \sum_{\alpha\in\mathcal{J}^+(n,k)} \sum_{\substack{\rho\in\mathcal{N}(n) \\ \rho_n\leq N}} |a_{\alpha,\rho}||x_{\rho_1}|^{\alpha_1} \cdots |x_{\rho_n}|^{\alpha_n} \\
&\leq C + C\sum_{k=1}^{d}\sum_{n=1}^{k} \sum_{\alpha\in\mathcal{J}^+(n,k)} \left(\sum_{\substack{\rho\in\mathcal{N}(n) \\ \rho_1=1,\ \rho_n\leq N}} |x_{\rho_2}|^p \cdots |x_{\rho_n}|^p + \sum_{\substack{\rho\in\mathcal{N}(n) \\ \rho_1>1,\ \rho_n\leq N}} |x_{\rho_1}|^p \cdots |x_{\rho_n}|^p \right) \\
&\leq C + C\sum_{k=1}^{d}\sum_{n=1}^{k} \sum_{\alpha\in\mathcal{J}^+(n,k)} \left(\frac{1}{(n-1)!} + \frac{1}{n!}\right) \leq C + 2C\sum_{k=1}^{d}\sum_{n=1}^{k} \binom{k-1}{n-1} \\
&= C + 2C\sum_{k=1}^{d} 2^{k-1} = C + 2C(2^d - 1) \leq 2^{d+1}C,
\end{aligned}$$

since $|\mathcal{J}^+(n,k)| = \binom{n+(k-n)-1}{n-1} = \binom{k-1}{n-1}$ (indeed, from the combinatorial point of view it represents the number of distributions of k identical balls into n distinct boxes such that each box is non-empty). □

By Fact 55 for every sub-symmetric polynomial $Q \in P^d(\ell_p^0)$ there is a collection $\{a_\alpha;\ \alpha \in \mathcal{J}^+(k), 1 \le k \le d\} \subset \mathbb{R}$ such that

$$Q(x) = Q(0) + \sum_{\substack{\alpha\in\mathcal{J}^+(k)\\ k=1,\dots,d}} a_\alpha P_\alpha(x) \tag{9}$$

for every $x \in \ell_p^0$.

Lemma 71. *Let $P \in P^d(\ell_p^0)$, $1 \le p < \infty$ be given by* (8)*, let $M \subset \mathbb{N}$ be an increasing sequence, and let $Q \in P^d(\ell_p^0)$ be sub-symmetric and given by* (9)*. Then* (7) *holds if and only if*

$$\lim_{N\to\infty} \sup_{\substack{(\alpha,\rho)\in\mathcal{K}(M,d)\\ \rho_1\ge N}} |a_{\alpha,\rho} - a_\alpha| = 0.$$

Proof. We may assume without loss of generality that $P(0) = Q(0) = 0$. For any $\rho \in \mathcal{N}(k)$ we define $T_\rho\colon \ell_p^k \to \ell_p^0$ by $T_\rho\big(\sum_{j=1}^k x_j e_j\big) = \sum_{j=1}^k x_j e_{\rho_j}$ and put $P_\rho = P \circ T_\rho$. Then $P_\rho \in \mathcal{P}^d(\ell_p^k)$. Note that (7) can be reformulated this way: for every $k \in \mathbb{N}$ and every $\varepsilon > 0$ there is $N \in \mathbb{N}$ such that $|Q(x) - P_\rho(x)| \le \varepsilon$ for every $x \in B_{\ell_p^k}$ and every $\rho \in \mathcal{N}(k)$ satisfying $\rho \subset M$ and $\rho_1 \ge N$. In other words, for every $k \in \mathbb{N}$

$$\lim_{N\to\infty} \sup_{\substack{\rho\in\mathcal{N}(k)\\ \rho\subset M,\ \rho_1\ge N}} \|Q\restriction_{\ell_p^k} - P_\rho\| = 0.$$

We have

$$P_\rho\Big(\sum_{j=1}^k x_j e_j\Big) = \sum_{\substack{(\alpha,\tau)\in\mathcal{K}(\mathbb{N},d)\\ \tau_n\le k}} a_{\alpha,(\rho_{\tau_1},\dots,\rho_{\tau_n})} x_{\tau_1}^{\alpha_1}\cdots x_{\tau_n}^{\alpha_n}.$$

Since the space $\mathcal{P}^d(\ell_p^k)$ is finite-dimensional, all norms on this space are equivalent. In particular, the canonical norm is equivalent to the norm that takes the maximum of absolute values of the corresponding coefficients. Hence there is $K_k \ge 1$ such that $\frac{1}{K_k}\|Q\restriction_{\ell_p^k} - P_\rho\| \le \max_{\substack{(\alpha,\sigma)\in\mathcal{K}(\mathbb{N},d)\\ \sigma\subset\rho}} |a_{\alpha,\sigma} - a_\alpha| \le K_k\|Q\restriction_{\ell_p^k} - P_\rho\|$ for every $\rho \in \mathcal{N}(k)$. As

$$\begin{aligned}\sup_{\substack{(\alpha,\rho)\in\mathcal{K}(M,d)\\ \rho_1\ge N}} |a_{\alpha,\rho} - a_\alpha| &= \max_{1\le k\le d} \sup_{\substack{\rho\in\mathcal{N}(k)\\ \rho\subset M,\ \rho_1\ge N}} \max_{\substack{(\alpha,\sigma)\in\mathcal{K}(\mathbb{N},d)\\ \sigma\subset\rho}} |a_{\alpha,\sigma} - a_\alpha|\\ &= \sup_{k\in\mathbb{N}} \sup_{\substack{\rho\in\mathcal{N}(k)\\ \rho\subset M,\ \rho_1\ge N}} \max_{\substack{(\alpha,\sigma)\in\mathcal{K}(\mathbb{N},d)\\ \sigma\subset\rho}} |a_{\alpha,\sigma} - a_\alpha|,\end{aligned}$$

the statement of the lemma follows. □

Lemma 72. *Let $P \in \mathcal{P}^d(\ell_p)$, $1 \le p < \infty$ be a polynomial consisting only of large powers. There are a subsequence $\{e_{n_j}\}$ of the canonical basis $\{e_j\}$ and a sub-symmetric polynomial $Q \in \mathcal{P}^d(\ell_p)$ such that*

$$\lim_{N\to\infty} \big\|(Q - P)\restriction_{\overline{\operatorname{span}}\{e_{n_j}\}_{j=N}^\infty}\big\| = 0.$$

Proof. By Theorem 26 there are an increasing sequence $M \subset \mathbb{N}$ and a sub-symmetric $Q \in \mathcal{P}^d(\ell_p)$ such that (7) holds. By Corollary 58 the polynomial Q consists only of large powers and so clearly the same holds also for $Q - P$. Let Q be given by (9) and $M = \{n_j\}$. Lemma 70 gives the estimate

$$\|(Q - P) \restriction_{\overline{\operatorname{span}}\{e_{n_j}\}_{j=N}^{\infty}}\| \le 2^{d+1} \sup_{\substack{(\alpha,\rho)\in\mathcal{K}(M,d)\\ \rho_1 \ge N}} |a_{\alpha,\rho} - a_\alpha|.$$

Lemma 71 now finishes the proof. □

Lemma 73. *Let $M \subset \mathbb{N}$ be an increasing sequence and suppose that $P \in \mathcal{P}^d(\ell_p(M))$, $1 \le p < \infty$ is given by (8). If P consists only of small powers and $P(0) = 0$, then there is a subsequence M_1 of M such that*

$$\lim_{N\to\infty} \sup_{\substack{(\alpha,\rho)\in\mathcal{K}(M_1,d)\\ \rho_1 \ge N}} |a_{\alpha,\rho}| = 0.$$

Proof. By Theorem 26 there are a subsequence M_1 of M and a sub-symmetric $Q \in \mathcal{P}^d(\ell_p)$ such that (7) holds on M_1. Let Q be given by (9). For any $\alpha \in \mathcal{I}^+(n,k)$ such that $\alpha_j \ge p$ for all $j \in \{1,\dots,n\}$ we have $a_{\alpha,\rho} = 0$ for all $\rho \in \mathcal{N}(n)$. From Lemma 71 it follows that $a_\alpha = 0$. Consequently $Q = 0$ by Corollary 58 and we appeal to Lemma 71 again. □

Proposition 74. *Let $M_0 \subset \mathbb{N}$ be an increasing sequence and $P \in \mathcal{P}^d(\ell_p(M_0))$, $1 \le p < \infty$ a polynomial consisting only of small powers with $P(0) = 0$. Then there is a subsequence $M = \{m_j\}$ of M_0 such that $\lim_{N\to\infty}\|P \restriction_{\overline{\operatorname{span}}\{e_{m_j}\}_{j=N}^{\infty}}\| = 0$.*

Proof. To simplify the notation we denote the restriction of a polynomial Q to $\operatorname{span}\{e_j\}_{j\in A}$ by $Q \restriction_A$. Using induction we construct M together with auxiliary nested subsequences $M_0 \supset M_1 \supset M_2 \supset \cdots$, $M_k = \{m_j^k\}_{j=1}^{\infty}$, and a sequence of polynomials $\{S_k\} \subset \mathcal{P}^d(\ell_p^0)$ with the following properties:

(i) $m_j^k = m_j$ whenever $j \le k$,

(ii) $\|S_k\| \le \frac{1}{2^k}$,

(iii) $S_k \restriction_{\{m_j^k;\ j>k\}} = 0$.

(iv) Denote $R_k = \big(P - \sum_{j=1}^k S_j\big) \restriction_{M_k}$. If $R_k(x) = \sum_{(\alpha,\rho)\in\mathcal{K}(M_k,d)} b_{\alpha,\rho} x_{\rho_1}^{\alpha_1}\cdots x_{\rho_n}^{\alpha_n}$ for all $x \in \ell_p^0(M_k)$, then each $(\alpha,\rho) \in \mathcal{K}(M_k,d)$ with $b_{\alpha,\rho} \neq 0$ satisfies $\max\{\rho_j;\ \alpha_j < p\} > m_k$.

Let the polynomial P be given by the formula (8). By Lemma 73 there exists a subsequence M_1 of M_0 such that $\sup_{(\alpha,\rho)\in\mathcal{K}(M_1,d)}|a_{\alpha,\rho}| \le \frac{1}{2^{d+2}}$. We define $m_1 = m_1^1$, $A = \{(\alpha,\rho) \in \mathcal{K}(M_1,d);\ \rho_1 = m_1, \alpha_1 < p,\ \alpha_j \ge p \text{ for } j > 1\}$, and $S_1(x) = \sum_{(\alpha,\rho)\in A} a_{\alpha,\rho} x_{\rho_1}^{\alpha_1}\cdots x_{\rho_n}^{\alpha_n}$ for $x \in \ell_p^0$. By Lemma 70 the polynomial S_1 is continuous and $\|S_1\| \le \frac{1}{2}$. Since P consist only of small powers, the properties (iii) and (iv) are clearly satisfied.

Now assume $m_1,\dots,m_k$, M_k, and $S_1,\dots,S_k$ are already defined for some $k \in \mathbb{N}$. Fix $y \in \ell_p$ with $\operatorname{supp} y \subset \{m_1,\dots,m_k\}$ and define a polynomial $R_k^y \in \mathcal{P}^d(\ell_p^0)$

by $R_k^y(\sum_{j=1}^{s} x_j e_j) = R_k(y + \sum_{j=m_k+1}^{s} x_j e_j)$. By the inductive hypothesis R_k^y consists only of small powers and $R_k^y(0) = 0$. After re-grouping the coefficients we obtain

$$R_k^y(x) = \sum_{\substack{(\beta,\sigma)\in\mathcal{K}(M_k,d)\\ \sigma_1>m_k}} c_{\beta,\sigma}(y) x_{\sigma_1}^{\beta_1}\cdots x_{\sigma_n}^{\beta_n},$$

for every $x \in \ell_p^0$, where

$$c_{\beta,\sigma}(y) = \sum_{s=1}^{d-|\beta|} \sum_{r=1}^{s} \sum_{\substack{\gamma\in\mathcal{J}^+(r,s)\\ \tau\in\mathcal{N}(r)\\ \tau_r\le m_k}} b_{(\gamma_1,\ldots,\gamma_r,\beta_1,\ldots,\beta_n),(\tau_1,\ldots,\tau_r,\sigma_1,\ldots,\sigma_n)} y_{\tau_1}^{\gamma_1}\cdots y_{\tau_r}^{\gamma_r}$$

if $\beta = (\beta_1,\ldots,\beta_n)$. Since the set of the coefficients $\{b_{\alpha,\rho}\}$ is bounded, it is not difficult to see that $\{y \mapsto c_{\beta,\sigma}(y);\ (\beta,\sigma) \in \mathcal{K}(M_k,d), \sigma_1 > m_k\}$ is an equi-Lipschitz system of functions on $B = \{y \in B_{\ell_p};\ \operatorname{supp} y \subset \{m_1,\ldots,m_k\}\}$. Since B is compact, there is a finite set $F \subset B$ such that each $c_{\beta,\sigma}(F)$ is $\frac{1}{2^{d+k+3}}$-dense in $c_{\beta,\sigma}(B)$. Therefore by using Lemma 73 for each R_k^y, $y \in F$, we can find a subsequence $W = \{m_j^{k+1}\}_{j=k+1}^{\infty}$ of $M_k \setminus \{m_1,\ldots,m_k\}$ such that $\sup_{(\beta,\sigma)\in\mathcal{K}(W,d)} |c_{\beta,\sigma}(y)| \le \frac{1}{2^{d+k+2}}$ for each $y \in B$. We put $m_{k+1} = m_{k+1}^{k+1}$ and $M_{k+1} = \{m_1,\ldots,m_k\} \cup W$.

Next, we set $A = \{(\alpha,\rho) \in \mathcal{K}(M_{k+1},d);\ \max\{\rho_j;\ \alpha_j < p\} = m_{k+1}\}$ and $S_{k+1}(x) = \sum_{(\alpha,\rho)\in A} b_{\alpha,\rho} x_{\rho_1}^{\alpha_1}\cdots x_{\rho_n}^{\alpha_n}$ for $x \in \ell_p^0$. Fix any $z \in B_{\ell_p^0}(M_{k+1})$. Then $z = y + x$ for some $y \in B$ and $x \in B_{\ell_p^0}$ with $\operatorname{supp} x > m_k$. Notice that

$$S_{k+1}(z) = S_{k+1}(y+x) = \sum_{\substack{(\beta,\sigma)\in\mathcal{K}(M_{k+1},d)\\ \sigma_1=m_{k+1},\ \beta_1<p\\ \beta_j\ge p \text{ for } j>1}} c_{\beta,\sigma}(y) x_{\sigma_1}^{\beta_1}\cdots x_{\sigma_n}^{\beta_n}.$$

Because the polynomial $x \mapsto S_{k+1}(y+x)$ consists almost only of large powers on $\{m_j^{k+1}\}_{j=k+1}^{\infty}$, Lemma 70 gives $|S_{k+1}(z)| \le \frac{1}{2^{k+1}}$, and so (ii) holds. The property (iii) is clearly satisfied. Finally, to see (iv) note that $R_{k+1} = (R_k - S_{k+1})\restriction_{M_{k+1}}$.

To finish the proof, let $x = \sum_{j=N}^{k} x_{m_j} e_{m_j}$ with $\|x\| \le 1$. Then $R_k(x) = 0$ by (iv) and hence $P(x) = \sum_{j=1}^{k} S_j(x) = \sum_{j=N}^{k} S_j(x)$ by (iii). It follows that $|P(x)| \le \sum_{j=N}^{k} |S_j(x)| \le \frac{1}{2^{N-1}}$ by (ii). □

Proof of Theorem 69. Suppose that the polynomial P is given by the formula (8). Put $\mathcal{K} = \{((\alpha_1,\ldots,\alpha_n),\rho) \in \mathcal{K}(\mathbb{N},d);\ \alpha_j \ge p, j = 1,\ldots,n\}$ and define $R \in P^d(\ell_p^0)$ by $R(x) = \sum_{(\alpha,\rho)\in\mathcal{K}} a_{\alpha,\rho} x_{\rho_1}^{\alpha_1}\cdots x_{\rho_n}^{\alpha_n}$, i.e. R consists of all the large powers of P. Then R is continuous by Lemma 70 and by Proposition 1.25 it can be extended to a continuous polynomial on ℓ_p. Hence $P - R$ is a continuous polynomial on ℓ_p that consists only of small powers. The proof now follows easily by combining Lemma 72 with Proposition 74. □

We continue by applying these results in the setting of general Banach spaces.

Definition 75. Let X be a real Banach space with a Schauder basis $\{e_j\}_{j=1}^\infty$ and let $f: X \to \mathbb{R}$ be a p-homogeneous function (i.e. $f(\lambda x) = \lambda^p f(x)$ for all $\lambda > 0$). We say that f stabilises at zero level if

$$\lim_{n\to\infty} \sup\left\{|f(x)|;\ x = \sum_{j=n}^{\infty} a_j e_j, x \in B_X\right\} = 0.$$

It would be more precise to say that f stabilises at zero level with respect to the basis $\{e_j\}$. Note that if f stabilises at zero level with respect to the basis $\{e_j\}$ and $\{u_j\}$ is a block basis of $\{e_j\}$, then $f\restriction_{\overline{\text{span}}\{u_j\}}$ stabilises at zero level with respect to the basis $\{u_j\}$.

Theorem 76 ([HaHá]). *Let X be a real infinite-dimensional Banach space and let $P \in \mathcal{P}(^dX)$. Then there exists a basic sequence $\{x_j\}_{j=1}^\infty \subset X$ such that $P\restriction_{\overline{\text{span}}\{x_j\}}$ is either separating or stabilises at zero level.*

Proof. If X does not contain ℓ_p, p even, then X does not have a separating family of polynomials (Fact 47 and Theorem 64). Thus by Lemma 67 there is a normalised basic sequence $\{x_j\}_{j=1}^\infty \subset X$ such that $P\restriction_{\overline{\text{span}}\{x_j\}}$ stabilises at zero level. For the rest of the proof we may assume that $X = \ell_p$, p even, and thanks to Theorem 69 that P is sub-symmetric.

We put $u_n = \sum_{j=t_n}^{t_n+n-1} \frac{1}{n^{1/p}} e_j$, where $\{t_n\} \subset \mathbb{N}$ is a suitable sequence so that $\{u_n\}$ is a block basis of the canonical basis $\{e_n\}$. Then $\|u_n\|_{\ell_p}=1$, while $\lim_{n\to\infty}\|u_n\|_{\ell_q}=0$ for any $q > p$. Denote $\mathcal{J} = \{\alpha = (\alpha_1, \dots, \alpha_k) \in \mathcal{I}^+(d);\ \alpha_j \ge p, j = 1, \dots, k\}$. If $\alpha = (\alpha_1, \dots, \alpha_k) \in \mathcal{J}$, then $|P_\alpha(x)| \le \prod_{j=1}^k \|x\|_{\ell_{\alpha_j}}^{\alpha_j}$ for every $x \in X$ (see Fact 54). It follows that $P_\alpha\restriction_{\overline{\text{span}}\{u_n\}}$ stabilises at zero level whenever there is $j \in \{1, \dots, k\}$ with $\alpha_j > p$.

By Fact 55 and Corollary 58 there are $b_\alpha \in \mathbb{R}$ such that $P = \sum_{\alpha\in\mathcal{J}} b_\alpha P_\alpha$. There are two possibilities. Either $d = kp$ for some $k \in \mathbb{N}$ and $b_{(p,\dots,p)} \ne 0$, or the opposite holds. In the latter case P stabilises at zero level on $\overline{\text{span}}\{u_n\}$ by the previous paragraph. In the former case, consider the polynomial s_p^k. This is a sub-symmetric polynomial and so there are $c_\beta \in \mathbb{R}$ such that $s_p^k = \sum_{\beta\in\mathcal{I}^+(d)} c_\beta P_\beta$. From the multinomial formula we get

$$\|x\|^d = s_p(x)^k = k!P_{(p,\dots,p)}(x) + \sum_{\substack{\beta\in\mathcal{J}\\ \beta\ne(p,\dots,p)}} c_\beta P_\beta(x).$$

Since all P_β in the sum on the right stabilise at zero level on $\overline{\text{span}}\{u_n\}$, it follows that $P_{(p,\dots,p)}$ is separating on $\overline{\text{span}}\{u_n\}_{n\ge n_0}$ for a suitable $n_0 \in \mathbb{N}$. Consequently, P is separating on $\overline{\text{span}}\{u_n\}_{n\ge n_1}$ for some $n_1 \in \mathbb{N}$. □

Theorem 76 implies that odd degree homogeneous polynomials stabilise at zero level on infinite-dimensional subspaces of X. A similar result in the finite-dimensional case is the quantitative Theorem 2.45. We show that these two results cannot be combined in the infinite-dimensional spaces, in a rather strong and quantitative way. In particular, it follows from the result below that on every separable real Banach space X there is a homogeneous polynomial P, of any prescribed odd degree greater than one, which has no infinite-dimensional null space (i.e. a subspace $Y \subset X$ such that $P(Y) = \{0\}$).

Theorem 77 ([ArHá2]). *Let X be a real Banach space with w^*-separable dual and $n > 1$ an odd integer. Then there exists $P \in \mathcal{P}(^nX)$ without any infinite-dimensional null space. More precisely, given any $x \in X \setminus \{0\}$, $P(x) = 0$, there exists an $N \in \mathbb{N}$ such that every null space $Y \subset X$ containing x satisfies* $\dim Y \leq N$.

Proof. Suppose that we have already proved the statement of the theorem for $X = c_0$ and $n = 3$. Let $P \in \mathcal{P}(^3c_0)$ be the polynomial with the required property. Given any Banach space X with w^*-separable dual and $n = 3 + 2l$, $l \in \mathbb{N}_0$, we construct the desired $Q \in \mathcal{P}(^nX)$ as follows: Let $T \in \mathcal{L}(X; c_0)$ be a one-to-one operator (take for example $T(x) = \left(\frac{1}{j} f_j(x)\right)_{j=1}^{\infty}$, where $\{f_j\}_{j=1}^{\infty} \subset B_{X^*}$ is w^*-dense and hence it separates the points of X), and put $Q(x) = \left(\sum_{j=1}^{\infty} \frac{1}{2^j} f_j(x)^{2l}\right) P \circ T(x)$. Clearly, any subspace $Y \subset X$ where Q vanishes translates via T onto a subspace of c_0 of the same dimension where P vanishes, which concludes the proof of the general case.

It remains to produce P on c_0. We put

$$P(x) = \sum_{1 \leq j < k} \alpha_{jk} x_j x_k^2$$

where $\alpha_{jk} = \frac{1}{2^k 8^j k^j}$, $j, k \in \mathbb{N}$, $j < k$. Further, let $\beta_{jk}^l = \frac{1}{2^{l-j}} \frac{1}{k} \alpha_{jk}$, $j, k, l \in \mathbb{N}$, $j < k$, $j < l$. It is straightforward to check that the systems $\{\alpha_{jk}\}$ and $\{\beta_{jk}^l\}$ satisfy the following conditions:

(i) $\sum_{1 \leq j < k} \alpha_{jk} < +\infty$,

(ii) $\sum_{l=j+1}^{k-1} \alpha_{lk} \leq \frac{1}{k} \alpha_{jk}$ for each $j, k \in \mathbb{N}$, $j < k$,

(iii) $\sum_{l=j+1}^{\infty} \beta_{jk}^l \leq \frac{1}{k} \alpha_{jk}$ for each $j, k \in \mathbb{N}$, $j < k$,

(iv) $(\alpha_{jk})^2 \leq \frac{1}{16} \beta_{lj}^k \beta_{lk}^j$ for each $j, k, l \in \mathbb{N}$, $l < j < k$.

Note that (i) guarantees that the generalised series in the formula for P converges absolutely for every $x \in c_0$ and so indeed defines a 3-homogeneous polynomial on c_0. We claim that P satisfies the statement of the theorem. So let $0 \neq x \in X$, $P(x) = 0$. We may without loss of generality assume that $\|x\| = 1$. Consider the polynomial $R(y) = P(x + y)$ of degree 3. Then $R(y) = \sum_{1 \leq j < k} \alpha_{jk}(x_j + y_j)(x_k + y_k)^2$. Writing $R = R_0 + R_1 + R_2 + R_3$, where R_m is m-homogeneous, we obtain in particular

$$\begin{aligned} R_2(y) &= \sum_{1 \leq j < k} \alpha_{jk}(x_j y_k^2 + 2y_j x_k y_k) \\ &= \sum_{k=2}^{\infty} \left(\sum_{j=1}^{k-1} \alpha_{jk} x_j \right) y_k^2 + \sum_{1 \leq j < k} 2\alpha_{jk} x_k y_j y_k. \end{aligned}$$

To prove the claim it suffices to find $N \in \mathbb{N}$ such that R_2 restricted to the subspace $Z = \overline{\operatorname{span}}\{e_j;\ j > N\} \subset c_0$ of codimension N is non-zero away from the origin.

Indeed, if so, then $\lambda \mapsto R(\lambda z) = \sum_{m=0}^{3} \lambda^m R_m(z)$ is a non-zero polynomial in λ for every $0 \neq z \in Z$, and in particular for every $0 \neq z \in Z$ there exists some $\lambda \in \mathbb{R}$ such that $P(x + \lambda z) = R(\lambda z) \neq 0$. Now if $x \in Y \subset c_0$ is a null space, then $Z \cap Y = \{0\}$ and so $\dim Y \leq N$ as stated.

Let $r = \min\{j;\ x_j \neq 0\}$ and assume without loss of generality that $x_r > 0$. We claim that $N \geq \max\{r, \frac{2}{x_r}\}$ is sufficient. First note that by (ii) and (iii) for $k > N$

$$\begin{aligned}\sum_{j=r}^{k-1} \alpha_{jk} x_j &\geq \alpha_{rk} x_r - \sum_{j=r+1}^{k-1} \alpha_{jk} |x_j| > \frac{2}{k}\alpha_{rk} - \sum_{j=r+1}^{k-1} \alpha_{jk} \\ &\geq \frac{1}{k}\alpha_{rk} \geq \sum_{l=r+1}^{\infty} \beta_{rk}^{l} \geq \sum_{l=N+1}^{\infty} \beta_{rk}^{l}.\end{aligned}$$

Thus if $y \in Z$, then

$$\begin{aligned}R_2(y) &= \sum_{k=N+1}^{\infty} \left(\sum_{j=r}^{k-1} \alpha_{jk} x_j \right) y_k^2 + \sum_{N+1 \leq j < k} 2\alpha_{jk} x_k y_j y_k \\ &\geq \sum_{k=N+1}^{\infty} \left(\sum_{l=N+1}^{\infty} \beta_{rk}^{l} \right) y_k^2 + \sum_{N+1 \leq j < k} 2\alpha_{jk} x_k y_j y_k \\ &= \frac{1}{2} \sum_{k=N+1}^{\infty} \sum_{l=N+1}^{\infty} \beta_{rk}^{l} y_k^2 + \frac{1}{2} \sum_{k=N+1}^{\infty} \sum_{l=N+1}^{\infty} \beta_{rl}^{k} y_l^2 + \sum_{N+1 \leq j < k} 2\alpha_{jk} x_k y_j y_k \\ &\geq \sum_{j=N+1}^{\infty} \sum_{k=j+1}^{\infty} \frac{1}{2}(\beta_{rj}^{k} y_j^2 + \beta_{rk}^{j} y_k^2) + 2\alpha_{jk} x_k y_j y_k.\end{aligned}$$

But using (iv) we get that for $r < j < k$

$$\begin{aligned}\frac{1}{2}(\beta_{rj}^{k} y_j^2 + \beta_{rk}^{j} y_k^2) + 2\alpha_{jk} x_k y_j y_k &\geq \frac{1}{2}(\beta_{rj}^{k} y_j^2 + \beta_{rk}^{j} y_k^2) - 2\alpha_{jk} |y_j y_k| \\ &\geq \frac{1}{4}(\beta_{rj}^{k} y_j^2 + \beta_{rk}^{j} y_k^2) + \frac{1}{2}\sqrt{\beta_{rj}^{k}}|y_j|\sqrt{\beta_{rk}^{j}}|y_k| - 2\alpha_{jk}|y_j y_k| \\ &\geq \frac{1}{4}(\beta_{rj}^{k} y_j^2 + \beta_{rk}^{j} y_k^2).\end{aligned}$$

It follows that $R_2(y) > 0$ for every $0 \neq y \in Z$. □

The statement of the theorem applies to all separable Banach spaces, ℓ_∞, or $C(K)$, where K is a separable compact (not necessarily metrisable). The w^*-separability of the dual is inherited by the subspaces, so since $\ell_1(\mathrm{c}) \subset \ell_\infty$, the theorem applies also to $\ell_1(\mathrm{c})$. In order to obtain infinite-dimensional null spaces, we need to pass to spaces with large density character, as described by the next result, whose proof is omitted. Denote $\exp \alpha = 2^\alpha$ and $\exp^{n+1} \alpha = \exp(\exp^n \alpha)$, where α is a cardinal.

Theorem 78 ([ArHá2]). *Assume the generalised continuum hypothesis. Let $n \in \mathbb{N}$ be odd and let X be a real Banach space of* $\operatorname{dens} X \geq \exp^{n+1} \alpha$, *where* $\alpha \geq \mathfrak{c}$. *Then every n-homogeneous polynomial on X has a null space of density at least α^+.*

6. Sub-symmetric polynomials on $\mathbb{R}^n$

Let $N \in \mathbb{N}$ and $e_1, \ldots, e_N$ be the canonical basis of $\mathbb{R}^N$. Analogously as in the infinite-dimensional case we say that a polynomial $P \in \mathcal{P}(\mathbb{R}^N)$ is sub-symmetric if

$$P\Big(\sum_{j=1}^{k} x_j e_{n_j}\Big) = P\Big(\sum_{j=1}^{k} x_j e_j\Big)$$

whenever $1 \leq k < N$, $x_1, \ldots, x_k \in \mathbb{R}$, and $1 \leq n_1 < \cdots < n_k \leq N$. Clearly, if $P \in \mathcal{P}(\mathbb{R}^N)$ is sub-symmetric and $1 \leq n_1 < \cdots < n_k \leq N$, then the polynomial $Q \in \mathcal{P}(\mathbb{R}^k)$ given by $Q\big(\sum_{j=1}^k x_j e_j\big) = P\big(\sum_{j=1}^k x_j e_{n_j}\big)$ is also sub-symmetric.

We set $\mathcal{N}(k, N) = \{\rho \in \mathcal{N}(k);\ \rho_k \leq N\} = \{\rho \in \{1, \ldots, N\}^k;\ \rho_1 < \cdots < \rho_k\}$ and define the elementary sub-symmetric polynomials $P_\alpha^N \in \mathcal{P}(^d\mathbb{R}^N)$ for $\alpha \in \mathcal{I}^+(k, d)$, $k \leq N$, by

$$P_\alpha^N(x) = \sum_{\rho \in \mathcal{N}(k,N)} x_{\rho_1}^{\alpha_1} \cdots x_{\rho_k}^{\alpha_k}. \tag{10}$$

Similarly as in the infinite-dimensional case (Fact 55) we can see that the polynomials P_α^N, $\alpha \in \mathcal{I}^+(k, d)$, $k = 1, \ldots, \min\{d, N\}$ form a basis of the space of sub-symmetric d-homogeneous polynomials on $\mathbb{R}^N$.

The concept of sub-symmetric polynomials on $\mathbb{R}^N$ can be used to capture the essential information on the behaviour of a given general polynomial, similarly to the infinite-dimensional situation.

Theorem 79. *Let $n, d \in \mathbb{N}$ and $\varepsilon > 0$. There exists $N = N(n, d, \varepsilon)$ such that for every polynomial $P \in \mathcal{P}(^d\ell_1^N)$, $\|P\| \leq 1$, there exist $A \subset \{1, \ldots, N\}$, $|A| = n$, and a sub-symmetric polynomial $Q \in \mathcal{P}(^dY)$ such that $\|P\restriction_Y - Q\| < \varepsilon$, where* $Y = \operatorname{span}\{e_k;\ k \in A\}$ *and $\{e_k\}_{k=1}^N$ is the canonical basis of ℓ_1^N.*

Proof. For a given $P \in \mathcal{P}(^d\ell_1^N)$ satisfying $\|P\| \leq 1$ there are $a_{\alpha,\rho} \in \mathbb{R}$ such that $P = \sum_{\alpha \in \mathcal{I}^+(d)} R_\alpha$, where

$$R_\alpha(x) = \sum_{\rho \in \mathcal{N}(k,N)} a_{\alpha,\rho} x_{\rho_1}^{\alpha_1} \cdots x_{\rho_k}^{\alpha_k} \tag{11}$$

for $\alpha \in \mathcal{I}^+(k, d)$. Using (3) and the Polarisation formula we can see that each $|a_{\alpha,\rho}| < \binom{d}{\alpha}\|\check{P}\| \leq d^d$. We show that for any $n \in \mathbb{N}$, $\varepsilon > 0$, $K > 0$, and $\alpha \in \mathcal{I}^+(d)$ there is $N = N_\alpha(n, \varepsilon, K)$ such that for any polynomial $R \in \mathcal{P}(^d\ell_1^N)$ of the form (11) with $|a_{\alpha,\rho}| \leq K$ for all $\rho \in \mathcal{N}(k, N)$ there are $A \subset \{1, \ldots, N\}$, $|A| = n$, and $c \in \mathbb{R}$

such that $\|R\restriction_Y - cP^n_\alpha\| < \varepsilon$, where $Y = \operatorname{span}\{e_k;\ k \in A\}$. It is then clear that we may take

$$N(n,d,\varepsilon) = N_{\alpha^v}\big(\dots N_{\alpha^2}\big(N_{\alpha^1}(n, \tfrac{\varepsilon}{v}, d^d), \tfrac{\varepsilon}{v}, d^d\big)\dots, \tfrac{\varepsilon}{v}, d^d\big),$$

where $\alpha^1, \dots, \alpha^v$ is an enumeration of $\mathcal{I}^+(d)$.

So fix $\alpha \in \mathcal{I}^+(k,d)$, $n \in \mathbb{N}$, $\varepsilon > 0$, and $K > 0$. Let $\delta = \frac{\varepsilon}{2n!}$ and $M = \left[\frac{K}{\delta}\right]$. By Ramsey's theorem there is $N \in \mathbb{N}$ such that for every $2(M+1)$-colouring of k-subsets (i.e. subsets of cardinality k) of $\{1,\dots,N\}$ there is $A \subset \{1,\dots,N\}$, $|A| = n$, such that all k-subsets of A have the same colour. Now given $R \in \mathcal{P}({}^d\ell_1^N)$ of the form (11) with $|a_{\alpha,\rho}| \le K$ for all $\rho \in \mathcal{N}(k,N)$ we put $m(\rho) = \left[\frac{a_{\alpha,\rho}}{\delta}\right] \in \{-M-1, -M, \dots, M\}$. Note that $|a_{\alpha,\rho} - \delta m(\rho)| < \delta$. Each $\rho \in \mathcal{N}(k,N)$ uniquely determines a k-subset of $\{1,\dots,N\}$ and vice versa and so the function m induces a $2(M+1)$-colouring of the k-subsets of $\{1,\dots,N\}$. Let $A \subset \{1,\dots,N\}$, $|A| = n$, be such that there is $m_0 \in \mathbb{N}$ satisfying $m(\rho) = m_0$ for all $\rho \subset A$. Then

$$\left|R\Big(\sum_{j\in A} x_j e_j\Big) - \delta m_0 P^n_\alpha\Big(\sum_{j\in A} x_j e_j\Big)\right| \le \delta \sum_{\rho\subset A} |x_{\rho_1}^{\alpha_1}\cdots x_{\rho_k}^{\alpha_k}| \le \delta\binom{n}{k} < \varepsilon$$

whenever $\left\|\sum_{j\in A} x_j e_j\right\| \le 1$. □

The main result of this section (Corollary 86) concerns the properties of subsymmetric polynomials, however in its proof we need to work also with partial derivatives of the polynomials P^N_α and for this reason we consider also the polynomials P^N_α given by the formula (10) where $\alpha \in \mathcal{I}(k,d)$, $k \le N$. We denote by $H^{d,K}(\mathbb{R}^N)$ the subspace of $\mathcal{P}^d(\mathbb{R}^N)$ generated by the polynomials P^N_α, $\alpha \in \bigcup_{k=1}^K \mathcal{I}(k,d)$. For formal reasons we also put $P^N_\alpha = 0$ if $k > N$ and $P^N_{()} = 1$, both even for $N = 0$, further $\mathcal{I}(0,0) = \{()\}$, and $\mathbb{R}^0 = \{0\}$. Note that these definitions are consistent with (10).

The following fact describes an important relation between the restriction of P^M_α to the first N coordinates and P^N_α. Note that for $M > N$ we consider canonically $\mathbb{R}^N$ as a subspace of $\mathbb{R}^M$.

Fact 80. *Let $M, N, k, d \in \mathbb{N}_0$, $N < M$, and $\alpha \in \mathcal{I}(k,d)$ be such that $\alpha_m > 0$ and $\alpha_{m+1} = \dots = \alpha_k = 0$ for some $0 \le m \le k$. Then*

$$P^M_\alpha(x) = \sum_{j=m}^{k} \binom{M-N}{k-j} P^N_{(\alpha_1,\dots,\alpha_j)}(x)$$

for every $x \in \mathbb{R}^N$. Conversely,

$$P^N_\alpha(x) = \sum_{j=m}^{k} (-1)^{k-j}\binom{M-N+k-j-1}{k-j} P^M_{(\alpha_1,\dots,\alpha_j)}(x)$$

for every $x \in \mathbb{R}^N$.

Proof. The first relation follows from the following (recall that $x \in \mathbb{R}^N$, that is $x_{N+1} = \cdots = x_M = 0$ as per the aforementioned convention):

$$P_\alpha^M(x) = \sum_{\rho \in \mathcal{N}(k,M)} x_{\rho_1}^{\alpha_1} \cdots x_{\rho_m}^{\alpha_m} = \sum_{\substack{\rho \in \mathcal{N}(k,M) \\ \rho_m \le N}} x_{\rho_1}^{\alpha_1} \cdots x_{\rho_m}^{\alpha_m}$$

$$= \sum_{j=m}^{k} \sum_{\substack{1 \le \rho_1 < \cdots < \rho_k \le M \\ \rho_j \le N < \rho_{j+1}}} x_{\rho_1}^{\alpha_1} \cdots x_{\rho_m}^{\alpha_m} = \sum_{j=m}^{k} \binom{M-N}{k-j} P_{(\alpha_1,\dots,\alpha_j)}^N(x).$$

The second relation can be proved by induction on $k-m$. For $k-m=0$ it follows immediately from the first one. For the induction step we use the first relation together with the inductive hypothesis to obtain

$$P_\alpha^N(x) = P_\alpha^M(x) - \sum_{j=m}^{k-1} \binom{M-N}{k-j} P_{(\alpha_1,\dots,\alpha_j)}^N(x)$$

$$= P_\alpha^M(x) - \sum_{j=m}^{k-1} \binom{M-N}{k-j} \sum_{l=m}^{j} (-1)^{j-l} \binom{M-N+j-l-1}{j-l} P_{(\alpha_1,\dots,\alpha_l)}^M(x)$$

$$= P_\alpha^M(x) - \sum_{l=m}^{k-1} \left(\sum_{j=l}^{k-1} (-1)^{j-l} \binom{M-N}{k-j} \binom{M-N+j-l-1}{j-l} \right) P_{(\alpha_1,\dots,\alpha_l)}^M(x)$$

and the result now follows from the identity $\sum_{j=l}^{k} (-1)^{j-l} \binom{M-N}{k-j} \binom{M-N+j-l-1}{j-l} = 0$. Adding or removing a couple of zero summands, this is equivalent to

$$\sum_{p=0}^{M-N} (-1)^{k-l-p} \binom{M-N}{p} \binom{M-N+k-l-p-1}{M-N-1} = 0,$$

which is the Fréchet formula for the polynomial $t \mapsto \binom{M-N+k-l-t-1}{M-N-1}$ of degree $M-N-1$ (Theorem 2.50). □

It is very important to notice that the previous fact covers all the special cases like $N < k \le M$, $k > M$, $N = 0$, $m = 0$, or $k = 0$. Observe also that in particular in the sub-symmetric case (i.e. $\alpha \in \mathcal{J}^+(d)$) we have $P_\alpha^M \restriction_{\mathbb{R}^N} = P_\alpha^N$. Hence for the elementary sub-symmetric polynomials the superscript N can be dropped. We will use this simplification for the power sum symmetric polynomials $s_n^N = P_{(n)}^N = s_n$.

The next fact deals with the situation when we fix the first N coordinates of P_α^M.

Fact 81. *Let $N, d \in \mathbb{N}_0$, $M, k \in \mathbb{N}$, $N < M$, $k \le M$, $\alpha \in \mathcal{J}(k,d)$, and $y \in \mathbb{R}^N$. Then the polynomial $(x_1,\dots,x_{M-N}) \mapsto P_\alpha^M(y_1,\dots,y_N,x_1,\dots,x_{M-N})$ belongs to $H^{d,\min\{k,M-N\}}(\mathbb{R}^{M-N})$.*

Proof.

$$P_\alpha^M(y_1,\dots,y_N,x_1,\dots,x_{M-N}) = \sum_{j=0}^{k} \sum_{\substack{1\le\rho_1<\dots<\rho_k\le M\\ \rho_j\le N<\rho_{j+1}}} y_{\rho_1}^{\alpha_1}\cdots y_{\rho_j}^{\alpha_j} x_{\rho_{j+1}-N}^{\alpha_{j+1}}\cdots x_{\rho_k-N}^{\alpha_k}$$
$$= \sum_{\substack{0\le j\le k\\ k-(M-N)\le j\le N}} P^N_{(\alpha_1,\dots,\alpha_j)}(y) P^{M-N}_{(\alpha_{j+1},\dots,\alpha_k)}(x_1,\dots,x_{M-N}).$$

□

Let $k,d\in\mathbb{N}$, $\alpha\in\mathcal{J}(k,d)$, $k\le N$, $x\in\mathbb{R}^N$, and $1\le l\le N$. Then

$$\frac{\partial P_\alpha^N}{\partial x_l}(x) = \frac{\partial}{\partial x_l}\left(\sum_{j=1}^{k} \sum_{\substack{\rho\in\mathcal{N}(k,N)\\ \rho_j=l}} x_{\rho_1}^{\alpha_1}\cdots x_{\rho_k}^{\alpha_k}\right)$$
$$= \sum_{\substack{j=1\\ \alpha_j>0}}^{k} \alpha_j \sum_{\substack{1\le\rho_1<\dots<\rho_{j-1}<l\\ l<\rho_{j+1}<\dots<\rho_k\le N}} x_{\rho_1}^{\alpha_1}\cdots x_{\rho_{j-1}}^{\alpha_{j-1}} x_l^{\alpha_j-1} x_{\rho_{j+1}}^{\alpha_{j+1}}\cdots x_{\rho_k}^{\alpha_k} \tag{12}$$
$$= \sum_{\substack{j=1\\ \alpha_j>0}}^{k} \alpha_j P^{l-1}_{(\alpha_1,\dots,\alpha_{j-1})}(x_1,\dots,x_{l-1}) x_l^{\alpha_j-1} P^{N-l}_{(\alpha_{j+1},\dots,\alpha_k)}(x_{l+1},\dots,x_N).$$

These partial derivatives have the following useful property:

Fact 82. *Let $k,d,N\in\mathbb{N}$, $\alpha\in\mathcal{J}(k,d)$, $k\le N$. Then $\sum_{l=1}^N \frac{\partial P_\alpha^N}{\partial x_l}\in H^{d-1,k}(\mathbb{R}^N)$.*

Proof.

$$\sum_{l=1}^{N}\frac{\partial P_\alpha^N}{\partial x_l}(x) = \sum_{l=1}^{N}\sum_{\substack{j=1\\ \alpha_j>0}}^{k} \alpha_j \sum_{\substack{1\le\rho_1<\dots<\rho_{j-1}<l\\ l<\rho_{j+1}<\dots<\rho_k\le N}} x_{\rho_1}^{\alpha_1}\cdots x_{\rho_{j-1}}^{\alpha_{j-1}} x_l^{\alpha_j-1} x_{\rho_{j+1}}^{\alpha_{j+1}}\cdots x_{\rho_k}^{\alpha_k}$$
$$= \sum_{\substack{j=1\\ \alpha_j>0}}^{k} \alpha_j \sum_{l=1}^{N} \sum_{\substack{\rho\in\mathcal{N}(k,N)\\ \rho_j=l}} x_{\rho_1}^{\alpha_1}\cdots x_{\rho_{j-1}}^{\alpha_{j-1}} x_{\rho_j}^{\alpha_j-1} x_{\rho_{j+1}}^{\alpha_{j+1}}\cdots x_{\rho_k}^{\alpha_k}$$
$$= \sum_{\substack{j=1\\ \alpha_j>0}}^{k} \alpha_j P^N_{(\alpha_1,\dots,\alpha_{j-1},\alpha_j-1,\alpha_{j+1},\alpha_k)}(x).$$

□

We note that this fact does not hold with $\mathcal{J}^+(k,d)$ and the space of sub-symmetric polynomials in place of $\mathcal{J}(k,d)$ and $H^{d-1,k}(\mathbb{R}^N)$, and this is the sole reason for considering the spaces $H^{d,K}(\mathbb{R}^N)$.

For each $x \in \mathbb{R}^N$ we naturally identify the derivative $DP_\alpha^N(x)$ with the vector $\left(\frac{\partial P_\alpha^N}{\partial x_1}(x), \dots, \frac{\partial P_\alpha^N}{\partial x_N}(x)\right) \in \mathbb{R}^N$.

Fact 83. *Let $M, N, k, d \in \mathbb{N}$, where $M > N$ and $k \leq N$, let $\alpha \in \mathcal{I}(k,d)$, and let $x \in \mathbb{R}^N$. Then the derivative $DP_\alpha^N(x)$ is a linear combination of the vectors $DP_\beta^M(x)\restriction_N = \left(\frac{\partial P_\beta^M}{\partial x_1}(x), \dots, \frac{\partial P_\beta^M}{\partial x_N}(x)\right) \in \mathbb{R}^N$, $\beta \in \bigcup_{m=1}^k \mathcal{I}(m,d)$.*

Proof. Let $1 \leq m \leq k$ be such that $\alpha_m > 0$ and $\alpha_{m+1} = \dots = \alpha_k = 0$. Fix $1 \leq l \leq N$. If $\alpha_j > 0$, then $m \geq j$ and hence by Fact 80

$$P^{N-l}_{(\alpha_{j+1},\dots,\alpha_k)}(x_{l+1},\dots,x_N) = \sum_{s=m}^{k} c_s P^{M-l}_{(\alpha_{j+1},\dots,\alpha_s)}(x_{l+1},\dots,x_N,0,\dots,0),$$

where $c_s = (-1)^{k-s}\binom{M-N+k-s-1}{k-s}$. Therefore using the formula (12) and the fact that $\alpha_{s+1} = \dots = \alpha_k = 0$ if $m \leq s \leq k$ we obtain

$$\begin{aligned}\frac{\partial P_\alpha^N}{\partial x_l}(x) &= \sum_{\substack{j=1\\ \alpha_j>0}}^{k} \Bigg(\alpha_j P^{l-1}_{(\alpha_1,\dots,\alpha_{j-1})}(x_1,\dots,x_{l-1})x_l^{\alpha_j-1} \\ &\qquad\qquad \cdot \sum_{s=m}^{k} c_s P^{M-l}_{(\alpha_{j+1},\dots,\alpha_s)}(x_{l+1},\dots,x_N,0,\dots,0)\Bigg) \\ &= \sum_{s=m}^{k} c_s \frac{\partial P^M_{(\alpha_1,\dots,\alpha_s)}}{\partial x_l}(x),\end{aligned}$$

from which the statement follows. □

We will also make use of the following version of the Lagrange multipliers theorem.

Theorem 84. *Let $G \subset \mathbb{R}^n$ be an open set, $f \in C^1(G)$, $F \in C^1(G;\mathbb{R}^m)$, and assume that F has a constant rank. If the function f has a local extremum with respect to $M = \{x \in G;\ F(x) = 0\}$ at $a \in M$, then $Df(a)$ is a linear combination of $DF_1(a), \dots, DF_m(a)$, where $F_1, \dots, F_m$ are the components of the mapping F.*

Proof. Let $k = \operatorname{rank} F(x)$ for $x \in G$. Since DF is continuous, we may without loss of generality assume that $DF_1(x), \dots, DF_k(x)$ are linearly independent for each $x \in G$. From the Rank theorem it follows that there are C^1-smooth functions g_j of k variables, $j = k+1, \dots, m$, and a neighbourhood U of a such that $F_j(x) = g_j\big(F_1(x), \dots, F_k(x)\big)$ for each $x \in U$, $j = k+1, \dots, m$ (see e.g. [Zo, Proposition 8.6.3.1]). Notice that $g_j(0,\dots,0) = g_j\big(F_1(a),\dots,F_k(a)\big) = F_j(a) = 0$, $j = k+1, \dots, m$. Therefore $M \cap U = \{x \in U;\ F_1(x) = 0, \dots, F_k(x) = 0\}$ and we may use the classical version of the Lagrange multipliers theorem □

Now we are ready to prove the key lemma.

Lemma 85. *For every $n, K \in \mathbb{N}$ there exist $N \in \mathbb{N}$ and $u, v \in \mathbb{R}^N$ such that $P(u) = P(v)$ for every $P \in H^{n,K}(\mathbb{R}^N)$ but $s_{n+1}(u) \neq s_{n+1}(v)$.*

Proof. The proof is based on the observation that $\sum_{l=1}^{N} \frac{\partial s_{n+1}}{\partial x_l}(x) = (n+1)s_n(x)$, which together with Fact 82 leads to an inductive proof. For each fixed $K \in \mathbb{N}$ we prove the statement by induction on n. So fix an arbitrary $K \in \mathbb{N}$ and denote $\mathcal{M}(n) = \bigcup_{1\le d\le n} \bigcup_{1\le k\le K} \mathcal{I}(k,d)$. The space $H^{n,K}(\mathbb{R}^N)$ is generated by a constant function and polynomials P_α^N, $\alpha \in \mathcal{M}(n)$. For $n = 1$ the functions P_α^N, $\alpha \in \mathcal{M}(n)$ are linear and so there is $N \in \mathbb{N}$ large enough such that $\bigcap_{\alpha\in\mathcal{M}(n)} \ker P_\alpha^N$ contains a non-zero element u. Then it suffices to take $v = 2u$.

The inductive step from $n-1$ to n will be proved by contradiction. So assume that for each $N \ge K$ and each $u, v \in \mathbb{R}^N$ satisfying $P_\alpha^N(u) = P_\alpha^N(v)$ for all $\alpha \in \mathcal{M}(n)$ we have $s_{n+1}(u) = s_{n+1}(v)$. Now let $F^N : \mathbb{R}^N \to \mathbb{R}^{|\mathcal{M}(n)|}$ be the mapping whose components are the polynomials P_α^N, $\alpha \in \mathcal{M}(n)$ in some fixed order and let $A_N(x)$ be its Jacobi matrix at $x \in \mathbb{R}^N$, i.e. $A_N(x) = \left(\frac{\partial P_\alpha^N}{\partial x_l}(x)\right)_{\substack{\alpha\in\mathcal{M}(n)\\ l=1,\dots,M}}$. Note that the number of rows of the matrix of functions A_N does not depend on N. Thus there are $N \ge K$ and $y \in \mathbb{R}^N$ such that $\operatorname{rank} A_N(y) = r = \max_{M\ge K, x\in\mathbb{R}^M} \operatorname{rank} A_M(x)$.

By the inductive hypothesis there are $M > N$ and $g, h \in \mathbb{R}^{M-N}$ such that $P(g) = P(h)$ for all $P \in H^{n-1,K}(\mathbb{R}^{M-N})$ but $s_n(g) \ne s_n(h)$. If we denote by $A_M(x)\restriction_N$ the matrix consisting of the first N columns of the matrix $A_M(x)$, then $r = \operatorname{rank} A_N(y) \le \operatorname{rank} A_M(y)\restriction_N \le \operatorname{rank} A_M(y) \le r$, where the first inequality follows from Fact 83. Let $w_1^M, \dots, w_r^M$ be the rows of A_M such that $w_1^M(y)\restriction_N, \dots, w_r^M(y)\restriction_N$ are linearly independent. Using the continuity of the entries of A_M it is easy to see that there is a neighbourhood $U \subset \mathbb{R}^M$ of y such that for each $x \in U$ the vectors $w_1^M(x)\restriction_N, \dots, w_r^M(x)\restriction_N$ are linearly independent and so they form a basis of the space spanned by the rows of $A_M(x)\restriction_N$. Clearly the same holds for $w_1^M(x), \dots, w_r^M(x)$ and $A_M(x)$.

Fix an arbitrary $z \in U$ and put $S = \{x \in U;\ P_\alpha^M(x) = P_\alpha^M(z), \alpha \in \mathcal{M}(n)\}$. By our assumption s_{n+1} is constant on S and so Theorem 84 implies that $Ds_{n+1}(z)$ is a linear combination of the rows of $A_M(z)$. It follows that for each $z \in U$ the vector $Ds_{n+1}(z)$ is a linear combination of $w_1^M(z), \dots, w_r^M(z)$.

Next, we put $u = y + c\sum_{j=1}^{M-N} g_j e_{N+j}$, $v = y + c\sum_{j=1}^{M-N} h_j e_{N+j}$ for some suitable $c \ne 0$ so that $u, v \in U$. Notice that since $H^{n-1,K}(\mathbb{R}^{M-N})$ is generated by homogeneous polynomials, we still have $P(cg) = P(ch)$ for all $P \in H^{n-1,K}(\mathbb{R}^{M-N})$ but $s_n(cg) \ne s_n(ch)$. For a fixed $\alpha \in \mathcal{M}(n)$ and $1 \le l \le N$ consider the polynomial $P(x) = \frac{\partial P_\alpha^M}{\partial x_l}(y_1, \dots, y_N, x_1, \dots, x_{M-N})$. Then by (12) and Fact 81 we have $P \in H^{n-1,K}(\mathbb{R}^{M-N})$ and so $P(cg) = P(ch)$. Therefore

$$w_j^M(u)\restriction_N = w_j^M(v)\restriction_N, \quad j = 1, \dots, r. \tag{13}$$

We have $Ds_{n+1}(u) = \sum_{j=1}^{r} \lambda_j w_j^M(u)$ and $Ds_{n+1}(v) = \sum_{j=1}^{r} \mu_j w_j^M(v)$ for some $\lambda_j, \mu_j \in \mathbb{R}$ and clearly the same holds when we restrict to the first N coordinates of all these vectors. But since $Ds_{n+1}(u)\restriction_N = (n+1)(y_1^n, \dots, y_N^n) = Ds_{n+1}(v)\restriction_N$, combining this with (13) and the fact that $w_1^M(u)\restriction_N, \dots, w_r^M(u)\restriction_N$ are linearly independent we obtain $\mu_j = \lambda_j$, $j = 1,\dots,r$. Finally, from Fact 82 and Fact 81 it follows that $x \mapsto \sum_{l=1}^{M} w_j^M \big(y + \sum_{j=1}^{M-N} x_j e_{N+j}\big)_l \in H^{n-1,K}(\mathbb{R}^{M-N})$ for $j = 1,\dots,r$. Therefore

$$\begin{aligned}(n+1)s_n(u) &= \sum_{l=1}^{M} \frac{\partial s_{n+1}}{\partial x_l}(u) = \sum_{j=1}^{r} \lambda_j \sum_{l=1}^{M} w_j^M(u)_l \\ &= \sum_{j=1}^{r} \lambda_j \sum_{l=1}^{M} w_j^M(v)_l = \sum_{l=1}^{M} \frac{\partial s_{n+1}}{\partial x_l}(v) = (n+1)s_n(v).\end{aligned}$$

Since $s_n(u) = s_n(y) + s_n(cg)$ and $s_n(v) = s_n(y) + s_n(ch)$, we get $s_n(cg) = s_n(ch)$, which is a contradiction. □

Let X be a real Banach space. We are going to work with algebras of polynomials on X, i.e. linear subspaces of $\mathcal{P}(X)$ that are closed with respect to pointwise multiplication. Given a set $B \subset \mathcal{P}(X)$ we say that B generates an algebra $\mathcal{A} \subset \mathcal{P}(X)$ if $\mathcal{A}$ is the smallest algebra containing B, i.e. it is the intersection of all algebras containing B. It is easy to see that B generates $\mathcal{A}$ if and only if for every $p \in \mathcal{A}$ there are a subset $\{b_1,\dots,b_k\} \subset B$ and a polynomial $P \in \mathcal{P}(\mathbb{R}^k)$ such that $p = P \circ (b_1,\dots,b_k)$.

Corollary 86. *For every $n \in \mathbb{N}$ there are $N \in \mathbb{N}$ and $\varepsilon > 0$ such that if $M \geq N$, then*

$$\sup_{x \in B_{\ell_1^M}} |p(x) - s_{n+1}(x)| \geq \varepsilon$$

for every p from the algebra generated by the sub-symmetric polynomials on $\mathbb{R}^M$ of degree at most n.

Proof. Applying Lemma 85 to $K = n$ we obtain $N \in \mathbb{N}$ and $u, v \in B_{\ell_1^N}$ such that $P(u) = P(v)$ for every $P \in H^{n,n}(\mathbb{R}^N)$ but $s_{n+1}(u) \neq s_{n+1}(v)$. We put $\varepsilon = \frac{1}{2}|s_{n+1}(u) - s_{n+1}(v)|$. Let $M \geq N$. Since all sub-symmetric polynomials from $\mathcal{P}^n(\mathbb{R}^N)$ are contained in $H^{n,n}(\mathbb{R}^N)$, from the remark after Fact 80 it follows that in particular $P(u) = P(v)$ for every sub-symmetric $P \in \mathcal{P}^n(\mathbb{R}^M)$. We conclude that $p(u) = p(v)$ for every p from the algebra generated by the sub-symmetric polynomials from $\mathcal{P}^n(\mathbb{R}^M)$. The statement now easily follows. □

7. Polynomial algebras on Banach spaces

All spaces in this section are real. Let X be a Banach space. Observe that if $\mathcal{A}$ is a subalgebra of $C(X)$ such that each member of $\mathcal{A}$ is bounded on CCB subsets of X, then $\overline{\mathcal{A}}^{\tau_b}$ is also a subalgebra of $C(X)$. For every Banach space X we denote by $\mathcal{A}_n(X)$ the algebra generated by $\mathcal{P}^n(X)$. Generally speaking, with a single exception when $\mathcal{P}(X) = \mathcal{P}_{wu}(X)$, there are no results giving a characterisation of the uniform closure $\overline{\mathcal{P}(X)}^{\tau_b}$ in any infinite-dimensional Banach space. The refinement of the problem is finding the characterisation of $\overline{\mathcal{A}_n(X)}$, and this is wide open as well. The results in this section focus on the natural question when $\overline{\mathcal{A}_n(X)} = \overline{\mathcal{A}_{n+1}(X)}$ (of course, the inclusion $\subset$ always holds). More precisely, we are going to use the theory of sub-symmetric polynomials developed in the previous section together with the asymptotic approach to polynomial behaviour to obtain rather general results showing that the inclusion $\supset$ is almost never satisfied. We begin by formulating a positive result, which easily follows from results proved later on in Chapters 6 and 7.

Proposition 87. *Let X be a Banach space such that it does not contain ℓ_1 and $\mathcal{P}(^nX) = \mathcal{P}_{\mathrm{wsc}}(^nX)$. Then*

$$\overline{\mathcal{A}_1(X)} = \overline{\mathcal{A}_2(X)} = \cdots = \overline{\mathcal{A}_n(X)}.$$

Proof. It is clear that $\mathcal{P}_{\mathrm{f}}(X) \subset \mathcal{A}_1(X)$ (see Definition 7.10). Therefore we have $\mathcal{P}^n(X) = \mathcal{P}^n_{\mathrm{wu}}(X) \subset \mathcal{C}_{\mathrm{wu}}(X) = \overline{\mathcal{P}_{\mathrm{f}}(X)}^{\tau_b} \subset \overline{\mathcal{A}_1(X)}$, where the first equality follows from Proposition 3.29 and Corollary 3.80, and the second equality is Theorem 7.17. Since $\overline{\mathcal{A}_1(X)}$ is an algebra, $\mathcal{A}_n(X) \subset \overline{\mathcal{A}_1(X)}$, from which the result follows. □

By Corollary 3.69 and the classical Theorem 5.125, the above proposition can be applied to all $C(K)$ spaces, K scattered compact, and all $n \in \mathbb{N}$. All our results going in the opposite direction will rely on the following criterion.

Theorem 88. *Let X be a Banach space and $n \in \mathbb{N}$ such that for each $\varepsilon > 0$ there is $P \in \mathcal{P}(^nX)$ with the following property: For every $N \in \mathbb{N}$ there exists a linearly independent set $\{e_j\}_{j=1}^N \subset S_X$ such that*

$$\sup_{\sum_{j=1}^N |a_j| \le 1} \left| P\left(\sum_{j=1}^N a_j e_j\right) - \sum_{j=1}^N a_j^n \right| \le \varepsilon.$$

Then $\mathcal{P}(^nX) \not\subset \overline{\mathcal{A}_{n-1}(X)}$.

Proof. Denote by $S^n(\ell_1^m)$ the algebra generated by all sub-symmetric polynomials on ℓ_1^m of degree at most n. By Corollary 86 there is $m \in \mathbb{N}$ and $\varepsilon > 0$ such that $\|Q - s_n\| \ge 3\varepsilon$ for all $Q \in S^{n-1}(\ell_1^m)$. Let $P \in \mathcal{P}(^nX)$ be the polynomial from the assumptions of the theorem. We claim that $P \notin \overline{\mathcal{A}_{n-1}(X)}$.

By contradiction, suppose that there exist $P_1, \dots, P_k \in \mathcal{P}^{n-1}(X)$ and $r \in \mathcal{P}(\mathbb{R}^k)$ such that for $R = r \circ (P_1, \dots, P_k)$ we have $\|P - R\| < \varepsilon$. Put $K = 1 + \max_j \|P_j\|$ and let $0 < \eta \le 1$ be such that $|r(u) - r(v)| < \varepsilon$ whenever $u, v \in KB_{\ell_\infty^k}$, $\|u - v\|_{\ell_\infty^k} < \eta$. Using Theorem 79 recursively $k(n-1)$ times we find $N \in \mathbb{N}$ such that for any linearly independent $\{e_j\}_{j=1}^N \subset S_X$ there exist $A \subset \{1, \dots, N\}$, $|A| = m$, and sub-symmetric polynomials $Q_1, \dots, Q_k \in \mathcal{P}^{n-1}(Y)$ such that $\|P_j \restriction_Y - Q_j\| < \eta$, $j = 1, \dots, k$, where $Y = \operatorname{span}\{e_j;\ j \in A\}$ with ℓ_1-norm.

Let $\{e_j\}_{j=1}^N$ be the linearly independent set from the assumptions of the theorem and $A \subset \{1, \dots, N\}$, $Q_1, \dots, Q_k \in \mathcal{P}^{n-1}(Y)$ as above. Note that as $\{e_j\}$ is normalised, $\|R\restriction_Y - P\restriction_Y\|_Y \le \|R\restriction_Y - P\restriction_Y\|_X < \varepsilon$. Put $Q = r \circ (Q_1, \dots, Q_k)$. Then $Q \in S^{n-1}(\ell_1^m)$ and $\|Q - s_n\| \le \|Q - R\restriction_Y\| + \|R\restriction_Y - P\restriction_Y\| + \|P\restriction_Y - s_n\| < 3\varepsilon$, a contradiction. □

Theorem 89. *Let X be a Banach space that contains ℓ_1. Then*

$$\overline{\mathcal{A}_1(X)} \subsetneq \overline{\mathcal{A}_2(X)} \subsetneq \overline{\mathcal{A}_3(X)} \subsetneq \cdots$$

Proof. Combine Proposition 3.91 and Theorem 88. □

The following easy fact will be used later in the proofs.

Fact 90. *Let X be a Banach space and $\{x_n\} \subset X$. If there is a w^*-null sequence $\{\phi_n\} \subset X^*$ (in particular if X has a Schauder basis $\{(e_n; f_n)\}$ satisfying $\inf_n \|e_n\| > 0$ and $\{\phi_n\}$ is a subsequence of $\{f_n\}$) such that $\varepsilon = \inf_n |\phi_n(x_n)| > 0$, then $\{x_n;\ n \in \mathbb{N}\}$ is not relatively compact.*

Proof. We construct a subsequence $\{x_{n_k}\}$ of $\{x_n\}$ by induction. We put $n_1 = 1$ and if n_k is already defined for some $k \in \mathbb{N}$, then we find $n_{k+1} > n_k$ such that $|\phi_j(x_{n_k})| < \frac{\varepsilon}{2}$ whenever $j \geq n_{k+1}$. By our assumption $K = \sup_n \|\phi_n\| < +\infty$. The set $\{x_{n_k};\ k \in \mathbb{N}\}$ is then an $\frac{\varepsilon}{2K}$-separated set. Indeed, let $k, l \in \mathbb{N}$, $k < l$. Then $\|x_{n_l} - x_{n_k}\| \geq \frac{1}{K}\left|\phi_{n_l}(x_{n_l} - x_{n_k})\right| \geq \frac{1}{K}\left(|\phi_{n_l}(x_{n_l})| - |\phi_{n_l}(x_{n_k})|\right) \geq \frac{\varepsilon}{2K}$. □

Now we prove another type of a diagonalisation result (see also Lemma 67 and Lemma 6.72).

Lemma 91. *Let X be a Banach space with a Schauder basis $\{(x_n; f_n)\}$, Y a Banach space with a bounded shrinking Schauder basis, and let $m \in \mathbb{N}$ be such that $\mathcal{P}^{m-1}(X; Y^*) = \mathcal{P}_{\mathrm{K}}^{m-1}(X; Y^*)$. Let $P \in \mathcal{P}(^m X; Y^*)$ and denote by P_k the kth component of P, i.e. the composition of P with the kth coordinate functional on Y^*. Then for each $\varepsilon > 0$ there is a subsequence $\{x_{n_k}\}$ such that for every $k \in \mathbb{N}$*

$$\left|P_{n_k}(x) - a_k^m P_{n_k}(x_{n_k})\right| \leq \frac{\varepsilon}{2^k}$$

whenever $x = \sum_{j=1}^{\infty} a_j x_{n_j} \in B_X$ is such that $\min\{j;\ a_j \neq 0\} \geq |\operatorname{supp} x|$.

Proof. Note that we may assume without loss of generality that $\{x_n\}$ is normalised. For any polynomial $Q \in \mathcal{P}(X; Y^*)$ we will use the convention that Q_k denotes the kth component of Q, i.e. the composition of Q with the kth coordinate functional on Y^*.

For $\alpha \in \mathcal{J}(l, m)$, $q \in \mathbb{N}_0$ satisfying $q + |\alpha| \leq m$, and $k \in \mathbb{N}$ we define polynomials $R_k^{\alpha,q} \in \mathcal{P}(^{m-|\alpha|-q}X)$ by

$$R_k^{\alpha,q}(x) = \binom{m}{\alpha, q, m - |\alpha| - q} \check{P}_k(^{\alpha_1}x_1, \dots, {}^{\alpha_l}x_l, {}^q x_k, {}^{m-|\alpha|-q}x).$$

Note that using Proposition 1.22 we obtain

$$P_k\left(\sum_{j=1}^{\infty} a_j x_j\right) = \sum_{\substack{\alpha \in \mathcal{J}(k-1,m) \\ 0 \leq q \leq m-|\alpha|}} a_1^{\alpha_1} \cdots a_{k-1}^{\alpha_{k-1}} a_k^q R_k^{\alpha,q}\left(\sum_{j=k+1}^{\infty} a_j x_j\right).$$

The proof of the lemma is two-stage. In the first step we discard the coordinates before k, in the second step we get rid of the coordinates after k (i.e. the polynomials $R_k^{\alpha,q}$).

Clearly $\{R_k^{\alpha,q};\ k \in \mathbb{N}, \alpha \in \mathcal{J}(l,m), 0 \le q \le m-|\alpha|, l \in \mathbb{N}\}$ is bounded in $\mathcal{P}^m(X)$. We claim that for fixed $\alpha \in \mathcal{J}(l,m)$ and $q \in \mathbb{N}_0$ with $q+|\alpha| \le m$ and $|\alpha| > 0$ we have $\lim_{k\to\infty} \|R_k^{\alpha,q}\| = 0$. Indeed, if this is not the case, then there are a sequence $\{v_j\} \subset B_X$ and a subsequence $\{R_{k_j}^{\alpha,q}\}$ such that $\inf_j |R_{k_j}^{\alpha,q}(v_j)| > 0$. By passing to further subsequences we may also assume that $b_p = \lim_{j\to\infty} R_{k_j}^{\alpha,p}(v_j)$ exists finite for all $0 \le p \le m-|\alpha|$. Obviously $b_q \ne 0$ and so there is $t \in \mathbb{R}$ such that $\sum_{p=0}^{m-|\alpha|} b_p t^p \ne 0$.

Put $Q(x) = \check{P}({}^{\alpha_1}x_1, \dots, {}^{\alpha_l}x_l, {}^{m-|\alpha|}x)$. Then $Q \in \mathcal{P}({}^{m-|\alpha|}X; Y^*)$ and so by our assumption the polynomial Q is compact. By the Polarisation formula $(\check{P})_k = \check{P_k}$ and so $Q_k(tx_k + x) = \binom{m}{\alpha, m-|\alpha|}^{-1} \sum_{p=0}^{m-|\alpha|} t^p R_k^{\alpha,p}(x)$ by Lemma 1.21. Hence $\lim_{j\to\infty} Q_{k_j}(tx_{k_j} + v_j) = \binom{m}{\alpha, m-|\alpha|}^{-1} \sum_{p=0}^{m-|\alpha|} b_p t^p \ne 0$. This however contradicts the compactness of Q by Fact 90.

Now we construct a subsequence $\{x_{m_k}\}$ so that

$$\left| P_{m_k}\left(\sum_{j=1}^{\infty} a_j x_{m_j}\right) - \sum_{q=0}^{m} a_k^q R_{m_k}^q \left(\sum_{j=k+1}^{\infty} a_j x_{m_j}\right)\right| \le \frac{\varepsilon}{3\cdot 2^k} \tag{14}$$

for any $\sum_{j=1}^{\infty} a_j x_{m_j} \in B_X$, where $R_k^q = R_k^{(0,\dots,0),q}$ for short. Let $K \ge 1$ be the basis constant of $\{x_n\}$. We set $m_1 = 1$. If $m_1, \dots, m_k$ are already defined, then we find $m_{k+1} > m_k$ so that $m|\mathcal{J}(k,m)|(2K)^m \|R_{m_{k+1}}^{\alpha,q}\| < \frac{\varepsilon}{3\cdot 2^{k+1}}$ for all $\alpha \in \mathcal{J}(m_k, m)$ such that $\alpha_j = 0$ if $j \notin \{m_1, \dots, m_k\}$ and $|\alpha| > 0$, and all $0 \le q \le m-|\alpha|$. It is easily checked that the sequence $\{m_k\}$ satisfies (14).

In the second step we show that the polynomials $R_{m_k}^q$, $q < m$ are asymptotically zero on a suitable subspace. Using induction we construct a subsequence $\{n_k\}$ of $\{m_k\}$ along with auxiliary nested subsequences $\{m_k\} \supset \{m_k^1\} \supset \{m_k^2\} \supset \cdots$. Set $n_1 = m_1$. By Theorem 26 there are a subsequence $\{m_k^1\}$ of $\{m_k\}$, a Banach space E with a Schauder basis $\{e_n\}$ that is a spreading model of $\{x_{m_k^1}\}$, and sub-symmetric polynomials $S_{n_1}^q \in \mathcal{P}({}^{m-q}E)$, $q = 0, \dots, m-1$, satisfying the following for $r = 1$:

$$\left| S_{n_r}^q\left(\sum_{j=1}^{N} a_j e_j\right) - R_{n_r}^q\left(\sum_{j=1}^{N} a_j x_{m_{l_j}^r}\right)\right| \le \frac{\varepsilon}{3m(2K)^m 2^r} \tag{15}$$

for all $N \le l_1 < l_2 < \cdots < l_N$, all scalars $a_1, \dots, a_N$ with $\sum_{j=1}^{N} a_j x_{m_{l_j}^r} \in B_X$, and all $0 \le q < m$. Now assume that $\{m_k^{r-1}\}$ is already defined for some $r \in \mathbb{N}$, $r > 1$. Then we put $n_r = m_r^{r-1}$ and again by Theorem 26 there are a subsequence $\{m_k^r\}$ of $\{m_k^{r-1}\}$ and sub-symmetric polynomials $S_{n_r}^q \in \mathcal{P}({}^{m-q}E)$, $q = 0, \dots, m-1$, satisfying (15). (Recall that subsequences of a sequence with a spreading model have the same spreading model.) Note that for each $k \in \mathbb{N}$ the sequence $\{n_j\}_{j=k+1}^{\infty}$ is a subsequence of $\{m_j^k\}_{j=1}^{\infty}$. It follows that

$$\left| S_{n_k}^q\left(\sum_{j=1}^{N} a_j e_j\right) - R_{n_k}^q\left(\sum_{j=1}^{N} a_j x_{n_{l_j}}\right)\right| \le \frac{\varepsilon}{3m(2K)^m 2^k}$$

for all $\max\{k+1, N\} \le l_1 < l_2 < \dots < l_N$, all scalars $a_1, \dots, a_N \in \mathbb{R}$ satisfying $\sum_{j=1}^N a_j x_{n_{l_j}} \in B_X$, all $0 \le q < m$, and $k \in \mathbb{N}$. Combining these estimates with (14) and the fact that $P_k(x_k) = \sum_{q=0}^m R_k^q(0) = R_k^m(0)$ we obtain

$$\begin{aligned} |P_{n_k}(x) - f_{n_k}(x)^m P_{n_k}(x_{n_k})| &\le \frac{2}{3}\frac{\varepsilon}{2^k} + \sum_{q=0}^{m-1} (2K)^q \left| S_{n_k}^q \left(\sum_{\substack{j=1 \\ l_j \ge k+1}}^N a_j e_j \right) \right| \\ &\le \frac{2}{3}\frac{\varepsilon}{2^k} + (4K)^m \sum_{q=0}^{m-1} \|S_{n_k}^q\| \end{aligned}$$

whenever $x = \sum_{j=1}^N a_j x_{n_{l_j}} \in B_X$ with $N \le l_1 < \dots < l_N$ (assuming, as we may, that (1) in Proposition 25 holds with $\varepsilon_k \le \frac{1}{2}$).

We claim that $\lim_{k\to\infty} \|S_{n_k}^q\| = 0$ for each $0 \le q < m$. This finishes the proof by passing to a suitable subsequence of $\{n_k\}$ in the estimate above. By contradiction, assume that this does not hold for some $0 \le q < m$. Since $\{R_k^q;\ k \in \mathbb{N}\}$ is a bounded set, it follows that $\{S_{n_k}^q\}$ is a bounded sequence in a finite-dimensional space of sub-symmetric $(m-q)$-homogeneous polynomials on E. Thus $\{S_{n_k}^q\}$ has a non-zero cluster point S. In particular, there is a finitely supported $w = \sum_{j=1}^r w_j e_j \in \frac{1}{2} B_E$ such that $S(w) \ne 0$, which means that $\inf_k |S_{p_k}^q(w)| = 2\delta > 0$ for some subsequence $\{p_k\}$ of $\{n_k\}$. Put $w_l = \sum_{j=1}^r w_j x_{n_{l+j}}$ and note that $w_l \in B_X$ for l large enough. Then $\liminf_{l\to\infty} |R_{p_k}^q(w_l)| \ge \delta$ for each $k \in \mathbb{N}$ (dropping finitely many members of $\{p_k\}$ if necessary).

Now let $W^l(x) = \binom{m}{q} \check{P}({}^q x, {}^{m-q} w_l)$. Then $W^l \in \mathcal{P}({}^q X; Y^*) = \left((\otimes_{\pi,s}^q X) \otimes_\pi Y\right)^*$ (Corollary 3.14). Since $\{w_l\}$ is bounded, so is $\{W^l\}$. Since $(\otimes_{\pi,s}^q X) \otimes_\pi Y$ is separable, there is a subsequence $\{W^{l_j}\}$ such that $W^{l_j} \overset{w^*}{\to} W \in \mathcal{P}({}^q X; Y^*)$, and in particular $w^*\text{-}\lim_{j\to\infty} W^{l_j}(x_k) = W(x_k)$ for each $k \in \mathbb{N}$ (Corollary 3.14). Hence $|W_{p_k}(x_{p_k})| = \lim_{j\to\infty} |W_{p_k}^{l_j}(x_{p_k})| = \lim_{j\to\infty} |R_{p_k}^q(w_{l_j})| \ge \delta$ for each $k \in \mathbb{N}$. But according to our assumption the polynomial W is compact, which contradicts Fact 90. □

Theorem 92. *Let X be a Banach space and let $m \in \mathbb{N}$ be such that there is a non-compact $P \in \mathcal{P}({}^m X; \ell_1)$. Then $\mathcal{P}({}^n X) \not\subset \overline{\mathcal{A}_{n-1}(X)}$ for every $n \ge m$.*

Proof. If X has a subspace isomorphic to ℓ_1, then the result follows from Theorem 89, so for the rest of the proof we assume that X does not contain ℓ_1. We may assume that m is minimal such that there is a non-compact $P \in \mathcal{P}({}^m X; \ell_1)$. Denote by $\{e_k\}$ the canonical basis of ℓ_1 and by P_k the kth component of P, i.e. the composition of P with the kth coordinate functional on ℓ_1. We may assume that there is a basic sequence $\{x_k\} \subset X$ such that $P(x_k) = e_k$ (although in the proof it suffices to have $\limsup |P_k(x_k)| > 0$). Indeed, by Lemma 3.31 there is a weakly null sequence $\{x_k\} \subset S_X$ such that $\{P(x_k);\ k \in \mathbb{N}\}$ is not relatively compact. By passing to a subsequence we may assume that $\{x_k\}$ is a normalised basic sequence such that $\{P(x_k)\}$ is not relatively compact. Then we may use Proposition 3.36.

Denote by $\{f_k\}$ the functionals biorthogonal to $\{x_k\}$. Let $\varepsilon > 0$ and let $\{x_{n_k}\}$ be the subsequence from Lemma 91. Given $n \geq m$ we put

$$Q(x) = \sum_{k=1}^{\infty} f_{n_k}(x)^{n-m} P_{n_k}(x).$$

Then $Q \in \mathcal{P}(^n X)$ (Theorem 1.29) and it satisfies the condition laid out in Theorem 88. Indeed, given $N \in \mathbb{N}$ we put $y_j = x_{n_{N+j}}$, $j = 1, \dots, N$. If $x = \sum_{j=1}^{N} b_j y_j$ is such that $\sum_{j=1}^{N} |b_j| \leq 1$, then $x \in B_X$. We put $a_{N+j} = b_j$, $j = 1, \dots, N$ and $a_k = 0$ for $k \notin [N+1, N+N]$. Then $x = \sum_{k=1}^{\infty} a_k x_{n_k}$ and so

$$\begin{aligned}\left| Q(x) - \sum_{j=1}^{N} b_j^n \right| &= \left| \sum_{k=1}^{\infty} a_k^{n-m} P_{n_k}(x) - \sum_{k=1}^{\infty} a_k^n P_{n_k}(x_{n_k}) \right| \\ &\leq \sum_{k=1}^{\infty} \left| P_{n_k}(x) - a_k^m P_{n_k}(x_{n_k}) \right| \leq \varepsilon. \end{aligned}$$

□

Corollary 93. *Let X be a Banach space admitting a non-compact $T \in \mathcal{L}(X; \ell_p)$, $1 \leq p < \infty$. Put $n = \lceil p \rceil$. Then*

$$\overline{\mathcal{A}_{n-1}(X)} \subsetneq \overline{\mathcal{A}_n(X)} \subsetneq \overline{\mathcal{A}_{n+1}(X)} \subsetneq \cdots. \tag{16}$$

Proof. By Proposition 3.33 we may assume that $T(B_X)$ contains the canonical basis of ℓ_p. It then suffices to compose T with the polynomial $P \in \mathcal{P}(^n\ell_p; \ell_1)$ given by $P((x_j)_{j=1}^{\infty}) = (x_j^n)_{j=1}^{\infty}$ to obtain a non-compact n-homogeneous polynomial from X into ℓ_1. It remains to apply Theorem 92. □

Corollary 94. *Let $X = L_p([0,1])$, $1 \leq p \leq \infty$, or $X = \ell_\infty$, or $X = C(K)$, K a non-scattered compact. Then*

$$\overline{\mathcal{A}_1(X)} \subsetneq \overline{\mathcal{A}_2(X)} \subsetneq \overline{\mathcal{A}_3(X)} \subsetneq \cdots$$

Proof. The spaces $L_1([0,1])$, ℓ_∞, $L_\infty([0,1])$, and $C(K)$, K non-scattered, contain ℓ_1 ([FHHMZ, Proposition 5.4], Theorem 5.125) and so Theorem 89 applies. The spaces $L_p([0,1])$, $1 < p < \infty$ contain a complemented subspace isomorphic to ℓ_2 ([FHHMZ, Theorem 4.53]) and we may use Corollary 93. □

Corollary 95. *Let $1 \leq p < \infty$ and $n = \lceil p \rceil$. Then*

$$\overline{\mathcal{A}_1(\ell_p)} = \cdots = \overline{\mathcal{A}_{n-1}(\ell_p)} \subsetneq \overline{\mathcal{A}_n(\ell_p)} \subsetneq \overline{\mathcal{A}_{n+1}(\ell_p)} \subsetneq \cdots.$$

Proof. By Proposition 87 and Corollary 3.59 we obtain $\overline{\mathcal{A}_{n-1}(\ell_p)} = \overline{\mathcal{A}_1(\ell_p)}$. The rest follows readily from Corollary 93. □

Corollary 96. *Let X be a Banach space such that X^* is of type $p > 1$. Then*

$$\overline{\mathcal{A}_{n-1}(X)} \subsetneq \overline{\mathcal{A}_n(X)} \subsetneq \overline{\mathcal{A}_{n+1}(X)} \subsetneq \cdots,$$

where $n = [q] + 1$ and $\frac{1}{p} + \frac{1}{q} = 1$.

Proof. Let $s \in (q, n)$. By Corollary 43 there are a normalised basic sequence $\{x_k\}_{k=1}^{\infty}$ in X and $T \in \mathcal{L}(X; \ell_s)$ such that $T(x_k) = e_k$, where $\{e_k\}$ is the canonical basis of ℓ_s. Thus T is not compact and an appeal to Corollary 93 finishes the argument. □

Corollary 97 ([DiGo]). *Let X be a Banach space with an unconditional basis that does not contain ℓ_1, and suppose that n is the smallest integer such that there exists $P \in \mathcal{P}(^nX)$ which is not weakly sequentially continuous. Then*

$$\overline{\mathcal{A}_1(X)} = \cdots = \overline{\mathcal{A}_{n-1}(X)} \subsetneqq \overline{\mathcal{A}_n(X)} \subsetneqq \overline{\mathcal{A}_{n+1}(X)} \subsetneqq \cdots$$

Proof. The initial sequence of equalities follows from Proposition 87. On the other hand, by Lemma 3.30 we may assume that there are $a > 0$ and a normalised weakly null block sequence $\{x_j\}_{j=1}^{\infty}$ of the given unconditional basis such that $|P(x_j)| \geq a$, $j \in \mathbb{N}$. By Proposition 2.79 the space $\mathcal{P}(^nX)$ has a subspace isomorphic to ℓ_∞. To finish, we combine Theorems 3.37 and 92. □

We remark that by only formal changes in the argument the result holds for any Banach space with an unconditional FDD (this is the formulation in [DiGo]).

8. Notes and remarks

Section 1. The important concept of finite representability was introduced by Robert Clarke James [Ja1] and has become one of the key notions in modern Banach space theory. Local theory of Banach spaces is studied in a number of monographs, e.g. [Pis4], [Pis5], [DJT], [Tom2], [MS]. Ultraproduct techniques are nicely exposed in [He].

Section 2. The notion of a spreading model was introduced by Antoine Brunel and Louis Sucheston [BrSu1], [BrSu2], and thoroughly studied in [BeaLa] or [Gue]. Basic facts can also be found in [FHHMZ]. The Banach Saks property was first investigated by Stefan Banach and Stanisław Saks [BaSa] where it is shown that L_p-spaces, $1 < p < \infty$ have this property (the super-reflexive case can be deduced from this using the James-Gurarii result on the lower estimates for weakly null sequences in super-reflexive spaces). A reflexive counterexample is provided by the Baernstein space [Bae], [CasaSh]. William Buhmann Johnson [Johns2] introduced the notion of the weak p-Banach-Saks property, see also [CastSá1], [CastSá2]. It follows from [JohnsZip1] that quotients of ℓ_p have the weak p-Banach-Saks property. By [Johns2], for quotients of L_p, $2 < p < \infty$, this condition is equivalent to being a quotient of ℓ_p. These properties were also investigated by Aleksander Pełczyński in [Peł1] under the name of τ_s-convergence, and used to study continuity of multilinear mappings under this topology. His notion of a $\tau_{\frac{1}{p}}$-null sequence is equivalent to the hereditarily p-Banach-Saks sequence. For more information see [Rak], [Bea], [BeaLa].

Versions of Lemma 32 can be found in many places, e.g. [Rak], [GR], [CastSá2], [FaJ]. We reproduced the proof given in the last reference.

Section 3. Versions of the key Lemma 39 were proved independently in [Dev1], [GJ2], [Hál1], and [FaJ]. The important concept of $\mathcal{P}^n$-null sequences appears naturally in the study of polynomial versions of the Schur and Dunford-Pettis properties [Ry2], [CCG], [FaJ], see also [GuJL] for a nice survey of this area. The important Proposition 44 comes from [GJ2] and provides an essential ingredient to the proof of Theorem 64 in the next section.

Section 4. Robert Deville's Theorem 51 is a generalisation of results by Victor Zakharovich Meshkov [Me] in case $n = 2$. This work, and its methods of proofs based on averaging and Kahane's inequality, is based on previous results of Tadeusz Figiel, and of Stanisław Kwapień in [Kw2].

In the works [Gon2], [GJ3], and [GGo] the Rademacher averaging technique from [ALRT] has been systematically used to obtain new interesting results as well as to considerably simplify the proofs (and often generalise the statements) of the results of R. Deville [Dev1], Jaroslav Kurzweil [Kur1], and [HaHá].

Corollary 63 has inspired many similar characterisations within classes of Banach spaces equipped with some invariant conditions on the domain: Let X be a separable rearrangement invariant function space on $[0, 1]$. Then X admits a separating polynomial if and only if up to an equivalent renorming X coincides with $L_{2k}([0, 1])$ for some $k \in \mathbb{N}$ ([GJ3]). Similarly ([GGJ]), let X be a separable rearrangement invariant function space on $[0, +\infty)$. Then X admits a separating polynomial if and only if X is isomorphic to $L_{2n}([0, +\infty)) \cap L_{2m}([0, +\infty))$.

The proof of Theorem 64 from [GGo] follows a similar pattern as R. Deville's argument, but it relies explicitly on the spreading models techniques. Of course, the additional important point obtained by Manuel González and Raquel Gonzalo is that it locates the canonical basis of ℓ_p within any weakly null sequence. The same techniques imply, [GGo], that if X is a Banach lattice with an exact cotype p, then X contains a complemented subspace isomorphic to ℓ_p. If moreover X is ℓ_p saturated, then every normalised weakly null sequence contains a subsequence spanning a complemented subspace isomorphic to ℓ_p.

For examples showing that a Banach space with a separating polynomial X need not have a homogeneous separating polynomial of degree $q(X)$ see the discussion after Fact 45.

It seems to be open if Theorem 51 can be strengthened as follows (for Banach lattices the answer is positive by the general theory):

Problem 98. If X is a Banach space with a separating polynomial, does it have a renorming with modulus of convexity of power type $q(X)$?

In connection with the previous problem we remark that there is a super-reflexive Banach space X such that $q(X) < \inf\{q;\ X$ has an equivalent norm with modulus of convexity of power type $q\}$, [Pis2].

Problem 99. Assume that a Banach space X admits a separating polynomial. Does it admit a separating polynomial that is convex?

Problem 100. Is there a separable universal space for separable spaces with a separating homogeneous polynomial of degree n?

Similarly, we may ask if the analogue of Mordecay Zippin's result for reflexive (or Asplund) spaces holds. Namely:

Problem 101. If X is a separable Banach space with a separating polynomial, is X isomorphic to a subspace of a Banach space with a Schauder basis that admits a separating polynomial?

Section 5. All the proofs in this section are simplifications of the results first shown in an unpublished preprint, whose main results were listed in the communication [HaHá].

Given an equivalent norm $|\cdot|$ on an infinite-dimensional Banach space $(X, \|\cdot\|)$ we can define a distortion of $|\cdot|$ on X by $\lambda(|\cdot|) = \inf_Y \sup\{\frac{|x|}{|y|};\ x, y \in S_{(Y,\|\cdot\|)}\}$, where Y runs over all infinite-dimensional subspaces of X. A norm $|\cdot|$ is called distorted if $\lambda(|\cdot|) > 1$. Solving a longstanding open problem, it was shown in [OS1] that on ℓ_p, $1 < p < \infty$ we can find an equivalent distorted norm. For more information on the distortion property see e.g. [OS2]. The results on polynomials in our section can be combined with this theory to show that polynomials are non-distorted, unless the Banach space is saturated with distorted copies of ℓ_p.

Linear sets of zeros of polynomials under real or complex scalars have been studied in [PZ], [AroRu], [AGZ], [Fern], [Ferr], and [AT]. The main positive result for real Banach spaces is Theorem 78, whose proof relies on the use of the Erdős-Rado infinite combinatorial lemma.

Section 6. The main part of the theory presented in Sections 6 and 7 was originally built in [Há4], but the proof of the key finite dimensional ingredient (Corollary 86) in this paper contains a serious flaw. A different proof was given in [DHJ], which forms the content of the present Section 6. A precursor to this work was the seminal paper [NS], where the authors show that the chain of closures of the algebras $\mathcal{A}_n(\ell_p)$ does not become eventually constant by using a similar, but much easier observation based on the dimensions of the algebras of symmetric polynomials on $\mathbb{R}^n$.

Section 7. The theory presented in this section comes from [D'H]. The natural problem of finding the closure of the space of polynomials in the uniform topology goes back at least to Georgiy Evgenievich Shilov [Shi]. Under the assumption $\mathcal{P}(X) = \mathcal{P}_{\mathrm{wu}}(X)$ it was solved in [AP], [Aro1], and this is the only case when the solution is known (see also [FGL] for related results). There seems to be no hope for a topological approach to the polynomial approximation problem. In fact, the natural candidate, the topology generated by polynomials, introduced and studied in [ACL], [HL], [GaJL], [GL], does not seem to be appropriate.

Sufficient conditions for the polynomial approximability were found by Arkadij Semenovich Nemirovskij and S. M. Semenov, [NS]. They introduce a regular topology on ℓ_p generated by semi-norms $\{\|U(x)\|_\infty;\ U \in \mathcal{L}(\ell_p; \ell_p)\}$. A mapping $f : \ell_p \to Y$ is said to be regular if it is uniformly continuous in the regular topology on every bounded set. Every regular mapping $f\colon \ell_p \to Y$ is shown to be uniformly approximable by polynomials on bounded sets. As a corollary, every symmetric $C^{2,+}$-smooth function on ℓ_2 is uniformly approximable by polynomials. On the other hand, Jon Vanderwerff, [Va1], showed that there exist (non-symmetric, of course) $C^{2,+}$-smooth renormings of ℓ_2 which cannot be uniformly approximated by polynomials.

Problem 102. Give a description of $\overline{\mathcal{P}(X)}$ for a general separable Banach space X.

This problem is open even for $X = \ell_2$.

In view of Corollary 93 we remark the following: If X contains ℓ_1, then X^* contains ℓ_2 (see Theorem 3.88 and the remark after). In particular, X^* contains a weakly null sequence with an upper 2-estimate, although $\mathcal{P}(\ell_1) = \mathcal{P}_{\mathrm{wsc}}(\ell_1)$. On the other hand, if X does not contain ℓ_1, then the existence of upper estimates in X^* immediately leads to a construction of non-wsc polynomials, as was indicated in Section 3. Also, non-wsc polynomials give rise to non-compact operators into ℓ_p from some subspaces of X.

Problem 103. Suppose that X does not contain ℓ_1 and $\mathcal{P}(X) \neq \mathcal{P}_{\mathrm{wsc}}(X)$. Is there a non-compact bounded linear operator from X into ℓ_p for some $1 \le p < \infty$? Equivalently, is there a weakly null sequence in X^* with an upper q-estimate for $\frac{1}{p} + \frac{1}{q} = 1$?

An important remaining open problem is the following.

Problem 104. Let X be a separable Banach space not containing ℓ_1 and let $n \in \mathbb{N}$ be the smallest integer such that $\mathcal{P}(^nX) \neq \mathcal{P}_{\mathrm{wsc}}(^nX)$. Is then

$$\overline{\mathcal{A}_1(X)} = \cdots = \overline{\mathcal{A}_{n-1}(X)} \subsetneq \overline{\mathcal{A}_n(X)} \subsetneq \overline{\mathcal{A}_{n+1}(X)} \subsetneq \cdots ?$$

By combining Theorems 3.37 and 92 the answer to this problem is positive provided the following problem has a positive answer (note that the opposite implication in the statement below is true using Theorem 3.37 and a simple construction).

Problem 105. Let X be a separable Banach space not containing ℓ_1 and let $n \ge 2$ be an integer such that $\mathcal{P}(^nX) \neq \mathcal{P}_{\mathrm{wsc}}(^nX)$. Does then the space $\mathcal{P}(^nX)$ contain c_0?

In particular, if the dual X^* has a subspace isomorphic to c_0, then Theorem 92 implies that (16) holds for $n = 2$. If the dual X^* has a super-reflexive subspace, then we can conclude that (16) holds for some n. Indeed, by the James-Gurarii theorem [FHHMZ, Theorem 9.25], X admits a non-compact linear operator into some ℓ_p. This leaves us with two possibilities. If X fails (16) for every $n \in \mathbb{N}$, then either X^* is ℓ_1-saturated, or it contains a Tsirelson-like subspace Y, i.e. Y does not contain ℓ_1, c_0, or a super-reflexive space.

Problem 106. Suppose that X is a Banach space such that X^* is ℓ_1-saturated. Is then $\mathcal{P}(X) = \mathcal{P}_{\mathrm{wsc}}(X)$?

Chapter 5

Smoothness and structure

In this chapter we are going to study the impact of higher smoothness on the structure and geometry of real Banach spaces. It turns out that even very mild smoothness assumptions, such as the existence of a separating twice Gâteaux differentiable function, have serious structural consequences for the space in question.

We begin Section 1 by recalling some of the equivalent conditions for a Banach space to be an Asplund space. In particular, a separable Banach space is an Asplund space if and only if it has a C^1-smooth renorming. We proceed by studying some basic properties of Legendre-Fenchel duality related to smoothness.

Unfortunately, since all of these results are deeply intertwined, we have split the section on impact of smoothness on the structure into two parts and inserted a section on variational principles in between. In Section 2 we pass to the basic study of Banach spaces which admit a locally uniformly smooth bump function. We prove a version of the compact variational principle, which in the absence of c_0 subspaces allows one to pass to bumps with uniformly continuous derivatives. This is shown to imply one of the fundamental results of our theory: if a Banach space X admits a bump with a locally uniformly continuous derivative, then either X contains a subspace isomorphic to c_0 or X is super-reflexive.

Section 3 begins with the celebrated Ekeland variational principle. This is followed by the first order Deville-Godefroy-Zizler variational principle and Stegall's variational principle for spaces with the RNP, which will be our principal tool. Generally speaking, variational principles claim the existence of a point where a given lower semi-continuous function attains its (strong) minimum. These principles have striking consequences for Banach spaces with uniformly rotund norms and admitting bumps with higher derivatives. For example, if a space with a norm with modulus of convexity of power type q has a T^q-smooth bump, then it admits a separating polynomial.

In Section 4 we show that the modulus of convexity of power type q in Theorem 62 can be relaxed to cotype q, provided we upgrade the smoothness assumption on the bump to be UT_c^q-smooth. We prove the fundamental result of Robert Deville concerning the structure of the spaces with C^∞-smooth bumps. We show that X is an Asplund space provided that it admits a bump with some rather weak differentiability properties. Similarly, under very weak convexity and differentiability assumptions we can already construct a norm with modulus of smoothness of power type 2 (and hence the space is super-reflexive). Finally, improving on the fundamental work of Victor Zakharovich Meshkov we show that X is isomorphic to a Hilbert space provided that both X and X^* admit a bump with certain smoothness properties.

Sections 5–6 are devoted to separable polyhedral Banach spaces. These spaces are characterised by admitting an equivalent renorming which locally depends on finitely many coordinates (LFC). The space c_0 is the main polyhedral space, and every polyhedral space is an Asplund space and contains c_0. These spaces admit many characterisations, e.g. they can be renormed by means of C^∞-smooth and simultaneously LFC norms.

In the remaining, rather computational, Sections 7–9 we find the best order of smoothness for the classical Banach spaces L_p, $C(K)$, and the Orlicz spaces, respectively.

In this chapter every vector space is real.

1. Convex functions

This section deals with various topics concerning the smoothness of convex functions, including norms. We first review (without proofs) some fundamental results about Asplund, weak Asplund, and weak* Asplund spaces, which will we used throughout, and which are covered in detail elsewhere, notably in [DGZ].

Further, we define the Moreau-Rockafellar subdifferential, and a directional/pointwise modulus of smoothness for a general mapping. We show how these notions relate to the differentiability of convex functions, and how the power type of the moduli relate to the T^p-smoothness. Using a Baire category argument we show that for convex functions on Banach spaces a directional pointwise modulus of smoothness of power type implies a pointwise modulus of smoothness of power type.

Next, we prove some basic facts about convex functions and norms with uniformly continuous derivatives (also of higher order). We also show that the Minkowski functionals of sub-level sets of various smooth convex functions share the same smoothness properties and thus give smooth norms.

We proceed by introducing the powerful concept of Legendre-Fenchel duality and show some elementary properties of this notion. We define a (pointwise) modulus of exposedness, show some of its basic properties and prove that it is dual to the pointwise modulus of smoothness. Finally, we prove that a Lipschitz convex function which on a large set has a modulus of exposedness of power type gives rise to an equivalent norm with modulus of convexity of some power type. This allows us to prove via duality various smoothness results in later sections.

We end this section by showing that subdifferentials of weakly sequentially continuous convex functions exhibit some compactness, provided that the domain space does not contain ℓ_1.

The following concept plays an important role both in the theory of differentiability and in an investigation of the structure of Banach spaces.

Definition 1. A Banach space X is called an Asplund space if for every continuous convex function f defined on an open convex subset $U \subset X$ the set of points at which f is Fréchet differentiable contains a dense G_δ subset of U. A Banach space X

is called a weak Asplund space if for every continuous convex function f defined on an open convex subset $U \subset X$ the set of points at which f is Gâteaux differentiable contains a dense G_δ subset of U.

Let us remark that the set of points of Fréchet differentiability of a continuous convex function defined on an open convex set is always a G_δ set (Fact 13), while this is no longer true for the set of points of Gâteaux differentiability – examples can be found in [Tal1] or [HŠZ].

The norm of a normed linear space X is said to be C^k-smooth if it is a C^k-smooth function on $X \setminus \{0\}$.

Theorem 2. *Let X be a Banach space. The following statements are equivalent:*

(i) *X is an Asplund space.*
(ii) *Every continuous convex function on X is Fréchet differentiable on a dense G_δ subset of X.*
(iii) *X^* has the Radon-Nikodým property (RNP).*
(iv) *Every equivalent norm on X is Fréchet differentiable at some point.*
(v) *Every separable subspace of X is an Asplund space.*
(vi) *Every separable subspace of X has separable dual.*
(vii) *Every separable subspace of X admits an equivalent C^1-smooth norm.*
(viii) *Every separable subspace of X admits a Fréchet differentiable bump.*

For the proof see [DGZ] and [FHHMZ, Theorem 11.8, Theorem 11.15].

From the equivalence of (i) and (viii) we immediately obtain the following corollary.

Corollary 3. *If a Banach space admits a Fréchet differentiable bump, then it is an Asplund space.*

We will show much stronger result later on in Theorem 72. We note that it is not known whether the converse holds.

Theorem 4 (Stanisław Mazur). *Let X be a separable Banach space and let $f : X \to \mathbb{R}$ be a continuous convex function. Then the set of points of Gâteaux differentiability of f is a dense G_δ subset of X. In particular, X is a weak Asplund space.*

For the proof see [FHHMZ, Theorem 8.2].

Fact 5. *Let X be a Banach space, $U \subset X$ an open convex set, and $f : U \to \mathbb{R}$ a lower semi-continuous convex function. Then f is continuous.*

Proof. Let $F_n = \{x \in U;\ f(x) \le n\}$, $n \in \mathbb{N}$. Then each F_n is relatively closed in U and $U = \bigcup_{n\in\mathbb{N}} F_n$. Since U is a Baire space, there is $n \in \mathbb{N}$ such that F_n has a non-empty interior. Hence f is bounded above on some open set and so f is continuous on U. □

Fact 6. *Let X be a normed linear space, $U \subset X^*$ open, and let $f : U \to \mathbb{R}$ be w^*-lower semi-continuous. If f is Fréchet differentiable at $x \in U$, then $Df(x) \in X$.*

Proof. Denote $F = Df(x) \in X^{**}$. The functions $f_n(h) = n\left(f(x + \frac{1}{n}h) - f(x)\right)$, $h \in B_{X^*}$ are w^*-lower semi-continuous for n sufficiently large and $f_n \to F$ uniformly on B_{X^*}. Hence F is w^*-lower semi-continuous on B_{X^*}. As F is linear, it is w^*-continuous on B_{X^*}. Therefore F is w^*-continuous on X^* by the Banach-Dieudonné theorem, which implies that $F \in X$. □

Definition 7. A dual Banach space X^* is called a weak* Asplund space if for every w^*-lower semi-continuous convex function f defined on an open convex subset $U \subset X^*$ the set of points at which f is Fréchet differentiable contains a dense G_δ subset of U.

Note that by Fact 5 each w^*-lower semi-continuous convex function f defined on an open convex set is continuous and hence each dual Asplund space is a weak* Asplund space.

Theorem 8 ([Col]). *Let X be a Banach space. Then X^* is a weak* Asplund space if and only if X has the Radon-Nikodým property.*

For the proof see [BourgR, Theorem 5.7.4].

Let A be a set. For $f: A \to \mathbb{R} \cup \{+\infty\}$ we denote by $\operatorname{dom} f$ the effective domain of f, i.e. the set $f^{-1}(\mathbb{R})$. The function f is called proper if $\operatorname{dom} f$ is non-empty. The (closed) epigraph of $f: A \to \overline{\mathbb{R}}$ is defined by $\operatorname{epi} f = \{(x,t) \in A \times \mathbb{R};\ t \geq f(x)\}$.

Definition 9. Let X be a normed linear space, $\varphi: X \to \mathbb{R} \cup \{+\infty\}$, and let $x \in \operatorname{dom} \varphi$. We define the (Moreau-Rockafellar) subdifferential of φ at x by

$$\partial\varphi(x) = \{f \in X^*;\ f(h) + \varphi(x) \leq \varphi(x+h) \text{ for all } h \in X\}.$$

Note that the subdifferential $\partial\varphi(x)$ is always a w^*-closed convex subset of X^*. If $\varphi: X \to \mathbb{R} \cup \{+\infty\}$ is a convex function continuous on $\operatorname{Int}\operatorname{dom}\varphi$ and $x \in \operatorname{Int}\operatorname{dom}\varphi$, then the subdifferential $\partial\varphi(x)$ is non-empty and bounded. This follows from the Hahn-Banach theorem used on the zero functional on $\{0\}$ and the dominating convex function $h \mapsto \varphi(x+h) - \varphi(x)$. The boundedness is a consequence of the fact that φ is locally Lipschitz.

Note also that every convex function $\varphi: C \to \mathbb{R}$ defined on a convex subset C of a vector space can be extended to a convex function defined on the whole space by setting $\varphi(x) = +\infty$ for $x \notin C$.

Definition 10. Let X, Y be normed linear spaces, $U \subset X$ open, and $f: U \to Y$. For $x \in U$, $h \in X$ we define a directional pointwise modulus of smoothness of f at x in the direction h by

$$\rho^f_{x,h}(t) = \sup_{0 \leq s \leq t} \|f(x+sh) + f(x-sh) - 2f(x)\| \quad \text{for } t \geq 0 \text{ such that } x \pm th \in U.$$

Further, we define a pointwise modulus of smoothness of f at x by

$$\rho^f_x(t) = \sup_{h \in S_X} \rho^f_{x,h}(t) \quad \text{for } t \geq 0 \text{ such that } B(x,t) \subset U.$$

Let X be a normed linear space and $f: X \to \mathbb{R} \cup \{+\infty\}$. Then for $x \in \operatorname{dom} f$ the moduli $\rho^f_{x,h}$ and ρ^f_x are well-defined on $[0, +\infty)$ by the formulae above. If f is moreover convex, then it is easy to check that $\rho^f_{x,h}(t) = f(x+th)+f(x-th)-2f(x)$ and $\rho^f_x(t) = \sup_{h\in S_X}\big(f(x+th)+f(x-th)-2f(x)\big)$.

Lemma 11. *Let $f: U \to \mathbb{R}$ be a continuous convex function defined on an open convex subset U of a normed linear space X and $x \in U$. Then the following statements are equivalent:*

(i) *The function f is Gâteaux differentiable at x.*
(ii) *The directional derivative $\frac{\partial f}{\partial h}(x)$ exists for every $h \in X$.*
(iii) *The subdifferential $\partial f(x)$ is a singleton.*
(iv) *$\rho^f_{x,h}(t) = o(t)$, $t \to 0+$ for every $h \in S_X$.*
(v) *$\displaystyle\lim_{t\to 0}\frac{f(x+th)+f(x-th)-2f(x)}{t} = 0$ for every $h \in X$.*

If f is Gâteaux differentiable at x, then $\partial f(x) = \{\delta f(x)\}$.

Proof. (i)⇒(v) is obvious and (iv) is just a reformulation of (v).

(v)⇒(iii) We already know that the subdifferential $\partial f(x)$ is non-empty. Suppose that there are functionals $g_1, g_2 \in \partial f(x)$ and $h \in X$ such that $g_1(h) > g_2(h)$. Then $f(x+th)+f(x-th)-2f(x) \ge g_1(th)+g_2(-th) = t(g_1(h)-g_2(h))$ for every sufficiently small $t > 0$, which contradicts (v).

(iii)⇒(ii) Extend f to the whole of X with the value $+\infty$ outside U. Fix $h \in X\setminus\{0\}$. From the convexity of f it follows that for all $t \in \mathbb{R}\setminus\{0\}$

$$t \mapsto \frac{f(x+th)-f(x)}{t} \quad \text{is non-decreasing.} \tag{1}$$

Thus $\frac{\partial f}{\partial h^+}(x) = \lim\limits_{t\to 0+}\frac{1}{t}\big(f(x+th)-f(x)\big)$ and $\frac{\partial f}{\partial h^-}(x) = \lim\limits_{t\to 0-}\frac{1}{t}\big(f(x+th)-f(x)\big)$ always exist (finite) and $\frac{\partial f}{\partial h^+}(x) \ge \frac{\partial f}{\partial h^-}(x)$. Next, we set $g_1(th) = t\frac{\partial f}{\partial h^+}(x)$ and $g_2(th) = t\frac{\partial f}{\partial h^-}(x)$. Obviously $g_j(th) \le f(x+th)-f(x)$ for all $t \in \mathbb{R}$, $j = 1, 2$. By the Hahn-Banach theorem there is an extension of g_j to the whole of X such that $g_j \in X^*$ and $g_j(u) \le f(x+u)-f(x)$ for all $u \in X$, $j = 1, 2$. By (iii) it follows that $g_1 = g_2$ and hence $\frac{\partial f}{\partial h^+}(x) = \frac{\partial f}{\partial h^-}(x)$, which means that $\frac{\partial f}{\partial h}(x)$ exists.

(ii)⇒(i) Using the convexity of f we obtain for every $h, k \in X$ and sufficiently small $t > 0$

$$\frac{f\big(x+t(h+k)\big)-f(x)}{t} \le \frac{f(x+2th)-f(x)}{2t} + \frac{f(x+2tk)-f(x)}{2t},$$

from which it follows that the function $h \mapsto \frac{\partial f}{\partial h^+}(x)$ is sub-additive. Similarly we obtain that the function $h \mapsto \frac{\partial f}{\partial h^-}(x)$ is super additive. Hence $h \mapsto \frac{\partial f}{\partial h}(x)$ is linear (the homogeneity is obvious). Further, from (1) and the continuity of f we obtain that $h \mapsto \frac{\partial f}{\partial h^+}(x)$ is upper semi-continuous, while $h \mapsto \frac{\partial f}{\partial h^-}(x)$ is lower semi-continuous. Hence $h \mapsto \frac{\partial f}{\partial h}(x)$ is continuous.

The last statement follows from (iii) and (1). □

Note that from (1) it follows that if f is convex, then both $t \mapsto \frac{1}{t}\rho^f_{x,h}(t)$ and $t \mapsto \frac{1}{t}\rho^f_x(t)$ are non-decreasing.

Lemma 12. *Let $f: U \to \mathbb{R}$ be a continuous convex function defined on an open convex subset U of a normed linear space X and $x \in U$. Then the following statements are equivalent:*

(i) *The function f is Fréchet differentiable at x.*

(ii) $\rho^f_x(t) = o(t)$, $t \to 0+$.

(iii) $\lim_{h\to 0} \frac{f(x+h)+f(x-h)-2f(x)}{\|h\|} = 0$ *for every $h \in X$.*

Proof. (i)⇒(iii)⇒(ii) is obvious.

(ii)⇒(i) By Lemma 11 we know that f is Gâteaux differentiable at x. So for small $h \in X$ we have

$$\begin{aligned} 0 \le f(x+h) - f(x) - \delta f(x)[h] &= f(x+h) - f(x) + \delta f(x)[-h] \\ &\le f(x+h) + f(x-h) - 2f(x) \le \rho^f_x(\|h\|) = o(\|h\|),\ h \to 0. \end{aligned}$$

□

We note that if f is not convex, then $\rho^f_x(t) = o(t)$, $t \to 0+$ does not imply the smoothness of f at x: for the function $f(x,y) = \frac{x^2y}{x^2+y^2}$, $f(0,0) = 0$ we have $\rho^f_{(0,0)}(t) = 0$, but f is not even Gâteaux differentiable at the origin.

Fact 13. *Let $f: U \to \mathbb{R}$ be a continuous convex function defined on an open convex subset U of a normed linear space X. Then the set of points of Fréchet differentiability of f is a G_δ set.*

Proof. By Lemma 12 the set of points of Fréchet differentiability of f is equal to $\bigcap_{n\in\mathbb{N}} G_n$, where

$$G_n = \bigcup_{\delta>0} \left\{x \in U;\ \tfrac{1}{t}\rho^f_x(t) < \tfrac{1}{n} \text{ for } t \in (0,\delta]\right\}.$$

To see that each G_n is open fix $x \in G_n$ and let $\delta > 0$ be such that f is Lipschitz on $B(x, 2\delta)$ and $\frac{1}{\delta}\rho^f_x(\delta) < \frac{1}{n}$. The family of functions $\{y \mapsto \rho^f_{y,h}(\delta)\}_{h\in S_X}$ is equi-Lipschitz on $B(x,\delta)$, and hence $y \mapsto \rho^f_y(\delta)$ is Lipschitz on $B(x,\delta)$. Thus there is a neighbourhood V of x such that $\frac{1}{\delta}\rho^f_y(\delta) < \frac{1}{n}$ for all $y \in V$. Since for f convex $t \mapsto \frac{1}{t}\rho^f_x(t)$ is non-decreasing, it follows that $\frac{1}{t}\rho^f_y(t) < \frac{1}{n}$ for all $t \in (0,\delta]$ and $y \in V$, which shows that $V \subset G_n$. □

Definition 14. Let X, Y be normed linear spaces, $U \subset X$ an open set, and $f: U \to Y$. We say that $\rho^f_{x,h}$ (or ρ^f_x) is of power type $p \in (1, +\infty)$, if $\rho^f_{x,h}(t) = O(t^p)$, $t \to 0+$ (or $\rho^f_x(t) = O(t^p)$, $t \to 0+$).

Clearly, if the modulus $\rho^f_{x,h}$ (or ρ^f_x) is of power type p, then it is also of power type q for every $q \in (1,p)$. It is obvious that $\rho^f_{x,h}$ is of power type p if and only if $\|f(x+th) + f(x-th) - 2f(x)\| = O(t^p)$, $t \to 0+$. Similarly, ρ^f_x is of power type p if and only if $\|f(x+h)+f(x-h)-2f(x)\| = O(\|h\|^p)$, $h \to 0$. Also, $\rho^f_{x,ch}(t) = \rho^f_{x,h}(ct)$ for $c \geq 0$ and small t.

Observe that if f is T^p-smooth at x for $p \in (1,2]$, then ρ^f_x is of power type p. Similarly, if f is directionally T^p-smooth at x for $p \in (1,2]$, then $\rho^f_{x,h}$ is of power type p for each $h \in S_X$.

Fact 15. *Let X be a normed linear space, $U \subset X$ an open convex set, $f: U \to \mathbb{R}$ a convex continuous function, $x \in U$, and $p \in (1,2]$. If $\rho^f_{x,h}$ is of power type p for each $h \in S_X$, then f is weakly T^q-smooth at x for every $q \in (1,p)$. If ρ^f_x is of power type p, then f is T^q-smooth at x for every $q \in (1,p)$.*

Proof. In both cases the function f is Gâteaux differentiable at x by Lemma 11. By the convexity $\delta f(x)[h] \leq f(x+h) - f(x)$ for every $h \in X$ satisfying $x+h \in U$. The assertion now follows from the inequalities $0 \leq f(x+h) - f(x) - \delta f(x)[h] = f(x+h) - f(x) + \delta f(x)[-h] \leq f(x+h) + f(x-h) - 2f(x)$. □

Proposition 16. *Let X be a Banach space, $U \subset X$ an open convex set, $f: U \to \mathbb{R}$ a convex continuous function, and $x \in U$. If for each $h \in S_X$ the directional pointwise modulus $\rho^f_{x,h}$ is of some power type p_h, then the pointwise modulus ρ^f_x is of some power type. Moreover, if $p = \inf\{p_h;\ h \in S_X\} > 1$, then ρ^f_x is of power type p.*

Proof. Let $\delta > 0$ be such that $B(x,\delta) \subset U$ and f is bounded on $B(x,\delta)$. We put

$$F_n = \big\{h \in B_X;\ f(x+th) + f(x-th) - 2f(x) \leq nt^{1+\frac{1}{n}} \text{ for } 0 \leq t \leq \delta\big\}.$$

From the assumption on the modulus $\rho^f_{x,h}$ and from the boundedness of f on the ball $B(x,\delta)$ it follows that $B_X = \bigcup_{n\in\mathbb{N}} F_n$. Each of the sets F_n is closed because $F_n = \bigcap_{t\in[0,\delta]}\{h \in B_X;\ f(x+th) + f(x-th) - 2f(x) \leq nt^{1+\frac{1}{n}}\}$ and f is continuous. Further, each F_n is absolutely convex. Indeed, let $u, v \in F_n$ and $\lambda \in (0,1)$. Then

$$\begin{aligned}
&f\big(x + t(\lambda u + (1-\lambda)v)\big) + f\big(x - t(\lambda u + (1-\lambda)v)\big) - 2f(x)\\
&\leq \lambda f(x+tu) + (1-\lambda)f(x+tv) + \lambda f(x-tu) + (1-\lambda)f(x-tv) - 2f(x)\\
&= \lambda\big(f(x+tu)+f(x-tu)-2f(x)\big) + (1-\lambda)\big(f(x+tv)+f(x-tv)-2f(x)\big)\\
&\leq nt^{1+\frac{1}{n}}
\end{aligned}$$

for $0 \leq t \leq \delta$. Obviously, if $h \in F_n$, then also $-h \in F_n$.

Hence by the Baire category theorem there is $n \in \mathbb{N}$ such that F_n has a non-empty interior. It follows from the absolute convexity that there is $\eta > 0$ such that $B(0,\eta) \subset F_n$. Therefore $0 \leq f(x+h) + f(x-h) - 2f(x) \leq n\big(\frac{\|h\|}{\eta}\big)^{1+\frac{1}{n}}$ for every $h \in B(0,\delta\eta)$, which means that ρ^f_x is of power type $1+\frac{1}{n}$.

Finally, notice that if $p > 1$, then instead of the sets F_n as above we may consider the sets

$$F_n = \big\{h \in B_X;\ f(x+th) + f(x-th) - 2f(x) \leq nt^p \text{ for } 0 \leq t \leq \delta\big\}. \qquad \square$$

Combining Proposition 16 with Fact 15 we obtain the following corollary.

Corollary 17. *Let X be a Banach space, $U \subset X$ an open convex set, $f : U \to \mathbb{R}$ a convex continuous function, $x \in U$, and $p \in (1, 2]$. If f is directionally T^p-smooth at x, then f is T^q-smooth at x for every $q \in (1, p)$.*

Corollary 18. *Let X be a Banach space that admits a separating continuous directionally T^p-smooth convex function f for some $p > 1$. Then X is an Asplund space.*

Proof. By Corollary 17 f is Fréchet differentiable. Hence by composing f with a suitable smooth real function we obtain a Fréchet differentiable bump (Theorem 1.69). Thus X is an Asplund space by Corollary 3. □

Next, we turn our attention to convex functions with uniformly continuous derivatives.

Lemma 19. *Let X be a normed linear space, $U \subset X$ an open convex set, and let $\Omega \subset \mathcal{M}_s$ be the cone generated by $\omega \in \mathcal{M}_s$.*

(i) *If Y is a Banach space and $f \in C^{1,\Omega}(U; Y)$, then there is $C > 0$ such that*

$$\|f(x+h) + f(x-h) - 2f(x)\| \le C\omega(\|h\|)\|h\|$$

for every $x, h \in X$ satisfying $x \pm h \in U$.

(ii) *Suppose that $f : U \to \mathbb{R}$ is a continuous convex function. If there is $C > 0$ such that*

$$f(x+h) + f(x-h) - 2f(x) \le C\omega(\|h\|)\|h\|$$

for every $x, h \in X$ satisfying $x \pm h \in U$, then $f \in C^{1,\Omega}(V)$ for each non-empty bounded open $V \subset U$ satisfying $\operatorname{dist}(V, X \setminus U) > 0$.

Proof. (i) Suppose that $f \in C^1(U; Y)$ and Df is uniformly continuous on X with modulus $K\omega$ for some $K > 0$. By Theorem 1.89 for any $x, h \in X$ satisfying $x \pm h \in U$ we get

$$\begin{aligned}\|f(x+h) + f(x-h) - 2f(x)\| &= \left\| \int_0^1 Df(x+th)[h]\,\mathrm{d}t + \int_0^1 Df(x-th)[-h]\,\mathrm{d}t \right\| \\ &\le \int_0^1 \|Df(x+th) - Df(x-th)\|\,\|h\|\,\mathrm{d}t \\ &\le K\omega(2\|h\|)\|h\| \le 2K\omega(\|h\|)\|h\|.\end{aligned}$$

(ii) First, notice that f is Fréchet differentiable on U by Lemma 12. Let $V \subset U$ be a non-empty bounded open set for which $\operatorname{dist}(V, X \setminus U) > 0$. Put $d = \operatorname{diam} V$ and $\varepsilon = \min\{\frac{d}{2}, \operatorname{dist}(V, X \setminus U)\}$. Fix any $x, y \in V$. For an arbitrary $h \in X$ satisfying $\|h\| = \frac{\varepsilon}{d}\|x-y\|$ (note that $x+h \in U$, $y-h \in U$) we use the fact that $Df(x) \in \partial f(x)$

and $Df(y) \in \partial f(y)$ to obtain

$$\begin{aligned}
Df(x)[h] - Df(y)[h] &= Df(x)[h] + Df(y)[-h] \\
&\le f(x+h) - f(x) + f(y-h) - f(y) \\
&= f(x+h) + f(y-h) - 2f\Big(\frac{x+y}{2}\Big) - \Big(f(x) + f(y) - 2f\Big(\frac{x+y}{2}\Big)\Big) \\
&\le f(x+h) + f(y-h) - 2f\Big(\frac{x+y}{2}\Big) \\
&= f\Big(\frac{x+y}{2} + \frac{x-y}{2} + h\Big) + f\Big(\frac{x+y}{2} - \frac{x-y}{2} - h\Big) - 2f\Big(\frac{x+y}{2}\Big) \\
&\le C\omega\Big(\Big\|\frac{x-y}{2} + h\Big\|\Big)\Big\|\frac{x-y}{2} + h\Big\| \\
&\le \Big(\frac{1}{2} + \frac{\varepsilon}{d}\Big)C\omega\Big(\Big(\frac{1}{2} + \frac{\varepsilon}{d}\Big)\|x-y\|\Big)\|x-y\| \\
&\le C\omega(\|x-y\|)\|x-y\|.
\end{aligned}$$

Taking supremum over $h \in X$, $\|h\| = \frac{\varepsilon}{d}\|x-y\|$, we get

$$\frac{\varepsilon}{d}\|x-y\|\|Df(x) - Df(y)\| \le C\omega(\|x-y\|)\|x-y\|,$$

which shows that Df is uniformly continuous on V with modulus $\frac{d}{\varepsilon}C\omega$. □

We note that for $\omega(t) = t$ (the Lipschitz case) it can be shown that for continuous convex functions the condition $f(x+h) + f(x-h) - 2f(x) \le C\|h\|^2$ for $x, h \in X$, $x \pm h \in U$, is equivalent to $f \in C^{1,1}(U)$. (We restrict $\|x-y\|$ to be small and use the fact that mappings on normed linear spaces that are Lipschitz for small distances are globally Lipschitz with the same constant.)

Further, for convex functions that are defined on the whole space X we can take $\|h\| = \frac{1}{2}\|x-y\|$ in the previous proof, which gives the following version.

Lemma 20. *Let $f\colon X \to \mathbb{R}$ be a continuous convex function on a normed linear space X and let $\Omega \subset \mathcal{M}_s$ be the cone generated by $\omega \in \mathcal{M}_s$. Then $f \in C^{1,\Omega}(X)$ if and only if there is $C > 0$ such that for every $x, h \in X$*

$$f(x+h) + f(x-h) - 2f(x) \le C\omega(\|h\|)\|h\|.$$

Fact 21. *Let (X, ν) be a normed linear space such that the norm ν is C^k-smooth. Then $d^j\nu(cx) = \frac{1}{c^{j-1}}d^j\nu(x)$ for every $x \in X \setminus \{0\}$, $c > 0$, and every $j \in \{1, \dots, k\}$.*

Proof. Just differentiate the equality $\nu(cx) = c\nu(x)$. □

Let (X, ν) be a normed linear space. The norm ν is said to be $C^{k,\Omega}$-smooth if it is C^k-smooth on $X \setminus \{0\}$ and $d^k\nu$ is uniformly continuous on S_X with modulus ω for some $\omega \in \Omega$.

Fact 22. *Let (X,ν) be a normed linear space and let $\Omega\subset\mathcal{M}_s$ be a convex cone. If the norm ν is $C^{1,\Omega}$-smooth, then for each $\delta>0$ there is $\omega\in\Omega$ such that $D\nu$ is uniformly continuous on $X\setminus B(0,\delta)$ with modulus ω. If the norm ν is $C^{k,\Omega}$-smooth, $k>1$, then for each $0<\delta<1$ there is $\omega\in\Omega$ such that $d^k\nu$ is uniformly continuous on $B(0,\frac{1}{\delta})\setminus B(0,\delta)$ with modulus ω.*

Proof. For convenience we will also denote $\|x\|=\nu(x)$. Suppose $k>1$. Let $\psi\in\Omega$ be such that $d^k\nu$ is uniformly continuous on S_X with modulus ψ. Fix $x_0\in S_X$ and denote $C=\|d^k\nu(x_0)\|+\psi(2)$. Then $\|d^k\nu(x)\|\le\|d^k\nu(x_0)\|+\|d^k\nu(x)-d^k\nu(x_0)\|\le\|d^k\nu(x_0)\|+\psi(\|x-x_0\|)\le C$ for any $x\in S_X$. Using Fact 21 we can estimate for $x,y\in B(0,\frac{1}{\delta})\setminus B(0,\delta)$

$$\begin{aligned}
\|d^k\nu(x)-d^k\nu(y)\| &= \left\|\frac{1}{\|x\|^{k-1}}d^k\nu\Big(\frac{x}{\|x\|}\Big)-\frac{1}{\|y\|^{k-1}}d^k\nu\Big(\frac{y}{\|y\|}\Big)\right\|\\
&\le\frac{1}{\|x\|^{k-1}}\left\|d^k\nu\Big(\frac{x}{\|x\|}\Big)-d^k\nu\Big(\frac{y}{\|y\|}\Big)\right\|+\left\|d^k\nu\Big(\frac{y}{\|y\|}\Big)\right\|\cdot\left|\frac{1}{\|x\|^{k-1}}-\frac{1}{\|y\|^{k-1}}\right|\\
&\le\frac{1}{\delta^{k-1}}\psi\Big(\Big\|\frac{x}{\|x\|}-\frac{y}{\|y\|}\Big\|\Big)+\frac{C}{\|x\|^{k-1}\|y\|^{k-1}}\big|\|x\|^{k-1}-\|y\|^{k-1}\big|\\
&\le\frac{1}{\delta^{k-1}}\psi\Big(\frac{2}{\delta}\|x-y\|\Big)+\frac{C}{\delta^{2k-2}}\big(\|x\|^{k-2}+\|x\|^{k-3}\|y\|+\dots+\|y\|^{k-2}\big)\|x-y\|\\
&\le\frac{1}{\delta^{k-1}}\left\lceil\frac{2}{\delta}\right\rceil\psi(\|x-y\|)+\frac{C(k-1)}{\delta^{3k-4}}\|x-y\|.
\end{aligned}$$

The assertion now follows from Fact 1.120. The proof for $k=1$ is analogous. □

Lemma 23. *Let X be a normed linear space, $C\subset X$ an open convex set, and let $f\colon C\to\mathbb{R}$ be a continuous convex function. Suppose that there are $x_0\in C$ and $a\in\mathbb{R}$, $a>f(x_0)$ such that the set $\{x\in C;\ f(x)\le a\}$ is a closed bounded subset of X. (This in particular holds if f is separating.) Assume further that there is an open $\Omega\subset C$ such that $f\in C^k(\Omega)$, where $k\in\mathbb{N}\cup\{\infty,\omega\}$, and that $S=\{x\in C;\ f(x)=a\}\subset\Omega$. Then X admits an equivalent C^k-smooth norm β.*

If we suppose further that d^jf is bounded and uniformly continuous on S with modulus $\omega\in\mathcal{M}_s$ for each $j\in\{1,\dots,k\}$, then there is $K>0$ such that $d^j\beta$ is bounded and uniformly continuous on S with modulus $K\omega$ for each $j\in\{1,\dots,k\}$.

Proof. Assume without loss of generality that $x_0=0$. Put $B=\{x\in C;\ f(x)\le a\}$. Since by the assumption $f(0)<a$, the set B is a closed convex bounded neighbourhood of 0 and hence the Minkowski functional of B, denoted by ν, is a continuous function and $\partial B=\{x\in X;\ \nu(x)=1\}$. On the other hand, using the facts that B is closed, $f(0)<a$, and f is convex, we obtain $\partial B=S$. It follows that for every $x\in X\setminus\{0\}$ and $y\in\mathbb{R}^+$ we have $f\big(\frac{x}{y}\big)=a$ if and only if $y=\nu(x)$.

Let $H\colon X\times\mathbb{R}^+\to X$ be given by $H(x,y)=\frac{x}{y}$. The set $G=H^{-1}(\Omega)$ is open by the continuity of H. Define $F\colon G\to\mathbb{R}$ by $F(x,y)=f\big(\frac{x}{y}\big)$. Then $F\in C^k(G)$ (Corollary 1.84 and Theorem 1.83, or Proposition 1.173 and Theorem 1.172). Note that by the convexity of f and the fact that $f(0)<a$ there is $\delta>0$ such that $Df(u)[u]\ge\delta$

for every $u \in S$. It follows that $D_2F(x, \nu(x)) = Df(\frac{x}{\nu(x)})[-\frac{x}{\nu(x)^2}] \neq 0$ for every $x \in X \setminus \{0\}$. Therefore the Implicit function theorem (Theorem 1.87, Theorem 1.174) implies that $\nu \in C^k(X \setminus \{0\})$. Moreover,

$$\begin{aligned} D\nu(x) &= -\left(Df\left(\frac{x}{\nu(x)}\right)\left[-\frac{x}{\nu(x)^2}\right]\right)^{-1} \cdot \left(\frac{1}{\nu(x)}Df\left(\frac{x}{\nu(x)}\right)\right) \\ &= \frac{1}{Df\left(\frac{x}{\nu(x)}\right)\left[\frac{x}{\nu(x)}\right]} Df\left(\frac{x}{\nu(x)}\right). \end{aligned} \tag{2}$$

We show the last statement of the theorem (with β replaced by ν) by induction on k. Suppose first that $k = 1$. Since $D\nu(x) = \frac{1}{Df(x)[x]} Df(x)$ for $x \in S$, it is easily checked that $D\nu$ is bounded and uniformly continuous with modulus $K\omega$ on S for some $K > 0$ (recall that $Df(u)[u] \geq \delta$ for $u \in S$ and use also Fact 1.120).

Now suppose that the statement holds for $k - 1$. Let $\varepsilon\colon X^* \times X \to \mathbb{R}$ be the evaluation operator, i.e. $\varepsilon(g, x) = g(x)$. Then $\varepsilon \in \mathcal{L}(X^*, X; \mathbb{R})$. By Proposition 1.79 and Proposition 1.71 we can see that $d^j\varepsilon$ is bounded and Lipschitz on bounded sets for all $j \in \mathbb{N}$. Notice that (2) can be rewritten as $D\nu = p \circ q$, where $p = \frac{1}{\varepsilon\circ(Df, Id)} \cdot Df$ and $q(x) = \frac{x}{\nu(x)}$. By the inductive hypothesis and Proposition 1.128 together with Proposition 1.129 there is $L > 0$ such that both d^jp and d^jq are bounded and uniformly continuous with modulus $L\omega$ on S for all $j \in \{1, \dots, k-1\}$. So, using Proposition 1.128 once again, we conclude that there is $K > 0$ such that $d^j(D\nu)$ is bounded and uniformly continuous with modulus $K\omega$ on S for all $j \in \{1, \dots, k-1\}$, which together with the case $k = 1$ gives that $d^j\nu$ is bounded and uniformly continuous with modulus $K\omega$ on S for all $j \in \{1, \dots, k\}$.

Finally, $\beta(x) = \nu(x) + \nu(-x)$ is the desired equivalent norm. □

We note that a function which is not separating but satisfies the weaker assumptions of the previous lemma is used e.g. in the proof of Theorem 104.

The following concept is very important in convex analysis.

Definition 24. Let X be a normed linear space and $\varphi\colon X \to \overline{\mathbb{R}}$. We define the (Legendre-Fenchel) conjugate function $\varphi^*\colon X^* \to \overline{\mathbb{R}}$ by

$$\varphi^*(f) = \sup\{f(x) - \varphi(x);\ x \in X\}.$$

Further, for $\psi\colon X^* \to \overline{\mathbb{R}}$ we define $\psi_* = \psi^* \restriction_X$.

Note that for a given $f \in X^*$ we have

$$\begin{aligned} -\varphi^*(f) &= \inf\{\varphi(x) - f(x);\ x \in X\} \\ &= \sup\{c \in \mathbb{R};\ f(x) + c \leq \varphi(x) \text{ for all } x \in X\} \end{aligned}$$

and if $\varphi^*(f) \in \mathbb{R}$, then the last supremum is attained. Further, φ^* is always a convex w^*-lower semi-continuous function.

Fact 25. *Let X be a normed linear space and $\varphi\colon X \to \overline{\mathbb{R}}$. Then $(\varphi^*)_* \le \varphi$ and*

$$(\varphi^*)_*(x) = \sup\big\{f(x) + c;\ f \in X^*, c \in \mathbb{R}, f(y) + c \le \varphi(y) \textit{ for all } y \in X\big\}. \quad (3)$$

In particular, if φ is convex and lower semi-continuous, then $(\varphi^*)_* = \varphi$.

Dually, if $\psi\colon X^* \to \overline{\mathbb{R}}$, then $(\psi_*)^* \le \psi$ and

$$(\psi_*)^*(f) = \sup\big\{f(x) + c;\ x \in X, c \in \mathbb{R}, g(x) + c \le \psi(g) \textit{ for all } g \in X^*\big\}.$$

If the function ψ is convex and w^*-lower semi-continuous, then $(\psi_*)^* = \psi$. In particular, $((\varphi^*)_*)^* = \varphi^*$ for any $\varphi\colon X \to \overline{\mathbb{R}}$.

Proof. For any $x \in X$ we have

$$\begin{aligned}\varphi(x) &\ge \sup\big\{f(x) + c;\ f \in X^*, c \in \mathbb{R}, f(y) + c \le \varphi(y) \text{ for all } y \in X\big\}\\ &= \sup\big\{f(x) + c;\ f \in X^*, c \in \mathbb{R}, c \le \inf\{\varphi(y) - f(y);\ y \in X\}\big\}\\ &= \sup\big\{f(x) + c;\ f \in X^*, c \in \mathbb{R}, c \le -\varphi^*(f)\big\} = (\varphi^*)_*(x).\end{aligned}$$

Now suppose that φ is convex and lower semi-continuous and there is $x \in X$ such that $\varphi(x) > (\varphi^*)_*(x)$. Let $\alpha \in \mathbb{R}$ be such that $\varphi(x) > \alpha > (\varphi^*)_*(x)$. By the separation theorem used on the closed convex set $\operatorname{epi}\varphi$ and a point (x, α) there are $f \in X^*$ and $c \in \mathbb{R}$ such that $f(x) + c \ge \alpha$ and $f(y) + c < \varphi(y)$ for every $y \in X$. (Some extra care is needed if $x \notin \operatorname{dom}\varphi$.) This contradicts the formula (3).

The dual statement is proved in the same way. We only use the separation theorem in the locally convex space $(X^*, w^*) \times \mathbb{R}$. □

Lemma 26. *Let X be a normed linear space, $\varphi\colon X \to \mathbb{R} \cup \{+\infty\}$, $x \in \operatorname{dom}\varphi$, and $f \in X^*$. The following statements are equivalent:*

(i) *$\varphi(x) + \varphi^*(f) = f(x)$.*
(ii) *$f \in \partial\varphi(x)$.*
(iii) *$x \in \partial\varphi^*(f)$ and $(\varphi^*)_*(x) = \varphi(x)$.*

Proof. (i)⇒(ii) For any $y \in X$ we get $\varphi(y) - \varphi(x) = \varphi(y) + \varphi^*(f) - f(x) \ge \varphi(y) + f(y) - \varphi(y) - f(x) = f(y - x)$ and so $f \in \partial\varphi(x)$.

(ii)⇒(i) We have $f(y) - \varphi(y) \le f(x) - \varphi(x)$ for every $y \in X$. Thus we obtain $\varphi^*(f) = \sup\{f(y) - \varphi(y);\ y \in X\} = f(x) - \varphi(x)$.

(i)⇒(iii) For any $g \in X^*$ we get $\varphi^*(g) - \varphi^*(f) = \varphi^*(g) + \varphi(x) - f(x) \ge g(x) - \varphi(x) + \varphi(x) - f(x) = x(g - f)$ and therefore $x \in \partial\varphi^*(f)$. Moreover, $\varphi(x) \ge (\varphi^*)_*(x) \ge f(x) - \varphi^*(f) = \varphi(x)$.

(iii)⇒(i) From (ii)⇒(i) used on φ^* we obtain $\varphi^*(f) + (\varphi^*)_*(x) = f(x)$ and so (i) follows. □

From the definition we immediately obtain that if $\varphi_1 \le \varphi_2$ on X, then $\varphi_1^* \ge \varphi_2^*$ on X^*. The next lemma shows that in some circumstances this dual behaviour holds even locally.

Lemma 27. *Let X be a normed linear space, $\varphi_1, \varphi_2 \colon X \to \mathbb{R} \cup \{+\infty\}$ where φ_2 is convex, $x \in \operatorname{dom} \varphi_2$, and $f \in \partial\varphi_2(x)$. If there is $r > 0$ such that $\varphi_1 \le \varphi_2$ on $B(x,r)$ and*

$$\eta = \inf\bigl\{\varphi_2(y) - f(y) - \bigl(\varphi_2(x) - f(x)\bigr);\ y \in X, \|y - x\| = r\bigr\} > 0,$$

then $\varphi_1^* \ge \varphi_2^*$ on $B(f, \frac{\eta}{r})$.

Dually, if φ_2 is moreover lower semi-continuous and there is $r > 0$ such that $\varphi_1^* \le \varphi_2^*$ on $B(f,r)$ and

$$\eta = \inf\bigl\{\varphi_2^*(g) - g(x) - \bigl(\varphi_2^*(f) - f(x)\bigr);\ g \in X^*, \|g - f\| = r\bigr\} > 0,$$

then $\varphi_1 \ge \varphi_2$ on $B(x, \frac{\eta}{r})$.

Proof. Fix $g \in B(f, \frac{\eta}{r})$. Let $z \in X$ be such that $\|z - x\| > r$. Set $y = x + r\frac{z-x}{\|z-x\|}$. By the convexity $\varphi_2(y) - f(y) \le \frac{r}{\|z-x\|}\bigl(\varphi_2(z) - f(z)\bigr) + \bigl(1 - \frac{r}{\|z-x\|}\bigr)\bigl(\varphi_2(x) - f(x)\bigr)$. Therefore

$$\begin{aligned}\varphi_2(z) - g(z) - \bigl(\varphi_2(x) - g(x)\bigr)&\\ &\ge \varphi_2(z) - f(z) - \bigl(\varphi_2(x) - f(x)\bigr) - \|g - f\|\|z - x\|\\ &\ge \frac{\|z - x\|}{r}\Bigl(\varphi_2(y) - f(y) - \bigl(\varphi_2(x) - f(x)\bigr)\Bigr) - \frac{\eta}{r}\|z - x\| \ge 0,\end{aligned}$$

which implies that $\sup\{g(z) - \varphi_2(z);\ z \in X\} = \sup\{g(z) - \varphi_2(z);\ z \in B(x,r)\}$. Thus $\varphi_1^*(g) \ge \sup\{g(z) - \varphi_1(z);\ z \in B(x,r)\} \ge \sup\{g(z) - \varphi_2(z);\ z \in B(x,r)\} = \varphi_2^*(g)$.

To prove the dual statement, note that $x \in \partial\varphi_2^*(f)$ by Lemma 26. Thus by the first statement used on φ_1^* and φ_2^* we obtain $(\varphi_1^*)_* \ge (\varphi_2^*)_*$ on $B(x, \frac{\eta}{r})$. Hence $\varphi_1 \ge (\varphi_1^*)_* \ge (\varphi_2^*)_* = \varphi_2$ on $B(x, \frac{\eta}{r})$ by Fact 25. □

The following notion turns out to be a dual notion to the pointwise modulus of smoothness.

Definition 28. Let X be a normed linear space, $\varphi \colon X \to \mathbb{R} \cup \{+\infty\}$, $x \in \operatorname{dom} \varphi$, and $f \in \partial\varphi(x)$. We define a modulus of exposedness of φ at x by f by

$$\Delta^{\varphi}_{x,f}(\varepsilon) = \inf\bigl\{\varphi(y) - f(y) + \varphi(z) - f(z) - 2\bigl(\varphi(x) - f(x)\bigr);\ y, z \in X, \|y - z\| \ge \varepsilon\bigr\}$$

for $\varepsilon \ge 0$.

Note that $\Delta^{\varphi}_{x,f}$ is a non-decreasing non-negative function, $\Delta^{\varphi}_{x,f}(0) = 0$, and $\Delta^{\varphi}_{x,f}(\varepsilon) = +\infty$ for $\varepsilon > \operatorname{diam} \operatorname{dom} \varphi$. Further, $\varphi(x + h) \ge \varphi(x) + f(h) + \Delta^{\varphi}_{x,f}(\|h\|)$ for all $h \in X$. This in turn implies that if $\varphi(x + h) < \varphi(x) + f(h) + \Delta^{\varphi}_{x,f}(\varepsilon)$, then $\|h\| < \varepsilon$. On the other hand the following fact holds.

Fact 29. *Let X be a normed linear space, $\varphi \colon X \to \mathbb{R} \cup \{+\infty\}$, $x \in \operatorname{dom} \varphi$, and let $f \in \partial\varphi(x)$. If there is a non-decreasing function $\theta \colon [0, +\infty) \to \mathbb{R}$ such that $\varphi(x + h) \ge \varphi(x) + f(h) + \theta(\|h\|)$ for all $h \in X$, then $\Delta^{\varphi}_{x,f}(\varepsilon) \ge \theta(\frac{\varepsilon}{2})$.*

Proof. It suffices to notice that if $y, z \in X$ are such that $\|y - z\| \ge \varepsilon$, then either $\|x - y\| \ge \frac{\varepsilon}{2}$ or $\|x - z\| \ge \frac{\varepsilon}{2}$. □

The next fact justifies the name of the modulus.

Fact 30. *Let X be a normed linear space, $\varphi\colon X \to \mathbb{R} \cup \{+\infty\}$, $x \in \operatorname{dom}\varphi$, and $f \in \partial\varphi(x)$. Then $(x, \varphi(x))$ is a strongly exposed point of the set* epi φ *exposed by $(f, -1) \in X^* \oplus \mathbb{R}$ if and only if $\Delta^{\varphi}_{x,f}(\varepsilon) > 0$ for all $\varepsilon > 0$.*

Proof. Let $F = (f, -1) \in (X \oplus \mathbb{R})^*$, i.e. $F(y,t) = f(y) - t$. As $f \in \partial\varphi(x)$, we have $F(y,t) \le f(y) - \varphi(y) \le f(x) - \varphi(x) = F(x, \varphi(x))$ for any $(y,t) \in \operatorname{epi}\varphi$.

$\Leftarrow$ Suppose that $F(x_n, t_n) \to F(x, \varphi(x))$ for $x_n \in \operatorname{dom}\varphi$, $t_n \ge \varphi(x_n)$. Then $\Delta^{\varphi}_{x,f}(\|x_n - x\|) \le \varphi(x_n) - f(x_n) - \big(\varphi(x) - f(x)\big) \le t_n - f(x_n) - \big(\varphi(x) - f(x)\big) = F(x, \varphi(x)) - F(x_n, t_n) \to 0$. Since $\Delta^{\varphi}_{x,f}$ is non-decreasing, this implies that $x_n \to x$, which together with $F(x_n, t_n) \to F(x, \varphi(x))$ gives $(x_n, t_n) \to (x, \varphi(x))$.

$\Rightarrow$ By contrary, let $\varepsilon > 0$ be such that $\Delta^{\varphi}_{x,f}(\varepsilon) = 0$. Then for each $n \in \mathbb{N}$ there are $y_n, z_n \in X$ satisfying $0 \le \varphi(y_n) - f(y_n) + \varphi(z_n) - f(z_n) - 2\big(\varphi(x) - f(x)\big) \le \frac{1}{n}$ and $\|y_n - z_n\| \ge \varepsilon$. It follows that $0 \le \varphi(y_n) - f(y_n) - \big(\varphi(x) - f(x)\big) \le \frac{1}{n}$ and also $0 \le \varphi(z_n) - f(z_n) - \big(\varphi(x) - f(x)\big) \le \frac{1}{n}$, which can be rewritten as $0 \le F(x, \varphi(x)) - F(y_n, \varphi(y_n)) \le \frac{1}{n}$ and $0 \le F(x, \varphi(x)) - F(z_n, \varphi(z_n)) \le \frac{1}{n}$. The strong exposedness now gives $y_n \to x$ and $z_n \to x$, a contradiction. □

Lemma 31. *Let X be a normed linear space, $\varphi\colon X \to \mathbb{R} \cup \{+\infty\}$, $x \in \operatorname{dom}(\varphi^*)_*$, and let φ be lower semi-continuous at x. Suppose that there is $f \in \partial(\varphi^*)_*(x)$ such that $\Delta^{(\varphi^*)_*}_{x,f}(\varepsilon) > 0$ for every $\varepsilon > 0$. Then $x \in \operatorname{dom}\varphi$ and $(\varphi^*)_*(x) = \varphi(x)$.*

Proof. Assume that the statement does not hold. Let $\delta \in \mathbb{R}$, $\delta > 0$ be such that $(\varphi^*)_*(x) + \delta \le \varphi(x)$. From the lower semi-continuity of φ at x there is $\varepsilon > 0$ such that $\varphi(y) > (\varphi^*)_*(x) + \frac{\delta}{2}$ for each $y \in U(x, \varepsilon)$. Without loss of generality we may assume that $\|f\|\varepsilon \le \frac{\delta}{4}$. Put $\eta = \min\{\Delta^{(\varphi^*)_*}_{x,f}(\varepsilon), \frac{\delta}{4}\}$ and note that $\eta > 0$. Now for $y \in U(x, \varepsilon)$ we have $f(y) + (\varphi^*)_*(x) - f(x) + \eta < \|f\|\varepsilon + (\varphi^*)_*(x) + \eta \le (\varphi^*)_*(x) + \frac{\delta}{2} < \varphi(y)$, while for $y \in X$ with $\|y - x\| \ge \varepsilon$ we have $f(y) + (\varphi^*)_*(x) - f(x) + \eta \le f(y - x) + (\varphi^*)_*(x) + \Delta^{(\varphi^*)_*}_{x,f}(\|y - x\|) \le (\varphi^*)_*(y) \le \varphi(y)$. Thus $f(y) + c \le \varphi(y)$ for all $y \in X$, where $c > (\varphi^*)_*(x) - f(x)$, a contradiction with Fact 25. □

For a convex function φ the infimum in the definition of the modulus $\Delta^{\varphi}_{x,f}$ can be taken only over $\|y - z\| = \varepsilon$:

Lemma 32. *Let X be a normed linear space, $\varphi\colon X \to \mathbb{R} \cup \{+\infty\}$ a convex function, $x \in \operatorname{dom}\varphi$, and $f \in \partial\varphi(x)$. Then for $\varepsilon \ge 0$ we have*

$$\Delta^{\varphi}_{x,f}(\varepsilon) = \inf\big\{\varphi(y) - f(y) + \varphi(z) - f(z) - 2\big(\varphi(x) - f(x)\big);\ y, z \in X, \|y - z\| = \varepsilon\big\}.$$

Proof. Set $\psi(u) = \varphi(u) - f(u) - \big(\varphi(x) - f(x)\big)$ for $u \in X$ and note that ψ is convex. Also note that $\Delta^{\varphi}_{x,f}(\varepsilon) = \inf\{\psi(y) + \psi(z);\ y, z \in X, \|y - z\| \ge \varepsilon\}$. Let $\varepsilon > 0$ and pick any $y, z \in X$ satisfying $\|y - z\| \ge \varepsilon$. Without loss of generality we may assume that $\psi(y) \le \psi(z)$. Put $v = y + \varepsilon\frac{z - y}{\|z - y\|}$. Then $\|y - v\| = \varepsilon$ and $\psi(v) \le \frac{\varepsilon}{\|z - y\|}\psi(z) + \big(1 - \frac{\varepsilon}{\|z - y\|}\big)\psi(y) \le \psi(z)$. Thus $\psi(y) + \psi(v) \le \psi(y) + \psi(z)$, from which the statement of the lemma follows. □

For formal reasons we extend the moduli ρ_x^φ and $\Delta_{x,f}^\varphi$ to the whole $\mathbb{R}$ to be even functions, i.e. $\rho_x^\varphi(t) = \rho_x^\varphi(-t)$ for $t < 0$, $\Delta_{x,f}^\varphi(\varepsilon) = \Delta_{x,f}^\varphi(-\varepsilon)$ for $\varepsilon < 0$. Now we come to our main duality tool.

Proposition 33. *Let X be a normed linear space, $\varphi\colon X \to \mathbb{R} \cup \{+\infty\}$ a convex function, $x \in \operatorname{dom}\varphi$, and $f \in \partial\varphi(x)$. Then $\rho_f^{\varphi^*} = (\Delta_{x,f}^\varphi)^*$. If φ is moreover lower semi-continuous, then $\rho_x^\varphi = (\Delta_{f,x}^{\varphi^*})^*$.*

Proof. By Lemma 26, $f \in \operatorname{dom}\varphi^*$ and so $\rho_f^{\varphi^*}$ is well-defined. For any $t \geq 0$ we have

$$\begin{aligned}
\rho_f^{\varphi^*}(t) &= \sup\{\varphi^*(f+tg) + \varphi^*(f-tg) - 2\varphi^*(f);\ g \in S_{X^*}\}\\
&= \sup\{f(y) + tg(y) - \varphi(y) + f(z) - tg(z) - \varphi(z) - 2\varphi^*(f);\ g \in S_{X^*}, y, z \in X\}\\
&= \sup\{t\|y-z\| + f(y) - \varphi(y) + f(z) - \varphi(z) - 2\varphi^*(f);\ y, z \in X\}\\
&= \sup\{t\varepsilon + f(y) - \varphi(y) + f(z) - \varphi(z) - 2\varphi^*(f);\ y, z \in X, \|y-z\| = \varepsilon, \varepsilon \geq 0\}\\
&= \sup\{t\varepsilon + \sup\{f(y) - \varphi(y) + f(z) - \varphi(z) - 2(f(x) - \varphi(x));\\
&\qquad\qquad y, z \in X, \|y-z\| = \varepsilon\};\ \varepsilon \geq 0\}\\
&= \sup\{t\varepsilon - \Delta_{x,f}^\varphi(\varepsilon);\ \varepsilon \geq 0\} = \sup\{t\varepsilon - \Delta_{x,f}^\varphi(\varepsilon);\ \varepsilon \in \mathbb{R}\} = (\Delta_{x,f}^\varphi)^*(t),
\end{aligned}$$

where we used Lemma 26 and Lemma 32. For $t < 0$ the formula follows from the obvious fact that if $\psi\colon \mathbb{R} \to \mathbb{R} \cup \{+\infty\}$ is even, then so is ψ^*.

Further, by Lemma 26, $x \in \partial\varphi^*(f)$ and so $\Delta_{f,x}^{\varphi^*}$ is well-defined. If φ is moreover lower semi-continuous, then similarly as above and using moreover Fact 25 we obtain

$$\begin{aligned}
\rho_x^\varphi(t) &= \sup\{(\varphi^*)_*(x+th) + (\varphi^*)_*(x-th) - 2\varphi(x);\ h \in S_X\}\\
&= \sup\{u(x) + tu(h) - \varphi^*(u) + v(x) - tv(h) - \varphi^*(v) - 2\varphi(x);\ h \in S_X, u, v \in X^*\}\\
&= \sup\{t\|u-v\| + u(x) - \varphi^*(u) + v(x) - \varphi^*(v) - 2\varphi(x);\ u, v \in X^*\}\\
&= \sup\{t\varepsilon + u(x) - \varphi^*(u) + v(x) - \varphi^*(v) - 2\varphi(x);\ u, v \in X^*, \|u-v\| = \varepsilon, \varepsilon \geq 0\}\\
&= \sup\{t\varepsilon - \Delta_{f,x}^{\varphi^*}(\varepsilon);\ \varepsilon \geq 0\} = \sup\{t\varepsilon - \Delta_{f,x}^{\varphi^*}(\varepsilon);\ \varepsilon \in \mathbb{R}\} = (\Delta_{f,x}^{\varphi^*})^*(t).
\end{aligned}$$

□

Corollary 34. *Let X be a normed linear space and let $\varphi\colon X \to \mathbb{R} \cup \{+\infty\}$ be lower semi-continuous. Suppose that φ^* is Fréchet differentiable at $f \in X^*$ and put $x = D\varphi^*(f)$. Then $x \in \operatorname{dom}\varphi$, $(\varphi^*)_*(x) = \varphi(x)$, and $f \in \partial(\varphi^*)_*(x) \cap \partial\varphi(x)$.*

Proof. Fact 6 implies $x \in X$. Denote $\psi = (\varphi^*)_*$ and note that $\psi^* = \varphi^*$ by Fact 25. By Lemma 26 we have $x \in \operatorname{dom}\psi$ and $f \in \partial\psi(x)$, and so $\Delta_{x,f}^\psi$ is well-defined. Suppose that there is $\varepsilon_0 > 0$ such that $\Delta_{x,f}^\psi(\varepsilon_0) = 0$. Proposition 33 implies that $\rho_f^{\varphi^*}(t) = \rho_f^{\psi^*}(t) \geq \varepsilon_0 t$ for all $t > 0$, which contradicts Lemma 12. Thus $\Delta_{x,f}^\psi(\varepsilon) > 0$ for every $\varepsilon > 0$, which in turn implies that $x \in \operatorname{dom}\varphi$ and $(\varphi^*)_*(x) = \varphi(x)$ by Lemma 31. Since $f(h) + \varphi(x) = f(h) + \psi(x) \leq \psi(x+h) \leq \varphi(x+h)$ for every $h \in X$, we get $f \in \partial\varphi(x)$. □

Definition 35. Let X be a normed linear space, $\varphi\colon X \to \mathbb{R} \cup \{+\infty\}$, $x \in \operatorname{dom}\varphi$, and $f \in \partial\varphi(x)$. We say that $\Delta^{\varphi}_{x,f}$ is of power type $q \in (1, +\infty)$ if

$$\liminf_{\varepsilon \to 0+} \frac{\Delta^{\varphi}_{x,f}(\varepsilon)}{\varepsilon^q} > 0.$$

Note that since $\Delta^{\varphi}_{x,f}$ is non-decreasing, it is of power type q if and only if for each $r > 0$ there is $K > 0$ such that $\Delta^{\varphi}_{x,f}(\varepsilon) \geq K\varepsilon^q$ for all $\varepsilon \in [0, r]$.

The following fact is proved by an elementary analysis.

Fact 36. *Let $p \in (1, +\infty)$ and $\varphi(t) = \frac{1}{p}|t|^p$, $t \in \mathbb{R}$. Then $\varphi^*(u) = \frac{1}{q}|u|^q$, $u \in \mathbb{R}$, where $\frac{1}{p} + \frac{1}{q} = 1$.*

Corollary 37. *Let X be a normed linear space, $\varphi\colon X \to \mathbb{R} \cup \{+\infty\}$ a convex function, $x \in \operatorname{dom}\varphi$, and $f \in \partial\varphi(x)$. If $\rho^{\varphi^*}_f$ is of power type p, then $\Delta^{\varphi}_{x,f}$ is of power type q, where $\frac{1}{p} + \frac{1}{q} = 1$. Similarly, if φ is moreover lower semi-continuous and if ρ^{φ}_x is of power type p, then $\Delta^{\varphi^*}_{f,x}$ is of power type q.*

Proof. Combine Proposition 33, Lemma 27, and Fact 36 together with the formula $(C\psi)^*(u) = C\psi^*(\frac{u}{C})$ for any $C > 0$. □

Lemma 38. *Let X be a normed linear space such that its norm has modulus of convexity of power type q. Set $\varphi(x) = \|x\|^2$. Then for each $x \in X$ there is a constant $C > 0$ such that for every $f \in \partial\varphi(x)$ we have $\Delta^{\varphi}_{x,f}(\varepsilon) \geq C \min\{\varepsilon^q, \varepsilon^2\}$ for all $\varepsilon \geq 0$.*

Proof. Denote by δ the modulus of convexity of the norm of X and let $K > 0$ be such that $\delta(\varepsilon) \geq K\varepsilon^q$ for $\varepsilon \in [0, 2]$. Note that $q \geq 2$. By Fact 29 it suffices to show that for each $x \in X$ there is a constant $C > 0$ such that for every $f \in \partial\varphi(x)$ we have $\varphi(x+h) - \varphi(x) - f(h) \geq C\min\{\|h\|^q, \|h\|^2\}$. For $x = 0$ we have $\partial\varphi(0) = \{0\}$ and the inequality obviously holds with $C = 1$. Now fix $x \in X$, $x \neq 0$. Put $L = \frac{K}{(9\|x\|)^{q-2}}$ and $C = \min\{\frac{1}{4}, \frac{1}{2^{q-1}}, \frac{L}{2^{q-1}}\}$. Suppose that $f \in \partial\varphi(x)$. It is not difficult to check that $\|f\| = 2\|x\|$ and $f(x) = 2\|x\|^2$. Put $g = \frac{1}{2\|x\|}f$. Then $g \in S_{X^*}$ and $g(x) = \|x\|$. Hence $|g(tx)| = \|tx\|$ for all $t \in \mathbb{R}$.

Choose any $h \in X$. If $\|h\| \geq 8\|x\|$, then

$$\|x+h\|^2 - \|x\|^2 - f(h) \geq \big(\|h\| - \|x\|\big)^2 - \|x\|^2 - \|f\|\|h\| = \|h\|^2 - 4\|x\|\|h\| \geq \frac{1}{2}\|h\|^2.$$

So now suppose that $\|h\| < 8\|x\|$. Put $h_1 = \frac{g(h)}{\|x\|}x$ and $h_2 = h - h_1$. Note that h_1 is a multiple of x and $h_2 \in \ker g$. We will use the fact that on $\operatorname{span}\{h_1\}$ the square of the norm is just a quadratic function, and on $\ker g$, which gives a supporting hyperplane to B_X at $\frac{x}{\|x\|}$, we can use the modulus of convexity. So, we have

$$\|x+h\|^2 - \|x\|^2 - f(h) = \|x+h_1+h_2\|^2 - \|x+h_1\|^2 + \|x+h_1\|^2 - \|x\|^2 - f(h_1). \quad (4)$$

First we deal with the increment in the direction h_1:

$$\|x+h_1\|^2 - \|x\|^2 - f(h_1) = \|x\|^2\Big(1+\frac{g(h)}{\|x\|}\Big)^2 - \|x\|^2 - \frac{g(h)}{\|x\|}f(x) \qquad (5)$$
$$= g(h_1)^2 = \|h_1\|^2.$$

Next, if $x+h_1=0$, then from (4), (5), and $\|h_1\|^2+\|h_2\|^2 \geq \frac{1}{2}\|h\|^2$ the required inequality immediately follows. Otherwise we put $y_1 = \frac{x+h}{\|x+h\|}$ and $y_2 = \frac{x+h_1}{\|x+h\|}$. Notice that $\|x+h_1\| = |g(x+h_1)| = |g(x+h)| \leq \|x+h\|$, and so $\|y_1\|=1$ and $\|y_2\| \leq 1$. Therefore $1-\|\frac{y_1+y_2}{2}\| \geq \delta(\|y_1-y_2\|) = \delta\big(\frac{\|h_2\|}{\|x+h\|}\big)$. On the other hand,

$$\Big\|\frac{y_1+y_2}{2}\Big\| \geq \Big|g\Big(\frac{y_1+y_2}{2}\Big)\Big| = \frac{|g(x+h_1)|}{\|x+h\|} = \frac{\|x+h_1\|}{\|x+h\|}.$$

Putting the last two estimates together with the fact that δ is of power type q yields

$$K\frac{\|h_2\|^q}{\|x+h\|^q} \leq \delta\Big(\frac{\|h_2\|}{\|x+h\|}\Big) \leq 1-\frac{\|x+h_1\|}{\|x+h\|} \leq 1-\frac{\|x+h_1\|^2}{\|x+h\|^2},$$

which gives

$$K\|h_2\|^q \leq \|x+h\|^{q-2}\big(\|x+h\|^2-\|x+h_1\|^2\big)$$
$$\leq (9\|x\|)^{q-2}\big(\|x+h_1+h_2\|^2-\|x+h_1\|^2\big).$$

If we now get back to (4) and use the last inequality together with (5), we obtain

$$\|x+h\|^2-\|x\|^2-f(h) \geq L\|h_2\|^q+\|h_1\|^2$$
$$\geq \begin{cases} L\|h_2\|^q+\|h_1\|^q \geq \min\{L,1\}\frac{1}{2^{q-1}}\|h\|^q & \text{if } \|h_1\|\leq 1,\\ L\|h_2\|^2+\|h_1\|^2 \geq \min\{L,1\}\frac{1}{2}\|h\|^2 & \text{if } \|h_2\|\geq 1,\\ \|h_1\|^2 > \frac{1}{4}\|h\|^2 & \text{if } \|h_1\|>1 \text{ and } \|h_2\|<1.\end{cases}$$

To see the last inequality, note that in this case $\|h\| \leq \|h_1\|+\|h_2\| < \|h_1\|+1 < 2\|h_1\|$. This finishes the proof. □

Theorem 39. *Let X be a Banach space which admits a Lipschitz convex function $\varphi\colon U\to\mathbb{R}$ defined on a non-empty open convex set $U\subset X$, with the following property: There is a set Ω residual in U such that for each $x\in\Omega$ there is $f\in\partial\varphi(x)$ for which $\Delta^{\varphi}_{x,f}$ is of some power type q_x. Then X admits an equivalent norm with modulus of convexity of some power type. If $q=\sup_{x\in\Omega}q_x<+\infty$, then X admits an equivalent norm with modulus of convexity of power type q.*

Proof. Without loss of generality we may assume that φ is 1-Lipschitz on U. For a convex function $\psi\colon X\to\mathbb{R}\cup\{+\infty\}$ which is 1-Lipschitz on $\operatorname{Int}\operatorname{dom}\psi$ and for $x\in\operatorname{Int}\operatorname{dom}\psi$ we define

$$\delta^{\psi}_x(\varepsilon) = \inf\big\{\psi(y)+\psi(z)-2\psi(x)+\|2x-(y+z)\|;\ y,z\in X, \|y-z\|\geq\varepsilon\big\},\ \varepsilon\geq 0.$$

Note that δ^{ψ}_x is non-decreasing and for any $f\in\partial\psi(x)$ we have $\Delta^{\psi}_{x,f}\leq\delta^{\psi}_x$.

For each $n \in \mathbb{N}$ set

$$F_n = \{x \in U;\ \delta_x^{\varphi}(\varepsilon) \geq \tfrac{1}{n}\varepsilon^n \text{ for all } \varepsilon \in [0,1]\}.$$

Each F_n is closed in U. Indeed, let $\{x_k\} \subset F_n$, $x_k \to x \in U$. Then for any $\varepsilon \in [0,1]$ and $y, z \in X$, $\|y-z\| \geq \varepsilon$ we have $\varphi(y)+\varphi(z)-2\varphi(x_k)+\|2x_k-(y+z)\| \geq \frac{1}{n}\varepsilon^n$ and hence also $\varphi(y)+\varphi(z)-2\varphi(x)+\|2x-(y+z)\| \geq \frac{1}{n}\varepsilon^n$.

By our assumption $\Omega \subset \bigcup_{n\geq 2} F_n$ and since U is a Baire space, it follows that $\bigcup_{n\geq 2} F_n$ is of the second category in U. Therefore there are $n \in \mathbb{N}$, $n \geq 2$, $x \in U$, and $0 < r < \frac{1}{2}$ such that $B(x,r) \subset F_n$. Define $\psi(h) = \frac{1}{2}\varphi(x+h)+\frac{1}{2}\varphi(x-h)-\varphi(x)$ for all $h \in X$. Then ψ is an even convex function, 1-Lipschitz on $\operatorname{dom}\psi$, with $\psi(0)=0$ and $0 \in \partial\psi(0)$.

For every $u \in B(0,r)$ and $\varepsilon \in [0,1]$ we have

$$\begin{aligned}\delta_u^{\psi}(\varepsilon) &= \inf\{\psi(y)+\psi(z)-2\psi(u)+\|2u-(y+z)\|;\ y,z \in X, \|y-z\| \geq \varepsilon\}\\ &\geq \inf\{\tfrac{1}{2}\varphi(x+y)+\tfrac{1}{2}\varphi(x+z)-\varphi(x+u)+\tfrac{1}{2}\|2u-(y+z)\|;\\ &\qquad\qquad y,z \in X, \|y-z\| \geq \varepsilon\}\\ &\quad + \inf\{\tfrac{1}{2}\varphi(x-y)+\tfrac{1}{2}\varphi(x-z)-\varphi(x-u)+\tfrac{1}{2}\|2u-(y+z)\|;\\ &\qquad\qquad y,z \in X, \|y-z\| \geq \varepsilon\}\\ &= \tfrac{1}{2}\delta_{x+u}^{\varphi}(\varepsilon)+\tfrac{1}{2}\delta_{x-u}^{\varphi}(\varepsilon) \geq \tfrac{1}{n}\varepsilon^n.\end{aligned}$$

In particular, for any $y \in B(0,\frac{1}{2})$ we can use $z=-y$ in the definition of δ_0^{ψ} to obtain $2\psi(y) \geq \delta_0^{\psi}(2\|y\|) \geq \frac{1}{n}2^n\|y\|^n$. Next, we put $a = \frac{1}{2n}2^n r^n$. Then the set $B = \{x \in X;\ \psi(x) \leq a\} \subset B(0,r)$ is an absolutely convex bounded neighbourhood of 0. Hence the Minkowski functional ν of B is an equivalent norm on X.

We claim that the modulus of convexity of ν is of power type n. Let $0 < C \leq 1$ be such that $\|x\| \geq C\nu(x)$ for all $x \in X$. Fix any $\varepsilon \in [0,1]$. Let $y, z \in X$ be such that $\nu(y)=\nu(z)=1$ and $\nu(y-z)=\varepsilon$. Note that $\|y-z\| \geq C\varepsilon$ and $\nu(\frac{y+z}{2}) \geq \frac{1}{2}$. Put $x = \frac{y+z}{2}/\nu(\frac{y+z}{2})$. Then $\psi(x)=\psi(y)=\psi(z)=a$. Further,

$$\begin{aligned}\tfrac{1}{n}C^n\varepsilon^n &\leq \delta_x^{\psi}(C\varepsilon) \leq \psi(y)+\psi(z)-2\psi(x)+\|2x-(y+z)\| = \|2x-(y+z)\|\\ &= \|y+z\|\Big(\frac{1}{\nu(\frac{y+z}{2})}-1\Big) \leq \frac{2r}{\nu(\frac{y+z}{2})}\Big(1-\nu\big(\tfrac{y+z}{2}\big)\Big) \leq 4r\Big(1-\nu\big(\tfrac{y+z}{2}\big)\Big).\end{aligned}$$

This shows that the modulus of convexity of ν is of power type n.

To see the last statement, instead of the sets F_n as above we consider the sets $F_n = \{x \in U;\ \delta_x^{\varphi}(\varepsilon) \geq \frac{1}{n}\varepsilon^q \text{ for all } \varepsilon \in [0,1]\}$. □

At the end of this section we look at subdifferentials of weakly sequentially continuous convex functions on spaces which do not contain ℓ_1.

Fact 40. *Let X be a Banach space which does not contain ℓ_1, $U \subset X$ an open convex set, and let $\varphi\colon U \to \mathbb{R}$ be a weakly sequentially continuous convex function. Then $\partial\varphi(x)$ is compact for each $x \in U$.*

Proof. Assume the contrary. Let $x \in U$ be such that $\partial\varphi(x)$ is not compact. Since $\partial\varphi(x)$ is closed (even w^*) and bounded, there is a bounded sequence $\{f_n\} \subset \partial\varphi(x)$ which is not relatively compact. Let $\delta > 0$ be such that $B(x, \delta) \subset U$. By Theorem 3.56 there are a subsequence $\{f_{n_k}\}$, a weakly null sequence $\{h_k\} \subset B(0, \delta)$, and $\varepsilon > 0$ such that $f_{n_k}(h_k) \geq \varepsilon$ for each $k \in \mathbb{N}$. Then $\varphi(x + h_k) \geq \varphi(x) + f_{n_k}(h_k) \geq \varphi(x) + \varepsilon$ for all $k \in \mathbb{N}$, which contradicts the fact that $\varphi(x + h_k) \to \varphi(x)$. □

Proposition 41. *Let X be a Banach space which does not contain ℓ_1, $U \subset X$ an open convex set, and let $\varphi\colon U \to \mathbb{R}$ be a weakly sequentially Cauchy-continuous convex function. Then $\bigcup_{x \in V} \partial\varphi(x)$ is relatively compact for each bounded $V \subset U$ satisfying* $\operatorname{dist}(V, X \setminus U) > 0$.

Proof. Assume the contrary. Let $V \subset U$ be bounded such that $\bigcup_{x \in V} \partial\varphi(x)$ is not relatively compact and $\delta = \frac{1}{2}\operatorname{dist}(V, X \setminus U) > 0$. There are sequences $\{x_n\} \subset V$, $\{f_n\} \subset X^*$ such that $f_n \in \partial\varphi(x_n)$ and $\{f_n\}$ is not relatively compact. By Rosenthal's ℓ_1-theorem we may assume that $\{x_n\}$ is weakly Cauchy. There are a subsequence $\{f_{n_k}\}$, a weakly null sequence $\{h_k\} \subset B(0, \delta)$, and $\varepsilon > 0$ such that $f_{n_k}(h_k) \geq \varepsilon$ for each $k \in \mathbb{N}$ (if $\{f_n\}$ is bounded, then we use Theorem 3.56; otherwise it is obvious). Then $\varphi(x_k + h_k) \geq \varphi(x_k) + f_{n_k}(h_k) \geq \varphi(x_k) + \varepsilon$ for all $k \in \mathbb{N}$. On the other hand, since the sequence $\{y_k\}$, where $y_{2k-1} = x_k$, $y_{2k} = x_k + h_k$, is weakly Cauchy, it follows that $\lim \varphi(x_k + h_k) = \lim \varphi(x_k)$, which is a contradiction. □

2. Smooth bumps and structure I

In this section we give an hors d'oeuvre of the results that show how the existence of bump functions (or norms) with certain kind of smoothness impacts the structure of the space. First we show a few results related to separating polynomials. Then we show how local uniform smoothness together with convexity combine to produce a global uniform smoothness. Next we introduce a lemma by Marián Fabian and Robert Deville that constructs a "strongly separating" function from a single bump on spaces that do not contain c_0. This will be used to prove that on such spaces $C^{k,\Omega}_{\mathrm{loc}}$-smoothness leads to a global $C^{k,\Omega}$-smoothness. Further, in case of the first order smoothness we construct a $C^{1,\Omega}$-smooth norm from a $C^{1,\Omega}$-smooth bump. Combining these two results we obtain that if X admits a $C^{1,+}_{\mathrm{loc}}$-smooth bump, then it is either super-reflexive, or it contains c_0.

We start with rather simple observation about the smoothness of (powers of) norms at the origin. Note that for every normed linear space and every $1 \leq q < p < +\infty$ the function $x \mapsto \|x\|^p$ is T^q-smooth at 0 with the polynomial P from the definition of T^q-smoothness being zero.

Proposition 42. *Let X be a normed linear space such that the function $x \mapsto \|x\|^p$ is weakly T^p-smooth at 0. Then p is an even integer and the norm of X is p-polynomial.*

In particular, if the square of the norm of X is G^2-smooth at 0, then X is a Hilbert space.

Proof. The weak T^p-smoothness implies that there is a polynomial $Q \in \mathcal{P}^{[p]}(X)$ such that $\|th\|^p - Q(th) = o(|t|^p)$, $t \to 0$ for each $h \in X$. However, since $\|th\|^p = o(|t|^q)$, $t \to 0$ for $q < p$, by the uniqueness of the polynomial expansion each k-homogeneous summand of Q is zero for $k < p$. It follows that $p \in \mathbb{N}$ and $Q \in \mathcal{P}(^pX)$. So for a given $h \in X$ we have $\|h\|^p - Q(h) = \frac{o(t^p)}{t^p}$, $t \to 0+$, whence $\|h\|^p = Q(h)$. From this we also get $Q(-h) = Q(h)$, thus p is even. □

Corollary 43. *Let X be a normed linear space such that the norm of $X \oplus_p \mathbb{R}$ is weakly T^p-smooth at $[0, w]$ for some $w \in \mathbb{R}$. Then p is an even integer and the norm of X is p-polynomial.*

Proof. By the assumption the function $\nu(x, y) = \|x\|^p + |y|^p$ is weakly T^p-smooth at $[0, w]$ (remark after Proposition 1.115). Thus also the restriction of ν to $X + [0, w]$ is weakly T^p-smooth at $[0, w]$, i.e. $x \mapsto x^p + |w|^p$ is weakly T^p-smooth at 0 and we may use Proposition 42. □

The following theorem shows that the existence of a separating polynomial (of a given degree) is separably determined. This will later allow us to use certain smoothing techniques that are inherently separable in a general setting.

Theorem 44. *Let X be a normed linear space and $n \in \mathbb{N}$. If every separable closed subspace of X admits a separating polynomial of degree at most n (resp. n-homogeneous), then X admits a separating polynomial of degree at most n (resp. n-homogeneous).*

Proof. Let us define $c_Y = \sup\{\inf_{S_Y} P;\ P \in \mathcal{P}^n(Y), \|P\| \le 1, P(0) = 0\}$ for a subspace $Y \subset X$. Notice that by the assumption $c_Y > 0$ for every separable closed subspace Y of X. Then

$$c = \inf\{c_Y;\ Y \text{ is a separable closed subspace of } X\} > 0.$$

Indeed, otherwise there would be a sequence of separable closed subspaces $\{Y_n\}$ of X with $c_{Y_n} < \frac{1}{n}$. Set $Y = \overline{\operatorname{span}} \bigcup_{n=1}^{\infty} Y_n$. Then Y is a separable closed subspace of X and $0 < c_Y \le c_{Y_n} < \frac{1}{n}$ for every $n \in \mathbb{N}$, a contradiction.

Now for every separable closed subspace Y of X we find a polynomial $P^Y \in \mathcal{P}^n(Y)$ such that $\|P^Y\| \le 1$, $P(0) = 0$, and $\inf_{S_Y} P^Y \ge \frac{c}{2}$. Denote by P_j^Y, $j = 1, \dots, n$, the j-homogeneous terms of P^Y and note that by Fact 1.42 there is $K > 0$ such that $\|P_j^Y\| \le K$ for every separable closed subspace Y of X and every $j \in \{1, \dots, n\}$. Define $Q_j^Y : X \to \mathbb{R}$ by $Q_j^Y(x) = P_j^Y(x)$ for $x \in Y$, $Q_j^Y(x) = 0$ otherwise.

Let Λ be a family of all separable closed subspaces of X ordered by inclusion. Then $\{Q_j^Y \restriction_{B_X}\}_{Y \in \Lambda}$, $j = 1, \dots, n$, are nets in the compact space $[-K, K]^{B_X}$. Thus we can extract convergent subnets $\{Q_j^{Y_\gamma} \restriction_{B_X}\}_{\gamma \in \Gamma}$, $j = 1, \dots, n$. Since each of the functions Q_j^Y is j-homogeneous, it follows that the nets $\{Q_j^{Y_\gamma}\}_{\gamma \in \Gamma}$ converge pointwise on the whole of X. Define $Q_j : X \to \mathbb{R}$ by $Q_j(x) = \lim_\Gamma Q_j^{Y_\gamma}(x)$. It is not difficult to check using the Polarisation formula that $Q_j \in P(^jX)$.

Put $Q = \sum_{j=1}^{n} Q_j$. Then $Q \in P^n(X)$ and $Q(0) = 0$. It is easy to see that $\|Q\| \le 1$ and since for every $x \in S_X$ there is $\beta \in \Gamma$ such that $x \in Y_\gamma$ for $\gamma \ge \beta$, we have $\inf_{S_X} Q \ge \frac{c}{2}$. The homogeneous case is clear from the above. □

Corollary 45 ([Li2]). *Assume that every separable subspace of a Banach space X is isomorphic to a Hilbert space. Then X is isomorphic to a Hilbert space.*

The next theorem shows that local uniform smoothness of the norm turns to global uniform smoothness under additional convexity assumptions.

Theorem 46 ([FWZ]). *Let $\Omega \subset \mathcal{M}_s$ be a convex cone and let X be a Banach space with a $C^{k,\Omega}_{\mathrm{loc}}$-smooth norm such that its unit ball has at least one strongly exposed point. Then X admits an equivalent $C^{k,\Omega}$-smooth norm.*

Proof. Let $w \in S_X$ be a strongly exposed point of B_X exposed by $f \in S_{X^*}$ and let $\delta > 0$ be such that $\|\cdot\|$ is $C^{k,\Omega}$-smooth on $U(w, 2\delta)$. Denote $H = \ker f$. By the strong exposedness there is $\eta > 0$ such that $\|w + h\| \ge 1 + \eta$ whenever $h \in H$, $\|h\| \ge \delta$. Indeed, otherwise there would be a sequence $\{h_n\} \subset H$, $\|h_n\| \ge \delta$, such that $\|w+h_n\| \le 1+\frac{1}{n}$. This means that $1 = f(w) = f(w+h_n) \le \|w+h_n\| \le 1+\frac{1}{n} \to 1$ and hence $h_n \to 0$, a contradiction. Thus $h \mapsto \|w + h\|$ is separating on $H \cap U(0, 2\delta)$ and by Lemma 23 there is a $C^{k,\Omega}$-smooth equivalent norm ν on H. Now it suffices to take the norm $x \mapsto \sqrt{\nu(x - f(x)w)^2 + f(x)^2}$. □

The following lemma is an important tool for constructing a "strongly separating" function from a single bump without any additional structure.

Lemma 47. *Let X be a Banach space which does not contain c_0. Let U be a bounded symmetric (i.e. $U = -U$) open neighbourhood of the origin and let $f : \overline{U} \to \mathbb{R}$ be an even lower semi-continuous function such that $f(0) \le 0$ and $\inf f(\partial U) > 0$. Then there exist a symmetric compact set $K \subset U$ and a neighbourhood V of 0 such that $K + V \subset U$ and for each $\delta > 0$ there are $\varepsilon > 0$ and a finite subset $F \subset K$ such that for each $x \in V$, $\|x\| \ge \delta$ there is $z \in F$ satisfying $f(z + x) > f(z) + \varepsilon$.*

Proof. Let $m = \inf f(\partial U)$. We will inductively construct a sequence $\{x_j\}_{j=0}^{\infty} \subset X$ such that for each $n \in \mathbb{N}_0$ it satisfies

(i) $K_n \subset U$, where $K_n = \{\sum_{j=0}^{n} \varepsilon_j x_j;\ (\varepsilon_j) \in \{-1,1\}^{n+1}\}$,

(ii) $f(z) \le \frac{m}{2}\left(1 - \frac{1}{2^n}\right)$ for all $z \in K_n$.

Set $x_0 = 0$. If $x_0, x_1, \dots, x_{n-1}$ have already been defined, we put

$$E_n = \left\{x \in X;\ z + x \in U,\ f(z + x) \le \frac{m}{2}\left(1 - \frac{1}{2^n}\right) \text{ for all } z \in K_{n-1}\right\}.$$

From the inductive hypothesis it follows that at least $0 \in E_n$. We put $\alpha_n = \sup_{x \in E_n} \|x\|$ and choose $x_n \in E_n$ such that $\|x_n\| \ge \alpha_n/2$. Using the symmetry of K_{n-1} and U, the fact that f is even, and the inductive hypothesis we check that the properties (i) and (ii) are satisfied.

Let $K = \overline{\bigcup_{n=0}^{\infty} K_n}$. Obviously K is symmetric. Further, $K \subset \overline{U}$, hence K is bounded. From the Bessaga-Pełczyński theorem (Theorem 3.46) it follows that K

is compact and $x_n \to 0$, hence also $\alpha_n \to 0$. Since f is lower semi-continuous, $f(x) \le \frac{m}{2}$ for each $x \in K$ and thus $K \cap \partial U = \emptyset$. Hence there is a neighbourhood V of 0 satisfying $K + V \subset U$.

Pick any $\delta > 0$. Let $n \in \mathbb{N}$ be such that $\alpha_n < \delta$ and put $\varepsilon = \frac{m}{2^{n+1}}$ and $F = K_{n-1}$. If $x \in V$ and $\|x\| \ge \delta$, then $\|x\| > \alpha_n$ and hence $x \notin E_n$. Thus there is $z \in F$ satisfying $f(z+x) > \frac{m}{2}\left(1 - \frac{1}{2^n}\right)$. On the other hand from (ii) it follows that $f(z) \le \frac{m}{2}\left(1 - \frac{1}{2^{n-1}}\right)$ and so $f(z + x) > f(z) + \varepsilon$. □

Theorem 48. *Let X be a Banach space which does not contain c_0. Let $\Omega \subset \mathcal{M}$ be a convex cone. If X admits a $C^{k,\Omega}_{\mathrm{loc}}$-smooth bump, then X admits $C^{k,\Omega}$-smooth bump.*

Proof. By the assumption there is a function $f \in C^{k,\Omega}_{\mathrm{loc}}(X)$ satisfying $f(0) = 0$ and $f(x) = c > 0$ for x outside some ball. Replacing f with $x \mapsto f(x) + f(-x)$ we may additionally assume that f is even. Let $K \subset X$ be the compact set and V the neighbourhood of 0 from Lemma 47 used on f. Let $r > 0$ be such that $U(0, r) \subset V$. There is a covering of K by open balls $U(x_j, \delta_j)$, $j = 1, \dots, n$, such that f is $C^{k,\Omega}$-smooth on each $U(x_j, 2\delta_j)$.

Put $\delta = \frac{1}{2}\min\{r, \delta_1, \dots, \delta_n\}$. For this δ we find $\varepsilon > 0$ and $F \subset K$ finite from Lemma 47 such that for each $x \in X$, $\delta \le \|x\| < r$ there is $z \in F$ satisfying $f(z + x) > f(z) + \varepsilon$. Define $g\colon U(0, 2\delta) \to \mathbb{R}$ by

$$g(x) = \sum_{z \in F} \big(f(z + x) - f(z)\big)^2.$$

Since for each $z \in F$ there is $j \in \{1, \dots, n\}$ such that $U(z, 2\delta) \subset U(x_j, 2\delta_j)$, it follows from Proposition 1.128 that $g \in C^{k,\Omega}\big(U(0, 2\delta)\big)$.

Further, $g(0) = 0$ and $g(x) > \varepsilon^2$ for each $x \in X$ satisfying $\delta \le \|x\| < 2\delta$. Let $\theta \in C^\infty(\mathbb{R}; [0, 1])$ be a function with all derivatives bounded and such that $\theta(0) = 1$ and $\theta(t) = 0$ for $t \ge \varepsilon^2$. Put $h(x) = \theta\big(g(x)\big)$ for $x \in U(0, 2\delta)$ and $h(x) = 0$ elsewhere. Then it is easy to check again using Proposition 1.128 that $h \in C^{k,\Omega}(X)$. Finally, observe that $h(0) = 1$ and $h(x) = 0$ for $x \in X$, $\|x\| \ge \delta$. □

As the next proposition shows, the assumption that X does not contain c_0 is necessary in the previous theorem (note that there is a $C^{1,1}_{\mathrm{loc}}$-smooth bump on c_0, see Theorem 94).

Proposition 49 ([Wel1]). *There is no $C^{1,+}$-smooth bump on c_0.*

Proof. Assume such a bump f exists. Without loss of generality we may assume that $f(0) = 1$ and $f(x) = 0$ whenever $\|x\| = 1$. Let $n \in \mathbb{N}$ be such that

$$\|Df(x) - Df(y)\| \le \frac{1}{2} \quad \text{for any } x, y \in c_0,\ \|x - y\| \le \frac{1}{n}. \tag{6}$$

Let A be the set of all edges of a cube in the first 2^n coordinates of c_0:

$$A = \bigcup_{j_0=1}^{2^n} \left\{x \in c_0;\ |x_j| = \tfrac{1}{n} \text{ if } j \in \{1, \dots, 2^n\} \setminus \{j_0\},\ |x_{j_0}| \le \tfrac{1}{n},\ x_j = 0 \text{ if } j > 2^n\right\}.$$

Since A is symmetric and connected, and $Df(0)$ is even and continuous, there is $h_1 \in A$ with $Df(0)[h_1] = 0$. Moreover by changing the sign of h_1 if necessary we may assume that at least 2^{n-1} coordinates of h_1 is equal to $\frac{1}{n}$. Similarly there is $h_2 \in A$ with $Df(h_1)[h_2] = 0$ and such that at least 2^{n-2} coordinates of $h_1 + h_2$ equals $\frac{2}{n}$. By induction we construct $h_k \in A$ for $k = 3, \dots, n$ such that $Df(h_1 + \dots + h_{k-1})[h_k] = 0$ and $h_1 + \dots + h_k$ has at least 2^{n-k} coordinates equal to $\frac{k}{n}$. Then $\|h_1 + \dots + h_n\| = 1$ and so

$$\begin{aligned}
1 = \left|f(h_1 + \dots + h_n) - f(0)\right| &\le \sum_{k=1}^{n} \left|f(h_1 + \dots + h_k) - f(h_1 + \dots + h_{k-1})\right| \\
&= \sum_{k=1}^{n} \left|Df(h_1 + \dots + h_{k-1} + t_k h_k)[h_k]\right| \qquad \text{for some } 0 < t_k < 1 \\
&= \sum_{k=1}^{n} \left|Df(h_1 + \dots + h_{k-1} + t_k h_k)[h_k] - Df(h_1 + \dots + h_{k-1})[h_k]\right| \\
&\le \sum_{k=1}^{n} \frac{1}{2}\|h_k\| \le n\frac{1}{2}\frac{1}{n} = \frac{1}{2},
\end{aligned}$$

which is a contradiction. (The second equality follows from the Mean value theorem and the second inequality is assured by (6), since $\|t_k h_k\| \le \frac{1}{n}$.) □

Theorem 50 ([FWZ]). *Let $\Omega \subset \mathcal{M}_s$ be a convex cone. Let X be a Banach space that admits a $C^{1,\Omega}$-smooth bump. Then X admits an equivalent $C^{1,\Omega}$-smooth norm.*

Proof. By shifting and composing with a suitable smooth function on $\mathbb{R}$ we construct from a $C^{1,\Omega}$-smooth bump a $C^{1,\Omega}$-smooth function $\varphi\colon X \to [0, 6]$ such that $\varphi(0) = 0$ and $\varphi(x) = 6$ for $\|x\| \ge 1$ (Proposition 1.128).

Define a function $\psi\colon B(0, 6) \to \mathbb{R}$ by

$$\psi(x) = \inf\left\{\sum_{j=1}^{n} \alpha_j \varphi(x_j);\ x = \sum_{j=1}^{n} \alpha_j x_j, \sum_{j=1}^{n} \alpha_j = 1, \alpha_j \ge 0, x_j \in B(0, 6), j = 1, \dots, n,\ n \in \mathbb{N}\right\}.$$

It is easy to check that ψ is convex. Clearly, $\psi(x) \le \varphi(x)$ for any $x \in B(0, 6)$. Hence ψ is convex and bounded above, so it is continuous on $U(0, 6)$. Note also that $\psi(x) \le \frac{\|x\|}{6}\varphi\big(\frac{6}{\|x\|}x\big) + \big(1 - \frac{\|x\|}{6}\big)\varphi(0) = \|x\|$ for all $x \in B(0, 6)$.

By Lemma 19(i) there is $\omega \in \Omega$ such that $\varphi(x+h)+\varphi(x-h)-2\varphi(x) \le \omega(\|h\|)\|h\|$ for $x, h \in X$. We show that this implies $\psi(x+h)+\psi(x-h)-2\psi(x) \le 2\omega(\|h\|)\|h\|$ for $x, h \in X$, $x \pm h \in U(0, 2)$. So fix $x, h \in X$, $x \pm h \in U(0, 2)$ and notice that $\|h\| < 2$. Choose any $\varepsilon > 0$ satisfying $\|x\| + \varepsilon < 2$. Let $n \in \mathbb{N}$, $x_j \in B(0, 6)$, and $\alpha_j \ge 0$, $j \in \{1, \dots, n\}$, be such that $\sum_{j=1}^{n} \alpha_j = 1$, $x = \sum_{j=1}^{n} \alpha_j x_j$, and

$\sum_{j=1}^{n} \alpha_j \varphi(x_j) < \psi(x) + \varepsilon$. To estimate the values of $\psi(x \pm h)$ we need to distribute the h only to those x_js that lie in $B(0, 2)$, otherwise $x_j + h$ may not lie in $B(0, 6)$. Without loss of generality we may assume that there is $k \in \mathbb{N}_0$ such that $x_1, \dots, x_k \in B(0, 2)$, while $x_{k+1}, \dots, x_n \notin B(0, 2)$. We have

$$2 > \|x\| + \varepsilon \geq \psi(x) + \varepsilon > \sum_{j=1}^{n} \alpha_j \varphi(x_j) \geq 6 \sum_{j=k+1}^{n} \alpha_j = 6\left(1 - \sum_{j=1}^{k} \alpha_j\right).$$

Put $\alpha = \sum_{j=1}^{k} \alpha_j$. The above inequality gives $\frac{2}{3} < \alpha \leq 1$. Thus $\|x_j + \frac{1}{\alpha} h\| < 5$ for $j \in \{1, \dots, k\}$. So we can estimate

$$\begin{aligned}
&\psi(x+h) + \psi(x-h) - 2\psi(x) \\
&\quad < \sum_{j=1}^{k} \alpha_j \varphi\big(x_j + \tfrac{1}{\alpha} h\big) + \sum_{j=k+1}^{n} \alpha_j \varphi(x_j) \\
&\qquad\qquad + \sum_{j=1}^{k} \alpha_j \varphi\big(x_j - \tfrac{1}{\alpha} h\big) + \sum_{j=k+1}^{n} \alpha_j \varphi(x_j) - 2 \sum_{j=1}^{n} \alpha_j \varphi(x_j) + 2\varepsilon \\
&\quad = \sum_{j=1}^{k} \alpha_j \varphi\big(x_j + \tfrac{1}{\alpha} h\big) + \sum_{j=1}^{k} \alpha_j \varphi\big(x_j - \tfrac{1}{\alpha} h\big) - 2 \sum_{j=1}^{k} \alpha_j \varphi(x_j) + 2\varepsilon \\
&\quad = \sum_{j=1}^{k} \alpha_j \Big(\varphi\big(x_j + \tfrac{1}{\alpha} h\big) + \varphi\big(x_j - \tfrac{1}{\alpha} h\big) - 2\varphi(x_j)\Big) + 2\varepsilon \\
&\quad \leq \sum_{j=1}^{k} \alpha_j \omega\big(\|\tfrac{1}{\alpha} h\|\big) \|\tfrac{1}{\alpha} h\| + 2\varepsilon = \omega\big(\|\tfrac{1}{\alpha} h\|\big) \|h\| + 2\varepsilon.
\end{aligned}$$

Since this inequality holds for arbitrarily small ε and ω is sub-additive, we conclude that $\psi(x+h) + \psi(x-h) - 2\psi(x) \leq 2\omega(\|h\|)\|h\|$. By Lemma 19(ii) this means that $\psi \in C^{1,\Omega}\big(U(0, \frac{7}{4})\big)$.

Further, we have $\psi(x) \geq \frac{1}{2}$ whenever $x \in B(0, 6)$ is such that $\|x\| \geq \frac{3}{2}$. Indeed, let $x = \sum_{j=1}^{n} \alpha_j x_j$ be a convex combination of points in $B(0, 6)$. Similarly as above let us suppose that $x_1, \dots, x_k \in B(0, 1)$, while $x_{k+1}, \dots, x_n \notin B(0, 1)$. Then $\frac{3}{2} \leq \|x\| \leq \sum_{j=1}^{k} \alpha_j \|x_j\| + \sum_{j=k+1}^{n} \alpha_j \|x_j\| \leq 1 + 6 \sum_{j=k+1}^{n} \alpha_j$. Therefore $\sum_{j=k+1}^{n} \alpha_j \geq \frac{1}{12}$, which in turn implies $\sum_{j=1}^{n} \alpha_j \varphi(x_j) \geq \sum_{j=k+1}^{n} 6\alpha_j \geq \frac{1}{2}$. Since $\psi(0) = 0$, the function ψ is separating on $U(0, \frac{7}{4})$. So, by Lemma 23 there is an equivalent $C^{1,\Omega}$-smooth norm on X. □

Corollary 51. *Let X be a Banach space that admits a $C^{1,+}_{\mathrm{loc}}$-smooth (resp. $C^{1,\alpha}_{\mathrm{loc}}$-smooth) bump. Then either X contains c_0, or X is super-reflexive (and admits an equivalent norm with modulus of smoothness of power type $1 + \alpha$).*

Combining Theorem 46 with Corollary 51 and Proposition 49 we immediately obtain the following:

Corollary 52 ([FWZ]). *If X is a Banach space with a $C^{1,+}_{\mathrm{loc}}$-smooth LUR norm, then X is super-reflexive.*

3. Smooth variational principles

This sections deals with various variational principles, concentrating on the higher order smoothness. The general idea of variational principles is that given a function we can find a perturbation of certain class such that the perturbed function attains (perhaps even strong) minimum. An alternative point of view is that given a bounded below function φ there is a function of certain class that "touches" φ from below.

First we present the Ekeland variational principle, and we state the first order Deville-Godefroy-Zizler variational principle and Stegall's variational principle (which is of course as smooth as it gets, but works only on linearly coercive functions), which will be our principal tool. We give a quantitative version of Stegall's variational principle for super-reflexive spaces, which states how fast the perturbed function rises away from its minimum.

Next, given a class of functions $\mathcal{S}$ we introduce a general idea of $\mathcal{S}$-variational principle. Using this, we prove a combined $C^{k,\alpha}$-smooth variational principle and structural theorem.

In the last part we revisit the separating polynomials. We use our variational approach to prove that if a space with a norm with modulus of convexity of power type q has a T^q-smooth bump, then it admits a separating polynomial. We end this section with a combined polynomial variational principle and structural theorem.

Let (P, ρ) be a metric space. We say that a function $f\colon P \to \mathbb{R} \cup \{+\infty\}$ attains a strong minimum at a point $x \in P$ if $f(x) = \inf_P f$ and $\rho(x_n, x) \to 0$ whenever $f(x_n) \to f(x)$.

The following classical theorem has a wide applicability in non-linear analysis.

Theorem 53 (Ekeland variational principle). *Let (P, ρ) be a complete metric space and let $\varphi\colon P \to \mathbb{R} \cup \{+\infty\}$ be a proper lower semi-continuous bounded below function. Then for every $\varepsilon > 0$ there is $x_0 \in P$ such that $\varphi(x) \geq \varphi(x_0) - \varepsilon\rho(x, x_0)$ for all $x \in P$. Moreover, the function $x \mapsto \varphi(x) + 2\varepsilon\rho(x, x_0)$ attains a strong minimum at x_0.*

Proof. Choose any $y_1 \in \operatorname{dom}\varphi$ and define inductively $y_{n+1} \in P$ so that

$$\varphi(y_{n+1}) + \varepsilon\rho(y_{n+1}, y_n) \leq \min\left\{\inf\{\varphi(x) + \varepsilon\rho(x, y_n);\ x \in P\} + \frac{1}{2^n},\ \varphi(y_n)\right\}. \tag{7}$$

This is always possible as we may take $y_{n+1} = y_n$ if the infimum above is attained in y_n, otherwise the infimum is strictly smaller than $\varphi(y_n)$ and we can choose a suitable y_{n+1} by the definition of infimum.

The sequence $\{\varphi(y_n)\}$ is non-increasing and bounded below, thus it is convergent in $\mathbb{R}$. Moreover, $\{y_n\}$ is Cauchy. Indeed, $\rho(y_m, y_n) \leq \sum_{k=n}^{m-1} \rho(y_{k+1}, y_k) \leq \sum_{k=n}^{m-1}(\varphi(y_k) - \varphi(y_{k+1}))/\varepsilon = (\varphi(y_n) - \varphi(y_m))/\varepsilon$ for $m > n$, and since $\{\varphi(y_n)\}$ is Cauchy, so is $\{y_n\}$. As P is complete, $y_n \to x_0 \in P$.

Finally, (7) implies $\varphi(y_{n+1}) \leq \varphi(x) + \varepsilon\rho(x, y_n) + \frac{1}{2^n} - \varepsilon\rho(y_{n+1}, y_n)$ for every $x \in P$ and $n \in \mathbb{N}$. From the lower semi-continuity of φ it follows that

$$\begin{aligned}\varphi(x_0) &\leq \lim_{n\to\infty} \varphi(y_n) \leq \lim_{n\to\infty} \left(\varphi(x) + \varepsilon\rho(x, y_n) + \frac{1}{2^n} - \varepsilon\rho(y_{n+1}, y_n)\right) \\ &= \varphi(x) + \varepsilon\rho(x, x_0).\end{aligned}$$

To show the statement about the strong minimum, let $\{x_n\}_{n\in\mathbb{N}} \subset P$ be such that $\varphi(x_n) + 2\varepsilon\rho(x_n, x_0) \to \varphi(x_0)$. Then $\varepsilon\rho(x_n, x_0) = \varepsilon\rho(x_n, x_0) + \varphi(x_0) - \varphi(x_0) \leq \varphi(x_n) + 2\varepsilon\rho(x_n, x_0) - \varphi(x_0) \to 0$. □

The following variational principle deals with the first order smoothness. See [DGZ, Theorem I.2.3].

Theorem 54. *Let X be a Banach space that admits a Fréchet (resp. Gâteaux) differentiable Lipschitz bump. Let $\varphi\colon X \to \mathbb{R} \cup \{+\infty\}$ be a bounded below proper lower semi-continuous function. Then for every $\varepsilon > 0$ there exists a Lipschitz function g which is Fréchet (resp. Gâteaux) differentiable on X such that $\sup_{x\in X}|g(x)| \leq \varepsilon$, $\sup_{x\in X}\|\delta g(x)\| \leq \varepsilon$, and $\varphi - g$ attains a strong minimum on X.*

Theorem 55 (Stegall's variational principle, [Steg2], [Steg3]). *Let X be a Banach space with the RNP and let A be a non-empty closed bounded subset of X. Let $\varphi\colon A \to \mathbb{R} \cup \{+\infty\}$ be a bounded below proper lower semi-continuous function. Then the set of $f \in X^*$ such that the function $\varphi - f$ attains a strong minimum on A is a dense G_δ subset of X^*.*

Note that the fact that $\varphi - f$ attains a strong minimum at x_0 can be reformulated as $f \in \partial\varphi(x_0)$ and $\Delta^{\varphi}_{x_0,f}(\varepsilon) > 0$ for all $\varepsilon > 0$ (Fact 29). For the proof of the theorem see [BorVa, Corollary 6.6.17].

Let X be a normed linear space. We say that a function $f\colon X \to \mathbb{R} \cup \{+\infty\}$ is linearly coercive if $\liminf_{\|x\|\to+\infty} \frac{\varphi(x)}{\|x\|} > 0$. Similarly we say that f is quadratically coercive if $\liminf_{\|x\|\to+\infty} \frac{\varphi(x)}{\|x\|^2} > 0$.

Corollary 56 ([Fab1]). *Let X be a Banach space with the RNP and assume that $\varphi\colon X \to \mathbb{R} \cup \{+\infty\}$ is a bounded below proper lower semi-continuous linearly coercive function. Put $r = \liminf_{\|x\|\to+\infty} \frac{\varphi(x)}{\|x\|}$. Then the set of $f \in X^*$ such that the function $\varphi - f$ attains a strong minimum is residual in $U_{X^*}(0, r)$.*

Proof. Denote by M the set of $f \in X^*$ such that the function $\varphi - f$ attains a strong minimum. It suffices to show that M is residual in $U_{X^*}(0, r - \delta)$ for every $0 < \delta < r$. So fix $0 < \delta < r$. Find $R_1 > 0$ such that $\frac{\varphi(x)}{\|x\|} > r - \frac{\delta}{2}$ for $\|x\| \geq R_1$. Put $R = \max\{R_1, \frac{2(\varphi(0)+1)}{\delta}\}$. Denote by M_δ the set of $f \in X^*$ such that $(\varphi - f)\restriction_{B(0,R)}$ attains a strong minimum. By Theorem 55 the set M_δ is residual in X^*.

Fix $f \in M_\delta \cap U(0, r - \delta)$. Let $x_0 \in B(0, R)$ be such that $(\varphi - f)\restriction_{B(0,R)}$ attains a strong minimum at x_0. For every $x \in X \setminus B(0, R)$ we have

$$\begin{aligned}\varphi(x) - f(x) &\geq \varphi(x) - \|f\|\|x\| > \left(r - \frac{\delta}{2}\right)\|x\| - (r - \delta)\|x\| \\ &= \frac{\delta}{2}\|x\| > R\frac{\delta}{2} \geq \varphi(0) + 1 = \varphi(0) - f(0) + 1.\end{aligned} \tag{8}$$

It follows that the function $\varphi - f$ attains its minimum only at the point x_0. Further, if $\varphi(x_n) - f(x_n) \to \varphi(x_0) - f(x_0)$, then eventually $\varphi(x_n) - f(x_n) < \varphi(0) + 1$, and hence by (8) eventually $x_n \in B(0, R)$. Therefore $\varphi - f$ attains a strong minimum at x_0, which implies that $f \in M$.

Finally, since $M \cap U(0, r - \delta) \supset M_\delta \cap U(0, r - \delta)$, the set M is residual in $U(0, r - \delta)$. □

The following observation shows that the perturbed function can be chosen so that its strong minimum is arbitrarily close to the infimum of the original function.

Corollary 57. *Let X be a Banach space with the RNP and let $\varphi\colon X \to \mathbb{R} \cup \{+\infty\}$ be a bounded below proper lower semi-continuous linearly coercive function. Then for every $\varepsilon > 0$ there is $f \in X^*$ such that $\|f\| < \varepsilon$, the function $\varphi - f$ attains a strong minimum at $x_0 \in X$, and $\varphi(x_0) < \inf_X \varphi + \varepsilon$.*

Proof. Without loss of generality we may assume that $\inf_X \varphi = 0$. Let $\delta > 0$ and $R > 0$ be such that $\varphi(x) \geq \delta\|x\|$ whenever $\|x\| > R$. By Corollary 56 there is $f \in X^*$ such that $\|f\| < \min\{\varepsilon, \frac{\delta}{2}, \frac{\varepsilon}{4R}\}$ and $\varphi - f$ attains a strong minimum at $x_0 \in X$. We need to show that $\varphi(x_0) < \varepsilon$. First observe that $\|x_0\| \leq R$. Indeed, otherwise for every $x \in B(0, R)$ we can estimate

$$\begin{aligned}\varphi(x) &= \varphi(x) - f(x) + f(x) \geq \varphi(x_0) - f(x_0) + f(x) \\ &> \delta\|x_0\| - \frac{\delta}{2}\|x_0 - x\| \geq \frac{\delta}{2}\big(\|x_0\| - \|x\|\big) \geq \frac{\delta}{2}\big(\|x_0\| - R\big) > 0,\end{aligned}$$

which together with the fact that $\varphi(x) > \delta R$ for $\|x\| > R$ contradicts the fact that $\inf_X \varphi = 0$.

Suppose that $\varphi(x_0) \geq \varepsilon$. Then for every $x \in B(0, R)$ we can estimate

$$\varphi(x) \geq \varphi(x_0) - f(x_0) + f(x) > \varepsilon - \frac{\varepsilon}{4R}\|x_0 - x\| \geq \varepsilon - \frac{\varepsilon}{4R}2R = \frac{\varepsilon}{2},$$

which again contradicts the fact that $\inf_X \varphi = 0$. □

For super-reflexive spaces we have the following quantitative version.

Theorem 58. *Let X be a Banach space such that its norm has modulus of convexity of power type q. Let $\varphi\colon X \to \mathbb{R} \cup \{+\infty\}$ be a bounded below proper lower semi-continuous quadratically coercive function. Then there are $f \in X^*$, $x_0 \in X$, and $C > 0$ such that for every $x \in X$*

$$\varphi(x) - f(x) \geq \varphi(x_0) - f(x_0) + C \min\{\|x - x_0\|^q, \|x - x_0\|^2\}.$$

Proof. Let us put $\alpha = \frac{1}{2} \liminf_{\|x\| \to +\infty} \frac{\varphi(x)}{\|x\|^2}$ and notice that $x \mapsto \varphi(x) - \alpha\|x\|^2$ is a bounded below proper lower semi-continuous quadratically coercive function. Therefore Corollary 56 implies that there exist $f_1 \in X^*$ and $x_0 \in X$ such that $\varphi(x_0) - \alpha\|x_0\|^2 - f_1(x_0) = \inf_{x \in X}\big(\varphi(x) - \alpha\|x\|^2 - f_1(x)\big)$. By Lemma 38 there are $f_2 \in X^*$ and $C > 0$ such that $\alpha\|x_0 + h\|^2 - \alpha\|x_0\|^2 - f_2(h) \geq C \min\{\|h\|^q, \|h\|^2\}$ for all $h \in X$. Put $f = f_1 + f_2$. Then for every $x \in X$

$$\begin{aligned}
&\varphi(x) - f(x)\\
&\quad= \varphi(x) - \alpha\|x\|^2 - f_1(x) + \alpha\|x\|^2 - \alpha\|x_0\|^2 - f_2(x - x_0) + \alpha\|x_0\|^2 - f_2(x_0)\\
&\quad\geq \varphi(x_0) - \alpha\|x_0\|^2 - f_1(x_0) + C \min\{\|x - x_0\|^q, \|x - x_0\|^2\} + \alpha\|x_0\|^2 - f_2(x_0)\\
&\quad= \varphi(x_0) - f(x_0) + C \min\{\|x - x_0\|^q, \|x - x_0\|^2\}.
\end{aligned}$$

□

Let $\mathcal{S}$ be a class of functions. We say that a Banach space X admits an $\mathcal{S}$-variational principle provided that for every bounded below proper lower semi-continuous function $\varphi\colon X \to \mathbb{R} \cup \{+\infty\}$ there is $f \in \mathcal{S}$ defined on X such that the function $\varphi - f$ attains its minimum. We say that a Banach space X admits a strong $\mathcal{S}$-variational principle if for every bounded below proper lower semi-continuous function $\varphi\colon X \to \mathbb{R} \cup \{+\infty\}$ and for every $\varepsilon > 0$ there is $f \in \mathcal{S}$ defined on X such that the function $\varphi - f$ attains a strong minimum at $x_0 \in X$ and $\varphi(x_0) < \inf_X \varphi + \varepsilon$.

Corollary 59. *Let X be a Banach space with the RNP. Let $S \subset C(X)$ be such that $g + f \in S$, $-g \in S$, and $x \mapsto g(x + a) \in S$ for every $g \in S$, $f \in X^*$, and $a \in X$. If S contains a linearly coercive bounded below function, then X admits a strong S-variational principle.*

Proof. Let $g \in S$ be linearly coercive and bounded below. Let $\varphi\colon X \to \mathbb{R} \cup \{+\infty\}$ be a bounded below proper lower semi-continuous function and $\varepsilon > 0$. Find $a, b \in X$ such that $\varphi(a) < \inf_X \varphi + \frac{\varepsilon}{4}$ and $g(b) < \inf_X g + \frac{\varepsilon}{4}$. Put $h(x) = g(x - a + b)$. The function $\varphi + h$ is a bounded below proper lower semi-continuous function which is linearly coercive. By Corollary 57 there is $f \in X^*$ such that the function $\varphi + h - f$ attains a strong minimum at $x_0 \in X$ and $\varphi(x_0) + h(x_0) < \inf_X(\varphi + h) + \frac{\varepsilon}{2}$. Then $-h + f \in S$ and $\varphi - (-h + f)$ attains a strong minimum at x_0. Finally, $\varphi(x_0) + h(x_0) < \inf_X(\varphi + h) + \frac{\varepsilon}{2} \leq \varphi(a) + h(a) + \frac{\varepsilon}{2} = \varphi(a) + g(b) + \frac{\varepsilon}{2} < \inf_X \varphi + \inf_X g + \varepsilon$, and so $\varphi(x_0) < \inf_X \varphi + \inf_X g + \varepsilon - h(x_0) = \inf_X \varphi + \inf_X h - h(x_0) + \varepsilon \leq \inf_X \varphi + \varepsilon$. □

Lemma 60. *Let X be a normed linear space and let $S \subset C(X)$ be a set of functions that is closed under shifting and scaling of the domain and under compositions with functions from $C^\infty(\mathbb{R})$ with all derivatives bounded, and closed under sums that are finite on bounded sets. If X admits a bump function from S, then X admits a linearly coercive function from S.*

Proof. By the assumptions there is $f_1 \in S$ such that $\operatorname{supp} f_1 \in B_X$ and $f_1(0) \neq 0$. By composing with a suitable C^∞-smooth function with all derivatives bounded we produce a non-negative function $f \in S$ for which $f(x) = 1$ for $\|x\| \geq 1$ and there is $r > 0$ such that $f(x) = 0$ for $x \in B(0, r)$. Put $g(x) = \sum_{n=1}^{\infty} f(\frac{1}{n}x)$. On bounded sets this is a finite sum and hence $g \in S$. Since $g(x) \geq [\|x\|] > \|x\| - 1$, it follows that $\liminf_{\|x\|\to+\infty} \frac{g(x)}{\|x\|} \geq 1$. □

Notice that if X admits a (strong) $\mathcal{S}$-variational principle and $\mathcal{S}$ is closed under taking restrictions to subspaces of X, then also every closed subspace of X admits a (strong) $\mathcal{S}$-variational principle (for $Y \subset X$ it suffices to extend the function φ from Y by putting $\varphi(x) = +\infty$ for $x \in X \setminus Y$.)

As the next theorem shows, in case of the higher order smoothness the assumption of the Radon-Nikodým property does not lead to a loss of generality. Note that it also covers the C^k-smooth case (when $\alpha = 0$) and the C^∞-smooth case (just consider the class $\mathcal{C}^\infty$ in all the statements).

Theorem 61. *Let X be a Banach space, $k \in \mathbb{N} \cup \{\infty\}$, and $\alpha \in [0, 1]$ such that $k + \alpha > 1$. The following statements are equivalent:*

(i) *X admits a strong $\mathcal{C}^{k,\alpha}$-variational principle.*
(ii) *X admits a $\mathcal{C}^{k,\alpha}_{\mathrm{loc}}$-variational principle.*
(iii) *X is super-reflexive, admits an equivalent norm with modulus of smoothness of power type $\min\{k + \alpha, 2\}$, and admits a $\mathcal{C}^{k,\alpha}$ bump.*
(iv) *X has the RNP and admits a $\mathcal{C}^{k,\alpha}$ bump.*
(v) *X does not contain c_0 and admits a $\mathcal{C}^{k,\alpha}_{\mathrm{loc}}$ bump.*

Proof. (i)⇒(ii) and (iii)⇒(iv)⇒(v) are clear.

(ii)⇒(v) First we show that X admits a $\mathcal{C}^{k,\alpha}_{\mathrm{loc}}$ bump. Let $\varphi\colon X \to \mathbb{R} \cup \{+\infty\}$ be defined as $\varphi(x) = \frac{1}{\|x\|}$ for $x \neq 0$ and $\varphi(0) = +\infty$. As X admits a $\mathcal{C}^{k,\alpha}_{\mathrm{loc}}$-variational principle, there is $f \in C^{k,\alpha}_{\mathrm{loc}}(X)$ such that $f(x_0) = \varphi(x_0)$ and $f(x) \leq \varphi(x)$ for every $x \in X$. Thus $x_0 \neq 0$ and $f(x) \leq \frac{1}{\|x\|} < \frac{1}{2\|x_0\|} = \frac{1}{2} f(x_0)$ whenever $\|x\| > 2\|x_0\|$. Since $f(x_0) > 0$, by composing f with a suitable smooth function we obtain a $C^{k,\alpha}_{\mathrm{loc}}$-smooth bump (see the end of the proof of Theorem 48).

Next, we show by contradiction that the space c_0 does not admit a $\mathcal{C}^{1,+}_{\mathrm{loc}}$-variational principle. Let $\|\cdot\|$ be a LUR norm on c_0 and let $\varphi\colon c_0 \to \mathbb{R} \cup \{+\infty\}$ be defined as $\varphi(x) = \frac{1}{\|x\|}$ for $x \neq 0$ and $\varphi(0) = +\infty$. Suppose that c_0 admits a $\mathcal{C}^{1,+}_{\mathrm{loc}}$-variational principle. Then there are $g \in C^{1,+}_{\mathrm{loc}}(c_0)$ and $x_0 \in c_0$ such that $g(x_0) = \varphi(x_0)$ and $g(x) \leq \varphi(x)$ for every $x \in c_0$. It follows that $x_0 \neq 0$ and there is $f \in c_0^*$, $\|f\| = 1$, such that $f(x_0) = \|x_0\|$.

Put $H = \ker f$. First we show that $\|\cdot\|$ attains a strong minimum at x_0 on the hyperplane $x_0 + H$. For any $h \in H$ we have $\|x_0\| = f(x_0) = f(x_0 + h) \leq \|x_0 + h\|$. Suppose that vectors $h_n \in H$ are such that $\|x_0 + h_n\| \to \|x_0\|$. Then we have $2\|x_0\| = f(2x_0 + h_n) \leq \|2x_0 + h_n\| \leq \|x_0\| + \|x_0 + h_n\| \to 2\|x_0\|$ and so

$2\|x_0\|^2+2\|x_0+h_n\|^2-\|x_0+x_0+h_n\|^2 \to 0$. Since the norm $\|\cdot\|$ is LUR, it follows that $h_n \to 0$.

Now let $\delta > 0$ be such that Dg is uniformly continuous on $U(x_0, 6\delta)$. From the above it follows that there is $\varepsilon > 0$ such that $\|x_0+h\| \geq \|x_0\|+\varepsilon$ for $h \in H$, $\|h\| \geq \delta$. Hence

$$g(x_0+h) \leq \frac{1}{\|x_0+h\|} \leq \frac{1}{\|x_0\|+\varepsilon} \quad \text{for all } h \in H,\ \|h\| \geq \delta. \tag{9}$$

Let $P \in \mathcal{L}(c_0; H)$ be the projection of c_0 onto H defined by $P(x) = x - f(x)\frac{x_0}{\|x_0\|}$. Notice that $\|P\| \leq 2$. Define a function $G\colon c_0 \to \mathbb{R}$ by

$$G(x) = g\big(x_0 + P(x)\big) - \frac{1}{\|x_0\|} - f(x)^2.$$

It is easy to check that DG is uniformly continuous on $U(0, 3\delta)$. We show that the function G is separating on $U(0, 3\delta)$. Let $x \in c_0$, $\|x\| \geq 2\delta$. If $\|P(x)\| \geq \delta$, then $G(x) \leq \frac{1}{\|x_0\|+\varepsilon} - \frac{1}{\|x_0\|}$ by (9). Otherwise $\delta > \|P(x)\| \geq \|x\| - |f(x)| \geq 2\delta - |f(x)|$ and hence $G(x) \leq \frac{1}{\|x_0+P(x)\|} - \frac{1}{\|x_0\|} - f(x)^2 \leq -f(x)^2 < -\delta^2$. Since $G(0) = 0$, by composing G with a suitable smooth function on $\mathbb{R}$ we obtain a $C^{1,+}$-smooth bump on c_0 (see the end of the proof of Theorem 48). This however contradicts Proposition 49.

(iv)$\Rightarrow$(i) We check that the construction in Lemma 60 works also for $\mathcal{C}^{k,\alpha}$ bumps. Indeed, $\|d^k g(x) - d^k g(y)\| \leq \sum_{n=1}^{\infty} \frac{1}{n^k} C \frac{1}{n^\alpha}\|x-y\|^\alpha = C\|x-y\|^\alpha \sum_{n=1}^{\infty} \frac{1}{n^{k+\alpha}}$. Thus we can use Corollary 59.

(v)$\Rightarrow$(iii) By Theorem 48 the space X admits a $\mathcal{C}^{k,\alpha}$ bump (note that it works also in the case $\alpha = 0$ by considering the $C^{k-1,1}$-smoothness). The rest follows from Corollary 51. □

We note that the same proof gives the statement of the previous theorem with $C^{k,\alpha}$ replaced by $C^{k,\Omega}$ with a general convex cone $\Omega \subset \mathcal{M}$, see also the formulation of Theorem 48. We only need to take extra care when $k = 1$ – when checking that the construction in Lemma 60 works we need to employ some careful estimates of the partial sums of $\sum \frac{1}{n}$.

The next theorem is a further example of an interplay between the geometry and smoothness of the space.

Theorem 62. *Let X be a Banach space such that its norm has modulus of convexity of power type q. Then the following statements are equivalent:*

(i) *X admits a T^q-smooth bump.*
(ii) *X admits a separating polynomial.*
(iii) *X admits a separating polynomial of degree at most q.*

Proof. (iii)$\Rightarrow$(ii) is obvious. (ii)$\Rightarrow$(i) follows from the fact that a polynomial is clearly T^p-smooth for every $p \in [0, +\infty)$, so a composition with a suitable smooth real function gives the desired bump (Proposition 1.115).

(i)⇒(iii) Let g be a T^q-smooth bump on X. Define a function $\varphi\colon X \to \mathbb{R} \cup \{+\infty\}$ by $\varphi(x) = g(x)^{-2}$ for $x \in \operatorname{supp}_\circ g$ and $\varphi(x) = +\infty$ otherwise. By Proposition 1.115 the function φ is T^q-smooth on $\operatorname{dom}\varphi$. Obviously φ is a bounded below proper lower semi-continuous quadratically coercive function. Thus by Theorem 58 there are $f \in X^*$, $x \in \operatorname{dom}\varphi$, and $C > 0$ such that $\varphi(x+h)-f(x+h) \geq \varphi(x)-f(x)+C\|h\|^q$ for every $h \in B_X$. Since φ is T^q-smooth at x, there is $P \in \mathcal{P}^{[q]}(X)$ satisfying $P(0) = \varphi(x)$ and $\varphi(x+h) - P(h) = o(\|h\|^q)$, $h \to 0$. Now put $Q = P - f - \varphi(x)$. Notice that $Q \in \mathcal{P}^{[q]}(X)$ because $q \geq 2$. Then $Q(0) = 0$ and

$$\begin{aligned} Q(h) &= P(h) - f(h) - \varphi(x) = \varphi(x+h) - f(h) - \varphi(x) + o(\|h\|^q) \\ &= \varphi(x+h) - f(x+h) - \big(\varphi(x) - f(x)\big) + o(\|h\|^q) \\ &\geq C\|h\|^q + o(\|h\|^q) \geq \frac{C}{2}\|h\|^q \end{aligned}$$

for all h from a suitably small neighbourhood of 0. Therefore Q is a separating polynomial. □

Theorem 63. *Let X be a Banach space such that its norm has modulus of convexity of power type 2. If X admits a weakly T^2-smooth bump, then X is isomorphic to a Hilbert space.*

Proof. Similarly as in the proof of Theorem 62 we obtain a function φ, $x \in X$, a functional $f \in X^*$, and a constant $C > 0$ such that φ is weakly T^2-smooth at x and $\varphi(x+h) - f(x+h) \geq \varphi(x) - f(x) + C\|h\|^2$ for every $h \in B_X$. Let $P \in \mathcal{P}^2(X)$ satisfy $P(0) = \varphi(x)$ and $\varphi(x+th) - P(th) = o(t^2)$, $t \to 0$ for each $h \in X$. Now put $Q(y) = P(y) + P(-y) - 2\varphi(x)$. Notice that $Q \in \mathcal{P}(^2X)$. Fix $h \in S_X$. Then

$$\begin{aligned} t^2 Q(h) &= Q(th) = P(th) + P(-th) - 2\varphi(x) \\ &= \varphi(x+th) - f(x+th) - \big(\varphi(x) - f(x)\big) \\ &\qquad + \varphi(x-th) - f(x-th) - \big(\varphi(x) - f(x)\big) + o(t^2) \\ &\geq 2Ct^2 + o(t^2) \end{aligned}$$

for all suitably small t. It follows that $Q(h) \geq 2C$ and therefore Q is a 2-homogeneous separating polynomial. □

Theorem 64. *Let X be a Banach space. The following statements are equivalent:*

(i) *X admits a strong polynomial variational principle.*
(ii) *X admits a strong $\mathcal{C}^\infty$-variational principle.*
(iii) *X admits $\mathcal{C}^k$-variational principle for every $k \in \mathbb{N}$.*
(iv) *X admits a separating polynomial.*
(v) *X is super-reflexive, admits an equivalent norm with modulus of smoothness of power type 2, and admits a $\mathcal{C}^\infty$ bump.*
(vi) *X does not contain c_0 and admits a $\mathcal{C}^k$ bump for every $k \in \mathbb{N}$.*

Proof. (i)⇒(ii)⇒(iii) and (v)⇒(vi) are obvious.

(ii)⇔(v) and (iii)⇔(vi) follow from Theorem 61.

(iv)⇒(i) follows from Corollary 59 and Fact 4.45.

(vi)⇒(iv) By Corollary 51 the space X is super-reflexive. Thus X has an equivalent norm with modulus of convexity of some power type, see [DGZ, Theorem IV.4.8]. Then X admits a separating polynomial by Theorem 62. □

4. Smooth bumps and structure II

We continue with our quest of investigating how the existence of smooth bumps (or norms) affects the structure of the underlying space. We show that we can replace the modulus of convexity in Theorem 62 by cotype q, but we need to assume a UT_c^q-smooth bump. Then we deal with the structure of spaces that admit C^∞-smooth bumps.

Next, we prove some basic arithmetic rules for the pointwise modulus of smoothness and we show that X is an Asplund space provided that it admits a bump with some rather weak differentiability properties. The main result here is Theorem 73, which says that under very weak convexity and differentiability assumptions we can already construct a norm with modulus of smoothness of power type (and hence the space is super-reflexive). We present some corollaries stating that X is isomorphic to a Hilbert space provided that both X and X^* admit a bump with certain smoothness properties. These vastly generalise for example the classical result of Victor Zakharovich Meshkov, which requires C^2-smoothness in both X and X^*.

We finish with a theorem stating that in some cases a weakly T^p-smooth bump suffices to give a polynomial with a certain kind of separation property. This will be used later in Section 7 when considering the smoothness of L_p spaces.

We start with an auxiliary observation.

Fact 65. *Let X be a normed linear space that does not admit a separating family of polynomials of degree at most $n \in \mathbb{N}$. Let $\mathcal{F} \subset \mathcal{P}^n(X)$ be a finite family satisfying $P(0) = 0$ for each $P \in \mathcal{F}$. Then for every $r > 0$ and $\varepsilon > 0$ there is $x \in S_X$ such that $\sup_{|\lambda|\le r}|P(\lambda x)| < \varepsilon$ for each $P \in \mathcal{F}$.*

Proof. Assume moreover that $r \ge 1$. For a polynomial $P \in \mathcal{P}(X)$ denote by P_j its j-homogeneous summand. Since the family $\{P_j;\ P \in \mathcal{F}, j \in \{1, \dots, n\}\}$ is not separating, there is $x \in S_X$ such that $|P_j(x)| < \frac{1}{nr^n}\varepsilon$ for each $P \in \mathcal{F}$, $j \in \{1, \dots, n\}$. Fix $P \in \mathcal{F}$. Then $|P(\lambda x)| \le \sum_{j=1}^n |\lambda|^j |P_j(x)| < \sum_{j=1}^n r^j \frac{1}{nr^n}\varepsilon \le \varepsilon$ for each $\lambda \in [-r, r]$. □

Theorem 62 has a counterpart in which we weaken the convexity assumption, but strengthen the smoothness assumption, and also the conclusion is slightly weaker.

Theorem 66. *Let X be a Banach space saturated with spaces of cotype q. If X admits a UT_c^q-smooth bump, then X admits a separating family of polynomials of degree at most q.*

Proof. By contradiction, suppose there is no separating family of polynomials of degree at most q on X. By Fact 4.48 no finite-codimensional subspace of X admits a separating family of polynomials of degree at most q. By the assumption there is a UT_c^q-smooth function $f: X \to \mathbb{R}$ satisfying $f(0) = 0$ and $f(x) \geq 2^q + 1$ for $x \in X$, $\|x\| \geq 1$. For each $x \in X$ let us denote by $P_x \in \mathcal{P}^{[q]}(X)$ the polynomial satisfying $P_x(0) = 0$ and $f(x+h) - f(x) - P_x(h) = o(\|h\|^q)$, $h \to 0$. We construct a normalised basic sequence $\{x_n\}_{n=1}^{\infty}$ in X such that

$$\sup_{\substack{x \in \operatorname{span}\{x_1,\dots,x_n\} \\ h \in \overline{\operatorname{span}}\{x_j;\, j>n\} \\ x,h \in B_X}} |P_x(h)| \leq \delta_n \tag{10}$$

for each $n \in \mathbb{N}$, where $\{\delta_n\}_{n=1}^{\infty} \subset \mathbb{R}^+$ satisfies $\sum_{n=1}^{\infty} \delta_n \leq 1$.

To this end put $\delta_n = \frac{1}{2^n}$ and let $\{\zeta_n\} \subset \mathbb{R}^+$ be such that $\prod_{n=1}^{\infty}(1+\zeta_n) \leq 2$. Put $\eta_n = \prod_{j=n}^{\infty}(1+\zeta_j)$. We construct the sequence $\{x_n\}$ by induction so that for every $n \in \mathbb{N}$ it satisfies

$$\sup_{\substack{x \in \operatorname{span}\{x_1,\dots,x_j\} \\ h \in \operatorname{span}\{x_{j+1},\dots,x_n\} \\ x,h \in \eta_n B_X}} |P_x(h)| \leq \sum_{k=j+1}^{n} \frac{1}{2^k} \quad \text{for each } 1 \leq j < n, \text{ and} \tag{11}$$

$$\|y\| \leq (1+\zeta_{n-1})\|y + \lambda x_n\| \quad \text{for all } y \in \operatorname{span}\{x_1,\dots,x_{n-1}\} \text{ and } \lambda \in \mathbb{R}. \tag{12}$$

This clearly implies (10).

Choose $x_1 \in S_X$ arbitrarily. Let $n \in \mathbb{N}$ and assume that $x_1,\dots,x_n$ are already defined such that (11) and (12) hold for this n. We find x_{n+1} such that (11) and (12) hold for $n+1$. Consider the family of polynomials

$$\mathcal{A} = \Big\{P;\ P(h) = P_x(y+h) - P_x(y),\\ x \in \operatorname{span}\{x_1,\dots,x_j\}, y \in \operatorname{span}\{x_{j+1},\dots,x_n\}, x, y \in \eta_n B_X, j = 1,\dots,n\Big\}.$$

An inspection of the proof of Theorem 1.110 reveals that y in the statement of that theorem can be restricted to a subspace. Thus from the proof we obtain that the UT_c^q-smoothness of f implies continuity of the mapping $x \mapsto P_x$ on the subspace $\operatorname{span}\{x_1,\dots,x_n\}$. Therefore the mapping $(x,y) \mapsto P_x(y+\cdot) - P_x(y)$ is continuous on $\operatorname{span}\{x_1,\dots,x_n\} \times X$. It follows that the set $\mathcal{A}$ is compact in $\mathcal{P}^{[q]}(X)$ and so it contains a finite subset $\mathcal{F}$ such that for each $P \in \mathcal{A}$ there is $Q \in \mathcal{F}$ satisfying $\sup_{h \in 4B_X}|P(h) - Q(h)| < \frac{1}{2^{n+2}}$.

Let Z be the finite-codimensional subspace of X from Lemma 4.66 used on $Y = \operatorname{span}\{x_1,\dots,x_n\}$ and $\varepsilon = \zeta_n$. By Fact 65 there is $x_{n+1} \in S_Z$ such that $\sup_{|\lambda|\leq 4}|P(\lambda x_{n+1})| < \frac{1}{2^{n+2}}$ for each $P \in \mathcal{F}$ and hence $\sup_{|\lambda|\leq 4}|P(\lambda x_{n+1})| < \frac{1}{2^{n+1}}$ for each $P \in \mathcal{A}$. The property (12) is clearly satisfied and so $\|y\| \leq (1+\zeta_n)\eta_{n+1} = \eta_n$ and $|\lambda| \leq \|y + \lambda x_{n+1}\| + \|y\| \leq \eta_{n+1} + \eta_n < 4$ whenever $y \in \operatorname{span}\{x_1,\dots,x_n\}$ and $\|y + \lambda x_{n+1}\| \leq \eta_{n+1}$. Using this together with the definition of $\mathcal{A}$ and the inductive hypothesis we obtain (11). This concludes the construction of $\{x_n\}$. Moreover, by (12), $\big\|\sum_{j=1}^{n} a_j x_j\big\| \leq (1+\zeta_n)\cdots(1+\zeta_{m-1})\big\|\sum_{j=1}^{m} a_j x_j\big\| \leq 2\big\|\sum_{j=1}^{m} a_j x_j\big\|$ for all $n < m$ and all $a_1,\dots,a_m \in \mathbb{R}$. Thus $\{x_n\}$ is a basic sequence in X.

By the assumption there is an infinite-dimensional subspace of $\overline{\operatorname{span}}\{x_n\}$ which is of cotype q. So there is a normalised block basic sequence $\{y_n\}$ of $\{x_n\}$ such that $Y = \overline{\operatorname{span}}\{y_n\}$ is of cotype q with the cotype constant C ([FHHMZ, Theorem 4.26]). Observe that (10) remains true when passing to a block sequence of $\{x_n\}$, but we also have to pass to a subsequence of $\{\delta_n\}$. Next, we will inductively construct a block sequence $\{u_n\}$ of $\{y_n\}$ and sets $K_n = \left\{\sum_{j=0}^{n} \varepsilon_j u_j;\ (\varepsilon_j) \in \{-1,1\}^{n+1}\right\}$. Set $u_0 = 0$. If $u_0, u_1, \dots, u_{n-1}$ have already been defined, we put

$$E_n = \Big\{h \in \operatorname{span}\{y_j;\ j > \max\operatorname{supp} u_{n-1}\} \cap B_X;$$
$$\big|f(x \pm h) - f(x) - P_x(\pm h)\big| \le \tfrac{1}{C^q}\|h\|^q \text{ for all } x \in K_{n-1}\Big\}.$$

It is clear that E_n contains some non-zero vectors. We put $\alpha_n = \sup_{h\in E_n}\|h\|$ and choose $u_n \in E_n$ such that $\|u_n\| > \alpha_n/2$.

There are two possibilities. Either $K = \overline{\bigcup_{n=1}^{\infty} K_n} \subset B_X$, or $K_n \setminus B_X \neq \emptyset$ for some $n \in \mathbb{N}$. In the first case, since Y is of a non-trivial cotype, it does not contain c_0 (check the canonical basis of c_0). From the Bessaga-Pełczyński theorem (Theorem 3.46) it follows that K is compact and $u_n \to 0$, hence also $\alpha_n \to 0$. However, since f is UT_c^q-smooth, there is $\delta > 0$ such that $|f(x+h) - f(x) - P_x(h)| \le \frac{1}{C^q}\|h\|^q$ for all $h \in X$, $\|h\| \le \delta$, and all $x \in K$. This contradicts the fact that $\alpha_n \to 0$.

In the latter case, let $n \in \mathbb{N}$ be the smallest such that $K_n \setminus B_X \neq \emptyset$. Then $\left\|\sum_{j=1}^{n} \sigma_j u_j\right\| > 1$ for some choice of signs $(\sigma_j) \in \{-1,1\}^n$. On the other hand, $\left\|\sum_{j=1}^{n} \varepsilon_j u_j\right\| \le \left\|\sum_{j=1}^{n-1} \varepsilon_j u_j\right\| + \|u_n\| \le 2$ for every $(\varepsilon_j) \in \{-1,1\}^n$. Hence using the definition of E_k and (10) we obtain

$$\begin{aligned} f\Big(\sum_{j=1}^{n} \sigma_j u_j\Big) &= \sum_{k=1}^{n} f\Big(\sum_{j=1}^{k} \sigma_j u_j\Big) - f\Big(\sum_{j=1}^{k-1} \sigma_j u_j\Big) \\ &\le \sum_{k=1}^{n} P_{\sum_{j=1}^{k-1}\sigma_j u_j}(\sigma_k u_k) + \frac{1}{C^q}\|u_k\|^q < \sum_{k=1}^{\infty} \delta_k + \frac{1}{C^q}\sum_{k=1}^{n}\|u_k\|^q \\ &\le 1 + \frac{1}{2^n}\sum_{(\varepsilon_j)\in\{-1,1\}^n}\Big\|\sum_{j=1}^{n}\varepsilon_j u_j\Big\|^q \le 1 + 2^q, \end{aligned}$$

which is a contradiction. □

The next result contains the main structural information concerning C^∞-smooth Banach spaces.

Theorem 67 ([Dev1]). *Let X be an infinite-dimensional Banach space which does not contain c_0 and admits a C^k-smooth bump for every $k \in \mathbb{N}$. Then X is of type 2, of exact cotype $q = q(X)$ which is an even integer, and admits a separating polynomial of degree d, where $q \le d \le 2q$ but not less. Further, every homogeneous separating polynomial on X is of degree mq for some $m \in \mathbb{N}$ and X has a subspace isomorphic to ℓ_q.*

Proof. The space X has a separating polynomial and is of type 2 according to Theorem 64. We know from Corollary 4.62 that $q = q(X)$ is an even integer. By Theorem 66 the space X has a separating family of polynomials of degree at most q. By Theorem 4.51 we conclude that X is of cotype q and has no separating polynomial of degree less that q. On the other hand by Fact 4.47 it has a separating polynomial of degree at most $2q$. By Lemma 4.68 there exists a subspace of X isomorphic to ℓ_q. The statement about the degrees of homogeneous separating polynomials then follows from Corollary 4.61. □

Corollary 68. *Let X be an infinite-dimensional Banach space that admits a C^k-smooth bump for every $k \in \mathbb{N}$. Then X is saturated by spaces from $\{\ell_p;\ p \text{ even}\} \cup \{c_0\}$.*

By combining Theorem 66, Theorem 64, and Theorem 4.51 we obtain another corollary.

Corollary 69 ([Dev1]). *Let X be a Banach space that admits a C^k-smooth bump for some $k \in \mathbb{N}$, $k > 1$, and suppose that X is saturated with spaces of cotype k. Then X admits a separating polynomial and is of type 2 and cotype k.*

The following lemma shows that the range of the derivative of a linearly coercive function is dense in a certain ball around origin.

Lemma 70. *Let X be a Banach space and let $\varphi\colon X \to \mathbb{R} \cup \{+\infty\}$ be a proper linearly coercive lower semi-continuous bounded below function that is Gâteaux differentiable on $\operatorname{dom}\varphi$. Then $\delta\varphi(\operatorname{dom}\varphi) \cap B_{X^*}(0,r)$ is dense in $B_{X^*}(0,r)$, where $r = \liminf\limits_{\|x\|\to+\infty} \frac{\varphi(x)}{\|x\|}$.*

Proof. Fix a functional $f \in U_{X^*}(0,r)$ and $0 < \varepsilon \le r - \|f\|$. Since we have $\liminf_{\|x\|\to+\infty} \frac{\varphi(x)-f(x)}{\|x\|} \ge r - \|f\| > 0$, the function $\varphi - f$ is lower semi-continuous and bounded below. So by the Ekeland variational principle (Theorem 53) there is $x \in X$ such that $\varphi(x+h) - f(x+h) \ge \varphi(x) - f(x) - \varepsilon\|h\|$ for all $h \in X$. It follows that $x \in \operatorname{dom}\varphi$ and $\varphi(x+h) - \varphi(x) \ge f(h) - \varepsilon\|h\|$ for all $h \in X$. This implies that $\delta g(x)[h] = \lim_{t\to 0+} \frac{1}{t}\big(\varphi(x+th) - \varphi(x)\big) \ge f(h) - \varepsilon\|h\|$ for every $h \in X$. Thus $|\delta g(x)[h] - f(h)| \le \varepsilon\|h\|$ for every $h \in X$ and consequently $\|\delta g(x) - f\| \le \varepsilon$. □

Let X, Y be normed linear spaces, $A \subset X$, and $f\colon A \to Y$. We say that f is α-Hölder at $x \in A$ if there are $\delta > 0$ and $L \ge 0$ such that $\|f(y) - f(x)\| \le L\|y - x\|^\alpha$ for all $y \in U(x,\delta) \cap A$. We say that f is pointwise α-Hölder on A if it is α-Hölder at every $x \in A$. We say that f is α-Hölder at $x \in A$ in the direction $h \in X$ if $f\restriction_{x+\operatorname{span}\{h\}}$ is α-Hölder at x. We say that f is directionally α-Hölder at x if for each $h \in S_X$ it is α-Hölder at x in the direction h. We say that f is directionally pointwise α-Hölder on A if it is directionally α-Hölder at every $x \in A$.

Note that it follows directly from the definitions that if f is Fréchet differentiable at x, then it is Lipschitz at x, and if f has directional derivatives in all directions at x (in particular if f is Gâteaux differentiable at x), then it is directionally Lipschitz at x.

Fact 71. *Let X, Y be normed linear spaces, $U \subset X$ open, $f : U \to \mathbb{R}$, $g : U \to Y$, $x \in U$, and $h \in X$. If f, g are α-Hölder at x in the direction h, resp. α-Hölder at x, then*

$$\rho_{x,h}^{fg}(t) \le O(1)\rho_{x,h}^{f}(t) + O(1)\rho_{x,h}^{g}(t) + O(t^{2\alpha}),\ t \to 0+, \qquad \textit{resp.}$$
$$\rho_{x}^{fg}(t) \le O(1)\rho_{x}^{f}(t) + O(1)\rho_{x}^{g}(t) + O(t^{2\alpha}),\ t \to 0+.$$

If f is α-Hölder at x in the direction h, resp. α-Hölder at x, and $f(x) \neq 0$, then

$$\rho_{x,h}^{1/f}(t) \le O(1)\rho_{x,h}^{f}(t) + O(t^{2\alpha}),\ t \to 0+, \qquad \textit{resp.}$$
$$\rho_{x}^{1/f}(t) \le O(1)\rho_{x}^{f}(t) + O(t^{2\alpha}),\ t \to 0+.$$

Proof. Everything follows from the following expansions:

$$\frac{1}{f(x+th)} + \frac{1}{f(x-th)} - \frac{2}{f(x)} = \frac{2f(x) - f(x+th) - f(x-th)}{f(x+th)f(x)} + \frac{\big(f(x+th) - f(x-th)\big)\big(f(x) - f(x-th)\big)}{f(x+th)f(x-th)f(x)},$$

$$\begin{aligned} 2\big(fg(x+th) + fg(x-th) - 2fg(x)\big) &= \big(f(x+th) + f(x-th) - 2f(x)\big)\big(g(x+th) + g(x)\big) \\ &\quad + \big(g(x+th) + g(x-th) - 2g(x)\big)\big(f(x+th) + f(x)\big) \\ &\quad + \big(g(x-th) - g(x)\big)\big(f(x-th) - f(x+th)\big) \\ &\quad + \big(f(x-th) - f(x)\big)\big(g(x-th) - g(x+th)\big). \end{aligned}$$

□

Theorem 72 ([DGZ2], [Va2]). *Let X be a Banach space that admits a continuous bump φ with one of the following properties:*

(i) *For each $x \in X$ there exists $\alpha > \frac{1}{2}$ such that the bump φ is α-Hölder at x and $\rho_x^{\varphi}(t) = o(t)$, $t \to 0+$.*

(ii) *φ is Gâteaux differentiable and for each $x \in X$ and $h \in S_X$ there is $p > 1$ such that $\varphi(x+th) - \varphi(x) - \delta\varphi(x)[th] = o(|t|^p)$, $t \to 0$.*

Then $\operatorname{dens} Y^* = \operatorname{dens} Y$ *for every subspace $Y \subset X$. In particular X is an Asplund space.*

Note that the condition (ii) is satisfied in particular if for each $x \in X$ there is $p > 1$ such that φ is weakly T^p-smooth at x.

Proof. Since both properties pass to subspaces, we may assume that $Y = X$. Let Q be a dense subset of X with $|Q| = \operatorname{dens} X$. Define a function $\psi : X \to \mathbb{R} \cup \{+\infty\}$ by $\psi(x) = \frac{1}{\varphi(x)^2}$ for $x \in \operatorname{supp}_o \varphi$ and $\psi(x) = +\infty$ otherwise.

Assume first that (i) holds. From Fact 71 we obtain $\rho_x^{\psi}(t) = o(t)$, $t \to 0+$ for each $x \in \operatorname{dom} \psi$. Fix $f \in X^*$ and $\varepsilon > 0$. By the Ekeland variational principle (Theorem 53) used on $\psi - f$ there is $x \in X$ such that $\psi(x+h) - f(x+h) \ge \psi(x) - f(x) - \frac{\varepsilon}{10}\|h\|$ for each $h \in X$. There is $0 < r \le 1$ such that $|\psi(x+h) + \psi(x-h) - 2\psi(x)| \le \frac{\varepsilon}{10}\|h\|$ for

$h \in B(0,2r)$. So $-\frac{\varepsilon}{10}\|h\| \le \psi(x+h)-\psi(x)-f(h) = \psi(x+h)-\psi(x)+f(-h) \le \psi(x+h)-\psi(x)+\psi(x-h)-\psi(x)+\frac{\varepsilon}{10}\|h\| \le \frac{\varepsilon}{5}\|h\|$ for all $h \in B(0,2r)$.

Let $z \in Q \cap B(x,r)$, $z \neq x$, and find $n \in \mathbb{N}$ such that $\frac{1}{2}\|z-x\| \le \frac{1}{n} \le \|z-x\|$. Then $z+\frac{1}{n}h \in B(x,2r)$ for any $h \in B_X$. Hence

$$\begin{aligned}
&\big|\psi(z+\tfrac{1}{n}h)-\psi(z)-\tfrac{1}{n}f(h)\big| \\
&\quad= \big|\psi(z+\tfrac{1}{n}h)-\psi(x)-f(z+\tfrac{1}{n}h-x)-\psi(z)+\psi(x)+f(z-x)\big| \\
&\quad\le \big|\psi(z+\tfrac{1}{n}h)-\psi(x)-f(z+\tfrac{1}{n}h-x)\big| + |\psi(z)-\psi(x)-f(z-x)| \\
&\quad\le \frac{\varepsilon}{5}\|z-x+\tfrac{1}{n}h\| + \frac{\varepsilon}{5}\|z-x\| \le \frac{\varepsilon}{5}(2\|z-x\|+\tfrac{1}{n})
\end{aligned}$$

for every $h \in B_X$. It follows that

$$\big|n\big(\psi(z+\tfrac{1}{n}h)-\psi(z)\big)-f(h)\big| \le \frac{\varepsilon}{5}(2n\|z-x\|+1) \le \varepsilon.$$

Now consider the function $g_{z,n}\colon B_X \to \mathbb{R}$, $g_{z,n}(h) = n\big(\psi(z+\frac{1}{n}h)-\psi(z)\big)$. We have just shown that $g_{z,n} \in \ell_\infty(B_X)$ and $\|g_{z,n}-f\| \le \varepsilon$ when considering f as a member of $\ell_\infty(B_X)$.

So setting $M = \{g_{z,n};\ z \in Q, n \in \mathbb{N}, g_{z,n} \in \ell_\infty(B_X)\}$ we have $X^* \subset \overline{M}$ in $\ell_\infty(B_X)$ and hence $\operatorname{dens} X^* \le |M| \le |Q| = \operatorname{dens} X$.

Now assume that (ii) holds. The function ψ is Gâteaux differentiable on $\operatorname{dom}\psi$ by Theorem 1.70. Proposition 1.115 implies that for each $x \in \operatorname{dom}\psi$ and $h \in S_X$ there is $1 < p < 2$ such that $\psi(x+th)-\psi(x)-\delta\psi(x)[th] = o(|t|^p)$, $t \to 0$.

Fix $f \in X^*$ and $\varepsilon > 0$. Since $\lim_{\|x\|\to+\infty}\frac{\psi(x)}{\|x\|} = +\infty$, by Lemma 70 there is $x \in \operatorname{dom}\psi$ such that $\|f-\delta\psi(x)\| < \frac{\varepsilon}{2}$. Let γ be such that ψ is bounded on $B(x,\gamma)$. Set

$$F_n = \big\{h \in B_X;\ |\psi(x+th)-\psi(x)-\delta\psi(x)[th]| \le n\|th\|^{1+\frac{1}{n}}, 0 \le t \le \gamma\big\}.$$

By the continuity of ψ each F_n is closed. From the properties of the function ψ it follows that $B_X = \bigcup_{n\in\mathbb{N}} F_n$. By the Baire category theorem there is $K \in \mathbb{N}$ such that F_K has a non-empty interior in B_X. Thus there are $y \in B_X$ and $r > 0$ such that $|\psi(x+th)-\psi(x)-\delta\psi(x)[th]| \le K\|th\|^{1+\alpha}$ for all $h \in B(y,r)$ and $0 \le t \le \gamma$, where $\alpha = \frac{1}{K}$. In other words, $|\psi(x+u)-\psi(x)-\delta\psi(x)[u]| \le K\|u\|^{1+\alpha}$ for all $u \in D = \{th;\ h \in B(y,r), 0 \le t \le \gamma\}$.

Put $D_0 = \{x+th;\ h \in B(y,\frac{r}{2}), 0 \le t \le \gamma\}$. Let $\eta = \big(\min\{\frac{\varepsilon}{20K}, \frac{r}{2\|y\|+r}, 1\}\big)^{\frac{2}{\alpha}}$ and choose $z \in Q \cap D_0 \cap B(x,\eta)$, $z \neq x$. Further, let $n \in \mathbb{N}$ be such that $\frac{1}{2}\|z-x\|^{1+\frac{\alpha}{2}} \le \frac{1}{n} \le \|z-x\|^{1+\frac{\alpha}{2}} \le \|z-x\|$. Note that $z-x \in D$ and also $z-x+\frac{1}{n}h \in D$ for each $h \in B_X$. Indeed, $z-x = tv$ for some $v \in B(y,\frac{r}{2})$ and $0 < t \le \gamma$. We have $z-x+\frac{1}{n}h = t(v+\frac{1}{t}\frac{1}{n}h)$, and since $t \ge \frac{\|z-x\|}{\|y\|+\frac{r}{2}}$, we get $\|v+\frac{1}{t}\frac{1}{n}h-y\| \le \frac{r}{2}+\frac{\|y\|+\frac{r}{2}}{\|z-x\|}\|z-x\|^{1+\frac{\alpha}{2}} \le \frac{r}{2}+(\|y\|+\frac{r}{2})\eta^{\frac{\alpha}{2}} \le r$.

Hence

$$\begin{aligned}&\left|\psi(z+\tfrac{1}{n}h)-\psi(z)-\delta\psi(x)[\tfrac{1}{n}h]\right|\\&\quad\le\left|\psi(z+\tfrac{1}{n}h)-\psi(x)-\delta\psi(x)[z-x+\tfrac{1}{n}h]\right|+\left|\psi(z)-\psi(x)-\delta\psi(x)[z-x]\right|\\&\quad\le K\|z-x+\tfrac{1}{n}h\|^{1+\alpha}+K\|z-x\|^{1+\alpha}\le K(2^{1+\alpha}+1)\|z-x\|^{1+\alpha}\end{aligned}$$

for every $h\in B_X$. It follows that

$$\begin{aligned}\left|n\big(\psi(z+\tfrac{1}{n}h)-\psi(z)\big)-f(h)\right|&\le\left|n\big(\psi(z+\tfrac{1}{n}h)-\psi(z)\big)-\delta\psi(x)[h]\right|+\frac{\varepsilon}{2}\\&\le 5Kn\|z-x\|^{1+\alpha}+\frac{\varepsilon}{2}\le 10K\|z-x\|^{\frac{\alpha}{2}}+\frac{\varepsilon}{2}\le\varepsilon.\end{aligned}$$

The proof can now be finished in the same way as in the case (i). □

Theorem 73 ([DGZ2], [MPVZ]). *Let X be a Banach space such that X^* is a weak* Asplund space. If X admits a continuous directionally pointwise Lipschitz bump β with $\rho^{\beta}_{x,h}$ of some power type for each $h\in S_X$, $x\in X$, then X is super-reflexive. If moreover there is $p\in(1,2]$ such that each $\rho^{\beta}_{x,h}$ is of power type p, then X admits an equivalent norm with modulus of smoothness of power type p.*

We remark that some kind of convexity condition on X (such as the RNP, see Theorem 8) is necessary in the previous theorem: the space c_0 admits a $C^{1,1}_{\mathrm{loc}}$-smooth norm (Theorem 94), but it is not super-reflexive.

Proof. For $x\in X$ put $\varphi(x)=\frac{1}{\beta(x)^2}$ if $\beta(x)\neq 0$ and $\varphi(x)=+\infty$ otherwise. Note that φ is continuous on X and $\operatorname{dom}\varphi$ is an open set. By Fact 71 for each $x\in\operatorname{dom}\varphi$, $h\in S_X$ if $\rho^{\beta}_{x,h}$ is of power type $p_{x,h}$, then $\rho^{\varphi}_{x,h}$ is of power type $\min\{p_{x,h},2\}$.

Since $\varphi(x)=+\infty$ for x outside some ball and $\varphi\ge 0$, it follows that $\operatorname{dom}\varphi^*=X^*$ and φ^* is Lipschitz, as it is a supremum of a family of Lipschitz functions with their Lipschitz constant bounded. Since X^* is a weak* Asplund space, there is a dense G_δ subset Ω of X^* such that φ^* is Fréchet differentiable at every point of Ω. Denote $\psi=(\varphi^*)_*$ and note that $\psi\le\varphi$ and $\psi^*=\varphi^*$ by Fact 25.

Fix $f\in\Omega$ and denote $x=D\varphi^*(f)$. By Corollary 34 we have $x\in\operatorname{dom}\varphi$, $\psi(x)=\varphi(x)$, and $f\in\partial\psi(x)$. Thus $\rho^{\psi}_{x,h}(t)=\psi(x+th)+\psi(x-th)-2\psi(x)\le\varphi(x+th)+\varphi(x-th)-2\varphi(x)\le\rho^{\varphi}_{x,h}(t)$ for all $t\ge 0$, $h\in X$. Hence $\rho^{\psi}_{x,h}$ is of the same power type as $\rho^{\varphi}_{x,h}$. The function ψ is continuous on $\operatorname{Int}\operatorname{dom}\psi\supset\operatorname{dom}\varphi$, since ψ is dominated by a continuous function φ. Therefore ρ^{ψ}_{x} is of some power type (resp. of power type p) by Proposition 16. This in turn implies that $\Delta^{\psi^*}_{f,x}$ is of some power type (resp. of power type q, where $\frac{1}{p}+\frac{1}{q}=1$) by Corollary 37. Now Theorem 39 and the standard duality [DGZ, Proposition IV.1.12] finishes the proof. □

Note that the above theorem applies for example if X has the RNP and admits a continuous directionally T^p-smooth bump (and in particular G^2-smooth bump, Proposition 1.112).

Corollary 74. *Let X be a Banach space such that either X^* or X^{**} is a weak* Asplund space, or equivalently such that either X or X^* has the RNP. If both X and X^* admit a continuous directionally pointwise Lipschitz bump with directional pointwise modulus of smoothness of power type 2, then X is isomorphic to a Hilbert space.*

Proof. The equivalence of the assumption of the RNP follows from Theorem 8.

First assume that X^* is a weak* Asplund space. Then X admits an equivalent norm with modulus of smoothness of power type 2 by Theorem 73. In particular, X is reflexive and so X^{**} is an Asplund space. Thus also X^* admits an equivalent norm with modulus of smoothness of power type 2 by Theorem 73. Hence X admits an equivalent norm with modulus of convexity of power type 2 ([DGZ, Proposition IV.1.12]). If Y is a separable subspace of X, then Y admits a G^2-smooth bump by Corollary 7.72 (together with a version of Lemma 20 for norms), and hence Y is isomorphic to a Hilbert space by Theorem 63. So the space X is isomorphic to a Hilbert space by Corollary 45.

If we assume that X^{**} is a weak* Asplund space, then X^* admits an equivalent norm with modulus of smoothness of power type 2 by Theorem 73. In particular, X^* is an Asplund space, and so X is isomorphic to a Hilbert space by the first part of the proof. □

Corollary 75. *Let X be a Banach space such that X admits a continuous directionally pointwise Lipschitz bump with directional pointwise modulus of smoothness of power type 2 and X^* admits a continuous bump φ with one of the following properties:*

(i) *φ is pointwise Lipschitz with pointwise modulus of smoothness of power type 2,*
(ii) *φ is Gâteaux differentiable and $\varphi(x+th)-\varphi(x)-\delta\varphi(x)[th] = O(|t|^2)$, $t \to 0$ for each $h \in S_{X^*}$, $x \in X^*$.*

Then X is isomorphic to a Hilbert space. The same holds if we switch the roles of X and X^*.

Note that condition (ii) is in particular satisfied if φ is weakly T^2-smooth.

Proof. Since X^* is an Asplund space by Theorem 72, we may use Corollary 74. If we switch the roles of X and X^*, then X is an Asplund space by Theorem 72, hence X^* has the RNP by Theorem 2, and we may again use Corollary 74. □

The following theorem says that in the non-separable setting a *weakly T^q-smooth* bump together with the modulus of convexity of power type q is sufficient for the existence of a polynomial with certain separating properties. Compare this with Theorem 62.

Theorem 76 ([MV]). *Let X be a non-separable Banach space such that its norm has modulus of convexity of power type q, where q is not an even integer. If X admits a continuous weakly T^q-smooth bump, then there is a polynomial on X of degree less than q which is not weakly sequentially continuous.*

Proof. Let $g\colon X \to \mathbb{R}$ be the continuous weakly T^q-smooth bump. We define a function $\varphi\colon X \to \mathbb{R} \cup \{+\infty\}$ by $\varphi(x) = \frac{1}{g(x)^2}$ if $g(x) \neq 0$ and $g(x) = +\infty$ otherwise. Then φ is a bounded below proper continuous quadratically coercive function and hence by Theorem 58 there are $f \in X^*$, $x \in \operatorname{dom}\varphi$, and $C > 0$ such that $\varphi(x+h) - \varphi(x) - f(h) \geq C\|h\|^q$ for every $h \in B_X$. As φ is weakly T^q-smooth at x (Proposition 1.115), there is $P \in \mathcal{P}^{[q]}(X)$ satisfying $P(0) = 0$ and such that $\varphi(x+th) - \varphi(x) - P(th) = o(|t|^q)$, $t \to 0$ for each $h \in X$. Put $Q(h) = P(h) + P(-h)$. Since q is not an even integer, $\deg Q < q$.

As X is a reflexive non-separable space, X has an uncountable total biorthogonal system $\{(x_\alpha; f_\alpha)\}_{\alpha\in\Lambda}$ with $\{x_\alpha\} \subset B_X$ (see e.g. [HMVZ, Corollary 4.11]). Assume that Q is weakly sequentially continuous. We claim that there is $\gamma \in \Lambda$ such that $Q(tx_\gamma) = 0$ for all $t \in [-1,1]$. If this is not the case, then (since Λ is uncountable) there are a sequence $\{\alpha_n\}$ of distinct elements of Λ and a sequence $\{t_n\} \subset [-1,1]$ such that $|Q(t_n x_{\alpha_n})| > \delta$ for some $\delta > 0$ and all $n \in \mathbb{N}$. By the weak sequential compactness of B_X there is a subsequence $\{t_{n_k} x_{\alpha_{n_k}}\}$ of $\{t_n x_{\alpha_n}\}$ which is weakly convergent; and in fact weakly null by the totality of $\{x_\alpha; f_\alpha\}$. This contradicts the fact that $Q(0) = 0$.

Put $h = x_\gamma$. Then

$$\begin{aligned}2C\|th\|^q &\leq \varphi(x+th) - \varphi(x) - f(th) + \varphi(x-th) - \varphi(x) - f(-th)\\ &= P(th) + P(-th) + o(|t|^q) = Q(th) + o(|t|^q) = o(|t|^q),\ t \to 0,\end{aligned}$$

which is a contradiction. It means that Q is not weakly sequentially continuous. □

Proposition 77. *Let X be a Banach space with an analytic norm such that there is a non-empty open $U \subset X$ on which the norm is weakly sequentially Cauchy-continuous. Then X^* is separable.*

Proof. First notice that X is an Asplund space (Corollary 3) and so X does not contain ℓ_1. Let V be a non-empty open ball in U with $\operatorname{dist}(V, X \setminus U) > 0$. Then $D\|\cdot\|(V)$ is relatively compact by Proposition 41. Further, since $D\|\cdot\| \in C^\omega(X \setminus \{0\}; X^*)$ and $\overline{\operatorname{span}}\, D\|\cdot\|(X \setminus \{0\}) = X^*$ by the Bishop-Phelps theorem, it follows from Corollary 1.159 that $\overline{\operatorname{span}}\, D\|\cdot\|(V) = X^*$ and so X^* is separable. □

5. Local dependence on finitely many coordinates

We start by introducing the concept of local dependence on finitely many coordinates and show that the existence of an LFC bump has a strong impact on structure, namely the space is then a c_0-saturated Asplund space. We show how the LFC property is used to construct a C^∞-smooth norm on $c_0(\Gamma)$. We prove that on a separable Asplund space a continuous LFC bump (or just a locally wsc bump) implies already the existence of a C^∞-smooth (and LFC) bump. Finally, we show how to construct a smooth approximation of a norm using a smooth generalised James boundary and the techniques of this section.

Definition 78. Let X be a topological vector space, $\Omega \subset X$ an open subset, E an arbitrary set, $M \subset X^*$, and $g: \Omega \to E$. We say that g *depends only on* M on a set $U \subset \Omega$ if $g(x) = g(y)$ whenever $x, y \in U$ are such that $f(x) = f(y)$ for all $f \in M$. We say that g *depends locally on finitely many coordinates from* M (g is LFC-M for short) if for each $x \in \Omega$ there are a neighbourhood $U \subset \Omega$ of x and a finite subset $F \subset M$ such that g depends only on F on U. We say that g depends locally on finitely many coordinates (g is LFC for short) if it is LFC-X^*.

A canonical example of a non-trivial LFC function is the supremum norm on c_0, which is LFC-$\{e_j^*\}$ away from the origin. Indeed, take any $x = (x_j) \in c_0$, $x \neq 0$. Let $n \in \mathbb{N}$ be such that $|x_j| < \frac{1}{2}\|x\|_\infty$ for $j > n$. Then $\|\cdot\|_\infty$ depends only on $\{e_1^*, \dots, e_n^*\}$ on $U(x, \frac{1}{4}\|x\|_\infty)$.

Fact 79. *Let X be a topological vector space, $\Omega \subset X$ an open subset, E an arbitrary set, $M \subset X^*$, and $g: \Omega \to E$. The mapping g is LFC-M if and only if for each $x \in \Omega$ there are an open neighbourhood $U \subset \Omega$ of x, $n \in \mathbb{N}$, a biorthogonal system $\{(e_j; f_j)\}_{j=1}^n \subset X \times M$, an open set $V \subset \mathbb{R}^n$, and a mapping $G: V \to E$, such that $g(y) = G\big(f_1(y), \dots, f_n(y)\big)$ for all $y \in U$, where $G(w) = g\big(x + \sum_{j=1}^n (w_j - f_j(x))e_j\big)$ for each $w \in V$.*

Proof. Let U_0 be an open neighbourhood of x and $n \in \mathbb{N}$ such that g depends only on $\{f_1, \dots, f_n\} \subset M$ on U_0. Without loss of generality we may assume that the functionals $f_1, \dots, f_n$ are linearly independent. Hence there are vectors $e_1, \dots, e_n \in X$ such that $\{(e_j; f_j)\}_{j=1}^n$ is a biorthogonal system (since $\bigcap_{j \neq k} \ker f_j \not\subset \ker f_k$ for each $1 \leq k \leq n$). Let $\Phi: \mathbb{R}^n \to X$ be defined as $\Phi(w) = x + \sum_{j=1}^n \big(w_j - f_j(x)\big)e_j$. This is a continuous mapping and so the set $V = \Phi^{-1}(U_0)$ is an open subset of $\mathbb{R}^n$. Notice that $G(w) = g(\Phi(w))$ for each $w \in V$. Let $\Psi: X \to \mathbb{R}^n$ be defined as $\Psi(y) = (f_1(y), \dots, f_n(y))$. This is a continuous mapping and so the set $U = \Psi^{-1}(V) \cap U_0$ is open. Moreover, $\Phi(\Psi(x)) = x$, hence U is an open neighbourhood of x.

Now choose any $y \in U$. Since $\Psi(y) \in V$, the composition $G(\Psi(y))$ is well-defined. Further, using the facts that $y \in U_0$, $\Phi(\Psi(y)) \in U_0$, g depends only on $\{f_1, \dots, f_n\}$ on U_0, and $f_k(\Phi(\Psi(y))) = f_k(x) + \sum_{j=1}^n \big(f_j(y) - f_j(x)\big) f_k(e_j) = f_k(y)$ for each $1 \leq k \leq n$, we may conclude that $G(\Psi(y)) = g\big(\Phi(\Psi(y))\big) = g(y)$.

The other implication is obvious. □

From the previous fact it follows that the mappings G through which g is "locally factorised" share some properties of g – for example they are continuous (or smooth) if and only if g is continuous (or smooth).

Notice further that if $g: \Omega \to E$ is LFC and $h: E \to F$ is any mapping, then the mapping $h \circ g$ is also LFC. The following fact is also immediate.

Fact 80. *Let X, Y be normed linear spaces, E an arbitrary set, $M \subset Y^*$, $\Omega \subset Y$ an open set, $g: \Omega \to E$ an LFC-M mapping, and $T \in \mathcal{L}(X; Y)$. Then the mapping $g \circ T: T^{-1}(\Omega) \to E$ is LFC-$T^*(M)$.*

A typical example of the use of the LFC notion is the following.

Lemma 81. *Let X, Y, Z be normed linear spaces, $U \subset X$, $V \subset Y$ open sets, $M \subset Y^*$, $k \in \mathbb{N} \cup \{\infty\}$, $\Phi\colon U \to V$ a continuous mapping such that $f \circ \Phi$ is C^k-smooth for each $f \in M$, and let $g \in C^k(V; Z)$ be a LFC-M mapping. Then $g \circ \Phi \in C^k(U; Z)$.*

Proof. Fix $x \in U$. By Fact 79 there are $f_1, \dots, f_n \in M$, an open neighbourhood W of $\Phi(x)$, an open $\Omega \subset \mathbb{R}^n$, and $G \in C^k(\Omega; Z)$ such that $g(y) = G\big(f_1(y), \dots, f_n(y)\big)$ for all $y \in W$. Then $g(\Phi(z)) = G\big(f_1 \circ \Phi(z), \dots, f_n \circ \Phi(z)\big)$ for each $z \in \Phi^{-1}(W)$, which is an open neighbourhood of x, and so the assertion follows. □

Let (X, τ) be a topological vector space, $(Y, \mathcal{U})$ a uniform space, $\Omega \subset X$ open, and $f\colon \Omega \to Y$. We say that f is locally weakly uniformly continuous if for each $x \in \Omega$ there is a τ-neighbourhood U of x such that $f \restriction_U$ is weakly uniformly continuous (or more precisely w–$\mathcal{U}$ uniformly continuous).

Fact 82. *Let X be a topological vector space, $(Y, \mathcal{U})$ a uniform space, $\Omega \subset X$ open, and let $g\colon \Omega \to Y$ be a continuous LFC mapping. Then g is locally weakly uniformly continuous.*

Proof. We use the notation from the proof of Fact 79. Let $K \subset V$ be a compact neighbourhood $\Psi(x)$. Since G is clearly continuous, it is uniformly continuous on K. The mapping Ψ is obviously w–$\|\cdot\|$ uniformly continuous and hence $g = G \circ \Psi$ is w–$\mathcal{U}$ uniformly continuous on $\Psi^{-1}(K) \cap U_0$. □

The next lemma serves as an addition to the Implicit function theorem.

Lemma 83. *Let X, Y, Z be normed linear spaces, $U \subset X$ and $V \subset Y$ open sets, and let $f\colon U \times V \to Z$ depend only on $\{f_1, \dots, f_n\} \subset (X \oplus Y)^*$ on $U \times V$. Further, assume that there is a unique mapping $u\colon U \to V$ satisfying $f\big(x, u(x)\big) = 0$ for all $x \in U$. Then u depends only on $\{f_1 \restriction_X, \dots, f_n \restriction_X\} \subset X^*$ on U.*

Proof. Let $x, y \in U$ be such that $f_j \restriction_X(x) = f_j \restriction_X(y)$ for $j = 1, \dots, n$. Then $(x, u(x)) \in U \times V$, $(y, u(x)) \in U \times V$, and

$$f_j\big((x,u(x))\big) = f_j \restriction_X(x) + f_j\big((0,u(x))\big) = f_j \restriction_X(y) + f_j\big((0,u(x))\big) = f_j\big((y,u(x))\big)$$

for $j = 1, \dots, n$. It follows that $0 = f\big(x, u(x)\big) = f\big(y, u(x)\big)$, which by the uniqueness of u gives $u(x) = u(y)$. □

The word "coordinate" in the term LFC originates of course from spaces with bases, where LFC was first defined using the coordinate functionals. In order to apply the LFC techniques to spaces without a Schauder basis, the notion was generalised using arbitrary functionals from the dual. However, as we will show, the generalisation does not substantially increase the supply of LFC mappings on Banach spaces with a Schauder basis. Interestingly, up to a small perturbation we can always in addition assume that the given LFC mapping in fact depends on the coordinate functionals. We begin with a simple related result for Markushevich bases.

Proposition 84. *Let X be a separable Banach space, E a set, and $g: X \to E$ an LFC mapping. Then there is a Markushevich basis $\{(e_j; f_j)\}_{j=1}^{\infty} \subset X \times X^*$ such that g is LFC-$\{f_j\}$.*

Proof. By the Lindelöf property of X we can choose a countable $\{h_j\} \subset X^*$ such that it separates the points of X and g is LFC-$\{h_j\}$. Then we can use the Markushevich theorem ([FHHMZ, Theorem 4.59]) to construct a Markushevich basis $\{(e_j; f_j)\}$ such that $\operatorname{span}\{f_j\} = \operatorname{span}\{h_j\}$. Now let $x \in X$ and $U \subset X$ be a neighbourhood of x such that g depends only on $M = \{h_1, \dots, h_n\}$ on U. Let $M \subset \operatorname{span}\{f_1, \dots, f_m\}$. Then for any $y, z \in U$ satisfying $f_k(y) = f_k(z)$ for $k = 1, \dots, m$ we also have $h_j(y) = h_j(z)$ for all $j = 1, \dots, n$ and hence $g(y) = g(z)$. Thus g depends only on $\{f_1, \dots, f_m\}$ on U. □

We would like to establish a similar result for Schauder bases. In this context shrinking Schauder bases emerge quite naturally, taking into account Theorem 91.

Fact 85. *Let X and Y be Banach spaces with equivalent Schauder bases $\{x_j\}$ and $\{y_j\}$ respectively. Then $\{x_j\}$ is shrinking if and only if $\{y_j\}$ is shrinking.*

Proof. Let $\{x_j\}$ be a shrinking basis and $T: Y \to X$ be an isomorphism of Y onto X such that $Ty_j = x_j$. Then $T^*: X^* \to Y^*$ is an isomorphism of X^* onto Y^* such that $T^* x_j^* = y_j^*$ and thus

$$Y^* = T^*(X^*) = T^*(\overline{\operatorname{span}}\{x_j^*\}) \subset \overline{T^*(\operatorname{span}\{x_j^*\})} = \overline{\operatorname{span}}\, T^*(\{x_j^*\}) = \overline{\operatorname{span}}\{y_j^*\}.$$

□

Lemma 86. *Let X be a Banach space with a shrinking Schauder basis $\{e_j\}$. Let $f \in X^*$, $\varepsilon > 0$, and $n \in \mathbb{N}$. Then there are a (shrinking) Schauder basis $\{x_j\}$ of X and $N \in \mathbb{N}$, $N > n$, such that $x_j = e_j$ for $1 \le j < N$, $\{x_j\}$ is $(1+\varepsilon)$-equivalent to $\{e_j\}$, $\operatorname{span}\{x_j\}_{j=k}^m = \operatorname{span}\{e_j\}_{j=k}^m$ for all $1 \le k \le n$ and $m \ge k$, and $\overline{\operatorname{span}}\{x_j;\ j \ge N\} \subset \ker f$.*

Proof. If $f(e_j) = 0$ for all $j > n$, then we may put $N = n+1$ and $x_j = e_j$ for all $j \in \mathbb{N}$. Otherwise by passing to a multiple of f we may without loss of generality assume that there is $K \ge n$ such that $f(e_K) = 1$. Let us denote $f_k = f - P_{k-1}^*(f)$, $k \in \mathbb{N}, k > 1$, where P_j are the projections associated with the basis $\{e_j\}$. As $\{e_j\}$ is shrinking, $\|f_k\| \to 0$ and hence we can find $N \in \mathbb{N}$ such that $N > K$ and $\|f_N\| \le \frac{\varepsilon}{(2+\varepsilon)\|e_K\|}$. Put $x_j = e_j$ for $1 \le j < N$ and $x_j = e_j - f(e_j)e_K$ for $j \ge N$. For any $m_1, m_2 \in \mathbb{N}$, $m_1 \le m_2$, and any sequence $\{a_j\}$ of scalars we have

$$\begin{aligned}\left\|\sum_{j=m_1}^{m_2} a_j x_j\right\| &= \left\|\sum_{j=m_1}^{m_2} a_j e_j - e_K \sum_{j=\max\{m_1,N\}}^{m_2} a_j f(e_j)\right\| \\ &\le \left\|\sum_{j=m_1}^{m_2} a_j e_j\right\| + \left\|e_K f_N\left(\sum_{j=m_1}^{m_2} a_j e_j\right)\right\| \\ &\le \left(1 + \|e_K\|\|f_N\|\right)\left\|\sum_{j=m_1}^{m_2} a_j e_j\right\| \le \left(1 + \frac{\varepsilon}{2+\varepsilon}\right)\left\|\sum_{j=m_1}^{m_2} a_j e_j\right\|\end{aligned}$$

and

$$\left\|\sum_{j=m_1}^{m_2} a_j x_j\right\| \ge \left\|\sum_{j=m_1}^{m_2} a_j e_j\right\| - \left\|e_K f_N\left(\sum_{j=m_1}^{m_2} a_j e_j\right)\right\| \ge \left(1-\frac{\varepsilon}{2+\varepsilon}\right)\left\|\sum_{j=m_1}^{m_2} a_j e_j\right\|.$$

This implies that $\{x_j\}$ is a basic sequence $(1+\varepsilon)$-equivalent to $\{e_j\}$. Since $x_K = e_K$ and $n \le K < N$, we have $\operatorname{span}\{x_j\}_{j=k}^m = \operatorname{span}\{e_j\}_{j=k}^m$ for all $1 \le k \le n$ and $m \ge k$, and therefore $\operatorname{span}\{x_j\} = \operatorname{span}\{e_j\}$, which implies that $\{x_j\}$ is a basis of X. Clearly, $f(x_j) = 0$ for $j \ge N$. □

It is certainly worth noticing that the method used in the previous lemma (and the next theorem) does not rely on the classical argument of perturbation by the norm-summable sequence. In fact our new basis is "far" away from the original one.

Theorem 87. *Let X be a Banach space with a shrinking Schauder basis $\{e_j\}$, let $\{f_j\} \subset X^*$ be a countable set, and $\varepsilon > 0$. Then there is a (shrinking) Schauder basis $\{x_j\}$ of X such that it is $(1+\varepsilon)$-equivalent to $\{e_j\}$, $\operatorname{span}\{x_j\}_{j=1}^m = \operatorname{span}\{e_j\}_{j=1}^m$ for all $m \in \mathbb{N}$, and $\operatorname{span}\{f_j\} \subset \operatorname{span}\{x_j^*\}$.*

Proof. Choose a sequence of $\varepsilon_j > 0$ such that $\prod_j (1+\varepsilon_j) \le 1+\varepsilon$ and put $N_0 = 1$. We apply Lemma 86 to $\{e_j\}$, f_1, ε_1, and $n = 1$. We obtain a basis $\{x_j^1\}$ which is $(1+\varepsilon_1)$-equivalent to $\{e_j\}$ and $N_1 \in \mathbb{N}$ such that $\overline{\operatorname{span}}\{x_j^1;\ j \ge N_1\} \subset \ker f_1$. Moreover, $x_j^1 = e_j$ for $j < N_1$ and $\operatorname{span}\{x_j^1\}_{j=1}^m = \operatorname{span}\{e_j\}_{j=1}^m$ for all $m \in \mathbb{N}$.

We proceed by induction. Suppose the basis $\{x_j^k\}$ and $N_k \in \mathbb{N}$ have already been defined in such a way that the basis $\{x_j^k\}$ is $\prod_{i \le k}(1+\varepsilon_i)$-equivalent to the basis $\{e_j\}$, $x_j^k = x_j^{k-1}$ for $j < N_k$, $\operatorname{span}\{x_j^k\}_{j=1}^m = \operatorname{span}\{e_j\}_{j=1}^m$ for all $m \in \mathbb{N}$, and finally $\overline{\operatorname{span}}\{x_j^k;\ j \ge N_i\} \subset \ker f_i$ for each $1 \le i \le k$. We apply Lemma 86 to $\{x_j^k\}$, f_{k+1}, ε_{k+1}, and $n = N_k$ in order to obtain a basis $\{x_j^{k+1}\}$ which is $\prod_{i \le k+1}(1+\varepsilon_i)$-equivalent to the basis $\{e_j\}$, and a number $N_{k+1} \in \mathbb{N}$ satisfying $N_{k+1} > N_k$, such that $\overline{\operatorname{span}}\{x_j^{k+1};\ j \ge N_{k+1}\} \subset \ker f_{k+1}$. Moreover, $x_j^{k+1} = x_j^k$ for $j < N_{k+1}$ and $\operatorname{span}\{x_j^{k+1}\}_{j=1}^m = \operatorname{span}\{x_j^k\}_{j=1}^m = \operatorname{span}\{e_j\}_{j=1}^m$ for all $m \in \mathbb{N}$. Since also $\operatorname{span}\{x_j^{k+1}\}_{j=N_i}^m = \operatorname{span}\{x_j^k\}_{j=N_i}^m$ for all $1 \le i \le k$ and $m \ge N_i$, we have $\overline{\operatorname{span}}\{x_j^{k+1};\ j \ge N_i\} \subset \ker f_i$ for $1 \le i \le k+1$.

Clearly, there is a sequence $\{x_j\}$ such that $\lim_{k\to\infty} x_j^k = x_j$ for all $j \in \mathbb{N}$. (This is because the sequence $\{N_k\}$ is increasing and thus $\{x_j^k\}_{k=1}^\infty$ is eventually constant.) It is straightforward to check that $\operatorname{span}\{x_j\}_{j=1}^m = \operatorname{span}\{e_j\}_{j=1}^m$ for all $m \in \mathbb{N}$, $\{x_j\}$ is a basis of X which is $(1+\varepsilon)$-equivalent to $\{e_j\}$, and $\overline{\operatorname{span}}\{x_j;\ j \ge N_i\} \subset \ker f_i$ (which means that $f_i \in \operatorname{span}\{x_j^*;\ j < N_i\}$) for any $i \in \mathbb{N}$. □

If a Banach space X has a shrinking Schauder basis, then using the Lindelöf property of X and Theorem 87 we obtain the following corollary, which usually allows us to work only with LFC-$\{e_j^*\}$ mappings.

Corollary 88. *Let X be a Banach space with a shrinking Schauder basis $\{e_j\}$, E any set, $g\colon X \to E$ an LFC mapping, and $\varepsilon > 0$. Then there is a (shrinking) Schauder basis $\{x_j\}$ of X, $(1+\varepsilon)$-equivalent to $\{e_j\}$, such that g is LFC-$\{x_j^*\}$.*

As we shall see, the existence of an LFC bump function on a Banach space has a strong impact on its structure. The next lemma shows that if a space admits an arbitrary (even non-measurable) LFC bump, then it admits also an upper semi-continuous LFC bump.

Lemma 89. *Let X be a topological vector space, $M \subset X^*$, and let $g\colon X \to \mathbb{R}$ be an LFC-M function. Then $\chi_{\operatorname{supp} g}$ is an upper semi-continuous LFC-M function.*

Proof. Since $\operatorname{supp} g$ is closed, $\chi_{\operatorname{supp} g}$ is upper semi-continuous. Fix $x \in X$. There is an open neighbourhood U of X such that g depends only on $\{f_1, \dots, f_n\} \subset M$ on U. We claim that $\chi_{\operatorname{supp} g}$ also depends only on $\{f_1, \dots, f_n\}$ on U. Suppose that this is not the case. Then there are $y, z \in U$ such that $z - y \in \bigcap_{j=1}^n \ker f_j$, $y \in \operatorname{supp} g$, and $z \notin \operatorname{supp} g$. There is a net $\{y_\gamma\}_{\gamma\in\Gamma} \subset X$ such that $g(y_\gamma) \neq 0$ for all $\gamma \in \Gamma$ and $y_\gamma \to y$. Since $y_\gamma + z - y \to z$ and U is open, we way suppose that both $\{y_\gamma\}_{\gamma\in\Gamma} \subset U$ and $\{y_\gamma + z - y\}_{\gamma\in\Gamma} \subset U$. Hence $g(y_\gamma + z - y) = g(y_\gamma) \neq 0$ for all $\gamma \in \Gamma$ and so $z \in \operatorname{supp} g$, a contradiction. □

Theorem 90 ([PWZ]). *An infinite-dimensional Banach space X that admits an (arbitrary) LFC bump contains c_0.*

Proof. By Lemma 89 there is a non-empty bounded closed set $A \subset X$ such that χ_A is LFC. Without loss of generality we may assume that A contains the origin. Put $g(x) = 1 - \chi_A(x) + 1 - \chi_A(-x)$. Then g is an even lower semi-continuous LFC function with $g(0) = 0$ and $g(x) = 2$ for all x outside some ball. Assume that X does not contain c_0. We apply Lemma 47 on the function g. Let K be the compact set and V the neighbourhood of 0 from Lemma 47. Let $r > 0$ be such that $U(0, r) \subset V$.

Since g is LFC, there are a covering of K by open balls $U(x_j, \delta_j)$, $j = 1, \dots, n$, and a finite set $M \subset X^*$ such that g depends only on M on each $U(x_j, 2\delta_j)$. Put $\delta = \min\{\frac{r}{2}, \delta_1, \dots, \delta_n\}$ and pick any $x \in \bigcap_{f\in M} \ker f$ with $\|x\| = \delta$. Since $x \in V$, by Lemma 47 there is $z \in K$ such that $g(z + x) > g(z)$. Let $k \in \mathbb{N}$ be such that $z \in U(x_k, \delta_k)$. Since $z + x \in U(x_k, 2\delta_k)$ and g depends only on M on $U(x_k, 2\delta_k)$, we have $g(z + x) = g(z)$, a contradiction. □

Theorem 91 ([FZ1]). *Let X be a Banach space, $M \subset X^*$, and suppose that X admits an (arbitrary) LFC-M bump. Then $\overline{\operatorname{span}}\, M = X^*$.*

Proof. By Lemma 89 there is a non-empty bounded closed set $A \subset X$ such that χ_A is LFC-M. Pick any $f \in X^*$ and $\varepsilon > 0$ and notice that f is bounded on A. Let $\mathfrak{I}_A\colon X \to \mathbb{R} \cup \{+\infty\}$ be the indicator function of the set A, i.e. $\mathfrak{I}_A(x) = 0$ for $x \in A$ and $\mathfrak{I}_A = +\infty$ for $x \in X \setminus A$. Put $\varphi = \mathfrak{I}_A - f$. Then φ is a lower semi-continuous bounded below function and so by the Ekeland variational principle (Theorem 53) there is $x_0 \in X$ such that $\varphi(x) \geq \varphi(x_0) - \varepsilon\|x - x_0\|$ for every $x \in X$. Obviously $x_0 \in A$ and for every $x \in A$ we have $-f(x) \geq -f(x_0) - \varepsilon\|x - x_0\|$ from which it follows that

$$f(x - x_0) \le \varepsilon \|x - x_0\| \quad \text{for every } x \in A. \tag{13}$$

Let $\delta > 0$ and $f_1, \dots, f_n \in M$ be such that χ_A depends only on $\{f_1, \dots, f_n\}$ on $U(x_0, \delta)$. Put $Z = \bigcap_{j=1}^n \ker f_j$. For any $z \in Z$, $\|z\| < \delta$ we have $x_0 + z \in A$ and hence $f(z) \le \varepsilon\|z\|$ by (13). This means that $\|f \restriction_Z\| \le \varepsilon$. By the Hahn-Banach theorem we can find a functional $g \in X^*$ such that $g = f$ on Z and $\|g\| \le \varepsilon$. Clearly, $f - g \in \operatorname{span} M$ and $\|f - (f - g)\| \le \varepsilon$. □

Corollary 92. *An infinite-dimensional Banach space X which admits an LFC bump is a c_0-saturated Asplund space.*

Proof. Since the existence of an LFC bump passes to subspaces, X is c_0-saturated by Theorem 90 and it is enough to show that X^* is separable provided that X is separable.

Let g be an LFC bump on X. By the Lindelöf property of X there exists a countable set $M \subset X^*$ such that g is LFC-M. Then $X^* = \overline{\operatorname{span}}\, M$ by Theorem 91, hence X^* is separable. □

We say that a norm on X is LFC-M it is LFC-M on the set $X \setminus \{0\}$.

Lemma 93. *If in Lemma 23 we assume additionally that f is LFC-M on Ω for some $M \subset X^*$, then the norm β is also LFC-M.*

Proof. We use the same notation as in the proof of Lemma 23. It is easy to check that if the function f is LFC-M on Ω, then the function F is LFC-N on G, where $N = \{g \circ P_1;\ g \in M\} \cup \{P_2\} \subset (X \oplus \mathbb{R})^*$, and P_1, P_2 denote the canonical projections of $X \oplus \mathbb{R}$ onto $X, \mathbb{R}$. Thus in this case ν is LFC-M by Lemma 83. □

The space c_0 plays a prominent role in this theory both as a canonical example of a space with LFC structure and through Corollary 92. The LFC property usually allows for certain smooth constructions. A prime example is the next theorem.

Theorem 94. *For any set Γ the space $c_0(\Gamma)$ admits an equivalent C^∞-smooth LFC-$\{e_\gamma^*\}$ norm.*

Proof. Let $\varphi \in C^\infty(\mathbb{R})$ be a convex non-negative function such that $\varphi(t) = 0$ for $t \in [-\frac{1}{2}, \frac{1}{2}]$ and $\varphi(1) = 1$. Define a function $\Phi\colon c_0(\Gamma) \to \mathbb{R}$ by

$$\Phi(x) = \sum_{\gamma \in \Gamma} \varphi\big(x(\gamma)\big).$$

Since the sum is locally finite, the function Φ is C^∞-smooth and LFC-$\{e_\gamma^*\}$. Obviously Φ is convex and separating, since $\Phi(0) = 0$, while $\Phi(x) \ge 1$ whenever $\|x\| \ge 1$. Now we apply Lemma 93. □

Of course by composing the norm from Theorem 94 with a suitable C^∞-smooth function we can produce a C^∞-smooth LFC bump on $c_0(\Gamma)$. Next, we show that on a separable Banach space the existence of a continuous LFC bump already implies the existence of a smooth bump. But first we need a few technical results.

Fact 95. *Let $f:\mathbb{R}\to\mathbb{R}$ be an even function that is non-decreasing on $[0,+\infty)$ and let $\varphi:\mathbb{R}\to\mathbb{R}$ be an even function with bounded support that is non-increasing on $[0,+\infty)$. Then the convolution $f*\varphi(x)=\int_{\mathbb{R}}f(x-t)\varphi(t)\,\mathrm{d}t$ is an even function that is non-decreasing on $[0,+\infty)$.*

Proof. Note that $f*\varphi$ is well-defined as the functions f and φ are clearly Borel measurable and bounded on bounded sets. Further, $f*\varphi(-x)=\int_{\mathbb{R}}f(-x-t)\varphi(t)\,\mathrm{d}t=\int_{\mathbb{R}}f(x+t)\varphi(t)\,\mathrm{d}t=\int_{\mathbb{R}}f(x-t)\varphi(t)\,\mathrm{d}t=f*\varphi(x)$, using first the fact that f is even and then the fact that φ is even.

Now pick any $0\le x<y<+\infty$. The function $\psi(t)=\varphi\big(\frac{y-x}{2}-t\big)-\varphi\big(\frac{x-y}{2}-t\big)$ is an odd function (this is obvious), such that $\psi(t)\ge 0$ for $t\ge 0$. Indeed, either we have $0\ge\frac{y-x}{2}-t\ge\frac{x-y}{2}-t$, or $0<\frac{y-x}{2}-t\le t-\frac{x-y}{2}$ and in both cases we use the properties of φ. Similarly we obtain that the function $t\mapsto f\big(\frac{x+y}{2}+t\big)-f\big(\frac{x+y}{2}-t\big)$ is non-negative for $t\ge 0$. Therefore

$$\begin{aligned}
f*\varphi(y)-f*\varphi(x)&=\int_{\mathbb{R}}f(t)\big(\varphi(y-t)-\varphi(x-t)\big)\,\mathrm{d}t=\int_{\mathbb{R}}f\big(\tfrac{x+y}{2}+t\big)\psi(t)\,\mathrm{d}t\\
&=\int_{(-\infty,0)}f\big(\tfrac{x+y}{2}+t\big)\psi(t)\,\mathrm{d}t+\int_{(0,+\infty)}f\big(\tfrac{x+y}{2}+t\big)\psi(t)\,\mathrm{d}t\\
&=-\int_{(0,+\infty)}f\big(\tfrac{x+y}{2}-t\big)\psi(t)\,\mathrm{d}t+\int_{(0,+\infty)}f\big(\tfrac{x+y}{2}+t\big)\psi(t)\,\mathrm{d}t\\
&=\int_{(0,+\infty)}\Big(f\big(\tfrac{x+y}{2}+t\big)-f\big(\tfrac{x+y}{2}-t\big)\Big)\psi(t)\,\mathrm{d}t\ge 0.
\end{aligned}$$

□

Let X be a Banach lattice. We say that a function $f:X\to\mathbb{R}$ is a *strongly lattice* function if $f(x)\le f(y)$ whenever $|x|\le|y|$. Recall that a Banach space X with an unconditional basis $\{e_j\}$ has a natural lattice structure defined by $\sum_{j=1}^{\infty}a_je_j\ge 0$ if and only if $a_j\ge 0$ for all $j\in\mathbb{N}$, and similarly for ℓ_∞.

Lemma 96. *Let $\varepsilon>0$ and a sequence $\Delta=\{\delta_n\}_{n=1}^{\infty}$, $\delta_n>0$, be given. Denote*

$$V_\Delta=\Big\{x\in\ell_\infty;\ |x(n_0)|-\delta_{n_0}>\sup_{n>n_0}|x(n)|+\delta_{n_0}\quad\textit{for some } n_0\in\mathbb{N}\Big\},$$

which is an open subset of ℓ_∞. There is a convex strongly lattice 1-Lipschitz function $F:\ell_\infty\to[0,+\infty)$ such that $\big|F(x)-\|x\|\big|\le\varepsilon$ for any $x\in\ell_\infty$ and F is LFC-$\{e_j^*\}$ and C^∞-smooth on V_Δ, where e_j^* are the coordinate functionals on ℓ_∞. Moreover, for each $x\in V_\Delta$ satisfying $\|x\|\ge 4\varepsilon$ there is a neighbourhood V_x of x such that F depends only on a finite subset of $\{e_j^*;\ j\in\operatorname{supp}x\}$ on V_x.

Proof. Let $\{e_j\}\subset\ell_\infty$ be the canonical basic sequence biorthogonal to $\{e_j^*\}$ and P_n the projection $P_n(x)=\sum_{j=1}^{n}e_j^*(x)e_j$. Set $\varepsilon_1=\min\{\delta_1,\varepsilon\}$ and $\varepsilon_n=\min\{\delta_n,\varepsilon_{n-1},\frac{1}{n}\}$ for $n>1$. Pick a sequence $\{\varphi_n\}_{n=1}^{\infty}$ of C^∞-smooth even functions $\varphi_n:\mathbb{R}\to[0,+\infty)$ such that $\operatorname{supp}\varphi_n\subset[-\varepsilon_n,\varepsilon_n]$, φ_n is non-increasing on $[0,+\infty)$, and $\int_{\mathbb{R}}\varphi_n(t)\,\mathrm{d}t=1$.

Define a sequence $\{F_n\}_{n=1}^{\infty}$ of functions $F_n\colon \ell_\infty \to [0,\infty)$ by

$$F_n(x) = \int_{\mathbb{R}^n} \Big\| x - \sum_{j=1}^{n} t_j e_j \Big\| \prod_{j=1}^{n} \varphi_j(t_j)\,\mathrm{d}\lambda_n.$$

It is easily checked that each F_n is convex, 1-Lipschitz and $\big|F_n(x) - \|x\|\big| \le \varepsilon$ for any $x \in \ell_\infty$.

We show that F_n is strongly lattice by induction. Put $F_0 = \|\cdot\|$ and note that F_0 is strongly lattice. Assume that F_{n-1} is strongly lattice for some $n \in \mathbb{N}$. Pick any $x, y \in \ell_\infty$, $x = (x_j)$, $y = (y_j)$, satisfying $|x| \le |y|$. Define $g\colon \mathbb{R} \to \mathbb{R}$ by $g(u) = F_{n-1}(x + (u - x_n)e_n)$. Then by the inductive hypothesis g is an even function non-decreasing on $[0, +\infty)$. Using the Fubini theorem we obtain

$$\begin{aligned}
F_n(y) - F_n(x) &= \int_{\mathbb{R}} (F_{n-1}(y - te_n) - F_{n-1}(x - te_n))\varphi_n(t)\,\mathrm{d}t \\
&= \int_{\mathbb{R}} \Big(F_{n-1}(y - te_n) - F_{n-1}\big(x + (y_n - x_n - t)e_n\big)\Big)\varphi_n(t)\,\mathrm{d}t \\
&\quad + \int_{\mathbb{R}} \Big(F_{n-1}\big(x + (y_n - x_n - t)e_n\big) - F_{n-1}(x - te_n)\Big)\varphi_n(t)\,\mathrm{d}t \\
&= \int_{\mathbb{R}} \Big(F_{n-1}(y - te_n) - F_{n-1}\big(x + (y_n - x_n - t)e_n\big)\Big)\varphi_n(t)\,\mathrm{d}t \\
&\quad + g * \varphi_n(y_n) - g * \varphi_n(x_n) \ge 0,
\end{aligned}$$

because $F_{n-1}(y - te_n) \ge F_{n-1}(x + (y_n - x_n - t)e_n)$, which follows from the inductive hypothesis (notice that $|y - te_n| \ge |x + (y_n - x_n - t)e_n|$ in the lattice sense), and also using Fact 95.

Using the Fubini theorem and the fact that $\int_{\mathbb{R}} \varphi_j = 1$ for any $j \in \mathbb{N}$ we obtain for $m > n$ and any $x \in \ell_\infty$

$$\begin{aligned}
|F_m(x) - F_n(x)| &= \left| \int_{\mathbb{R}^m} \left(\Big\| x - \sum_{j=1}^{m} t_j e_j \Big\| - \Big\| x - \sum_{j=1}^{n} t_j e_j \Big\| \right) \prod_{j=1}^{m} \varphi_j(t_j)\,\mathrm{d}\lambda_m \right| \\
&\le \int_{\mathbb{R}^m} \Big\| \sum_{j=n+1}^{m} t_j e_j \Big\| \prod_{j=1}^{m} \varphi_j(t_j)\,\mathrm{d}\lambda_m \le \varepsilon_{n+1}.
\end{aligned}$$

It follows that the sequence $\{F_n\}$ converges uniformly on ℓ_∞ to some function $F\colon \ell_\infty \to [0, +\infty)$. Consequently F is convex, strongly lattice, 1-Lipschitz, and $\big|F(x) - \|x\|\big| \le \varepsilon$ for any $x \in \ell_\infty$.

Fix any $x \in V_\Delta$ and let $n_0 \in \mathbb{N}$ be such that $|x(n_0)| - \delta_{n_0} > \sup_{n>n_0} |x(n)| + \delta_{n_0}$. Note that the set $U_x = \{y \in \ell_\infty;\ |y(n_0)| - \delta_{n_0} > \sup_{n>n_0} |y(n)| + \delta_{n_0}\} \subset V_\Delta$ is an open neighbourhood of x. Pick any $y \in U_x$ and $k > n_0$. Then

$$\|y - t_1e_1 - \cdots - t_ke_k\| = \|y - t_1e_1 - \cdots - t_{n_0}e_{n_0}\| = \|P_{n_0}(y) - t_1e_1 - \cdots - t_{n_0}e_{n_0}\|,$$

as long as $|t_j| \le \delta_{n_0}$ for $n_0 \le j \le k$. Since $\varepsilon_n \le \delta_{n_0}$ for $n \ge n_0$ and $\int_{\mathbb{R}} \varphi_n = 1$, it follows that $F_k(y) = F_{n_0}(y) = F_{n_0}(P_{n_0}(y))$. This means that $F(y) = F_{n_0}(P_{n_0}(y))$

and so F depends only on $\{e_j^*\}_{j=1}^{n_0}$ on U_x. Using substitution and Corollary 1.91 (similarly as in the proof of Lemma 7.69) it is easy to see that F is C^∞-smooth.

Finally, let $x \in V_\Delta$ be such that $\|x\| \geq 4\varepsilon$ and let U_x be as above. Further, put $V_x = U_x \cap U(x, \varepsilon)$ and $M = \{j \in \operatorname{supp} x;\ j \leq n_0\}$. Pick any $y \in U_x$ and $k > n_0$, and denote $P_M(y) = \sum_{j \in M} e_j^*(y) e_j$. Then

$$2\varepsilon < \|y - t_1 e_1 - \cdots - t_k e_k\| = \|P_M(y) - t_1 e_1 - \cdots - t_{n_0} e_{n_0}\|$$

whenever $|t_j| \leq \varepsilon_j$ for each $1 \leq j \leq k$. This follows from the discussion above combined with the fact that $|y(j) - t_j| < 2\varepsilon$ for $1 \leq j \leq n_0$, $j \notin \operatorname{supp} x$. Therefore $F(y) = F_{n_0}(P_M(y))$ and so F depends only on $\{e_j^*;\ j \in M\}$ on V_x. □

Let X be a normed linear space, $\Omega \subset X$ open, and $f \colon \Omega \to \mathbb{R}$. We say that f is locally weakly sequentially continuous (locally wsc for short) if for each $x \in \Omega$ there is a (norm) neighbourhood U of x such that $f \restriction_U$ is weakly sequentially continuous.

Theorem 97. *Let X be a separable Banach space. The following statements are equivalent:*

(i) *X admits a continuous LFC bump.*
(ii) *X^* is separable and X admits a bump that is locally weakly sequentially continuous.*
(iii) *X admits a C^∞-smooth LFC bump.*

Proof. (iii)⇒(i) is clear and (i)⇒(ii) follows from Corollary 92 and Fact 82.

(ii)⇒(iii) Let $\{f_n\}_{n\in\mathbb{N}}$ be a dense subset of S_{X^*} and define a bounded linear operator $T \colon X \to c_0$ by $T(x) = \big(2^{-n} f_n(x)\big)_{n\in\mathbb{N}}$. Denote $Y = T(X)$, which is a linear subspace of c_0. Notice that T is one-to-one and the linear operator $T^{-1} \colon Y \to X$ has the following property: $T^{-1} \restriction_{T(M)}$ is $\|\cdot\|$–w continuous whenever $M \subset X$ is bounded. It follows that if $A \subset X$ is a closed convex bounded set, then so is $T(A)$ as a subset of a normed linear space Y. Indeed, if $\{y_n\} \subset T(A)$ is such that $y_n \to y \in Y$, then $T^{-1}(\{y_n\}) \cup \{T^{-1}(y)\}$ is bounded and hence $T^{-1}(y_n) \xrightarrow{w} T^{-1}(y)$. As A is closed and convex, it is weakly closed and hence $T^{-1}(y) \in A$, which in turn implies $y \in T(A)$.

As X admits a locally wsc bump, by shifting, scaling, and a composition with a suitable continuous function we produce a locally wsc function $g \colon X \to [1, 2]$ such that $g(0) = 1$ and $g(x) = 2$ whenever $\|x\| \geq 1$. By the Lindelöf property of X there exists a countable covering of X by open balls $\{U_n\}_{n\in\mathbb{N}}$ such that $g \restriction_{\overline{U_n}}$ is wsc for each $n \in \mathbb{N}$. Without loss of generality we may assume that $0 \in U_1$. Put $A_n = \bigcup_{j=1}^n \overline{U_j}$, $n \in \mathbb{N}$. Since each $\overline{U_j}$, $j \in \mathbb{N}$, is weakly closed, it is easily checked that $g \restriction_{A_n}$ is wsc for each $n \in \mathbb{N}$.

Now for each $n \in \mathbb{N}$ let us define $G_n(y) = g\big(T^{-1}(y)\big)$ for $y \in T(A_n)$. From the discussion above it follows that $T(A_n)$ is closed in Y and G_n is continuous on $T(A_n)$, hence by Tietze's theorem G_n can be extended to a continuous function $G_n \colon Y \to \mathbb{R}$. Note that $g(x) = G_n\big(T(x)\big)$ for each $x \in A_n$.

Choose a sequence of real numbers $\{\eta_n\}$ decreasing to 1 such that $\eta_1 \leq 1 + \frac{1}{4}$ and a decreasing sequence $\Delta = \{\delta_n\}$ such that $0 < \delta_n < \frac{1}{4}(\eta_n - \eta_{n+1})$ and $\delta_1 \leq \frac{1}{8}$. By Theorem 94 the space Y admits a C^∞-smooth LFC bump. The class of C^∞-smooth LFC functions is an approximation class and hence by Corollary 7.50 and Theorem 7.52 for each $n \in \mathbb{N}$ there is $H_n \in C^\infty(Y)$ which is LFC and satisfies $\sup_{y\in Y}|H_n(y) - \eta_n G_n(y)| < \delta_n$. Set $h_n = H_n \circ T$. Then $h_n \in C^\infty(X)$, h_n is LFC by Fact 80, and $\sup_{x\in A_n}|h_n(x) - \eta_n g(x)| < \delta_n$. It follows that $h_n \to g$ locally uniformly.

Further, let us define $\Phi\colon X \to \ell_\infty$ by $\Phi(x) = \big(h_n(x)\big)_n$. First we show that Φ is continuous. Pick any $x \in X$ and $\varepsilon > 0$. Let $k \in \mathbb{N}$ be such that $x \in U_k$ and $\delta_k < \frac{\varepsilon}{4}$. There is a neighbourhood $U \subset U_k$ of x such that $|g(x) - g(y)| < \frac{\varepsilon}{4}$ whenever $y \in U$. Then

$$\begin{aligned}|h_n(x) - h_n(y)| &\leq |h_n(x) - \eta_n g(x)| + \eta_n|g(x) - g(y)| + |\eta_n g(y) - h_n(y)|\\ &< 2\delta_n + \eta_n\frac{\varepsilon}{4} < \varepsilon\end{aligned}$$

whenever $y \in U$ and $n \geq k$. This, together with the fact that $h_1, \dots, h_{k-1}$ are continuous, implies that Φ is continuous at x.

Next, consider the set V_Δ from Lemma 96. We claim that $\Phi(X) \subset V_\Delta$. Indeed, let $x \in X$ and $n_0 \in \mathbb{N}$ be such that $x \in U_{n_0}$. Then for $n > n_0$ we have

$$\begin{aligned}h_{n_0}(x) - \delta_{n_0} &> \eta_{n_0} g(x) - 2\delta_{n_0} > \eta_{n_0+1} g(x) + 2\delta_{n_0}\\ &> \eta_n g(x) + \delta_n + \delta_{n_0} > h_n(x) + \delta_{n_0},\end{aligned}$$

where the second inequality follows from the definition of δ_{n_0}.

We now apply Lemma 96 to the sequence Δ and $\varepsilon = \frac{1}{8}$ in order to obtain the corresponding function F, and we set $f = F \circ \Phi$. The properties of F and h_n together with the continuity of Φ and Fact 79 imply that f is a C^∞-smooth LFC function (see also Lemma 81).

Finally, since $0 \in U_1$,

$$\begin{aligned}f(0) = F\big(\Phi(0)\big) &\leq \|\Phi(0)\| + \frac{1}{8} = \sup_n h_n(0) + \frac{1}{8}\\ &\leq \sup_n\big(\eta_n g(0) + \delta_n\big) + \frac{1}{8} = \eta_1 + \delta_1 + \frac{1}{8} \leq \frac{3}{2}.\end{aligned}$$

On the other hand, if $\|x\| \geq 1$, then we find n_0 such that $x \in U_{n_0}$ and we get

$$\begin{aligned}f(x) &\geq \|\Phi(x)\| - \frac{1}{8} = \sup_n h_n(x) - \frac{1}{8} \geq h_{n_0}(x) - \frac{1}{8}\\ &> \eta_{n_0} g(x) - \delta_{n_0} - \frac{1}{8} > 2 - \delta_1 - \frac{1}{8} \geq \frac{7}{4}.\end{aligned}$$

Therefore f is a separating function on X and we obtain the desired bump by composing f with a suitable C^∞-smooth real function. □

Let X be a normed linear space. Recall that a set $B \subset B_{X^*}$ is called a James boundary of X if for each $x \in X$ there is $f \in B$ such that $\|x\| = f(x)$.

Definition 98. Let X be a normed linear space and let B be a set of functions on X. We say that B is a generalised James boundary of X if $\|x\| = \max_{f \in B} f(x)$ for each $x \in X$.

The following approximation result is another corollary of Lemma 96.

Theorem 99. *Let $(X, \|\cdot\|)$ be a normed linear space and $k \in \mathbb{N} \cup \{\infty\}$. If X has a countable James boundary, then the norm $\|\cdot\|$ can be approximated by a C^∞-smooth LFC norm uniformly on bounded sets. If X has a countable equi-continuous generalised James boundary $\{g_n\}_{n\in\mathbb{N}}$ such that each g_n is a convex function C^k-smooth on* $\operatorname{supp}_o g_n$, *then the norm $\|\cdot\|$ can be approximated by a C^k-smooth norm uniformly on bounded sets.*

Proof. Let $\{f_n;\ n \in \mathbb{N}\} \subset S_{X^*}$ be a countable James boundary of $(X, \|\cdot\|)$. Choose $\varepsilon > 0$, a sequence of real numbers $\{\eta_n\}$ decreasing to 1 such that $\eta_1 < 1 + \varepsilon$, and a sequence $\Delta = \{\delta_n\}$ such that $0 < \delta_n < \frac{1}{2}(\eta_n - \eta_{n+1})$. Let us define $T \in \mathcal{L}(X; \ell_\infty)$ by $T(x) = \big(\eta_n f_n(x)\big)_n$. Consider the set V_Δ from Lemma 96. We claim that $T(\Omega) \subset V_\Delta$, where $\Omega = \{x \in X;\ \|x\| > 1\}$. Indeed, let $x \in \Omega$ and $n_0 \in \mathbb{N}$ be such that $f_{n_0}(x) = \|x\|$. Then for $n > n_0$ we have

$$\begin{aligned} |\eta_{n_0} f_{n_0}(x)| - \delta_{n_0} &= \eta_{n_0}\|x\| - \delta_{n_0} > \eta_{n_0+1}\|x\| + \delta_{n_0} \\ &\geq \eta_n\|x\| + \delta_{n_0} \geq |\eta_n f_n(x)| + \delta_{n_0}, \end{aligned}$$

where the first inequality follows from the definition of δ_{n_0}.

We now apply Lemma 96 to the sequence Δ and ε in order to obtain the corresponding function F, and we set $f = F \circ T$. Then f is a continuous convex function that is C^∞-smooth and LFC on Ω (Fact 80).

We have $f(0) = F(0) \leq \varepsilon$. Notice that $\|x\| < \|T(x)\| \leq \eta_1\|x\| < (1+\varepsilon)\|x\|$ for every $x \in X \setminus \{0\}$ (the first inequality follows from the fact that $\{f_n\}$ is a James boundary). Set $A = \{x \in X;\ f(x) \leq 1 + 2\varepsilon\}$. Then for every $x \in A$ we have $\|x\| \leq \|T(x)\| \leq F(T(x)) + \varepsilon \leq 1 + 3\varepsilon$, hence A is a CCB set. Moreover, for $x \in X$ satisfying $f(x) = 1 + 2\varepsilon$ we have $(1+\varepsilon)\|x\| > \|T(x)\| \geq F(T(x)) - \varepsilon = 1 + \varepsilon$, in particular $x \in \Omega$. Therefore by Lemma 93 there is an equivalent C^∞ smooth and LFC norm on X. Notice also that through the choice of ε this norm can be made arbitrarily close to the norm $\|\cdot\|$.

To prove the case of the generalised boundary we choose any $0 < \varepsilon \leq \frac{1}{4}$, put $f_n = \max\{g_n, 0\}$, and notice that each f_n is a non-negative convex function C^k-smooth on $\operatorname{supp}_o f_n$ and $\{f_n\}_{n\in\mathbb{N}}$ is still an equi-continuous generalised James boundary of X. The proof is then almost identical as above with the following changes: T is not a bounded linear operator, but it is a continuous mapping (this follows from the equi-continuity of $\{f_n\}$). The convexity of f follows easily from the fact that F is convex and strongly lattice, and each f_n is convex and non-negative. Finally, for any $x \in \Omega$ there are a neighbourhood V of $T(x)$ and a finite $M \subset \operatorname{supp} T(x)$ such that F is C^∞-smooth and depends only on $\{e_n^*;\ n \in M\}$ on V (Lemma 96). There is a neighbourhood U of x in Ω such that $T(U) \subset V$ and $\operatorname{supp} T(y) \cap M = M$ for each $y \in U$, i.e. $U \subset \operatorname{supp}_o f_n$ for each $n \in M$. The C^k-smoothness of f on U then follows from Fact 79. We finish by using Lemma 23. $\square$

6. Isomorphically polyhedral spaces

This section deals with spaces isomorphic to polyhedral spaces. The main structural theorem gives several characterisations of separable isomorphically polyhedral Banach spaces, for example the existence of a renorming with a countable James boundary or the existence of a (smooth) LFC renorming. We also show that every separable polyhedral Banach space admits an equivalent analytic norm and conversely if a Banach space with an analytic norm has all polynomials weakly sequentially continuous, then it is a separable isomorphically polyhedral space.

A closed convex bounded subset of a finite-dimensional normed linear space is called a (convex) polytope if it has only finitely many extreme points, or equivalently if it is a convex hull of finitely many points, or equivalently if it is a bounded intersection of finitely many closed half-spaces (see e.g. [Zie, Theorem 1.1]).

We say that a normed linear space X is polyhedral, if the unit ball of every finite-dimensional subspace of X is a polytope.

Proposition 100 (Victor Klee). *Let X be a normed linear space. The following statements are equivalent:*

(i) *X is a finite-dimensional polyhedral space.*
(ii) *X^* is a finite-dimensional polyhedral space.*
(iii) Ext B_{X^*} *is finite.*
(iv) *X has a finite James boundary.*

Proof. (ii)⇒(iii)⇒(iv) are obvious.

(iv)⇒(i) Let B be a finite James boundary of the space X. It is easy to see that $B_X = B_\circ = \bigcap_{f \in B}\{x \in X;\ f(x) \leq 1\}$. In particular, $\bigcap_{f \in B} \ker f = \{0\}$, hence X must be finite-dimensional. Since B_X is an intersection of finitely many half-spaces, it is a polytope.

(i)⇒(ii) Using the reflexivity of X we obtain that X^{**} is a finite-dimensional polyhedral space. Now we can use the already proved implication (ii)⇒(i). □

Lemma 101 ([GlMc]). *Let X be a normed linear space. If there is a 1-norming set $B \subset B_{X^*}$ such that no w^*-accumulation point of B with norm 1 attains its norm, then B is a James boundary of X and X is polyhedral.*

Proof. To see that B is a James boundary of X, choose $x \in S_X$. There is $\{f_n\} \subset B$ such that $f_n(x) \to 1$. By the w^*-compactness of B_{X^*} the sequence $\{f_n\}$ has a w^*-cluster point $f \in B_{X^*}$. Then $f(x) = 1$ and consequently $\|f\| = 1$. It follows that f is not a w^*-accumulation point of B, hence $f \in B$.

Suppose further that X is not polyhedral. Then there is a finite-dimensional subspace $Y \subset X$ that is not polyhedral. Since $M = \{f\restriction_Y;\ f \in B\} \cap S_{Y^*}$ is a James boundary of Y, the set M is infinite by Proposition 100. By the compactness of B_{Y^*} there is a sequence $\{f_n\}$ of distinct elements of M such that $f_n \to g \in Y^*$. Hence $g \in S_{Y^*}$ and g attains its norm (by the compactness of B_Y). For each $n \in \mathbb{N}$ there is $h_n \in B$ such that $f_n = h_n\restriction_Y$. By the w^*-compactness of B_{X^*} the sequence $\{h_n\}$ has a w^*-cluster point $h \in B_{X^*}$, which is also a w^*-accumulation point of B, as all h_ns are distinct. Clearly $h \restriction_Y = g$. Therefore $\|h\| = 1$ and h attains its norm, a contradiction with the assumptions. □

Let X be a normed linear space. We say that $F \subset S_X$ is a facet of B_X if there is a closed hyperplane $H \subset X$ supporting B_X such that $F = B_X \cap H$ and F has a non-empty relative interior in H.

Lemma 102 ([Ve]). *Let X be a polyhedral Banach space and $x \in S_X$. The following statements are equivalent:*

(i) *The point x is an interior point of a facet of B_X relative to the hyperplane defining the facet.*
(ii) *There is $f \in S_{X^*}$ such that $\|y\| = f(y)$ for y from a neighbourhood of x.*
(iii) *The norm is Gâteaux differentiable at x.*

Proof. (i)⇒(ii)⇒(iii) are easy to see.

(iii)⇒(i) Put $f = \delta\|\cdot\|(x)$ and $Z = \ker f$. Then $H = x + Z$ is a hyperplane supporting B_X. We show that x is an interior point of $F = B_X \cap H$ in H.

Put $C = F - x$, which is a closed convex subset of Z. Fix $h \in S_Z$. The space $Y = \operatorname{span}\{x, h\}$ is a two-dimensional polyhedral space and so B_Y is a polygon. Since $\|\cdot\| \restriction_Y$ is Gâteaux differentiable at x, S_Y contains a line segment L such that x is its interior point (Lemma 11). It follows that $L \subset F$ and so $L - x \subset C$, in particular $th \in C$ for all t sufficiently small.

This shows that C is absorbing in Z and so it has a non-empty interior in Z by Lemma 1.134. Now suppose that 0 is not an interior point of C in Z. Then by the separation theorem there is $g \in Z^*$ such that $g(y) \geq 0$ for each $y \in C$. This however contradicts the fact that C is absorbing. □

Note that if $x \in X$ and $f \in B_{X^*}$, then $f(x) = \|x\|$ if and only if $f \in \partial\|\cdot\|(x)$. Therefore $B \subset B_{X^*}$ is a James boundary of X if and only if $\partial\|\cdot\|(x) \cap B \neq \emptyset$ for each $x \in S_X$. In particular, $\bigcup_{x \in S_X} \partial\|\cdot\|(x)$ is a James boundary of X.

Let X be a normed linear space, $\Omega \subset X$ open, and $f : \Omega \to \mathbb{R}$. We say that f is locally weakly sequentially Cauchy-continuous (locally wsC for short) if for each $x \in \Omega$ there is a (norm) neighbourhood U of x such that $f \restriction_U$ is weakly sequentially Cauchy-continuous.

Theorem 103. *Let X be a Banach space. The following statements are equivalent:*

(i) *X is a separable isomorphically polyhedral space.*
(ii) *X admits an equivalent norm with a countable James boundary.*
(iii) *X admits an equivalent norm with a K_σ James boundary.*
(iv) *X is separable and admits an equivalent LFC norm.*
(v) *X is separable and admits an equivalent norm which is C^∞-smooth and LFC.*
(vi) *X is separable, does not contain ℓ_1, and admits an equivalent norm that is locally weakly sequentially Cauchy-continuous on $X \setminus \{0\}$.*

Proof. (v)⇒(iv) is obvious.

(iv)⇒(iii) Let $\|\cdot\|$ be an equivalent LFC norm on X. By the Lindelöf property of S_X there exist a countable family of finite sets $\{M_n\}_{n\in\mathbb{N}}$, $M_n \subset X^*$, and a countable covering $\{U_n\}_{n\in\mathbb{N}}$ of S_X by open balls such that $\|\cdot\|$ depends only on M_n on U_n. For each $n \in \mathbb{N}$ set $K_n = (\operatorname{span} M_n) \cap S_{X^*}$, which is clearly a compact set.

Now choose any $x \in S_X$. Let $g \in S_{X^*}$ be such that $g(x) = 1$ and let $n \in \mathbb{N}$ be such that $x \in U_n$. For any $h \in \bigcap_{f\in M_n} \ker f$ sufficiently small so that $x \pm h \in U_n$ we have $\|x \pm h\| = \|x\| = 1$. Hence $\pm g(h) = g(x \pm h) - g(x) \le \|x \pm h\| - 1 = 0$. It follows that $h \in \ker g$. Thus $\bigcap_{f\in M_n} \ker f \subset \ker g$, which in turn implies that $g \in K_n$. Therefore $\bigcup_{n=1}^{\infty} K_n$ is a K_σ James boundary of $(X, \|\cdot\|)$.

(iii)⇒(ii) Let $\|\cdot\|$ be an equivalent norm on X with a K_σ James boundary and let $K_n \subset S_{X^*}$ be compact sets such that $\bigcup_{n=1}^{\infty} K_n$ is a James boundary of $(X, \|\cdot\|)$.

Take a sequence of real numbers $\{\varepsilon_n\} \subset (0, 1]$ such that $\varepsilon_n \to 0$. Let M_n be a finite $\frac{\varepsilon_n}{4}$-dense subset of K_n. We put $B = \bigcup_{n=1}^{\infty}\{\pm(1 + \varepsilon_n)f;\ f \in M_n\}$ and define

$$|||x||| = \sup_{f\in B} f(x) \tag{14}$$

for $x \in X$. We claim that $|||\cdot|||$ is an equivalent norm on X such that B is a James boundary of $(X, |||\cdot|||)$.

Choose any $x \in S_X$. Clearly, $|||x||| \le 2$. On the other hand, there are $j \in \mathbb{N}$ and $g \in K_j$ such that $g(x) = 1$. Let $h \in M_j$ be such that $\|g - h\| < \frac{\varepsilon_j}{4}$. Then $h(x) > 1 - \frac{\varepsilon_j}{4}$ and consequently

$$(1 + \varepsilon_j)h(x) > (1 + \varepsilon_j)\left(1 - \frac{\varepsilon_j}{4}\right) \ge 1 + \frac{\varepsilon_j}{2}.$$

It follows that $|||x||| > 1 + \frac{\varepsilon_j}{2}$ and so $|||\cdot|||$ is an equivalent norm on X. Furthermore, since $\varepsilon_n \to 0$, there is $n_0 \in \mathbb{N}$ such that $(1 + \varepsilon_n)f(x) \le 1 + \frac{\varepsilon_j}{2}$ for $n \ge n_0$ and any $f \in S_{X^*}$. Thus the supremum in (14) is actually over a finite set and hence it is attained. This proves our claim.

(ii)⇒(i) Let $\{f_n;\ n \in \mathbb{N}\} \subset S_{X^*}$ be a countable James boundary of $(X, \|\cdot\|)$ and put $Q = \{\pm f_n;\ n \in \mathbb{N}\}$. By Rodé's theorem (see [FHHMZ, Theorem 3.122]) $B_{X^*} = \overline{\operatorname{conv}}\, Q$, hence both X^* and X are separable.

Choose a sequence of real numbers $\{\eta_n\}$ decreasing to 1 and put $g_n = \eta_n f_n$. Let $B = \{\pm g_n;\ n \in \mathbb{N}\}$. Define a new norm on X by the formula $|||x||| = \sup_{n\in\mathbb{N}} |g_n(x)|$. Then $\|x\| \le |||x||| \le \eta_1\|x\|$ for each $x \in X$, so $|||\cdot|||$ is an equivalent norm on X. Clearly B is a 1-norming set for $(X, |||\cdot|||)$.

Suppose that $g \in X^*$ is a w^*-accumulation point of B. We claim that g does not attain its norm $|||g|||$. It is easy to see that g is also a w^*-accumulation point of Q and therefore $\|g\| \le 1$. Choose any $x \in X$ with $|||x||| = 1$. There is $k \in \mathbb{N}$ such that $f_k(x) = \|x\|$. Thus

$$g(x) \le \|g\|\|x\| \le \|x\| = f_k(x) < g_k(x) \le |||x||| = 1.$$

So the space $(X, |||\cdot|||)$ is polyhedral by Lemma 101.

(i)⇒(ii) Suppose that $(X, \|\cdot\|)$ is a polyhedral space. We show that S_X can be covered by facets of B_X. The set $B = \{f \in S_{X^*};\ f^{-1}(1) \cap B_X \text{ is a facet of } B_X\}$ is then a (minimal) James boundary of X. From the separability of S_X it follows that it can be covered only by countably many facets, hence B is countable.

Let S be the union of all the facets of B_X and $Q = \{x \in X \setminus \{0\};\ \frac{x}{\|x\|} \notin S\}$. Obviously Q is a cone, i.e. $ty \in Q$ whenever $y \in Q$ and $t > 0$. We claim that for any $u \in Q$ the set $Q - u$ is absorbing. So fix $u \in Q$ and put $v = \frac{u}{\|u\|}$. Choose any $z \in S_X$, $z \neq \pm v$, and put $Y = \operatorname{span}\{v, z\}$. Since Y is two-dimensional and X is polyhedral, B_Y is a polygon. As $v \in S_Y$, it follows that there is $w \in S_Y$, $w \neq v$, such that the line segment $[v, w]$ lies in S_Y, and w and z lie in the same half-plane in Y relative to $\operatorname{span}\{v\}$. Now it is not difficult to see that the line segment $[v, w) \in Q$. Indeed if some $y \in (v, w)$ belongs to a facet of B_X, then the hyperplane defining this facet must contain the line segment $[v, w]$, a contradiction with $v \in Q$. From this and the fact that Q is a cone we easily obtain that there is $\delta > 0$ such that $u + tz \in Q$ for $t \in [0, \delta]$. This shows that $Q - u$ is absorbing.

The norm $\|\cdot\|$ is not Gâteaux differentiable at any point of Q (Lemma 102). So by Theorem 4 the set Q is contained in an F_σ set with an empty interior, which means according to Lemma 1.134 that there is no $u \in Q$ such that $Q - u$ is absorbing. Therefore Q is empty.

(ii)$\Rightarrow$(v) The separability of X was already discussed above. The rest follows from Theorem 99.

(iv)$\Rightarrow$(vi) follows from Corollary 92 and Fact 82.

(vi)$\Rightarrow$(iii) Proposition 41 combined with the Lindelöf property of S_X implies that $\bigcup_{x \in S_X} \partial\|\cdot\|(x)$ is a countable union of relatively compact sets. Thus $(X, \|\cdot\|)$ has a K_σ James boundary. □

Theorem 104 ([DFH2]). *Every separable isomorphically polyhedral Banach space X admits an equivalent analytic norm.*

Proof. By Theorem 103 we can consider X with an equivalent norm for which there is $\{f_n;\ n \in \mathbb{N}\} \subset S_{X^*}$ a countable James boundary of X. Define a function $\Phi\colon U(0,1) \to \mathbb{R}$ by

$$\Phi(x) = \sum_{n=1}^{\infty} f_n(x)^{2n}.$$

Obviously this is a convex function that is analytic (Theorem 1.168, see also Example 1.137). Let $a \in (0,1)$ and set $A = \{x \in U(0,1);\ \Phi(x) \leq a\}$. The set A is closed. Indeed, let $\{x_n\} \subset A$ be such that $x_n \to x \in X$. Then obviously $x \in B_X$. Suppose that $x \in S_X$. Let $n \in \mathbb{N}$ be such that $f_n(x) = 1$. There is a neighbourhood V of x such that $f_n(y) > \sqrt[2n]{a}$ for $y \in V$. Hence $\Phi(y) \geq f_n(y)^{2n} > a$ for $y \in V \cap U(0,1)$, which contradicts the fact that $\Phi(x_n) \leq a$ for all $n \in \mathbb{N}$. It follows that $x \in U(0,1)$ and so $x \in A$ by the continuity of Φ. Now we apply Lemma 23. □

The following is a corollary of Theorem 103.

Corollary 105 ([HáTr]). *Let X be a Banach space with an analytic norm. If all polynomials on X are weakly sequentially continuous, then X is separable and isomorphically polyhedral.*

Proof. For every $x \in X$, $x \neq 0$ there exists a neighbourhood U of x such that $\|y\| = \sum_{n=0}^{\infty} P_n(y - x)$ for all $y \in U$, where $P_n \in \mathcal{P}(^n X)$, and the sum converges uniformly on U. Since each P_n is wsC (Corollary 3.27), it follows that $\|\cdot\|$ is locally wsC on $X \setminus \{0\}$. Thus X^* is separable by Proposition 77 and we may apply Theorem 103. □

7. L_p spaces

In this section we study in detail the smoothness of L_p spaces. We begin with a classical positive result of Robert Bonic and John Frampton on the smoothness of $L_p(\mu)$ spaces. Then we prove some negative results on smoothness of ℓ_p spaces (which of course translate immediately to general L_p spaces). These show that the results of Bonic and Frampton are sharp in a strong sense. Then we proceed with the study of the smoothness of norms and bumps in the Lebesgue-Bochner spaces $L_p(\mu; X)$, proving some basic results on the differentiability of homogeneous mappings along the way. Finally we show some sharper results on smoothness in $L_p(\mu; L_q(\nu))$ spaces.

Theorem 106 ([BF]). *Let μ be a measure and $1 < p < +\infty$. Let $\Phi = \|\cdot\|^p$, where $\|\cdot\|$ is the canonical norm of $X = L_p(\mu)$.*

(i) *If p is an even integer, then Φ is a p-homogeneous polynomial.*
(ii) *If p is an odd integer, then $\Phi \in C^{p-1,1}(X)$.*
(iii) *If p is not an integer, then $\Phi \in C^{[p],p-[p]}(X)$.*

Proof. (i) For any $f_1, \dots, f_p \in X$ we put $M(f_1, \dots, f_p) = \int f_1 \cdots f_p \,d\mu$. By repeated use of the Hölder inequality we can check that the integral exists. Obviously $M \in L({}^pX)$ and $\Phi = \widehat{M}$. Hence $\Phi \in \mathcal{P}({}^pX)$.

(ii), (iii) Define $\varphi\colon \mathbb{R} \to \mathbb{R}$ by $\varphi(t) = |t|^p$. Let $k = p - 1$ in the case (ii) and $k = [p]$ in the case (iii). The function φ is $C^{k,p-k}$-smooth on $\mathbb{R}$, i.e. there is $C > 0$ such that

$$\left|\varphi^{(k)}(x) - \varphi^{(k)}(y)\right| \le C|x - y|^{p-k} \quad \text{for all } x, y \in \mathbb{R}. \tag{15}$$

This follows from the fact that $\left||u|^\alpha - |v|^\alpha\right| \le |u - v|^\alpha$ and $\left||u|^\alpha \operatorname{sgn} u - |v|^\alpha \operatorname{sgn} v\right| \le 2|u - v|^\alpha$ for all $u, v \in \mathbb{R}$ and $\alpha \in (0, 1]$, which can be proved by standard analysis. By Corollary 1.108

$$\left|\varphi(x + h) - \sum_{j=0}^{k} \frac{1}{j!}\varphi^{(j)}(x)h^j\right| \le \frac{1}{k!}C|h|^p \quad \text{for all } x, h \in \mathbb{R}. \tag{16}$$

For $x, h \in X$ and $j \in \{0, \dots, k\}$ we have by the Hölder inequality

$$\int \left|\varphi^{(j)}\big(x(t)\big)h(t)^j\right| d\mu(t) \le \left(\int \left|\varphi^{(j)}\big(x(t)\big)\right|^{\frac{p}{p-j}} d\mu(t)\right)^{\frac{p-j}{p}} \left(\int |h(t)|^p \,d\mu(t)\right)^{\frac{j}{p}}$$
$$= p(p-1)\cdots(p-j+1)\|x\|^{p-j}\|h\|^j.$$

Thus for any $x \in X$ and $j \in \{0, \dots, k\}$ we may set $P_j^x(h) = \int \varphi^{(j)}\big(x(t)\big)h(t)^j \,d\mu(t)$ for $h \in X$. Similarly as in (i) and together with the above estimate we can see that

$P_j^x \in \mathcal{P}({}^jX)$. Using (16) we obtain

$$\left|\Phi(x+h) - \sum_{j=0}^{k} \frac{1}{j!} P_j^x(h)\right| \le \int \left|\varphi\big(x(t)+h(t)\big) - \sum_{j=0}^{k} \frac{1}{j!} \varphi^{(j)}\big(x(t)\big)h(t)^j\right| \mathrm{d}\mu(t)$$
$$\le \frac{1}{k!} C \int |h(t)|^p \,\mathrm{d}\mu(t) = \frac{1}{k!} C \|h\|^p.$$

Now Theorem 1.125 implies that $\Phi \in C^{k,p-k}(X)$. □

Theorem 107.

(i) *If $1 \le p < +\infty$ and p is not an even integer, then ℓ_p does not admit a T^p-smooth bump.*

(ii) *If $1 \le p < 2$, then ℓ_p does not admit a continuous bump β with the property that for each $x \in \ell_p$ there is $q > p$ such that β is directionally T^q-smooth at x.*

(iii) *If $1 \le p < +\infty$, p is not an even integer, and Γ is an uncountable set, then $\ell_p(\Gamma)$ does not admit a continuous weakly T^p-smooth bump.*

Note that it means in particular that ℓ_p, $1 \le p < 2$, does not admit a continuous G^2-smooth bump, and $\ell_p(\Gamma)$, Γ uncountable, $1 \le p < +\infty$, p not an even integer, does not admit a continuous $G^{\lceil p \rceil}$-smooth bump.

Proof. The first part of the proof is the same for all three statements (i), (ii), and (iii). Denote by X the space ℓ_p in the case (i) or (ii), or $\ell_p(\Gamma)$ in the case (iii). By S-smoothness we denote the smoothness of the type given in (i), (ii), or (iii) respectively. Let $\varphi\colon X \to \mathbb{R} \cup \{+\infty\}$ be defined as $\varphi(x) = \frac{1}{\|x\|^p}$ for $x \ne 0$ and $\varphi(0) = +\infty$. Assuming that X admits an S-smooth bump, by Corollary 59 (note that $\ell_1(\Gamma)$ has the RNP by Theorem 2) and Lemma 60 the space X admits also an S-variational principle. Thus there are an S-smooth function $f\colon X \to \mathbb{R}$ and $x_0 \in X$ such that $f(x_0) = \varphi(x_0)$ and $f(x) \le \varphi(x)$ for every $x \in X$. It follows that $x_0 \ne 0$ and $f(x_0) > 0$, and so there is a neighbourhood U of x_0 such that $f > 0$ on U. Hence $g = \frac{1}{f}$ is S-smooth on U (Proposition 1.115), $g(x_0) = \|x_0\|^p$ and $g(x) \ge \|x\|^p$ for every $x \in U$.

(i) By the T^p-smoothness there are $P \in \mathcal{P}^{[p]}(X)$ satisfying $P(0) = 0$, and $\delta > 0$ such that $g(x_0+h) - g(x_0) - P(h) \le \frac{1}{2}\|h\|^p$ for $\|h\| \le \delta$. Put $Q(h) = P(h) + P(-h)$. Since p is not an even integer, $\deg Q < p$. Further,

$$Q(h) \ge g(x_0+h) + g(x_0-h) - 2g(x_0) - \|h\|^p$$
$$\ge \|x_0+h\|^p + \|x_0-h\|^p - 2\|x_0\|^p - \|h\|^p$$

for $\|h\| \le \delta$. The polynomial Q is weakly sequentially continuous (Corollary 3.59), and so $\lim_{n\to\infty} Q(\delta e_n) = 0$. (Note also that in case that $p < 2$ simply $Q = 0$). By the properties of the ℓ_p-norm we have $\lim_{n\to\infty}\|x_0 \pm \delta e_n\|^p = \|x_0\|^p + \delta^p$. Putting everything together we obtain

$$0 = \lim_{n\to\infty} Q(\delta e_n) \ge \lim_{n\to\infty} \big(\|x_0+\delta e_n\|^p + \|x_0-\delta e_n\|^p\big) - 2\|x_0\|^p - \delta^p = \delta^p,$$

which is a contradiction.

(ii) Let $q > p$ be such that g is directionally T^q-smooth at x_0. Without loss of generality we may assume that $q < 2$. For a given $h \in X$ there is $a \in \mathbb{R}$ such that $g(x_0 + th) = g(x_0) + at + o(|t|^q)$, $t \to 0$. Since

$$\begin{aligned}\|x_0 + th\|^p + \|x_0 - th\|^p - 2\|x_0\|^p &\le g(x_0 + th) + g(x_0 - th) - 2g(x_0)\\ &= o(|t|^q),\ t \to 0,\end{aligned}$$

it follows that there is $\eta > 0$ such that

$$\frac{1}{t^q}\left(\|x_0 + th\|^p + \|x_0 - th\|^p - 2\|x_0\|^p\right) \le 1$$

for $t \in (0, \eta)$. Since the expression on the left-hand side is obviously bounded for $t \in [\eta, 1]$, we can conclude that

$$\psi(h) = \sup_{t\in(0,1]} \frac{1}{t^q}\left(\|x_0 + th\|^p + \|x_0 - th\|^p - 2\|x_0\|^p\right) < +\infty$$

for each $h \in X$. Therefore $X = \bigcup_{n=1}^{\infty}\{h \in X;\ \psi(h) \le n\}$. Since ψ is lower semicontinuous, each of the sets $\{h \in X;\ \psi(h) \le n\}$ is closed and so by the Baire category theorem there are $C > 0$ and an open set $V \subset X$ such that $\psi(h) \le C$ for $h \in V$. Since ψ is even, $\psi(h) \le C$ also for $h \in -V$. The convexity of ψ implies that there is $\delta > 0$ such that $\psi(h) \le C$ for $h \in B(0, \delta)$. Now similarly as in the proof of (i) we obtain

$$2t^p\delta^p = \lim_{n\to\infty}\left(\|x_0 + t\delta e_n\|^p + \|x_0 - t\delta e_n\|^p\right) - 2\|x_0\|^p \le Ct^q$$

for all $t \in (0, 1]$, which is a contradiction.

(iii) By the weak T^p-smoothness there is a polynomial $P \in \mathcal{P}^{[p]}(X)$ satisfying $P(0) = 0$ and $g(x_0 + th) - g(x_0) - P(th) = o(|t|^p)$, $t \to 0$ for each $h \in X$. Put $Q(h) = P(h) + P(-h)$. Since p is not an even integer, $\deg Q < p$, and hence each homogeneous summand of Q is countably supported (Corollary 3.62). Since also $\operatorname{supp} x_0$ is countable, there is $\gamma \in \Gamma \setminus \operatorname{supp} x_0$ such that $Q(te_\gamma) = 0$ for all $t \in \mathbb{R}$. Therefore,

$$\begin{aligned}2|t|^p &= \|x_0 + te_\gamma\|^p + \|x_0 - te_\gamma\|^p - 2\|x_0\|^p\\ &\le g(x_0 + te_\gamma) + g(x_0 - te_\gamma) - 2g(x_0)\\ &= Q(te_\gamma) + o(|t|^p) = o(|t|^p),\ t \to 0,\end{aligned}$$

which is a contradiction. □

We note that the statement (i) for $p > 2$ follows also from Theorem 62 and Corollary 4.46, and the statement (iii) for $p > 2$ follows also from Theorem 76 and Corollary 3.59.

Notice that for $p \in \mathbb{N}$ odd there is an equivalent norm on ℓ_p that is $C^{p-1,1}$-smooth and G^p-smooth (variants of Corollary 7.72, Corollary 7.70), while there is no continuous G^p-smooth bump on $\ell_p(\Gamma)$ for Γ uncountable (Theorem 107(iii)).

Theorem 108 ([LeSu]). *Let (Ω, μ) be a measure space with an infinite family of pairwise disjoint measurable subsets of Ω of finite positive measure. Further, let X be a Banach space, $1 < p < +\infty$, and $k \in \mathbb{N}$, $k < p$. The following statements are equivalent:*

(i) *The norm of X is C^k-smooth and its kth differential is bounded on S_X.*
(ii) *The canonical norm of the space $L_p(\mu; X)$ is C^k-smooth.*
(iii) *The canonical norm of the space $L_p(\mu; X)$ is C^k-smooth with $d^j\|\cdot\|$ bounded on $L_p(\mu; X) \setminus B(0, r)$ for each $r > 0$, $j \in \{1, \dots, k\}$.*

We note that the assumption on the measure space is equivalent to $L_p(\mu)$ (the scalar Lebesgue space) being infinite-dimensional and also equivalent to the fact that $L_p(\mu; X)$ contains a subspace isometric to $\ell_p(\mathbb{N}; X)$.

Before proving the theorem we gather a few facts about the differentiability of p-homogeneous mappings, i.e. the mappings f satisfying $f(\lambda x) = \lambda^p f(x)$ for all $\lambda > 0$.

Fact 109. *Let X, Y be normed linear spaces, $p \in \mathbb{R}$, and let $f : X \setminus \{0\} \to Y$ be a p-homogeneous mapping that is k-times Fréchet differentiable at every $x \in S_X$. Then f is k-times Fréchet differentiable on $X \setminus \{0\}$ and $d^j f(\lambda x) = \lambda^{p-j} d^j f(x)$ for each $x \in X \setminus \{0\}$, $\lambda > 0$, and $j \in \{1, \dots, k\}$.*

Proof. Given $x \in X$, $x \neq 0$, we have $f(\lambda x + h) - f(\lambda x) - \lambda^{p-1} Df(x)[h] = \lambda^p\big(f(x + \frac{1}{\lambda}h) - f(x) - Df(x)[\frac{1}{\lambda}h]\big) = o(\|h\|)$, $h \to 0$. The rest follows by induction. □

Fact 110. *Let X, Y be normed linear spaces, $p \in (1, +\infty)$, let $f : X \to Y$ be a p-homogeneous mapping with $f(0) = 0$, and $k \in \mathbb{N}$, $k < p$.*

(i) *For each $1 \le q < p$ the mapping f is weakly T^q-smooth at 0 with the approximating polynomial being zero.*
(ii) *For each $1 \le q < p$ the mapping f is T^q-smooth at 0 if and only if f is bounded on S_X.*
(iii) *If f is G^{k-1}-smooth on $X \setminus \{0\}$, then f is G^k-smooth at 0 with $\delta^j f(0) = 0$, $j = 1, \dots, k$.*

Proof. (i) and (ii) follow directly from the definitions. (iii) follows by induction: $\lim_{t \to 0} \frac{1}{t}\delta^{j-1} f(th_1)[h_2, \dots, h_j] = \lim_{t \to 0} t^{p-j}\delta^{j-1} f(h_1)[h_2, \dots, h_j] = 0$, using an analogue of Fact 109 for Gâteaux derivatives. □

Fact 111. *Let X, Y be normed linear spaces, $p \in (1, +\infty)$, $f : X \to Y$ a p-homogeneous mapping, and $k \in \mathbb{N}$, $k < p$. The following statements are equivalent:*

(i) *$f \in C^k(X; Y)$.*
(ii) *$f \in C^k(X; Y)$ and $d^j f$ is bounded on bounded sets for all $j \in \{1, \dots, k\}$.*
(iii) *The mapping f is k-times Fréchet differentiable at the points of S_X, $d^k f$ is continuous and bounded on S_X, and $f(0) = 0$.*

Proof. (i)⇒(ii) Let $j \in \{1, \dots, k\}$. Since $d^j f$ is continuous at 0, it is bounded on a neighbourhood of 0. By Fact 109 it is bounded on any bounded set. Note also that from Fact 109 it follows that $d^j f(0) = 0$.

(ii)⇒(iii) Since f is continuous, the p-homogeneity implies $f(0) = 0$. The rest is obvious.

(iii)⇒(i) From Fact 109 it follows that $f \in C^k(X \setminus \{0\}; Y)$ and $d^k f$ is bounded on $B_X \setminus \{0\}$. Using repeatedly Proposition 1.72 we conclude that the mappings $d^{k-1}f, \dots, df, f$ are Lipschitz (and therefore bounded) on $B_X \setminus \{0\}$. It follows that $f(h) = \|h\|^p f\big(\frac{h}{\|h\|}\big) = o(\|h\|)$, $h \to 0$, which means that $df(0) = 0$. Since by Fact 109 the mapping df is $(p-1)$-homogeneous, the rest follows by induction. □

Lemma 112. *Let X be a normed linear space,* $\dim X \geq 2$, *and $x, y \in X$, $\|x\| \leq \|y\|$. There are points $u, v \in X$ such that the line segments $[x, u]$, $[u, v]$, and $[v, y]$ lie in $X \setminus U(0, \frac{1}{3}\|x\|)$ and $\|u - x\| + \|v - u\| + \|y - v\| \leq 3\|y - x\|$.*

Proof. If $[x, y] \cap U(0, \frac{1}{3}\|x\|) = \emptyset$, then the assertion is trivial, so assume the contrary. Let $Z \subset X$ be a two-dimensional subspace containing x and y. Let $v \in [x, y]$ be such that $\|v\| = \|x\|$ and $v \neq x$. From the convexity and the choice of v we get $\|z\| \geq \|x\|$ for all $z \in [v, y]$. Let $f \in S_{Z^*}$ be such that $f(v - x) = 0$ and $f(x) \geq 0$. By the compactness of $B_Z(0, \|x\|)$ there is $u \in Z$ with $f(u) = \|u\| = \|x\|$. For any $t \in [0, \frac{1}{3}]$ we have $\|tu + (1-t)x\| \geq (1-t)\|x\| - t\|u\| = (1-2t)\|x\| \geq \frac{1}{3}\|x\|$, while for $t \in [\frac{1}{3}, 1]$ we have $\|tu + (1-t)x\| \geq f(tu + (1-t)x) = t\|x\| + (1-t)f(x) \geq t\|x\| \geq \frac{1}{3}\|x\|$. Since $f(v) = f(x) \geq 0$, we can show analogously that $[u, v] \cap U(0, \frac{1}{3}\|x\|) = \emptyset$.

Because $[x, v] \cap U(0, \frac{1}{3}\|x\|) \neq \emptyset$, it follows that $\|v - x\| \geq \frac{4}{3}\|x\|$. Therefore $\|u - x\| + \|v - u\| + \|y - v\| \leq 4\|x\| + \|y - v\| \leq 3\|v - x\| + \|y - v\| \leq 3\|y - x\|$. □

Fact 113. *Let X be a normed linear space, $p \in (1, +\infty)$, $\varphi(x) = \|x\|^p$, and $k \in \mathbb{N}$, $k < p$. Then $\varphi \in C^k(X)$ if and only if the norm is C^k-smooth and its kth differential is bounded on S_X.*

Proof. ⇒ Follows from Fact 111 and Corollary 1.117.

⇐ Using repeatedly Fact 109 and Proposition 1.71 together with Lemma 112 we conclude that $d^{k-1}\|\cdot\|, \dots, d\|\cdot\|$ are Lipschitz (and therefore bounded) on $X \setminus B(0, \frac{1}{2})$. So $\varphi \in C^k(X \setminus \{0\})$ and $d^k\varphi$ is bounded on S_X by Corollary 1.117. Hence $\varphi \in C^k(X)$ by Fact 111. □

The proof of Theorem 108 is based on the following observation.

Lemma 114. *Let (Ω, μ) be a finite measure space, X a normed linear space, and let $f\colon \Omega \to X$ be a measurable mapping. Then for each $\varepsilon > 0$ there is a measurable set $Q \subset \Omega$ such that $f(Q)$ is totally bounded and $\mu(\Omega \setminus Q) < \varepsilon$.*

Proof. Let $\varphi_n\colon \Omega \to X$ be simple measurable mappings such that $\varphi_n \to f$ almost everywhere. Let $\varepsilon > 0$ be given. By Egorov's theorem there is a measurable set $Q \subset \Omega$ such that $\varphi_n \to f$ uniformly on Q and $\mu(\Omega \setminus Q) < \varepsilon$. Therefore for every $\delta > 0$ there is $n \in \mathbb{N}$ such that the set $\varphi_n(Q)$ is a finite δ-dense set for $f(Q)$. □

Proof of Theorem 108. (i)$\Rightarrow$(iii) We will show that the function $\Phi(x) = \|x\|^p$ is C^k-smooth on $L_p(\mu; X)$ using Theorem 1.110. Let $\varphi\colon X \to \mathbb{R}$ be defined as $\varphi(u) = \|u\|^p$. Then $\varphi \in C^k(X)$ and $d^j\varphi$ are bounded by $M > 0$ on S_X for all $j \in \{1,\dots,k\}$ by Fact 113 and Fact 111. Hence $d^j\varphi(u) \le M\|u\|^{p-j}$ for each $u \in X$, $j \in \{1,\dots,k\}$ by Fact 109. For $x, h \in L_p(\mu; X)$ and $j \in \{1,\dots,k\}$ we have by the Hölder inequality

$$\int \big\|d^j\varphi\big(x(t)\big)[h(t)]\big\|\,\mathrm{d}\mu(t) \le \Big(\int \big\|d^j\varphi\big(x(t)\big)\big\|^{\frac{p}{p-j}}\,\mathrm{d}\mu(t)\Big)^{\frac{p-j}{p}}\Big(\int \|h(t)\|^p\,\mathrm{d}\mu(t)\Big)^{\frac{j}{p}}$$
$$\le \Big(\int \big(M\|x(t)\|^{p-j}\big)^{\frac{p}{p-j}}\,\mathrm{d}\mu(t)\Big)^{\frac{p-j}{p}}\|h\|^j = M\|x\|^{p-j}\|h\|^j.$$

It follows that for a given $x \in L_p(\mu; X)$ and $j \in \{0,\dots,k\}$ we may define $P_j^x(h) = \int d^j\varphi\big(x(t)\big)[h(t)]\,\mathrm{d}\mu(t)$ for every $h \in L_p(\mu; X)$. Similarly as in the proof of Theorem 106 and together with the above estimate it is easily seen that $P_j^x \in \mathcal{P}({}^jL_p(\mu; X))$. To estimate the remainder of the approximation by the polynomials as required in Theorem 1.110 we first use Corollary 1.108 and the Hölder inequality to obtain

$$\begin{aligned}
|R(y,h)| &= \Big|\Phi(y+h) - \sum_{j=0}^{k}\frac{1}{j!}P_j^y(h)\Big| \\
&\le \int \Big|\varphi\big(y(t)+h(t)\big) - \sum_{j=0}^{k}\frac{1}{j!}d^j\varphi\big(y(t)\big)[h(t)]\Big|\,\mathrm{d}\mu(t) \qquad (17)\\
&\le \frac{1}{k!}\int\Big(\sup_{\xi\in[0,1]}\big\|d^k\varphi\big(y(t)+\xi h(t)\big) - d^k\varphi\big(y(t)\big)\big\|\Big)\|h(t)\|^k\,\mathrm{d}\mu(t) \\
&\le \frac{1}{k!}\Big(\int\Big(\sup_{\xi\in[0,1]}\big\|d^k\varphi\big(y(t)+\xi h(t)\big) - d^k\varphi\big(y(t)\big)\big\|\Big)^{\frac{p}{p-k}}\,\mathrm{d}\mu(t)\Big)^{\frac{p-k}{p}}\|h\|^k
\end{aligned}$$

for any $y, h \in L_p(\mu; X)$.

Now fix $x \in L_p(\mu; X)$ (or more precisely fix a concrete representative of the equivalence class x, which will be denoted by x again) and choose an arbitrary $\varepsilon > 0$. Set $\varepsilon_1 = \frac{1}{6\cdot 2^p}\big(\frac{\varepsilon}{2M}\big)^{\frac{p}{p-k}}$. There is a measurable set $A_1 \subset \Omega$ such that $\int_{A_1}\|x(t)\|^p\,\mathrm{d}\mu(t) < \varepsilon_1$ and $\mu(\Omega_1) < +\infty$, where $\Omega_1 = \Omega \setminus A_1$. (Take a simple non-negative function $\psi(t) \le \|x(t)\|^p$ for which $\int\|x(t)\|^p\,\mathrm{d}\mu - \varepsilon_1 < \int\psi\,\mathrm{d}\mu$ and set $A_1 = \psi^{-1}(0)$.)

Further, there is $\eta > 0$ such that $\int_A\|x(t)\|^p\,\mathrm{d}\mu < \varepsilon_1$ whenever $A \subset \Omega$ is measurable with $\mu(A) < \eta$. This follows from the fact that $A \mapsto \int_A\|x(t)\|^p\,\mathrm{d}\mu$ is a finite measure absolutely continuous with respect to μ. By Lemma 114 there is a measurable set $Q \subset \Omega_1$ such that $K = \overline{x(Q)}$ is compact and $\mu(A_2) < \eta$, where $A_2 = \Omega_1 \setminus Q$. This means that $\int_{A_2}\|x(t)\|^p\,\mathrm{d}\mu(t) < \varepsilon_1$. By the compactness of K there is $\Delta > 0$ such that $\|d^k\varphi(u) - d^k\varphi(v)\| < \big(2\mu(Q)\big)^{-\frac{p-k}{p}}\varepsilon$ whenever $u, v \in X$, $\operatorname{dist}(u, K) < 2\Delta$, $\operatorname{dist}(v, K) < 2\Delta$, and $\|u - v\| < \Delta$. (Here we consider $\frac{1}{0} = +\infty$.) Let $\delta > 0$ be such that $2(\frac{\delta}{\Delta})^p < \eta$ and $2\delta < (3\varepsilon_1)^{\frac{1}{p}}$.

Now let $y, h \in L_p(\mu; X)$ be such that $\|y - x\| < \delta$ and $\|h\| < \delta$ (here we again choose concrete representatives of the equivalence classes for y and h). Set $Q_1 = \{t \in Q;\ \|y(t) - x(t)\| < \Delta, \|h(t)\| < \Delta\}$ and $A_3 = Q \setminus Q_1$. It is easy to see that $\mu(A_3) \le \mu(\{t \in Q;\ \|y(t) - x(t)\| \ge \Delta\}) + \mu(\{t \in Q;\ \|h(t)\| \ge \Delta\}) \le \frac{\|y-x\|^p}{\Delta^p} + \frac{\|h\|^p}{\Delta^p} < 2(\frac{\delta}{\Delta})^p < \eta$. Therefore $\int_{A_3} \|x(t)\|^p \,\mathrm{d}\mu(t) < \varepsilon_1$.

We can finally estimate the remainder (17). Note that for $t \in Q_1$ and $\xi \in [0, 1]$ we have $\operatorname{dist}(y(t), K) \le \|y(t) - x(t)\| < \Delta$ and $\operatorname{dist}(y(t) + \xi h(t), K) < 2\Delta$. Hence

$$\begin{aligned}\int_{Q_1} \Big(\sup_{\xi\in[0,1]} \big\|d^k\varphi\big(y(t) + \xi h(t)\big) - d^k\varphi\big(y(t)\big)\big\| \Big)^{\frac{p}{p-k}} \,\mathrm{d}\mu(t) \\ \le \varepsilon^{\frac{p}{p-k}} \frac{\mu(Q_1)}{2\mu(Q)} \le \frac{1}{2}\varepsilon^{\frac{p}{p-k}}.\end{aligned} \tag{18}$$

Next, for any $t \in \Omega$ and $\xi \in [0, 1]$ we have by Fact 109

$$\begin{aligned}\big\|d^k\varphi\big(y(t) + \xi h(t)\big) - d^k\varphi\big(y(t)\big)\big\| &\le \big\|d^k\varphi\big(y(t) + \xi h(t)\big)\big\| + \big\|d^k\varphi\big(y(t)\big)\big\| \\ &\le M\big(\|y(t) + \xi h(t)\|^{p-k} + \|y(t)\|^{p-k}\big) \le 2M\big(\|y(t)\| + \|h(t)\|\big)^{p-k}.\end{aligned}$$

So, setting $B = A_1 \cup A_2 \cup A_3$ and using the Minkowski inequality and the fact that $\int_{A_i} \|x(t)\|^p \,\mathrm{d}\mu(t) < \varepsilon_1$, $i = 1, 2, 3$, we obtain

$$\begin{aligned}&\int_{B} \Big(\sup_{\xi\in[0,1]} \big\|d^k\varphi\big(y(t) + \xi h(t)\big) - d^k\varphi\big(y(t)\big)\big\| \Big)^{\frac{p}{p-k}} \,\mathrm{d}\mu(t) \\ &\qquad \le (2M)^{\frac{p}{p-k}} \int_B \big(\|y(t)\| + \|h(t)\|\big)^p \,\mathrm{d}\mu(t) \\ &\qquad \le (2M)^{\frac{p}{p-k}} \bigg(\Big(\int_B \|x(t)\|^p \,\mathrm{d}\mu(t) \Big)^{\frac{1}{p}} + \|y - x\| + \|h\| \bigg)^p \\ &\qquad < (2M)^{\frac{p}{p-k}} \Big((3\varepsilon_1)^{\frac{1}{p}} + 2\delta \Big)^p < (2M)^{\frac{p}{p-k}} 2^p \cdot 3\varepsilon_1 = \frac{1}{2}\varepsilon^{\frac{p}{p-k}}.\end{aligned}$$

This, together with (18) and (17) gives $|R(y, h)| < \frac{1}{k!}\varepsilon\|h\|^k$ for any $y, h \in L_p(\mu; X)$ satisfying $\|y - x\| < \delta$ and $\|h\| < \delta$. Therefore by Theorem 1.110 the function Φ is C^k-smooth on $L_p(\mu; X)$ and $d^j\Phi$, $j = 1, \dots, k$, are bounded on bounded sets by Fact 111. Hence by Corollary 1.117 the canonical norm on $L_p(\mu; X)$ is C^k-smooth with $d^j\|\cdot\|$ bounded on $S_{L_p(\mu;X)}$, $j = 1, \dots, k$. Finally, $d^j\|\cdot\|$ are bounded on $L_p(\mu; X) \setminus B(0, r)$ for each $r > 0$, $j \in \{1, \dots, k\}$ by Fact 109.

(iii)⇒(ii) is obvious.

(ii)⇒(i) By the assumption the spaces X and $\ell_p(\mathbb{N}; X)$ are isometric to certain subspaces of $L_p(\mu; X)$, hence their norms are C^k-smooth. From now on we will work in the space $\ell_p(\mathbb{N};X)$. Let $\varphi\colon X\to\mathbb{R}$ be defined as $\varphi(x)=\|x\|^p$ and similarly $\Phi\colon \ell_p(\mathbb{N};X)\to\mathbb{R}$ as $\Phi(x)=\|x\|^p=\sum_{n=1}^{\infty}\varphi(x_n)$. Note that $\varphi\in C^k(X\setminus\{0\})$, $\Phi\in C^k(\ell_p(\mathbb{N};X)\setminus\{0\})$, and both φ and Φ are G^k-smooth and T^k-smooth at 0 (Fact 110).

Fix $x,h\in\ell_p(\mathbb{N};X)$ such that $x\neq 0$ and $\operatorname{supp} h\subset\{1,\dots,N\}$ for some $N\in\mathbb{N}$. Then

$$\begin{aligned}\Phi(x+th)&=\sum_{n=1}^{N}\varphi(x_n+th_n)+\sum_{n=N+1}^{\infty}\varphi(x_n)\\&=\sum_{n=1}^{N}\left(\sum_{j=0}^{k}\frac{t^j}{j!}\widehat{\delta^j\varphi(x_n)}[h_n]+o(t^k)\right)+\sum_{n=N+1}^{\infty}\varphi(x_n)\\&=\sum_{j=0}^{k}\frac{t^j}{j!}\sum_{n=1}^{N}\widehat{\delta^j\varphi(x_n)}[h_n]+\sum_{n=N+1}^{\infty}\varphi(x_n)+o(t^k),\ t\to 0.\end{aligned}$$

Note that the Gâteaux derivatives are there to cater for the case that some $x_n=0$. Since $\Phi(x+th)=\sum_{j=0}^{k}\frac{t^j}{j!}d^j\Phi(x)[h]+o(t^k)$, $t\to 0$, from the uniqueness of the approximating polynomials we obtain $d^j\Phi(x)[h]=\sum_{n=1}^{N}\widehat{\delta^j\varphi(x_n)}[h_n]$, $j=1,\dots,k$. The continuity of $d^j\Phi(x)$ and the density of finitely supported vectors in $\ell_p(\mathbb{N};X)$ imply that $d^j\Phi(x)[h]=\sum_{n=1}^{\infty}\widehat{\delta^j\varphi(x_n)}[h_n]$ for every $x,h\in\ell_p(\mathbb{N};X)$, $x\neq 0$, and $j\in\{1,\dots,k\}$.

Now assume that $d^k\varphi$ is not bounded on S_X. Then there are vectors $x_n,h_n\in S_X$ such that $d^k\varphi(x_n)[h_n]\geq 2^n$. Let $\tilde{x}=(x_n/2^{\frac{n}{p}})_{n=1}^{\infty}$ and $\tilde{h}=(h_n/2^{\frac{n}{p}})_{n=1}^{\infty}$. Then $\tilde{x},\tilde{h}\in\ell_p(\mathbb{N};X)$, $\|\tilde{x}\|=\|\tilde{h}\|=1$, and

$$\begin{aligned}d^k\Phi(\tilde{x})[\tilde{h}]&=\sum_{n=1}^{\infty}d^k\varphi\left(\frac{x_n}{2^{\frac{n}{p}}}\right)\left[\frac{h_n}{2^{\frac{n}{p}}}\right]\\&=\sum_{n=1}^{\infty}\left(\frac{1}{2^{\frac{n}{p}}}\right)^{p-k}\left(\frac{1}{2^{\frac{n}{p}}}\right)^{k}d^k\varphi(x_n)[h_n]\geq\sum_{n=1}^{\infty}\frac{1}{2^n}2^n=+\infty,\end{aligned}$$

a contradiction. Thus $d^k\varphi$ is bounded on S_X and by Fact 111 and Corollary 1.117 also $d^k\|\cdot\|_X$ is bounded on S_X. □

Note that if X is the space c_0 with a C^∞-smooth norm, then by Theorem 108 and Proposition 49 the canonical norm of $\ell_2(\mathbb{N};X)$ is not even C^2-smooth.

Proposition 115. *Let μ be a measure with at least two disjoint sets of finite non-zero measure, X a Banach space, and $1\leq p<+\infty$. The following statements are equivalent:*

(i) *The canonical norm of $L_p(\mu;X)$ is weakly T^p-smooth.*
(ii) *p is an even integer and the norm of X is p-polynomial.*
(iii) *p is an even integer and the canonical norm of $L_p(\mu;X)$ is p-polynomial.*

Proof. (i)⇒(ii) By the assumption on μ the space $L_p(\mu; X)$ contains a subspace isometric to $X \oplus_p \mathbb{R}$ and so Corollary 43 applies.

(ii)⇒(iii) For $y \in X$ denote $Q(y) = \|y\|^p$. By the Polarisation formula we have

$$\int \left|\check{Q}(x_1(t), \dots, x_p(t))\right| \mathrm{d}\mu(t) = \int \left| \frac{1}{2^p p!} \sum_{\varepsilon_j = \pm 1} \varepsilon_1 \cdots \varepsilon_p Q\left(\sum_{j=1}^p \varepsilon_j x_j(t) \right) \right| \mathrm{d}\mu(t)$$

$$\le \frac{1}{2^p p!} \sum_{\varepsilon_j = \pm 1} \int \left| Q\left(\sum_{j=1}^p \varepsilon_j x_j(t) \right) \right| \mathrm{d}\mu(t) = \frac{1}{2^p p!} \sum_{\varepsilon_j = \pm 1} \left\| \sum_{j=1}^p \varepsilon_j x_j \right\|^p$$

for $x_1, \dots, x_p \in L_p(\mu; X)$. Thus $M(x_1, \dots, x_p) = \int \check{Q}(x_1(t), \dots, x_p(t)) \, \mathrm{d}\mu(t)$ exists and obviously $M \in \mathcal{L}({}^p L_p(\mu; X))$. It follows that the canonical norm of $L_p(\mu; X)$ is p-polynomial.

(iii)⇒(i) is clear. □

Corollary 116. *Let X be a Banach space. If the canonical norm of $\ell_2(\mathbb{N}; X)$ is weakly T^2-smooth, then X is a Hilbert space.*

Almost the same proof as that of Theorem 106 gives the following proposition.

Proposition 117 ([DGJ]). *Let μ be a measure, X, Y Banach spaces, $1 < p < +\infty$, and let $f \in C^k(X; Y)$ be a p-homogeneous mapping such that $d^k f$ is α-Hölder on S_X and $k + \alpha \le p$. Then the mapping $F \colon L_p(\mu; X) \to Y$ defined by*

$$F(x) = \int f\left(x(t)\right) \mathrm{d}\mu(t)$$

is a C^k-smooth p-homogeneous mapping with $d^k F$ being α-Hölder on bounded sets.

Corollary 118 ([DGJ]). *Let μ be a measure with an infinite family of pairwise disjoint measurable sets of finite positive measure. Let X be a Banach space, $1 < p < +\infty$, $k \in \mathbb{N}$, and $\alpha \in (0, 1]$ such that $k + \alpha \le p$. If the norm of X is C^k-smooth with its kth differential α-Hölder on S_X, then the canonical norm of $L_p(\mu; X)$ has the same property. Further, the following statements are equivalent:*

(i) *X admits a $C^{k,\alpha}$-smooth bump.*
(ii) *$L_p(\mu; X)$ admits a $C^{k,\alpha}$-smooth bump.*
(iii) *$L_p(\mu; X)$ admits a $C^{k,\alpha}_{\mathrm{loc}}$-smooth bump.*

Proof. The statement about norms follows directly from Proposition 117. For bumps, (i)⇒(ii) follows from Proposition 117 after constructing a suitable p-homogeneous function from the bump using a variant of [DGZ, Proposition II.5.1]. (ii)⇒(iii) is obvious.

(iii)⇒(ii) Since X is isometric to a subspace of $L_p(\mu; X)$, it admits a $C^{k,\alpha}_{\mathrm{loc}}$-smooth bump. Assume first that X contains c_0. Then $L_p(\mu; X)$ contains $\left(\bigoplus_{n=1}^\infty \ell_\infty^n\right)_p$, which is a reflexive space that is not super-reflexive, as every Banach space is finitely representable in it. This contradicts Corollary 51. Thus X does not contain c_0, and so it admits a $C^{k,\alpha}$-smooth bump by Theorem 48. □

Corollary 119 ([DGJ]). *Let μ, ν be measures, $1 < p, q < +\infty$, and $X = L_p(\mu; L_q(\nu))$.*

(i) *If q is an even integer and p is a multiple of q, then the norm of X is p-polynomial.*

(ii) *If q is an even integer, then the norm of X is $C^{k,p-k}$-smooth, where $k = \lceil p \rceil - 1$.*

(iii) *If q is not an even integer, then the norm of the space X is $C^{k,r-k}$-smooth, where $r = \min\{p, q\}$ and $k = \lceil r \rceil - 1$.*

Proof. (i) follows from Proposition 115.

(ii), (iii) We use Proposition 117 on the function $f : L_q(\nu) \to \mathbb{R}$, $f(x) = \|x\|^p$, which is p-homogeneous. If q is an even integer, then the norm of $L_q(\nu)$ is polynomial, and so f is C^k-smooth with $d^k f$ being $(p-k)$-Hölder on $S_{L_q(\nu)}$ (Proposition 1.128). If q is not an even integer, then f is C^k-smooth with $d^k f$ being $(r-k)$-Hölder on $S_{L_q(\nu)}$ (Theorem 106, the fact that $r \le q$, and Proposition 1.128). □

Proposition 120 ([DGJ]). *If the space $X = \left(\bigoplus_{n=1}^{\infty} \ell_q^n\right)_p$ admits a separating polynomial, then p and q are even integers and p is an integer multiple of q.*

Proof. As ℓ_p is isometric to a subspace of X, p is an even integer by Corollary 4.61. Further, by the sliding hump argument, every normalised weakly null sequence in X has a subsequence equivalent to a block basis such that no two supports of different blocks share the same ℓ_q^n, which means that the subsequence is equivalent to the canonical basis of ℓ_p. Therefore X is saturated by ℓ_p and has property S_p. By Corollary 69 the space X is of cotype p. Since ℓ_q is finitely representable in X, it follows that $p \ge q$. By [Fi, Corollary 22] the norm of X has modulus of convexity of power type p. Thus X admits a p-homogeneous separating polynomial (Theorem 62 and Proposition 4.49). By Corollary 4.11, ℓ_q also admits a p-homogeneous separating polynomial, so finally by Corollary 4.61 we get the conclusion. □

Theorem 121 ([DGJ]). *Let $1 < p, q < +\infty$. If $X = \left(\bigoplus_{n=1}^{\infty} \ell_q^n\right)_p$ admits a T^p-smooth bump, then p and q are even integers and p is a multiple of q (and hence the norm of X is p-polynomial).*

Proof. We show that if X admits a T^p-smooth bump, then it admits a separating polynomial. The rest is Proposition 120 and Corollary 119. Exactly as in the proof of Theorem 107 we find $x_0 \in X$, $x_0 \ne 0$, $P \in \mathcal{P}^{\lfloor p \rfloor}(X)$, and $\delta > 0$ such that

$$P(h) \ge \|x_0 + h\|^p - \|x_0\|^p - \frac{1}{2}\|h\|^p \tag{19}$$

for all $h \in X$, $\|h\| \le \delta$. Since $\|\cdot\|^p$ is Lipschitz on bounded sets, there is $\eta > 0$ such that $\|y + h\|^p < \|x_0 + h\|^p + \frac{1}{8}\delta^p$ and $\|y\|^p > \|x_0\|^p - \frac{1}{8}\delta^p$ whenever $y \in X$, $\|y - x_0\| < \eta$, and $h \in B(0, \delta)$.

Denote by $T_m : X \to X$ the projection $T_m\big((z_1, z_2, \dots)\big) = (z_1, \dots, z_m, 0, 0, \dots)$, where $z_n \in \ell_q^n$. Find $N \in \mathbb{N}$ such that $\|x_0 - T_N(x_0)\| < \eta$ and put $H = (Id - T_N)(X)$. Note that if $u \in T_N(X)$ and $v \in H$, then $\|u + v\|^p = \|u\|^p + \|v\|^p$. Therefore for any $h \in H$, $\|h\| = \delta$ the estimate (19) implies

$$P(h) > \|T_N(x_0) + h\|^p - \|T_N(x_0)\|^p - \frac{1}{2}\|h\|^p - \frac{1}{4}\delta^p = \frac{1}{2}\|h\|^p - \frac{1}{4}\delta^p = \frac{1}{4}\delta^p.$$

It follows that H admits a non-negative homogeneous separating polynomial Q_1 (Fact 4.45). As $T_N(X)$ is finite-dimensional, it admits a non-negative homogeneous separating polynomial Q_2. The polynomial $Q_1 \circ (Id - T_N) + Q_2 \circ T_N$ is then a separating polynomial on X. □

Corollary 122 ([DGJ]). *Let μ, ν be measures with infinite families of pairwise disjoint measurable sets of finite positive measure, $1 < p, q < +\infty$, and $X = L_p(\mu; L_q(\nu))$.*

(i) *If q is an even integer and X admits a T^p-smooth bump, then p is a multiple of q (and hence the norm of X is p-polynomial).*

(ii) *If q is not an even integer, then X does not admit a T^r-smooth bump, where $r = \min\{p, q\}$.*

Proof. (i) follows from Theorem 121, as $\left(\bigoplus_{n=1}^{\infty} \ell_q^n\right)_p$ is isometric to a subspace of X.

(ii) The case $p < q$ follows also from Theorem 121. The case $q \leq p$ follows from Theorem 107, as ℓ_q is isometric to a subspace of X. □

Note that ℓ_3 admits an equivalent norm that is $C^{2,1}$-smooth and G^3-smooth, but does not admit a T^3-smooth bump (Theorem 106, variants of Corollary 7.72 and Corollary 7.70, Theorem 107). The space $\ell_2(\mathbb{N}; \ell_4)$ admits an equivalent norm that is $C^{1,1}$-smooth and G^2-smooth, but does not admit a T^2-smooth bump (Corollary 119, variants of Corollary 7.72 and Corollary 7.70, Corollary 122).

8. $C(K)$ spaces

In the present section we collect, mostly without proof, some results on higher smoothness of $C(K)$ spaces.

Fact 123. *Let T be a topological space and X a Banach space of all bounded Borel functions on T with the supremum norm. Consider the space $[-1, 1]^{\mathbb{N}}$ with the product topology. Let $\Phi\colon T \to [-1, 1]^{\mathbb{N}}$ be a Borel mapping and $L = \overline{\Phi(T)}$ (this is a metrisable compact space). Then the linear operator $T\colon C(L) \to X$ given by $T(f) = f \circ \Phi$ is an isometry into. The mapping Φ is Borel if and only if each $\pi_n \circ \Phi$, $n \in \mathbb{N}$ is Borel, where π_n is the projection onto the nth coordinate.*

Proof. T is an isometry because $\Phi(T)$ is dense in L. The last statement follows from the fact that the topology of $[-1, 1]$ has a countable basis. The last statement follows from the fact that the topology of $[-1, 1]$ has a countable basis of the form $I_1 \times \cdots \times I_n \times [-1, 1]^{\mathbb{N}}$, I_j intervals. □

Proposition 124. *Let K be a compact space and $X \subset C(K)$ a separable subspace. Then there are a metrisable compact L which is a continuous image of K and a subspace $Y \subset C(K)$ such that Y is isometric to $C(L)$ and $X \subset Y$. In particular, if K is scattered, then L is countable, and so Y is isometric to $C([0, \alpha])$ for some countable ordinal α.*

Proof. Let $\{f_n\}$ be a dense subset of B_X and let us define $\Phi\colon K \to [-1,1]^{\mathbb{N}}$ by $\Phi(x)(n) = f_n(x)$. Then Φ is continuous and hence the operator T from Fact 123 is a linear isometry onto some $Y \subset C(K)$. Note that $f_n = \pi_n \circ \Phi$ and hence $T(\pi_n \restriction_L) = f_n$. Thus $f_n \in Y$ and consequently $X \subset Y$.

If K is scattered, then by [FHHMZ, Lemma 14.20] and [FHHMZ, Lemma 14.21] the space $L = \Phi(K)$ is countable. So $C(L)$ is isometric to $C([0,\alpha])$ for some countable ordinal α by the Mazurkiewicz-Sierpiński theorem ([HMVZ, Theorem 2.56]). □

Theorem 125. *Let K be a compact space. The following statements are equivalent:*

(i) *K is scattered.*
(ii) *$C(K)$ is an Asplund space.*
(iii) *$C(K)^*$ is isometric to $\ell_1(\Gamma)$ for some Γ (in fact, we can choose $\Gamma = K$).*
(iv) *$C(K)$ does not contain ℓ_1.*
(v) *$C(K)$ is c_0-saturated.*

(i)⇒(iii) is [FHHMZ, Theorem 14.24]. (i)⇒(ii) follows from Theorem 2, Proposition 124, and the implication (i)⇒(iii). (ii)⇒(i) can be found in [FHHMZ, Theorem 14.25]. (iii)⇒(ii) follows from Theorem 2 and the fact that $\ell_1(\Gamma)$ has the RNP. (ii)⇒(iv) follows from Theorem 2. (i)⇔(iv)⇔(v) are results of Aleksander Pełczyński and Zbigniew Semadeni, [PS].

Since every Banach space admitting a C^1-smooth bump function is an Asplund space (Corollary 3), we see that smooth renormings of $C(K)$ spaces may exist only for scattered K.

Theorem 126. *For any set Γ the space $c_0(\Gamma)$ admits an equivalent norm which is simultaneously LFC-$\{e^*_\gamma\}$, C^∞-smooth, and uniformly Gâteaux differentiable.*

We postpone the proof until Chapter 7, page 437.

The following result improves an earlier result of Michel Talagrand which considers only C^1-smoothness.

Theorem 127 ([Hay1]). *For every ordinal μ the space $C([0,\mu])$ admits an equivalent C^∞-smooth LFC norm.*

Theorem 128 ([Hay3]). *Let $\widehat{T}$ be a one-point compactification of a locally compact tree T.*

(i) *The space $C(\widehat{T})$ admits a C^∞-smooth LFC bump.*
(ii) *The space $C(\widehat{T})$ admits an equivalent C^∞-smooth LFC norm if and only if it admits an equivalent C^1-smooth norm.*

Theorem 129 ([HáHa]). *Let K be a σ-discrete compact space. Then $C(K)$ admits an equivalent C^∞ smooth LFC-K norm.*

Proof. Let $K = \bigcup_{n\in\mathbb{N}} D_n$, where the sets D_n are pairwise disjoint and relatively discrete. First note that K is scattered. Indeed, by the Baire category theorem there is $k \in \mathbb{N}$ such that $\overline{D_k}$ has a non-empty interior. Thus there are an open $U \subset \overline{D_k}$ and $x \in D_k \cap U$. Since D_k is relatively discrete, there is an open neighbourhood

V of x such that $V \cap D_k = \{x\}$. Thus also $V \cap \overline{D_k} = \{x\}$, and consequently $V \cap U = V \cap U \cap \overline{D_k} = \{x\}$, i.e. x is an isolated point of K. Since the σ-discreteness is hereditary, the scatteredness follows. Therefore K is sequentially compact ([Fab3, Lemma 2.1.1]).

We define the new norm as a Musielak-Orlicz norm on $\ell_\infty(K)$ and show that it is smooth on its subspace $C(K)$. To this end we first define two functions $\alpha, \beta\colon K \to (0,1]$ that will provide endpoints of certain intervals determining the Orlicz functions $t \mapsto \varphi(x,t)$, $x \in K$. So fix $r_n \in (0,1)$, $n \in \mathbb{N}$ such that $a = \prod_{n\in\mathbb{N}} r_n > 0$ and for $x \in K$ set

$$\alpha(x) = \prod_{\{n;\ x\in\overline{D_n}\}} r_n, \qquad \beta(x) = \prod_{\{n;\ x\in D'_n\}} r_n.$$

Notice that $\alpha(x) = r_k\beta(x)$, where $k \in \mathbb{N}$ is the unique index such that $x \in D_k$. This follows from the fact that $D_k \cap D'_k = \emptyset$ and D_n are pairwise disjoint. Therefore $0 < a \leq \alpha(x) < \beta(x) \leq 1$.

Further, we claim that the functions α, β have the following property:

$$\beta(x) \leq \liminf_{n\to\infty} \alpha(x_n) \quad \text{whenever } x_n \to x \in K \text{ and } x_n \in K \text{ are distinct.} \tag{20}$$

To prove this, let $\{y_n\}$ be a subsequence of $\{x_n\}$ such that $\lim \alpha(y_n) = \liminf \alpha(x_n)$. Next, by passing to further subsequences and then diagonalising we may assume that for each $k \in \mathbb{N}$ the set $\{y_n;\ n \geq k\}$ is either a subset of $\overline{D_k}$ or disjoint with $\overline{D_k}$. Let $M = \{k \in \mathbb{N};\ \{y_n;\ n \geq k\} \subset \overline{D_k}\}$. Then

$$\prod_{\substack{k\leq n\\ k\in M}} r_k \cdot \prod_{k>n} r_k \leq \alpha(y_n) \leq \prod_{\substack{k\leq n\\ k\in M}} r_k,$$

and therefore $\lim_{n\to\infty} \alpha(y_n) = \prod_{k\in M} r_k$. On the other hand, if $k \in M$, then $\lim y_n = x \in D'_k$, as y_n are distinct and $\{y_n;\ n \geq k\} \subset \overline{D_k}$. Hence $\beta(x) \leq \prod_{k\in M} r_k$, which proves the claim.

Let $\varphi\colon K \times \mathbb{R} \to [0,+\infty)$ be such that for each $x \in K$ the function $t \mapsto \varphi(x,t)$ is a C^∞-smooth even convex function non-decreasing on $[0,+\infty)$ and satisfying $\varphi(x,t) = 0$ for $|t| \leq \alpha(x)$ and $\varphi(x,t) > 1$ for $|t| \geq \beta(x)$. Now define a function $\Phi\colon \ell_\infty(K) \to [0,+\infty]$ by

$$\Phi(f) = \sum_{x\in K} \varphi\big(x, f(x)\big).$$

Note that the function Φ is convex, even, lower semi-continuous, $\Phi = 0$ on the ball $B_{\ell_\infty(K)}(0,a)$, and $\Phi(f) > 1$ whenever $\|f\|_\infty > 1$. Since $B(0,a) \subset B_1 \subset B(0,1)$, where $B_1 = \{f \in \ell_\infty(K);\ \Phi(f) \leq 1\}$, the Minkowski functional of B_1 is an equivalent norm on $\ell_\infty(K)$.

Put $B = \{f \in C(K);\ \Phi(f) \leq 1\}$. We show that Φ is LFC-K and C^∞-smooth on a neighbourhood of B in $C(K)$. Let $f \in B$. We claim that there are $\delta > 0$ and $F \subset K$ finite such that $\varphi(x, g(x)) = 0$ whenever $\|f - g\|_\infty < \delta$ and $x \in K \setminus F$. Indeed, if this is not the case, then there are sequences $\{f_n\} \subset C(K)$ and $\{x_n\} \subset K$ such that

x_n are distinct, $f_n \to f$ uniformly on K, and $\varphi(x_n, f_n(x_n)) > 0$ for each $n \in \mathbb{N}$. It follows that $|f_n(x_n)| > \alpha(x_n)$. By passing to a subsequence we may assume that $x_n \to x \in K$. Since $f_n \to f$ uniformly on K and f is continuous, $f_n(x_n) \to f(x)$. Thus $|f(x)| \geq \liminf \alpha(x_n) \geq \beta(x)$ by (20). This implies that $\varphi(x, f(x)) > 1$, a contradiction with $\Phi(f) \leq 1$. Clearly, Φ depends only on F on $U(f, \delta) \cap C(K)$ and is C^∞-smooth there as a finite sum of C^∞-smooth functions.

Finally, set $C = \operatorname{Int}\{f \in C(K);\ \Phi(f) \leq 2\}$. Then C is a non-empty open convex set and $B \subset C$, since Φ is continuous on a neighbourhood of B. Further, B is closed in $C(K)$, as Φ is lower semi-continuous. So, Lemma 23 together with Lemma 93 finishes the proof. □

Corollary 130 ([Há8]). *Let K be a compact space with $K^{(\omega_1)} = \emptyset$. Then $C(K)$ admits an equivalent C^∞-smooth LFC-K norm.*

Recall that in general a dual LUR norm on X^* implies C^1-smoothness of the norm on X, and this is often used to produce a C^1-smooth renorming. The following corollary shows that in case of $C(K)$ spaces, this approach already gives a C^∞-smooth renorming.

Corollary 131 ([HáHa]). *Let K be a compact space. If $C(K)^*$ admits an equivalent dual LUR norm, then $C(K)$ admits an equivalent C^∞-smooth LFC-K norm.*

Proof. This follows from Theorem 129 combined with the result of Matías Raja [Raj] that $C(K)^*$ admits an equivalent dual LUR norm if and only if K is σ-discrete. □

Theorem 132 ([Há3]). *Let K be a compact space. Then $C(K)$ has an equivalent analytic norm if and only if K is countable.*

Proof. $\Leftarrow$ follows from Theorem 103 and Theorem 104.

$\Rightarrow$ Since $C(K)$ has an equivalent analytic norm $\|\cdot\|$, it is an Asplund space (Corollary 3) and so $C(K)^*$ is isometric to $\ell_1(K)$ (Theorem 125). Since $C(K)$ spaces enjoy the DPP, by Theorem 3.67 we get $\mathcal{P}(C(K)) = \mathcal{P}_{\mathrm{wsC}}(C(K))$. For each $x \in C(K) \setminus \{0\}$ there is a neighbourhood U of x such that $\|y\| = \sum_{n=0}^{\infty} P_n(y - x)$ for all $y \in U$, where $P_n \in \mathcal{P}(^n X)$, and the sum converges uniformly on U. It follows that $\|\cdot\|$ is locally wsC on $C(K) \setminus \{0\}$. Thus $C(K)^*$ is separable by Proposition 77, which in turn implies that K is countable. □

9. Orlicz spaces

We briefly review some properties of Orlicz sequence spaces and then proceed to study how their smoothness depends on the behaviour of the Orlicz function. We then turn to polyhedrality of Orlicz sequence spaces which leads to an example of a separable c_0-saturated Asplund space that is not isomorphically polyhedral. This serves also as an example of a separable space that has C^∞-smooth norm but does not admit an equivalent analytic norm.

A function $M: \mathbb{R} \to [0, +\infty)$ is called an Orlicz function if it is even, convex, non-decreasing on $[0, +\infty)$, $M(0) = 0$, and if M is not constant. We define $\Phi_M: \ell_\infty \to [0, +\infty]$ by $\Phi_M(x) = \sum_{n=1}^{\infty} M(x_n)$, which is an even convex function. An Orlicz sequence space ℓ_M is a linear subspace of ℓ_∞ consisting of $x \in \ell_\infty$ satisfying $\Phi_M(x/\rho) < +\infty$ for some $\rho > 0$, equipped with the norm given by the Minkowski functional of $\{x \in \ell_M;\ \Phi_M(x) \leq 1\}$. With this norm the space ℓ_M is a Banach space. It is easily seen that the canonical coordinate projections in ℓ_M are continuous and so Φ_M is lower semi-continuous on ℓ_M and it is continuous on $U_{\ell_M}(0, 1)$. From the convexity we obtain $\Phi_M(x) \leq \|x\|$ for $x \in B_{\ell_M}$, while $\Phi_M(x) \geq \|x\|$ for $x \in \ell_M$, $\|x\| > 1$.

An important subspace of ℓ_M is a space h_M consisting of $x \in \ell_M$ satisfying $\Phi_M(x/\rho) < +\infty$ for all $\rho > 0$. The space h_M is a closed subspace of ℓ_M and the canonical basis vectors $\{e_n\}_{n=1}^{\infty}$ form a symmetric basis of h_M. Further, Φ_M is continuous on h_M and for $x \in h_M$ we have $\Phi_M(x) = 1$ if and only if $\|x\| = 1$.

If M is such that $M(t) = 0$ for some $t > 0$, then it is called a degenerate Orlicz function. The associated space ℓ_M is then isomorphic to ℓ_∞ and h_M is isomorphic to c_0.

For Orlicz functions M_1, M_2 the spaces ℓ_{M_1}, ℓ_{M_2} are isomorphic if and only if the canonical bases of h_{M_1}, h_{M_2} are equivalent, if and only if the functions M_1, M_2 are equivalent at 0, i.e. there is $C \geq 1$ such that $\frac{1}{C} M_1\big(\frac{1}{C}t\big) \leq M_2(t) \leq C M_1(Ct)$ for each $t \in [0, 1]$.

A non-degenerate Orlicz function M is said to satisfy the Δ_2-condition at 0 (at $+\infty$), if there is $K > 0$ such that $M(2t) \leq K M(t)$ for $t \in [0, 1]$ ($t \in [1, +\infty)$). An Orlicz function M satisfies the Δ_2-condition at 0 if and only if $h_M = \ell_M$. For an introduction to Orlicz sequence spaces we refer the reader to [LiTz2].

To a non-degenerate Orlicz function M we associate the following numbers, called the Boyd indices:

$$\alpha_M = \sup\Big\{q \in (0, +\infty);\quad \sup_{\lambda, t \in (0,1]} \frac{M(\lambda t)}{\lambda^q M(t)} < +\infty\Big\},$$

$$\beta_M = \inf\Big\{q \in (0, +\infty);\quad \inf_{\lambda, t \in (0,1]} \frac{M(\lambda t)}{\lambda^q M(t)} > 0\Big\}.$$

It is not difficult to check that $1 \leq \alpha_M \leq \beta_M \leq +\infty$ and that $\beta_M < +\infty$ if and only if M satisfies the Δ_2-condition at 0.

Theorem 133 ([LiTz2, Theorem 4.a.9]). *Let M be a non-degenerate Orlicz function. The space ℓ_p, or c_0 if $p = +\infty$, is isomorphic to a subspace of an Orlicz space h_M if and only if $p \in [\alpha_M, \beta_M]$.*

Corollary 134. *Let M be a non-degenerate Orlicz function. The Orlicz space h_M is c_0-saturated if and only if $\alpha_M = +\infty$.*

Proof. $\Rightarrow$ follows directly from Theorem 133.

$\Leftarrow$ If the space h_M is not c_0-saturated, then there is an infinite-dimensional subspace $X \subset h_M$ that does not contain c_0. By [LiTz2, Proposition 4.a.7] there is a subspace

Y of X isomorphic to h_N for some Orlicz function N. Since h_N does not contain c_0, from Theorem 133 it follows that $\beta_N < +\infty$ and h_N contains ℓ_{β_N}. Thus also h_M contains ℓ_{β_N}, a contradiction with $\alpha_M = +\infty$. □

Proposition 135 ([Kn]). *Let M be a non-degenerate Orlicz function and $1 < p < \alpha_M$. Then every weakly null sequence in h_M has a subsequence with an upper p-estimate. In particular, every polynomial on h_M with degree smaller than α_M is weakly sequentially continuous.*

Proof. Let $\{x_n\} \subset h_M$ be a weakly null sequence. Without loss of generality we may assume that $M(1) \geq 1$ and $\{x_n\} \subset B_{h_M}$. We can find a subsequence $\{x_{n_k}\}$ and a block basic sequence $y_k = \sum_{j=l_k}^{l_{k+1}-1} c_j e_j \in B_{h_M}$ such that $\|x_{n_k} - y_k\| \leq \frac{1}{2^k}$. Note that $\sum_{j=l_k}^{l_{k+1}-1} M(c_j) \leq 1$, which also implies that $|c_j| \leq 1$ for all $j \in \mathbb{N}$. By the definition of α_M there is $C \geq 1$ such that $M(\lambda t) \leq C^p \lambda^p M(t)$ for all $\lambda, t \in [0,1]$. Let $a_1, \dots, a_N$ be any scalars and denote $\nu = \big(\sum_{k=1}^N |a_k|^p\big)^{\frac{1}{p}}$. Then

$$\begin{aligned}\sum_{k=1}^{N} \sum_{j=l_k}^{l_{k+1}-1} M\Big(\frac{a_k c_j}{C\nu}\Big) &\leq \sum_{k=1}^{N} \sum_{j=l_k}^{l_{k+1}-1} C^p \Big(\frac{|a_k|}{C\nu}\Big)^p M(c_j) \\ &= \frac{1}{\nu^p} \sum_{k=1}^{N} |a_k|^p \sum_{j=l_k}^{l_{k+1}-1} M(c_j) \leq 1,\end{aligned}$$

which means that $\big\|\sum_{k=1}^N a_k y_k\big\| \leq C\nu$. Hence

$$\Big\|\sum_{k=1}^{N} a_k x_{n_k}\Big\| \leq \Big\|\sum_{k=1}^{N} a_k y_k\Big\| + \sum_{k=1}^{N} |a_k| \frac{1}{2^k} \leq (C+1)\nu.$$

The last statement follows from Corollary 3.58. □

Somewhat weaker statement is obtained for polynomials of degree $k \geq \alpha_M$ when M satisfies certain condition. Note that an example of such function (where $k > \alpha_M$) is $M(t) = t^{p+\sin\log|\log t|}$, see also [LiTz2, Example 4.c.2].

Proposition 136. *Let M be a non-degenerate Orlicz function such that it satisfies $\liminf_{t\to 0+} \frac{M(t)}{t^k} = 0$ for some $k \in \mathbb{N}$. Then $\{e_n\}$ does not have a lower k-estimate. Consequently, for each finitely supported $x \in X$ the sequence $\{S^n(x)\}_{n=1}^\infty$ is $\mathcal{P}^k$-null, where $S \in \mathcal{L}(h_M; h_M)$ is the right shift operator.*

Proof. For $n \in \mathbb{N}$ denote $\mu_n = \|e_1 + \dots + e_n\|$. Suppose that $\{e_n\}$ has a lower k-estimate. Then there is $C > 0$ such that $Cn^{\frac{1}{k}} \leq \mu_n$ for each $n \in \mathbb{N}$. Hence

$$1 = \Phi\Big(\frac{e_1 + \dots + e_n}{\mu_n}\Big) = nM\Big(\frac{1}{\mu_n}\Big) \leq nM\Big(\frac{1}{Cn^{\frac{1}{k}}}\Big)$$

for each $n \in \mathbb{N}$. Let $\{t_n\} \subset (0, \frac{1}{C}]$ be such that $\lim_{n\to\infty} \frac{M(t_n)}{t_n^k} = 0$. For each $n \in \mathbb{N}$ find $j_n \in \mathbb{N}$ such that $\frac{1}{C}(j_n + 1)^{-\frac{1}{k}} < t_n \leq \frac{1}{C} j_n^{-\frac{1}{k}}$. Then

$$1 \leq (j_n+1)M\Big(\frac{1}{C(j_n+1)^{\frac{1}{k}}}\Big) < (j_n+1)M(t_n) \leq \frac{j_n+1}{C^k j_n}\cdot\frac{M(t_n)}{t_n^k} \leq \frac{2}{C^k}\cdot\frac{M(t_n)}{t_n^k},$$

which is a contradiction. Thus we may apply Corollary 4.41. □

Theorem 137 ([MaT2]). *Let M be a non-degenerate Orlicz function and the Orlicz space h_M is not isomorphic to ℓ_p, p an even integer. If $\alpha_M < +\infty$, then h_M does not admit a T^{α_M}-smooth bump.*

Proof. If α_M is not an even integer, then the result follows from Theorem 133 and Theorem 107. So suppose that $\alpha_M = k$ is an even integer and h_M admits a T^k-smooth bump. Without loss of generality we may assume that $M(1) = 1$. Denote $\Phi = \Phi_M$. Analogously as in the proof of Theorem 107 (this time using the function $\varphi = \frac{1}{\Phi}$) we obtain $x \in h_M$ and $P \in \mathcal{P}^k(h_M)$ satisfying $P(0) = 0$ such that for each $\varepsilon > 0$ there is $\delta > 0$ such that

$$\Phi(x+h) - \Phi(x) - P(h) \leq \varepsilon\|h\|^k \quad \text{for all } \|h\| \leq \delta. \tag{21}$$

Since h_M is not isomorphic to ℓ_k, we have either $\limsup_{t\to 0+} \frac{M(t)}{t^k} = +\infty$, or $\liminf_{t\to 0+} \frac{M(t)}{t^k} = 0$. In the first case let $P = Q + P_k$, where $Q \in \mathcal{P}^{k-1}(h_M)$ and $P_k \in \mathcal{P}(^k h_M)$. Let $\delta > 0$ be such that (21) holds with $\varepsilon = 1$ and let $0 < \eta \leq \delta$ be such that $\frac{M(\eta)}{\eta^k} > \|P_k\| + 1$. Then

$$\begin{aligned} Q(\eta e_n) &\geq \Phi(x+\eta e_n) - \Phi(x) - P_k(\eta e_n) - \eta^k \\ &\geq M\big(e_n^*(x)+\eta\big) - M\big(e_n^*(x)\big) - \|P_k\|\eta^k - \eta^k. \end{aligned}$$

Note that by the definition of α_M we have $\lim_{t\to 0+} \frac{M(t)}{t^q} = 0$ for all $q < \alpha_M = k$ and thus $\lim_{n\to\infty} Q(\eta e_n) = 0$ by Proposition 136. So by passing to a limit as $n \to \infty$ we obtain $0 \geq M(\eta) - \eta^k\big(\|P_k\| + 1\big) > 0$, a contradiction.

Now we deal with the second case. By Theorem 133 there is a basic sequence $\{y_n\} \subset h_M$ which is equivalent to the canonical basis of ℓ_k. Since $k > 1$, $\{y_n\}$ is weakly null and hence there is a subsequence of $\{y_n\}$ equivalent to a block basic sequence of $\{e_n\}$ (Proposition 3.32). Therefore there is a normalised block basic sequence $\{u_n\}$ of $\{e_n\}$ which is equivalent to the canonical basis of ℓ_k. For $n \in \mathbb{N}$ denote $\mu_n = \|u_1 + \dots + u_n\|$. Since $\{u_n\}$ is equivalent to the canonical basis of ℓ_k, there is $a > 0$ such that $an^{\frac{1}{k}} \leq \mu_n$ for each $n \in \mathbb{N}$. Let $\delta > 0$ be such that (21) holds with $\varepsilon = \frac{1}{2}a^k$ and find $m \in \mathbb{N}$ such that $\frac{1}{\mu_m} \leq \delta$. As $\Phi\big(\frac{u_1+\dots+u_m}{\mu_m}\big) = 1$, we obtain

$$m\Big(\frac{a}{\mu_m}\Big)^k \leq 1 = \Phi\Big(\frac{u_1+\dots+u_m}{\mu_m}\Big) = \sum_{j=1}^{m} \Phi\Big(\frac{u_j}{\mu_m}\Big).$$

It follows that there is $j \in \{1,\dots,m\}$ such that $\Phi\big(\frac{u_j}{\mu_m}\big) \geq \big(\frac{a}{\mu_m}\big)^k$. Set $v = \frac{u_j}{\mu_m}$ and note that $\|v\| = \frac{1}{\mu_m} \leq \delta$.

Let $S \in \mathcal{L}(h_M; h_M)$ be the right shift operator and let us denote $v_n = S^n(v)$. Then $\Phi(v_n) = \Phi(v)$ and $\|v_n\| = \|v\|$ for each $n \in \mathbb{N}$. Proposition 136 implies that $P(v_n) \to 0$, while obviously $\Phi(x + v_n) \to \Phi(x) + \Phi(v)$. Therefore from $P(v_n) \geq \Phi(x+v_n) - \Phi(x) - \frac{1}{2}a^k\|v_n\|^k$ we get $0 \geq \Phi(v) - \frac{1}{2}a^k\|v\|^k \geq \frac{1}{2}\left(\frac{a}{\mu_m}\right)^k > 0$, a contradiction. □

Similarly as the Orlicz sequence spaces we may also define Orlicz spaces using an arbitrary measure μ. Let $\mathfrak{M}$ be the vector space of all of μ-measurable functions factorised by equality almost everywhere. We define $\Phi_M : \mathfrak{M} \to [0, +\infty]$ by $\Phi_M(x) = \int M(x(t))\,\mathrm{d}\mu(t)$, which is an even convex function. An Orlicz space $L_M(\mu)$ is a linear subspace of $\mathfrak{M}$ consisting of $x \in \mathfrak{M}$ satisfying $\Phi_M(x/\rho) < +\infty$ for some $\rho > 0$, equipped with the norm given by the Minkowski functional of $\{x \in L_M(\mu);\ \Phi_M(x) \leq 1\}$. With this norm the space $L_M(\mu)$ is a Banach space. Obviously Φ_M is continuous on $U_{L_M(\mu)}(0,1)$ and using Fatou's lemma it can be shown that Φ_M is lower semi-continuous on $L_M(\mu)$. From the convexity we obtain $\Phi_M(x) \leq \|x\|$ for $x \in B_{L_M(\mu)}$, while $\Phi_M(x) \geq \|x\|$ for $x \in L_M(\mu)$, $\|x\| > 1$. The space $H_M(\mu)$ consisting of $x \in L_M(\mu)$ satisfying $\Phi_M(x/\rho) < +\infty$ for all $\rho > 0$ is a closed subspace of $L_M(\mu)$. Further, Φ_M is continuous on $H_M(\mu)$ and for $x \in H_M(\mu)$ we have $\Phi_M(x) = 1$ if and only if $\|x\| = 1$.

Recall that for a function $\varphi\colon [0, +\infty) \to \overline{\mathbb{R}}$ its non-decreasing envelope is given by $\bar{\varphi}(t) = \sup_{s \in [0,t]} \varphi(s)$.

Theorem 138 ([Mal1]). *Let $k \in \mathbb{N}$ and let $M \in C^k(\mathbb{R})$ be a non-degenerate Orlicz function which is C^{k+1}-smooth on $\mathbb{R}^+$ and has the following property: There is $p > k$ and $C \geq 2$ such that*

$$M(\lambda t) \leq C\lambda^p M(t) \quad \text{for all } \lambda \in [0,1],\ t \in \mathbb{R}^+, \text{ and} \tag{22}$$

$$t^j \left|M^{(j)}(t)\right| \leq CM(Ct) \quad \text{for all } t \in \mathbb{R}^+,\ j = 1, \dots, k+1. \tag{23}$$

Let

$$\omega_0(t) = t \cdot \sup_{\substack{\lambda \in [t,1] \\ s \in \mathbb{R}^+}} \frac{M(\lambda s)}{\lambda^{k+1} M(s)} \quad \text{for } t \in (0,1],$$

$\omega_0(0) = 0$, and $\omega_0(t) = t$ for $t \in (1, +\infty)$. Then $\omega_M = \bar{\omega}_0 \in \mathcal{M}$. Further, let $\Omega \subset \mathcal{M}$ be the cone generated by ω_M. Then for any measure μ the function Φ_M is C^k-smooth on $U_{L_M(\mu)}(0, \frac{1}{2C})$, $C^{k,\Omega}$-smooth on $U_{L_M(\mu)}(0, \frac{1}{4C})$, and $C^{k,\Omega}_{\mathrm{loc}}$-smooth on $H_M(\mu)$. In particular, the norm of $H_M(\mu)$ is $C^{k,\Omega}_{\mathrm{loc}}$-smooth and if Φ_M is separating on $U_{L_M(\mu)}(0, \frac{1}{4C})$ (e.g. if M satisfies the Δ_2-condition at 0 and $+\infty$), then there is an equivalent $C^{k,\Omega}$-smooth norm on $L_M(\mu)$.

We remark that the assumptions of the theorem can be somewhat relaxed to not require the C^k-smoothness on $\mathbb{R}$ and to require (23) only for $j = k+1$. One can then painfully prove using several inductions back and forth that our assumptions are actually satisfied. Since it turns out that this formal generalisation is not useful for us, we chose to trade it for slightly shorter proof.

The proof of this theorem is analogous to the proof of Theorem 106, but instead of the inequality (15) we use the following lemma.

Lemma 139. *Let M be an Orlicz function as in Theorem 138 and $\omega = \omega_M$. Then there is $K > 0$ such that*

$$\left|M^{(k)}(x+tz) - M^{(k)}(x)\right|v^k \le K\big(M(2Cx) + M(z) + M(v)\big)\omega(t)$$

for all $x, z \in \mathbb{R}$, $v \in [0, +\infty)$, $t \in [0, \frac{1}{2C}]$.

The proof of the lemma is based mainly on the estimate (iii) in the following fact. It seems that it would be natural to define ω as the non-decreasing envelope of $\lambda \mapsto \sup_{s \in \mathbb{R}^+} \frac{M(\lambda s)}{\lambda^k M(s)}$. This does not work however, and the particular formula for ω is chosen exactly so that the estimate (iii) below holds.

Fact 140. *Let M be a non-degenerate Orlicz function and let $\omega = \omega_M$ as defined in Theorem 138. Then*

(i) $\omega(t) \ge t$ *for each* $t \in [0, +\infty)$,
(ii) $M(\lambda t) \le \omega(\lambda)\lambda^k M(t)$ *for each* $\lambda \in [0, 1]$, $t \in [0, +\infty)$,
(iii) $\lambda^{k+1} M(a) \le \big(M(\frac{\lambda}{t}a) + M(a)\big)\omega(t)t^k$ *for each* $\lambda \in [0, 1]$, $a \in [0, +\infty)$, *and* $t \in \mathbb{R}^+$.

Proof. (i) and (ii) are clear from the definition of ω.

(iii) If $\lambda \le t$, then $\lambda^{k+1} M(a) \le t^{k+1} M(a) \le M(a)\omega(t)t^k$ by (i). If $\lambda > t$, then $t \le \frac{t}{\lambda} < 1$ and hence

$$\omega(t) \ge t \frac{M(\frac{t}{\lambda}\frac{\lambda}{t}a)}{(\frac{t}{\lambda})^{k+1} M(\frac{\lambda}{t}a)},$$

from which the inequality in (iii) follows. □

Proof of Lemma 139. First we prove that there is $C_1 > 0$ such that for each $x, y \in \mathbb{R}$ not both zero

$$\left|M^{(k)}(y) - M^{(k)}(x)\right| \le C_1 \frac{M\big(C \max\{|x|, |y|\}\big)}{\max\{|x|, |y|\}^{k+1}} |y - x|. \tag{24}$$

Without loss of generality we may assume that $|x| \le |y|$. If $|y - x| \le \frac{1}{2}|y|$, then either $0 < x \le y$, or $y \le x < 0$, and so using (23) we obtain

$$\begin{aligned}\left|M^{(k)}(y) - M^{(k)}(x)\right| &\le \int_{|x|}^{|y|} \left|M^{(k+1)}(t)\right| \mathrm{d}t \le C \int_{|x|}^{|y|} \frac{M(Ct)}{t^{k+1}} \mathrm{d}t \\ &\le C \frac{M\big(C|y|\big)}{|x|^{k+1}} |y - x| \le 2^{k+1} C \frac{M\big(C|y|\big)}{|y|^{k+1}} |y - x|.\end{aligned}$$

On the other hand, if $|y-x| > \frac{1}{2}|y|$ and $x \neq 0$, then (23) and (22) imply

$$\begin{aligned}\big|M^{(k)}(y) - M^{(k)}(x)\big| &\leq \big|M^{(k)}(y)\big| + \big|M^{(k)}(x)\big| \leq C\frac{M\big(C|y|\big)}{|y|^k} + C\frac{M\big(C|x|\big)}{|x|^k} \\ &= C\frac{M\big(C|y|\big)}{|y|^k} + C\frac{M\big(\frac{|x|}{|y|}C|y|\big)}{|x|^k} \leq C\frac{M\big(C|y|\big)}{|y|^k} + C^2\frac{M\big(C|y|\big)}{|y|^k} \\ &< 2(C+C^2)\frac{M\big(C|y|\big)}{|y|^{k+1}}|y-x|\end{aligned}$$

If $x = 0$, then it suffices to notice that (22) implies that M is T^k-smooth at 0 with the approximating polynomial being zero and hence $M^{(k)}(0) = 0$. Then we repeat the estimate above.

Now fix $x, z \in \mathbb{R}$, $v \in [0, +\infty)$, and $t \in [0, \frac{1}{2C}]$. Clearly we may also assume that $tz \neq 0$. Set $\zeta = \max\{|x|, |x+tz|\}$. Obviously, $t|z| \leq 2\zeta$. Also, since M is non-decreasing on $[0, +\infty)$,

$$M(C\zeta) \leq \max\{M(2C|x|), M(2Ct|z|)\} \leq M(2Cx) + M(z). \tag{25}$$

The estimate (24) implies

$$\big|M^{(k)}(x+tz) - M^{(k)}(x)\big|v^k \leq C_1 t|z|\frac{M(C\zeta)}{\zeta^{k+1}}v^k.$$

If $v \leq \frac{C\zeta}{t}$, then by Fact 140(iii) and (25)

$$\begin{aligned}t|z|\frac{M(C\zeta)}{\zeta^{k+1}}v^k &\leq \frac{C^{k+1}}{t^k}\left(\left(\frac{t|z|}{C\zeta}\right)^{k+1} + \left(\frac{tv}{C\zeta}\right)^{k+1}\right)M(C\zeta) \\ &\leq C^{k+1}\big(M(|z|) + M(v) + 2M(C\zeta)\big)\omega(t) \\ &\leq C^{k+1}\big(3M(z) + M(v) + 2M(2Cx)\big)\omega(t).\end{aligned}$$

Finally, in case that $v > \frac{C\zeta}{t}$ we use the inequality $t|z| \leq 2\zeta$, (22), and Fact 140(ii) to obtain

$$t|z|\frac{M(C\zeta)}{\zeta^{k+1}}v^k \leq 2\frac{v^k}{\zeta^k}M\left(\frac{C\zeta}{tv}tv\right) \leq 2C\frac{v^k}{\zeta^k}\left(\frac{C\zeta}{tv}\right)^k M(tv) \leq 2C^{k+1}M(v)\omega(t).$$

□

Proof of Theorem 138. Let us denote $X = L_M(\mu)$, $H = H_M(\mu)$, $\Phi = \Phi_M$, and $\omega = \omega_M$. From (22) it easily follows that $\omega(t) \leq C\max\{t^{p-k}, t\}$, $t \in [0,1]$, and hence $\omega \in \mathcal{M}$. By Corollary 1.108 and Lemma 139

$$\left|M(x+th) - \sum_{j=0}^{k}\frac{1}{j!}M^{(j)}(x)t^j h^j\right| \leq \frac{1}{k!}K\big(M(2Cx) + 2M(h)\big)\omega(t)t^k \tag{26}$$

for all $x, h \in \mathbb{R}$, $t \in [0, \frac{1}{2C}]$.

Let $x \in U_X(0, \frac{1}{C}) \cup H$, $h \in U_X(0, \frac{1}{C})$, and $j \in \{1, \dots, k\}$. Fix concrete representatives for x, h and let (S, μ) be the measure space we are working on. Put $T = \{s \in S;\ x(s)h(s) \neq 0\}$ and $A = \big\{s \in T;\ |x(s)| \leq |h(s)|\big\}$. Then by (23) and (22)

$$\begin{aligned}\int \big|M^{(j)}\big(x(s)\big)h(s)^j\big|\,\mathrm{d}\mu(s) &\leq C\int_T M\big(Cx(s)\big)\left|\frac{h(s)}{x(s)}\right|^j \mathrm{d}\mu(s)\\ &= C\int_A M\big(Cx(s)\big)\left|\frac{h(s)}{x(s)}\right|^j \mathrm{d}\mu(s) + C\int_{T\setminus A} M\big(Cx(s)\big)\left|\frac{h(s)}{x(s)}\right|^j \mathrm{d}\mu(s)\\ &\leq C^2\int_A \left|\frac{x(s)}{h(s)}\right|^{p-j} M\big(Ch(s)\big)\,\mathrm{d}\mu(s) + C\Phi(Cx) \leq C^2\Phi(Ch) + C\Phi(Cx).\end{aligned}$$

Therefore for any given $x \in U_X(0, \frac{1}{C}) \cup H$ and $j \in \{0, \dots, k\}$ we may define $P_j^x(h) = \int M^{(j)}\big(x(s)\big)h(s)^j\,\mathrm{d}\mu(s)$ for $h \in X$.

For a fixed μ-measurable function $f \in \mathfrak{M}$ the mapping $(h_1, \dots, h_j) \mapsto f \cdot h_1 \cdot h_2 \cdots h_j$ clearly belongs to $L({}^j\mathfrak{M}; \mathfrak{M})$. Thus the estimate above gives that $h \mapsto (M^{(j)} \circ x) \cdot h^j$ belongs to $\mathcal{P}({}^jX; L_1(\mu))$. Since $g \mapsto \int g\,\mathrm{d}\mu$ is a bounded linear functional on $L_1(\mu)$, it follows that $P_j^x \in \mathcal{P}({}^jX)$.

Using (26) we obtain

$$\begin{aligned}\left|\Phi(x+h) - \sum_{j=0}^{k}\frac{1}{j!}P_j^x(h)\right| &\leq \int\left|M\big(x(s)+h(s)\big) - \sum_{j=0}^{k}\frac{1}{j!}M^{(j)}\big(x(s)\big)h(s)^j\right|\mathrm{d}\mu(s)\\ &\leq \frac{1}{k!}K\omega\big(\|h\|\big)\|h\|^k\int\left(M\big(2Cx(s)\big) + 2M\Big(\frac{h(s)}{\|h\|}\Big)\right)\mathrm{d}\mu(s)\\ &= \frac{1}{k!}K\left(\Phi(2Cx) + 2\Phi\Big(\frac{h}{\|h\|}\Big)\right)\omega\big(\|h\|\big)\|h\|^k\\ &\leq \frac{1}{k!}K\big(\Phi(2Cx) + 2\big)\omega\big(\|h\|\big)\|h\|^k\end{aligned}$$

for any $x, h \in X$, $\|h\| \leq \frac{1}{2C}$. Now Theorem 1.110 together with the continuity of Φ implies that Φ is C^k-smooth on $U_X(0, \frac{1}{2C})$ and also C^k-smooth on H. Finally, using Lemma 139 again, we get

$$\begin{aligned}&\big\|d^k\Phi(x) - d^k\Phi(y)\big\| = \|P_k^x - P_k^y\| = \sup_{\|h\|\leq 1}|P_k^x(h) - P_k^y(h)|\\ &\leq \sup_{\|h\|\leq 1}\int\big|M^{(k)}\big(x(s)\big) - M^{(k)}\big(y(s)\big)\big||h(s)|^k\,\mathrm{d}\mu(s)\\ &\leq K\omega\big(\|x-y\|\big)\sup_{\|h\|\leq 1}\int\left(M\big(2Cx(s)\big) + M\Big(\frac{x(s)-y(s)}{\|x-y\|}\Big) + M\big(h(s)\big)\right)\mathrm{d}\mu(s)\\ &\leq K\omega\big(\|x-y\|\big)\big(\Phi(2Cx) + 2\big)\end{aligned}$$

whenever $x, y \in U_X(0, \frac{1}{2C}) \cup H$ and $\|x - y\| \leq \frac{1}{2C}$. In particular, it follows that $\Phi \in C^{k,\Omega}(U_X(0, \frac{1}{4C}))$ and $\Phi \in C^{k,\Omega}_{\mathrm{loc}}(H)$. The statements about the norms follow from Lemma 23, noting that the last part of the proof of Lemma 23 works also locally and that the minimal modulus of continuity of $d^k\Phi$ is sub-additive due to the convexity of the domains. □

Corollary 141 ([Mal1]). *Let M be a non-degenerate Orlicz function such that h_M is not isomorphic to any ℓ_p. If $1 < \alpha_M < +\infty$, then set $k = \lceil \alpha_M \rceil - 1$. In this case h_M admits an equivalent norm that is $C^{k,q-k}_{\mathrm{loc}}$-smooth for any $q \in (k, \alpha_M)$, but does not admit a T^{α_M}-smooth bump. If moreover $\beta_M < +\infty$, then h_M admits an equivalent norm that is $C^{k,q-k}$-smooth for any $q \in (k, \alpha_M)$.*

If $\alpha_M = +\infty$, then h_M admits an equivalent C^∞-smooth norm.

Proof. We construct an Orlicz function N which is equivalent to M at 0 and satisfies the assumptions of Theorem 138. Without loss of generality we may assume that $M(1) = e - 1$. We put $M_1(t) = M(t)$ for $t \in [0, 1]$ and $M_1(t) = \exp t - 1$ for $t \in (1, +\infty)$. Further, set $\varphi(t, s) = \exp \frac{s}{s-t}$ for $t > s \geq 0$ and $\varphi(t, s) = 0$ for $0 < t \leq s$. We define

$$N(t) = \int_0^{|t|} \frac{M_1(s)}{s} \varphi(|t|, s)\,\mathrm{d}s$$

for $t \in \mathbb{R}$. By the convexity, $M_1(s) \leq (e-1)s$ for $s \in [0, 1]$ and hence N is real-valued. Since $s \mapsto \frac{M_1(s)}{s}$ is non-decreasing on $(0, +\infty)$ (this follows from the convexity of M and $t \mapsto \exp t - 1$),

$$N(t) \geq \int_{\frac{t}{2}}^{\frac{3t}{4}} \frac{M_1(s)}{s} \varphi(t, s)\,\mathrm{d}s \geq e^{-3} \frac{M_1(\frac{t}{2})}{\frac{t}{2}} \int_{\frac{t}{2}}^{\frac{3t}{4}} 1\,\mathrm{d}s = \frac{1}{2e^3} M_1\Big(\frac{1}{2}t\Big)$$

for $t \in \mathbb{R}^+$. On the other hand, $N(t) \leq \frac{M_1(t)}{t} \int_0^t 1\,\mathrm{d}s = M_1(t)$ for $t \in \mathbb{R}^+$. Therefore M and N are equivalent at 0.

It is easy to show by induction that there are constants $c_{n,j} \in \mathbb{N}$, $0 \leq j < n$, $n \in \mathbb{N}$, such that

$$\frac{\partial^n \varphi}{\partial t^n}(t, s) = s \exp \frac{s}{s-t} \sum_{j=0}^{n-1} \frac{c_{n,j} s^{n-1-j}}{(s-t)^{2n-j}}$$

for all $t > s \geq 0$, $n \in \mathbb{N}$. Put $f(t, s) = \frac{M_1(s)}{s} \varphi(t, s)$ for $s, t \in \mathbb{R}^+$, $f(t, 0) = 0$. Then the mappings $(t, s) \mapsto \frac{\partial^n f}{\partial t^n}(t, s)$ are continuous on $\mathbb{R}^+ \times \mathbb{R}^+_0$ and hence by Corollary 1.91 the function N is C^∞-smooth on $\mathbb{R}^+$ and

$$N^{(n)}(t) = \int_0^t \frac{M_1(s)}{s} \frac{\partial^n \varphi}{\partial t^n}(t, s)\,\mathrm{d}s = \frac{1}{t^n} \int_0^1 M_1(ut) \exp \frac{u}{u-1} \sum_{j=0}^{n-1} \frac{c_{n,j} u^{n-1-j}}{(u-1)^{2n-j}}\,\mathrm{d}u$$

for $t \in \mathbb{R}^+$ and $n \in \mathbb{N}$. It follows that N' is positive and non-decreasing on $(0, +\infty)$ and thus N is an Orlicz function.

Next,

$$t^n|N^{(n)}(t)| \le M_1(t)\int_0^1 \exp\frac{u}{u-1}\left|\sum_{j=0}^{n-1}\frac{c_{n,j}u^{n-1-j}}{(u-1)^{2n-j}}\right| du = C_n M_1(t) \le 2e^3 C_n N(2t)$$

for each $t \in \mathbb{R}^+$, $n \in \mathbb{N}$, where C_n are appropriate constants. This means that N satisfies (23) in Theorem 138 for any $k \in \mathbb{N}$. Since for each $q < \alpha_M$ there is a constant A such that $M(t) \le At^q$ for $t \in [0,1]$, it follows from the above inequality that $\lim_{t\to 0+} N^{(n)}(t) = 0$ whenever $n < \alpha_M$. Using the continuity at 0 it follows inductively that $N \in C^n(\mathbb{R})$ for all $n < \alpha_M$.

It remains to show that for any $1 < p < \alpha_M$ the function N satisfies (22) in Theorem 138. First, we show this statement for M_1. Let $K_1 = \sup_{\lambda,t\in(0,1]} \frac{M(\lambda t)}{\lambda^p M(t)} < +\infty$, $K_2 = \sup_{\lambda\in(0,1]} \frac{e}{\lambda^p(\exp\frac{1}{\lambda}-1)} < +\infty$, and put $K_0 = \max\{K_1, K_2, 1\}$. For $\lambda, t \in [0,1]$ we have $M_1(\lambda t) = M(\lambda t) \le K_1\lambda^p M(t) \le K_0\lambda^p M_1(t)$. If $\lambda \in (0,1]$ and $\lambda t \ge 1$, then $M_1(\lambda t) = \exp(\lambda t) - 1 < \exp(\lambda t) \le K_2\lambda^p(\exp t - 1) \le K_0\lambda^p M_1(t)$. And finally, if $t > 1$ and $\lambda t < 1$, then we use both of the previous estimates to obtain $M_1(\lambda t) \le K_0(\lambda t)^p M_1(1) = K_0\lambda^p t^p M_1(\frac{1}{t}t) \le K_0\lambda^p t^p K_0\frac{1}{t^p}M_1(t) = K_0^2\lambda^p M_1(t)$.

Now for $\lambda \in [0, \frac{1}{2}]$ and $t \in \mathbb{R}^+$ we obtain $N(\lambda t) = N(2\lambda\frac{t}{2}) \le M_1(2\lambda\frac{t}{2}) \le K_0^2 2^p\lambda^p M_1(\frac{t}{2}) \le 2^{p+1}e^3 K_0^2\lambda^p N(t)$, while of course $N(\lambda t) \le N(t) \le 2^p\lambda^p N(t)$ for $\lambda \in [\frac{1}{2}, 1]$, $t \in \mathbb{R}^+$.

To sum up what we have proved so far: There is an Orlicz function N equivalent to M at 0 which is C^n-smooth on $\mathbb{R}$ for each $n < \alpha_M$, it is C^∞-smooth on $\mathbb{R}^+$, satisfies (22) for any $1 < p < \alpha_M$, and satisfies (23) for any $k \in \mathbb{N}$. Thus in case $\alpha_M = +\infty$ Theorem 138 implies that there is an equivalent C^∞-smooth norm on h_M. If $\alpha_M < +\infty$, we put $k = \lceil\alpha_M\rceil - 1$ and choose any $p \in (k, \alpha_M)$. It is easily checked, that for each $q \in (k, \alpha_M)$ there is $L > 0$ such that $\omega_N(t) \le Lt^{q-k}$ for $t \in [0,1]$. Hence the equivalent norm on h_M given by Theorem 138 is $C^{k,q-k}_{\mathrm{loc}}$-smooth, or even $C^{k,q-k}$-smooth in case $\beta_M < +\infty$ (in this case N satisfies the Δ_2-condition at 0 and so Φ_N is separating in any ball around 0).

The negative statement is Theorem 137. □

We remark that the Orlicz spaces for a general measure μ can be treated in a similar way.

Next we turn to study of polyhedrality in Orlicz sequence spaces.

Theorem 142 ([Leu3]). *The following statements are equivalent for every non-degenerate Orlicz function M:*

(i) *There exists a constant $K > 0$ such that* $\lim_{t\to 0+} \frac{M(Kt)}{M(t)} = +\infty$.

(ii) *The Orlicz space h_M is isomorphic to a subspace of $C([0,\omega^\omega])$.*

(iii) *The Orlicz space h_M is isomorphic to a subspace of $C(K)$ for some scattered compact K.*

All spaces satisfying the condition (ii) are isomorphically polyhedral (Theorem 103), and D. H. Leung in [Leu1] conjectured that conversely all polyhedral Orlicz sequence spaces fall under this description. There is a strong evidence supporting this idea. First, it is shown in [Leu1] and [HJ3] that the naturally defined LFC renormings exist precisely for those spaces. Second, negating the condition in (i) we obtain the following statement

$$\forall K > 0 \quad \exists \{t_n\}_{n=1}^{\infty}, t_n \to 0+: \sup_{n\in\mathbb{N}} \frac{M(Kt_n)}{M(t_n)} < +\infty.$$

Reversing the order of the quantifiers leads to the following stronger (less general) condition

$$\exists \{t_n\}_{n=1}^{\infty}, t_n \to 0+ \quad \forall K > 0: \sup_{n\in\mathbb{N}} \frac{M(Kt_n)}{M(t_n)} < +\infty.$$

By Theorem 144 below, the Orlicz sequence spaces satisfying the last condition are not isomorphically polyhedral (although they may be c_0-saturated). Thus Leung's theorem above is a near characterisation of polyhedrality for Orlicz sequence spaces, the gap lying in the exchange of quantifiers.

The following theorem follows of course from Theorem 142, nevertheless we give a short direct proof.

Theorem 143 ([Leu1]). *Let M be a non-degenerate Orlicz function such that there is a $K > 1$ for which*

$$\lim_{t\to 0+} \frac{M(Kt)}{M(t)} = +\infty.$$

Then the Orlicz space h_M is isomorphically polyhedral.

Proof. For $k \in \mathbb{N}$ let $b_k = \inf_{0<t\le M^{-1}(\frac{1}{k})} \frac{M(Kt)}{M(t)}$. By our assumption $b_k \to +\infty$. Also notice that by the continuity of $t \mapsto \frac{M(Kt)}{M(t)}$ we have $b_k > 1$ for all $k \in \mathbb{N}$ and so we may set $\eta_k = \frac{b_k}{b_k - 1}$, $k \in \mathbb{N}$. Note that $\eta_k > 1$ and $\eta_k \to 1$. Define a new norm on h_M by

$$|||x||| = \sup_{k\in\mathbb{N}} \eta_k \|P_k(x^*)\|, \tag{27}$$

where $\|\cdot\|$ is the canonical norm on h_M, P_k are the projections associated with the canonical basis, and if $x = (x_n) \in h_M$, then $x^* = (x_n^*)$ is the non-increasing reordering of $(|x_n|)$. It is easy to check that $|||\cdot|||$ is an equivalent norm on h_M and the canonical basis is monotone with respect to this norm. We prove that for each $x \in h_M$ there is $m \in \mathbb{N}$ such that

$$|||x||| = |||P_m x|||. \tag{28}$$

First we show that if $x = (x_n)$ is such that the sequence (x_n) is non-negative and non-increasing, then there is $k \in \mathbb{N}$ such that $\|x\| \le \eta_k \|P_k x\|$. Without loss of generality we may assume that $\|x\| = 1$. Let $k \in \mathbb{N}$ be such that $\|x - P_k x\| < \frac{1}{K}$. Then $\sum_{n=k+1}^{\infty} M(Kx_n) < 1$. Suppose that $\eta_k \|P_k x\| < \|x\| = 1$. Since (x_n) is non-increasing, from $1 > \Phi_M(\eta_k P_k x) \ge \sum_{n=1}^{k} M(\eta_k x_k) \ge kM(x_k)$ it follows that $M(x_k) < \frac{1}{k}$. For all $n \geq k$ we have $M(x_n) \leq M(x_k) < \frac{1}{k}$ and therefore $M(Kx_n) \geq b_k M(x_n)$. Thus

$$\begin{aligned} 1 = \Phi_M(x) &= \Phi_M(P_k x) + \sum_{n=k+1}^{\infty} M(x_n) \\ &\leq \frac{1}{\eta_k}\Phi_M(\eta_k P_k x) + \frac{1}{b_k}\sum_{n=k+1}^{\infty} M(Kx_n) < \frac{1}{\eta_k} + \frac{1}{b_k} = 1, \end{aligned}$$

a contradiction. (We note that in the first inequality we used the convexity of Φ_M.)

Now let $x \in h_M$ be arbitrary. We have shown that there exists $k \in \mathbb{N}$ such that $\|x^*\| \leq \eta_k \|P_k(x^*)\|$. Since $\lim_{n\to\infty} \eta_n \|P_n x^*\| = \|x^*\|$, the supremum in the formula (27) is attained, say at $j \in \mathbb{N}$. Let $\sigma\colon \mathbb{N} \to \mathbb{N}$ be a one-to-one mapping such that $x_n^* = |x_{\sigma(n)}|$, $n \in \mathbb{N}$, and let $m = \max_{n\leq j} \sigma(n)$. Then

$$|||P_m x||| \leq |||x||| = \eta_j \|P_j(x^*)\| = \eta_j \|P_j\big((P_m x)^*\big)\| \leq |||P_m x|||.$$

Finally, denote $X = (h_M, |||\cdot|||)$ and let $\{(e_j; f_j)\}$ be the canonical basis of X. For each $n \in \mathbb{N}$ define $K_n = \operatorname{span}\{f_1, \dots, f_n\} \cap B_{X^*}$, which is a compact set, and put $B = \bigcup_{n=1}^{\infty} K_n$. We claim that B is a James boundary for X, which in view of Theorem 103 will finish the proof. Fix $x \in X$. By (28) there is $m \in \mathbb{N}$ such that $|||x||| = |||P_m x|||$. Given a functional $f \in B_{X^*}$ we let $g = \sum_{j=1}^{m} f(e_j) f_j$. Then $g(y) = g(P_m y) = \sum_{j=1}^{m} f(e_j) y_j = f(P_m y)$ for any $y \in h_M$. Since $g(y) = f(P_m y) \leq |||P_m y||| \leq |||y|||$, it follows that $g \in K_m$. This shows that $\sup_{f\in B_{X^*}} f(P_m x) = \sup_{g\in K_m} g(P_m x)$. Now by the compactness of K_m there is $g \in K_m$ such that $g(x) = g(P_m x) = \sup_{f\in K_m} f(P_m x) = \sup_{f\in B_{X^*}} f(P_m x) = |||P_m x||| = |||x|||$. □

Theorem 144. *Let M be a non-degenerate Orlicz function for which there exists a sequence $\{t_n\}_{n=1}^{\infty} \subset \mathbb{R}$ decreasing to 0 such that*

$$\sup_{n\in\mathbb{N}} \frac{M(Kt_n)}{M(t_n)} < +\infty \quad \textit{for all } K > 0.$$

Then the Orlicz space h_M does not admit any (even non-continuous) LFC bump function.

Proof. Suppose that h_M admits some LFC bump g. Without loss of generality we may assume that $g = \chi_A$ for some set $0 \in A \subset B_X$ (by shifting, scaling, and composing with a suitable function) and that g is LFC-$\{e_i^*\}$. (Since h_M is c_0-saturated by Theorem 90, it does not contain ℓ_1. As $\{e_i\}$ is unconditional, it is shrinking by James's theorem. Now consider $g \circ T$, where $T\colon h_M \to h_M$ is an equivalence isomorphism of the bases $\{e_i\}$ and $\{x_i\}$ from Corollary 88.)

Note that by our assumption the vectors with coordinates in the set $\{t_n;\ n \in \mathbb{N}\} \cup \{0\}$ have the property of "bounded completeness": If $\left\|\sum_{i=1}^{k} t_{m_i} e_i\right\| \leq 1$ for all $k \in \mathbb{N}$, where $m_i \in \mathbb{N}_0$ are not necessarily distinct (we put $t_0 = 0$), then $\sum_{i=1}^{\infty} t_{m_i} e_i$ converges in h_M. Indeed, it follows that $\sum_{i=1}^{k} M(t_{m_i}) \leq 1$ for all $k \in \mathbb{N}$. Further,

$$\sum_{i=1}^{k} M(Kt_{m_i}) \leq \sup_{n\in\mathbb{N}} \frac{M(Kt_n)}{M(t_n)} \sum_{i=1}^{k} M(t_{m_i}) \leq \sup_{n\in\mathbb{N}} \frac{M(Kt_n)}{M(t_n)}$$

for all $K > 0$ and all $k \in \mathbb{N}$. Consequently, $\sum_{i=1}^{\infty} M(Kt_{m_i}) < +\infty$ for all $K > 0$ and hence the sum $\sum_{i=1}^{\infty} t_{m_i} e_i$ converges in h_M.

We construct a sequence $\{x_k\}_{k=1}^{\infty} \subset A$ by induction. Put $x_0 = 0 \in A$ and define natural numbers $m_0 = n_0 = 1$. If $m_{k-1} \in \mathbb{N}$, $n_{k-1} \in \mathbb{N}$, and $x_{k-1} \in A$ are already defined, we put

$$M_k = \{(m,n) \in \mathbb{N}^2;\ m \geq m_{k-1}, n > n_{k-1}, \text{ and } x_{k-1} + t_m e_n \in A\}.$$

As g depends only on some finite subset of $\{e_i^*\}$ on a neighbourhood of x_{k-1}, and $t_m \to 0$, we can see that $M_k \neq \emptyset$. Let $(m_k, n_k) = \min M_k$ in the lexicographic ordering of $\mathbb{N}^2$ and put $x_k = x_{k-1} + t_{m_k} e_{n_k}$.

As $\{x_k\} \subset A \subset B_X$ and $x_k = \sum_{i=1}^{k} t_{m_i} e_{n_i}$, by the above argument $x_k \to x \in h_M$. We can find $\delta > 0$ and $N \in \mathbb{N}$ such that g depends only on $\{e_i^*\}_{i<N}$ on $U(x,\delta)$. Since $\{x_k\}$ converges, $m_k \to \infty$ and so there is $j \in \mathbb{N}$ such that $x_j \in U(x, \frac{\delta}{2})$, $\|t_{m_j} e_1\| < \frac{\delta}{2}$, $m_j < m_{j+1}$, and $n_j > N$. Then $x_j + t_{m_j} e_{n_j+1} \in A$ and therefore $(m_j, n_j+1) \in M_{j+1}$. But $(m_j, n_j+1) < (m_{j+1}, n_{j+1})$, which is a contradiction. □

Theorem 145 ([Leu1]). *There is an Orlicz function M such that $\alpha_M = +\infty$, but the Orlicz space h_M does not admit an LFC bump. In particular, h_M is a separable c_0-saturated Asplund space that is not isomorphically polyhedral, and h_M admits an equivalent C^∞-smooth norm, but not an LFC bump.*

In the proof we will use the following lemma.

Lemma 146. *Let $\{b_n\}_{n=0}^{\infty} \subset (0,+\infty)$ be a non-increasing sequence such that*

$$\sup_{m,n\in\mathbb{N}_0} \frac{b_{n+m}}{b_n} K^m < +\infty$$

for every $K > 0$. Let M be the even, continuous, piecewise affine function such that $M(0) = 0$ and $M'(t) = b_n$ for $t \in (2^{-n-1}, 2^{-n})$, $n \in \mathbb{N}_0$, $M'(t) = b_0$ for $t \geq 1$. Then M is a non-degenerate Orlicz function with $\alpha_M = +\infty$.

Proof. Clearly M is a non-degenerate Orlicz function. Fix any $q \in (0,+\infty)$. Since $2^{-n-1} b_n \leq M(2^{-n}) \leq 2^{-n} b_n$ for each $n \in \mathbb{N}_0$,

$$C = \sup_{m,n\in\mathbb{N}_0} \frac{M(2^{-(n+m)})}{M(2^{-n})} 2^{mq} \leq 2 \sup_{m,n\in\mathbb{N}_0} \frac{b_{n+m}}{b_n} (2^{q-1})^m < +\infty.$$

Given any $\lambda, t \in (0,1]$ we can find $m, n \in \mathbb{N}_0$ such that $\lambda \in (2^{-m-1}, 2^{-m}]$ and $t \in (2^{-n-1}, 2^{-n}]$. Then $\lambda t \leq 2^{-(n+m)}$ and hence

$$\frac{M(\lambda t)}{\lambda^q M(t)} < \frac{M(2^{-(n+m)})}{2^{-(m+1)q} M(2^{-n-1})} = 4^q \frac{M(2^{-(n+1+m-1)})}{M(2^{-(n+1)})} 2^{(m-1)q} \leq 4^q C$$

whenever $m \geq 1$. In case that $m = 0$ we have $M(\lambda t) \leq M(t) < 2^q \lambda^q M(t)$. □

Proof of Theorem 145. We construct a sequence $\{b_n\}_{n=0}^{\infty} \subset (0,+\infty)$ which satisfies the assumptions of Lemma 146 and the resulting Orlicz function M satisfies the assumptions of Theorem 144. The rest follows from Corollary 134, Theorem 103, and Corollary 141.

Set $\alpha_n = \frac{1}{n!}$ and note the following properties of the sequence $\{\alpha_n\}_{n=0}^{\infty}$: it is non-increasing, $\alpha_0 = 1$, $\alpha_{n+m} \le \alpha_n \alpha_m$ for all $m,n \in \mathbb{N}_0$, and $\alpha_n K^n \to 0$ for each $K > 0$. Further, put $s_n = \sum_{j=0}^{n} j$ for $n \in \mathbb{N}_0$, and define a sequence $\{c_n\}_{n=0}^{\infty}$ by $c_0 = 1$ and $c_{n+1} = \alpha_{n+1}\alpha_{s_{n+2}} c_n$. Finally, let $b_{s_n+k} = \frac{c_n}{\alpha_{n-k}}$ for $n,k \in \mathbb{N}_0$, $k \le n$.

The sequence $\{b_n\}_{n=1}^{\infty}$ is non-increasing. Indeed, let $n,k,l \in \mathbb{N}_0$ with $k < l \le n$. Then $b_{s_n+k} = \frac{c_n}{\alpha_{n-k}} \ge \frac{c_n}{\alpha_{n-l}} = b_{s_n+l}$ and $b_{s_{n+1}} = \frac{c_{n+1}}{\alpha_{n+1}} = \alpha_{s_{n+2}} c_n \le c_n = b_{s_n+n}$.

Next, we show that $b_{n+m} \le \alpha_m b_n$ for all $m,n \in \mathbb{N}_0$. Fix $m,n \in \mathbb{N}_0$ and let $i,j,k,l \in \mathbb{N}_0$ be such that $n = s_i + k$, $k \le i$, and $n+m = s_j + l$, $l \le j$. If $i = j$, then $m = l - k$ and hence $b_{n+m} = b_{s_j+l} = \frac{c_j}{\alpha_{j-l}} \le \alpha_m \frac{c_j}{\alpha_{j-k}} = \alpha_m b_n$. Otherwise $i < j$ and $m < s_{j+1}$, and therefore $b_{n+m} \le b_{s_j} = \frac{c_j}{\alpha_j} = \alpha_{s_{j+1}} c_{j-1} \le \alpha_m c_i = \alpha_m \alpha_0 b_{s_i+i} \le \alpha_m b_n$. It follows that for any $K > 0$

$$\sup_{m,n\in\mathbb{N}_0} \frac{b_{n+m}}{b_n} K^m \le \sup_{m\in\mathbb{N}_0} \alpha_m K^m < +\infty.$$

Thus the requirements of Lemma 146 are satisfied.

Now let M be the Orlicz function defined in Lemma 146. Set $t_n = 2^{-(s_n+n)}$. We claim that the function M and the sequence $\{t_n\}_{n=1}^{\infty}$ satisfies the assumptions of Theorem 144. This sequence is clearly decreasing to 0. Fix $m \in \mathbb{N}$. Then for each $n \in \mathbb{N}$, $n \ge m$,

$$\begin{aligned} M(2^m t_n) &= M(2^{-s_n-n+m}) \le \frac{b_{s_n+n-m}}{2^{s_n+n-m}} = \frac{c_n}{2^{s_n+n-m}\alpha_m} \\ &= \frac{2^{m+1}}{\alpha_m}\frac{\alpha_0 b_{s_n+n}}{2^{s_n+n+1}} \le \frac{2^{m+1}}{\alpha_m} M(2^{-s_n-n}) = \frac{2^{m+1}}{\alpha_m} M(t_n), \end{aligned}$$

which means that $\sup_{n\in\mathbb{N}} \frac{M(2^m t_n)}{M(t_n)} < +\infty$. □

Theorem 147 ([HáTr]). *Let M be an Orlicz function. Then h_M admits an equivalent analytic norm if and only if either h_M is isomorphic to ℓ_{2n}, $n \in \mathbb{N}$, or h_M is isomorphically polyhedral. In particular,*

(i) *if there is $K > 0$ such that $\lim_{t\to 0+} \frac{M(Kt)}{M(t)} = +\infty$, then h_M has an equivalent analytic norm;*

(ii) *if $\alpha_M = +\infty$ and there exists a sequence $\{t_n\}_{n=1}^{\infty}$ decreasing to 0 such that $\sup_{n\in\mathbb{N}} \frac{M(Kt_n)}{M(t_n)} < +\infty$ for all $K > 0$, then h_M does not admit an equivalent analytic norm.*

Proof. $\Leftarrow$ follows from Theorem 106 and Theorem 104.

$\Rightarrow$ If M is degenerate, then h_M is isomorphic to c_0 and so it is isomorphically polyhedral. Otherwise, if h_M admits an analytic norm, then either h_M is isomorphic to some ℓ_{2n}, or $\alpha_M = +\infty$ (Theorem 137). In the latter case, by Proposition 135 every polynomial on h_M is weakly sequentially continuous. Thus by Corollary 105 the space h_M is isomorphically polyhedral.

(i) follows from Theorem 143, (ii) follows from Theorem 133, Theorem 144, and Theorem 103. □

From Theorem 145 and Theorem 147 we immediately obtain the following corollary.

Corollary 148. *There is a separable c_0-saturated Asplund space with a C^∞-smooth norm that does not admit an equivalent analytic norm.*

10. Notes and remarks

Our approach to parts of Sections 3, 4, and 7 was inspired by [FZ2].

Section 1. The theory of convex functions with respect to Legendre-Fenchel duality and differentiability is thoroughly studied in [BorVa], where many additional results can be found. Proposition 16 is essentially contained in [MPVZ] and is based on the idea in [BN]. The definition of the modulus of exposedness and the duality results are motivated by [BGV]. Theorem 39 is essentially implicitly contained in [DGZ2].

For the theory of moduli of smoothness and convexity of Banach spaces we refer to [DGZ], [LiTz3], or [God]. The fundamental result in this area is the Enflo-Pisier renorming theorem [Enf], [Pis3], which states that a Banach space has an equivalent norm with a modulus of convexity (resp. smoothness) of power type if and only if the space is super-reflexive. There is a vast number of papers studying these moduli for concrete classes of Banach spaces e.g. [Fi], [MaT1], [IT].

In connection with finite-dimensional convex analysis, recall that Alexandrov's theorem states that a convex function on $\mathbb{R}^n$ is T^2-smooth at almost every point in $\mathbb{R}^n$. We refer to [BN] for some, mostly negative, results concerning possible generalisations to infinite-dimensional spaces. However, the following seems to be still open.

Problem 149. Suppose that f is a convex continuous function on ℓ_2. Is f weakly T^2-smooth at some point?

A very similar problem is open for convex functions on c_0.

Problem 150 (M. Fabian)**.** Let f be a convex continuous function on c_0. Are there $x_0 \in c_0$ and $K, \delta > 0$ such that $f(x_0 + h) + f(x_0 - h) - 2f(x_0) \leq K\|h\|^2$ for all $h \in U(0, \delta)$?

Section 2. The idea of Theorem 44 comes from [Li2], where Corollary 45 is proved. The idea of Lemma 47 comes from [Fab2] and it is explicitly formulated in [DevFa]. Theorem 48 is essentially from [Fab2] (where only first order smoothness is considered), an earlier version with additional assumptions is in [FWZ].

Section 3. Variational principles have been studied and applied extensively in various parts of functional analysis. Their systematic use in problems on higher smoothness in Banach spaces was spurred by the paper of Jonathan M. Borwein and David Preiss [BorPr].

Theorem 61 for $\alpha = 0$ is proved in [FHV]. Theorem 62 appears in [DGJ] and some earlier versions in [FPWZ]. Theorem 64 (without variational principles) is proved in [FPWZ].

For a sample of other applications of variational principles see e.g. [DGZ3], [DeGh].

Section 4. Variants of Theorem 66 appear in [Dev1] and [Dev2]. The main structural Theorem 67, due to Robert Deville (except for the degrees of the homogeneous polynomials), is a culmination of the work of many mathematicians. The proof relies on ideas and techniques which originated in the papers of Jaroslav Kurzweil, Victor Zakharovich Meshkov, Boris Mikhailovich Makarov, Stanisław Kwapień, Tadeusz Figiel, and others. Regarding the assumptions of this theorem we pose the following well-known problem.

Problem 151. Let X be a Banach space which admits a C^k-smooth bump (resp. norm) for every $k \in \mathbb{N}$. Does X admit a C^∞-smooth bump (norm)?

Regarding the lowest degree for a single homogeneous separating polynomial we pose the next problem.

Problem 152. Let X be a Banach space admitting a separating polynomial. Consider the set $S = \{q;$ there is a subspace $Y \subset X$ such that $q(Y) = q\}$. Let m be the least common integer multiple of the elements in S. Does there exist an m-homogeneous separating polynomial on X?

By our results we know that m is the lowest value for which a separating polynomial might exist. If S is a singleton, then every normalised weakly null sequence in X contains a subsequence equivalent to the canonical basis of ℓ_q. An example of such space is any subspace of ℓ_p.

In [JO] it is shown that if a subspace X of $L_p(\mu)$, $2 < p < \infty$, does not contain ℓ_2, then X is isomorphic to a subspace of $\ell_p(\Gamma)$. Hence there are spaces with separating polynomials which are not isomorphic to a subspace of $L_p(\mu)$ for some p even. For example $\ell_r \oplus \ell_q$ for distinct even integers $r, q > 2$ has a separating polynomial, but does not embed into any $L_p(\mu)$, p even. Indeed, according to the mentioned result and the well-known structural properties of ℓ_p-spaces ([FHHMZ, p. 209]), both spaces ℓ_r and ℓ_q would have to be isomorphic to a subspace of ℓ_p, which is a contradiction.

Corollary 69 together with S. Kwapień's characterisation implies the following result: If X is a Banach space that admits a C^2-smooth bump and is saturated with spaces of cotype 2, then X is isomorphic to a Hilbert space.

By Corollary 68 Tsirelson-like spaces (i.e. spaces not containing a subspace isomorphic to c_0 or ℓ_p) cannot have a C^∞-smooth bump. The next problem represents a natural generalisation of this result.

Problem 153. If X has a $C^{2,+}$-smooth norm, does there exist a subspace of X isomorphic to ℓ_p for some $p \in (1, \infty)$?

In a similar vein, it seems to be unknown if the 2-convexified Tsirelson space $T^{(2)}$ constructed in [FiJ] has a C^2-smooth bump (or norm). This space is of type 2 and weak cotype 2 (which implies that it is of cotype q for every $q > 2$), [Johns1]. On the other hand, [DeZi], if $T^{(2)}$ had a $C^{2,\alpha}$-smooth bump, then by Theorem 66 and Fact 4.47 it would have a separating polynomial, and hence it would have cotype $2 = q(T^{(2)})$. By S. Kwapień's result again it would have to be isomorphic to ℓ_2, a contradiction. We refer to [Pis5] for the theory of weak cotype 2 spaces, which is strongly related to this subject.

Problem 154. If a separable Banach space X admits a C^k-smooth bump, does it admit a C^k-smooth norm?

In particular,

Problem 155. If a separable Banach space X admits a separating polynomial, does it admit a C^∞-smooth norm?

According to the remark following the proof of Theorem 73, if X has the RNP and a G^2-smooth bump, then X is super-reflexive. The same conclusion holds, due to Corollary 51, provided that X does not contain c_0 and admits a $C^{1,+}_{\mathrm{loc}}$-smooth bump. We do not know if these two sets of assumptions can be meshed together into a single and stronger one. Namely:

Problem 156. If a Banach space X has a G^2-smooth norm and does not contain c_0, is then X super-reflexive?

Corollary 74 and Corollary 75 have a long history, starting with results of [Me] and subsequently gradually improved by a number of authors, e.g. [Fab2], [DGZ2], [Va2], and [MPVZ].

Theorem 76 has an analogue in [GJT]. Let X be a Banach space, $1 < p < \infty$, and suppose that X admits a UT^p-smooth bump and does not contain ℓ_k, k even integer, $k < p$. Then every polynomial on X of degree less than p is weakly sequentially continuous.

Section 5. The concept of local dependence on finitely many coordinates goes back to Nicolaas Hendrik Kuiper. The main structural results on Banach spaces admitting a bump function which depends locally on finitely many coordinates are Theorem 90 and Theorem 91 due to Jaroslav Pechanec, John H. M. Whitfield and Václav Zizler, [PWZ], resp. Marián Fabian and V. Zizler, [FZ1]. Independently, similar results were obtained by Vladimir P. Fonf, [Fo2] under the somewhat weaker assumption of having a LFC norm rather than just a bump.

Proposition 84 and Theorem 87 are from [HJ4]. Theorem 94 is due to N. Kuiper and appears (for Γ countable) in [BF]. Lemma 96 is essentially implicitly contained in [Há2]. Equivalence of (i) and (iii) in Theorem 97 is proved in [JS] based on some earlier results from [HJ4].

Section 6. Polyhedral spaces were introduced by Victor Klee, as a purely isometric concept of finite-dimensional character. They were thoroughly investigated by V. P. Fonf who studied among other things their isomorphic and topological aspects as well as renorming properties, e.g. [Fo1], [Fo3], [Fo4].

Parts of Theorem 103 appear in [Fo2] and [Há2]. The proof of (i)⇒(ii) that we present comes from [Ve]. A version of Theorem 104 for the space c_0 appeared in [FPWZ]. It is clear that polyhedral spaces, as well as spaces admitting an LFC bump are closed with respect to taking subspaces, and polyhedral spaces admit a LFC bump.

Problem 157. If a separable Banach space X admits a LFC bump, is then X isomorphic to a polyhedral space?

This problem has been addressed in [HJ3], where a tentative counterexample has been constructed, but the proof of non-polyhedrality of the candidate space is still missing.

Polyhedral spaces, or more generally spaces admitting a LFC bump are c_0-saturated Asplund spaces. The converse fails, as in [CKT] there is a first example of a c_0-saturated space (in fact an Orlicz space) which is not isomorphically polyhedral and its quotient is isomorphic to ℓ_2. Denny Ho-Hon Leung [Leu2] constructed first such example with an unconditional basis. Ioannis Gasparis improved this result in [GaspaI1], [GaspaI2] showing that every ℓ_p, $1 < p < \infty$, is isomorphic to a quotient of a polyhedral Banach space with an unconditional basis. Spiros A. Argyros, Vaggelis Felouzis, and Theocharis Raikoftsalis [ArgFe], [ArgRa] showed that every reflexive Banach space is a quotient of a c_0-saturated Banach space.

Considering the relationship between polyhedral spaces and L_1-preduals, by the results in [JohnsZip2] if X is separable and X^* is isometric to $L_1(\mu)$, then X is isometric to a quotient of $C(\Delta)$. The isomorphic version is false, as there exist separable isomorphic preduals of ℓ_1 not containing a subspace isomorphic to c_0. First such spaces were constructed in [BD].

Problem 158. Is every separable polyhedral Banach space isomorphic to a subspace of an isometric predual of ℓ_1?

Problem 159. Is every quotient of $C(K)$, K scattered, c_0-saturated?

It is known ([Al]) that a quotient of $C([0, \omega^\omega])$, which is an isometric predual of ℓ_1, need not be isomorphic to a subspace of any $C(K)$, K scattered.

Section 7. The Banach spaces of the type L_p form the first class of spaces whose higher smoothness properties were investigated in quantitative detail. The upper estimate of the best smoothness of the classical L_p spaces summarised in Theorem 107 has its origins in the work of J. Kurzweil [Kur1], and also [BF]. The results are essentially from [DGJ], [FZ2], [MV], and [HeTr].

The closely related spaces $L_p(\mu; X)$ were treated by Isaac Edward Leonard and Kondagunta Sundaresan in [LeSu], and in the special case $X = L_q$ also in [DGJ].

The following result from [DGJ] has a proof similar to those in Section 7: Let $1 < p, q < +\infty$ and q is not an even integer. Then the space $\left(\bigoplus_{n=1}^{\infty} \ell_q^n\right)_p$ does not admit a $C_{\text{loc}}^{k,\alpha}$-smooth bump for $k + \alpha > q$.

Among the classes of Banach spaces that are close to L_p-spaces and for which the best smoothness renormings have been found there are also the trace-class spaces $c_p(H)$ (the non-commutative ℓ_p-spaces). The results are similar to their classical ℓ_p-counterparts (this is not very surprising, since by [Fri] every subspace of $c_p(H)$ which is not isomorphic to ℓ_2 contains ℓ_p). In particular, in [Tom1] and [GGo] the following is shown: Let H be a real infinite-dimensional Hilbert space and $1 < p < +\infty$.

(i) If p is an even integer, then the norm of $c_p(H)$ is p-polynomial.
(ii) The norm of $c_p(H)$ is $C^{\lceil p \rceil - 1}$-smooth.
(iii) If p is not an even integer, then $c_p(H)$ does not admit a T^p-smooth bump.

The space $c_4(\ell_2)$ is an example of a space with a separating 4-homogeneous polynomial which is not isomorphic to a subspace of $L_4(\mu)$. Indeed, Charles McCarthy in [McC] proved that $c_p(H)$ does not embed into $L_p(\mu)$, $2 < p < +\infty$.

To complete the picture, there is an equivalent C^∞-smooth norm on $\mathcal{L}_K(H; H)$, the space of compact operators on H [Tom1]. However, the following is open.

Problem 160 (V. Zizler). Does the space of compact operators $\mathcal{L}_K(\ell_2; \ell_2)$ admit a real analytic norm?

Section 8. By the results in Chapter 6, every C^2-smooth function on $c_0(\Gamma)$ depends locally on countably many coordinates. In particular we conclude the next result.

Theorem 161 ([Há5]). *If Γ is uncountable, then there is no C^2-smooth function on $c_0(\Gamma)$ which attains its minimum at one point. In particular, there is no equivalent rotund C^2-smooth norm on $c_0(\Gamma)$.*

There is a large number of papers devoted to renorming theory and smoothness of $C(K)$ spaces, apart from the results presented in this section. To begin with, by Theorem 125 we may restrict our attention to scattered compact sets K only. A testing ground for this theory is provided by a special class of scattered compacts – the one point compactifications of trees considered by Richard Haydon in his seminal paper [Hay3]. R. Haydon proved that $C_0(T)$ has a C^∞-smooth bump for every tree T. He also characterised those trees for which $X = C_0(T)$ admits a C^∞-smooth renorming by the existence of the so called Talagrand operator, i.e. a bounded linear operator $T\colon X \to c_0(S_{X^*} \times \Gamma)$ such that for every $x \in X$ there exists $(f, m) \in S_{X^*} \times \Gamma$ with $f(x) = \|x\|$ and $T(x)(f, m) \neq 0$ (this formulation in fact comes from the subsequent paper [FPST], and appears to be more general). In both cases mentioned above, the C^∞-smooth bumps and norms are LFC.

By intricate analysis of the branching properties of trees, R. Haydon constructed examples of trees such that $C_0(T)$ does not have even Gâteaux differentiable equivalent renorming.

By contrast to Theorem 126 it is shown in [MoT] that $C(K)$ admits no equivalent uniformly Gâteaux differentiable renorming provided K is a separable and uncountable scattered compact. In this context we point out that UG renormability for any Banach space was characterised in [FGZ].

In light of the above results and the results in this section, we pose the following well-known problem.

Problem 162. If K is a scattered compact, does $C(K)$ admit a C^∞-smooth bump?

In [FPST] an analogous theory has been built for polyhedral renormings. Namely, it is shown there that $C(K)$ admits a polyhedral renorming if and only if $C(K)$ admits a Talagrand operator. Apart from the trees such that $C_0(T)$ has a C^∞-smooth renorming, this condition is also satisfied for all σ-discrete compacts K. It should be noted that in these cases one can even construct C^1-smooth and LUR norms which are uniform limits of C^∞-smooth norms, [HP]. For more information on the subject of smooth renormings of non-separable Banach spaces we refer especially to [DGZ] and the survey paper [ST].

On the other hand, the following remains to be one of the main open problems in renorming theory.

Problem 163. Does every Asplund space admit a Fréchet differentiable bump?

Section 9. The best smoothness of Orlicz-type Banach spaces has been initially studied in particular by Rumen Petrov Maleev and Stanimir L. Troyanski in several papers [MaT3], [Mal1], [Mal2]. The polyhedral case received particular attention in the works of D. H. Leung, [Leu1], [Leu3].

Theorem 144 is from [Leu1] and [HJ3]. In [MZ] (and partially in [GJT]) the best smoothness of Orlicz-Musielak spaces has been studied. In connection with Theorem 143 there is a result of S. L. Troyanski: Let $\{p_n\} \subset [1, +\infty)$ and $p_n \to +\infty$. Then the Nakano space $h_{\{p_n\}}$ is isomorphically polyhedral.

Results concerning the best Gâteaux smoothness for separable Banach spaces are quite rare. It is proved in [Tr] that $L_p(\mu)$ for a σ-finite μ admits an equivalent G^p-smooth norm for p an odd integer. (For separable L_p-spaces this result can be gleaned out by combining Theorem 106 and Corollary 7.72.) There are somewhat strengthened results in [Mal2]. The assumption on the measure space cannot be relaxed (see Theorem 107). This leaves us with the following natural problem.

Problem 164. Let $p > 2$, p not an even integer. Does ℓ_p admit a $G^{[p]+1}$-smooth bump or an equivalent norm?

Table 1. Smoothness in various spaces

space		admits	does not admit	remark
$L_p(\mu)$	p even	p-polynomial norm		Th. 106
	p not even	$C^{k,p-k}$ norm, $k = \lceil p \rceil - 1$	T^p bump	Th. 106, Th. 107
$\ell_p(\Gamma)$, Γ uncountable, p not even			continuous weakly T^p bump	Th. 107
$L_p(\mu; X)$	X has C^k norm with $d^k\|\cdot\|$ bounded, $k < p$	C^k norm		Th. 108
	X has $C^{k,\alpha}$ bump, $k + \alpha \le p$	$C^{k,\alpha}$ bump		Cor. 118
$L_p(\mu; L_q(\nu))$	q even, $p = kq$	p-polynomial norm		Cor. 119, Cor. 122
	q even, $p \ne kq$	$C^{k,p-k}$ norm, $k = \lceil p \rceil - 1$	T^p bump	
	q not even	$C^{k,r-k}$ norm, $r = \min\{p,q\}$, $k = \lceil r \rceil - 1$	T^r bump, $r = \min\{p,q\}$	
$\left(\bigoplus_{n=1}^{\infty} \ell_q^n\right)_p$	q not even or $p \ne kq$		T^p bump	Th. 121
c_0		analytic norm;	$C^{1,+}$ bump	Th. 94, Pr. 49,
separable polyhedral		C^∞ LFC norm		Th. 104
$c_0(\Gamma)$, Γ uncountable		C^∞ LFC and UG norm	analytic norm	T. 126; C. 3.59, C. 105
$C(K)$	$K = [0, \mu]$	C^∞ LFC norm		Th. 127
	$K = \widehat{T}$	C^∞ LFC bump		Th. 128
	K is σ-discrete	C^∞ LFC norm		Th. 129
	K uncountable		analytic norm	Th. 132
$h_M \not\cong \ell_p$	$1 < \alpha_M < +\infty$	$C^{k,q-k}_{\mathrm{loc}}$ norm, $k = \lceil \alpha_M \rceil - 1, q \in (k, \alpha_M)$	T^{α_M} bump	Cor. 141
	$1 < \alpha_M, \beta_M < +\infty$	$C^{k,q-k}$ norm, $k = \lceil \alpha_M \rceil - 1, q \in (k, \alpha_M)$		
	$\alpha_M = +\infty$	C^∞ norm		
	$\lim\limits_{t \to 0+} \frac{M(Kt)}{M(t)} = +\infty$	polyhedral		Th. 143
	$\sup\limits_{n \in \mathbb{N}} \frac{M(Kt_n)}{M(t_n)} < +\infty$		LFC bump	Th. 144
	$\sup\limits_{n \in \mathbb{N}} \frac{M(Kt_n)}{M(t_n)} < +\infty$, $\alpha_M = +\infty$		analytic norm	Th. 147
$\mathcal{L}_K(H; H)$		C^∞ norm		[Tom1]
$c_p(H)$	p not even	$C^{\lceil p \rceil - 1}$ norm	T^p bump	
	p even	p-polynomial norm		[GGo]
h_φ		$C^{\lceil \alpha_\varphi \rceil - 1}$ norm		[MZ]
$h_{\{p_n\}}$	$p = \liminf p_n < +\infty$	$C^{\lceil p \rceil - 1}$ norm		
	$p_n \to +\infty$	polyhedral		Troyanski, unpubl.
$\ell_{\{p_n\}} \not\cong \ell_p$	$p_n \to q \ne 2k$		T^q bump	[MZ]
	$p_n \to 2k$, $p_n \ge 2k$		C^{2k} bump	
	$p_n \to 2k$, $p_n \le 2k$		T^{2k} bump	[GJT]

Chapter 6

Structural behaviour of smooth mappings

In the present chapter we develop the theory of smooth mappings between Banach spaces with emphasis laid on the preservation of the structure of the initial space. In the first section we study general properties of weakly continuous mappings with regards to higher smoothness. The important Theorem 8 bootstraps the weak uniform continuity into the higher differentials provided that $d^k f$ is uniformly continuous. If X does not contain ℓ_1, then these results combine into a very efficient tool for handling uniformly smooth mappings.

In Section 2 we study the bidual extensions of mappings $f \in \mathcal{C}^{k,+}(B_X; Y)$ into the bidual denoted by $f^{**} \in \mathcal{C}^{k,+}(B_{X^{**}}; Y^{**})$. The construction requires several ingredients, namely the Converse Taylor theorem and the powerful ultrapower construction based on the principle of local reflexivity. The extension will be a key tool later on for generalising some results concerning $C(K)$ for K scattered to the general compact spaces K.

In Section 3 we start developing the theory of $\mathcal{W}_\lambda$-spaces, i.e. all those X such that $\mathcal{C}^{1,+}(B_X) \subset \mathcal{C}_{\mathrm{wsC}}(\lambda B_X)$ holds for some $\lambda \in (0, 1]$. The main goal of this theory is to generalise Theorem 3.47 into the setting of uniformly smooth mappings. This objective is achieved in the main Theorem 57 after having completed several important steps. The first step is the quantitative finite-dimensional Lemma 27, which shows the rigidity of ℓ_∞^n with respect to uniformly smooth functions. Using this essential observation, the second step consists of showing that $C(K)$ for K scattered are $\mathcal{W}_1$-spaces. The third step is completed in Theorem 46. In Section 4 uniformly smooth non-compact mappings from $\mathcal{C}^{1,+}(B_{c_0}; X)$ are studied in detail. They are characterised by the fact that their bidual extension has a point where the derivative is non-compact, and hence fixes a copy of c_0. This implies that X^{**} contains c_0. Moreover, non-compact uniformly smooth mappings from $C(K)$, K scattered compact space, can always be reduced to a suitable subspace isomorphic to c_0 where the restriction remains non-compact and this wraps up the scattered case. The case of a general $C(K)$ space is taken up in Section 5. The new ingredient, Theorem 56, states that weakly Cauchy sequences in $C(K)$ spaces are uniformly close to weakly Cauchy sequences in some $C([1, \alpha])$ after passing to the bidual.

In Section 6 we show that for non-compact polynomials from $C(K)$, K scattered, fixing a copy of c_0 remains true, but we give an example of a uniformly smooth mapping from B_{c_0} where the fixing property fails.

In Section 7 we give some rather general results on the ranges of smooth mappings (and derivatives of smooth functions) which illustrate that the theory of the class $\mathcal{W}$ works under nearly optimal assumptions. Indeed, according to Theorem 69 no structure is preserved by surjective C^∞-smooth mappings from spaces with property $\mathcal{B}$. In order to obtain positive results we introduce the notion of separating mappings. In the rest of the chapter our attention moves to smooth separating mappings from ℓ_p-spaces, proving a strong structural result in this case.

In the whole chapter the scalars are real unless specified otherwise.

1. Weak uniform continuity and higher smoothness

We continue with the study of higher smoothness in connection with the weak continuity.

Definition 1. Let X, Y be normed linear spaces, $U \subset X$ a convex set with non-empty interior, and $k \in \mathbb{N}_0$ (resp. $k = \infty$). By $\mathcal{C}^k(U;Y)$ we denote the locally convex space of C^k-smooth mappings $f\colon \operatorname{Int} U \to Y$ such that every $d^j f$, $j = 0, \dots, k$ (resp. $j \in \mathbb{N}_0$), has a continuous extension to the whole U, and such that all $d^j f$ are bounded on CCB subsets of U. We will use the expression $d^j f(x)$ for all $x \in U$ in the sense of these extensions. The space $\mathcal{C}^k(U;Y)$ is endowed with the locally convex topology τ_b^k of uniform convergence of $d^j f$, $j = 0, \dots, k$ (resp. $j \in \mathbb{N}_0$) on CCB subsets of U.

It is easy to see using Theorem 1.85 that if $U = X$ (resp. if U is a CCB set) and Y is a Banach space, then $(\mathcal{C}^k(U;Y), \tau_b^k)$ is a Fréchet locally convex space (resp. a Banach space, provided that $k < \infty$).

Let X, Y be normed linear spaces, $U \subset X$ a convex set with non-empty interior, $k \in \mathbb{N}$ (resp. $k = \infty$), and $f \in \mathcal{C}^k(U;Y)$. If $V \subset U$ is a CCB set, then there is a CCB set $W \subset U$ satisfying $V \subset W = \overline{W \cap \operatorname{Int} U}$. Indeed, assuming without loss of generality that $0 \in \operatorname{Int} U$, we may take $W = \bigcup_{t \in [0,1]} tV$. Thus applying Proposition 1.71 to $W \cap \operatorname{Int} U$ we san see that all $d^j f$, $j = 0, \dots, k-1$ (resp. all $j \in \mathbb{N}_0$ if $k = \infty$) are Lipschitz on V. In the next we are interested in general theory of weakly continuous mappings (in several senses, as defined in Section 3.2) which have uniformly continuous derivatives.

Definition 2. Let X, Y be normed linear spaces, $U \subset X$ a convex set with non-empty interior, and $k \in \mathbb{N}_0$. By $\mathcal{C}_u(U;Y)$ we denote the subspace of $\mathcal{C}(U;Y)$ consisting of mappings that are uniformly continuous on CCB subsets of U. By $\mathcal{C}^{k,+}(U;Y)$ we denote the subspace of $\mathcal{C}^k(U;Y)$ consisting of mappings f such that $d^k f$ is uniformly continuous on CCB subsets of U. As usual, in the scalar case we use the notation $\mathcal{C}^{k,+}(U)$.

Notice that by the remarks above if $f \in \mathcal{C}^{k,+}(U;Y)$, then all $d^j f$, $j = 0, \dots, k$ are uniformly continuous on CCB subsets of U. In view of Theorem 1.125 we will also call the mappings in $\mathcal{C}^{k,+}(U;Y)$ uniformly smooth.

Fact 3. *Let X, Y be normed linear spaces, $U \subset X$ a convex set with non-empty interior, $k \in \mathbb{N}$, $f \in \mathcal{C}^{k,+}(U)$, $C \subset U$ a CCB set satisfying $C \subset \overline{C \cap \operatorname{Int} U}$, and let $\omega \in \mathcal{M}$ be a modulus of continuity of $d^k f$ on C. Then for any $x, y \in C$*

$$\left\| f(y) - \sum_{j=0}^{k} \frac{1}{j!} d^j f(x)[y-x] \right\| \le \frac{1}{k!} \omega(\|y-x\|)\|y-x\|^k.$$

Proof. By considering the minimal modulus of continuity of $d^k f$ we may assume without loss of generality that $\omega \in \mathcal{M}_s$. For $x, y \in C \cap \operatorname{Int} U$ the estimate follows from Corollary 1.108. Then we extend the estimate to the boundary by the continuity. □

Proposition 4. *Let X, Y, Z be normed linear spaces, $U \subset X$ and $V \subset Y$ convex sets with non-empty interior, $g \in \mathcal{C}^1(U;V)$, and $f \in \mathcal{C}^{1,+}(V;Z)$. Suppose further that for any CCB subset C of U the set $g(C)$ is contained in a CCB subset of V (this holds in particular if the set V is closed). Then $f \circ g \in \mathcal{C}^1(U;Z)$ and $D(f \circ g)(x) = Df(g(x)) \circ Dg(x)$ for every $x \in U$. If moreover $g \in \mathcal{C}^{1,+}(U;Y)$, then $f \circ g \in \mathcal{C}^{1,+}(U;Z)$.*

Proof. Fix an arbitrary $x \in \operatorname{Int} U$ and let $\delta > 0$ be such that $x + h \in \operatorname{Int} U$ and $\|g(x+h) - g(x) - Dg(x)[h]\| \le \|h\|$ for $h \in B(0,\delta)$. Then $\|g(x+h) - g(x)\| \le \|g(x+h) - g(x) - Dg(x)[h]\| + \|Dg(x)\|\|h\| \le (\|Dg(x)\| + 1)\|h\|$ for $h \in B(0,\delta)$. Let $W \subset V$ be a CCB set satisfying $g(B(0,\delta)) \subset W \subset \overline{W \cap \operatorname{Int} V}$ and let ω be a modulus of continuity of Df on W. Using Fact 3 we obtain

$$\begin{aligned}
&\big\| f\big(g(x+h)\big) - f\big(g(x)\big) - Df\big(g(x)\big)\big[Dg(x)[h]\big]\big\| \\
&\quad \le \big\| f\big(g(x+h)\big) - f\big(g(x)\big) - Df\big(g(x)\big)[g(x+h) - g(x)]\big\| \\
&\qquad + \big\| Df\big(g(x)\big)\big[g(x+h) - g(x) - Dg(x)[h]\big]\big\| \\
&\quad \le \omega\big(\|g(x+h) - g(x)\|\big)\|g(x+h) - g(x)\| \\
&\qquad + \big\| Df\big(g(x)\big)\big\| \, \|g(x+h) - g(x) - Dg(x)[h]\| \\
&\quad = o(\|h\|), \; h \to 0.
\end{aligned}$$

Thus $f \circ g$ is Fréchet differentiable at x and its derivative satisfies the Chain rule formula.

Now for any $x, y \in U$ and $h \in B_X$ we have

$$\begin{aligned}
&\big\| Df\big(g(y)\big)\big[Dg(y)[h]\big] - Df\big(g(x)\big)\big[Dg(x)[h]\big]\big\| \\
&\quad \le \big\| Df\big(g(y)\big)\big[Dg(y)[h]\big] - Df\big(g(x)\big)\big[Dg(y)[h]\big]\big\| \\
&\qquad + \big\| Df\big(g(x)\big)\big[Dg(y)[h]\big] - Df\big(g(x)\big)\big[Dg(x)[h]\big]\big\| \\
&\quad \le \big\| Df\big(g(y)\big) - Df\big(g(x)\big)\big\| \, \|Dg(y)\| + \big\| Df\big(g(x)\big)\big\| \, \|Dg(y) - Dg(x)\|.
\end{aligned}$$

Thus the mapping $x \mapsto Df(g(x)) \circ Dg(x)$ is continuous on U and consequently $f \circ g \in \mathcal{C}^1(U;Z)$ and the Chain rule formula holds on the whole of U. If moreover $g \in \mathcal{C}^{1,+}(U;Y)$, then the estimate above implies that $f \circ g \in \mathcal{C}^{1,+}(U;Z)$. □

Definition 5 ([AP]). Let X be a normed linear space, Y a Banach space, $U \subset X$ a convex set with non-empty interior, and $k \in \mathbb{N}_0$. By $\mathcal{C}^k_{\mathrm{wu}}(U;Y)$ we denote the subspace of $\mathcal{C}^k(U;Y)$ consisting of all f that satisfy $d^j f \in \mathcal{C}_{\mathrm{wu}}\big(U;\mathcal{P}_{\mathrm{w}}({}^jX;Y)\big)$, $j = 0,\dots,k$.

We remind that $f \in \mathcal{C}^k_{\mathrm{wu}}(U;Y)$ if and only if $d^j f \in \mathcal{C}_{\mathrm{wu}}\big(U;\mathcal{P}_{\mathrm{wu}}({}^jX;Y)\big)$ for $j = 0,\dots,k$ (Corollary 3.27).

Lemma 6. *Let X be a normed linear space, Y a Banach space, and let $U \subset X$ be a convex set with non-empty interior. If $f \in \mathcal{C}^{1,+}(U;Y) \cap \mathcal{C}_{\mathrm{wu}}(U;Y)$, then $Df \in \mathcal{C}_{\mathrm{wu}}\big(U;\mathcal{L}_{\mathrm{wu}}(X;Y)\big)$.*

Proof. The fact that $Df(x) \in \mathcal{L}_{\mathrm{wu}}(X;Y)$ for every $x \in \operatorname{Int} U$ follows at once combining Proposition 3.70, Lemma 3.18, and Corollary 3.24. For non-interior points we use the fact that $\mathcal{L}_{\mathrm{wu}}(X;Y)$ is closed in $\mathcal{L}(X;Y)$.

Now let $V \subset U$ be a CCB set and $\varepsilon > 0$. First assume that $\rho = \operatorname{dist}(V, X \setminus U) > 0$. Put $W = V + U(0,\frac{\rho}{2})$. Then $\overline{W}$ is a CCB set with $\overline{W} \subset \operatorname{Int} U$. Since f is UT^1-smooth on W, there is $0 < \delta < \frac{\rho}{2}$ such that $\big\|f(x+h) - f(x) - Df(x)[h]\big\| \le \frac{\varepsilon}{4}\|h\|$ whenever $x \in V$, $h \in B(0,\delta)$. By the assumptions there exist $\eta > 0$ and $\phi_1,\dots,\phi_n \in X^*$ such that $\|f(x) - f(y)\| < \frac{\varepsilon\delta}{4}$ whenever $x, y \in \overline{W}$ and $\max_l |\phi_l(x-y)| < \eta$. Thus for $x, y \in V$ such that $\max_l |\phi_l(x-y)| < \eta$ and $h \in B_X$ we have

$$\begin{aligned}
\big\|Df(x)[h] - Df(y)[h]\big\| &= \frac{1}{\delta}\big\|Df(x)[\delta h] - Df(y)[\delta h]\big\| \\
&\le \frac{1}{\delta}\Big(\big\|f(x+\delta h) - f(x) - Df(x)[\delta h]\big\| + \big\|f(y+\delta h) - f(y) - Df(y)[\delta h]\big\| \\
&\qquad + \|f(y+\delta h) - f(x+\delta h)\| + \|f(y) - f(x)\|\Big) < \varepsilon.
\end{aligned}$$

Now assume that V is a general CCB subset of U. Without loss of generality we may assume that $0 \in \operatorname{Int} U$. Let $r = \operatorname{dist}(0, X \setminus U) > 0$ and let $R > 0$ be such that $V \subset B(0,R)$. Put $A = \bigcup_{t\in[0,1]} tV$. Then A is a CCB subset of U and so there is $\delta > 0$ such that $\|Df(x) - Df(y)\| < \frac{\varepsilon}{3}$ whenever $x, y \in A$, $\|x - y\| < \delta$. Choose $\lambda \in (0,1)$ so that $(1-\lambda)R < \delta$. Since $U(0,r) \subset U$, from the convexity we get $U(\lambda x, (1-\lambda)r) \subset U$ for any $x \in U$. Consequently $\operatorname{dist}(\lambda V, X \setminus U) \ge (1-\lambda)r > 0$. By the previous part of the proof applied to the CCB set λV there are $\eta > 0$ and $\phi_1,\dots,\phi_n \in B_{X^*}$ such that $\|Df(u) - Df(v)\| < \frac{\varepsilon}{3}$ whenever $u, v \in \lambda V$ and $\max_l |\phi_l(u-v)| < \eta$.

Now pick any $x, y \in V$ such that $\max_l |\phi_l(x-y)| < \eta$. Put $u = \lambda x$ and $v = \lambda y$. Then $\|x - u\| = (1-\lambda)\|x\| \le (1-\lambda)R < \delta$ and similarly $\|y - v\| < \delta$. Clearly $u, v \in \lambda V$ and since $\max_l |\phi_l(u-v)| = \lambda \max_l |\phi_l(x-y)| < \eta$, it follows that

$$\|Df(x) - Df(y)\| \le \|Df(x) - Df(u)\| + \|Df(u) - Df(v)\| + \|Df(v) - Df(y)\| < \varepsilon.$$

□

Fact 7. *Let X be a normed linear space, Y a Banach space, $n \in \mathbb{N}$, and further let $I: \mathcal{L}(X;\mathcal{L}(^nX;Y)) \to \mathcal{L}(^{n+1}X;Y)$ be the canonical isometry from Fact 1.9. Then*

$$\begin{aligned}\mathcal{L}^{\mathrm{s}}_{\mathrm{wu}}(^{n+1}X;Y) &= I\big(\mathcal{L}_{\mathrm{wu}}(X;\mathcal{L}^{\mathrm{s}}_{\mathrm{wu}}(^nX;Y))\big) \cap \mathcal{L}^{\mathrm{s}}(^{n+1}X;Y)\\ &= I\big(\mathcal{L}_{\mathrm{K}}(X;\mathcal{L}(^nX;Y))\big) \cap \mathcal{L}^{\mathrm{s}}(^{n+1}X;Y).\end{aligned}$$

Proof. Let $M \in \mathcal{L}_{\mathrm{wu}}(^{n+1}X;Y)$ be given. Denote $T = I^{-1}(M)$ and choose an arbitrary $\varepsilon > 0$. There exist weak neighbourhoods $W_0,\dots,W_n$ of the origin in X such that we have $\|M(x_0,\dots,x_n) - M(y_0,\dots,y_n)\| < \varepsilon$ whenever $x_j, y_j \in B_X$ and $x_j - y_j \in W_j$ for $j = 0,\dots,n$. Therefore for every fixed $x \in B_X$ we have $\|T(x)(x_1,\dots,x_n) - T(x)(y_1,\dots,y_n)\| = \|M(x,x_1,\dots,x_n) - M(x,y_1,\dots,y_n)\| < \varepsilon$ whenever $x_j, y_j \in B_X$ and $x_j - y_j \in W_j$, $j = 1,\dots,n$. So $T \in \mathcal{L}(X;\mathcal{L}_{\mathrm{wu}}(^nX;Y))$. Further, if $x, y \in B_X$ are such that $x - y \in W_0$, then

$$\begin{aligned}\|T(x) - T(y)\| &= \sup_{h_1,\dots,h_n \in B_X} \|T(x)(h_1,\dots,h_n) - T(y)(h_1,\dots,h_n)\|\\ &= \sup_{h_1,\dots,h_n \in B_X} \|M(x,h_1,\dots,h_n) - M(y,h_1,\dots,h_n)\| \le \varepsilon.\end{aligned}$$

Thus the inclusion $\mathcal{L}_{\mathrm{wu}}(^{n+1}X;Y) \subset I\big(\mathcal{L}_{\mathrm{wu}}(X;\mathcal{L}_{\mathrm{wu}}(^nX;Y))\big)$ holds even in the non-symmetric case.

The inclusion $\subset$ in the second equality holds trivially. To finish the proof, let $T \in \mathcal{L}_{\mathrm{K}}(X;\mathcal{L}(^nX;Y))$ be such that the $(n+1)$-linear mapping $I(T)$ is symmetric. Note that $T \in \mathcal{L}_{\mathrm{wu}}(X;\mathcal{L}(^nX;Y))$ by Corollary 3.24. Let $\{x^j_\gamma\}_{\gamma\in\Gamma} \subset B_X$, $j = 0,\dots,n$ be arbitrary weakly convergent nets such that $\{x^0_\gamma\}$ is weakly null. Then $\|I(T)(x^0_\gamma,\dots,x^n_\gamma)\| = \|T(x^0_\gamma)(x^1_\gamma,\dots,x^n_\gamma)\| \le \|T(x^0_\gamma)\| \to 0$. Therefore from the symmetry of $I(T)$, Theorem 3.26, and Proposition 3.15 it follows that $I(T) \in \mathcal{L}_{\mathrm{wu}}(^{n+1}X;Y)$. □

We note that if we replace the property wu by property K in the previous fact, then only the inclusion $\supset$ holds in general: Consider $X = \ell_2$ and the bilinear form $M \in \mathcal{L}_{\mathrm{K}}(^2X;\mathbb{R})$, $M(x,y) = \langle x,y\rangle$. Then $I^{-1}(M) = \big(x \mapsto \langle x,\cdot\rangle\big)$ which is clearly not a compact linear operator.

Theorem 8. *Let X be a normed linear space, Y a Banach space, $U \subset X$ a convex set with non-empty interior, and $k \in \mathbb{N}$. Then $\mathcal{C}^{k,+}(U;Y) \cap \mathcal{C}_{\mathrm{wu}}(U;Y) = \mathcal{C}^k_{\mathrm{wu}}(U;Y)$.*

Proof. The inclusion $\subset$ follows immediately by induction, working with D^jf instead of d^jf, using Lemma 6, Fact 3.21, and Fact 7. The inclusion $\supset$ is trivial. □

Theorem 9. *Let X be a normed linear space, Y a Banach space, $U \subset X$ a convex set with non-empty interior, and $f \in \mathcal{C}^k(U;Y)$. Assume that $d^kf \in \mathcal{C}_{\mathrm{K}}(U;\mathcal{P}_{\mathrm{wu}}(^kX;Y))$ and $d^jf(x) \in \mathcal{P}_{\mathrm{wu}}(^jX;Y)$ for each $x \in U$ and $j \in \{1,\dots,k\}$. Then $f \in \mathcal{C}_{\mathrm{wu}}(U;Y)$.*

Proof. First assume that $k = 1$. Let $V \subset U$ be a CCB subset and $\varepsilon > 0$. Without loss of generality we may assume that $V \subset \overline{V \cap \operatorname{Int} U}$ (see the remark after Definition 1). Let $R > 0$ be such that $V \subset B(0,R)$. Since $Df(V)$ is relatively compact, there is a finite set $\{x_1,\dots,x_m\} \subset V$ such that the set $\{Df(x_j)\}_{j=1}^m$ is $\frac{\varepsilon}{4R}$-dense in $Df(V)$.

Further, there exist $\delta > 0$ and a finite set $\Phi \subset B_{X^*}$ such that $\|Df(x_j)[h]\| < \frac{\varepsilon}{2}$, $j = 1, \dots, m$ whenever $h \in B(0, 2R)$ satisfies $\max_{\phi\in\Phi} |\phi(h)| < \delta$. Now if $x \in V$ is arbitrary, then there is $j \in \{1, \dots, m\}$ such that $\|Df(x) - Df(x_j)\| < \frac{\varepsilon}{4R}$ and hence $\|Df(x)[h]\| \le \|Df(x_j)[h]\| + 2R\|Df(x) - Df(x_j)\| < \varepsilon$ whenever $h \in B(0, 2R)$ satisfies $\max_{\phi\in\Phi} |\phi(h)| < \delta$. So if $x, y \in V \cap \operatorname{Int} U$ are such that $\max_{\phi\in\Phi} |\phi(x-y)| < \delta$, then Theorem 1.107 yields

$$\|f(y) - f(x)\| = \left\| \int_0^1 Df\big(x + t(y - x)\big)[y - x]\,\mathrm{d}t \right\| \le \varepsilon.$$

The same inequality for points in the closure is now immediate.

The case of general $k > 1$ follows by induction: Assume that the statement holds for $k - 1$ and set $g = D^{k-1}f$ and $Z = \mathcal{L}^{s}_{wu}({}^{k-1}X; Y)$. Then by the assumption and Fact 3.21 we have $g\colon U \to Z$ and $D^k f \in \mathcal{C}_K\big(U; \mathcal{L}^{s}_{wu}({}^{k}X; Y)\big)$. By Fact 7, $Dg \in \mathcal{C}_K(U; \mathcal{L}_{wu}(X; Z))$ and so by the first part of the proof $g \in \mathcal{C}_{wu}(U; Z)$. Consequently f satisfies the assumptions of the inductive hypothesis and the proof is finished. □

The previous theorem in general does not hold without the assumption that all the differentials at each point are w–$\|\cdot\|$ uniformly continuous polynomials, as the identity mapping on ℓ_2 or c_0 clearly shows (its second derivative is a constant zero mapping, but its first derivative at each point is the identity and hence not compact). However, in the special case of polynomials (of degree at least 2) it does hold.

Theorem 10. *Let X be a normed linear space, Y a Banach space, $n \in \mathbb{N}$, $n \ge 2$, and $P \in \mathcal{P}({}^nX; Y)$. The following statements are equivalent:*

(i) $P \in \mathcal{P}_{wu}({}^nX; Y)$.

(ii) $d^kP \in \mathcal{P}_K({}^{n-k}X; \mathcal{P}({}^kX; Y))$ *for some* $1 \le k < n$.

(iii) $P \in \mathcal{C}^k_{wu}(X; Y)$ *for every* $k \in \mathbb{N}$.

Proof. (i)⇔(iii) follows from Theorem 8, (iii)⇒(ii) follows from Lemma 3.18.

(ii)⇒(i) It suffices to prove this implication for $k = 1$ and then use induction. So assume that $DP \in \mathcal{P}_K({}^{n-1}X; \mathcal{L}(X; Y))$. We claim that this in fact means that $DP \in \mathcal{P}_K({}^{n-1}X; \mathcal{L}_K(X; Y))$ and so we can apply Theorem 9. By Fact 3.21 we have $\check{DP} \in \mathcal{L}_K({}^{n-1}X; \mathcal{L}(X; Y))$. Fix $x \in X$. Using Lemma 1.99 and the symmetry of $\check{P}$ we obtain $DP(x)[B_X] = n\check{P}({}^{n-1}x, B_X) = n\check{P}(B_X, {}^{n-1}x) = \check{DP}(B_X, {}^{n-2}x)[x] = \varepsilon_x \circ \check{DP}(B_X, {}^{n-2}x)$, where $\varepsilon_x \in \mathcal{L}(\mathcal{L}(X; Y); Y)$ is the operator of evaluation at x. It follows that $DP(x)[B_X]$ is relatively compact in Y. □

2. Bidual extensions

In this section we study further the extension of continuous mappings with domain in X to mappings with domain in X^{**} initiated in Section 4.1. In particular we concentrate on the behaviour of the derivative of the extension.

Let X, Y be Banach spaces and $f: B_X \to Y$ a uniformly continuous mapping. Recall that we can define an extension $E_{\mathcal{U},T}(f): B_{X^{**}} \to Y^{**}$ by

$$E_{\mathcal{U},T}(f)(x) = w^*\text{-}\lim_{\mathcal{U}} f(T(x)_\alpha)$$

for any $x \in B_{X^{**}}$, where $\mathcal{U}$ is the ultrafilter and $T \in \mathcal{L}(X^{**}; (X)_\mathcal{U})$ is the isometric embedding from Corollary 4.7. (The representation $T(x) = (T(x)_\alpha)_\mathcal{U}$ is chosen so that $T(x)_\alpha \in B_X$ for all $\alpha \in I$, Proposition 4.8.) In general the extension depends on $\mathcal{U}$ and T, but as mentioned in Section 4.1 if e.g. f is w–w uniformly continuous, then the extension is uniquely determined. This is the case of an extension of a linear operator $L \in \mathcal{L}(X; Y)$, and $E_{\mathcal{U},T}(L) = L^{**}$ (the double adjoint operator, in the sense of a restriction), independently of $\mathcal{U}$ and T.

Our interest lies mostly in some applications of the bidual extension which are independent of the concrete choice of $\mathcal{U}$ and T. We are going to assume from now on that $\mathcal{U}$ and T have been fixed. In order to simplify the notation we will use the notation $E(f) = E_{\mathcal{U},T}(f)$, or the more suggestive $f^{**} = E(f)$ (in accordance with the extension of linear operators), and in this section also $\overline{f} = E(f)$ for better readability of some formulae.

We remark that if Y is a dual space (e.g. if $Y = \mathcal{L}(X^{**}; Z^{**})$), then we can define an extension with range in Y using directly the w^* topology of Y (see Section 4.1). This extension is in general different from the extension considered above (using the w^* topology in Y^{**}) and for our applications (mainly Theorem 18) we always need to extend into Y^{**}, as the w^* topology on Y is too weak here.

Fact 11. *Let X, Y be Banach spaces and $f: B_X \to Y$ a uniformly continuous mapping. If $f \in \mathcal{C}_K(B_X; Y)$, then $\overline{f} \in \mathcal{C}_K(B_{X^{**}}; Y)$. Similarly, if $f \in \mathcal{C}_{wK}(B_X; Y)$, then $\overline{f} \in \mathcal{C}_{wK}(B_{X^{**}}; Y)$.*

Proof. This is immediate from the fact that $\overline{f}(B_{X^{**}}) \subset \overline{f(B_X)}^{w^*} = \overline{f(B_X)}^{\|\cdot\|} \subset Y$, resp. $\overline{f}(B_{X^{**}}) \subset \overline{f(B_X)}^{w^*} = \overline{f(B_X)}^{w} \subset Y$, since the compact topology is the weakest Hausdorff topology. □

Theorem 4.9 has the following corollary.

Corollary 12. *Let X and Y be Banach spaces and let $f \in \mathcal{C}^{k,+}(B_X; Y)$. Then $\overline{f} \in \mathcal{C}^{k,+}(B_{X^{**}}; Y^{**})$ and $d^j\overline{f}(x)[h] = w^*\text{-}\lim_\mathcal{U} d^j f(T(x)_\alpha)[T(h)_\alpha]$ for any $j = 1, \dots, k$, $x \in B_{X^{**}}$, and $h \in X^{**}$.*

Proof. Theorem 4.9 implies $\overline{f} \in C^{k,+}(\operatorname{Int} B_{X^{**}}; Y^{**})$ and the validity of the formula on $\operatorname{Int} B_{X^{**}}$. All the differentials can be continuously extended to $B_{X^{**}}$ by the uniform continuity and so $\overline{f} \in \mathcal{C}^{k,+}(B_{X^{**}}; Y^{**})$. To prove the formula on the whole $B_{X^{**}}$ fix $j \in \{1, \dots, k\}$, $x \in S_{X^{**}}$, $h \in B_{X^{**}}$, $\phi \in B_{X^*}$, and $\varepsilon > 0$. There is $\delta > 0$ such that $\|d^j f(u) - d^j f(v)\| < \varepsilon$ whenever $u, v \in B_X$ satisfy $\|u - v\| \le \delta$. Find $y \in \operatorname{Int} B_{X^{**}}$ such that $\|x - y\| < \delta$ and $\|d^j\overline{f}(x)[h] - d^j\overline{f}(y)[h]\| < \varepsilon$. There is $U \in \mathcal{U}$ such that $\|T(x)_\alpha - T(y)_\alpha\| < \delta$ and $\left|\langle d^j\overline{f}(y)[h] - d^j f(T(y)_\alpha)[T(h)_\alpha], \phi\rangle\right| < \varepsilon$ for all $\alpha \in U$.

Then for any $\alpha \in U$ we obtain

$$\begin{aligned}&\left|\langle d^j\overline{f}(x)[h] - d^j f(T(x)_\alpha)[T(h)_\alpha], \phi\rangle\right| \\ &\quad \le \left\|d^j\overline{f}(x)[h] - d^j\overline{f}(y)[h]\right\| + \left|\langle d^j\overline{f}(y)[h] - d^j f(T(y)_\alpha)[T(h)_\alpha], \phi\rangle\right| \\ &\qquad + \left\|d^j f(T(y)_\alpha)[T(h)_\alpha] - d^j f(T(x)_\alpha)[T(h)_\alpha]\right\| < 3\varepsilon.\end{aligned}$$

□

The following fact is immediate using the w^*–w^* continuity of adjoint operators.

Fact 13. *Let X, Y, Z be Banach spaces, $f: B_X \to Y$ a uniformly continuous mapping, and $L \in \mathcal{L}(Y; Z)$. Then $\overline{L \circ f} = L^{**} \circ \overline{f}$.*

Theorem 14 ([CHL]). *Let X be a $\mathcal{L}_{\infty,\lambda}$-space, $\lambda \ge 1$, and let $i: X \to Z$ be an isometric embedding into some Banach space Z. Then for every Banach space Y and $f \in \mathcal{C}^{k,+}(B_X; Y)$ there exists a "partial extension" $g \in \mathcal{C}^{k,+}(\frac{1}{\lambda} B_Z; Y^{**})$ such that $g(i(x)) = f(x)$ for $x \in \frac{1}{\lambda} B_X$.*

Proof. The operator $i^{**}: X^{**} \to Z^{**}$ is again an isometry into. Since X^{**} is a $\mathcal{P}_\lambda$-space (Theorem 4.15), there is a projection $P: Z^{**} \to i^{**}(X^{**})$ of norm at most λ. We define the extension by $g = \overline{f} \circ (i^{**})^{-1} \circ P$. □

We now focus on the relationship between the operations of taking the extension E and taking the derivative D, chiefly as a tool for studying the properties of the derivative of the extension (Theorem 18). Let $f \in \mathcal{C}^{1,+}(B_X; Y)$. Then

$$D \circ E(f) = D\overline{f}: B_{X^{**}} \to \mathcal{L}(X^{**}; Y^{**}),$$
$$E \circ D(f) = \overline{Df}: B_{X^{**}} \to \mathcal{L}(X; Y)^{**}.$$

The ranges of the respective operations are different spaces, so strictly speaking E and D cannot commute. Moreover, the values of the latter composition are not even linear operators. However, there is a very natural way to consider their commutation in a weaker sense which in some sense reflects that both of these compositions somehow coincide as "operators" with respect to evaluations on elements in X^{**}. In fact, and somewhat more precisely, we will consider the situation when these two objects coincide with respect to certain duality evaluations involving the tensor product $X^{**} \otimes_\pi Y^*$, which can be paired naturally with both of the spaces in question.

Denote by $\mathcal{L}^{**}(X^{**}; Y^{**})$ the subspace of $\mathcal{L}(X^{**}; Y^{**})$ consisting of all double adjoint bounded linear operators. The operation E restricted to bounded linear operators is an isometry $E: \mathcal{L}(X; Y) \to \mathcal{L}^{**}(X^{**}; Y^{**})$ onto. Further, the space $\mathcal{L}(X; Y)$ is canonically isometric to a subspace of its double dual space $\mathcal{L}(X; Y)^{**}$ via $i: \mathcal{L}(X; Y) \to \mathcal{L}(X; Y)^{**}$ such that $\langle i(L), \phi\rangle = \langle \phi, L\rangle$ for $L \in \mathcal{L}(X; Y)$, $\phi \in \mathcal{L}(X; Y)^*$. By Fact 3.8, $\mathcal{L}(X^{**}; Y^{**}) = (X^{**} \otimes_\pi Y^*)^*$, hence there is a canonical isometric embedding into the double dual space $j: X^{**} \otimes_\pi Y^* \to \mathcal{L}(X^{**}; Y^{**})^*$ such that $\langle j(\phi), S\rangle = \langle S, \phi\rangle$ for $\phi \in X^{**} \otimes_\pi Y^*$, $S \in \mathcal{L}(X^{**}; Y^{**})$. We consider the following restriction of the quotient operator E^* adjoint to the embedding $E: \mathcal{L}(X; Y) \to \mathcal{L}(X^{**}; Y^{**})$:

$$J = E^* \circ j: X^{**} \otimes_\pi Y^* \to \mathcal{L}(X; Y)^*.$$

Then $J^*\colon \mathcal{L}(X;Y)^{**} \to \mathcal{L}(X^{**};Y^{**})$ satisfies $E = J^* \circ i$. Indeed,

$$\begin{aligned} \langle J^* \circ i(L), \phi\rangle &= \langle i(L), J(\phi)\rangle = \langle J(\phi), L\rangle \\ &= \langle E^* \circ j(\phi), L\rangle = \langle j(\phi), E(L)\rangle = \langle E(L), \phi\rangle \end{aligned} \tag{1}$$

for all $\phi \in X^{**} \otimes_\pi Y^*$, $L \in \mathcal{L}(X;Y)$. Hence the following diagram commutes:

$$\begin{array}{ccc} \mathcal{L}(X;Y) & \xrightarrow{\;E\;} & \mathcal{L}^{**}(X^{**};Y^{**}) \\ \Big\downarrow i & & \Big\downarrow Id \\ \mathcal{L}(X;Y)^{**} & \xrightarrow{\;J^*\;} & \mathcal{L}(X^{**};Y^{**}) \end{array} \tag{2}$$

Definition 15. Let X, Y be Banach spaces and $f \in \mathcal{C}^{1,+}(B_X;Y)$. We say that the extension and differentiation operations E and D are weakly commuting for f at $z \in B_{X^{**}}$ if

$$D\overline{f}(z) = J^*(\overline{Df}(z)).$$

We say that the operations E and D are weakly commuting for f on a set $V \subset B_{X^{**}}$ provided they are weakly commuting at every point $z \in V$.

We note that if $Y = \mathbb{K}$, then the extension operator restricted to X^* is an identity $E\colon X^* \to X^* \subset X^{***}$ and $J^*\colon X^{***} \to X^{***}$ is also an identity. Thus the weak commutation in this case has a natural form $D\overline{f}(z) = \overline{Df}(z)$.

Recall that $w^*\text{-}\lim_{\mathcal{U}} T(z)_\alpha = z$ for every $z \in X^{**}$.

Proposition 16. *Let X, Y be Banach spaces, $f \in \mathcal{C}^{1,+}(B_X;Y)$, and $z \in B_{X^{**}}$. The following statements are equivalent:*

(i) *The operations E and D are weakly commuting for f at z.*
(ii) $\langle D\overline{f}(z), h \otimes \phi\rangle = \langle J^*(\overline{Df}(z)), h \otimes \phi\rangle$ *for all $h \in X^{**}$ and $\phi \in Y^*$.*
(iii) $w^*\text{-}\lim_{\mathcal{U}} Df(T(z)_\alpha)^{**}(h) = w^*\text{-}\lim_{\mathcal{U}} Df(T(z)_\alpha)[T(h)_\alpha]$ *for all $h \in X^{**}$.*
(iv) *The operations E and D are weakly commuting for $\phi \circ f$ at z for all $\phi \in Y^*$.*

Proof. We have $D\overline{f}(z) = J^*(\overline{Df}(z))$ if and only if $\langle D\overline{f}(z), \psi\rangle = \langle J^*(\overline{Df}(z)), \psi\rangle$ for all $\psi \in X^{**} \otimes Y^*$ (Fact 3.8). It follows that (i) and (ii) are equivalent, because the span of all elementary tensors is norm dense in $X^{**} \otimes_\pi Y^*$.

To show (ii)$\Leftrightarrow$(iii) fix $h \in X^{**}$ and $\phi \in Y^*$. Then, using Corollary 12,

$$\langle D\overline{f}(z), h \otimes \phi\rangle = \langle D\overline{f}(z)[h], \phi\rangle = \lim_{\mathcal{U}} \big\langle Df(T(z)_\alpha)[T(h)_\alpha], \phi\big\rangle,$$

while

$$\begin{aligned} \langle J^*(\overline{Df}(z)), h \otimes \phi\rangle &= \langle \overline{Df}(z), J(h \otimes \phi)\rangle = \lim_{\mathcal{U}} \big\langle Df(T(z)_\alpha), E^* \circ j(h \otimes \phi)\big\rangle \\ &= \lim_{\mathcal{U}} \big\langle j(h \otimes \phi), E(Df(T(z)_\alpha))\big\rangle = \lim_{\mathcal{U}} \big\langle j(h \otimes \phi), Df(T(z)_\alpha)^{**}\big\rangle \\ &= \lim_{\mathcal{U}} \big\langle Df(T(z)_\alpha)^{**}, h \otimes \phi\big\rangle = \lim_{\mathcal{U}} \big\langle Df(T(z)_\alpha)^{**}(h), \phi\big\rangle. \end{aligned}$$

If $L \in \mathcal{L}(X;Y)$, $\phi \in Y^*$, and $h \in X^{**}$, then $L^*(\phi) = \phi \circ L$ and so

$$\langle L^{**}(h), \phi\rangle = \langle h, L^*(\phi)\rangle = \langle h, \phi \circ L\rangle. \tag{3}$$

Also, for $\psi \in X^*$ we have $\psi^{**}(h) = h(\psi)$. Combining this with the Chain rule we get

$$\begin{aligned}\langle Df(T(z)_\alpha)^{**}(h), \phi\rangle &= \langle h, \phi \circ Df(T(z)_\alpha)\rangle = \langle h, D(\phi \circ f)(T(z)_\alpha)\rangle \\ &= D(\phi \circ f)(T(z)_\alpha)^{**}(h).\end{aligned}$$

On the other hand, we have $\langle \phi, Df(T(z)_\alpha)[T(h)_\alpha]\rangle = \big(\phi \circ Df(T(z)_\alpha)\big)(T(h)_\alpha) = D(\phi \circ f)(T(z)_\alpha)[T(h)_\alpha]$. Thus (iii) is equivalent to

$$\lim_{\mathcal{U}} D(\phi \circ f)(T(z)_\alpha)^{**}(h) = \lim_{\mathcal{U}} D(\phi \circ f)(T(z)_\alpha)[T(h)_\alpha]$$

for all $h \in X^{**}$ and $\phi \in Y^*$. So the equivalence of (i) and (iii) applied to the functions $\phi \circ f$ gives (i)$\Leftrightarrow$(iv). □

The operations E and D always weakly commute on B_X. Indeed, let $x \in B_X$. Since $T(x)_\alpha = x$ for all $\alpha \in I$, the condition (iii) in Proposition 16 transforms to $Df(x)^{**}(h) = w^*\text{-}\lim_{\mathcal{U}} Df(x)[T(h)_\alpha]$, which holds by the w^*–w^* continuity of adjoint operators. Moreover, $D\overline{f}(x) = Df(x)^{**} \in \mathcal{L}^{**}(X^{**}; Y^{**})$ by Theorem 4.9(iii). In general, however, $D\overline{f}(z)$ need not be a double adjoint operator if $z \in B_{X^{**}}$. Indeed, let $f : c_0 \to c_0$ be given by $f(x_0, x_1, \dots) = (x_0, x_0x_1, x_0x_2, \dots)$. Then $\overline{f} : \ell_\infty \to \ell_\infty$ is given by the same formula and $D\overline{f}(1, 1, \dots)[h] = (h_0, h_0 + h_1, h_0 + h_2, \dots)$. So $D\overline{f}(1, 1, \dots) \notin \mathcal{L}^{**}(\ell_\infty; \ell_\infty)$, since it does not map c_0 into c_0, although it is a w^*–w^* continuous operator and the operations E and D weakly commute on B_{ℓ_∞} in this case (Theorem 23 and Theorem 30).

Theorem 17 ([CHL]). *Let X, Y be Banach spaces, $f \in \mathcal{C}^{1,+}(B_X; Y)$, and suppose that $Df \in \mathcal{C}_K\big(B_X; \mathcal{L}(X; Y)\big)$. Then the operations E and D are weakly commuting for f on $B_{X^{**}}$ and $D\overline{f}(z) \in \mathcal{L}^{**}(X^{**}; Y^{**})$ for every $z \in B_{X^{**}}$.*

Proof. Fix $z \in B_{X^{**}}$. Since $Df \in \mathcal{C}_K\big(B_X; \mathcal{L}(X; Y)\big)$ and the compact topology is the weakest Hausdorff topology, $\overline{Df}(z) = \lim_{\mathcal{U}} Df(T(z)_\alpha)$ in the norm topology of $\mathcal{L}(X; Y)^{**}$. Hence $\overline{Df}(z) \in i(\mathcal{L}(X; Y))$, i.e. there is $L \in \mathcal{L}(X; Y)$ such that $i(L) = \overline{Df}(z)$. Thus $\lim_{\mathcal{U}} \|Df(T(z)_\alpha)^{**} - L^{**}\| = \lim_{\mathcal{U}} \|Df(T(z)_\alpha) - L\| = 0$. Since

$$\begin{aligned}&\big|\langle Df(T(z)_\alpha)^{**}(h), \phi\rangle - \langle \phi, Df(T(z)_\alpha)[T(h)_\alpha]\rangle\big| \\ &\quad\le \big\|Df(T(z)_\alpha)^{**} - L^{**}\big\| \|\phi\| \|h\| + \big|\langle L^{**}(h) - L(T(h)_\alpha), \phi\rangle\big| \\ &\qquad + \|L - Df(T(z)_\alpha)\| \|\phi\| \|T(h)_\alpha\|\end{aligned}$$

for $h \in X^{**}$ and $\phi \in Y^*$, using the w^{**}–w^{**} continuity of L^{**} we conclude that (iii) in Proposition 16 holds. Finally, by combining the established weak commutation property with the commutative diagram (2) we obtain

$$\langle D\overline{f}(z)[h], \phi\rangle = \langle J^*(\overline{Df}(z))(h), \phi\rangle = \langle L^{**}[h], \phi\rangle$$

for each $\phi \in Y^*$ and $h \in X^{**}$, and so $D\overline{f}(z) = L^{**} \in \mathcal{L}^{**}(X^{**}; Y^{**})$. □

Let Y, Z be Banach spaces. We consider two natural topologies on $\mathcal{L}(Z;Y^{**})$. The first one is the weak* topology coming from the duality $(Z \otimes_\pi Y^*)^* = \mathcal{L}(Z;Y^{**})$ of Fact 3.8. The second one is the weak* operator topology (W*OT for short), i.e. the locally convex topology generated by the semi-norms $L \mapsto |\langle L(x), \phi\rangle|$, $x \in Z$, $\phi \in Y^*$. Since the elementary tensors are linearly dense in $Z \otimes_\pi Y^*$, the W*OT and w^* topologies coincide on bounded subsets of $\mathcal{L}(Z;Y^{**})$.

Theorem 18. *Let X, Y be Banach spaces and let $f \in \mathcal{C}^{1,+}(B_X;Y)$ be w–w uniformly continuous. Then the bidual extension $\overline{f} \in \mathcal{C}^{1,+}(B_{X^{**}};Y^{**})$ is w^*–w^* uniformly continuous, and in particular independent of the choice of $\mathcal{U}$ and T. Thus for any subspace $Z \subset X$ we have $\overline{f\restriction_{B_Z}} = \overline{f}\restriction_{B_{Z^{**}}}$.*

Moreover, the mapping $D\overline{f} \in \mathcal{C}_{\mathrm{u}}(B_{X^{**}};\mathcal{L}(X^{**};Y^{**}))$ is also w^*–w^* uniformly continuous.

Proof. The first statement is clear (see the discussion preceding Theorem 4.9). If $Z \subset X$ is a subspace, then $f\restriction_{B_Z}$ is also w–w uniformly continuous. Since the w^*–w^* continuous mappings $\overline{f\restriction_{B_Z}}$ and $\overline{f}$ agree on B_Z which is w^*-dense in $B_{Z^{**}}$, they also agree on $B_{Z^{**}}$.

To prove the last assertion, fix arbitrary $F \in X^{**}$ and $\phi \in Y^*$. We claim that

$$J(F \otimes \phi) \circ Df = F \circ D(\phi \circ f). \tag{4}$$

Indeed, let $\psi = F \otimes \phi$. For any $x \in B_X$ we have $(J(\psi)\circ Df)(x) = \langle J(\psi), Df(x)\rangle = \langle Df(x)^{**}, \psi\rangle = \langle Df(x)^{**}(F), \phi\rangle = \langle F, \phi \circ Df(x)\rangle = \langle F, D(\phi \circ f)(x)\rangle = (F \circ D(\phi \circ f))(x)$, where we successively used (1), Fact 3.8, (3), and the Chain rule.

Because by our assumption $\phi \circ f \in \mathcal{C}_{\mathrm{wu}}(B_X)$, it follows from Lemma 6 that $D(\phi \circ f) \in \mathcal{C}_{\mathrm{wu}}(B_X;X^*)$. This combined with (4) gives $J(\psi) \circ Df \in \mathcal{C}_{\mathrm{wu}}(B_X)$. It also implies using Lemma 3.18 and Theorem 17 that the operations E and D are weakly commuting for $\phi \circ f$ on $B_{X^{**}}$. Hence by Proposition 16 the operations E and D are weakly commuting for f on $B_{X^{**}}$. So for any $z \in B_{X^{**}}$ we have

$$\begin{aligned} \langle D\overline{f}(z), \psi\rangle &= \langle J^*(\overline{Df}(z)), \psi\rangle = \overline{Df}(z)(J(\psi)) \\ &= \big(J(\psi)^{**} \circ \overline{Df}\big)(z) = \overline{J(\psi) \circ Df}(z), \end{aligned}$$

where the last equality follows from Fact 13. Because the weak uniform continuity of $J(\psi) \circ Df$ implies the w^*-uniform continuity of $\overline{J(\psi) \circ Df}$, it follows that the mapping $D\overline{f}\colon B_{X^{**}} \to \mathcal{L}(X^{**};Y^{**})$ is w^*–W*OT uniformly continuous. Since its range is bounded, the uniformities W*OT and w^* on this range coincide and so $D\overline{f}$ is w^*–w^* uniformly continuous. □

3. Class $\mathcal{W}$

The class $\mathcal{W}$ of Banach spaces lies at the core of the theory of smooth mappings between Banach spaces. It provides a link between uniform smoothness and weak continuity leading to strong structural results for smooth mappings in the following sections. In some sense, these spaces are the closest to finite-dimensional ones, from the point of view of smoothness. Class $\mathcal{W}$ appears naturally in several contexts, including the polynomial approximations studied in Chapter 7, and also the weak commuting of bidual extensions. The pivotal result in this section is the finite-dimensional quantitative Lemma 27, which leads eventually to the proof that c_0, and more generally $C(K)$ spaces for scattered compacts K, are $\mathcal{W}_1$-spaces.

Definition 19. Let $\lambda \in (0,1]$. We say that a Banach space X is a $\mathcal{W}_\lambda$-space (or that it belongs to the class $\mathcal{W}_\lambda$) if

$$\mathcal{C}^{1,+}(B_X) \subset \mathcal{C}_{\mathrm{wsC}}(\lambda B_X)$$

(in the sense of restriction). We say that a Banach space is a $\mathcal{W}$-space (or belongs to the class $\mathcal{W}$) if it is a $\mathcal{W}_\lambda$-space for some $\lambda \in (0,1]$.

Clearly if X is a $\mathcal{W}_\lambda$-space, then it is a $\mathcal{W}_\xi$-space for every $0 < \xi < \lambda$. Conversely, using the uniform continuity it is easy to check that if X is a $\mathcal{W}_\xi$-space for every $0 < \xi < \lambda$, then X is a $\mathcal{W}_\lambda$-space. Every Schur space is trivially a $\mathcal{W}_1$-space. Note also that if X is a $\mathcal{W}$-space, then $\mathcal{C}^{1,+}(X) \subset \mathcal{C}_{\mathrm{wsC}}(X)$.

Since the definition of a $\mathcal{W}$-space is sequential, the following observation is immediate:

Fact 20. *Let X be a Banach space. If for each separable subspace Y of X there is a subspace Z satisfying $Y \subset Z \subset X$ such that Z is a $\mathcal{W}_\lambda$-space, then X is a $\mathcal{W}_\lambda$-space.*

This fact is similar to the usual separable reduction principle, but a warning is in order: The class $\mathcal{W}$ is not closed with respect to subspaces as we shall see in Example 39. This is related to the extensions of $C^{1,+}$-smooth functions, see also Theorem 34. On the other hand, the class $\mathcal{W}$ behaves rather well with respect to quotients. Compare with Proposition 3.87 (especially in view of Fact 22) and Theorem 3.41(v).

Proposition 21 ([CHL]). *Let X be a $\mathcal{W}_\lambda$-space and $Y \subset X$ a separable subspace that does not contain ℓ_1. Then X/Y is a $\mathcal{W}_{\frac{\lambda}{3}}$-space. More generally, if Z is a Banach space such that $\mathcal{C}^{1,+}(B_X; Z) \subset \mathcal{C}_{\mathrm{wsC}}(\tau B_X; Z)$ for some $\tau \in (0,1]$, then $\mathcal{C}^{1,+}(B_{X/Y}; Z) \subset \mathcal{C}_{\mathrm{wsC}}(\frac{\tau}{3} B_{X/Y}; Z)$.*

Proof. Suppose by contradiction that there is $f \in \mathcal{C}^{1,+}(B_{X/Y}; Z) \setminus \mathcal{C}_{\mathrm{wsC}}(\frac{\tau}{3} B_{X/Y}; Z)$. By Fact 3.86 there exists a weakly Cauchy sequence $\{y_j\} \subset \frac{\tau}{3} B_{X/Y}$ such that $\varepsilon = \operatorname{dist}\big(\{f(y_{2k});\, k \in \mathbb{N}\}, \{f(y_{2k-1});\, k \in \mathbb{N}\}\big) > 0$. Since f is uniformly continuous on $B_{X/Y}$, there is $\delta > 0$ such that $\|f(x) - f(y)\| < \frac{\varepsilon}{2}$ whenever $x, y \in B_{X/Y}$ are such that $\|x - y\| < \delta$.

Denote by $Q: X \to X/Y$ the canonical quotient mapping. By Lemma 3.85 there is a weakly Cauchy sequence $\{x_j\} \subset X$ such that $\lim\|x_{2k-1}\| \le \frac{\tau}{3}$, $\lim\|x_{2k}\| = r \le \tau$, $\{Q(x_{2k-1})\}$ is a subsequence of $\{y_{2k-1}\}$, and $\{Q(x_{2k})\}$ is a subsequence of $\{y_{2k}\}$. By passing to subsequences we may assume that $\{x_{2k-1}\} \subset \tau B_X$ and $\big|\|x_{2k}\| - r\big| < \delta$ for all $k \in \mathbb{N}$. We set $z_{2k-1} = x_{2k-1}$ and $z_{2k} = \frac{r}{\|x_{2k}\|} x_{2k}$ for $k \in \mathbb{N}$. Then

$\{z_j\} \subset \tau B_X$ is weakly Cauchy and $\|z_{2k} - x_{2k}\| < \delta$ for all $k \in \mathbb{N}$. However, since by the assumption (and Proposition 4) $f \circ Q \in \mathcal{C}_{\mathrm{wsC}}(\tau B_X; Z)$, the sequence $\{f(Q(z_j))\}$ is Cauchy, but $\operatorname{dist}(\{f(Q(z_{2k-1}))\}, \{f(Q(z_{2k}))\}) \geq \frac{\varepsilon}{2}$, which is a contradiction. □

The assumption that Y does not contain ℓ_1 in the above proposition cannot be avoided: By Corollary 3.89 the space ℓ_1, which is a $\mathcal{W}_1$-space, has a quotient isomorphic to ℓ_2, which is not a $\mathcal{W}$-space (check the square of the norm). Note also that as a consequence of the bidual extensions (Corollary 12) we obtain that if X^{**} is a $\mathcal{W}_\lambda$-space, then X is a $\mathcal{W}_\lambda$-space too.

Proposition 3.19 and Theorem 3.81 immediately imply the following fact.

Fact 22. *Let X be a $\mathcal{W}_\lambda$-space and Y a Banach space. Then each $f \in \mathcal{C}^{1,+}(B_X; Y)$ is w–w sequentially Cauchy-continuous on λB_X. In particular, $\mathcal{P}(X) = \mathcal{P}_{\mathrm{wsC}}(X)$ and $\mathcal{P}_{\mathrm{K}}(X; Y) \subset \mathcal{P}_{\mathrm{wsC}}(X; Y)$.*

If moreover the space X does not contain ℓ_1, then $\mathcal{C}^{1,+}(B_X) \subset \mathcal{C}_{\mathrm{wu}}(\lambda B_X)$. Consequently, each $f \in \mathcal{C}^{1,+}(B_X; Y)$ is w–w uniformly continuous on λB_X. In particular, $\mathcal{P}(X) = \mathcal{P}_{\mathrm{wu}}(X)$ and $\mathcal{P}_{\mathrm{K}}(X; Y) = \mathcal{P}_{\mathrm{wu}}(X; Y)$.

Theorem 23. *Let X be a $\mathcal{W}_\lambda$-space that does not contain ℓ_1, Y a Banach space, and let $f \in \mathcal{C}^{1,+}(B_X; Y)$. Then the operations E and D are weakly commuting for f on $\lambda B_{X^{**}}$.*

Proof. By Proposition 16 it suffices to show the result for $Y = \mathbb{R}$. In this case it follows by combining Fact 22, Lemma 6, Lemma 3.18, and Theorem 17. □

Theorem 24 ([Há6]). *Let X be a $\mathcal{W}_\lambda$-space that does not contain ℓ_1, Y a Banach space, $k \in \mathbb{N}$, and $f \in \mathcal{C}^{k,+}(B_X; Y)$. The following statements are equivalent:*

(i) $f \in \mathcal{C}_{\mathrm{K}}(\lambda B_X; Y)$.
(ii) $f \in \mathcal{C}_{\mathrm{wsC}}(\lambda B_X; Y)$.
(iii) $f \in \mathcal{C}^k_{\mathrm{wu}}(\lambda B_X; Y)$.
(iv) *There exists $m \in \{1, \dots, k\}$ such that $d^m f \in \mathcal{C}_{\mathrm{K}}(\lambda B_X; \mathcal{P}_{\mathrm{K}}(^m X; Y))$ and $d^j f(x) \in \mathcal{P}_{\mathrm{K}}(^j X; Y)$ for each $x \in \lambda B_X$ and each $j \in \{1, \dots, m\}$.*

Proof. (i)⇒(ii) Follows from Fact 22 and Proposition 3.19, (ii)⇒(iii) follows from Theorem 3.81 and Theorem 8, (iii)⇒(iv) follows from Lemma 3.18, and (iv)⇒(i) follows from Fact 22, Theorem 9, and Lemma 3.18. □

By combining Fact 22 with Theorem 18 we obtain immediately the following corollary:

Corollary 25. *Let X be a $\mathcal{W}_\lambda$-space that does not contain ℓ_1, Y a Banach space, and $f \in \mathcal{C}^{1,+}(B_X; Y)$. Then the bidual extension $f^{**} \in \mathcal{C}^{1,+}(B_{X^{**}}; Y^{**})$ restricted to $\lambda B_{X^{**}}$ is w^*–w^* uniformly continuous, and in particular this restriction is independent of the choice of $\mathcal{U}$ and T. Thus for any subspace $Z \subset X$ we have $(f\restriction_{\lambda B_Z})^{**} = f^{**}\restriction_{\lambda B_{Z^{**}}}$.*

Moreover, the mapping $D(f^{**}) \in \mathcal{C}_{\mathrm{u}}(\lambda B_{X^{**}}; \mathcal{L}(X^{**}; Y^{**}))$ is also w^*–w^* uniformly continuous.

Later in this section we are going to prove that the spaces $C(K)$, where K is a scattered compact space, belong to the class $\mathcal{W}_1$. At the core of our arguments lie some finite-dimensional estimates.

We denote by $\{e_j\}_{j=1}^n$ the canonical basis of the space ℓ_∞^n and by $B_{\ell_\infty^n}^+$ the subset of the unit ball $B_{\ell_\infty^n}$ consisting of vectors that have non-negative coordinates. The next lemma contains the main idea that is needed for the estimate of the values of uniformly smooth functions on the canonical basis vectors of ℓ_∞^n. It involves symmetric functions in the sense of Definition 4.20. Compare the lemma below also with Proposition 5.49.

Lemma 26 ([Há5]). *For every $\omega \in \mathcal{M}$ with $\omega(1) \in \mathbb{R}$ and every $\varepsilon > 0$ there is $n(\omega,\varepsilon) \in \mathbb{N}$ such that if $n \geq n(\omega,\varepsilon)$ and $f \in \mathcal{C}^{1,+}(B_{\ell_\infty^n}^+)$ is a symmetric function such that $f(0) = 0$, $Df(0) = 0$, and Df has modulus of continuity ω, then $|f(e_j)| < \varepsilon$ for $j = 1,\dots,n$.*

Proof. Put $\xi = \frac{\varepsilon}{5}$ and find $k \in \mathbb{N}$ so that $\omega(\frac{1}{k}) \leq \xi$ and $\frac{\omega(1)}{k} \leq \xi$. Further, set $m = \max\{\lceil \frac{\omega(1)}{\xi} \rceil, 2\}$ and $n(\omega,\varepsilon) = m^{2(k-1)} + 1$. Now assume that $n \geq n(\omega,\varepsilon)$ and $f \in \mathcal{C}^{1,+}(B_{\ell_\infty^n}^+)$ satisfies the assumptions of the lemma. For a vector v we denote its pth coordinate by $v(p)$ and for $A \subset \{1,\dots,n\}$ we denote by P_A the projection in ℓ_1^n associated with A, i.e. $P_A(v) = \sum_{j\in A} v(j)e_j$. We put $x_0 = 0$, $y_0 = e_1$, $A_0 = \{2,\dots,n(\omega,\varepsilon)\}$, and define inductively vectors $x_j, y_j \in B_{\ell_\infty^n}^+$ and sets $A_j \subset A_0$, $1 \leq j \leq k-1$, satisfying the following:

(i) $x_j(1) = 0$ and $y_j = e_1 + x_j$;

(ii) for each $0 \leq l \leq j$ there is $p > 1$ such that $x_j(p) = \frac{l}{k}$;

(iii) $|f(x_j) - f(x_{j-1})| \leq 2\frac{\xi}{k}$ and $|f(y_j) - f(y_{j-1})| \leq 2\frac{\xi}{k}$;

(iv) $|A_j| = m^{2(k-j-1)}$ and $x_j(p) = j/k$ for $p \in A_j$.

Assume that x_j, y_j, and A_j are already defined for some $0 \leq j < k-1$. Note that $Df(x_j) \in \ell_1^n$ and $\|P_{A_j}(Df(x_j))\| \leq \|Df(x_j)\| \leq \omega(1)$. Consider splitting the set A_j into m subsets of size $m^{2(k-j-1)-1}$. It follows that there is $A \subset A_j$ such that $|A| = m^{2(k-j-1)-1}$ and $\|P_A(Df(x_j))\| \leq \frac{\omega(1)}{m} \leq \xi$. Repeating this procedure for y_j we conclude that there is $B \subset A$ such that $|B| = m^{2(k-j-2)}$ and $\|P_B(Df(y_j))\| \leq \xi$. We set $x_{j+1} = x_j + \frac{1}{k}\chi_B$, $y_{j+1} = e_1 + x_{j+1}$, and $A_{j+1} = B$. Then (i) and (iv) clearly holds. Since $m \geq 2$, we have $A_j \setminus B \neq \emptyset$, and so (ii) holds too. To see (iii) note that $\big|Df(x_j)[x_{j+1} - x_j]\big| \leq \frac{1}{k}\|P_B(Df(x_j))\| \leq \frac{\xi}{k}$ and similarly $\big|Df(y_j)[y_{j+1} - y_j]\big| \leq \frac{\xi}{k}$, and use Fact 3.

To finish the proof we put $x = x_{k-1}$ and $y = y_{k-1}$. By (ii) there exists a set $C \subset \{2,\dots,n\}$ of cardinality k such that x attains the values $0, \frac{1}{k}, \frac{2}{k}, \dots, \frac{k-1}{k}$ on C. Note that $P_C(y) = P_C(x)$. It follows that the coordinates of the vector $x + \frac{1}{k}\chi_C$ are just a rearrangement of those of y, and so by the symmetry $f(x + \frac{1}{k}\chi_C) = f(y)$. In particular, $|f(y) - f(x)| = \big|f(x + \frac{1}{k}\chi_C) - f(x)\big| \leq \omega(1)\frac{1}{k} \leq \xi$, as the function f

is $\omega(1)$-Lipschitz (Proposition 1.71). Combining this with (iii) we obtain

$$|f(e_1)| = |f(e_1) - f(0)|$$
$$\leq \sum_{j=1}^{k-1} |f(y_j) - f(y_{j-1})| + |f(y) - f(x)| + \sum_{j=1}^{k-1} |f(x_j) - f(x_{j-1})| < 5\xi = \varepsilon.$$

□

Lemma 27 ([Há5]). *For each $\omega \in \mathcal{M}$ with $\omega(1) \in \mathbb{R}$, $L > 0$, and $\varepsilon > 0$ there is $N(\omega, L, \varepsilon) \in \mathbb{N}$ such that if $n \geq N(\omega, L, \varepsilon)$ and $f \in \mathcal{C}^{1,+}(B^+_{\ell^n_\infty})$ is an L-Lipschitz function such that Df has modulus of continuity ω, then $|f(e_j) - f(0)| < \varepsilon$ for some $j \in \{1, \dots, n\}$.*

Proof. We set $N(\omega, L, \varepsilon) = 2n(\omega, \frac{\varepsilon}{2}) + \lceil \frac{2L}{\varepsilon} \rceil$, where $n(\cdot, \cdot)$ is the function from Lemma 26. Assume that $n \geq N(\omega, L, \varepsilon)$ and $f \in \mathcal{C}^{1,+}(B^+_{\ell^n_\infty})$ satisfies the assumptions of the lemma. Suppose that $|f(e_j) - f(0)| \geq \varepsilon$ for each $j \in \{1, \dots, n\}$. Since $Df(0) \in \ell^n_1$ and $\|Df(0)\| \leq L$, setting $A = \{j;\ |Df(0)_j| \leq \frac{\varepsilon}{2}\}$ we have $|A| \geq n - \frac{2L}{\varepsilon} \geq 2n(\omega, \frac{\varepsilon}{2})$. Further, there is $B \subset A$ such that $|B| \geq \frac{1}{2}|A|$ and either $f(e_j) - f(0) \geq \varepsilon$ for all $j \in B$, or $f(e_j) - f(0) \leq -\varepsilon$ for all $j \in B$. Let $\sigma = \operatorname{sgn}(f(e_j) - f(0))$, $j \in B$, and put $g(x) = \sigma\big(f(x) - f(0) - Df(0)[x]\big)$. Then $g \in \mathcal{C}^{1,+}(B^+_{\ell^n_\infty})$, $g(0) = 0$, $Dg(0) = 0$, $g(e_j) \geq \varepsilon - |Df(0)[e_j]| \geq \frac{\varepsilon}{2}$ for each $j \in B$, and Dg has modulus ω.

Let $N = |B| \geq n(\omega, \frac{\varepsilon}{2})$. Finally, we define a symmetric function g_s on $B^+_{\ell^N_\infty}$ by

$$g_s\left(\sum_{j=1}^{N} a_j e_j\right) = \frac{1}{N!} \sum_{\pi \in S_N} g\left(\sum_{j=1}^{N} a_j e_{\pi(j)}\right),$$

where S_N is the set of all bijections $\{1, \dots, N\} \to B$. Then g_s has a uniformly continuous derivative with modulus ω, $g_s(0) = 0$, $Dg_s(0) = 0$, and $g_s(e_j) \geq \frac{\varepsilon}{2}$ for $1 \leq j \leq N$. (In fact, g_s is a convex combination of functions which satisfy these conditions.) This contradicts Lemma 26. □

The following slightly more general formulation is also useful. Given an ordinal α we denote by $C_0([1, \alpha])$ the subspace of $C([1, \alpha])$ consisting of functions that are zero at α. We say that the functions $f, g \in C([1, \alpha])$ are disjointly supported if $f(\gamma)g(\gamma) = 0$ for every $\gamma \in [1, \alpha]$.

Lemma 28 ([Há5]). *For each $\omega \in \mathcal{M}$ with $\omega(2) \in \mathbb{R}$, $L > 0$, and $\varepsilon > 0$ there is $M(\omega, L, \varepsilon) \in \mathbb{N}$ such that if α is an ordinal, $X = C([1, \alpha])$ or $X = C_0([1, \alpha])$, $f \in \mathcal{C}^{1,+}(B_X)$ is an L-Lipschitz function such that Df has modulus of continuity ω, $N \geq M(\omega, L, \varepsilon)$, and if $v \in B_X$ and $\{u_j\}_{j=1}^N \subset X$ is a disjointly supported sequence such that $\{v + u_j\} \subset B_X$, then $|f(v + u_j) - f(v)| < \varepsilon$ for some $j \in \{1, \dots, N\}$.*

Proof. We set $\omega_1(t) = 2\omega(2t)$ and $M(\omega, L, \varepsilon) = N(\omega_1, 2L, \varepsilon)$, where $N(\cdot, \cdot, \cdot)$ is the function from Lemma 27. Assume that $N \geq M(\omega, L, \varepsilon)$ and $f \in \mathcal{C}^{1,+}(B_X)$, $v \in B_X$, and $\{u_j\}_{j=1}^N \subset X$ satisfy the assumptions of the lemma. We define an affine operator $T: \ell_\infty^N \to X$ by $T\big(\sum_{j=1}^N a_j e_j\big) = v + \sum_{j=1}^N a_j u_j$. Then T is 2-Lipschitz, $T(B^+_{\ell_\infty^N}) \subset B_X$, $T(0) = v$, and $T(e_j) = v + u_j$. Further, $g = f \circ T \in \mathcal{C}^{1,+}(B^+_{\ell_\infty^N})$ (Proposition 4), g is $2L$-Lipschitz, and Dg has modulus of continuity ω_1. So by Lemma 27 there is $j \in \{1, \dots, N\}$ such that $|f(v+u_j) - f(v)| = |g(e_j) - g(0)| < \varepsilon$. □

Given $f \in \mathcal{C}^{1,+}(B_{C([1,\alpha])})$, the preceding lemmata allow us to perturb a vector in the domain of f by a vector with large norm essentially not affecting the value of f. Note that Lemma 27 does not have a general vector-valued analogue (consider the identity on c_0). Nevertheless, we can apply it at least "componentwise" as in the following result.

Lemma 29. *Let $X = C([1,\alpha])$ or $X = C_0([1,\alpha])$, where α is an infinite countable ordinal, and let $\{f_n\} \subset \mathcal{C}^{1,+}(B_X)$ be an equi-Lipschitz sequence such that all the mappings Df_n have the same finite modulus of continuity. Further, let $\{x_n\} \subset B_X$ be a weakly Cauchy sequence and $\varepsilon > 0$. Then there are an increasing sequence $\{n_k\} \subset \mathbb{N}$ such that n_k and k have the same parity for each $k \in \mathbb{N}$, a function $v \in B_X$ with a finite range, and a sequence $\{u_k\} \subset X$ of disjointly supported functions such that $\{v + u_k\} \subset B_X$ and $|f_{n_k}(v + u_k) - f_{n_k}(x_{n_k})| < \varepsilon$ for each $k \in \mathbb{N}$.*

Proof. Let $L > 0$ be the common Lipschitz constant of f_n, $n \in \mathbb{N}$. Since α is countable, we can fix a system $\{\varepsilon_\beta\}_{\beta \le \alpha}$ of positive numbers such that $\sum_{\beta \le \alpha} \varepsilon_\beta < \varepsilon$. The sequence $\{x_n\}$ converges pointwise to $F \in B_{\ell_\infty([1,\alpha])}$ which satisfies additionally $F(\alpha) = 0$ in case that $X = C_0([1,\alpha])$. Using a finite induction we are going to construct the following objects:

(i) a decreasing sequence of ordinals $\alpha = \beta_1 > \beta_2 > \dots > \beta_m = 0$;
(ii) sets $\mathbb{N} = M_1 \supset M_2 \supset \dots \supset M_m$ such that each M_j contains infinitely many odd and even numbers;
(iii) $v \in B_X$ defined as $v(\gamma) = F(\beta_j)$ for $\gamma \in [\beta_{j+1} + 1, \beta_j]$, $j = 1, \dots, m-1$;
(iv) a system of sequences $\{x_n^j\}_{n \in M_j} \subset B_X$, $j = 1, \dots, m$, satisfying
 (a) $x_n^1 = x_n$ for all $n \in \mathbb{N} = M_1$,
 (b) x_n^{j+1} may differ from x_n^j only on $[\beta_{j+1} + 1, \beta_j]$ (and consequently x_n^{j+1} agrees with x_n on $[1, \beta_{j+1}]$) for each $n \in M_{j+1}$ and $j = 1, \dots, m-1$,
 (c) for each $j \in \{1, \dots, m\}$ the system $\{\operatorname{supp}_o(x_n^j - v) \cap [\beta_j + 1, \alpha]\}_{n \in M_j}$ is pairwise disjoint,
 (d) $\big|f_n(x_n^{j+1}) - f_n(x_n^j)\big| < \varepsilon_{\beta_j}$ for all $n \in M_{j+1}$, $j = 1, \dots, m-1$.

Note that since the intervals $[\beta_{j+1} + 1, \beta_j]$ are clopen, (iii) defines a function in X once all the β_j are defined.

Set $\beta_1 = \alpha$, $M_1 = \mathbb{N}$, and $x_n^1 = x_n$ for $n \in \mathbb{N}$. We proceed by induction. Notice that since $[0, \alpha]$ is well-ordered, the induction will stop after finitely many steps reaching $\beta_m = 0$. Now assume that β_j, M_j, and $\{x_n^j\}_{n \in M_j}$ are already defined. If β_j is non-limit, then we set β_{j+1} to be the predecessor of β_j, find $M_{j+1} \subset M_j$ satisfying (ii) such that $|x_n(\beta_j) - F(\beta_j)| < \varepsilon_{\beta_j}/L$ for all $n \in M_{j+1}$, and finally set $x_n^{j+1}(\gamma) = x_n^j(\gamma)$

for $\gamma \neq \beta_j$, and $x_n^{j+1}(\beta_j) = v(\beta_j) = F(\beta_j)$. Condition (c) holds by the inductive hypothesis. Condition (d) holds, since $\|x_n^{j+1} - x_n^j\| = |x_n^{j+1}(\beta_j) - x_n^j(\beta_j)| = |F(\beta_j) - x_n(\beta_j)| < \varepsilon_{\beta_j}/L$ for each $n \in M_{j+1}$ by (b) of the inductive hypothesis.

Assume now that β_j is a limit ordinal. For $\zeta < \eta < \beta_j$ we define continuous affine operators $P_{\zeta,\eta}\colon X \to X$ by

$$P_{\zeta,\eta}(x) = (1 - \chi_{[\zeta+1,\eta]}) \cdot x + F(\beta_j)\chi_{[\zeta+1,\eta]}.$$

Note that $P_{\zeta,\eta}(B_X) \subset B_X$. We claim that there is $\beta_{j+1} < \beta_j$ such that for each $\zeta \in (\beta_{j+1}, \beta_j)$ the set

$$\{n \in M_j;\ \big|f_n\big(P_{\beta_{j+1},\zeta}(x_n^j)\big) - f_n(x_n^j)\big| < \varepsilon_{\beta_j}/2\}$$

contains infinitely many odd numbers and infinitely many even numbers. Assuming the contrary, fix an arbitrary $N \in \mathbb{N}$. Since β_j is a limit ordinal, there are ordinals $\zeta_1 < \eta_1 \le \zeta_2 < \eta_2 \le \cdots \le \zeta_{2N} < \eta_{2N} < \beta_j$ such that for each $l \in \{1, \dots, 2N\}$ the set $\{n \in M_j;\ \big|f_n\big(P_{\zeta_l,\eta_l}(x_n^j)\big) - f_n(x_n^j)\big| < \varepsilon_{\beta_j}/2\}$ contains only finitely many odd numbers or only finitely many even numbers. By discarding some of the ζ_l, η_l and relabelling we may assume without loss of generality that for each $l \in \{1, \dots, N\}$ the set $\{n \in M_j;\ \big|f_n\big(P_{\zeta_l,\eta_l}(x_n^j)\big) - f_n(x_n^j)\big| < \varepsilon_{\beta_j}/2\}$ contains only finitely many odd numbers. Thus, using (ii) of the inductive hypothesis, there is $n \in M_j$ such that $\big|f_n\big(P_{\zeta_l,\eta_l}(x_n^j)\big) - f_n(x_n^j)\big| \ge \varepsilon_{\beta_j}/2$ for each $l \in \{1, \dots, N\}$. Put $y = x_n^j$ and $w_l = \chi_{[\zeta_l+1,\eta_l]} \cdot (F(\beta_j) - y)$, $l = 1, \dots, N$. Then $\{w_l\}$ are disjointly supported, $y + w_l \in B_X$, and $|f_n(y + w_l) - f_n(y)| \ge \varepsilon_{\beta_j}/2$, $l = 1, \dots, N$. Consequently for N large enough this contradicts Lemma 28.

We continue by defining M_{j+1} and $\{x_n^{j+1}\}_{n \in M_{j+1}}$. To this end we construct increasing sequences $\{k_l\} \subset M_j$ and $\{\zeta_l\} \subset (\beta_{j+1}, \beta_j)$ by induction on l. Let $k_1 \in M_j$ be odd such that $|x_n^j(\beta_j) - F(\beta_j)| < \varepsilon_{\beta_j}/(2L)$ for all $n \in M_j$, $n \ge k_1$ (notice that $x_n^j(\beta_j) = x_n(\beta_j)$ by (b) of the inductive hypothesis). Using the continuity of $x_{k_1}^j$ at β_j we find $\zeta_1 \in (\beta_{j+1}, \beta_j)$ such that $x_{k_1}^{j+1} = P_{\zeta_1,\beta_j}(x_{k_1}^j)$ satisfies $\|x_{k_1}^{j+1} - x_{k_1}^j\| < \varepsilon_{\beta_j}/L$. (This is possible since β_j is a limit ordinal.) Then (b) and (d) clearly hold for $n = k_1$.

Now assume that k_l and ζ_l have been defined for some $l \in \mathbb{N}$. By the above claim there is $k_{l+1} \in M_j$, $k_{l+1} > k_l$ such that k_{l+1} has the same parity as $l + 1$ and

$$\big|f_{k_{l+1}}\big(P_{\beta_{j+1},\zeta_l}(x_{k_{l+1}}^j)\big) - f_{k_{l+1}}(x_{k_{l+1}}^j)\big| < \varepsilon_{\beta_j}/2.$$

Note that $\big|P_{\beta_{j+1},\zeta_l}(x_{k_{l+1}}^j)(\beta_j) - F(\beta_j)\big| < \varepsilon_{\beta_j}/(2L)$. Since $P_{\beta_{j+1},\zeta_l}(x_{k_{l+1}}^j)$ is continuous at β_j, we can find $\zeta_{l+1} \in (\zeta_l, \beta_j)$ such that $x_{k_{l+1}}^{j+1} = P_{\zeta_{l+1},\beta_j}\big(P_{\beta_{j+1},\zeta_l}(x_{k_{l+1}}^j)\big)$ satisfies $\big\|x_{k_{l+1}}^{j+1} - P_{\beta_{j+1},\zeta_l}(x_{k_{l+1}}^j)\big\| < \varepsilon_{\beta_j}/(2L)$. Then (b) and (d) clearly hold for $n = k_{l+1}$. Finally, we let $M_{j+1} = \{k_l\}_{l \in \mathbb{N}}$ and note that (c) holds by the construction above, since $\operatorname{supp}_{\text{o}}(x_{k_{l+1}}^{j+1} - v) \cap [\beta_{j+1} + 1, \beta_j] \subset [\zeta_l + 1, \zeta_{l+1}]$, and by (b) and (c) of the inductive hypothesis.

This completes the construction of all the aforementioned objects. To finish the proof we put $u_k = x^m_{n_k} - v$, where $\{n_k\} \subset M_m$ is an increasing sequence such that n_k and k have the same parity. □

Theorem 30 ([Há5], [Há6]). *Let α be a countable ordinal. Then the spaces $C([1,\alpha])$ and $C_0([1,\alpha])$ are $\mathcal{W}_1$-spaces. In particular, c_0 is a $\mathcal{W}_1$-space.*

Proof. Let $X = C([1,\alpha])$ or $X = C_0([1,\alpha])$, where α is an infinite countable ordinal. By contrary, assume that there is $f \in \mathcal{C}^{1,+}(B_X)\setminus\mathcal{C}_{\mathrm{wsC}}(B_X)$. Without loss of generality (using shifting and scaling in the range) we may assume that there is a weakly Cauchy sequence $\{x_n\} \subset B_X$ such that $f(x_{2k-1}) \le -2$ and $f(x_{2k}) \ge 2$ for all $k \in \mathbb{N}$. Put $f_n = f$, $n \in \mathbb{N}$. By Lemma 29 there are a function $v \in B_X$ and a sequence $\{u_k\} \subset X$ of disjointly supported functions such that $\{v + u_k\} \subset B_X$ and $f(v + u_k) < -1$ for k odd, $f(v + u_k) > 1$ for k even. If $f(v) < 0$, then we set $w_n = u_{2n}$, otherwise we set $w_n = u_{2n-1}$, $n \in \mathbb{N}$. Then $|f(v + w_n) - f(v)| > 1$ for each $n \in \mathbb{N}$, a contradiction with Lemma 28. □

Corollary 31 ([Há6]). *Let K be a scattered compact space. Then the space $C(K)$ is a $\mathcal{W}_1$-space.*

Proof. By Proposition 5.124 every separable subspace of $C(K)$, K scattered, is contained in a subspace of $C(K)$ isometric to $C([1,\alpha])$, where α is a countable ordinal. Fact 20 combined with Theorem 30 finishes the proof. □

Theorem 32 ([Há6]). *The (original) Tsirelson space T^* is a $\mathcal{W}_{\frac{1}{5}}$-space.*

Proof. Let $f \in \mathcal{C}^{1,+}(B_{T^*})$ and let $\{x_n\} \subset \frac{1}{5}B_{T^*}$ be weakly Cauchy. Since T^* is reflexive, there is $v \in \frac{1}{5}B_{T^*}$ such that $x_n \xrightarrow{w} v$. We show that $f(x_n) \to f(v)$. If this is not the case, then we may assume by passing to a subsequence that there is $\varepsilon > 0$ such that $|f(x_n) - f(v)| \ge 2\varepsilon$ for all $n \in \mathbb{N}$. Put $y_n = x_n - v$. Then $y_n \in \frac{2}{5}B_{T^*}$ and $y_n \xrightarrow{w} 0$. Let $L > 0$ be such that f is L-Lipschitz and put $\delta = \frac{\varepsilon}{2L}$. By Proposition 3.32 we may assume that there is a block basic sequence $\{u_n\}$ of the canonical basis of T^* such that $\|u_n - y_n\| < \delta$ for each $n \in \mathbb{N}$. By scaling the $\{u_n\}$ if necessary we get $\|u_n\| \le \frac{2}{5}$ and $\|u_n - y_n\| < 2\delta$. So $|f(v+u_n) - f(v)| \ge 2\varepsilon - 2L\delta = \varepsilon$ for all $n \in \mathbb{N}$. By [CasaSh, p. 16] we have

$$\left\|\sum_{j=1}^{N} a_j u_{N+j}\right\| \le 2 \max_{1\le j\le N} |a_j| \cdot \|u_{N+j}\| \le \frac{4}{5} \max_{1\le j\le N} |a_j|$$

for all scalars $a_1, \dots, a_N$ and for every $N \in \mathbb{N}$. In particular, the continuous affine operators $S_N \colon \ell_\infty^N \to T^*$ given by $S_N\big(\sum_{j=1}^N a_j e_j\big) = v + \sum_{j=1}^N a_j u_{N+j}$ are $\frac{4}{5}$-Lipschitz and satisfy $S_N(B_{\ell_\infty^N}) \subset B_{T^*}$, $S_N(0) = v$, and $S_N(e_j) = v + u_{N+j}$. Applying Lemma 27 to $f \circ S_N$ (using also Proposition 4) we arrive at a contradiction for N large enough. □

Being a $\mathcal{W}$-space is invariant under isomorphisms, but the precise value of λ (such that X is a $\mathcal{W}_\lambda$-space) may depend on the norm.

Proposition 33 ([D'H]). *For every $\lambda \in (0,1)$ there is an equivalent norm $\|\cdot\|_\lambda$ on c_0 such that $(c_0, \|\cdot\|_\lambda)$ is a $\mathcal{W}_\lambda$-space but it is not a $\mathcal{W}_\xi$-space for any $\xi > \lambda$. In fact if we denote by B_λ the unit ball of $(c_0, \|\cdot\|_\lambda)$, then there is a function $\Phi_{\lambda,\xi} \in \mathcal{C}^\infty(B_\lambda)$ such that $\Phi_{\lambda,\xi} \notin \mathcal{C}_{\mathrm{wsc}}(\xi B_\lambda)$.*

Proof. Let $\{(e_j; f_j)\}$ be the canonical basis of c_0 and let $\lambda \in (0,1)$. The renorming $\|\cdot\|_\lambda$ of c_0 is determined by its closed unit ball

$$B_\lambda = \overline{\mathrm{conv}}\big(B_{c_0} \cup \{\pm \tfrac{1}{\lambda} e_j;\ j \in \mathbb{N}\}\big).$$

Clearly, $B_{c_0} \subset B_\lambda \subset \frac{1}{\lambda} B_{c_0}$, and therefore $\|\cdot\|_\lambda$ is an equivalent norm on c_0 and $(c_0, \|\cdot\|_\lambda)$ is a $\mathcal{W}_\lambda$-space. Fix $\xi \in (\lambda, 1]$ and $\beta \in (1, \frac{\xi}{\lambda})$. For every $x = (x_j) \in B_\lambda$ we denote $A_x = \{j \in \mathbb{N};\ |x_j| \geq \beta\}$. We claim that $|A_x| \leq \frac{1}{(\beta-1)\lambda}$. Indeed, choose an arbitrary $\alpha \in (1, \beta)$ and let $z = (z_j) \in c_0$ be such that $\|x - z\| < \beta - \alpha$ and $z = a_0 y + \sum_{k=1}^n a_k \frac{1}{\lambda} e_k$, where $\sum_{k=0}^n |a_k| \leq 1$ and $y = (y_j) \in B_{c_0}$. Let $C = \{k;\ |a_k| \geq (\alpha - 1)\lambda\}$. Then $|C| \leq \frac{1}{(\alpha-1)\lambda}$. Clearly $|z_j| \leq 1$ for $j > n$. If $j \leq n$, then $|z_j| = |a_0 y_j + a_j \frac{1}{\lambda}| \leq |a_0| + |a_j| \frac{1}{\lambda} < \alpha$, unless $j \in C$. Therefore $A_x \subset C$, and since α can be chosen arbitrarily close to β, the estimate follows.

Let $\varphi \in C^\infty(\mathbb{R})$ be a function that has all derivatives bounded and satisfies $\varphi([-\beta, \beta]) = \{0\}$ and $\varphi(\frac{\xi}{\lambda}) > 0$. We set $\Phi_{\lambda,\xi} = \sum_{j=1}^\infty \varphi \circ f_j$. Let $x \in B_\lambda$ and $\gamma = \max_{j \notin A_x} |x_j| < \beta$. For any $y \in B_\lambda$ with $\|y - x\| < \beta - \gamma$ we have $A_y \subset A_x$ and so $\Phi_{\lambda,\xi}(y) = \sum_{j \in A_x} \varphi \circ f_j(y)$. It follows that $\Phi_{\lambda,\xi}$ is C^∞-smooth on $\operatorname{Int} B_\lambda$ and it has all derivatives bounded and thus also Lipschitz. Consequently, $\Phi_{\lambda,\xi} \in \mathcal{C}^\infty(B_\lambda)$. Finally, $\Phi_{\lambda,\xi}(\frac{\xi}{\lambda} e_k) = \varphi(\frac{\xi}{\lambda}) > 0$ for each $k \in \mathbb{N}$, while $\Phi_{\lambda,\xi}(0) = 0$, and so $\Phi_{\lambda,\xi} \notin \mathcal{C}_{\mathrm{wsc}}(\xi B_\lambda)$. □

Let X be a $\mathcal{W}_\lambda$-space. The change of the constant λ caused by the renorming of the space is intimately connected with the non-existence of extensions of functions with uniformly continuous derivatives from the unit ball into its larger multiple. It is easy to see that the functions $\Phi_{\lambda,\xi}$ from Proposition 33 cannot be extended to functions belonging to $\mathcal{C}^{1,+}(\frac{\xi}{\lambda} B_\lambda)$. More precisely, there is no function $g \in \mathcal{C}^{1,+}(\frac{\xi}{\lambda} B_\lambda)$ such that $g \restriction_{\xi B_\lambda} = \Phi_{\lambda,\xi}$, since $\mathcal{C}^{1,+}(\frac{\xi}{\lambda} B_\lambda) \subset \mathcal{C}_{\mathrm{wsC}}(\xi B_\lambda)$. We give an extreme example of this phenomenon next.

Theorem 34 ([D'H]). *There are a separable Banach space X and $f \in \mathcal{C}^\infty(B_X)$ such that for any $\lambda > 1$ no function $g \in \mathcal{C}^{1,+}(\lambda B_X)$ agrees with f on a neighbourhood of the origin.*

Proof. Let $\|\cdot\|_\lambda$ and $\Phi_{\lambda,\xi}$ be the norm and the functions from Proposition 33. We put $X = \big(\bigoplus_{n=2}^\infty (c_0, \|\cdot\|_{\frac{1}{n}})\big)_2$. Let $P_n \colon X \to (c_0, \|\cdot\|_{\frac{1}{n}})$ be the canonical projections onto the direct summands. Put $\Psi_n = c_n \Phi_{\frac{1}{n}, \frac{1}{n-1}}$, where the constants c_n are chosen so that $\|d^j \Psi_n(x)\|_{\frac{1}{n}} \leq \frac{1}{2^n}$ for all $x \in B_{\frac{1}{n}}$, $j \in \{1, \dots, n\}$. Finally, set $f = \sum_{n=2}^\infty \Psi_n \circ P_n$. Since $d^k(\Psi_n \circ P_n)(x) = d^k \Psi_n(P_n(x)) \circ P_n$ for $x \in \operatorname{Int} B_X$ (Corollary 1.117), Theorem 1.85 implies that $f \in C^\infty(\operatorname{Int} B_X)$ with all derivatives bounded and hence also Lipschitz. Thus $f \in \mathcal{C}^\infty(B_X)$. Now suppose that $\lambda > 1$ and $g \in \mathcal{C}^{1,+}(\lambda B_X)$ is such that $g = f$ on ηB_X for some $\eta < 1$. If $m \in \mathbb{N}$ is such that $\frac{1}{m-1} \leq \eta$ and $\frac{m}{m-1} \leq \lambda$, then $g \restriction_{\frac{m}{m-1} P_m(B_X)} \in \mathcal{C}^{1,+}(\frac{m}{m-1} B_{\frac{1}{m}})$ and $g \restriction_{\frac{1}{m-1} P_m(B_X)} = \Psi_m$, a contradiction with the remark preceding the theorem. □

There are strong structural limitations for Banach spaces to belong to the class $\mathcal{W}$, especially if they do not contain ℓ_1. The following property has been isolated by Sean Michael Bates in [Bat] in his work on C^∞-smooth surjective mappings between Banach spaces.

Definition 35. We say that a Banach space X has property $\mathcal{B}$ if X^* contains a normalised weakly null hereditarily Banach-Saks sequence.

Note that a Banach space contains a normalised weakly null sequence if and only if it is not a Schur space. In particular, if X is an infinite-dimensional Banach space such that X^* has the Banach-Saks property (or more generally if X^* is not a Schur space and has the weak Banach-Saks property), then X has property $\mathcal{B}$.

Using an argument due to E. Odell and H. Rosenthal, S. M. Bates has shown the following characterisation, whose proof is omitted.

Lemma 36 ([Bat]). *A Banach space X has property $\mathcal{B}$ if and only if X^* contains a normalised sequence $\{f_k\}_{k=1}^\infty$ such that for every $\varepsilon > 0$ there exists $k(\varepsilon) \in \mathbb{N}$ such that for each $x \in B_X$*

$$\left|\{k \in \mathbb{N};\ |f_k(x)| > \varepsilon\}\right| \leq k(\varepsilon).$$

Proposition 37 ([Bat]). *Let X be a Banach space with property $\mathcal{B}$ that does not contain ℓ_1. Then X does not belong to the class $\mathcal{W}$.*

Proof. Let $\{f_k\}_{k=1}^\infty \subset X^*$ be the sequence from Lemma 36. Then $\{f_k\}$ is not relatively compact. Indeed, assuming the contrary some subsequence of $\{f_k\}$ converges to $f \in S_{X^*}$. Choosing $y \in B_X$ with $f(y) > \frac{1}{2}$ and a subsequence $\{f_{n_k}\}$ such that $\|f_{n_k} - f\| < \frac{1}{4}$ for all $k \in \mathbb{N}$ we obtain $|f_{n_k}(y)| \geq |f(y)| - \|f - f_{n_k}\| > \frac{1}{4}$ for all $k \in \mathbb{N}$, a contradiction. Therefore by passing to a subsequence of $\{f_k\}$ using Theorem 3.56 we may assume that there exists a semi-normalised weakly null sequence $\{x_k\}_{k=1}^\infty \subset X$ biorthogonal to $\{f_k\}$.

Choose $\varphi \in C^\infty(\mathbb{R})$ satisfying $\varphi(t) = 0$ for $t \in [-1, 1]$ and $\varphi(t) = 1$ for $|t| \geq 2$, and let $f\colon X \to \mathbb{R}$ be defined by

$$f(x) = \sum_{k=1}^\infty \varphi\big(f_k(x)\big).$$

For a fixed $x \in B(0, R)$ put $A = \{k \in \mathbb{N};\ |f_k(x)| > \frac{1}{2}\}$. Then $|A| \leq k(\frac{1}{2R})$. For any $y \in B(x, \frac{1}{2})$ we have $\{k \in \mathbb{N};\ |f_k(y)| > 1\} \subset A$ and so $f(y) = \sum_{k\in A} \varphi(f_k(y))$. Consequently, f is C^∞-smooth on X and it has all derivatives bounded on bounded sets, i.e. $f \in \mathcal{C}^\infty(X)$. However, $\lim_{k\to\infty} f(2x_k) = 1 \neq 0 = f(0)$, so X is not a $\mathcal{W}$-space. □

In connection with the previous result note that the Tsirelson space T of T. Figiel and W. B. Johnson does not have the Banach-Saks property, but T^* (the original Tsirelson space) has the Banach-Saks property ([CasaSh, p. 18]) and so T has property $\mathcal{B}$, while T^* is a $\mathcal{W}$-space (Theorem 32).

Proposition 38. *Let X be a Banach space with a sub-symmetric Schauder basis. Then X has property $\mathcal{B}$ if and only if X is not isomorphic to c_0.*

Proof. $\Rightarrow$ If X is isomorphic to c_0, then X^* is a Schur space and so there is no normalised weakly null sequence in X^*.

$\Leftarrow$ Let $\{(e_k; f_k)\}$ be the sub-symmetric Schauder basis of X. Since a sub-symmetric basis is semi-normalised and unconditional, we may assume without loss of generality that $\{f_k\}$ is normalised. Suppose that X does not have property $\mathcal{B}$. By Lemma 36 there are $\varepsilon > 0$ and a sequence $\{u_n\} \subset B_X$ such that $|\{k \in \mathbb{N};\ |f_k(u_n)| > \varepsilon\}| > n$. Given $n \in \mathbb{N}$ we find $A = \{k \in \mathbb{N};\ |f_k(u_n)| > \varepsilon\}$ such that $|A| = n$ and put $v_n = \sum_{k \in A} e_k$. Then $\|v_n\| \le C\|\frac{1}{\varepsilon}u_n\| \le \frac{C}{\varepsilon}$, where C is the unconditional basis constant of $\{e_k\}$. Let $K \ge 1$ be such that $\{e_k\}$ is K-spreading. Then $\left\|\sum_{k=1}^n e_k\right\| \le K\|v_n\| \le K\frac{C}{\varepsilon}$ and therefore

$$\left\|\sum_{k=1}^n a_k e_k\right\| \le C \max_j |a_j| \left\|\sum_{k=1}^n e_k\right\| \le K\frac{C^2}{\varepsilon} \max_j |a_j|$$

for any $a_1, \dots, a_n \in \mathbb{R}$. The reverse inequality is automatic: for each $j \in \{1, \dots, n\}$ we have $\left\|\sum_{k=1}^n a_k e_k\right\| \ge \left|f_j\left(\sum_{k=1}^n a_k e_k\right)\right| = |a_j|$. Hence $\{e_k\}$ is equivalent to the canonical basis of c_0, a contradiction. □

Example 39 (Schreier's space)**.** Schreier's space S is defined as a completion of c_{00} under the norm

$$\|(x_j)\| = \sup_{A \text{ admissible}} \sum_{j \in A} |x_j|,$$

where $A \subset \mathbb{N}$ is admissible if $A = \{n_1, \dots, n_k\}$, where $k \le n_1 < n_2 < \dots < n_k$. It is obvious from the definition that the canonical basis vectors $\{e_j\}$ form a normalised 1-unconditional basis of S. The space S has many important properties (it was the first constructed example without the weak Banach-Saks property, [Schr]) of which we only mention the following ones: The space S is isometric to a subspace of some $C(K)$, K countable ([CaGo]), hence S is isomorphically polyhedral and it is a subspace of a $\mathcal{W}_1$-space (Corollary 31).

Denote by $\{f_k\} \subset S^*$ the functionals biorthogonal to $\{e_k\}$. Obviously $\{f_k\}$ is normalised. We claim that $\{f_k\}$ satisfies the condition in Lemma 36 with $k(\varepsilon) = 2\lceil\frac{1}{\varepsilon}\rceil$ and so S has property $\mathcal{B}$. Indeed, fix $\varepsilon > 0$, $x \in B_S$, and put $m = \lceil\frac{1}{\varepsilon}\rceil$. Then $\left|\{k \ge m;\ |x_k| > \varepsilon\}\right| < m$, otherwise $1 \ge \|x\| \ge \sum_{k \in A}|x_k| > m\varepsilon \ge 1$, where A consists of some m members of the set $\{k \ge m;\ |x_k| > \varepsilon\}$. Since S does not contain ℓ_1, [CasaSh, Theorem 0.5], S is not a $\mathcal{W}$-space (Proposition 37). Thus S is a witness to the fact that the class $\mathcal{W}$ is not closed with respect to subspaces. Similarly, as S is c_0-saturated (Theorem 5.103, Corollary 5.92), property $\mathcal{B}$ does not pass to subspaces.

Note also that S^* contains ℓ_1 by the lifting property of ℓ_1 and hence does not have the Banach-Saks property.

We close this section by an example of a Banach space that does not belong to the class $\mathcal{W}$, but such that all of its scalar valued polynomials are still weakly continuous (see Fact 22). Let T^* be the (original) Tsirelson space and let $\{f_k\}$ be its canonical basis. Recall that the symmetrised Tsirelson space $S(T^*)$ is a reflexive Banach space with a symmetric basis $\{e_k\}$ whose norm is given by the formula

$$\left\|\sum_{k=1}^{\infty} a_k e_k\right\| = \sup_{\pi \in S} \left\|\sum_{k=1}^{\infty} a_{\pi(k)} f_k\right\|_{T^*},$$

where S is the set of all permutations of $\mathbb{N}$.

Theorem 40 ([CHL]). *The symmetrised Tsirelson space $S(T^*)$ has property $\mathcal{B}$ (in particular, it is not a $\mathcal{W}$-space), although $\mathcal{P}(S(T^*)) = \mathcal{P}_{\mathrm{wu}}(S(T^*))$.*

Proof. The space $S(T^*)$ has property $\mathcal{B}$ by Proposition 38, so it is not a $\mathcal{W}$-space by Proposition 37.

By [CasaSh, Proposition V.10] the canonical basis of T^* has an upper p-estimate for every $p < \infty$ and so by the definition of the norm in $S(T^*)$ the same holds also for the canonical basis $\{e_k\}$ of $S(T^*)$. By [CasaSh, Proposition X.b.7] there is $K > 0$ such that

$$\left\|\sum_{k=1}^{n} a_k u_k\right\| \le K \left\|\sum_{k=1}^{n} a_k e_k\right\|$$

for every normalised sequence $\{u_k\} \subset S(T^*)$ of disjointly supported vectors and all scalars $a_1, \dots, a_n$, so each bounded block basis of $\{e_k\}$ has an upper p-estimate for every $p < \infty$. Thus $S(T^*)$ has property S_p for every $p < \infty$ by Proposition 3.32 and we may apply Corollary 3.58 (together with the separability of $S(T^*)$). □

4. Uniformly smooth mappings from $C(K)$, K scattered

In the present section we are going to investigate uniformly smooth mappings from $X = C(K)$ for K scattered. The results extend the scalar valued theory from the previous section. In particular, under the assumption that the range of $f \in \mathcal{C}^{1,+}(B_X; Y)$ is relatively weakly compact, or alternatively Y has property (WR), we prove that f maps weakly Cauchy sequences to norm convergent ones.

Let K be a compact space. Recall that from the Riesz representation theorem and the Lebesgue dominated convergence theorem it immediately follows that a sequence $\{\psi_n\} \subset C(K)$ is weakly convergent to $\psi \in C(K)$ if and only if $\{\psi_n\}$ is bounded and $\{\psi_n(x)\}$ converges to $\psi(x)$ for each $x \in K$, and $\{\psi_n\}$ is weakly Cauchy if and only if it is bounded and $\{\psi_n(x)\}$ is convergent for each $x \in K$.

Lemma 29 has the following direct consequence (compare also with Lemma 3.31):

Lemma 41 ([DeH2]). *Let $X = C([1,\alpha])$ or $X = C_0([1,\alpha])$, where α is a countable ordinal, and let Y be a Banach space. If $f \in \mathcal{C}^{1,+}(B_X;Y) \setminus \mathcal{C}_K(B_X;Y)$, then there exist a function $v \in B_X$ with a finite range and a weakly null sequence $\{u_n\} \subset X$ of disjointly supported functions such that $\{v+u_n\} \subset B_X$ and $\{f(v+u_n)\}$ is not relatively compact in Y.*

Proof. Since X is separable, by passing to $\overline{\operatorname{span}}\, f(B_X)$ we may assume without loss of generality that Y is separable as well. There is a sequence $\{x_n\} \subset B_X$ such that $\{f(x_n)\}$ is ε-separated for some $\varepsilon > 0$. By Rosenthal's ℓ_1-theorem we may assume that $\{x_n\}$ is weakly Cauchy. Using Theorem 3.56 (applied to $\{f(x_n)\} \subset Y^{**}$) and passing to a subsequence we find a bounded sequence $\{\phi_n\} \subset Y^*$ such that $\{(f(x_n);\phi_n)\}$ is a biorthogonal system. Moreover, as (B_{Y^*}, w^*) is metrisable, we may assume that $\{\phi_n\}$ is w^*-convergent. By passing to a biorthogonal system $\{(f(x_{2n});\phi_{2n}-\phi_{2n-1})\}$ we may finally assume that $\{\phi_n\}$ is w^*-null.

By Lemma 29 applied to functions $f_n = \phi_n \circ f$ there are an increasing sequence $\{n_k\} \subset \mathbb{N}$, a function $v \in B_X$ with a finite range, and a sequence $\{u_k\} \subset X$ of disjointly supported functions such that $\{v+u_k\} \subset B_X$ and $f_{n_k}(v+u_k) \geq \frac{1}{2}$ for each $k \in \mathbb{N}$. The sequence $\{u_k\}$ is clearly weakly null. Since $\phi_{n_k}(f(v+u_k)) \geq \frac{1}{2}$, the set $\{f(v+u_k)\}$ is not relatively compact (Fact 4.90). □

Compare the last statement of the next theorem with Corollary 3.27.

Theorem 42 ([DeH2]). *Let $X = c_0$ or $X = C(K)$, where K is a scattered compact space, let Y be a Banach space, and $f \in \mathcal{C}^{1,+}(B_X;Y)$. Then $f \in \mathcal{C}_{\mathrm{wsC}}(B_X;Y)$ if and only if $f \restriction_Z \in \mathcal{C}_{\mathrm{wsc}}(B_Z;Y)$ for every subspace $Z \subset X$ isomorphic to c_0. In particular, $f \in \mathcal{C}_{\mathrm{wsC}}(B_X;Y)$ if and only if $f \in \mathcal{C}_{\mathrm{wsc}}(B_X;Y)$.*

Proof. The implication $\Rightarrow$ is trivial. We prove $\Leftarrow$ by contradiction. As in the proof of Corollary 31 it suffices to consider only $K = [1,\alpha]$, where α is a countable ordinal. Assuming the contrary suppose by Theorem 24 that $f \notin \mathcal{C}_K(B_X;Y)$. By Lemma 41 there are $v \in B_X$ and a weakly null sequence $\{u_n\} \subset X$ of disjointly supported functions such that $\{v+u_n\} \subset B_X$ and $\{f(v+u_n)\}$ is not relatively compact. Consequently, $Z = \overline{\operatorname{span}}(\{v\} \cup \{u_n;\ n \in \mathbb{N}\})$ is isomorphic to c_0 and $f \notin \mathcal{C}_{\mathrm{wsc}}(B_Z;Y)$. □

Lemma 43. *Let X be a Banach space with a shrinking Schauder basis $\{e_n\}$, Y a Banach space, $U \subset X$ a CCB set, and $\{f_n\} \subset \mathcal{C}_{\mathrm{wu}}(U;Y)$. Then there is an increasing sequence $\{n_k\} \subset \mathbb{N}$ such that*

$$\left\| f_{n_k}(x) - f_{n_k}\left(\sum_{j=1}^{k} a_j e_{n_j}\right) \right\| < \frac{1}{k}$$

for every $k \in \mathbb{N}$ and every $x = \sum_{j=1}^{\infty} a_j e_{n_j} \in U$ such that $\sum_{j=1}^{k} a_j e_{n_j} \in U$.

Proof. We construct the sequence $\{n_k\}$ and an auxiliary sequence of finite subsets $\Phi_k \subset X^*$ by induction. Put $n_1 = 1$. Assume that n_k is defined for some $k \in \mathbb{N}$. By the assumption there is a weak neighbourhood W of the origin in X such that $\|f_{n_k}(x) - f_{n_k}(y)\| < \frac{1}{k}$ whenever $x, y \in U$, $x - y \in W$. Since $\{e_n\}$ is shrinking and U is bounded, there is a finite set $\Phi_k \subset X^*$ of finitely supported functionals such that $x, y \in U$ satisfy $x - y \in W$ whenever $\max_{\phi\in\Phi_k} |\phi(x-y)| < 1$. We find $n_{k+1} > n_k$ such that $n_{k+1} > \max_{\phi\in\Phi_k} \max \operatorname{supp} \phi$.

Now if $x = \sum_{j=1}^{\infty} a_j e_{n_j} \in U$ and $y = \sum_{j=1}^{k} a_j e_{n_j} \in U$, then $x - y \in \ker \phi$ for each $\phi \in \Phi_k$ and so the estimate follows. □

The following is another type of a diagonalisation lemma.

Lemma 44. *Let $X = C([1,\alpha])$ or $X = C_0([1,\alpha])$, where α is an ordinal, and let $\{f_n\} \subset \mathcal{C}^{1,+}(B_X)$ be an equi-Lipschitz sequence such that all the mappings Df_n have the same finite modulus of continuity. Further, let $v \in B_X$, let $\{u_n\}_{n=1}^{\infty} \subset X$ be a disjointly supported sequence such that $\{v + u_n\} \subset B_X$, and let $\varepsilon > 0$. Then there is an increasing sequence $\{n_k\} \subset \mathbb{N}$ such that*

$$\left| f_{n_k}\left(v + \sum_{j=1}^{k} a_j u_{n_j}\right) - f_{n_k}(v + a_k u_{n_k}) \right| < \varepsilon$$

for every $k \in \mathbb{N}$ and every $a_1, \dots, a_k \in [0,1]$.

Proof. Let $L > 0$ be the common Lipschitz constant of f_n, $n \in \mathbb{N}$, let $m \in \mathbb{N}$ satisfy $L\frac{1}{m} < \frac{\varepsilon}{4}$, and set $S = \left\{0, \frac{1}{m}, \dots, \frac{m-1}{m}, 1\right\}$. We will construct the sequence $\{n_k\}_{k=1}^{\infty}$ by induction along with an auxiliary sequence $\mathbb{N} = M_0 \supset M_1 \supset M_2 \supset \cdots$ of infinite sets so that $n_k \in M_{k-1}$ and

$$\left| f_n\left(v + \sum_{j=1}^{k} a_j u_{n_j} + a u_n\right) - f_n\left(v + \sum_{j=1}^{k-1} a_j u_{n_j} + a u_n\right) \right| < \frac{\varepsilon}{2^{k+1}}$$

for every $k \in \mathbb{N}$, every $a_1, \dots, a_k, a \in S$, and every $n \in M_k$.

Set $M_0 = \mathbb{N}$, $n_0 = 0$, and assume that M_j and n_j, $j = 1, \dots, k-1$, are already defined for some $k \in \mathbb{N}$. Let ω be the modulus of continuity of the mappings Df_n, put $N = M(\omega, L, \frac{\varepsilon}{2^{k+1}})$, where the function $M(\cdot,\cdot,\cdot)$ comes from Lemma 28, and $K = |S|^{k+1}(N-1) + 1$. Further, let $I \subset M_{k-1}$ be a set of cardinality K such that $\min I > n_{k-1}$. Then for each fixed $n \in \mathbb{N}, n > \max I$ and each fixed $a_1, \dots, a_k, a \in S$ the set

$$\left\{ i \in I;\ \left| f_n\left(v + \sum_{j=1}^{k-1} a_j u_{n_j} + a_k u_i + a u_n\right) - f_n\left(v + \sum_{j=1}^{k-1} a_j u_{n_j} + a u_n\right) \right| \geq \frac{\varepsilon}{2^{k+1}} \right\}$$

has cardinality at most $N - 1$. It follows that for each $n \in \mathbb{N}$, $n > \max I$ there is $i_n \in I$ such that

$$\left| f_n\left(v + \sum_{j=1}^{k-1} a_j u_{n_j} + a_k u_{i_n} + a u_n\right) - f_n\left(v + \sum_{j=1}^{k-1} a_j u_{n_j} + a u_n\right) \right| < \frac{\varepsilon}{2^{k+1}}$$

for each $a_1, \dots, a_k, a \in S$. Consequently, there are an infinite set $M_k \subset M_{k-1}$ and $n_k \in I$ such that $i_n = n_k$ for each $n \in M_k$.

For $k \in \mathbb{N}$ and $a_1, \dots, a_k \in S$ we can estimate

$$\begin{aligned}
&\left| f_{n_k}\left(v + \sum_{j=1}^{k} a_j u_{n_j}\right) - f_{n_k}(v + a_k u_{n_k}) \right| \\
&\quad \le \sum_{l=1}^{k-1} \left| f_{n_k}\left(v + \sum_{j=1}^{k-l} a_j u_{n_j} + a_k u_{n_k}\right) - f_{n_k}\left(v + \sum_{j=1}^{k-l-1} a_j u_{n_j} + a_k u_{n_k}\right) \right| \\
&\quad < \sum_{l=1}^{k-1} \frac{\varepsilon}{2^{k-l+1}} < \frac{\varepsilon}{2},
\end{aligned}$$

since $n_k \in M_{k-l}$ for $l = 1, \dots, k-1$. Finally, because f_{n_k} is L-Lipschitz, we can pass to arbitrary $a_1, \dots, a_k \in [0, 1]$. □

Theorem 45 ([DeH2]). *Let $X = c_0$ or $X = C(K)$, where K is a scattered compact space, and let Y be a Banach space. If $f \in \mathscr{C}^{1,+}(B_X; Y) \setminus \mathscr{C}_{\mathrm{K}}(B_X; Y)$, then there exist $z \in B_{X^{**}}$ and a subspace $Z \subset X$ isometric to c_0 such that*

$$D(f^{**})(z)\restriction_Z \in \mathscr{L}(Z; Y^{**}) \setminus \mathscr{L}_{\mathrm{wK}}(Z; Y^{**}).$$

Moreover, if the Banach space Y is weakly sequentially complete, then we obtain $D(f^{**})(z)\restriction_Z \in \mathscr{L}(Z; Y) \setminus \mathscr{L}_{\mathrm{wK}}(Z; Y)$. If Y is a dual space and $P : Y^{**} \to Y$ is Dixmier's projection, then $D(P \circ f^{**})(z)\restriction_Z \in \mathscr{L}(Z; Y) \setminus \mathscr{L}_{\mathrm{wK}}(Z; Y)$.

Proof. There is a sequence $\{x_n\} \subset B_X$ such that $\{f(x_n)\}$ is not relatively compact in Y. In the case $X = C(K)$ by Proposition 5.124 there is a subspace E of X isometric to $C([1, \alpha])$ for some countable ordinal α such that $\{x_n\} \subset E$ and so $f\restriction_{B_E} \notin \mathscr{C}_{\mathrm{K}}(B_E; Y)$. Further, $(f\restriction_{B_E})^{**} = f^{**}\restriction_{B_{E^{**}}}$ (Corollary 25) and consequently $D((f\restriction_{B_E})^{**})(z) = D(f^{**}\restriction_{B_{E^{**}}})(z) = D(f^{**})(z)\restriction_{E^{**}}$ for any $z \in B_{E^{**}}$. Therefore we may assume without loss of generality that $X = c_0$ or $X = C([1, \alpha])$, where α is a countable ordinal.

By Lemma 41 there are $v \in B_X$ and a disjointly supported weakly null sequence $\{u_n\} \subset X$ such that $\{v + u_n\} \subset B_X$ and $\{f(v + u_n)\}$ is not relatively compact in Y. Without loss of generality we may assume that $f(v) = 0$. By Theorem 3.56 passing to a subsequence we may assume that there is a bounded sequence $\{\phi_n\} \subset Y^*$ biorthogonal to $\{f(v + u_n)\}$. In case that Y is a dual space we may moreover assume that $\{\phi_n\} \subset Y_*$. Denote $f_k = \phi_k \circ f$. By Lemma 44 we may assume that

$$\left| f_k\left(v + \sum_{j=1}^{k} a_j u_j\right) - f_k(v + a_k u_k) \right| < \frac{1}{4} \tag{5}$$

for every $k \in \mathbb{N}$ and every $a_1, \dots, a_k \in [0, 1]$.

Set $g_k(x) = \phi_k\big(Df(x)[u_k]\big)$, $x \in B_X$. Since X is a $\mathcal{W}_1$-space with a separable dual, $f_k \in \mathscr{C}_{\mathrm{wu}}(B_X)$ and so by Lemma 6 also $Df_k \in \mathscr{C}_{\mathrm{wu}}(B_X; X^*)$. By the Chain rule $Df_k(x)[u_k] = g_k(x)$ for $x \in B_X$. Since the operator of evaluation at u_k is a bounded linear functional on X^*, we conclude that $g_k \in \mathscr{C}_{\mathrm{wu}}(B_X)$. Now consider the continuous affine operator $T: c_0 \to X$, $T\left(\sum_{n=1}^{\infty} a_n e_n\right) = v + \sum_{n=1}^{\infty} a_n u_n$. Then T is w–w uniformly continuous and $T(B_{c_0}^+) \subset B_X$, and so $g_k \circ T \in \mathcal{C}_{\mathrm{wu}}(B_{c_0}^+)$. Using Lemma 43 by passing to a further subsequence we may assume that

$$\begin{aligned}\left| g_k\left(v + \sum_{j=1}^{n} a_j u_j\right) - g_k\left(v + \sum_{j=1}^{k} a_j u_j\right)\right| \\ = \left| g_k \circ T\left(\sum_{j=1}^{n} a_j e_j\right) - g_k \circ T\left(\sum_{j=1}^{k} a_j e_j\right)\right| < \frac{1}{4}\end{aligned} \tag{6}$$

for every $k, n \in \mathbb{N}$, $n \geq k$ and every $a_1, \ldots, a_n \in [0,1]$.

Now we construct inductively a sequence $\{a_j\} \subset [0,1]$ so that

$$g_k\left(v + \sum_{j=1}^{n} a_j u_j\right) > \frac{1}{4} \tag{7}$$

for every $k, n \in \mathbb{N}$ with $n \geq k$. Assume that $a_1, \ldots, a_{k-1}$ are already defined for some fixed $k \in \mathbb{N}$. Recall that $f_k(v) = 0$ and $f_k(v + u_k) = 1$. Therefore (5) implies that $f_k\left(v + \sum_{j=1}^{k-1} a_j u_j\right) < \frac{1}{4}$ and $f_k\left(v + \sum_{j=1}^{k-1} a_j u_j + u_k\right) > \frac{3}{4}$. Put $h(t) = f_k\left(v + \sum_{j=1}^{k-1} a_j u_j + t u_k\right)$ for $t \in [0,1]$. Then $h'(t) = g_k\left(v + \sum_{j=1}^{k-1} a_j u_j + t u_k\right)$ for $t \in (0,1)$. Thus $\int_0^1 h'(t)\,\mathrm{d}t = h(1) - h(0) > \frac{1}{2}$, and so there exists $a_k \in (0,1)$ for which $g_k\left(v + \sum_{j=1}^{k} a_j u_j\right) = h'(a_k) > \frac{1}{2}$. Combining this with (6) we obtain (7) for every $n \geq k$.

Put $z_n = v + \sum_{j=1}^{n} a_j u_j$. Since the sequence $\{z_n\}$ is weakly Cauchy in X, it is w^*-convergent in the space X^{**}. Let $z \in B_{X^{**}}$ be its w^*-limit and let us denote $L = D(f^{**})(z) \in \mathcal{L}(X^{**}; Y^{**})$. By Corollary 25 and Fact 3.8

$$\langle L(w), \phi\rangle = \lim_{n\to\infty} \langle D(f^{**})(z_n)[w], \phi\rangle = \lim_{n\to\infty} \langle \phi, Df(z_n)[w]\rangle \tag{8}$$

for every $w \in X$ and $\phi \in Y^*$. In particular, $\langle L(u_k), \phi_k\rangle = \lim_{n\to\infty} g_k(z_n)$ for every $k \in \mathbb{N}$. Therefore $\langle L(u_k), \phi_k\rangle \geq \frac{1}{4}$ by (7). On the other hand, since $\{u_n\}$ is weakly null in X and thus also in X^{**}, the sequence $\{L(u_n)\}$ is weakly null in Y^{**}, and in particular $\lim_{n\to\infty}\langle L(u_n), \phi_k\rangle = 0$ for every $k \in \mathbb{N}$. It follows (similarly as in Fact 4.90) that $\{L(u_n)\}$ is not relatively compact in Y^{**}. The subspace $Z = \overline{\operatorname{span}}\{u_n\} \subset X$ is easily seen to be isometric to c_0. Then $L\restriction_Z \notin \mathcal{L}_{\mathrm{K}}(Z; Y^{**})$ and so $L\restriction_Z \notin \mathcal{L}_{\mathrm{wK}}(Z; Y^{**})$ by Theorem 3.47.

Suppose moreover that Y is weakly sequentially complete. For a fixed $w \in X$ by (8) the sequence $\{Df(z_n)[w]\}$ is weakly Cauchy and hence weakly convergent. It follows that $D(f^{**})(z)[w] = w\text{-}\lim_{n\to\infty} Df(z_n)[w] \in Y$.

Finally, assume that Y is a dual space and let $i: Y_* \to Y^*$ be the canonical embedding. Then $\langle P(L(w)), \phi\rangle = \langle L(w), i(\phi)\rangle$ for any $w \in X^{**}$ and $\phi \in Y_*$. Because $D(P \circ f^{**})(z)[w] = P\left(D(f^{**})(z)[w]\right) = P(L(w))$ for each $w \in X^{**}$, from the above we obtain $\left\langle D(P \circ f^{**})(z)[u_k], \phi_k\right\rangle \geq \frac{1}{4}$ and $\lim_{n\to\infty}\left\langle D(P \circ f^{**})(z)[u_n], \phi_k\right\rangle = 0$. So again, $D(P \circ f^{**})(z)\restriction_Z \notin \mathcal{L}_{\mathrm{wK}}(Z; Y)$. □

We note that passing to the bidual is necessary in the previous theorem: Consider the polynomial $P \in \mathcal{P}(^2c_0; c_0)$ given by $P((x_n)) = (x_n^2)$ (Theorem 1.29). Then P is clearly not compact, but $DP(x)[h] = 2(x_n h_n)_{n=1}^{\infty}$ by Lemma 1.99, so $D(P^{**})(x)\restriction_{c_0} = DP(x) \in \mathcal{L}_K(c_0; c_0)$ for each $x \in c_0$ (Theorem 4.9(iii)).

Let us now put together the previous results.

Theorem 46. *Let $X = c_0$ or $X = C(K)$, where K is a scattered compact space, let Y be a Banach space, and $f \in \mathcal{C}^{1,+}(B_X; Y)$. The following statements are equivalent:*

(i) $f \in \mathcal{C}_{\mathrm{wu}}(B_X; Y)$.
(ii) $f \in \mathcal{C}_{\mathrm{w}}(B_X; Y)$.
(iii) $f \in \mathcal{C}_{\mathrm{wsC}}(B_X; Y)$.
(iv) $f \in \mathcal{C}_{\mathrm{wsc}}(B_X; Y)$.
(v) $f \in \mathcal{C}_{\mathrm{K}}(B_X; Y)$.
(vi) $f \in \mathcal{C}_{\mathrm{wK}}(B_X; Y)$.
(vii) $D(f^{**})(z) \in \mathcal{L}_{\mathrm{K}}(X^{**}; Y^{**})$ *for all* $z \in B_{X^{**}}$.
(viii) $D(f^{**})(z) \in \mathcal{L}_{\mathrm{wK}}(X^{**}; Y^{**})$ *for all* $z \in B_{X^{**}}$.
(ix) $f^{**} \in \mathcal{C}_{\mathrm{K}}(B_{X^{**}}; Y^{**})$.
(x) $f^{**} \in \mathcal{C}_{\mathrm{wK}}(B_{X^{**}}; Y^{**})$.

Proof. (i)⇔(iii)⇔(v) by Theorem 24, (iii)⇔(iv) by Theorem 42, and (ii)⇔(iv) by Theorem 3.79,

(v)⇒(ix) by Fact 11, (ix)⇒(vii) by Proposition 3.70, since if (vii) fails, then by the continuity there is $z \in \operatorname{Int} B_{X^{**}}$ such that $D(f^{**})(z) \notin \mathcal{L}_K(X^{**}; Y^{**})$; (vii)⇒(viii) is trivial, and (viii)⇒(v) by Theorem 45.

(v)⇒(vi) is trivial, (vi)⇒(x) by Fact 11, and (x)⇒(viii) again by Proposition 3.70. □

In order to study when the properties of the class $\mathcal{W}$ pass to the vector-valued mappings we introduce the following property:

Definition 47. We say that a Banach space Y has property (WR) if

$$\mathcal{C}^{1,+}(B_{c_0}; Y) \subset \mathcal{C}_{\mathrm{wsC}}(B_{c_0}; Y).$$

Note that by Theorem 46 the property wsC in the definition can be equivalently replaced by any of the properties wu, w, wsc, K, wK. Similarly, we can equivalently replace the space c_0 in the definition by any $C(K)$ where K is a scattered compact:

Theorem 48 ([DeH2]). *Let $X = C(K)$, where K is a scattered compact space, and let Y be a Banach space with property (WR). Then $\mathcal{C}^{1,+}(B_X; Y) \subset \mathcal{C}_{\mathrm{wu}}(B_X; Y)$.*

Proof. Assuming the contrary, by Theorem 46 there is $f \in \mathcal{C}^{1,+}(B_X; Y) \setminus \mathcal{C}_K(B_X; Y)$. As in the proof of Theorem 45 we may assume that $K = [1, \alpha]$ for some countable ordinal α. By Lemma 41 there are $v \in B_X$ and a disjointly supported sequence $\{u_n\} \subset X$ such that $\{v + u_n\} \subset B_X$ and $\{f(v + u_n)\}$ is not relatively compact in Y. Define a polynomial $P \in \mathcal{P}^2(c_0; X)$ by $P\left(\sum_{n=1}^{\infty} a_n e_n\right) = v + \sum_{n=1}^{\infty} a_n^2 u_n$ and note that $P(B_{c_0}) \subset B_X$. Then $f \circ P \in \mathcal{C}^{1,+}(B_{c_0}; Y)$, but $f \circ P(e_n) = f(v + u_n)$ and hence $f \circ P \notin \mathcal{C}_K(B_{c_0}; Y)$. This means that Y fails property (WR), a contradiction. □

Let us now pass to the description of some of the properties of the class of Banach spaces with property (WR). Clearly, property (WR) passes to subspaces and is separably determined, i.e. Y has property (WR) if and only if every separable subspace of Y has property (WR). Moreover, if Y has property (WR), then it does not contain c_0, since every bounded linear operator $T\colon c_0 \to Y$ is compact by Theorem 3.47.

Proposition 49 ([DeH2]). *Let Y be a Banach space that does not have property (WR). Then Y has the following properties:*

(i) *Y^* has a complemented subspace isomorphic to ℓ_1 and so Y^{**} contains ℓ_∞.*
(ii) *Y is not of a non-trivial cotype.*
(iii) *Y is not weakly sequentially complete.*

Proof. By Theorem 46 there is $f \in \mathcal{C}^{1,+}(B_{c_0}; Y)\setminus\mathcal{C}_{\mathrm{K}}(B_{c_0}; Y)$. By Theorem 45 there exists a non-compact operator $L \in \mathcal{L}(c_0; Y^{**})$, so Y^{**} contains c_0 by Theorem 3.47. (i) then follows from [FHHMZ, Theorem 4.44].

(ii) If Y is of a non-trivial cotype, then the same holds for Y^{**} by the principle of local reflexivity. This is not possible, since Y^{**} contains c_0.

To prove (iii), suppose that Y is weakly sequentially complete. By Theorem 45 there exists a non-compact operator $L \in \mathcal{L}(c_0; Y)$, so Y contains c_0 by Theorem 3.47. This is a contradiction, since c_0 is not weakly sequentially complete. □

We say that a Banach space X has property (u) (of Aleksander Pełczyński) if for every weakly Cauchy sequence $\{x_n\} \subset X$ there exists a weakly unconditionally Cauchy sequence $\{u_n\} \subset X$ such that $\left\{x_n - \sum_{k=1}^{n} u_k\right\}_{n=1}^{\infty}$ is weakly null. Property (u) passes to subspaces, [AK, Proposition 3.5.2]. Banach spaces with an unconditional basis, or more generally order continuous Banach lattices, have property (u), [LiTz3, Proposition 1.c.2]. We point out that a Banach space with property (u) or a complemented subspace of a Banach lattice is weakly sequentially complete if and only if it does not contain c_0, Theorem 3.46 and [LiTz3, Theorem 1.c.7].

Corollary 50 ([DeH2]). *Let Y be a Banach space that does not have property (WR) and*

- *Y is a dual space, or*
- *Y is a complemented subspace of a Banach lattice, or*
- *Y has property (u).*

Then Y contains c_0.

Proof. By Theorem 46 there is $f \in \mathcal{C}^{1,+}(B_{c_0}; Y) \setminus \mathcal{C}_{\mathrm{K}}(B_{c_0}; Y)$. If Y is a dual space, then by Theorem 45 there exists a non-compact operator $L \in \mathcal{L}(c_0; Y)$, so Y contains c_0 by Theorem 3.47. The rest follows from Proposition 49(iii) and the classical results mentioned above. □

By combining the above with [FHHMZ, Theorem 4.44] we immediately conclude the following characterisation.

Corollary 51. *A dual Banach space has property (WR) if and only if it does not contain ℓ_∞.*

Recall that a Banach space X has the point of continuity property (PCP) if for every weakly closed bounded set $C \subset X$ the mapping $Id\colon (C, w) \to (C, \|\cdot\|)$ has a point of continuity. Spaces with the PCP have been extensively studied by many authors. In particular, all spaces with the RNP belong to this class – use [FHHMZ, Theorem 11.15, Corollary 11.4] and the fact that if C is closed, then any strongly exposed point of $\overline{\operatorname{conv}}\, C$ belongs to C.

Theorem 52 ([DeH2]). *A Banach space Y has property (WR) whenever*

- *Y^{**} does not contain ℓ_∞, or*
- *Y has property (u) and does not contain c_0, or*
- *Y is weakly sequentially complete, or*
- *Y has the PCP.*

Proof. The first three cases follow immediately from Proposition 49 and Corollary 50. We proceed with the proof of the last case. Since the PCP is clearly a hereditary property and property (WR) is separably determined, we may assume without loss of generality that Y is separable. By [GhMa, Theorem II.1] there is a finite-dimensional decomposition of Y such that each of its skipped blockings is boundedly complete, i.e. there exists a sequence $\{Y_n\}$ of finite-dimensional subspaces of Y satisfying

- $Y = \overline{\operatorname{span}} \bigcup_{n=1}^{\infty} Y_n$;
- $Y_k \cap \overline{\operatorname{span}} \bigcup_{n \neq k} Y_n = \{0\}$;
- if $\{m_k\} \subset \mathbb{N}$ and $\{n_k\} \subset \mathbb{N}$ are sequences satisfying $m_k < n_k + 1 < m_{k+1}$, $k \in \mathbb{N}$, and if $\{x_k\} \subset Y$ is a sequence satisfying $x_k \in Y_{m_k} \oplus Y_{m_k+1} \oplus \cdots \oplus Y_{n_k}$, $k \in \mathbb{N}$, and $\sup_n \left\| \sum_{k=1}^{n} x_k \right\| < +\infty$, then the series $\sum_{k=1}^{\infty} x_k$ converges.

(In fact the existence of such finite-dimensional decomposition is equivalent to the PCP.)

By contradiction, using again Theorem 46 there is $f \in \mathcal{C}^{1,+}(B_{c_0}; Y) \setminus \mathcal{C}_{\mathrm{K}}(B_{c_0}; Y)$. By Lemma 41 there are $v \in B_{c_0}$ and a disjointly supported weakly null sequence $\{u_n\} \subset c_0$ such that $\{v + u_n\} \subset B_{c_0}$ and $\{f(v + u_n)\}$ is not relatively compact in Y. Without loss of generality we may assume that $f(v) = 0$. By Theorem 3.56 passing to a subsequence we may assume that there is a bounded sequence $\{\phi_n\} \subset KB_{Y^*}$ biorthogonal to $\{f(v + u_n)\}$. By Lemma 44 we may assume that

$$\left| \phi_k \circ f\left(v + \sum_{j=1}^{k} a_j u_j \right) - \phi_k \circ f(v + a_k u_k) \right| < \frac{1}{4} \tag{9}$$

for every $k \in \mathbb{N}$ and each $a_1, \dots, a_k \in \{0, 1\}$.

We now proceed by constructing inductively increasing sequences of natural numbers $\{m_k\}$, $\{n_k\}$, $\{l_k\}$ so that $m_k < n_k + 1 < m_{k+1}$ and

$$\operatorname{dist}\left(f\left(v + \sum_{j=1}^{k} u_{l_j} \right) - f\left(v + \sum_{j=1}^{k-1} u_{l_j} \right), \operatorname{span} \bigcup_{j=m_k}^{n_k} Y_j \right) < \frac{1}{2^k} \tag{10}$$

for every $k \in \mathbb{N}$. We put $m_1 = 1$ and $l_1 = 1$, and we choose $n_1 \in \mathbb{N}$ such that $\operatorname{dist}\bigl(f(v+u_1) - f(v), \operatorname{span} \bigcup_{j=1}^{n_1} Y_j\bigr) < \frac{1}{2}$. Now assume m_j, l_j, and n_j, $1 \le j \le k-1$, are already defined for some $k \in \mathbb{N}, k > 1$. Put $m_k = n_{k-1} + 2$. Since f is w–w continuous (Fact 22), the sequence $\{f(v + \sum_{j=1}^{k-1} u_{l_j} + u_n) - f(v + \sum_{j=1}^{k-1} u_{l_j})\}_{n=1}^{\infty}$ is weakly null. The subspace $\overline{\operatorname{span}} \bigcup_{j=m_k}^{\infty} Y_j$ is finite-codimensional and hence there is $l_k > l_{k-1}$ such that

$$\operatorname{dist}\left(f\left(v + \sum_{j=1}^{k} u_{l_j}\right) - f\left(v + \sum_{j=1}^{k-1} u_{l_j}\right), \overline{\operatorname{span}} \bigcup_{j=m_k}^{\infty} Y_j\right) < \frac{1}{2^k}.$$

Finally, we find $n_k \geq m_k$ for which (10) holds.

Let us denote $y_k = f(v + \sum_{j=1}^{k} u_{l_j})$ for $k \in \mathbb{N}_0$. For every $k \in \mathbb{N}$ there is $w_k \in Y$, $\|w_k\| < \frac{1}{2^k}$ such that $z_k = y_k - y_{k-1} + w_k \in \operatorname{span} \bigcup_{j=m_k}^{n_k} Y_j$. Then $\sum_{k=1}^{n} z_k = y_n - y_0 + \sum_{k=1}^{n} w_k$ and since $\{y_n\}$ is bounded, by the bounded completeness the series $\sum_{k=1}^{\infty} z_k$ converges. It follows that $\{y_n\}$ is convergent. However, by (9), $\phi_{l_n}(y_n) > \phi_{l_n} \circ f(v + u_{l_n}) - \frac{1}{4} = \frac{3}{4}$ and $\phi_{l_n}(y_{n-1}) < \phi_{l_n} \circ f(v) + \frac{1}{4} = \frac{1}{4}$. Thus $\|y_n - y_{n-1}\| \geq \frac{1}{K}\phi_{l_n}(y_n - y_{n-1}) > \frac{1}{2K}$ for each $n \in \mathbb{N}$, a contradiction. □

In particular, Banach spaces of non-trivial cotype satisfy the first condition in Theorem 52 and hence they have property (WR).

5. Uniformly smooth mappings from $\mathcal{W}$-spaces

In this section we begin by showing that spaces with the hereditary DPP are $\mathcal{W}$-spaces. We proceed by proving that weakly Cauchy sequences in $C(K)$ spaces are arbitrarily close to weakly Cauchy sequences in $C(K)^{**}$ which are contained in a subspace isometric to some $C([1, \alpha])$ for a countable ordinal α. This result will ultimately allow us to bootstrap the results obtained for uniformly smooth mappings from $C(K)$, K scattered compact, into the case of an arbitrary compact space K. The main result of the theory of $\mathcal{W}$-spaces is Theorem 57, which gives a generalisation of the Dunford-Pettis property into the setting of uniformly smooth mappings from $\mathcal{L}_\infty$-spaces.

Theorem 53 ([CHL]). *Let X be a Banach space with the hereditary DPP. Then X is a $\mathcal{W}_\lambda$-space for some $\lambda > 0$ and moreover $\mathcal{C}^{1,+}(B_X; Y) \subset \mathcal{C}_{\mathrm{wsC}}(\lambda B_X; Y)$ for any Banach space Y with property (WR).*

Proof. The hereditary DPP is equivalent to property S_∞ ([Dies1], [Cem]). So by Theorem 3.55 there is $C > 0$ such that for every weakly null sequence $\{y_n\} \subset \lambda B_X$, $\lambda \geq 0$ there is a subsequence $\{y_{n_k}\}$ satisfying

$$\left\|\sum_{k=1}^{n} a_k y_{n_k}\right\| \leq \lambda C \max_{k=1,\dots,n} |a_k| \tag{11}$$

whenever $n \in \mathbb{N}$ and $a_1, \dots, a_n \in \mathbb{R}$. Hence by the proof of [KO1, Theorem 2.1] the space X has a uniform version of property (u): If $\{x_n\} \subset \lambda B_X$ is a weakly Cauchy sequence, then there are a wuC sequence $\{u_k\} \subset 2\lambda B_X$ and a weakly null sequence $\{y_n\} \subset 2\lambda B_X$ such that

$$x_n = y_n + \sum_{k=1}^{n} u_k \tag{12}$$

and

$$\sup_{\phi \in B_{X^*}} \sum_{k=1}^{\infty} |\phi(u_k)| \le \lambda(4C+1). \tag{13}$$

(In the notation of the proof of [KO1, Theorem 2.1] set $u_{m_1} = x_{m_1}, u_{m_j} = x_{m_j} - x_{m_{j-1}}$, and $u_k = 0$ otherwise.)

Put $\lambda = \frac{1}{8C+1}$. The first statement of the theorem follows from the second one for $Y = \mathbb{R}$. So let Y be a space with property (WR) and $f \in \mathcal{C}^{1,+}(B_X; Y)$. By contradiction, assume that $\{x_n\} \subset \lambda B_X$ is weakly Cauchy such that $\{f(x_n)\}$ is not convergent. By Fact 3.86 we may assume that $\operatorname{dist}\big(\{f(x_{2k-1});\ k \in \mathbb{N}\}, \{f(x_{2k});\ k \in \mathbb{N}\}\big) > 0$. Let $\{u_k\} \subset 2\lambda B_X$ be a wuC sequence and let $\{y_n\} \subset 2\lambda B_X$ be a weakly null sequence satisfying (12) and (13). By passing to further subsequences first in the odd indices and then in the even indices we may assume that (11) holds both for $\{y_{2k-1}\}$ and for $\{y_{2k}\}$. (Note that simultaneously we do not pass to a subsequence of $\{u_k\}$ but rather to a "subseries" $\{\sum_{j=n_{k-1}}^{n_k-1} u_j\}_{k=1}^{\infty}$, which is still wuC and satisfies (13).)

By (13) there is $S \in \mathcal{L}(c_0; X)$ satisfying $S(e_k) = u_k$ and $\|S\| \le \lambda(4C+1)$. Similarly, using (11) separately on the odd and even summands of $\sum_{k=1}^{n} a_k y_k$ and recalling that $\{y_k\} \subset 2\lambda B_X$ we conclude that there exists $T \in \mathcal{L}(c_0; X)$ satisfying $T(e_k) = y_k$ and $\|T\| \le 4\lambda C$. Finally, we define an operator $U \in \mathcal{L}(c_0 \oplus_\infty c_0; X)$ by $U(x,y) = S(x) + T(y)$. It follows that $\|U\| \le \|S\| + \|T\| \le \lambda(8C+1) = 1$, and consequently $U(B_{c_0 \oplus_\infty c_0}) \subset B_X$ and $f \circ U \in \mathcal{C}^{1,+}(B_{c_0 \oplus_\infty c_0}; Y)$. Let us put $z_n = \big(\sum_{k=1}^{n} e_k, e_n\big) \in c_0 \oplus_\infty c_0$. Then $U(z_n) = x_n$. Because $\{z_n\}$ is a weakly Cauchy sequence in the unit ball of the space $c_0 \oplus_\infty c_0$ which is isometric to c_0, and because Y has property (WR), the sequence $\{f(x_n)\} = \{f \circ U(z_n)\}$ is convergent, which is a contradiction. □

For the general $\mathcal{W}$-spaces not containing ℓ_1 we have a slightly more restrictive (in view of Theorem 52 and the remark after it) result.

Theorem 54 ([CHL]). *Let X be a $\mathcal{W}_\lambda$-space that does not contain ℓ_1 and Y a Banach space of a non-trivial type. Then $\mathcal{C}^{1,+}(B_X; Y) \subset \mathcal{C}_{\mathrm{wu}}(\lambda B_X; Y)$.*

Proof. By Theorem 24 it suffices to prove that $\mathcal{C}^{1,+}(B_X; Y) \subset \mathcal{C}_{\mathrm{K}}(\lambda B_X; Y)$. Let us first prove the theorem in the special case $Y = \ell_p$, $1 < p < \infty$. By contradiction, if $f \in \mathcal{C}^{1,+}(B_X; \ell_p)$ is such that $f(\lambda B_X)$ is not relatively compact, then using Proposition 3.34 and Rosenthal's ℓ_1-theorem we may assume that there exists a weakly Cauchy sequence $\{x_j\} \subset \lambda B_X$ such that $f(x_j) = e_j$, $j \in \mathbb{N}$, where $\{e_j\}$ is the canonical basis of ℓ_p. Define $P \in \mathcal{P}(^{\lceil p \rceil}\ell_p)$ by $P((a_j)) = \sum_{j=1}^{\infty} (-1)^j a_j^{\lceil p \rceil}$. Then $P(e_j) = (-1)^j$ and so $P \circ f \in \mathcal{C}^{1,+}(B_X) \setminus \mathcal{C}_{\mathrm{wsC}}(\lambda B_X)$, a contradiction.

Let now Y be a general Banach space of a non-trivial type and suppose by contradiction that there exists $f \in \mathcal{C}^{1,+}(B_X; Y)$ such that $S = f(\lambda B_X)$ is not relatively compact. Since Y is of a non-trivial type, it does not contain ℓ_1 and so by Rosenthal's ℓ_1-theorem S contains a weakly Cauchy δ-separated sequence $\{y_j\}$. Since Y^* is also of a non-trivial type, by passing to a subsequence using Corollary 4.43 there is $L \in \mathcal{L}(Y;\ell_p)$ for some $1 < p < \infty$ such that $L(y_{2j} - y_{2j-1}) = e_j$, where $\{e_j\}$ is the canonical basis of ℓ_p. In particular, $L(S)$ is not relatively compact in ℓ_p. Indeed, either $\{L(y_{2j-1})\}$ is not relatively compact or there is a convergent subsequence $\{L(y_{2k_j-1})\}$ and in this case $\{L(y_{2k_j})\} = \{e_{k_j} + L(y_{2k_j-1})\}$ is not relatively compact. Hence $L \circ f \in \mathcal{C}^{1,+}(B_X;\ell_p) \setminus \mathcal{C}_K(\lambda B_X;\ell_p)$, a contradiction with the first part of the proof. □

The assumption that X does not contain ℓ_1 cannot be removed from the statement of the result: By Corollary 3.89 there is a quotient mapping $T \in \mathcal{L}(\ell_1;\ell_2)$ which is obviously not compact.

Our aim now is to generalise our results on $C(K)$ spaces for scattered compacts K to the general setting of arbitrary compact spaces K. For a sequence $a = (a_k) \in \mathbb{R}^{\mathbb{N}}$ we define its oscillation on a set $I \subset \mathbb{N}$ by $\operatorname{osc}(a, I) = \sup_{j,k\in I} |a_k - a_j|$.

Lemma 55. *Let $\Delta \subset \mathbb{R}$ be a discrete set, $D = \Delta^{\mathbb{N}}$ with the product topology, and $\delta > 0$. There is a mapping $Q: D \to D$ with the following properties:*

(i) *If $a \in D$ is such that there is $m \in \mathbb{N}$ for which $\operatorname{osc}(a, [m, \infty)) \le \delta$, then the sequence $Q(a)$ is eventually constant.*
(ii) *Q is continuous.*
(iii) $\|Q(a) - a\|_\infty = \sup_{k\in\mathbb{N}} |Q(a)_k - a_k| \le \delta$ *for each $a \in D$.*

Proof. For every $a \in D$ there exist uniquely determined $k_0 \in \mathbb{N} \cup \{\infty\}$ and a (possibly finite) increasing sequence $n(a,k) \in \mathbb{N}$, $1 \le k < k_0$ such that $n(a,1) = 1$, $n(a,k_0) = \infty$ if $k_0 \in \mathbb{N}$, and

$$\operatorname{osc}\big(a, \big[n(a,k), n(a,k+1)\big)\big) \le \delta \quad \text{for } 1 \le k < k_0,$$
$$\operatorname{osc}\big(a, \big[n(a,k), n(a,k+1)\big]\big) > \delta \quad \text{for } 1 \le k < k_0 - 1.$$

It is important to notice that if $a = (a_j) \in D$, $b = (b_j) \in D$, and $a_j = b_j$ for $1 \le j \le n(a,k)$, then $n(a,j) = n(b,j)$ for $j = 1, \dots, k$. We define $Q: D \to D$ by

$$Q(a)_j = a_{n(a,k)} \quad \text{for } j \in \big[n(a,k), n(a,k+1)\big),\ 1 \le k < k_0.$$

Property (iii) is clear from the definition. If $a \in D$ satisfies the condition in (i), then clearly $k_0 \in \mathbb{N}$ and $Q(a)$ is constant on $[n(a,k_0-1),\infty)$. Finally, to prove (ii) suppose that $\{a^m\}_{m=1}^\infty \subset D$ converges pointwise to $a \in D$. Fix $j \in \mathbb{N}$. There is $1 \le k < k_0$ such that $j \in [n(a,k), n(a,k+1))$. Since Δ is discrete, there is $m_0 \in \mathbb{N}$ such that $a_l^m = a_l$ for each $l \in [1, j]$, $m \ge m_0$. Thus $n(a^m,k) = n(a,k)$ and $n(a^m,k+1) > j$ whenever $m \ge m_0$ and consequently $Q(a^m)_j = Q(a)_j$ for $m \ge m_0$. □

Theorem 56 ([CHL]). *Let $X = C(K)$, where K is a compact space. Let $\{x_k\} \subset B_X$ be a weakly Cauchy sequence and $\varepsilon > 0$. Then there exist a subspace $Z \subset X^{**}$ isometric to some $C([1,\alpha])$, α countable ordinal, and a weakly Cauchy sequence $\{z_k\} \subset B_Z$ such that $\|z_k - x_k\| < \varepsilon$ for every $k \in \mathbb{N}$.*

Proof. By considering $\overline{\operatorname{span}}\{x_k\}$ according to Proposition 5.124 we may without loss of generality assume that K is metrisable. Note that the Banach space Y of all bounded Borel functions on K with the supremum norm is isometric to a subspace of X^{**}. Indeed, the isometry $S\colon Y \to X^{**}$ is given by $S(f)(\mu) = \int_K f \,\mathrm{d}\mu$.

Let $x \in B_Y$ be a pointwise limit of the sequence $\{x_k\}$. Choose $n \in \mathbb{N}$ satisfying $\frac{3}{n} < \varepsilon$ and denote $\Delta = \left\{\frac{j}{n};\ j = -n, -n+1, \dots, n-1, n\right\}$. We define a simple Borel function $y\colon K \to \Delta$ approximating x within $\frac{1}{n}$ by $y(t) = \frac{j}{n}$ if and only if $\frac{j+1}{n} > x(t) \geq \frac{j}{n}$. Further, we construct a sequence $\{y_k\} \subset B_Y$ of Borel functions $y_k\colon K \to \Delta$ converging pointwise to y and satisfying $\|y_k - x_k\| \leq \frac{1}{n}$, $k \in \mathbb{N}$: We let

$$y_k(t) = \begin{cases} \frac{j}{n} & \text{if } |x_k(t) - y(t)| \geq \frac{1}{n} \text{ and } \frac{j+1}{n} > x_k(t) \geq \frac{j}{n} \text{ for some } -n \leq j \leq n, \\ y(t) & \text{otherwise.} \end{cases}$$

Fix $t \in K$. Since $|y(t) - x(t)| < \frac{1}{n}$ and $\{x_k(t)\}$ converges to $x(t)$, there is $m \in \mathbb{N}$ such that $|x_k(t) - y(t)| < \frac{1}{n}$ for all $k \geq m$ and so $y_k(t) = y(t)$ for all $k \geq m$.

Let $D = \Delta^{\mathbb{N}}$ with the product topology and define $\Phi\colon K \to D$ by $\Phi(t)_k = y_k(t)$. Note that D is a metrisable compact space. Put $H = \overline{\Phi(K)}$. We claim that for every $h = (h_k) \in H$ there is $m \in \mathbb{N}$ for which $\operatorname{osc}(h, [m, \infty)) \leq \frac{2}{n}$. Indeed, there is $\{t_j\} \subset K$ such that $\Phi(t_j) \to h$. By passing to a subsequence we may assume that $t_j \to t \in K$. Let $m \in \mathbb{N}$ be such that $|x_k(t) - y(t)| < \frac{1}{n}$ for every $k \geq m$. Fix $k \geq m$ and let $\eta > 0$ be such that $|x_k(t) - y(t)| < \frac{1}{n} - 2\eta$. Let $j_0 \in \mathbb{N}$ be such that $|h_k - y_k(t_j)| = |h_k - \Phi(t_j)_k| < \eta$ for every $j \geq j_0$. By the continuity of x_k there exists $j \geq j_0$ such that $|x_k(t_j) - x_k(t)| < \eta$. Then $|h_k - y(t)| \leq |h_k - y_k(t_j)| + |y_k(t_j) - x_k(t_j)| + |x_k(t_j) - x_k(t)| + |x_k(t) - y(t)| < \frac{2}{n}$. Since $y(t) \in \Delta$ and $h \in D$, the claim follows.

Let $Q\colon D \to D$ be the mapping from Lemma 55 for $\delta = \frac{2}{n}$ and set $\Psi = Q \circ \Phi$ and $L = \overline{\Psi(K)}$. Then L is a countable compact set. Indeed, if $g \in L$, then there is $\{t_j\} \subset K$ such that $\Psi(t_j) \to g$. By passing to a subsequence we may assume that $\Phi(t_j) \to h \in H$. Since Q is continuous, we have $\Psi(t_j) = Q(\Phi(t_j)) \to Q(h)$ and hence $g = Q(h)$. By (i) in Lemma 55 and the claim above we conclude that the sequence g is eventually constant. Thus L consists of sequences in D that are eventually constant and so it is countable. Therefore $C(L)$ is isometric to $C([1, \alpha])$ for some countable ordinal α by the Mazurkiewicz-Sierpiński theorem ([HMVZ, Theorem 2.56]).

Finally, by Fact 5.123 the mapping Φ is Borel, and hence so is Ψ, and further $T\colon C(L) \to Y$, $T(f) = f \circ \Psi$ is an isometry onto a subspace $Z \subset Y$. Note that $y_k = \pi_k \circ \Phi$, where π_k is the projection onto the kth coordinate. Put $z_k = T(\pi_k\restriction_L)$, $k \in \mathbb{N}$. Then $z_k = \pi_k \circ \Psi$ and so $\|z_k - y_k\| = \sup_{t \in K} |(Q \circ \Phi(t))_k - \Phi(t)_k| \leq \frac{2}{n}$ by (iii) in Lemma 55. Clearly, $z_k \in B_Z$ and $\|z_k - x_k\| \leq \|z_k - y_k\| + \|y_k - x_k\| \leq \frac{3}{n} < \varepsilon$ for each $k \in \mathbb{N}$. Since each $g \in L$ is eventually constant, $\{\pi_k(g)\}$ is convergent. Hence $\{\pi_k\restriction_L\}$ is weakly Cauchy in $C(L)$ and consequently $\{z_k\}$ is weakly Cauchy in Z. □

The next theorem generalises the Dunford-Pettis property of $\mathcal{L}_\infty$-spaces (Theorem 3.41, see also Theorem 3.47) to the case of uniformly smooth mappings. Recall that for any compact space K the space $C(K)$ is a $\mathcal{L}_{\infty,\lambda}$-space for every $\lambda > 1$ (Theorem 4.15).

Theorem 57 ([CHL]). *Let X be a $\mathcal{L}_{\infty,\lambda}$-space, $\lambda \geq 1$, and Y a Banach space. Then*

$$\mathcal{C}^{1,+}(B_X;Y) \cap \mathcal{C}_{\mathrm{wK}}(B_X;Y) \subset \mathcal{C}_{\mathrm{wsC}}(\tfrac{1}{\lambda}B_X;Y).$$

If Y^{**} has property (WR) or if Y has property (WR) and is isomorphic to a complemented subspace of a dual Banach space, then

$$\mathcal{C}^{1,+}(B_X;Y) \subset \mathcal{C}_{\mathrm{wsC}}(\tfrac{1}{\lambda}B_X;Y).$$

In particular, X is a $\mathcal{W}_{\frac{1}{\lambda}}$-space.

Proof. Embed X isometrically into $E = C\big((B_{X^*}, w^*)\big)$. By Theorem 4.15, X^{**} is a $\mathcal{P}_\lambda$-space. Hence there exists a projection $P\colon E^{**} \to X^{**}$ with $\|P\| \leq \lambda$. Assume that $f \in \mathcal{C}^{1,+}(B_X;Y) \setminus \mathcal{C}_{\mathrm{wsC}}(\frac{1}{\lambda}B_X;Y)$ and that $\{x_k\} \subset \frac{1}{\lambda}B_X$ is a weakly Cauchy sequence such that $\{f(x_k)\}$ is not convergent. Passing to a subsequence by Fact 3.86 we may assume that $\operatorname{dist}\big(\{f(x_{2k-1});\ k \in \mathbb{N}\}, \{f(x_{2k});\ k \in \mathbb{N}\}\big) > 0$. Put $g = f^{**} \circ P$ and note that $g(x_k) = f(x_k)$, $k \in \mathbb{N}$. Then g is uniformly continuous on $\frac{1}{\lambda}B_{E^{**}}$ and so by Theorem 56 there exist a subspace $Z \subset E^{**}$ isometric to $C([1,\alpha])$, α countable ordinal, and a weakly Cauchy sequence $\{z_k\} \subset \frac{1}{\lambda}B_Z$ such that $\|z_k - x_k\|$ is small enough so that $\operatorname{dist}\big(\{g(z_{2k-1});\ k \in \mathbb{N}\}, \{g(z_{2k});\ k \in \mathbb{N}\}\big) > 0$. Thus $g \in \mathcal{C}^{1,+}(\frac{1}{\lambda}B_{E^{**}};Y^{**}) \setminus \mathcal{C}_{\mathrm{wsC}}(\frac{1}{\lambda}B_Z;Y^{**})$. This contradicts Theorem 48 in case that Y^{**} has property (WR). To prove the first part of the theorem, note that by Theorem 46 the set $g(\frac{1}{\lambda}B_Z)$ is not relatively weakly compact and so $f^{**}(B_{X^{**}})$ is not relatively weakly compact either. Consequently, $f \notin \mathcal{C}_{\mathrm{wK}}(B_X;Y)$ by Fact 11.

Finally, suppose that Y has property (WR) and is isomorphic to a complemented subspace of a dual Banach space. Then there is a projection $Q\colon Y^{**} \to Y$ (use Dixmier's projection). We put $g = Q \circ f^{**} \circ P$ and arrive at a contradiction with Theorem 48 similarly as above. □

We remind that Y^{**} has property (WR) if and only if Y^{**} does not contain ℓ_∞ (Corollary 51).

Recall that the class $\mathcal{W}$ behaves rather well with respect to quotients (Proposition 21). Passing to non-linear smooth images, we have the following structure-preserving properties.

Proposition 58. *Let X be a $\mathcal{W}_\lambda$-space that does not contain ℓ_1, Y a Banach space, and let $f \in \mathcal{C}^{1,+}(B_X;Y)$ be such that* $\operatorname{Int} f(\lambda B_X) \neq \emptyset$. *Then Y does not contain ℓ_1.*

Proof. Suppose that Y contains ℓ_1. Then by Corollary 3.89 the space Y has a quotient isomorphic to ℓ_2. Hence there is $g \in \mathcal{C}^{1,+}(B_X;\ell_2)$ such that $\operatorname{Int} g(\lambda B_X) \neq \emptyset$. By Theorem 54, $g \in \mathcal{C}_{\mathrm{K}}(\lambda B_X;\ell_2)$, which is a contradiction. □

Proposition 59 ([CHL]). *Let X be a reflexive $\mathcal{W}_\lambda$-space, Y a Banach space, and let $f \in \mathcal{C}^{1,+}(B_X;Y)$ be such that* $\operatorname{Int} f(\lambda B_X) \neq \emptyset$. *Then Y is a reflexive $\mathcal{W}$-space.*

Proof. By shifting and scaling in the range we may suppose that there is $\xi > 0$ such that $\xi B_Y \subset f(\lambda B_X)$ and $f(B_X) \subset B_Y$. To prove that Y is reflexive, by the Eberlein-Šmulyan theorem it suffices to show that any sequence $\{y_k\} \subset \xi B_Y$ contains a weakly convergent subsequence. Let $\{x_k\} \subset \lambda B_X$ be such that $f(x_k) = y_k$. As X is reflexive, there is a subsequence $\{x_{n_k}\}$ such that $x_{n_k} \xrightarrow{w} x$. The mapping f is w–w sequentially continuous on λB_X (Fact 22) and hence $y_{n_k} = f(x_{n_k}) \xrightarrow{w} f(x)$.

We proceed by proving that Y is a $\mathcal{W}_\xi$-space. Suppose by contradiction that there exist a function $g \in \mathcal{C}^{1,+}(B_Y)$ and a weakly Cauchy sequence $\{y_k\} \subset \xi B_Y$ such that $\lim_{k\to\infty} g(y_{2k-1}) \neq \lim_{k\to\infty} g(y_{2k})$. By the weak compactness of ξB_Y there is $y \in \xi B_Y$ such that $y_k \xrightarrow{w} y$. There is a sequence $\{x_k\} \subset \lambda B_X$ satisfying $f(x_k) = y_k$. By passing to subsequences separately in odd and even indices we may assume that $x_{2k-1} \xrightarrow{w} u$ and $x_{2k} \xrightarrow{w} v$. Clearly, $f(u) = y = f(v)$, as f is sequentially w–w continuous on λB_X. Since $g \circ f \in \mathcal{C}^{1,+}(B_X) \subset \mathcal{C}_{\mathrm{wsC}}(\lambda B_X)$, we obtain $\lim g(y_{2k-1}) = \lim g \circ f(x_{2k-1}) = g \circ f(u) = g(y)$ and similarly $\lim g(y_{2k}) = \lim g \circ f(x_{2k}) = g \circ f(v) = g(y)$, a contradiction. □

6. Fixing the canonical basis of c_0

In the present section we are going to investigate the phenomenon of fixing the canonical basis of c_0, which holds for non-weakly compact linear operators from $C(K)$ spaces, in the non-linear setting. We start by showing that in a polynomial version of Theorem 3.47 the implication (iv)⇒(i) remains valid, provided that the compact space K is scattered. However, our main result is a counterexample to this implication for $C^{1,1}$-smooth mappings constructed below. This means that a qualitative change takes place when replacing polynomials with $C^{1,1}$-smooth mappings from c_0.

Theorem 60 ([DeH2]). *Let $X = c_0$ or $X = C(K)$, where K is a scattered compact space, let Y be a Banach space, and $P \in \mathcal{P}(X;Y)$ a non-compact polynomial. Then there exist $v \in B_X$ and a sequence $\{v_n\} \subset B_X$ such that both $\{v_n - v\}$ and $\{P(v_n) - P(v)\}$ are equivalent to the canonical basis of c_0.*

Proof. By Theorems 46 and 42 and Proposition 3.22 it suffices to prove the result for $X = c_0$. By Lemma 41, Proposition 3.32, and [FHHMZ, Proposition 4.45] there exist $v \in B_{c_0}$ and a weakly null sequence $\{u_n\} \subset c_0$ such that $\{v + u_n\} \subset B_{c_0}$, $\{u_n\}$ is equivalent to the canonical basis of c_0, and $\lim_{n\to\infty} P(v + u_n)$ does not exist. Put $v_n = v + u_n$ and $z_n = P(v_n) - P(v)$. Since P is w–w sequentially continuous (Corollary 3.59), $\{z_n\}$ is weakly null. Thus by passing to a subsequence we may assume that $\{z_n\}$ is a semi-normalised basic sequence. Define a polynomial Q by $Q(x) = P(v + x) - P(v)$. Then $z_n = Q(u_n)$. Since $\{u_n\}$ is a wuC sequence, by Theorem 3.53 the sequence $\{z_n\}$ is also wuC, i.e. it has an upper ∞-estimate (Fact 3.51). The lower ∞-estimate is automatic, since $\{z_n\}$ is a semi-normalised basic sequence: $\left\|\sum_{n=1}^m a_n z_n\right\| \geq \frac{1}{2K} \max_n \|a_n z_n\| \geq \frac{1}{2K} \inf_n \|z_n\| \max_n |a_n|$, where K is the basis constant of $\{z_n\}$. □

The next simple example shows that Theorem 60 is optimal in the sense that the shifting of the basis cannot be avoided, so the result is necessarily of an affine rather than linear nature. Define $P \in \mathcal{P}({}^4c_0; c_0)$ by $P(x) = x_1^2 \sum_{j=1}^{\infty} x_j^2 e_j$. Then P is non-compact, but $\lim_{n\to\infty} P(u_n) = 0$ for every weakly null sequence $\{u_n\} \subset c_0$. Since the image of P is contained in the non-negative cone of c_0, we also see that $T(B_{c_0}) \not\subset P(B_{c_0})$ for any $T \in \mathcal{L}(c_0; c_0)$.

In the rest of this section we will construct a $C^{1,1}$-smooth and non-compact mapping $f : B_{c_0} \to \ell_\infty$ which has the property that $f(B_{c_0})$ does not "contain" the canonical basis of c_0. To this end we first construct separating functions on ℓ_∞^n with an upper estimate on the second derivative that increases slower than the dimension of the domain.

Lemma 61. *The functions $p_n \in \mathcal{P}^{4n}(\ell_\infty^n)$ defined by*

$$p_n(x) = 1 - \prod_{k=1}^{n} (1 - x_k^4)$$

have the following properties:

(i) $0 \le p_n(x) \le 1$ *for every* $x \in B_{\ell_\infty^n}$ *and* $p_n(0) = 0$,

(ii) $p_n(x) \ge \delta^4$ *for each* $x \in B_{\ell_\infty^n}$ *with* $\|x\| \ge \delta \ge 0$,

(iii) $p_n(x) \le 1 - (1-\eta^4)^n$ *for each* $x \in \ell_\infty^n$ *with* $\|x\| \le \eta \le 1$,

(iv) $\|Dp_n(x)\| \le 4\sqrt[4]{n}$ *for every* $x \in B_{\ell_\infty^n}$,

(v) *there is a constant* $C > 0$ *such that* $\|D^2 p_n(x)\| \le C\sqrt{n}$ *for every* $x \in B_{\ell_\infty^n}$ *and every* $n \in \mathbb{N}$.

Proof. The first three properties are clear. We show the estimates on the derivatives by induction on n. Put $q_n(x) = \prod_{k=1}^n (1 - x_k^4)$ and note that $\|D^j q_n(x)\| = \|D^j p_n(x)\|$ for all $x \in \ell_\infty^n$, $j \in \mathbb{N}$. Clearly, $\sup_{x\in B_{\ell_\infty^1}} \|Dq_1(x)\| = \sup_{|x|\le 1} |4x^3| = 4$. Now assume that the estimate $\sup_{x\in B_{\ell_\infty^n}} \|Dq_n(x)\| \le 4\sqrt[4]{n}$ holds for some $n \in \mathbb{N}$. Since $q_{n+1}(x) = q_n\big(P_n(x)\big)(1 - x_{n+1}^4)$, where P_n is the projection to the first n coordinates, by the Leibniz formula (Corollary 1.116)

$$Dq_{n+1}(x)[h] = Dq_n\big(P_n(x)\big)[P_n(h)] \cdot (1 - x_{n+1}^4) + q_n\big(P_n(x)\big) \cdot (-4x_{n+1}^3) h_{n+1}$$

for every $x, h \in \ell_\infty^{n+1}$. Consider the function $\varphi(t) = 4\sqrt[4]{n}(1 - t^4) + 4t^3$, $t \in \mathbb{R}$. By the inductive hypothesis

$$\begin{aligned}\|Dq_{n+1}(x)\| &\le \|Dq_n(P_n(x))\| \cdot |1 - x_{n+1}^4| + |q_n(P_n(x))| \cdot |-4x_{n+1}^3| \\ &\le \varphi(|x_{n+1}|)\end{aligned}$$

for $x \in B_{\ell_\infty^{n+1}}$. Since $\varphi'(t) = -16\sqrt[4]{n}t^3 + 12t^2$, the function φ attains its maximum on $[0,1]$ at the point $\frac{3}{4\sqrt[4]{n}}$. So $\max_{[0,1]} \varphi = 4\sqrt[4]{n} + 4\frac{27}{256\sqrt[4]{n^3}} < 4\sqrt[4]{n+1}$, where the last inequality is easy to check. This finishes the proof of the estimate for Dq_n.

Similarly, we have $\|D^2 q_1(x)\| \le 12$ for $x \in B_{\ell_\infty^1}$. Now assume that the estimate $\sup_{x\in B_{\ell_\infty^n}} \|D^2 q_n(x)\| \le C\sqrt{n}$ holds for some $n \in \mathbb{N}$ and $C \ge 12$. By the Leibniz formula

$$D^2q_{n+1}(x)[h,k] = D^2q_n\big(P_n(x)\big)[P_n(h), P_n(k)]\cdot(1-x_{n+1}^4)$$
$$+ Dq_n\big(P_n(x)\big)[P_n(h)]\cdot(-4x_{n+1}^3)k_{n+1} + Dq_n\big(P_n(x)\big)[P_n(k)]\cdot(-4x_{n+1}^3)h_{n+1}$$
$$+ q_n\big(P_n(x)\big)\cdot(-12x_{n+1}^2)h_{n+1}k_{n+1}$$

for $x, h, k \in \ell_\infty^{n+1}$. Consider the function $\psi(t) = C\sqrt{n}(1-t^4)+32\sqrt{C}\sqrt[4]{n}t^3+12t^2$, $t \in \mathbb{R}$. By the inductive hypothesis $\|D^2q_{n+1}(x)\| \le \psi(|x_{n+1}|)$ for $x \in B_{\ell_\infty^{n+1}}$ (we note that the multiplicative constant $\sqrt{C}$ was added so that the analysis of the function ψ is much easier). We have $\psi'(t) = -4C\sqrt{n}t^3 + 96\sqrt{C}\sqrt[4]{n}t^2 + 24t$ and so $\psi'(t) = 0$ if and only if $t = 0$ or $t = \frac{12\pm5\sqrt{6}}{\sqrt{C}\sqrt[4]{n}}$. Thus ψ attains its maximum on $[0,+\infty)$ at $\frac{a}{\sqrt{C}\sqrt[4]{n}}$, where $a = 12+5\sqrt{6}$, and hence $\max_{[0,1]}\psi \le C\sqrt{n} + \frac{b}{C\sqrt{n}}$, where $b = -a^4 + 32a^3 + 12a^2$. It is not difficult to check that if $C^2 \ge (1+\sqrt{2})b$, then $\max_{[0,1]}\psi \le C\sqrt{n+1}$. This finishes the proof.

We remark that the estimate above gives $C \le 533$. Had we not artificially inflated the estimate by $\sqrt{C}$, a more tedious computation would have resulted in $C \le 32$. □

The next result shows among others that a uniformly smooth version of Theorem 3.53 is false.

Theorem 62 ([DeH2]). *There is a $C^{1,1}$-smooth mapping $f\colon B_{c_0} \to \ell_\infty$ that is not weakly compact and has the property that there is no sequence $\{y_n\}_{n=0}^\infty \subset f(B_{c_0})$ such that $\{y_n - y_0\}_{n=1}^\infty$ is equivalent to the canonical basis of c_0.*

Proof. Let Γ be the set of all finite subsets of $\mathbb{N}$. Then Γ is countable and so $\ell_\infty(\Gamma)$ is isometric to ℓ_∞. Let C be the constant and p_n the functions from Lemma 61. For each $A \in \Gamma$, $A = \{k_1,\dots,k_n\}$ with $k_1 < \dots < k_n$ we define $T_A \in \mathcal{L}(c_0;\ell_\infty^n)$ by $T_A(x) = (x_{k_1},\dots,x_{k_n})$ and set $\varphi_A = \frac{1}{C\sqrt{n}}p_n \circ T_A$. Then $\|D^2\varphi_A(x)\| \le 1$ for every $x \in B_{c_0}$. Finally we define $f\colon B_{c_0} \to \ell_\infty(\Gamma)$ by $f(x) = (\varphi_A(x))_{A\in\Gamma}$. Then f is $C^{1,1}$-smooth by Proposition 1.130. Also, $\|f(e_k) - f(e_j)\| \ge |\varphi_{\{k\}}(e_k) - \varphi_{\{k\}}(e_j)| = \frac{1}{C}|p_1(1) - p_1(0)| = \frac{1}{C}$ if $k \ne j$, where $\{e_k\}$ is the canonical basis of c_0. Thus f is not compact, and by Theorem 46 even not weakly compact.

We proceed by contradiction. Assume that $y_n = f(u_n)$, $n \in \mathbb{N}_0$, and $\{y_n - y_0\}_{n=1}^\infty$ is equivalent to the canonical basis of c_0. We claim that by passing to a subsequence of $\{u_n\}_{n=1}^\infty$ we may assume that there exist $\delta > 0$ and an increasing sequence $\{k_n\} \subset \mathbb{N}$ such that $|u_n(k_n)| \ge \delta$, $n \in \mathbb{N}$. Indeed, by passing to a subsequence we may assume that there is $u \in \ell_\infty$ such that $w^*\text{-}\lim u_n = u$. Since $u_n \not\to u$ in norm (otherwise $\{y_n - y_0\}$ would be convergent, which is not possible), we may assume that there is $\delta > 0$ such that $\|u_n - u\| > 2\delta$ for all $n \in \mathbb{N}$. Suppose that $m_0, n_0 \in \mathbb{N}$ are given. If there is $m \ge m_0$ such that $|u(m)| > \delta$, then $|u_n(m)| > \delta$ for sufficiently large $n \ge n_0$. Otherwise there is $n \ge n_0$ such that $|u_n(j) - u(j)| < \delta$ for $1 \le j \le m_0$. Since $\|u_n - u\| > 2\delta$, there is $m > m_0$ such that $|u_n(m) - u(m)| > 2\delta$. Consequently $|u_n(m)| \ge |u_n(m) - u(m)| - |u(m)| > 2\delta - \delta = \delta$.

To continue the proof assume that $K > 0$ is such that $\left\|\sum_{j=1}^{n}(y_{m+j} - y_0)\right\| \leq K$ for every $m, n \in \mathbb{N}$. Let $n \in \mathbb{N}$ be such that $\frac{\sqrt{n}}{2C}\delta^4 > K$. Further, let $\eta \in (0, 1]$ be such that $1 - (1 - \eta^4)^n < \frac{\delta^4}{2}$ and let $m \in \mathbb{N}$ be such that $|u_0(j)| \leq \eta$ for all $j \geq m$. Put $A = \{k_{m+1}, k_{m+2}, \dots, k_{m+n}\}$ and note that $k_{m+1} > m$. We have $\varphi_A(u_0) = \frac{1}{C\sqrt{n}} p_n\big(u_0(k_{m+1}), \dots, u_0(k_{m+n})\big) < \frac{1}{2C\sqrt{n}}\delta^4$ and $\varphi_A(u_{m+j}) \geq \frac{1}{C\sqrt{n}}\delta^4$ for each $j = 1, \dots, n$. So $\left\|\sum_{j=1}^{n}(y_{m+j} - y_0)\right\| \geq \sum_{j=1}^{n}\big(y_{m+j}(A) - y_0(A)\big) = \sum_{j=1}^{n}\big(\varphi_A(u_{m+j}) - \varphi_A(u_0)\big) > n\frac{1}{2C\sqrt{n}}\delta^4 > K$, a contradiction. □

7. Ranges of smooth mappings

In the present section we are going to investigate in some detail the necessity of the assumption on the initial space to belong to the class $\mathcal{W}$ in order to preserve some structural properties through uniformly smooth mappings. In particular, Theorem 69 below shows that property $\mathcal{B}$ (which appears to be rather close to the negation of $\mathcal{W}$) leads to a complete loss of any structural property via C^∞-smooth mappings (or even via homogeneous polynomials). In the case of C^1-smooth mappings the same is true even for every separable Banach space. These results are due to Sean Michael Bates, [Bat].

We begin with a lemma that allows us to give a joint proof of two types of results regarding the ranges of smooth mappings and the ranges of derivatives of smooth functions.

Lemma 63. *Let X, Y be Banach spaces, $B \subset X$ a closed convex set containing the origin, $0 < r < R \leq 1$, and let $\varphi \in \mathcal{C}^k(X; [0, 1])$ be a function satisfying* $\operatorname{supp}\varphi \subset RB$ *and $\varphi(x) = 1$ whenever $x \in rB$. Further, let $\{x_\gamma\}_{\gamma\in\Gamma} \subset X$ be bounded and such that $\{x_\gamma + RB\}_{\gamma\in\Gamma}$ is a δ-separated collection of subsets of B for some $\delta > 0$ and let $\{f_\gamma\}_{\gamma\in\Gamma} \subset \mathcal{C}^k(X; Y)$, $k \in \mathbb{N} \cup \{\infty\}$, be bounded (in the space $\mathcal{C}^k(X; Y)$). Then there are a mapping $\psi \in \mathcal{C}^k(X; Y)$ with* $\operatorname{supp}\psi \subset B$ *and a sequence $\{\delta_n\}_{n=0}^{\infty} \subset \mathbb{R}^+$ with $\delta_0 = 1$ and $\delta_n \leq \frac{1}{n!}$ such that for any $0 \leq j < k + 1$ and any sequence $\{\gamma_n\}_{n=0}^{\infty} \subset \Gamma$*

$$D^j\psi\left(\sum_{n=0}^{\infty} r^n x_{\gamma_n}\right) = \sum_{n=0}^{\infty} \frac{\delta_n}{r^{nj}} D^j f_{\gamma_n}\left(\frac{1}{r^n}\sum_{l=n+1}^{\infty} r^l x_{\gamma_l}\right).$$

Proof. Since $\{f_\gamma\}$ is bounded in $\mathcal{C}^k(X; Y)$, for each $\rho > 0$ and $0 \leq j < k + 1$ there is $K_j(\rho) \geq 1$ such that $\|D^l g_\gamma(x)\| \leq K_j(\rho)$ for each $x \in B(0, \rho)$, $\gamma \in \Gamma$, and $0 \leq l \leq j$, where $g_\gamma(x) = \varphi(x - x_\gamma) f_\gamma(x - x_\gamma)$. Put $K_0(0) = 1$ and $K_j(\rho) = K_k(\rho)$ if $k \in \mathbb{N}$ and $j > k$. We may assume without loss of generality that $j \mapsto K_j(\rho)$ and $\rho \mapsto K_j(\rho)$ are non-decreasing. We define inductively a sequence of mappings

$\{\psi_n\}_{n=0}^{\infty} \subset C^k(X;Y)$ by

$$\psi_0(x) = \sum_{\gamma\in\Gamma} \varphi(x - x_\gamma) f_\gamma(x - x_\gamma) \quad \text{and}$$

$$\psi_n(x) = \frac{1}{nK_n\left(\frac{2^n}{r^n}n\right)} \sum_{\gamma\in\Gamma} \psi_{n-1}\left(\tfrac{1}{r}(x - x_\gamma)\right).$$

First note that for each ψ_n the sum in the definition locally consists only of at most one non-zero summand (and consequently ψ_n is C^k-smooth) and $\operatorname{supp}\psi_n \subset B$. Indeed, for ψ_0 this is clear from the assumptions. Assuming that this statement holds for ψ_{n-1}, we get $\operatorname{supp}\left(x \mapsto \psi_{n-1}\left(\frac{1}{r}(x - x_\gamma)\right)\right) \subset x_\gamma + rB \subset x_\gamma + RB \subset B$, and so it holds also for ψ_n.

By induction on n we get $\|D^j\psi_n(x)\| \le \frac{1}{r^{jn}n!}K_n\left(\frac{2^n}{r^n}n\right)^{-1}K_j\left(\frac{2^n}{r^n}\rho\right) \le \frac{1}{r^{jn}n!}K_j\left(\frac{2^n}{r^n}\rho\right)$ for every $\rho > 0$, $x \in B(0,\rho)$, $n \in \mathbb{N}_0$, and $0 \le j < k+1$. Hence for a fixed $0 \le j < k+1$ and $\rho > 0$ we have $\|D^j\psi_n(x)\| \le \frac{1}{r^{jn}n!}$ for all $n \ge \max\{j,\rho\}$ and $x \in B(0,\rho)$. Therefore the mapping $\psi = \sum_{n=0}^{\infty}\psi_n$ belongs to $\mathcal{C}^k(X;Y)$, $\operatorname{supp}\psi \subset B$, and $D^j\psi = \sum_{n=0}^{\infty} D^j\psi_n$ (Theorem 1.85).

Next, set $\delta_n = \prod_{m=1}^{n}\frac{1}{m}K_m\left(\frac{2^m}{r^m}m\right)^{-1}$. We claim that $\psi_n(x) = \delta_n f_{\gamma_n}\left(\frac{1}{r^n}(x-y)\right)$ and $x \in B$ whenever $n \in \mathbb{N}_0$, $y = \sum_{j=0}^{n} r^j x_{\gamma_j}$ for some $\gamma_0,\dots,\gamma_n \in \Gamma$, and $x \in y + r^{n+1}B$. Indeed, for $n = 0$ this is clear. Assuming that the claim holds for $n-1$ we put $u = \frac{1}{r}(x - x_{\gamma_0})$ and $v = \sum_{j=0}^{n-1} r^j x_{\gamma_{j+1}}$. Then $u - v = \frac{1}{r}(x-y)$ and hence $u \in v + r^n B$. Therefore $\psi_{n-1}(u) = \delta_{n-1} f_{\gamma_n}\left(\frac{1}{r^{n-1}}(u-v)\right)$ and $u \in B$ by the inductive hypothesis. Since $x = x_{\gamma_0} + ru \in x_{\gamma_0} + rB \subset B$, it follows that $x \notin x_\gamma + rB$ for every $\gamma \in \Gamma \setminus \{\gamma_0\}$ and hence $\psi_n(x) = \frac{1}{n}K_n\left(\frac{2^n}{r^n}n\right)^{-1}\psi_{n-1}(u) = \delta_n f_{\gamma_n}\left(\frac{1}{r^n}(x-y)\right)$. Since by the assumption $\operatorname{Int} B \ne \emptyset$, we obtain also $D^j\psi_n(x) = \frac{1}{r^{nj}}\delta_n D^j f_{\gamma_n}\left(\frac{1}{r^n}(x-y)\right)$ for $0 \le j < k+1$.

We have shown above that $\sum_{n=0}^{m} r^n x_{\gamma_n} \in B$ for every $m \in \mathbb{N}$ and every sequence $\{\gamma_n\}_{n=0}^{\infty} \subset \Gamma$. Thus $\sum_{n=0}^{\infty} r^n x_{\gamma_n} \in B$. So finally if $x = \sum_{n=0}^{\infty} r^n x_{\gamma_n}$ for some sequence $\{\gamma_n\}_{n=0}^{\infty} \subset \Gamma$, then $x - \sum_{j=0}^{n} r^j x_{\gamma_j} = r^{n+1}\sum_{j=0}^{\infty} r^j x_{\gamma_{n+1+j}} \in r^{n+1}B$ for every $n \in \mathbb{N}_0$ and hence the statement of the lemma follows. □

Fact 64. *Let Y be a normed linear space, $\{y_\gamma\}_{\gamma\in\Gamma}$ a dense subset of B_Y, and let $\{\delta_n\}_{n=1}^{\infty} \subset \mathbb{R}^+$ be such that $\delta_n \to 0$. For each $y \in \delta_0 B_Y$ there is a sequence $\{\gamma_n\}_{n=0}^{\infty}$ of distinct elements of Γ such that $y = \sum_{n=0}^{\infty}\delta_n y_{\gamma_n}$. If $\Gamma = \mathbb{N}$, then $\{\gamma_n\}$ can be chosen to be increasing.*

Proof. We find a sequence $\{\gamma_n\}_{n=0}^{\infty} \subset \Gamma$ such that $\left\|y - \sum_{k=0}^{n}\delta_k y_{\gamma_k}\right\| < \delta_{n+1}$ for every $n \in \mathbb{N}_0$ using induction: Let $\gamma_0 \in \Gamma$ be such that $\left\|\frac{1}{\delta_0}y - y_{\gamma_0}\right\| < \frac{\delta_1}{\delta_0}$. By the inductive assumption $z = \frac{1}{\delta_n}\left(y - \sum_{k=0}^{n-1}\delta_k y_{\gamma_k}\right) \in B_Y$ and so there exists $\gamma_n \in \Gamma \setminus \{\gamma_0,\dots,\gamma_{n-1}\}$ such that $\|z - y_{\gamma_n}\| < \delta_{n+1}/\delta_n$, which finishes the construction. In case that $\Gamma = \mathbb{N}$ we can of course choose $\gamma_n > \gamma_{n-1}$. □

Fact 65. *Let X be an infinite-dimensional normed linear space. For every $\varepsilon \in (0,1)$ the sphere S_X contains an ε-separated set of cardinality* dens X.

Proof. Let $D \subset S_X$ be a maximal ε-separated set. By the Riesz lemma card $D \geq \omega$. Let E be the set of all linear combinations of vectors in D with rational coefficients. Then card $E =$ card D. Since the set D is maximal, $\overline{E} = X$, as otherwise the Riesz lemma would imply the existence of $x \in S_X$ with $\operatorname{dist}(x, \overline{E}) > \varepsilon$. It follows that dens $X \leq$ card $E =$ card $D \leq$ dens X. □

Theorem 66 ([AzDe]). *Let X be an infinite-dimensional Banach space that admits a C^1-smooth Lipschitz bump. Then there exists a Lipschitz bump $\psi \in C^1(X)$ such that $B_{X^*} \subset D\psi(X)$.*

Proof. By Theorem 5.72 we have dens $X^* =$ dens X. Take a dense $\{f_\gamma\}_{\gamma\in\Gamma} \subset B_{X^*}$ with card $\Gamma =$ dens X. By Fact 65 the ball B_X contains a $\frac{1}{2}$-separated set of cardinality card Γ. Thus the result follows from Lemma 63 for $Y = \mathbb{R}$ combined with Fact 64. □

Similarly we can obtain that if X is an infinite-dimensional Banach space that admits a C^1-smooth Lipschitz bump and Y is a Banach space with dens $Y \leq$ dens X, then there is a C^1-smooth Lipschitz mapping from X onto Y. For Y separable this holds even without the assumption of the existence of the smooth bump.

Theorem 67 ([Bat]). *Let X be an infinite-dimensional Banach space and Y a separable Banach space. Then there exists a C^1-smooth Lipschitz mapping from X onto Y.*

Proof. By the Josefson-Nissenzweig theorem ([Dies2, Chapter XIII]) there is a w^*-null sequence $\{g_n\} \subset S_{X^*}$. Define $T \in \mathcal{L}(X; c_0)$ by $T(x) = (g_n(x))_{n=1}^\infty$. For each $n \in \mathbb{N}$ there is $u_n \in B_X$ with $g_n(u_n) \geq \frac{1}{2}$ and so T is not compact by Fact 4.90. By Proposition 3.33 we may assume that there is a sequence $\{x_n\} \subset B_X$ such that $T(x_n) = e_n$, $n \in \mathbb{N}$, where $\{e_n\}$ is the canonical basis of c_0. Let $\{y_n\}_{n=1}^\infty$ be dense in B_Y and define constant functions $f_n(x) = y_n$ for $x \in c_0$, $n \in \mathbb{N}$. By Theorem 5.94 the space c_0 admits a C^1-smooth Lipschitz bump. By Lemma 63 and Fact 64 there are $r \in (0, \frac{1}{2})$ and a Lipschitz mapping $\psi \in C^1(c_0; Y)$ such that $\operatorname{supp}\psi \subset 2B_{c_0}$ and $B_Y \subset \psi(M)$, where $M = \bigl\{\sum_{k=0}^\infty r^k e_{n_k};\ \{n_k\} \subset \mathbb{N} \text{ increasing}\bigr\}$. Note that $M \subset T(2B_X)$.

Let us define $\Psi(x) = \sum_{n=1}^\infty n\psi\bigl(\frac{1}{n}T(x) - 3ne_1\bigr)$. Fix an arbitrary $x \in X$. Since $3n^2 + 2n < 3(n+1)^2 - 2(n+1)$ for each $n \in \mathbb{N}$, it is easy to see that there is at most one $n \in \mathbb{N}$ such that $|T(x)_1 - 3n^2| \leq 2n$. Thus there is a neighbourhood U of x such that $\bigl\|\frac{1}{m}T(y) - 3me_1\bigr\| \geq \frac{1}{m}|T(y)_1 - 3m^2| > 2$ for every $y \in U$ and $m \in \mathbb{N}$, $m \neq n$. Hence the sum in the definition of Ψ consists locally of at most one summand and consequently $\Psi \in C^1(X; Y)$ and it has a bounded derivative. For a fixed $n \in \mathbb{N}$ we have $\frac{1}{n}T(2nB_X + 3n^2x_1) - 3ne_1 = T(2B_X) \supset M$. Therefore $\Psi(X) = Y$. □

The following is a generalisation of the well-known fact that every separable Banach space is a quotient of ℓ_1, [FHHMZ, Theorem 5.1].

Theorem 68 ([Há7]). *Let X be a Banach space such that there exists a non-compact operator $T \in \mathcal{L}(X;\ell_p)$, $1 \le p < \infty$. Then for every separable Banach space Y there exists a surjection $P \in \mathcal{P}({}^{\lceil p\rceil}X;Y)$.*

Proof. By Proposition 3.33 we may assume that there is a sequence $\{x_n\} \subset B_X$ such that $T(x_n) = e_n$, $n \in \mathbb{N}$, where $\{e_n\}$ is the canonical basis of ℓ_p. Choose a dense sequence $\{y_n\} \subset B_Y$. Denote $m = \lceil p\rceil$ and define a polynomial $Q \in \mathcal{P}({}^m\ell_p;Y)$ by $Q\bigl(\sum_{n=1}^\infty a_n e_n\bigr) = \sum_{n=1}^\infty a_n^m y_n$. Put $P = Q \circ T$. If $y \in Y$, then by Fact 64 there is an increasing sequence $\{n_k\}_{k=0}^\infty \subset \mathbb{N}$ such that $y = \|y\|\sum_{k=0}^\infty 2^{-km} y_{n_k}$. Putting $x = \|y\|^{\frac{1}{m}} \sum_{k=0}^\infty 2^{-k} x_{n_k}$ we get $P(x) = y$. □

Theorem 69 ([Bat]). *If a Banach space X has property $\mathcal{B}$, then for any separable Banach space Y there exists a surjective mapping $f \in \mathcal{C}^\infty(X;Y)$.*

Proof. Let $\{g_n\} \subset S_{X^*}$ be the sequence from Lemma 36. As in the proof of Proposition 37 we may assume that there exists a semi-normalised basic sequence $\{z_n\} \subset X$ which is biorthogonal to $\{g_n\}$. Put $x_n = \frac{3}{4}z_n$ and $R = \frac{1}{4}$. Let $\theta \in C^\infty(\mathbb{R};[0,1])$ be such that $\theta(t) = 1$ for $t \in [-\frac{R}{2},\frac{R}{2}]$ and $\theta(t) = 0$ for $|t| \ge R$ and let $\varphi\colon X \to \mathbb{R}$ be defined by

$$\varphi(x) = \prod_{n=1}^\infty \theta\bigl(g_n(x)\bigr).$$

For a fixed $x \in B(0,\rho)$ put $A = \{n \in \mathbb{N};\ |g_n(x)| > \frac{R}{4}\}$. Then $|A| \le k(\frac{R}{4\rho})$. For any $y \in B(x,\frac{R}{4})$ we have $\{n \in \mathbb{N};\ |g_n(y)| > \frac{R}{2}\} \subset A$ and so $\varphi(y) = \prod_{n\in A}\theta(g_n(y))$. Consequently, φ is C^∞-smooth on X and it has all derivatives bounded on bounded sets, i.e. $\varphi \in \mathcal{C}^\infty(X;[0,1])$. Denote by B the cube $B = \{x \in X;\ |g_n(x)| \le 1, n \in \mathbb{N}\}$ and note that $rB = \{x \in X;\ |g_n(x)| \le r, n \in \mathbb{N}\}$ for $r > 0$. Put $r = \frac{R}{2}$. Then $\operatorname{supp}\varphi \subset RB$ and $\varphi(x) = 1$ for $x \in rB$.

Note that we have $x_n + RB \subset B$ for $n \in \mathbb{N}$ and $\|x - y\| \ge |g_k(x-y)| \ge |g_k(x_k - x_n)| - |g_k(x - x_k)| - |g_k(y - x_n)| \ge \frac{3}{4} - 2R = \frac{1}{4}$ whenever $x \in x_k + RB$, $y \in x_n + RB$, $k \ne n$. Choose a dense sequence $\{y_k\} \subset B_Y$ and define constant functions $f_k(x) = y_k$ for $x \in X$, $k \in \mathbb{N}$. By Lemma 63 and Fact 64 there is $\psi \in \mathcal{C}^\infty(X;Y)$ such that $\operatorname{supp}\psi \subset B$ and $B_Y \subset \psi(B)$.

Put $\Psi(x) = \sum_{n=1}^\infty n\psi\bigl(\frac{1}{n}x - 2nz_1\bigr)$. Fix $x \in X$. Since $2n^2+n < 2(n+1)^2-(n+1)$ for every $n \in \mathbb{N}$, it is easy to see that there exists at most one $n \in \mathbb{N}$ that satisfies $|g_1(x) - 2n^2| \le n$. Therefore there exists a neighbourhood U of x such that $\bigl|g_1\bigl(\frac{1}{m}y - 2mz_1\bigr)\bigr| > 1$ for every $y \in U$ and $m \in \mathbb{N}$, $m \ne n$. Hence the sum in the definition of Ψ consists locally of at most one summand and consequently $\Psi \in C^\infty(X;Y)$. Moreover, $\bigl|g_1\bigl(\frac{1}{n}x - 2nz_1\bigr)\bigr| \ge 2n - \frac{1}{n}\|x\| > 1$ for all $n > \|x\|$ and so $\frac{1}{n}x - 2nz_1 \notin B$ for $n > \|x\|$. Thus all the derivatives of Ψ are bounded on bounded sets. Finally, for a fixed $n \in \mathbb{N}$ we have $\frac{1}{n}(nB + 2n^2z_1) - 2nz_1 = B$ and hence $\Psi(X) = Y$. □

8. Harmonic behaviour of smooth mappings

As we have seen in the previous section any separable Banach space can be a smooth image (even by a homogeneous polynomial) of an ℓ_p. In order to preserve some structural properties of the initial space ℓ_p one needs to consider other properties of the mappings than just surjectivity. The notion which leads to strong properties is the harmonic behaviour of the mapping (more precisely, the lack of it). The harmonic behaviour itself can be viewed as a condition of smallness of the image, in a broad sense. We begin the section by proving a key result (Theorem 71) which properly belongs to the linear structural theory of Banach spaces.

Let Y be a Banach space and Z a Banach space with a Schauder basis $\{e_n\}$. Let us denote the positive cone of Z by $Z^+ = \{z \in Z;\ z = \sum_{n=1}^{\infty} a_n e_n, a_n \geq 0\}$. We say that Z^+ embeds into Y onto a basic sequence $\{y_n\} \subset Y$ if there are $A > 0$ and $C \geq 1$ such that

$$A\left\|\sum_{n=1}^{\infty} a_n e_n\right\| \leq \left\|\sum_{n=1}^{\infty} a_n y_n\right\| \leq CA\left\|\sum_{n=1}^{\infty} a_n e_n\right\| \quad \text{for any } \sum_{n=1}^{\infty} a_n e_n \in Z^+. \tag{14}$$

We say that C is an *embedding constant.*

We are interested in the problem whether the fact that Z^+ embeds into Y already implies that Y contains Z. Note that the summing basis $\{x_n\}$ of c_0 has the property that $\left\|\sum_{n=1}^{\infty} a_n x_n\right\| = \sum_{n=1}^{\infty} a_n$ provided that $a_n \geq 0$, which means that ℓ_1^+ embeds into c_0, although c_0 does not contain ℓ_1. On the other hand, it is rather easy to see that if c_0^+ embeds into Y onto a basic sequence $\{y_n\}$, then Y contains c_0, and in fact $\{y_n\}$ is equivalent to the canonical basis of c_0. Indeed, for $a \in \mathbb{R}$ denote $a^+ = \max\{a, 0\}$ and $a^- = \max\{-a, 0\}$. If $\sum_{n=1}^{k} a_n y_n \in Y$, then

$$\begin{aligned}\left\|\sum_{n=1}^{k} a_n y_n\right\| &= \left\|\sum_{n=1}^{k} a_n^+ y_n - \sum_{n=1}^{k} a_n^- y_n\right\| \leq \left\|\sum_{n=1}^{k} a_n^+ y_n\right\| + \left\|\sum_{n=1}^{k} a_n^- y_n\right\| \\ &\leq CA\max\{a_n^+\} + CA\max\{a_n^-\} \leq 2CA\max\{|a_n|\}.\end{aligned}$$

But since $\{y_n\}$ is a semi-normalised basic sequence, the reverse inequality is automatic (see the proof of Theorem 60).

In the case of unconditional basic sequences the problem is easy:

Fact 70. *Let Z be a Banach space with an unconditional Schauder basis $\{e_n\}$, Y a Banach space, and suppose that Z^+ embeds into Y onto an unconditional basic sequence $\{y_n\}$. Then Y contains Z and in fact $\{y_n\}$ is equivalent to $\{e_n\}$.*

Proof. There exists $K_1 \geq 1$ such that $K_1^{-1}\left\|\sum_{n=1}^{k}|a_n| y_n\right\| \leq \left\|\sum_{n=1}^{k} a_n y_n\right\| \leq K_1\left\|\sum_{n=1}^{k}|a_n| y_n\right\|$ for any $\sum_{n=1}^{k} a_n y_n \in Y, k \in \mathbb{N}$, and there exists $K_2 \geq 1$ such that $K_2^{-1}\left\|\sum_{n=1}^{k}|a_n| e_n\right\| \leq \left\|\sum_{n=1}^{k} a_n e_n\right\| \leq K_2\left\|\sum_{n=1}^{k}|a_n| e_n\right\|$ for any $\sum_{n=1}^{k} a_n e_n \in Z$, $k \in \mathbb{N}$. Thus $K_1^{-1} A K_2^{-1}\left\|\sum_{n=1}^{k} a_n e_n\right\| \leq \left\|\sum_{n=1}^{k} a_n y_n\right\| \leq K_1 CAK_2\left\|\sum_{n=1}^{k} a_n e_n\right\|$ for any $\sum_{n=1}^{k} a_n e_n \in Z, k \in \mathbb{N}$. □

Theorem 71 ([HJ2]). *Let Y be a Banach space and $1 < p < \infty$. If ℓ_p^+ embeds into Y, then Y contains ℓ_p.*

Proof. Let $\{y_n\} \subset Y$ be the basic sequence such that ℓ_p^+ embeds onto $\{y_n\}$. We claim that there is an unconditional normalised block basic sequence of $\{y_n\}$ such that all its vectors have non-negative coordinates with respect to $\{y_n\}$. Then it is easily seen by Fact 70 and [FHHMZ, Proposition 4.45] that this block basic sequence is equivalent to the canonical basis of ℓ_p.

For $x = \sum_{n=1}^k a_n y_n \in Y$ we denote $x^+ = \sum_{n=1}^k a_n^+ y_n$, $x^- = \sum_{n=1}^k a_n^- y_n$, and $\widehat{x} = \sum_{n=1}^k a_n e_n \in \ell_p$. Suppose that $\{y_n\}$ is not unconditional and ℓ_p^+ embeds onto $\{y_n\}$ with an embedding constant C. Then for any $\varepsilon > 0$ there is $y \in \operatorname{span}\{y_n\}$ such that $\|y^+\| = 1$ and $\|y\| < \varepsilon$. If this was not true for some $\varepsilon > 0$, then for any $x \in \operatorname{span}\{y_n\}$

$$\|x\| \geq \varepsilon \max\{\|x^+\|, \|x^-\|\} \geq \frac{\varepsilon}{2}(\|x^+\| + \|x^-\|) \geq \frac{\varepsilon}{2}\|x^+ + x^-\|.$$

On the other hand

$$\begin{aligned}\|x\| = \|x^+ - x^-\| &\leq \|x^+\| + \|x^-\| \leq CA\Big(\big\|\widehat{x^+}\big\| + \big\|\widehat{x^-}\big\|\Big)\\ &\leq CA2^{1-\frac{1}{p}}\big\|\widehat{x^+} + \widehat{x^-}\big\| \leq C2^{1-\frac{1}{p}}\|x^+ + x^-\|,\end{aligned}$$

which means that $\{y_n\}$ would be unconditional. Thus we can find a block basic sequence $\{v_n\}$ of $\{y_n\}$ such that $\|v_n\| < \frac{A}{2} \cdot \frac{1}{2^n}$ and $\big\|\widehat{v_n^+}\big\| = 1$.

Let $\{a_n\}_{n=1}^k$ be a finite sequence of non-negative real numbers not all zero. Then

$$\Big\|\sum_{n=1}^k a_n v_n\Big\| \leq \sum_{n=1}^k a_n\|v_n\| \leq \max\{a_n\}\sum_{n=1}^k \|v_n\| \leq \frac{A}{2}\Big(\sum_{n=1}^k a_n^p\Big)^{\frac{1}{p}} \quad \text{and} \tag{15}$$

$$\begin{aligned}\Big\|\sum_{n=1}^k a_n v_n\Big\| &= \Big\|\sum_{n=1}^k a_n v_n^+ - \sum_{n=1}^k a_n v_n^-\Big\| \geq \Big\|\sum_{n=1}^k a_n v_n^-\Big\| - \Big\|\sum_{n=1}^k a_n v_n^+\Big\|\\ &\geq A\Big\|\sum_{n=1}^k a_n \widehat{v_n^-}\Big\| - CA\Big\|\sum_{n=1}^k a_n \widehat{v_n^+}\Big\| = A\Big\|\sum_{n=1}^k a_n \widehat{v_n^-}\Big\| - CA\Big(\sum_{n=1}^k a_n^p\Big)^{\frac{1}{p}},\end{aligned}$$

which implies

$$\Big\|\sum_{n=1}^k a_n \widehat{v_n^-}\Big\| \leq \Big(C + \frac{1}{2}\Big)\Big(\sum_{n=1}^k a_n^p\Big)^{\frac{1}{p}}. \tag{16}$$

As $\big\|\widehat{v_n^+}\big\| = 1$, clearly

$$\Big\|\sum_{n=1}^k a_n v_n^+\Big\| \geq A\Big\|\sum_{n=1}^k a_n \widehat{v_n^+}\Big\| = A\Big(\sum_{n=1}^k a_n^p\Big)^{\frac{1}{p}}. \tag{17}$$

For an upper estimate take $f \in S_{(\overline{\operatorname{span}}\{y_n\})^*}$ so that $f\left(\sum_{n=1}^k a_n v_n^+\right) = \left\|\sum_{n=1}^k a_n v_n^+\right\|$. We will estimate the positive part of the functional f on $\operatorname{span}\{v_n^+\}$ using duality on ℓ_p. Let $b_n = f(y_n)$, $n \in \mathbb{N}$ and put $M = \bigcup_{n=1}^k \operatorname{supp} v_n$ (which is a finite set). Further, we define $g = \sum_{n\in M} b_n y_n^*$, $g^+ = \sum_{n\in M} b_n^+ y_n^*$, $\widehat{g} = \sum_{n\in M} b_n e_n^*$, and $\widehat{g^+} = \sum_{n\in M} b_n^+ e_n^*$, where $\{y_n^*\}$ and $\{e_n^*\}$ are the functionals biorthogonal to $\{y_n\}$ and $\{e_n\}$ respectively. Note that $f(x) = g(x)$ for every $x \in \operatorname{span}\{y_n\}$ with $\operatorname{supp} x \subset M$. Let $\frac{1}{p} + \frac{1}{q} = 1$ and put $y = \sum_{n\in M} (b_n^+)^{q-1} y_n$. Then

$$\begin{aligned}\|\widehat{g^+}\| &= \left(\sum_{n\in M} (b_n^+)^q\right)^{\frac{1}{q}} = \frac{\sum_{n\in M}(b_n^+)^q}{\left(\sum_{n\in M}(b_n^+)^q\right)^{\frac{1}{p}}} \\ &= \frac{g(y)}{\|\widehat{y}\|} \le CA\frac{g(y)}{\|y\|} = CA\frac{f(y)}{\|y\|} \le CA.\end{aligned} \tag{18}$$

Using (17) we have

$$\begin{aligned}\left\|\sum_{n=1}^k a_n v_n\right\| &\ge f\left(\sum_{n=1}^k a_n v_n\right) = f\left(\sum_{n=1}^k a_n v_n^+\right) - f\left(\sum_{n=1}^k a_n v_n^-\right) \\ &= \left\|\sum_{n=1}^k a_n v_n^+\right\| - g\left(\sum_{n=1}^k a_n v_n^-\right) \ge \left\|\sum_{n=1}^k a_n v_n^+\right\| - g^+\left(\sum_{n=1}^k a_n v_n^-\right) \\ &\ge A\left(\sum_{n=1}^k a_n^p\right)^{\frac{1}{p}} - g^+\left(\sum_{n=1}^k a_n v_n^-\right).\end{aligned}$$

Denote $M^+ = \bigcup_{n=1}^k \operatorname{supp} v_n^+$, $M^- = \bigcup_{n=1}^k \operatorname{supp} v_n^-$, $\widehat{g^+}\restriction_{M^+} = \sum_{n\in M^+} b_n^+ e_n^*$, and $\widehat{g^+}\restriction_{M^-}$ similarly. The last inequality together with (15), the Hölder inequality, and (16) gives

$$\begin{aligned}\frac{A}{2}\left(\sum_{n=1}^k a_n^p\right)^{\frac{1}{p}} &\le g^+\left(\sum_{n=1}^k a_n v_n^-\right) \le \|\widehat{g^+}\restriction_{M^-}\| \left\|\sum_{n=1}^k a_n \widehat{v_n^-}\right\| \\ &\le \|\widehat{g^+}\restriction_{M^-}\| \left(C + \frac{1}{2}\right)\left(\sum_{n=1}^k a_n^p\right)^{\frac{1}{p}},\end{aligned}$$

which means that $\|\widehat{g^+}\restriction_{M^-}\| \ge \frac{A}{2C+1}$. If we combine this inequality with (18), we obtain

$$\|\widehat{g^+}\restriction_{M^+}\| = \left(\|\widehat{g^+}\|^q - \|\widehat{g^+}\restriction_{M^-}\|^q\right)^{\frac{1}{q}} \le \left(C^q A^q - \frac{A^q}{(2C+1)^q}\right)^{\frac{1}{q}}.$$

This finally allows us to estimate

$$\begin{aligned}\left\|\sum_{n=1}^{k} a_n v_n^+\right\| &= f\left(\sum_{n=1}^{k} a_n v_n^+\right) = g\left(\sum_{n=1}^{k} a_n v_n^+\right)\\ &\le g^+\left(\sum_{n=1}^{k} a_n v_n^+\right) \le \|\widehat{g^+}\restriction_{M^+}\| \left\|\sum_{n=1}^{k} a_n \widehat{v_n^+}\right\|\\ &= \|\widehat{g^+}\restriction_{M^+}\| \left(\sum_{n=1}^{k} a_n^p\right)^{\frac{1}{p}}\\ &\le CA\left(1-\frac{1}{C^q(2C+1)^q}\right)^{\frac{1}{q}} \left(\sum_{n=1}^{k} a_n^p\right)^{\frac{1}{p}}.\end{aligned}$$

The last inequality and (17) shows that $\{v_n^+\}$ is a semi-normalised block basis such that ℓ_p^+ embeds onto $\{v_n^+\}$ with an embedding constant smaller than C. Now either $\{v_n^+\}$ is an unconditional basic sequence and we are done, or we can iterate the process to find another block basis. (Notice that in every iteration the constructed block basis is a block basis of $\{y_n\}$ such that all of its vectors have non-negative coordinates with respect to the previous basic sequence and hence with respect to $\{y_n\}$.) In every iteration the embedding constant drops at least by the factor of $\left(1-\frac{1}{C^q(2C+1)^q}\right)^{1/q} < 1$, where C is the initial embedding constant corresponding to the basic sequence $\{y_n\}$. Therefore after finitely many steps we obtain an unconditional block basic sequence as we claimed, otherwise the embedding constant would eventually drop below 1, which is impossible. □

Observe that in the case $Z = \ell_p$, $1 < p < \infty$, or $Z = c_0$ if there is an arbitrary sequence $\{y_n\} \subset Y$ (not necessarily basic) such that (14) holds, then Z^+ embeds into Y. This can be seen as follows: Let $f \in (\overline{\operatorname{span}}\{y_n\})^*$. Similarly as in (18) we can show that $\sum_{n=1}^{\infty}\left(f(y_n)^+\right)^q < +\infty$ and $\sum_{n=1}^{\infty}\left(f(y_n)^-\right)^q < +\infty$. This means that $\{y_n\}$ is weakly null, and thus some subsequence of $\{y_n\}$ is a basic sequence (Proposition 3.32).

We will make use of yet another diagonalisation result, this time for spaces with perfectly homogeneous basis; compare also with Lemma 4.67, Lemma 4.91, and Lemma 44. A Schauder basis is called perfectly homogeneous if it is equivalent to any of its normalised block bases. By the result of Mordecay Zippin ([Zip1], see also [LiTz2, Theorem 2.a.9]) a Banach space X has a normalised perfectly homogeneous basis if and only if the space X is isomorphic to c_0 or ℓ_p, $1 \le p < \infty$. Recall that a non-constant homogeneous polynomial $P: X \to Y$ separates 0 from S_X if $\inf_{x\in S_X} \|P(x)\| > 0$.

Lemma 72 ([HJ2]). *Let $X = c_0$ or $X = \ell_p$, $1 \le p < \infty$, let Y be a Banach space, and $k \in \mathbb{N}$. Suppose that no polynomial in $\mathcal{P}(^n X; Y)$, $1 \le n < k$, separates 0 from S_X. Let $P \in \mathcal{P}(^k X; Y)$ and $\varepsilon > 0$. Then there is a normalised block basis $\{z_j\}$ of the canonical basis $\{e_j\}$ of X such that if $\sum_{j=1}^{\infty} a_j z_j \in B_X$ and $m \in \mathbb{N}$, then*

$$\left\| P\left(\sum_{j=m}^{\infty} a_j z_j\right) - \sum_{j=m}^{\infty} a_j^k P(z_j) \right\| < \frac{\varepsilon}{2^m}.$$

If moreover no polynomial in $\mathcal{P}(^k X; Y)$ separates 0 from S_X, then there is a normalised block basis $\{u_j\}$ of $\{z_j\}$ such that $\| P \restriction_{\overline{\operatorname{span}}\{u_j\}} \| \le \varepsilon$.

Proof. We prove the lemma by induction on k. For $k = 1$ let $P \in \mathcal{L}(X; Y)$ and $\varepsilon > 0$. The diagonalisation is trivial: we put $z_j = e_j$, $j \in \mathbb{N}$. Now assume that there is no $T \in \mathcal{L}(X; Y)$ that separates 0 from S_X. As $\overline{\operatorname{span}}\{e_j\}_{j \ge m}$ is isometric to X and so $P \restriction_{\overline{\operatorname{span}}\{e_j\}_{j \ge m}}$ does not separate 0 from $S_{\overline{\operatorname{span}}\{e_j\}_{j \ge m}}$, we can inductively construct a normalised block basis $\{u_j\}$ of $\{e_j\}$ such that $\|P(u_j)\| < \frac{\varepsilon}{2^j}$. If $\left\|\sum_{j=1}^{\infty} a_j u_j\right\| \le 1$, then

$$\left\| P\left(\sum_{j=1}^{\infty} a_j u_j\right)\right\| \le \sum_{j=1}^{\infty} |a_j| \|P(u_j)\| \le \sum_{j=1}^{\infty} \|P(u_j)\| < \varepsilon.$$

Now suppose that the assertion holds for $k - 1$ and let $P \in \mathcal{P}(^k X; Y)$ and $\varepsilon > 0$. We define the block basis $\{z_j\}$ inductively along with auxiliary block bases $\{v_j^n\}_{j=1}^{\infty}$, $n \in \mathbb{N}$, such that $\{v_j^n\}_{j=1}^{\infty}$ is a normalised block basis of $\{v_j^{n-1}\}_{j=1}^{\infty}$, $z_n = v_n^n$, and

$$\|\check{P}(z_{j_1}, \dots, z_{j_k})\| < \frac{\varepsilon}{2^{n+2} n^k} \quad \text{whenever } \min\{j_1, \dots, j_k\} < \max\{j_1, \dots, j_k\} = n > 1. \tag{19}$$

Let us put $v_j^1 = e_j$ for $j \in \mathbb{N}$, and $z_1 = e_1$. Assume that $z_1, \dots, z_{n-1}$ and $\{v_j^{n-1}\}_{j=1}^{\infty}$ are already defined for some $n \in \mathbb{N}$, $n > 1$. Consider the set of polynomials $A \subset \mathcal{P}^{k-1}(X; Y)$,

$$A = \left\{x \mapsto \check{P}(z_{j_1}, \dots, z_{j_{k-l}}, {}^l x);\ 1 \le j_1 \le \dots \le j_{k-l} \le n-1, 1 \le l < k\right\}.$$

Since $Z = \overline{\operatorname{span}}\{v_j^{n-1}\}_{j=1}^{\infty}$ is isometric to X, by the assumption no homogeneous polynomial on Z of degree at most $k - 1$ separates 0 from S_Z. Thus by the inductive hypothesis (on $k - 1$) there is a normalised block basis $\{v_j^n\}_{j=1}^{\infty}$ of $\{v_j^{n-1}\}_{j=1}^{\infty}$ such that $\|Q \restriction_{\overline{\operatorname{span}}\{v_j^n\}}\| < \frac{\varepsilon}{2^{n+2} n^k}$ for all $Q \in A$. (Note that A is finite and so we can find successive block bases, starting with a normalised block basis of $\{v_j^{n-1}\}_{j=1}^{\infty}$, so that the above estimate finally holds for all polynomials in A.) Put $z_n = v_n^n$. Since $\check{P}$ is symmetric, (19) is obviously satisfied.

Clearly, $\{z_j\}$ is a normalised block basis of $\{e_j\}$. If $\sum_{j=1}^{\infty} a_j z_j \in B_X$ and $m \in \mathbb{N}$, then by (19)

$$\left\| P\left(\sum_{j=m}^{N} a_j z_j\right) - \sum_{j=m}^{N} a_j^k P(z_j) \right\| \leq \sum_{\substack{j_1,\dots,j_k \geq m \\ \min\{j_l\} < \max\{j_l\}}} |a_{j_1} \cdots a_{j_k}| \cdot \|\check{P}(z_{j_1}, \dots, z_{j_k})\|$$

$$\leq \sum_{\substack{j_1,\dots,j_k \geq m \\ \min\{j_l\} < \max\{j_l\}}} \|\check{P}(z_{j_1}, \dots, z_{j_k})\| < \sum_{n=m+1}^{\infty} \sum_{m \leq \min\{j_l\} < \max\{j_l\} = n} \frac{\varepsilon}{2^{n+2} n^k}$$

$$< \varepsilon \sum_{n=m+1}^{\infty} \frac{1}{2^{n+2}} \sum_{1 \leq j_1,\dots,j_k \leq n} \frac{1}{n^k} = \frac{\varepsilon}{2^{m+2}}$$

for any $N \in \mathbb{N}$. Consequently, the sum $\sum_{j=1}^{\infty} a_j^k P(z_j)$ converges and the estimate follows.

In the case that no polynomial in $\mathcal{P}(^k X; Y)$ separates 0 from S_X, we can (similarly as for $k = 1$) find a normalised block basis $\{u_j\}$ of $\{z_j\}$ such that $\|P(u_j)\| < \frac{\varepsilon}{2^{j+1}}$, $j \in \mathbb{N}$. Let $\{\alpha_j\} \subset \mathbb{N}$ be an increasing sequence such that $u_j = \sum_{n=\alpha_j}^{\alpha_{j+1}-1} b_n z_n$. Then $\|P(u_j) - \sum_{n=\alpha_j}^{\alpha_{j+1}-1} b_n^k P(z_n)\| < \frac{\varepsilon}{2^{\alpha_j+2}} \leq \frac{\varepsilon}{2^{j+2}}$. Thus if $\sum_{j=1}^{\infty} a_j u_j \in B_X$, then

$$\left\| P\left(\sum_{j=1}^{\infty} a_j u_j\right) - \sum_{j=1}^{\infty} a_j^k P(u_j) \right\|$$

$$\leq \left\| P\left(\sum_{j=1}^{\infty} a_j \sum_{n=\alpha_j}^{\alpha_{j+1}-1} b_n z_n\right) - \sum_{j=1}^{\infty} a_j^k \sum_{n=\alpha_j}^{\alpha_{j+1}-1} b_n^k P(z_n) \right\|$$

$$+ \left\| \sum_{j=1}^{\infty} a_j^k P(u_j) - \sum_{j=1}^{\infty} a_j^k \sum_{n=\alpha_j}^{\alpha_{j+1}-1} b_n^k P(z_n) \right\|$$

$$< \frac{\varepsilon}{4} + \sum_{j=1}^{\infty} |a_j|^k \left\| P(u_j) - \sum_{n=\alpha_j}^{\alpha_{j+1}-1} b_n^k P(z_n) \right\| < \frac{\varepsilon}{2},$$

and so

$$\left\| P\left(\sum_{j=1}^{\infty} a_j u_j\right) \right\| \leq \left\| P\left(\sum_{j=1}^{\infty} a_j u_j\right) - \sum_{j=1}^{\infty} a_j^k P(u_j) \right\| + \sum_{j=1}^{\infty} |a_j|^k \|P(u_j)\| < \varepsilon.$$

□

Compare the following lemma with Fact 4.45.

Lemma 73. *Let $X = c_0$ or $X = \ell_p$, $1 \leq p < \infty$, let Y be a Banach space, and $n \in \mathbb{N}$. If there is a polynomial $P \in \mathcal{P}^n(X; Y)$ that separates $z \in X$ from $\{x \in X;\ \|x - z\| = r\}$ for some $r > 0$, then there is a homogeneous $Q \in \mathcal{P}^n(X; Y)$ that separates 0 from $X \setminus B(0, s)$ for every $s > 0$.*

Proof. Let us put $R(x) = P(rx + z) - P(z)$. Then $\inf_{x \in S_X} \|R(x)\| = \varepsilon > 0$. Let $R_k \in \mathcal{P}(^k X; Y)$ be such that $R = \sum_{k=1}^{n} R_k$. Clearly it suffices to find a homogeneous $Q \in \mathcal{P}^n(X; Y)$ that separates 0 from S_X. Assume that no such homogeneous polynomial exists. Then by Lemma 72 there is a non-trivial subspace $Z \subset X$ such that $\|R_k \restriction_Z\| < \frac{\varepsilon}{n}$ for each $k = 1, \dots, n$, a contradiction. □

Recall that a harmonic function f on a bounded open $U \subset \mathbb{R}^n$ satisfies the Maximum principle: if f is continuous on $\overline{U}$, then f attains its maximum and minimum over $\overline{U}$ on ∂U. It follows that f is determined by its values on the boundary (see below). Also, if ∂U is connected, then $f(\overline{U}) = f(\partial U)$. This motivates the following definition.

Definition 74. Let X, Y be normed linear spaces and $U \subset X$ an open set. We say that a mapping $f : \overline{U} \to Y$ is harmonic-like on U if $f(\overline{V}) \subset \overline{f(\partial V)}$ for every bounded open $V \subset U$.

Note that if all mappings in some linear subspace $S \subset Y^{\overline{U}}$ are harmonic-like on U, then they are determined by their boundary values: if $f, g \in S$ are such that $f = g$ on ∂V for some bounded open $V \subset U$, then $f = g$ on $\overline{V}$. Indeed, $f - g \in S$ and so $(f - g)(\overline{V}) \subset \overline{(f - g)(\partial V)} = \{0\}$.

The following Proposition shows a prototype result concerning harmonic-like mappings.

Proposition 75 ([BF]). *If X is a non-Asplund Banach space, Y is an Asplund space, and $U \subset X$ is open, then every $f \in C(\overline{U}; Y)$ that is Fréchet differentiable on U is harmonic-like.*

Proof. Suppose that there is a bounded open set $V \subset U$ and $x \in V$ such that $\operatorname{dist}(f(x), f(\partial V)) = \delta > 0$. We may assume without loss of generality that $x = 0$. By Theorem 5.2 there is a separable subspace $Z \subset X$ that is not Asplund. Put $W = V \cap Z$ and $E = \overline{\operatorname{span}} f(\overline{W})$. Then E is a separable Asplund space and so there is $\varphi \in C^1(E)$ such that $\varphi(f(0)) = 1$ and $\varphi(y) = 0$ if $\|y - f(0)\| \geq \frac{\delta}{2}$ (Theorem 5.2). Define $\psi : Z \to \mathbb{R}$ by $\psi(z) = \varphi \circ f(z)$ for $z \in \overline{W}$ and $\psi(z) = 0$ otherwise. Since $\overline{W}$ is bounded, ψ is clearly a bump. Further, ψ is Fréchet differentiable on W and $\psi = 0$ on a neighbourhood of $Z \setminus W$, so ψ is Fréchet differentiable. This contradicts Corollary 5.3. □

The following variant is proved by an analogous argument.

Proposition 76 ([BF]). *Let X, Y be Banach spaces such that Y admits a $C^{k,\alpha}$-smooth bump but X does not and let $U \subset X$ be open. Then every $f \in C^{k,\alpha}(U; Y)$ continuously extended to $\overline{U}$ is harmonic-like.*

We are going to prove more results of this kind, concerning structural properties related to higher smoothness of the space. It is clear that if X admits a $C^{k,\alpha}$-smooth bump, then for every Banach space Y there exists $f \in C^{k,\alpha}(X; Y)$ that is not harmonic-like (as $\mathbb{R} \subset Y$). We investigate for a given $1 \leq p < \infty$ and $1 \geq \alpha > p - [p]$ the structural conditions on Y which imply that every $f \in C^{[p],\alpha}(U;Y)$, $U \subset \ell_p$ is harmonic-like. (Recall that ℓ_p has a $C^{[p],p-[p]}$-smooth bump if $p \notin \mathbb{N}$, Theorem 5.106.) In particular we show that every such mapping is harmonic-like unless Y contains $\ell_{\frac{p}{n}}$ for some integer $1 \leq n \leq [p]$. It should be noted that the image $f(\overline{U})$ may be large: If Y is separable, then there is even a homogeneous polynomial $P\colon \ell_p \to Y$ such that $B_Y \subset P(B_{\ell_p})$ (Theorem 68).

Proposition 77 ([HJ2]). *Let Y be a Banach space, $1 \leq p < \infty$, $U \subset \ell_p$ open, and $k \in \mathbb{N}$, $k \geq [p]$. Let $S = C(\overline{U};Y) \cap C^{k,\alpha}(U;Y)$ for some $\alpha \in (0,1]$ satisfying $k+\alpha > p$ or $S = C(\overline{U};Y) \cap C^{k,+}(U;Y)$ for $k = p$. All mappings in S are harmonic-like if and only if there is no polynomial in $\mathcal{P}({}^n\ell_p;Y)$, $n = 1,\dots,k$, that separates 0 from S_{ℓ_p}.*

Proof. Clearly, a polynomial that separates 0 from S_{ℓ_p} is not harmonic-like. Conversely, suppose that $f \in C(\overline{U};Y) \cap C^{k,\alpha}(U,Y)$ is not harmonic-like. There is a bounded open $V \subset U$ and $y \in V$ such that $\operatorname{dist}(f(y), f(\partial V)) = \varepsilon > 0$. We may assume without loss of generality that $y = 0$. Suppose that no polynomial in $\mathcal{P}({}^n\ell_p;Y)$, $n = 1,\dots,k$ separates 0 from S_{ℓ_p}.

Let $C > 0$ be such that $\|d^k f(u) - d^k f(v)\| \leq C\|u-v\|^\alpha$ for all $u,v \in V$, let $R > 0$ be such that $V \subset B(0,R)$, and find $N \in \mathbb{N}$ such that $\frac{1}{k!}CR^{k+\alpha}N^{1-\frac{k+\alpha}{p}} < \frac{\varepsilon}{2}$. By induction we find finitely supported vectors $x_0,\dots,x_m \in \ell_p$, $m \leq N$, such that $x_0 = 0$, $x_j \in V$ and $\|x_j\| \geq \left(\frac{j}{N}\right)^{\frac{1}{p}} R$ for $j = 0,\dots,m-1$, $x_m \in \partial V$, and $\|f(x_j) - f(x_{j-1})\| < \frac{\varepsilon}{N}$ for $j = 1,\dots,m$. Since $V \subset B(0,R)$, the requirements on x_j ensure that $m \leq N$.

Assume that the vectors $x_0,\dots,x_j$ are already defined for some $0 \leq j < N$. By Lemma 72 there is a normalised block basis $\{u_l\}$ of the canonical basis $\{e_l\}$ such that $\|d^n f(x_j)\restriction_{\overline{\operatorname{span}}\{u_l\}}\| < \frac{\varepsilon n! N^{n/p}}{2kNR^n}$ for $1 \leq n \leq k$. Thus there is a finitely supported $v \in S_{\ell_p}$ with $\max\operatorname{supp} x_j < \min\operatorname{supp} v$ such that $\frac{1}{n!}\frac{R^n}{N^{n/p}}\|d^n f(x_j)[v]\| < \frac{\varepsilon}{2kN}$ for all $1 \leq n \leq k$. If $x_j + tv \in V$ for all $t \in [0, \frac{R}{N^{1/p}}]$, then we set $x_{j+1} = x_j + \frac{R}{N^{1/p}}v$. Note that in this case $x_{j+1} \in V$ and $\|x_{j+1}\|^p = \|x_j\|^p + \frac{R^p}{N} \geq \frac{j+1}{N}R^p$. Otherwise we set $x_{j+1} = x_j + t_0 v$, where $t_0 = \inf\{t \geq 0;\ x_j + tv \notin V\}$, and further we set $m = j+1$ and finish the construction. In the latter case clearly $x_m \in \partial V$. In both cases by Corollary 1.108

$$\begin{aligned}\|f(x_{j+1}) - f(x_j)\| &\leq \sum_{n=1}^{k} \frac{1}{n!}\big\|d^n f(x_j)[x_{j+1}-x_j]\big\| + \frac{1}{k!}C\|x_{j+1}-x_j\|^{k+\alpha} \\ &\leq \sum_{n=1}^{k} \frac{\varepsilon}{2kN} + \frac{1}{k!}C\frac{R^{k+\alpha}}{N^{\frac{k+\alpha}{p}}} < \frac{\varepsilon}{N}.\end{aligned}$$

We have $\varepsilon = \operatorname{dist}(f(0), f(\partial V)) \leq \|f(x_m) - f(0)\| \leq \sum_{j=1}^m \|f(x_j) - f(x_{j-1})\| < \varepsilon$, which is a contradiction. The proof in the $C^{k,+}$-smooth case is analogous. □

The next theorem shows that separating vector-valued polynomials on ℓ_p or c_0 are analogues of linear isomorphisms.

Theorem 78 ([HJ2]). *Let Y be a Banach space, $1 \le p < \infty$, and $k \in \mathbb{N}$. Suppose that no polynomial in $\mathcal{P}({}^n\ell_p; Y)$, $1 \le n < k$, separates 0 from S_{ℓ_p}. If k is odd and $k \le p$, or if k is even and $k < p$, then there is $P \in \mathcal{P}({}^k\ell_p; Y)$ that separates 0 from S_{ℓ_p} if and only if Y contains $\ell_{\frac{p}{k}}$.*

There is a homogeneous $P \in \mathcal{P}(c_0; Y)$ that separates 0 from S_{c_0} if and only if Y contains c_0 if and only if for every $k \in \mathbb{N}$ there is $P \in \mathcal{P}({}^k c_0; Y)$ that separates 0 from S_{c_0}.

Proof. By $\{e_j\}$ we denote all the canonical bases in the respective spaces. First we prove the ℓ_p case.

$\Leftarrow$ Clearly, $Q\colon \ell_p \to \ell_{\frac{p}{k}}$ defined by $Q\big(\sum_{j=1}^{\infty} a_j e_j\big) = \sum_{j=1}^{\infty} a_j^k e_j$ is a k-homogeneous polynomial that separates 0 from S_{ℓ_p}. Hence if T is an isomorphism from $\ell_{\frac{p}{k}}$ into Y, then $T \circ Q \in \mathcal{P}({}^k\ell_p; Y)$ separates 0 from S_{ℓ_p}.

$\Rightarrow$ Let $\varepsilon = \inf_{x \in S_{\ell_p}} \|P(x)\| > 0$ and let $\{z_j\}$ be the normalised block basis of $\{e_j\}$ from Lemma 72. Put $y_j = P(z_j)$. If k is odd, then for any sequence $\{a_j\} \subset \mathbb{R}$ satisfying $\sum_{j=1}^{\infty} |a_j|^{\frac{p}{k}} = 1$ we have

$$\Big\|\sum_{j=1}^{\infty} a_j y_j\Big\| = \Big\|\sum_{j=1}^{\infty} \big(\sqrt[k]{a_j}\big)^k P(z_j)\Big\| < \Big\|P\Big(\sum_{j=1}^{\infty} \sqrt[k]{a_j} z_j\Big)\Big\| + \frac{\varepsilon}{2}$$
$$\le \|P\| \Big\|\sum_{j=1}^{\infty} \sqrt[k]{a_j} z_j\Big\|^k + \frac{\varepsilon}{2} = \|P\| + \frac{\varepsilon}{2}.$$

On the other hand,

$$\Big\|\sum_{j=1}^{\infty} a_j y_j\Big\| > \Big\|P\Big(\sum_{j=1}^{\infty} \sqrt[k]{a_j} z_j\Big)\Big\| - \frac{\varepsilon}{2} \ge \varepsilon \Big\|\sum_{j=1}^{\infty} \sqrt[k]{a_j} z_j\Big\|^k - \frac{\varepsilon}{2} = \frac{\varepsilon}{2}.$$

This implies that $\overline{\operatorname{span}}\{y_j\} \subset Y$ is a subspace isomorphic to $\ell_{\frac{p}{k}}$. If k is even, then $a_j = \big(\sqrt[k]{a_j}\big)^k$ only if $a_j \ge 0$ and therefore we obtain merely that $\ell^+_{\frac{p}{k}}$ embeds into Y. (In view of the remark after Theorem 71 we do not need $\{y_j\}$ to be a basic sequence.) Now Theorem 71 finishes the proof for k even.

In the case of c_0 we start by considering the homogeneous polynomial of the smallest degree that separates 0 from S_{c_0} and analogously as above we conclude that $c_0 \subset Y$. Then we use the fact that $Q\colon c_0 \to c_0$ defined by $Q\big(\sum_{j=1}^{\infty} a_j e_j\big) = \sum_{j=1}^{\infty} a_j^k e_j$ is a k-homogeneous polynomial that separates 0 from S_{c_0}. □

If $p \in \mathbb{N}$ is even, then $x \mapsto \|x\|^p$ is a separating polynomial on ℓ_p and so (using also Corollary 3.59) in this case the statement of Theorem 78 dos not hold for $k = p$. Putting together Proposition 77 and Theorem 78 we immediately obtain the following result.

Theorem 79 ([HJ2]). *Let Y be a Banach space, $1 \leq p < \infty$, p not an even integer, and $U \subset \ell_p$ open. Let $S = C(\overline{U}; Y) \cap C^{[p],\alpha}(U; Y)$ for some $\alpha \in (0, 1]$ satisfying $[p] + \alpha > p$ or $S = C(\overline{U}; Y) \cap C^{p,+}(U; Y)$ if $p \in \mathbb{N}$. Then either all mappings in S are harmonic-like or Y contains $\ell_{\frac{p}{k}}$ for some $1 \leq k \leq [p]$.*

9. Notes and remarks

Section 1. The results in this section are mostly improvements on the work concerning weakly uniformly continuous and simultaneously uniformly smooth mappings by [AP] and [CHL]. In particular, Theorems 8 and 9 are improvements of the results which appeared many times in the literature, e.g. in [AP], [CHL], [CiGu1], [CiGu2], [CGS]. As for the definition of $\mathcal{C}^k_{\mathrm{wu}}$, Guillermo Restrepo [Res] seems to be the earliest reference, in the case of reflexive X and $k = 1$.

Section 2. The bidual extension operation E introduced in this section had many precursors, all in the setting of polynomials or convergent power series. Perhaps one of the earliest versions appears for bilinear mappings in [Are]. Extensions of polynomial mappings between Banach spaces to their biduals have been considered by Aleksander Pełczyński in [Peł4], [Peł3], in the special case of $C(K)$ or spaces with the DPP. For a general Banach space they were described by Richard Martin Aron and Paul D. Berner [ArBer] in their seminal paper. The polynomial extension leads naturally to extensions of holomorphic functions via the Taylor formula. The topic received considerable attention in the literature, e.g. [DT2], [LindsRy], [GGMM], [Za1], [Vi], [Aro3], [Din, p. 79 and Chapter 6].

Our presentation is based on [CHL], where the extensions were obtained for the first time in the setting of uniformly k-times differentiable mappings, which are not necessarily analytic.

Problem 80. Find the most general conditions under which E and D are weakly commuting in $\lambda B_{X^{**}}$, $\lambda > 0$.

A natural and important question is whether the extensions exist without the uniformity assumption, namely:

Problem 81. Let $f \in \mathcal{C}^k(X; Y)$, $k \in \mathbb{N} \cup \{\infty\}$. Does there exist a bidual extension $f^{**} \in \mathcal{C}^k(X^{**}; Y^{**})$?

The special case with $X = c_0$, $Y = \mathbb{R}$ (a problem posed by R. M. Aron) seems to be also open. It was shown in [AFK1] that C^1-smooth functions can be extended (preserving C^1-smoothness) from subspaces of separable Asplund spaces. On the other hand, Václav Zizler [Ziz] proved that extensions preserving higher smoothness are rare. More precisely, let X be a separable Asplund space containing a subspace Y isomorphic to ℓ_2. Then the Hilbertian norm $\|\cdot\|$ of Y can be extended to a real valued function f on X that has Fréchet derivative on $B_Y \setminus \{0\}$ and such that Df is locally Lipschitz on S_Y if and only if Y is complemented.

Problem 82. Can C^1-smooth norms be extended from subspaces of a separable reflexive (or Asplund) space X to the whole space preserving the C^1-smoothness?

This result is false for Gâteaux smooth norms due to [Ziz]. Theorem 14 generalises the classical results from [Li1] concerning extensions of linear operators into the case of uniformly smooth mappings. For the general problem of extending mappings from subspaces to their superspaces preserving various properties see [BenLi].

Sections 3–6. The theory of the class $\mathcal{W}$ was developed in [Há5], [Há6], [DeH2], and [CHL].

In view of Proposition 33 we pose the next problem.

Problem 83. Assuming that X belongs to the class $\mathcal{W}$, is there an equivalent renorming of X such that X equipped with the new norm belongs to the class $\mathcal{W}_1$?

Belonging to the class $\mathcal{W}$ is not a three space property, as witnessed by the space $X = \ell_1 \oplus_1 S$, [CHL].

Problem 84. Let c_0 be equipped with the canonical supremum norm. Given a function $f \in \mathcal{C}^{1,+}(B_{c_0})$ and $\rho > 1$, is there an extension $F \in \mathcal{C}^{1,+}(\rho B_{c_0})$, $F = f$ on B_{c_0}?

Problem 85. If X is a $\mathcal{W}_1$-space, is then for some Banach space Y (or at least $Y = \mathbb{R}$)

$$\mathcal{C}^{1,+}(B_X; Y) \cap \mathcal{C}_{\mathrm{wsc}}(B_X; Y) = \mathcal{C}^{1,+}(B_X; Y) \cap \mathcal{C}_{\mathrm{wsC}}(B_X; Y)?$$

Perhaps the most interesting open problem of this theory is the following.

Problem 86. Is property (WR) for a Banach space Y is equivalent to Y not containing c_0?

Recall that if X is a $\mathcal{W}$-space, then $\mathcal{P}(X) = \mathcal{P}_{\mathrm{wsC}}(X)$. The following is a stronger version of Problem 4.103, since the assumptions imply that X does not admit a non-compact operator into ℓ_p, $1 < p < \infty$.

Problem 87. Let X be a Banach space failing property $\mathcal{B}$. Is then $\mathcal{P}(^nX) = \mathcal{P}_{\mathrm{wsC}}(^nX)$?

Theorem 56 was inspired by the results of Mordecay Zippin [Zip2]. However, our result does not follow from [Zip2] as there are examples of non-Schur Banach spaces that are ℓ_1-saturated, see [AzHa]. In particular, there is a Banach space with a weakly null Schauder basis which contains ℓ_1, and so its dual is non-separable.

Recall that X has the DPP and does not contain ℓ_1 if and only if X^* is a Schur space (Theorem 3.41).

Problem 88. If X^* is a Schur space, is then X a $\mathcal{W}$-space? More generally, if X has the DPP, is then X a $\mathcal{W}$-space?

In fact, we do not know if $L_1([0,1])$ is a $\mathcal{W}$-space.

The fixing of the canonical basis of c_0 by non-compact polynomials from $C(K)$, K scattered (Theorem 60) is a direct generalisation of the linear Theorem 3.47. Theorem 62 shows that a similar conclusion fails for uniformly smooth mappings from the unit ball. A more complicated variant of the construction in Theorem 62 leads to a somewhat stronger statement. Namely, the positive cone generated in ℓ_∞ by $f(B_{c_0})$ does not necessarily contain the canonical basis of c_0, [DeH2].

Let us mention for completeness that in the theory of linear operators on $C(K)$ spaces many results on the fixing of subspaces (preferably as large as possible) which generalise and strengthen Theorem 3.47 (the c_0-fixing property) have been obtained, see e.g. [BourgJ1], [AlBen], [Wol], [Ros1]. These results are connected with the still open problem of the isomorphic classification of complemented subspaces of $C[0,1]$. For more we refer e.g. to [Ros3].

In should be noted that the highly relevant theory of isomorphic ℓ_1-preduals is rapidly developing. It was shown in [FOS] that every Banach space with a separable dual can be isomorphically embedded into a $\mathcal{L}_\infty$-space whose dual is isomorphic to ℓ_1.

Section 7. The first example of the failure of Rolle's theorem (even for a polynomial of degree 4 on a Hilbert space) in an arbitrary infinite-dimensional space with a smooth norm is due to Stanislav A. Shkarin [Shk]. This seminal work has lead to a large number of papers studying the ranges of derivatives, e.g. [AzDe], [ADJ], [AFJ], [AJ1], [GHM], [BFKL], [BFL], the survey [Dev3], [DeH1], [GaspaT1], [GaspaT2], [HJ1], [FKK]. In [BenLi, p. 259] there is an example of a pair X, Y of non-isomorphic Banach spaces such that there exist real analytic (even polynomial of degree 3) surjections in both directions which are topological homeomorphisms.

Let us also mention a couple of results concerning C^k-diffeomorphisms. Parts of this theory have their roots in the work of Czesław Bessaga in [Bes]. It was shown in [Do3] and [Az] that if a Banach space X has a C^k-smooth norm, $k \in \mathbb{N} \cup \{\infty\}$, then any hyperplane in X is C^k-diffeomorphic with the unit sphere S_X. If X admits a C^k-smooth norm (which need not be equivalent to the original norm), then X is C^k-diffeomorphic with $X \setminus \{0\}$. These kinds of results play a role in the theory of C^k-smooth Banach manifolds. We refer to [Mo1], [BuKu], and [EE] for a sample of classical results in this vast field of research.

Section 8. The results in this section come from [HJ2]. In the same paper it was shown that if ℓ_p^+, $1 < p < \infty$ embeds into Y onto a basic sequence $\{y_n\}$ with the embedding constant $C < 2^{1-\frac{1}{p}}$, then $\{y_n\}$ is equivalent to the canonical basis of ℓ_p, and this inequality is sharp in the sense that for any $1 < p < \infty$ there is a space X isomorphic to $c_0 \oplus \ell_p$ with a Schauder basis $\{y_n\}$ such that ℓ_p^+ embeds into X onto $\{y_n\}$ with the embedding constant $2^{1-\frac{1}{p}}$. The paper [HJ2] also deals with the case of complex ℓ_p.

In [DM] the following result is proved:

Theorem 89. *Let X, Y be Banach spaces such that Y is of a non-trivial cotype, while X is not of a non-trivial cotype, and let $U \subset X$ be open. Then all mappings in $C(\overline{U}; Y) \cap C^{1,+}(U; Y)$ are harmonic-like.*

The C^k-smooth non-linear mappings are of course related with C^k-smooth vector fields in Banach spaces, in other words with ODEs corresponding to C^k-smooth right-hand side. This line of research was pursued in [HV2], and developed in a series of papers [HJ6], [HV1], and [HV3], where some of the problems regarding the local and global behaviour of abstract ODEs in Banach spaces, e.g. the local existence and global behaviour such as ω-limit sets and cross-sections of the solution funnels, have been solved. The general theory of ODEs in infinite-dimensional spaces is treated e.g. in [Dei] or [DMNZ].

Chapter 7

Smooth approximation

In the present chapter we are concerned with the general problem of approximating a given mapping from a subset of a Banach space X into a Banach space Y by means of polynomials or C^k-smooth mappings. The classical example is the Weierstraß-type Theorem 9, where arbitrary C^k-smooth mapping from a compact set $K \subset \mathbb{R}^n$ is shown to be approximable by polynomials, uniformly on K together with derivatives of order up to k. The best-known case is when $k = 0$.

Our interest lies primarily in the case when the unit ball B_X is non-compact and the Weierstraß theorem for uniformly continuous functions fails. It turns out, quite surprisingly, that in the special case of $\mathcal{W}$-spaces not containing ℓ_1, C^k-smooth mappings with uniformly continuous derivatives can be approximated together with their higher derivatives by polynomials uniformly on bounded sets. This result underlines the extremely poor supply of uniformly smooth functions on these spaces rather than the abundance of polynomials. Indeed, all polynomials are in this case weakly uniformly continuous on B_X.

In Section 3 we prove one of the highlights in the theory of smoothness, Theorem 29 of Jaroslav Kurzweil. This result claims that if a real separable Banach space X admits a separating polynomial, then every continuous mapping from X into a Banach space can be uniformly approximated by real analytic mappings. By adjusting the proof somewhat it can be shown that the result remains true for Banach spaces that admit a separating analytic function with uniform radii of convergence (e.g. for c_0), provided that the approximated mapping is uniformly continuous.

One of the principal tools for obtaining C^k-smooth approximations of continuous mappings on infinite-dimensional spaces are partitions of unity, which are studied in Section 5. This is a very powerful tool which leads to general positive results in separable spaces, as well as in non-separable WCG or $C(K)$ spaces, admitting a C^k-smooth bump.

The rest of the chapter is devoted to the study of smooth approximations preserving special properties of the approximated mapping. In order to study C^k-smooth approximations of Lipschitz mappings preserving the Lipschitz condition we introduce the concept of sup-partitions of unity and characterise it by means of componentwise C^k-smooth and bi-Lipschitz embeddings into $c_0(\Gamma)$. We show that every separable Banach space admitting a Lipschitz and C^k-smooth bump admits C^k-smooth sup-partitions of unity. This is applied to establish the existence of C^k-smooth and Lipschitz approximations of a given Lipschitz function in a separable Banach space X admitting a C^k-smooth and Lipschitz bump function. The real analytic case is also included, under the assumption of the existence of a separating polynomial. We also obtain results of this sort for vector valued Lipschitz mappings for certain types of the domain or range spaces.

In Section 8 we prove, again for certain types of the domain or range spaces, the existence of approximations of C^1-smooth mappings by C^k-smooth mappings for the mapping and its first derivative.

The last section is devoted to C^k-smooth (and convex) approximations of convex functions. It is rather easy to see that this problem is essentially equivalent to the statement that every equivalent norm on X can be approximated by C^k-smooth norms. A necessary condition for this result is clearly the existence of at least some equivalent C^k-smooth norm on the space X (similarly to the existence of a C^k-smooth bump in the general case). We prove that this condition is also sufficient for all separable Banach spaces.

The methods in this chapter can be essentially divided into two groups: Global methods, in which the approximating mapping is constructed on the whole space at once by a formula – these are represented by integral and infimal convolutions; and local methods, in which we approximate locally and then glue together the approximations using for example partitions of unity. Some of the methods are mixed, for example the real analytic approximations although using the partitions of unity are necessarily retaining the global flavour. A note on the hypotheses on the target space: if the method employed uses a limiting procedure (this includes the Bochner integral, or countable partitions of unity), we need the completeness of the space for the process to converge. This is not necessary if we use a finite procedure such as the locally finite partitions of unity. Likewise, the domain space needs to be complete only in some circumstances, for example when using Schauder bases, or when we are dealing with real analytic mappings.

In this chapter all the spaces are real unless stated otherwise.

1. Separation

As we shall see, a lot of the approximation methods ultimately boils down to the ability of separation of certain sets by smooth functions. In this section we present some of the separation results that will be used later in the chapter. Some of them are somewhat technical, as we require rather fine separation properties. Nevertheless, we begin with a lemma that serves as a prominent tool for smoothing up mappings on $\mathbb{R}^n$ and lies behind most of the approximation results. The method was used already by Karl Weierstraß.

Let X be a set, Y a normed linear space, $f: X \to Y$, and $S \subset X$. We denote $\|f\|_S = \sup_{x \in S} \|f(x)\|$. In the following lemma we consider the space $\mathbb{C}^n$ with the Euclidean norm.

Lemma 1. *Let Y be a Banach space, $C \subset Y$ a closed convex set, and let $f\colon \mathbb{R}^n \to C$ be strongly measurable (with respect to the Lebesgue measure) and bounded. Put $\Psi_\kappa(z) = \exp\bigl(-\kappa \sum_{j=1}^n z_j^2\bigr)$ for $z \in \mathbb{C}^n$ and $\kappa \in \mathbb{R}^+$, and define $g_\kappa\colon \mathbb{C}^n \to \tilde{Y}$ by the Bochner integral*

$$g_\kappa(z) = \frac{1}{c_\kappa} \int_{\mathbb{R}^n} \Psi_\kappa(z-y) f(y)\,\mathrm{d}y,$$

where $c_\kappa = \int_{\mathbb{R}^n} \Psi_\kappa(y)\,\mathrm{d}y = \bigl(\frac{\pi}{\kappa}\bigr)^{\frac{n}{2}}$. Then $g_\kappa \in H(\mathbb{C}^n;\tilde{Y})$ and $g_\kappa\restriction_{\mathbb{R}^n} \in C^\omega(\mathbb{R}^n;C)$ for every $\kappa \in \mathbb{R}^+$. Further, if f is Bochner integrable, then for every $\delta > 0$

$$\lim_{\kappa\to+\infty} \|g_\kappa\|_{G_\delta} = 0$$

(in case that $G_\delta \neq \emptyset$), where

$$G_\delta = \bigl\{z \in \mathbb{C}^n;\ \|\operatorname{Im} z\|^2 < \operatorname{dist}(\operatorname{Re} z, \operatorname{supp} f)^2 - \delta^2\bigr\}.$$

If f is L-Lipschitz, then so is each $g_\kappa\restriction_{\mathbb{R}^n}$, $\kappa \in \mathbb{R}^+$. Finally, if $f \in C^k(\mathbb{R}^n;Y)$ for some $k \in \mathbb{N}_0 \cup \{\infty\}$ with all $d^j f$, $j = 0,\dots,k$ bounded and uniformly continuous on $\mathbb{R}^n$, then for all $0 \le j \le k$

$$\lim_{\kappa\to+\infty} \bigl\|d^j g_\kappa - d^j f\bigr\|_{\mathbb{R}^n} = 0.$$

Proof. First note that

$$\begin{aligned}|\Psi_\kappa(z)| &= \exp\Bigl(-\kappa \sum_{j=1}^n \operatorname{Re}(z_j^2)\Bigr) = \exp\Bigl(-\kappa \sum_{j=1}^n \bigl((\operatorname{Re} z_j)^2 - (\operatorname{Im} z_j)^2\bigr)\Bigr)\\ &= \exp\Bigl(-\kappa\bigl(\|\operatorname{Re} z\|^2 - \|\operatorname{Im} z\|^2\bigr)\Bigr)\end{aligned}$$

and $D\Psi_\kappa(z)[h] = -2\kappa\Psi_\kappa(z)\sum_{j=1}^n z_j h_j$. Put $F(z,y) = \Psi_\kappa(z-y)f(y)$ and let $K > 0$ be such that $\|f\|_{\mathbb{R}^n} \le K$. Then for $z \in \mathbb{C}^n$, $\|z\| < r$, and $y \in \mathbb{R}^n$ we have

$$\begin{aligned}\|D_1 F(z,y)\| &= 2\kappa|\Psi_\kappa(z-y)|\|z-y\|\|f(y)\|\\ &= 2\kappa K \exp\bigl(\kappa\|\operatorname{Im} z\|^2\bigr)\exp\bigl(-\kappa\|\operatorname{Re} z - y\|^2\bigr)\|z-y\|\\ &\le 2\kappa K \exp(\kappa r^2)\exp\bigl(-\kappa(\max\{\|y\| - r, 0\})^2\bigr)(r + \|y\|).\end{aligned}$$

Thus we can apply Theorem 1.90 on bounded subsets of $\mathbb{C}^n$, which gives $g_\kappa \in H(\mathbb{C}^n;\tilde{Y})$ and hence also $g_\kappa\restriction_{\mathbb{R}^n} \in C^\omega(\mathbb{R}^n;Y)$. If $g_\kappa(x) \notin C$ for some $x \in \mathbb{R}^n$, then by the separation theorem there are $\phi \in Y^*$ and $\alpha \in \mathbb{R}$ such that $\phi\bigl(f(y)\bigr) < \alpha < \phi\bigl(g_\kappa(x)\bigr)$ for all $y \in \mathbb{R}^n$. But

$$\phi\bigl(g_\kappa(x)\bigr) = \frac{1}{c_\kappa}\int_{\mathbb{R}^n} \Psi_\kappa(x-y)\phi\bigl(f(y)\bigr)\,\mathrm{d}y < \frac{1}{c_\kappa}\int_{\mathbb{R}^n} \alpha\Psi_\kappa(x-y)\,\mathrm{d}y = \alpha,$$

which is a contradiction. Hence $g_\kappa\restriction_{\mathbb{R}^n} \in C^\omega(\mathbb{R}^n;C)$.

Further, if f is Bochner integrable, then for a fixed $\delta > 0$ and any $z \in G_\delta$ we can estimate

$$\begin{aligned}\|g_\kappa(z)\| &\le \frac{1}{c_\kappa}\int_{\mathbb{R}^n} |\Psi_\kappa(z-y)|\|f(y)\|\,\mathrm{d}y \\ &= \frac{1}{c_\kappa}\int_{\operatorname{supp} f} \exp\Big(-\kappa\big(\|\operatorname{Re} z - y\|^2 - \|\operatorname{Im} z\|^2\big)\Big)\|f(y)\|\,\mathrm{d}y \\ &\le \frac{1}{c_\kappa}\exp\left(-\kappa\delta^2\right)\int_{\operatorname{supp} f}\|f(y)\|\,\mathrm{d}y.\end{aligned}$$

Since $\lim_{\kappa\to+\infty}\frac{1}{c_\kappa}\exp(-\kappa\delta^2) = 0$, it follows that $\lim_{\kappa\to+\infty}\|g_\kappa\|_{G_\delta} = 0$.

For any $x \in \mathbb{R}^n$ we can use the substitution $y \to x - y$ to obtain

$$g_\kappa(x) = \frac{1}{c_\kappa}\int_{\mathbb{R}^n} \Psi_\kappa(y) f(x-y)\,\mathrm{d}y. \tag{1}$$

Thus if f is L-Lipschitz, then for any $u, v \in \mathbb{R}^n$ we have

$$\begin{aligned}\|g_\kappa(u) - g_\kappa(v)\| &= \frac{1}{c_\kappa}\left\|\int_{\mathbb{R}^n}\Psi_\kappa(y) f(u-y)\,\mathrm{d}y - \int_{\mathbb{R}^n}\Psi_\kappa(y) f(v-y)\,\mathrm{d}y\right\| \\ &\le \frac{1}{c_\kappa}\int_{\mathbb{R}^n}\Psi_\kappa(y)\|f(u-y) - f(v-y)\|\,\mathrm{d}y \\ &\le L\|u-v\|\frac{1}{c_\kappa}\int_{\mathbb{R}^n}\Psi_\kappa(y)\,\mathrm{d}y = L\|u-v\|.\end{aligned}$$

Now suppose that $f \in C^k(\mathbb{R}^n; Y)$ for some $k \in \mathbb{N}_0 \cup \{\infty\}$ with all the differentials $d^j f$, $j = 0, \dots, k$ bounded and uniformly continuous on $\mathbb{R}^n$. Using the formula (1), the boundedness of the differentials, Theorem 1.90, and induction we obtain

$$\begin{aligned}d^j g_\kappa(x) &= \frac{1}{c_\kappa}\int_{\mathbb{R}^n}\Psi_\kappa(y) d^j f(x-y)\,\mathrm{d}y \\ &= \frac{1}{c_\kappa}\int_{\mathbb{R}^n}\Psi_\kappa(x-y) d^j f(y)\,\mathrm{d}y\end{aligned}$$

for every $x \in \mathbb{R}^n$ and $1 \le j \le k$. Fix $j \in \{0, \dots, k\}$ and choose an arbitrary $\varepsilon > 0$. Consider the space $\mathbb{R}^n$ with the Euclidean norm. By the uniform continuity there exists $\delta > 0$ such that $\|d^j f(x) - d^j f(y)\| < \frac{\varepsilon}{2}$ whenever $x, y \in \mathbb{R}^n$, $\|x - y\| < \delta$. Moreover there is a constant $M > 0$ such that $\|d^j f(y)\| \le M$ for all $y \in \mathbb{R}^n$. For κ

large enough so that $2Me^{-\frac{1}{2}\kappa\delta^2}2^{\frac{n}{2}} < \frac{\varepsilon}{2}$ we then have

$$\begin{aligned}
&\|d^j g_\kappa(x) - d^j f(x)\| \\
&\quad = \left\| \frac{1}{c_\kappa}\int_{\mathbb{R}^n} \Psi_\kappa(x-y) d^j f(y)\,\mathrm{d}y - \frac{1}{c_\kappa}\left(\int_{\mathbb{R}^n} \Psi_\kappa(x-y)\,\mathrm{d}y\right) d^j f(x)\right\| \\
&\quad = \left\| \frac{1}{c_\kappa}\int_{\mathbb{R}^n} \Psi_\kappa(x-y)\left(d^j f(y) - d^j f(x)\right)\mathrm{d}y\right\| \\
&\quad \le \frac{1}{c_\kappa}\int_{\mathbb{R}^n} \Psi_\kappa(x-y)\|d^j f(y) - d^j f(x)\|\,\mathrm{d}y \\
&\quad \le \frac{1}{c_\kappa}\int_{\|x-y\|<\delta} \Psi_\kappa(x-y)\frac{\varepsilon}{2}\,\mathrm{d}y + \frac{1}{c_\kappa}\int_{\|x-y\|\ge\delta} \Psi_\kappa(x-y) 2M\,\mathrm{d}y \\
&\quad \le \frac{\varepsilon}{2} + \frac{2M}{c_\kappa}\int_{\|x-y\|\ge\delta} \Psi_\kappa(x-y)\,\mathrm{d}y = \frac{\varepsilon}{2} + \frac{2M}{c_\kappa}\int_{\|x-y\|\ge\delta} \exp\left(-\kappa\|x-y\|^2\right)\mathrm{d}y \\
&\quad \le \frac{\varepsilon}{2} + \frac{2M}{c_\kappa}e^{-\frac{1}{2}\kappa\delta^2}\int_{\mathbb{R}^n} \exp\left(-\frac{\kappa}{2}\|x-y\|^2\right)\mathrm{d}y \\
&\quad = \frac{\varepsilon}{2} + 2Me^{-\frac{1}{2}\kappa\delta^2}\frac{c_{\kappa/2}}{c_\kappa} = \frac{\varepsilon}{2} + 2Me^{-\frac{1}{2}\kappa\delta^2}2^{\frac{n}{2}} < \varepsilon
\end{aligned}$$

for every $x \in \mathbb{R}^n$. □

The next fact is used silently many times throughout this book.

Fact 2. *There exists a function $\theta \in C^\infty(\mathbb{R};[0,1])$ such that $\theta(t) = 0$ for $t \le 0$, $\theta(t) \in (0,1)$ for $t \in (0,1)$, and $\theta(t) = 1$ for $t \ge 1$.*

Proof. Define a function θ_0 by $\theta_0(t) = 0$ for $t \le 0$ and $\theta_0(t) = \exp\left(-\frac{1}{t}\right)$ for $t > 0$. It is standard to check that $\theta_0 \in C^\infty(\mathbb{R};[0,1))$. Put $\theta_1(t) = 1 - e^e\theta_0\left(\frac{1}{e} - t\right)$ and finally $\theta = \theta_1 \circ \theta_0$. □

Lemma 3. *Let $K \subset \mathbb{R}^n$ be a compact set and $U \subset \mathbb{R}^n$ an open neighbourhood of K. Then there is a function $\varphi \in C^\infty(\mathbb{R}^n;[0,1])$ such that $\operatorname{supp}\varphi \subset U$ and $\varphi = 1$ on some neighbourhood of K.*

Proof. Let θ be the function from Fact 2 and define the function $\psi \in C^\infty(\mathbb{R}^n)$ by $\psi(x) = \theta(1 - x_1^2 - \cdots - x_n^2)$. Notice that $\operatorname{supp}\psi \subset B(0,1)$ and $\psi(0) = 1$. Let $d = \frac{1}{2}\operatorname{dist}(K, \mathbb{R}^n \setminus U)$. For each $w \in K$ we define $\psi_w(x) = \psi\left(\frac{x-w}{d}\right)$ and $V_w = \{x \in \mathbb{R}^n;\ \psi_w(x) > \frac{1}{2}\}$. Since K is a compact set there are $w_1, \dots, w_k \in K$ such that $K \subset V_{w_1} \cup \cdots \cup V_{w_k}$. Put

$$\varphi(x) = \theta\left(2\sum_{j=1}^{k} \psi_{w_j}(x)\right).$$ □

Lemma 4. *There are functions $\theta_n \in H(\mathbb{C})$, $n \in \mathbb{N}$, with the following properties:*

(T1) $\theta_n \restriction_{\mathbb{R}}$ *maps into* $[0,1]$,
(T2) $\theta_n \restriction_{\mathbb{R}}$ *is* 4*-Lipschitz,*
(T3) $|\theta_n(z)| \le 2^{-n}$ *for every* $z \in \mathbb{C}$, $|z| \le \frac{1}{4}$,
(T4) $|\theta_n(x) - 1| \le 2^{-n}$ *for every* $x \in \mathbb{R}$, $x \ge 1$,
(T5) $|(\theta_n \restriction_{\mathbb{R}})'(x)| \le 2^{-n}$ *for every* $x \in \mathbb{R}$, $x \le \frac{1}{2}$ *or* $x \ge 1$.

Proof. Let $f : \mathbb{R} \to [0,1]$ be defined as $f(t) = 0$ for $t \le \frac{5}{8}$, $f(t) = 4t - \frac{5}{2}$ for $t \in (\frac{5}{8}, \frac{7}{8})$, and $f(t) = 1$ for $t \ge \frac{7}{8}$. Obviously f is a 4-Lipschitz function. We put $\theta_n = g_{\kappa_n}$ from Lemma 1, where $\kappa_n \in \mathbb{R}^+$ is chosen so that (T4) holds and

$$\sqrt{2}e^{-\kappa_n/128} \le 2^{-n}, \tag{2}$$

$$\frac{2\kappa_n}{c_{\kappa_n}} \int_{|t| \ge \frac{1}{8}} |t| e^{-\kappa_n t^2}\, dt = \frac{2e^{-\kappa_n/64}}{c_{\kappa_n}} \le 2^{-n}. \tag{3}$$

The function θ_n clearly has the properties (T1) and (T2). To prove (T3) we use successively the definition of f, the fact that $|\operatorname{Im} z| \le \frac{1}{4}$, $\operatorname{Re} z \le \frac{1}{4}$, and finally (2) to obtain

$$\begin{aligned}
|\theta_n(z)| &\le \frac{1}{c_{\kappa_n}} \int_{\mathbb{R}} f(t) e^{-\kappa_n \operatorname{Re}(z-t)^2}\, dt = \frac{e^{\kappa_n (\operatorname{Im} z)^2}}{c_{\kappa_n}} \int_{\mathbb{R}} f(t) e^{-\kappa_n (\operatorname{Re} z - t)^2}\, dt \\
&\le \frac{e^{\frac{1}{16}\kappa_n}}{c_{\kappa_n}} \int_{\frac{5}{8}}^{+\infty} e^{-\kappa_n (\operatorname{Re} z - t)^2}\, dt \\
&\le \frac{e^{\frac{1}{16}\kappa_n}}{c_{\kappa_n}} e^{-\frac{\kappa_n}{2}(\frac{3}{8})^2} \int_{\frac{5}{8}}^{+\infty} e^{-\frac{\kappa_n}{2}(\operatorname{Re} z - t)^2}\, dt \\
&\le \sqrt{2}e^{-\kappa_n/128} \le 2^{-n}.
\end{aligned}$$

Finally, we show (T5). Suppose that $x \le \frac{1}{2}$ or $x \ge 1$. Differentiating under the integral sign we obtain

$$\theta_n'(x) = \frac{2\kappa_n}{c_{\kappa_n}} \int_{\mathbb{R}} f(t)(t-x) e^{-\kappa_n (t-x)^2}\, dt.$$

Since $t \mapsto te^{-\kappa t^2}$ is odd and f is constant on $[x - \frac{1}{8}, x + \frac{1}{8}]$, we get (using also (3))

$$\begin{aligned}
|\theta_n'(x)| &= \frac{2\kappa_n}{c_{\kappa_n}} \left| \int_{|x-t| \ge \frac{1}{8}} f(t)(t-x) e^{-\kappa_n (t-x)^2}\, dt \right| \\
&\le \frac{2\kappa_n}{c_{\kappa_n}} \int_{|x-t| \ge \frac{1}{8}} |t-x| e^{-\kappa_n (t-x)^2}\, dt \le 2^{-n}.
\end{aligned}$$

□

Lemma 5. *Let $\{\varepsilon_n\}_{n=1}^{\infty}$ and $\{a_n\}_{n=1}^{\infty}$ be two sequences of positive real numbers. There are functions $\zeta_n \in H(\mathbb{C}^n)$, $n \in \mathbb{N}$, and a sequence $\{\delta_n\}_{n=1}^{\infty}$ of positive real numbers with the following properties:*

(Z1) *$\zeta_n \restriction_{\mathbb{R}^n}$ maps into $[0,1]$,*

(Z2) *$\zeta_n \restriction_{\mathbb{R}^n}$ is 2-Lipschitz with respect to the maximum norm,*

(Z3) *$|\zeta_n(z)| \le \varepsilon_n$ for every $z \in \mathbb{C}^n$ such that there is $k \in \{1,\dots,n-1\}$ for which $\operatorname{Re} z_k \le \frac{1}{4}$ and $\sum_{j=1}^n a_j(\operatorname{Im} z_j)^2 \le \delta_k$.*

(Z4) *$\zeta_n(x) \ge \frac{1}{2}$ for every $x \in \mathbb{R}^n$ for which $x_n \le 1$ and $x_j \ge 1$, $j = 1,\dots,n-1$,*

(Z5) *$\zeta_n(x) \le \varepsilon_n$ for $x \in \mathbb{R}^n$ satisfying $x_n \ge 2$.*

Proof. Let $f_n\colon \mathbb{R}^n \to [0,1]$ be a 2-Lipschitz function (with respect to the maximum norm) such that

$$f_n(x) = \begin{cases} 0 & \text{whenever } x_n \ge 2 \text{ or } \exists j \in \{1,\dots,n-1\}\colon x_j \le \frac{1}{2}, \\ 1 & \text{whenever } x_n \le 1 \text{ and } \forall j \in \{1,\dots,n-1\}\colon x_j \ge 1. \end{cases}$$

(See e.g. Lemma 39.) For each $n \in \mathbb{N}$ put $\delta_n = a_n/64$ and

$$\zeta_n(z) = \frac{1}{c_n} \int_{\mathbb{R}^n} f_n(t) \exp\left(-\kappa_n \sum_{j=1}^n a_j (z_j - t_j)^2\right) dt \quad \text{for } z \in \mathbb{C}^n,$$

where $c_n = \int_{\mathbb{R}^n} e^{-\kappa_n \sum_{j=1}^n a_j t_j^2}\, dt = \sqrt{\left(\frac{\pi}{\kappa_n}\right)^n \prod_{j=1}^n a_j^{-1}}$ and $\kappa_n \in \mathbb{R}^+$ is chosen so that (Z4) and (Z5) hold (analogously as in Lemma 1) and

$$e^{-\kappa_n a_j/64} \le 2^{-\frac{n}{2}} \varepsilon_n \text{ for } j = 1,\dots,n-1.$$

The function ζ_n belongs to $H(\mathbb{C}^n)$ and has the properties (Z1) and (Z2) (again similarly as in Lemma 1). To prove (Z3) we use successively use the definition of f_n, the fact that $\operatorname{Re} z_k \le \frac{1}{4}$, and the definition of δ_k to obtain

$$\begin{aligned}
|\zeta_n(z)| &\le \frac{1}{c_n} \int_{\mathbb{R}^n} f_n(t) e^{-\kappa_n \sum_{j=1}^n a_j \operatorname{Re}(z_j - t_j)^2}\, dt \\
&= \frac{e^{\kappa_n \sum_{j=1}^n a_j (\operatorname{Im} z_j)^2}}{c_n} \int_{\mathbb{R}^n} f_n(t) e^{-\kappa_n \sum_{j=1}^n a_j (\operatorname{Re} z_j - t_j)^2}\, dt \\
&\le \frac{e^{\kappa_n \delta_k}}{c_n} \int_{\substack{t \in \mathbb{R}^n \\ t_k > \frac{1}{2}}} e^{-\kappa_n \sum_{j=1}^n a_j (\operatorname{Re} z_j - t_j)^2}\, dt \\
&= \frac{e^{\kappa_n \delta_k}}{c_n} \int_{\substack{t \in \mathbb{R}^n \\ t_k > \frac{1}{2}}} e^{-\frac{\kappa_n}{2} \sum_{j=1}^n a_j (\operatorname{Re} z_j - t_j)^2} \cdot e^{-\frac{\kappa_n}{2} \sum_{j=1}^n a_j (\operatorname{Re} z_j - t_j)^2}\, dt \\
&\le \frac{e^{\kappa_n \delta_k}}{c_n} e^{-\frac{\kappa_n}{2} a_k \frac{1}{16}} \int_{\mathbb{R}^n} e^{-\frac{\kappa_n}{2} \sum_{j=1}^n a_j (\operatorname{Re} z_j - t_j)^2}\, dt = 2^{\frac{n}{2}} e^{-\kappa_n a_k/64} \le \varepsilon_n.
\end{aligned}$$

□

Lemma 6. *Let K be a compact space such that $C(K)$ admits a C^k-smooth bump function, $k \in \mathbb{N} \cup \{\infty\}$. Then for every $\zeta, \eta \in \mathbb{R}$, $0 < \zeta < \eta$, there is a function $\beta_{\zeta,\eta} \in C^k(C(K); [0,1])$ such that*

$$\beta_{\zeta,\eta}(f) = \begin{cases} 1 & \textit{when } \|f\|_\infty \le \zeta, \\ 0 & \textit{when } \|f\|_\infty \ge \eta. \end{cases}$$

Proof. By hypothesis there exists a function $\varphi \in C^k(C(K); [0,1])$ and $\alpha \in \mathbb{R}$, $\alpha > 0$, such that $\varphi(f) = 1$ for $\|f\|_\infty \le \alpha$, while $\varphi(f) = 0$ for $\|f\|_\infty \ge 1$. Choose $n \in \mathbb{N}$ so that $(\frac{\zeta}{\eta})^n \le \alpha$ and put

$$\beta_{\zeta,\eta}(f) = \varphi\Big(\frac{f^n}{\eta^n}\Big).$$

Since the mapping $f \mapsto f^n$ is a continuous n-homogeneous polynomial (and in particular it is C^∞-smooth), the function $\beta_{\zeta,\eta}$ has the required properties. □

We remark that the Taylor complexification $\tilde{c_0}$ of the real space c_0 is isometric to the complex space c_0.

Proposition 7. *Let $q \ge 1$. There are an open set $W \subset \tilde{c_0}$ and a function $\mu \in H(W)$ with the following properties:*

(M1) *For every $w \in c_0 \setminus \{0\}$ there is $\Delta_w > 0$ such that $U_{\tilde{c_0}}(y, \Delta_w) \subset W$ for every $y \in c_0$ satisfying $|w| \le |y| \le q|w|$, where the inequalities are understood in the lattice sense.*

(M2) *$\mu(w) \ge 8$ for $w \in c_0$, $\|w\| \ge 8$,*

(M3) *$|\mu(z)| < 2$ for $z \in U_{\tilde{c_0}}(y, \Delta_w)$, where $y \in c_0$, $\|y\| \le 1$, and $w \in c_0 \setminus \{0\}$, $|w| \le |y| \le q|w|$,*

(M4) *$\mu \restriction_{c_0}$ is $\sqrt{2}$-Lipschitz and maps into $\mathbb{R}$.*

Proof. Define μ on c_0 as the Minkowski functional of $\{x \in c_0;\ \sum_{n=1}^\infty (x_n)^{2n} \le 1\}$. Then μ is an equivalent norm on c_0 for which $\|x\| \le \mu(x) \le \sqrt{2}\|x\|$ (see also Theorem 5.104 and Example 1.137). This gives property (M2) and (M4).

Let $f : \tilde{c_0} \times (\mathbb{C} \setminus \{0\}) \to \mathbb{C}$ be defined as $f(z,u) = \sum_{n=1}^\infty (z_n/u)^{2n} - 1$. This function is holomorphic on $\tilde{c_0} \times (\mathbb{C} \setminus \{0\})$ and for every $x \in c_0 \setminus \{0\}$ we have $f(x, \mu(x)) = 0$.

Fix $w \in c_0 \setminus \{0\}$. Put $R = \frac{\|w\|}{2}$, $S = \frac{\|w\|}{4}$, $M = 1 + \sum_{n=1}^\infty \Big(\frac{1}{2} + \frac{2q}{\|w\|}|w_n|\Big)^{2n}$, $a = \frac{1}{\sqrt{2}q\|w\|}$, $r = \min\Big\{\frac{1}{2}\frac{aR^2}{aR+M}, 2 - \sqrt{2}\Big\}$, and $\Delta_w = s$ as defined in Theorem 1.176. Now choose any $y \in c_0$, $|w| \le |y| \le q|w|$. Then $R < \|w\| \le \|y\| \le \mu(y)$, thus $B(\mu(y), R) \subset V = \mathbb{C} \setminus \{0\}$. For any $z \in B(y, S)$, $u \in B(\mu(y), R)$ we have $|u| \ge \mu(y) - R \ge \|y\| - R \ge \|w\| - R = \frac{\|w\|}{2}$ and $|z_n| \le |y_n| + |z_n - y_n| \le q|w_n| + \|z - y\| \le q|w_n| + \frac{\|w\|}{4}$, and hence $|f(z,u)| \le 1 + \sum_{n=1}^\infty |\frac{z_n}{u}|^{2n} \le M$. Finally, $|D_2 f(y, \mu(y))| = \Big|-\frac{1}{\mu(y)}\sum_{n=1}^\infty 2n\big(\frac{y_n}{\mu(y)}\big)^{2n}\Big| \ge \frac{1}{\mu(y)}\sum_{n=1}^\infty \big(\frac{y_n}{\mu(y)}\big)^{2n} = \frac{1}{\mu(y)} \ge \frac{1}{\sqrt{2}\|y\|} \ge a$. Thus by Theorem 1.176 the equation $f(z,u) = 0$ uniquely determines a holomorphic function μ_y^w on $U_{\tilde{c_0}}(y, \Delta_w)$ with values in $U(\mu(y), r)$ and this holds for every $y \in c_0$, $|w| \le |y| \le q|w|$.

Take any two functions $\mu_1 = \mu_{y_1}^{w_1}$, $\mu_2 = \mu_{y_2}^{w_2}$ defined on open balls U_1 and U_2 respectively. If U_1 and U_2 intersect, then it is easy to check that $U_1 \cap U_2 \cap c_0$ is a non-empty set relatively open in c_0. Since $\mu_1 = \mu$ on $U_1 \cap c_0$ and $\mu_2 = \mu$ on $U_2 \cap c_0$, it follows that both holomorphic functions μ_1 and μ_2 agree on $U_1 \cap U_2 \cap c_0$ and therefore on the whole $U_1 \cap U_2$ (Corollary 1.158). This observation allows us to put $W = \bigcup\{U_{\tilde{c_0}}(y, \Delta_w);\ w \in c_0 \setminus \{0\}, y \in c_0, |w| \le |y| \le q|w|\}$ and define μ on W by $\mu(z) = \mu_y^w(z)$ whenever $z \in U(y, \Delta_w)$. This gives property (M1).

To prove (M3) let $w \in c_0 \setminus \{0\}$, $y \in c_0$, $|w| \le |y| \le q|w|$, $\|y\| \le 1$, and $z \in U_{\tilde{c_0}}(y, \Delta_w)$. Then by the choice of r above we have $\mu(z) \in U(\mu(y), 2 - \sqrt{2})$ and therefore $|\mu(z)| < |\mu(y)| + 2 - \sqrt{2} \le \sqrt{2}\|y\| + 2 - \sqrt{2} \le 2$. □

Lemma 8 ([RudinM]). *Let (P, ρ) be a metric space and $\mathcal{U} = \{U_\alpha\}_{\alpha \in \Lambda}$ an open covering of P. Then there are open refinements $\{V_{n\alpha}\}_{n \in \mathbb{N}, \alpha \in \Lambda}$, $\{W_{n\alpha}\}_{n \in \mathbb{N}, \alpha \in \Lambda}$ of $\mathcal{U}$ that satisfy the following:*

- *$V_{n\alpha} \subset W_{n\alpha} \subset U_\alpha$ for all $n \in \mathbb{N}$, $\alpha \in \Lambda$,*
- *$\operatorname{dist}(V_{n\alpha}, P \setminus W_{n\alpha}) \ge 2^{-n}$ for all $n \in \mathbb{N}$, $\alpha \in \Lambda$,*
- *$\operatorname{dist}(W_{n\alpha}, W_{n\beta}) \ge 2^{-n}$ for any $n \in \mathbb{N}$ and $\alpha, \beta \in \Lambda$, $\alpha \ne \beta$.*
- *for each $x \in P$ there are an open neighbourhood U_x of x and a number $n_x \in \mathbb{N}$ such that*
  - (i) *if $k > n_x$, then $U_x \cap W_{k\alpha} = \emptyset$ for any $\alpha \in \Lambda$,*
  - (ii) *if $k \le n_x$, then $U_x \cap W_{k\alpha} \ne \emptyset$ for at most one $\alpha \in \Lambda$.*

Proof. Choose some well-ordering of the set Λ. Define the sets $V_{n\alpha}$ by induction on $n \in \mathbb{N}$: If $V_{j\beta}$ are already defined for $j < n$ and all $\beta \in \Lambda$, let $V_{n\alpha}$ be the union of all $U(x, 2^{-n})$ such that

(a) α is the smallest with $x \in U_\alpha$,
(b) $x \notin V_{j\beta}$ for all $j < n$, $\beta \in \Lambda$,
(c) $U(x, 5 \cdot 2^{-n}) \subset U_\alpha$.

Further, let $W_{n\alpha} = \bigcup\{U(x, 2^{-n});\ x \in V_{n\alpha}\}$ for all $n \in \mathbb{N}$, $\alpha \in \Lambda$.

Certainly $V_{n\alpha} \subset W_{n\alpha} \subset U_\alpha$ and $\operatorname{dist}(V_{n\alpha}, P \setminus W_{n\alpha}) \ge 2^{-n}$ for all $n \in \mathbb{N}, \alpha \in \Lambda$. To see that $\{V_{n\alpha}\}$ covers P, observe that, for $x \in P$, there is a smallest $\alpha \in \Lambda$ such that $x \in U_\alpha$, and n so large that (c) holds. Then, by (b), $x \in V_{j\beta}$ for some $j \le n$, $\beta \in \Lambda$.

To prove the third property, suppose that $n \in \mathbb{N}$, $\alpha, \beta \in \Lambda$, $\alpha < \beta$, $p \in W_{n\alpha}$, and $q \in W_{n\beta}$. There is a ball $U(y, 2^{-n})$ in the definition of the set $V_{n\alpha}$ such that $\rho(p, y) < 2 \cdot 2^{-n}$, and a ball $U(z, 2^{-n})$ in the definition of the set $V_{n\beta}$ such that $\rho(q, z) < 2 \cdot 2^{-n}$. By (c), $U(y, 5 \cdot 2^{-n}) \subset U_\alpha$ but, by (a), $z \notin U_\alpha$. So $\rho(y, z) \ge 5 \cdot 2^{-n}$ and $\rho(p, q) \ge \rho(y, z) - \rho(p, y) - \rho(q, z) > 2^{-n}$.

Finally assume $x \in P$. Find some pair $n \in \mathbb{N}$, $\beta \in \Lambda$ such that $x \in V_{n\beta}$, and choose $j \in \mathbb{N}$ so that $U(x, 2^{-j+1}) \subset V_{n\beta}$. Put $n_x = n + j - 1$ and $U_x = U(x, 2^{-n-j})$. To show (i), suppose that $k > n_x$ and choose any $\alpha \in \Lambda$ and $z \in W_{k\alpha}$. It follows that there is a ball $U(y, 2^{-k})$ in the definition of $V_{k\alpha}$ such that $\rho(y, z) < 2 \cdot 2^{-k}$. Since

$k > n$, by (b), $y \notin V_{n\beta}$. And since $U(x, 2^{-j+1}) \subset V_{n\beta}$ and $k \geq j+1$,

$$\begin{aligned}\rho(x,z) &\geq \rho(x,y) - \rho(y,z) \geq 2^{-j+1} - \rho(y,z)\\ &> 2^{-j+1} - 2^{-k+1} \geq 2^{-j+1} - 2^{-j} = 2^{-j} > 2^{-n-j}.\end{aligned}$$

From the definition of U_x and n_x it is easy to see that (ii) follows from the third property. □

2. Approximation by polynomials

In this section we begin by proving the classical Weierstraß-type theorem on the density of polynomials among C^k-smooth functions in the uniform topology (together with its derivatives) on compact subsets of $\mathbb{R}^n$. Of course, the proof relies heavily on the compactness argument. The result can be extended into infinite-dimensional setting if we are interested in uniform topology on compact sets. We then deal with the approximation in uniform topology on bounded sets, which is not always possible. Applying the theory of $\mathcal{W}$-spaces we give a generalisation of the Weierstraß theorem in some special cases. We finish the section by showing that the assumptions used in order to get positive results are close to being optimal.

Theorem 9. *Let $\Omega \subset \mathbb{R}^n$ be an open set, Y a Banach space, and $f \in C^k(\Omega; Y)$, $k \in \mathbb{N}_0$. For every compact subset $K \subset \Omega$ and every $\varepsilon > 0$ there is a polynomial $p \in \mathcal{P}(\mathbb{R}^n; Y)$ such that $\|d^j f - d^j p\|_K \leq \varepsilon$ for $0 \leq j \leq k$.*

Proof. By Lemma 3 there is a function $\varphi \in C^\infty(\mathbb{R}^n)$ such that $\operatorname{supp}\varphi \subset \Omega$, $\operatorname{supp}\varphi$ is compact, and $\varphi = 1$ on a neighbourhood of K. Since $\varphi f \in C^k(\Omega; Y)$ (Corollary 1.84), φf has a compact support, and $d^j(\varphi f)(x) = d^j f(x)$ for $x \in K$, $0 \leq j \leq k$, replacing f by a function defined as φf on Ω and 0 on $\mathbb{R}^n \setminus \Omega$ we may suppose that $f \in C^k(\mathbb{R}^n; Y)$ and $S = \operatorname{supp} f$ is compact.

By Lemma 1 there is $\kappa \in \mathbb{R}^+$ such that $\|d^j g_\kappa - d^j f\|_{\mathbb{R}^n} < \frac{\varepsilon}{2}$ for $0 \leq j \leq k$. Put

$$Q_m(y) = \sum_{l=0}^{m} \frac{1}{l!}(-\kappa)^l \left(y_1^2 + \cdots + y_n^2\right)^l.$$

Then $Q_m \in \mathcal{P}(\mathbb{R}^n)$ and $\lim_{m\to\infty} d^j Q_m = d^j \Psi_\kappa$ locally uniformly on $\mathbb{R}^n$ for each $0 \leq j \leq k$ (Theorem 1.146). Put $M = \|f\|_{\mathbb{R}^n}$. The set $L = K - S$ is compact. Therefore there is $N \in \mathbb{N}$ such that $\|d^j Q_N - d^j \Psi_\kappa\|_L \leq \frac{c_\kappa}{M} \frac{1}{\lambda(L)} \frac{\varepsilon}{2}$ for each $0 \leq j \leq k$. Let

$$p(x) = \frac{1}{c_\kappa} \int_{\mathbb{R}^n} Q_N(x-y) f(y)\,\mathrm{d}y.$$

Then $p \in \mathcal{P}(\mathbb{R}^n; Y)$ (e.g. by Theorem 2.49). As

$$d^j g_\kappa(x) = \frac{1}{c_\kappa} \int_{\mathbb{R}^n} d^j \Psi_\kappa(x-y) f(y)\,\mathrm{d}y = \frac{1}{c_\kappa} \int_{\mathbb{R}^n} d^j \Psi_\kappa(y) f(x-y)\,\mathrm{d}y$$

and similarly for $d^j p$, we have for any $x \in K$

$$\begin{aligned}\|d^j g_\kappa(x) - d^j p(x)\| &= \left\| \frac{1}{c_\kappa} \int_{\mathbb{R}^n} \left(d^j \Psi_\kappa(y) - d^j Q_N(y)\right) f(x-y)\,\mathrm{d}y \right\| \\ &= \left\| \frac{1}{c_\kappa} \int_L \left(d^j \Psi_\kappa(y) - d^j Q_N(y)\right) f(x-y)\,\mathrm{d}y \right\| \\ &\le \frac{M}{c_\kappa} \int_L \left\|d^j \Psi_\kappa(y) - d^j Q_N(y)\right\| \mathrm{d}y \le \frac{M}{c_\kappa} \left\|d^j Q_N - d^j \Psi_\kappa\right\|_L \lambda(L) \le \frac{\varepsilon}{2}. \quad \square\end{aligned}$$

Definition 10. Let X and Y be normed linear spaces. By $\mathcal{P}_{\mathrm{f}}({}^nX;Y)$ we denote the linear subspace of $\mathcal{P}({}^nX;Y)$ consisting of all polynomials that can be written in the form $P(x) = \sum_{j=1}^k f_j(x)^n y_j$, where $f_j \in X^*$ and $y_j \in Y$. We also set $\mathcal{P}_{\mathrm{f}}(X;Y) = \operatorname{span} \bigcup_{n=0}^\infty \mathcal{P}_{\mathrm{f}}({}^nX;Y)$.

If X is finite-dimensional, then $\mathcal{P}_{\mathrm{f}}({}^nX;Y) = \mathcal{P}({}^nX;Y)$. This follows from Proposition 1.23 and the fact that $\operatorname{span}\{\langle y,\cdot\rangle^d;\ y \in \mathbb{R}^m\} = \mathcal{P}({}^d\mathbb{R}^m)$ (Section 2.4).

Fact 11. *Let X, Y be normed linear spaces, $P \in \mathcal{P}_{\mathrm{f}}({}^mX)$, and $Q \in \mathcal{P}_{\mathrm{f}}({}^nX;Y)$. Then $PQ \in \mathcal{P}_{\mathrm{f}}({}^{m+n}X;Y)$. In particular, $\mathcal{P}_{\mathrm{f}}(X)$ is a subalgebra of the algebra $\mathcal{P}(X)$ and $p \circ R \in \mathcal{P}_{\mathrm{f}}(X)$ whenever $R \in \mathcal{P}_{\mathrm{f}}(X)$ and $p \in \mathcal{P}(\mathbb{R})$.*

Proof. It is clear that it suffices to show that $x \mapsto f(x)^m g(x)^n \in \mathcal{P}_{\mathrm{f}}({}^{m+n}X)$ whenever $f, g \in X^*$. The polynomial $q \in \mathcal{P}({}^{m+n}\mathbb{R}^2)$ given by $q(u,v) = u^m v^n$ can be written as $q(u,v) = \sum_{j=1}^k c_j \langle (a_j,b_j),(u,v)\rangle^{m+n}$ (Section 2.4). Therefore $f(x)^m g(x)^n = \sum_{j=1}^k c_j h_j(x)^{m+n}$, where $h_j = a_j f + b_j g \in X^*$. $\square$

The following is an extension of the Weierstraß theorem into infinite-dimensional spaces. Of course the usefulness of this theorem is limited by the fact that here the compact sets are very small.

Theorem 12. *Let X, Y be normed linear spaces, $K \subset X$ compact, $f \in C(K;Y)$, and $\varepsilon > 0$. Then there is a polynomial $P \in \mathcal{P}_{\mathrm{f}}(X;Y)$ such that $\|f - P\|_K \le \varepsilon$.*

For the proof we need the following lemma on separation of sets by polynomials.

Lemma 13. *Let X be a normed linear space, $\Omega \subset X$ a bounded set, $C \subset \Omega$ a closed convex set, $K \subset \Omega$ a weakly compact set satisfying $C \cap K = \emptyset$, and $\delta > 0$. Then there is a polynomial $P \in \mathcal{P}_{\mathrm{f}}(X)$ such that $0 \le P(x) \le 1$ for $x \in \Omega$, $P(x) > 1-\delta$ for $x \in C$, and $P(x) < \delta$ for $x \in K$.*

Proof. Without loss of generality we may assume that the set C is non-empty. By the separation theorem for every $x \in K$ there are $f_x \in X^*$ and $b_x, c_x \in \mathbb{R}$ such that $f_x(y) < b_x < c_x < f_x(x)$ for $y \in C$. Since $\{f_x^{-1}((c_x,+\infty))\}_{x\in K}$ is a weakly open covering of the weak compact K, there are $x_1,\dots,x_n \in K$ such that $K \subset \bigcup_{k=1}^n f_{x_k}^{-1}((c_{x_k},+\infty))$. Denote $M_k = \max_{y\in K} f_{x_k}(y)$ and $a_k = \inf_{y\in C} f_{x_k}(y)$ for $k = 1,\dots,n$, and notice that $a_k \in \mathbb{R}$ by the boundedness of C. By Theorem 9 there are polynomials $q_k \in \mathcal{P}(\mathbb{R})$ such that $|q_k(t)| < \frac{1}{2n}$ for $t \in [a_k, b_{x_k}]$ and $q_k(t) > 1$ for $t \in [c_{x_k}, M_k]$, $k = 1, \dots, n$. Put

$$Q(x) = \sum_{k=1}^{n} q_k^2\big(f_{x_k}(x)\big).$$

Then $0 \le Q(x) < \frac{1}{4n} \le \frac{1}{4}$ for $x \in C$ and $Q(x) > 1$ for $x \in K$.

Denote $m = \inf_{y \in \Omega} Q(y)$ and $M = \sup_{y \in \Omega} Q(y)$. By the boundedness of Ω both m and M are finite. By Theorem 9 there is a polynomial $p \in \mathcal{P}(\mathbb{R})$ such that $0 \le q(t) \le 1$ for $t \in [m, M]$, $q(t) > 1 - \delta$ for $t \in [m, \frac{1}{4}]$, and $q(t) < \delta$ for $t \in [1, M]$. We obtain the desired polynomial P by setting $P = p \circ Q$, noting that $P \in \mathcal{P}_{\mathrm{f}}(X)$ by Fact 11. □

Definition 14. Let X be a set. A collection $\{\psi_\alpha\}_{\alpha \in \Lambda}$ of functions on X is called a partition of unity if

- $\psi_\alpha \colon X \to [0, 1]$ for all $\alpha \in \Lambda$,
- $\sum\limits_{\alpha \in \Lambda} \psi_\alpha(x) = 1$ for each $x \in X$.

Let $\mathcal{U}$ be a covering of X. We say that the partition of unity $\{\psi_\alpha\}_{\alpha \in \Lambda}$ is subordinated to $\mathcal{U}$ if $\{\operatorname{supp}_{\mathrm{o}} \psi_\alpha\}_{\alpha \in \Lambda}$ refines $\mathcal{U}$.

We note that from the second property it immediately follows that the collection $\{\operatorname{supp}_{\mathrm{o}} \psi_\alpha\}_{\alpha \in \Lambda}$ is point-countable, i.e. for every $x \in X$ the set $\{\alpha \in \Lambda;\ \psi_\alpha(x) \neq 0\}$ is countable. In applications often either the set Λ itself is countable, which allows for "global" constructions (e.g. analytic approximation, Section 3), or the collection $\{\operatorname{supp}_{\mathrm{o}} \psi_\alpha\}_{\alpha \in \Lambda}$ is locally finite, which then preserves local properties, like C^k-smoothness (Section 5).

Lemma 15. *Let X be a normed linear space, $K \subset X$ compact, $\{U(x_k, r_k)\}_{k=1}^n$ a covering of K, and $\delta > 0$. Then there exists a polynomial partition of unity $\{\psi_k\}_{k=1}^n \subset \mathcal{P}_{\mathrm{f}}(X)$ on K such that $\psi_k(x) < \delta$ whenever $x \in K \setminus U(x_k, 2r_k)$, $k = 1, \dots, n$.*

Proof. Using Lemma 13 we find $\varphi_1, \dots, \varphi_n \in \mathcal{P}_{\mathrm{f}}(X)$ satisfying $0 \le \varphi_k(x) \le 1$ for $x \in K$, $\varphi_k(x) > 1 - \delta$ for $x \in B(x_k, r_k)$, and $\varphi_k(x) < \delta$ for $x \in K \setminus U(x_k, 2r_k)$, $k = 1, \dots, n$. We construct inductively polynomials $\psi_1, \dots, \psi_n \in \mathcal{P}_{\mathrm{f}}(X)$ that will form a partition of unity on K. Put $\psi_1 = \varphi_1$ and $\psi_k = \varphi_k \cdot \big(1 - \sum_{j=1}^{k-1} \psi_j\big)$ for $k = 2, \dots, n-1$. Finally set $\psi_n = 1 - \sum_{j=1}^{n-1} \psi_j$. Notice that $\psi_k \in \mathcal{P}_{\mathrm{f}}(X)$ by Fact 11 and

$$\sum_{j=1}^{k} \psi_j = \left(\sum_{j=1}^{k-1} \psi_j\right) \cdot 1 + \left(1 - \sum_{j=1}^{k-1} \psi_j\right) \varphi_k, \quad k = 1, \dots, n-1.$$

Thus we can check by induction that

$$0 \le \varphi_k(x) \le \sum_{j=1}^{k} \psi_j(x) \le 1 \quad \text{for } x \in K, k = 1, \dots, n-1, \tag{4}$$

and consequently

$$0 \le \psi_k(x) \le \varphi_k(x) \le 1 \quad \text{for } x \in K, k = 1, \dots, n-1. \tag{5}$$

It follows that $\psi_1, \dots, \psi_n$ form a partition of unity on K.

Moreover this partition has the property that $\psi_k(x) < \delta$ whenever $x \in K \setminus U(x_k, 2r_k)$, $k = 1, \dots, n$. Indeed, for $k < n$ this follows from (5). If $k = n$ and $x \in K \setminus U(x_n, 2r_n)$, then there is $m \in \{1, \dots, n-1\}$ such that $x \in U(x_m, r_m)$. Thus, by (5) and (4), $\psi_n(x) \le 1 - \sum_{j=1}^{m} \psi_j(x) \le 1 - \varphi_m(x) < \delta$. □

Proof of Theorem 12. By the compactness there is a covering $\{U(x_k, r_k)\}_{k=1}^n$ of K such that $\|f(x) - f(x_k)\| < \frac{\varepsilon}{2}$ whenever $x \in U(x_k, 2r_k)$, $k = 1, \dots, n$. Let $M > 0$ be such that $\|f\|_K \le M$ and set $\delta = \varepsilon/(4nM)$. Let $\{\psi_k\}_{k=1}^n$ be the partition of unity from Lemma 15. Put

$$P(x) = \sum_{k=1}^{n} \psi_k(x) f(x_k), \qquad x \in X.$$

Obviously $P \in \mathcal{P}_{\mathrm{f}}(X;Y)$. To show that P approximates f on K fix any $x \in K$. Let $I = \{1 \le k \le n;\ x \in U(x_k, 2r_k)\}$ and $J = \{1, \dots, n\} \setminus I$. Then

$$\begin{aligned}
\|f(x) - P(x)\| &= \left\| \left(\sum_{k=1}^{n} \psi_k(x) \right) f(x) - \sum_{k=1}^{n} \psi_k(x) f(x_k) \right\| \\
&\le \sum_{k=1}^{n} \psi_k(x) \|f(x) - f(x_k)\| \\
&\le \sum_{k \in I} \psi_k(x) \|f(x) - f(x_k)\| + \sum_{k \in J} \psi_k(x) \big(\|f(x)\| + \|f(x_k)\| \big) \\
&\le \frac{\varepsilon}{2} + 2nM\delta = \varepsilon.
\end{aligned}$$

□

Note that from Theorem 9 it follows that for any mapping $f \in C(\Omega; Y)$, where $\Omega \subset \mathbb{R}^n$ is open and Y is a normed linear space, there is a sequence of polynomials $\{P_k\}_{k=1}^{\infty} \subset \mathcal{P}(\mathbb{R}^n; Y)$ such that $P_k \to f$ locally uniformly on Ω. If we are interested only in the pointwise convergence, then we have the following infinite-dimensional result:

Theorem 16. *Let X, Y be normed linear spaces, X separable, $\Omega \subset X$ open, and $f \in C(\Omega; Y)$. Then there is a sequence of polynomials $\{p_n\}_{n=1}^{\infty} \subset \mathcal{P}_{\mathrm{f}}(X; Y)$ such that $\lim_{n\to\infty} p_n(x) = f(x)$ for every $x \in \Omega$.*

Proof. As X is separable, there is a sequence $\{x_n\}_{n\in\mathbb{N}} \subset X$ such that $\overline{\operatorname{span}}\{x_n\} = X$. Put $K_n = \{x \in \operatorname{span}\{x_1, \dots, x_n\};\ \operatorname{dist}(x, X \setminus \Omega) \ge \frac{1}{n}, \|x\| \le n\}$. Then K_n is a compact subset of Ω. By Theorem 12 there are polynomials $p_n \in \mathcal{P}_{\mathrm{f}}(X; Y)$ such that $\|f - p_n\|_{K_n} \le \frac{1}{n}$ for every $n \in \mathbb{N}$. Choose any $x \in \Omega$ and $\varepsilon > 0$. There is $\delta > 0$ such that $U(x, 2\delta) \subset \Omega$ and $\|f(x) - f(y)\| < \frac{\varepsilon}{2}$ whenever $y \in U(x, \delta)$. Further, there is

$n_0 \in \mathbb{N}$ such that $\frac{1}{n_0} < \frac{\varepsilon}{2}, \delta > \frac{1}{n_0}, \|x\|+\delta \le n_0$, and $\operatorname{span}\{x_1,\dots,x_{n_0}\} \cap U(x,\delta) \neq \emptyset$. Choose $z \in \operatorname{span}\{x_1,\dots,x_{n_0}\} \cap U(x,\delta)$. It follows that $z \in K_n$ for every $n \ge n_0$ and hence

$$\|f(x)-p_n(x)\| \le \|f(x)-f(z)\|+\|f(z)-p_n(z)\| < \frac{\varepsilon}{2}+\frac{1}{n} < \varepsilon \quad \text{for every } n \ge n_0.$$

□

In contrast with that, the norm on $c_0(\Gamma)$, Γ uncountable, is not a pointwise limit of a sequence of polynomials on $S_{c_0(\Gamma)}$. Indeed, given any sequence of polynomials $\{P_n\}$ on $c_0(\Gamma)$, by Corollary 3.62 there is $\gamma \in \Gamma$ such that $P_n(e_\gamma) = P_n(0)$ for each $n \in \mathbb{N}$. This was first observed by Aleksander Pełczyński.

Next, we deal with the problem of uniform approximation by polynomials on bounded sets. The above methods relied on compactness, which is of course not available in the general infinite-dimensional setting. Nevertheless, we show that certain classes of spaces and mappings allow for some compactness argument, which then yields the uniform approximation on bounded sets.

Theorem 17 ([AP]). *Let X be a normed linear space, Y a Banach space, and $U \subset X$ a convex set. Then $\overline{\mathcal{P}_f(X;Y)}^{\tau_b} = \mathcal{C}_{wu}(U;Y)$.*

Proof. Let $f \in \mathcal{C}_{wu}(U;Y)$, let $V \subset U$ be a CCB set, and $\varepsilon > 0$. By the assumption there exist $\delta > 0$ and $\phi_1,\dots,\phi_n \in B_{X^*}$ such that $\|f(x) - f(y)\| < \frac{\varepsilon}{2}$ whenever $x, y \in V$, $|\phi_j(x-y)| < 2\delta$ for all $j \in \{1,\dots,n\}$. Since f is uniformly continuous on V, there is $M > 0$ such that $\|f\|_V \le M$. Let $\Phi\colon X \to \mathbb{R}^n$ be defined as $\Phi(x) = (\phi_1(x),\dots,\phi_n(x))$. We set $K = \overline{\Phi(V)}$. Consider $\mathbb{R}^n$ with the maximum norm. Since $\Phi(V)$ is relatively compact, there are points $y_1,\dots,y_m \in \Phi(V)$ such that $\{U(y_k,\delta)\}_{k=1}^m$ is a covering of K. By Lemma 15 there is a partition of unity $\{\psi_k\}_{k=1}^m \subset \mathcal{P}_f(\mathbb{R}^n)$ on K satisfying $\psi_k(y) < \frac{\varepsilon}{4mM}$ whenever $y \in K \setminus U(y_k, 2\delta)$, $k = 1,\dots,m$. Choose $x_k \in V$ such that $y_k = \Phi(x_k)$. Finally, put

$$P(x) = \sum_{k=1}^{m} \psi_k(\Phi(x)) f(x_k).$$

Obviously $P \in \mathcal{P}_f(X;Y)$. To show that P approximates f on V fix any $x \in V$. Let $I = \{1 \le k \le m;\ \Phi(x) \in U(y_k, 2\delta)\}$ and $J = \{1,\dots,m\} \setminus I$. Then

$$\begin{aligned}\|f(x) - P(x)\| &= \left\| \left(\sum_{k=1}^{m} \psi_k(\Phi(x)) \right) f(x) - \sum_{k=1}^{m} \psi_k(\Phi(x)) f(x_k) \right\| \\ &\le \sum_{k=1}^{m} \psi_k(\Phi(x)) \|f(x) - f(x_k)\| \\ &\le \sum_{k\in I} \psi_k(\Phi(x)) \|f(x) - f(x_k)\| + \sum_{k\in J} \psi_k(\Phi(x)) (\|f(x)\| + \|f(x_k)\|) \le \varepsilon.\end{aligned}$$

□

To prove the version of the above theorem for approximating also the derivatives we need the following notion.

Definition 18. A Banach space X is said to have the approximation property, if for every compact set $K \subset X$ and every $\varepsilon > 0$ there exists a finite rank operator $T \in \mathcal{L}(X;X)$ such that $\|T(x) - x\| < \varepsilon$ for all $x \in K$. Further, X is said to have the C-bounded approximation property, where $C \in \mathbb{R}$, $C \geq 1$, if the operator T can be always chosen so that $\|T\| \leq C$.

The approximation property, introduced by Alexander Grothendieck, is one of the central concepts in Banach space theory. For some basic properties and equivalent formulations see e.g. [FHHMZ]. We will rely on the following classical result concerning the approximation property in the dual.

Theorem 19 ([FHHMZ, Theorem 16.36]). *Let X be a Banach space. Then X^* has the approximation property if and only if for every Banach space Y, every $\varepsilon > 0$, and every $T \in \mathcal{L}_K(X;Y)$ there exists a finite rank operator $S \in \mathcal{L}(X;Y)$ such that $\|T - S\| < \varepsilon$.*

Using the principle of local reflexivity (Theorem 4.2) it is possible to obtain the following result (see also [Cas, Proposition 3.5]).

Proposition 20. *Let X be a Banach space such that X^* has a C-bounded approximation property. Then for every pair of compact sets $K \subset X$ and $L \subset X^*$, and for every $\varepsilon > 0$ there exists a finite rank operator $T \in \mathcal{L}(X;X)$ such that $\|T\| \leq C$, $\|T(x) - x\| < \varepsilon$ for each $x \in K$, and $\|\phi \circ T - \phi\| < \varepsilon$ for each $\phi \in L$.*

Theorem 21 ([AP]). *Let X be a Banach space. Then X^* has the approximation property if and only if $\mathcal{P}_{\mathrm{wu}}({}^nX;Y) = \overline{\mathcal{P}_{\mathrm{f}}({}^nX;Y)}$ for every Banach space Y and every $n \in \mathbb{N}$.*

Proof. $\Leftarrow$ follows from Corollary 3.24 and Theorem 19, which also implies the reverse implication for $n = 1$. We proceed with the proof of $\Rightarrow$ by induction on n. Let $P \in \mathcal{P}_{\mathrm{wu}}({}^nX;Y)$. Theorem 6.10 implies that $d^{n-1}P \in \mathcal{L}_K\big(X;\mathcal{P}_{\mathrm{wu}}({}^{n-1}X;Y)\big)$. By Lemma 1.99, $d^{n-1}P(x)[h] = n!\check{P}(x, {}^{n-1}h)$. By Theorem 19 for every $\varepsilon > 0$ there exist $\phi_1, \dots, \phi_k \in X^*$ and $P_1, \dots, P_k \in \mathcal{P}_{\mathrm{wu}}({}^{n-1}X;Y)$ such that

$$\sup_{x,h \in B_X} \left\| n!\check{P}(x, {}^{n-1}h) - \sum_{j=1}^{k} \phi_j(x) P_j(h) \right\| < \varepsilon.$$

We finish the proof by applying the inductive hypothesis to P_j, $j = 1, \dots, k$, and using Fact 11. □

Lemma 22 ([AP]). *Let X, Y be Banach spaces such that X^* has a C-bounded approximation property. Let $f \in \mathcal{C}^k_{\mathrm{wu}}(X;Y)$, $k \in \mathbb{N}$, and let $V \subset X$ be a CCB set and $\varepsilon > 0$. Then there exists a finite rank operator $T \in \mathcal{L}(X;X)$, $\|T\| \leq C$, such that*

$$\|d^j f - d^j(f \circ T)\|_V \leq \varepsilon, \quad j = 0, \dots, k.$$

Proof. Let $R > 0$ be such that $V \subset B(0, R)$. By the assumption there exist $\delta > 0$ and $\phi_1, \dots, \phi_n \in B_{X^*}$ such that $\|d^j f(x) - d^j f(y)\| < \frac{\varepsilon}{4C^k}$, $j = 0, \dots, k$, whenever $x, y \in B(0, CR)$, $|\phi_l(x - y)| < \delta$ for all $l \in \{1, \dots, n\}$. By Lemma 3.18 each set $d^j f(V)$, $j = 1, \dots, k$, is relatively compact in $\mathcal{P}_{\mathrm{wu}}({}^j X; Y)$ and so by Theorem 21 for each $j \in \{1, \dots, k\}$ there is a finite set $M_j \subset \mathcal{P}_{\mathrm{f}}({}^j X; Y)$ which is $\frac{\varepsilon}{4C^k}$-dense in $d^j f(V)$. Next, by Proposition 20 there exists a finite rank operator $T \in \mathcal{L}(X; X)$, $\|T\| \le C$ such that $\|\phi_l - \phi_l \circ T\| < \frac{\delta}{R}$, $l = 1, \dots, n$, and moreover $\|P - P \circ T\| < \frac{\varepsilon}{4}$ for all $P \in M_j$, $j = 1, \dots, k$.

Fix $x \in V$. We have $T(x) \in B(0, CR)$ and also $|\phi_l(x - T(x))| < \frac{\delta}{R} R = \delta$. Thus $\|f(x) - f \circ T(x)\| < \varepsilon$. Further, fix $j \in \{1, \dots, k\}$ and find $P \in M_j$ such that $\|d^j f(x) - P\| \le \frac{\varepsilon}{4C^k}$. Since T is a linear operator, the Chain rule (Corollary 1.117) implies that $d^j (f \circ T)(x) = d^j f(T(x)) \circ T$. Thus

$$
\begin{aligned}
&\|d^j f(x) - d^j (f \circ T)(x)\| \\
&\quad \le \|d^j f(x) - d^j f(x) \circ T\| + \|d^j f(x) \circ T - d^j f(T(x)) \circ T\| \\
&\quad \le \|d^j f(x) - P\| + \|P - P \circ T\| + \|P \circ T - d^j f(x) \circ T\| + \frac{\varepsilon}{4C^k}\|T\|^j \\
&\quad < \frac{\varepsilon}{4C^k} + \frac{\varepsilon}{4} + \frac{\varepsilon}{4C^k}\|T\|^j + \frac{\varepsilon}{4} \le \varepsilon.
\end{aligned}
$$

□

The next result is another generalisation of the Weierstraß theorem into infinite-dimensional spaces.

Theorem 23 ([AP]). *Let X, Y be Banach spaces and $k \in \mathbb{N}$. Suppose that X^* has the bounded approximation property. Then $\overline{\mathcal{P}_{\mathrm{f}}(X; Y)}^{\tau_b^k} = \mathcal{C}^k_{\mathrm{wu}}(X; Y)$.*

Proof. Let $f \in \mathcal{C}^k_{\mathrm{wu}}(X; Y)$, let $V \subset X$ be a CCB set, and $\varepsilon > 0$. By Lemma 22 there is a finite rank operator $T \in \mathcal{L}(X; X)$ such that $\|d^j f - d^j (f \circ T)\|_V \le \frac{\varepsilon}{2}$ for all $j \in \{0, \dots, k\}$. Put $g = f \restriction_{T(X)}$ and $C = \max\{\|T\|, 1\}$. Since $T(X)$ is finite-dimensional and $K = \overline{T(V)} \subset T(X)$ is compact, by Theorem 9 there is a polynomial $p \in \mathcal{P}(T(X); Y)$ such that $\|d^j g - d^j p\|_K \le \frac{\varepsilon}{2C^k}$, $j = 0, \dots, k$. Put $P = p \circ T$. Then $P \in \mathcal{P}_{\mathrm{f}}(X; Y)$ (see the remark after Definition 10). Now for any $x \in V$ and $j \in \{1, \dots, k\}$ we have

$$
\begin{aligned}
&\big\|d^j f(x) - d^j P(x)\big\| \\
&\quad \le \big\|d^j f(x) - d^j (f \circ T)(x)\big\| + \big\|d^j (f \circ T)(x) - d^j (p \circ T)(x)\big\| \\
&\quad \le \frac{\varepsilon}{2} + \big\|d^j f\big(T(x)\big) \circ T - d^j p\big(T(x)\big) \circ T\big\| \\
&\quad = \frac{\varepsilon}{2} + \big\|d^j g\big(T(x)\big) \circ T - d^j p\big(T(x)\big) \circ T\big\| \le \frac{\varepsilon}{2} + \frac{\varepsilon}{2C^k}\|T\|^j \le \varepsilon,
\end{aligned}
$$

and similarly for $j = 0$. □

The weak uniform topology from the previous result can sometimes be replaced by the uniform topology. The abstract form of this result is contained in the next theorem, which is followed by several concrete examples.

Theorem 24 ([CHL]). *Let X, Y be Banach spaces such that X does not contain ℓ_1 and X^* has the bounded approximation property. If $\mathcal{C}^{1,+}(X;Y) \subset \mathcal{C}_{\mathrm{wsC}}(X;Y)$, then for any $k \in \mathbb{N}$*

$$\overline{\mathcal{P}(X;Y)}^{\tau_b^k} = \overline{\mathcal{P}_{\mathrm{f}}(X;Y)}^{\tau_b^k} = \mathcal{C}^{k,+}(X;Y).$$

Proof. From Theorem 3.81 it follows that $\mathcal{C}^{k,+}(X;Y) \subset \mathcal{C}^{1,+}(X;Y) \subset \mathcal{C}_{\mathrm{wu}}(X;Y)$. Thus $\mathcal{C}^{k,+}(X;Y) = \mathcal{C}^k_{\mathrm{wu}}(X;Y)$ by Theorem 6.8 and the result follows from Theorem 23. □

Corollary 25 ([CHL]). *Let X, Y be Banach spaces such that X^* has the bounded approximation property. Then for any $k \in \mathbb{N}$*

$$\overline{\mathcal{P}(X;Y)}^{\tau_b^k} = \overline{\mathcal{P}_{\mathrm{f}}(X;Y)}^{\tau_b^k} = \mathcal{C}^{k,+}(X;Y)$$

provided that one of the following conditions is satisfied:

- *X is a $\mathcal{W}$-space which does not contain ℓ_1 and Y is of a non-trivial type, or*
- *X has the hereditary DPP and does not contain ℓ_1, and Y has property (WR), or*
- *X is a linear quotient of an isomorphic predual Z of ℓ_1, and Y^{**} does not contain ℓ_∞ or Y has property (WR) and is isomorphic to a complemented subspace of a dual Banach space.*

Proof. In the last case by Theorem 4.16 the space Z is a $\mathcal{L}_\infty$-space that does not contain ℓ_1. By the lifting property of ℓ_1 it follows that X does not contain ℓ_1 either. Thus in view of Theorem 24 it suffices to show that $\mathcal{C}^{1,+}(X;Y) \subset \mathcal{C}_{\mathrm{wsC}}(X;Y)$. This follows from Theorem 6.54, Theorem 6.53, or Theorem 6.57 and Proposition 6.21 respectively. □

The above results are quite special to the $\mathcal{W}$-spaces. In a general infinite-dimensional normed linear space X there are always continuous functions on X that cannot be uniformly approximated by polynomials on S_X. Indeed, while every polynomial is bounded on S_X, it is easy to construct a continuous function on X that is unbounded on S_X. But typically even much more regular functions cannot be uniformly approximated by polynomials.

Theorem 26 ([NS]). *Let X be an infinite-dimensional Banach space and let S be one of the spaces $C^k(X)$, $C^\infty(X)$, or $C^{k,\Omega}(X)$, where $k \in \mathbb{N}_0$ and $\Omega \subset \mathcal{M}$ is a convex cone. If X admits a bump function from S, then there is a bump function $f \in S$ such that it cannot be uniformly approximated on B_X by polynomials.*

Proof. The main ingredient of the proof is the fact that for every degree d there is $n \in \mathbb{N}$ such that the unit ball of $\mathbb{R}^n$ (with an arbitrary norm) contains a $\frac{1}{2}$-separated set of cardinality greater than the dimension of the space $\mathcal{P}^d(\mathbb{R}^n)$. Indeed let $A \subset B_{\mathbb{R}^n}$

be a maximal $\frac{1}{2}$-separated set, i.e. $\|x-y\| \geq \frac{1}{2}$ for every $x, y \in A$, $x \neq y$. By the maximality, $B_{\mathbb{R}^n} \subset \bigcup_{x\in A} U(x, \frac{1}{2})$. Therefore

$$\lambda(B_{\mathbb{R}^n}) \leq \sum_{x\in A} \lambda\big(U(x, \tfrac{1}{2})\big) \leq |A|\lambda\big(B(0, \tfrac{1}{2})\big) = |A|\frac{1}{2^n}\lambda(B_{\mathbb{R}^n}),$$

and hence $|A| \geq 2^n$. On the other hand, $\dim \mathcal{P}^d(\mathbb{R}^n) = \binom{n+d}{d}$ (Section 2.1). Since for every $d \in \mathbb{N}$ there is $n_d \in \mathbb{N}$ such that $2^{n_d} > \binom{n_d+d}{d}$, there is a $\frac{1}{2}$-separated set in $B_{\mathbb{R}^{n_d}}$ of cardinality greater that $\dim \mathcal{P}^d(\mathbb{R}^{n_d})$.

Put $\delta = \frac{3}{4}$. Since the space X is infinite-dimensional, there is a δ-separated set $\{z_d;\ d \in \mathbb{N}\} \subset B_X(0, \delta)$. Notice that $B\big(z_d, \frac{\delta}{3}\big) \subset B_X$ for every $d \in \mathbb{N}$. Let X_d be some n_d-dimensional subspace of X that contains z_d. By the discussion above there is a $\frac{\delta}{6}$-separated subset A_d of $B_{X_d}\big(z_d, \frac{\delta}{3}\big)$ satisfying $\dim \mathcal{P}^d(X_d) < |A_d| < \infty$. Consider the space $\mathbb{R}^{A_d}$ with the supremum norm. Define $R_d \in \mathcal{L}(\mathcal{P}^d(X_d); \mathbb{R}^{A_d})$ by $R_d(p) = p\restriction_{A_d}$. As $\dim R_d(\mathcal{P}^d(X_d)) \leq \dim \mathcal{P}^d(X_d) < |A_d| = \dim \mathbb{R}^{A_d}$, the space $R_d(\mathcal{P}^d(X_d))$ is a proper subspace of a finite-dimensional space $\mathbb{R}^{A_d}$ and so there is $f_d \in \mathbb{R}^{A_d}$ such that $\operatorname{dist}\big(f_d, R_d(\mathcal{P}^d(X_d))\big) = \|f_d\| = 1$.

Let $\varphi \in S$ be a bump function. By shifting and scaling we may suppose that $\varphi(0) = 1$ and $\operatorname{supp}\varphi \subset B_X\big(0, \frac{\delta}{18}\big)$. Define a function $f\colon X \to \mathbb{R}$ by

$$f(x) = \sum_{d\in\mathbb{N}} \sum_{y\in A_d} f_d(y)\varphi(x-y).$$

By the choice of the set $\{z_d\}$ the set $\bigcup_{d\in\mathbb{N}} A_d$ is a $\frac{\delta}{6}$-separated subset of B_X. Using this and the fact that $|f_d(y)| \leq 1$ for $y \in A_d$ it is easy to check that $f \in S$. Obviously $\operatorname{supp} f$ is bounded.

To see that f cannot be approximated on B_X by polynomials pick any $p \in P(X)$. Let $d \in \mathbb{N}$ be such that $p \in \mathcal{P}^d(X)$. Then $p\restriction_{X_d} \in \mathcal{P}^d(X_d)$ (Fact 1.35). Notice that $f\restriction_{A_d} = f_d$. Thus

$$\begin{aligned}\sup_{x\in B_X} |f(x) - p(x)| \geq \sup_{x\in A_d} |f(x) - p(x)| &= \|f_d - R_d(p\restriction_{X_d})\| \\ &\geq \operatorname{dist}\big(f_d, R_d(\mathcal{P}^d(X_d))\big) = 1.\end{aligned}$$

□

3. Approximation by real-analytic mappings

We begin with a Whitney-type approximation theorem stating that in a finite-dimensional case any mapping in $C^k(\Omega; Y)$, $k \in \mathbb{N}_0$, can be approximated on the whole Ω, in a fine topology, and together with its derivatives of order up to k by real analytic mappings. Then we present the famous result of Jaroslav Kurzweil which extends this result (for $k = 0$) to infinite-dimensional separable Banach spaces X that admit a separating polynomial. We also show that if we only require uniform approximations for uniformly continuous mappings, then it suffices that X admits a separating real analytic function with uniform radii of convergence.

Let X, Y be normed linear spaces, $\Omega \subset X$ open, and $f \in C^k(\Omega; Y)$ for some $k \in \mathbb{N}_0$. For $S \subset \Omega$ we define

$$\|f\|_{S,k} = \sum_{j=0}^{k} \sup_{x \in S} \|d^j f(x)\|.$$

Clearly $\|\cdot\|_{S,k}$ is a semi-norm on the subspace of $C^k(\Omega; Y)$ consisting of mappings with all derivatives up to k bounded on S.

Lemma 27. *Let X, Y be normed linear spaces over $\mathbb{K}$, $\Omega \subset X$ open, $k \in \mathbb{N}_0$, $\varphi \in C^k(\Omega)$, $f \in C^k(\Omega; Y)$, and $S \subset \Omega$. Then*

$$\|\varphi f\|_{S,k} \le \binom{k}{[\frac{k}{2}]} \|\varphi\|_{S,k} \|f\|_{S,k}.$$

Proof. Fix $x \in \Omega$ and $0 \le j \le k$. By the Leibniz formula (Corollary 1.116)

$$\begin{aligned} \|d^j(\varphi f)(x)\| &\le \sum_{l=0}^{j} \binom{j}{l} \|d^{j-l}\varphi(x) \cdot d^l f(x)\| \le \sum_{l=0}^{j} \binom{j}{l} \|d^{j-l}\varphi(x)\| \cdot \|d^l f(x)\| \\ &\le \binom{j}{[\frac{j}{2}]} \sum_{l=0}^{j} \|d^{j-l}\varphi(x)\| \cdot \|d^l f(x)\|. \end{aligned}$$

Therefore

$$\begin{aligned} \|\varphi f\|_{S,k} &\le \sum_{j=0}^{k} \binom{j}{[\frac{j}{2}]} \sum_{l=0}^{j} \|d^{j-l}\varphi\|_S \|d^l f\|_S \le \binom{k}{[\frac{k}{2}]} \sum_{j=0}^{k} \sum_{l=0}^{j} \|d^{j-l}\varphi\|_S \|d^l f\|_S \\ &\le \binom{k}{[\frac{k}{2}]} \sum_{j=0}^{k} \sum_{l=0}^{k} \|d^j\varphi\|_S \|d^l f\|_S = \binom{k}{[\frac{k}{2}]} \|\varphi\|_{S,k} \|f\|_{S,k}. \end{aligned}$$

□

In the next theorem we consider $\mathbb{C}^n$ with the Euclidean norm.

Theorem 28. *Let Y be a Banach space, $\Omega \subset \mathbb{R}^n$ an open set, $k \in \mathbb{N}_0 \cup \{\infty\}$, $f \in C^k(\Omega; Y)$, and $\varepsilon \in C(\Omega; \mathbb{R}^+)$. Put $G = \{z \in \mathbb{C}^n;\ \|\operatorname{Im} z\| < \operatorname{dist}(\operatorname{Re} z, \mathbb{R}^n \setminus \Omega)\}$. Then there exists a mapping $g \in H(G; \tilde{Y})$ such that $g \restriction_\Omega \in \mathbb{C}^\omega(\Omega; Y)$ and satisfies $\|d^j f(x) - d^j(g \restriction_\Omega)(x)\| < \varepsilon(x)$ for all $x \in \Omega$, $0 \le j \le \min\{k, 1/\varepsilon(x)\}$.*

Notice in particular, that if in the preceding theorem $\Omega = \mathbb{R}^n$, then the approximating mapping g will be an entire mapping, i.e. a mapping that is holomorphic on the whole $\mathbb{C}^n$.

Proof. Define $K_{-1} = K_0 = \emptyset$, $K_j = \{x \in \mathbb{R}^n;\ \operatorname{dist}(x, \mathbb{R}^n \setminus \Omega) \ge 2^{-j}\} \cap B(0, j)$, $L_j = K_j \setminus \operatorname{Int} K_{j-1}$, and $U_j = (\operatorname{Int} K_{j+1}) \setminus K_{j-2}$ for $j \in \mathbb{N}$. Note that $K_j \subset K_{j+1}$, L_j is compact, $U_j \subset \Omega$ is an open neighbourhood of L_j, $\Omega = \bigcup_{j=1}^{\infty} L_j$, and $L_j \cap U_l = \emptyset$ for $l > j+1$. By Lemma 3 there are functions $\varphi_j \in C^\infty(\mathbb{R}^n; [0,1])$, $j \in \mathbb{N}$, satisfying $\operatorname{supp} \varphi_j \subset U_j$ (hence $\operatorname{supp} \varphi_j$ is compact) and $\varphi_j = 1$ on a neighbourhood of L_j.

Further, we put $\varepsilon_0 = 1$, $\varepsilon_j = \min\{\varepsilon_{j-1}, \min_{x\in L_j} \varepsilon(x)\}$, $k_0 = 0$, $k_j = k$ if $k < \infty$, and finally $k_j = \max\{k_{j-1}, [\max_{x\in L_j} \frac{1}{\varepsilon(x)}]\}$ if $k = \infty$. Notice that the sequence $\{\varepsilon_j\}_{j=1}^{\infty}$ is non-increasing, while the sequence $\{k_j\}_{j=1}^{\infty}$ is non-decreasing. We set $M_j = \nu_{k_j} \|\varphi_j\|_{\mathbb{R}^n, k_j}$, where $\nu_l = \binom{l}{[\frac{l}{2}]}$. For each $j \in \mathbb{N}$ let $\delta_j > 0$ be such that

$$\delta_j(1 + M_{j+1}) < \frac{\varepsilon_j}{2^j}. \tag{6}$$

To slightly shorten our notation we denote $\bar{g} = g\restriction_{M\cap\mathbb{R}^n}$ for $g\colon M \to Y$, where $M \subset \mathbb{C}^n$. For each $j \in \mathbb{N}$ we define inductively mappings $f_j \in C^k(\mathbb{R}^n; Y)$ and $g_j \in H(\mathbb{C}^n; \tilde{Y})$ such that $\bar{g_j}$ maps into Y as follows: We put $f_j = 0$ on $\mathbb{R}^n \setminus \Omega$ and

$$f_j = \varphi_j \cdot \left(f - \sum_{l=1}^{j-1} \bar{g_l}\right) \tag{7}$$

on Ω. Then $f_j \in C^k(\mathbb{R}^n; Y)$ and since $\operatorname{supp}\varphi_j$ is compact, so is $\operatorname{supp} f_j$. By Lemma 1 there is a mapping $g_j \in H(\mathbb{C}^n; \tilde{Y})$ such that $\bar{g_j}$ maps into Y,

$$\|f_j - \bar{g_j}\|_{\mathbb{R}^n, k_j} < \delta_j, \tag{8}$$

and $\|g_j\|_{G_j} < \frac{1}{2^j}$, where $G_j = \{z \in \mathbb{C}^n;\ \|\operatorname{Im} z\|^2 < \operatorname{dist}(\operatorname{Re} z, \operatorname{supp} f_j)^2 - \frac{1}{4^j}\}$.

Put

$$g = \sum_{j=1}^{\infty} g_j. \tag{9}$$

Fix any $z \in G$ and put $\delta = \frac{1}{2}\min\{\operatorname{dist}(\operatorname{Re} z, \mathbb{R}^n \setminus \Omega) - \|\operatorname{Im} z\|, 1\}$. (We note that the minimum here is to cater for the case when $\operatorname{dist}(\operatorname{Re} z, \mathbb{R}^n \setminus \Omega) = +\infty$, i.e. $\Omega = \mathbb{R}^n$.) Further, put $V = \{w \in \mathbb{C}^n;\ \|\operatorname{Re} w - \operatorname{Re} z\| + \|\operatorname{Im} w - \operatorname{Im} z\| < \delta\}$, which is a neighbourhood of z. Let $j_0 \in \mathbb{N}$ be such that $2^{-j_0} < \frac{\delta}{2}$ and $\|\operatorname{Re} z\| + \|\operatorname{Im} z\| + \frac{3}{2}\delta \le j_0$. We claim that $V \subset G_j$ for all $j \ge j_0 + 2$. Indeed, pick any $w \in V$. Since j_0 is chosen so that $U_{\mathbb{R}^n}(\operatorname{Re} z, \|\operatorname{Im} z\| + \frac{3}{2}\delta) \subset K_{j_0}$, we have

$$\begin{aligned}
\|\operatorname{Im} w\| &\le \|\operatorname{Im} z\| + \|\operatorname{Im} w - \operatorname{Im} z\| < \|\operatorname{Im} z\| + \delta - \|\operatorname{Re} w - \operatorname{Re} z\| \\
&\le \|\operatorname{Im} z\| + \delta + \operatorname{dist}(\operatorname{Re} w, \mathbb{R}^n \setminus K_{j_0}) - \operatorname{dist}(\operatorname{Re} z, \mathbb{R}^n \setminus K_{j_0}) \\
&\le \operatorname{dist}(\operatorname{Re} w, \mathbb{R}^n \setminus K_{j_0}) - \frac{\delta}{2}.
\end{aligned}$$

Hence, using the fact that $(a - b)^2 \le a^2 - b^2$ for $a, b \in \mathbb{R}$, $a - b \ge 0$, and $b \ge 0$,

$$\begin{aligned}
\|\operatorname{Im} w\|^2 &\le \operatorname{dist}(\operatorname{Re} w, \mathbb{R}^n \setminus K_{j_0})^2 - \frac{\delta^2}{4} \le \operatorname{dist}(\operatorname{Re} w, \mathbb{R}^n \setminus K_{j-2})^2 - \frac{\delta^2}{4} \\
&\le \operatorname{dist}(\operatorname{Re} w, U_j)^2 - \frac{\delta^2}{4} \le \operatorname{dist}(\operatorname{Re} w, \operatorname{supp} f_j)^2 - \frac{\delta^2}{4} \\
&< \operatorname{dist}(\operatorname{Re} w, \operatorname{supp} f_j)^2 - \frac{1}{4^j}
\end{aligned}$$

and the claim follows. This means that $\|g_j\|_V < \frac{1}{2^j}$ for $j \geq j_0 + 2$. Therefore the series (9) converges absolutely locally uniformly on G and so $g \in H(G; \tilde{Y})$. Obviously since each $\bar{g_j}$ maps into Y, so does $\bar{g}$.

To show the approximation property of the mapping $\bar{g}$ fix an arbitrary $x \in \Omega$ and $0 \leq l \leq \min\{k, 1/\varepsilon(x)\}$. There is $p \in \mathbb{N}$ such that $x \in L_p$. Hence $l \leq k_p$ and $\varepsilon_p \leq \varepsilon(x)$. Since $\varphi_p = 1$ on a neighbourhood of L_p, by (7) and (8) we have

$$\left\| f - \sum_{j=1}^{p} \bar{g_j} \right\|_{L_p,k_p} = \|f_p - \bar{g_p}\|_{L_p,k_p} < \delta_p. \tag{10}$$

From Lemma 27, the fact that the sequences $\{k_j\}$ and $\{\nu_j\}$ are non-decreasing, (8), and (10) we obtain

$$\begin{aligned}
\|\overline{g_{p+1}}\|_{L_p,k_p} &\leq \|\overline{g_{p+1}} - f_{p+1}\|_{L_p,k_p} + \|f_{p+1}\|_{L_p,k_p} \\
&\leq \|\overline{g_{p+1}} - f_{p+1}\|_{L_p,k_p} + \nu_{k_p}\|\varphi_{p+1}\|_{L_p,k_p} \left\| f - \sum_{j=1}^{p} \bar{g_j} \right\|_{L_p,k_p} \\
&\leq \|\overline{g_{p+1}} - f_{p+1}\|_{\mathbb{R}^n,k_{p+1}} + \nu_{k_{p+1}}\|\varphi_{p+1}\|_{\mathbb{R}^n,k_{p+1}} \left\| f - \sum_{j=1}^{p} \bar{g_j} \right\|_{L_p,k_p} \\
&< \delta_{p+1} + M_{p+1}\delta_p.
\end{aligned}$$

Finally, for $j > p+1$ we have $U_j \cap L_p = \emptyset$ and since $\operatorname{supp} f_j \subset U_j$, $f_j = 0$ on a neighbourhood of L_p. Hence

$$\|\bar{g_j}\|_{L_p,k_p} = \|\bar{g_j} - f_j\|_{L_p,k_p} \leq \|\bar{g_j} - f_j\|_{\mathbb{R}^n,k_j} < \delta_j.$$

Putting all this together with (6) yields

$$\begin{aligned}
\|d^l f(x) - d^l \bar{g}(x)\| &= \left\| d^l f(x) - \sum_{j=1}^{\infty} d^l \bar{g_j}(x) \right\| \\
&\leq \left\| d^l f(x) - \sum_{j=1}^{p} d^l \bar{g_j}(x) \right\| + \sum_{j=p+1}^{\infty} \|d^l \bar{g_j}(x)\| \\
&\leq \left\| f - \sum_{j=1}^{p} \bar{g_j} \right\|_{L_p,k_p} + \sum_{j=p+1}^{\infty} \|\bar{g_j}\|_{L_p,k_p} < \delta_p(1 + M_{p+1}) + \sum_{j=p+1}^{\infty} \delta_j \\
&\leq \sum_{j=p}^{\infty} \delta_j(1 + M_{j+1}) < \sum_{j=p}^{\infty} \frac{\varepsilon_j}{2^j} \leq \sum_{j=p}^{\infty} \frac{\varepsilon_p}{2^j} \leq \varepsilon_p \leq \varepsilon(x).
\end{aligned}$$

We note that the first equality follows from the fact that the series (9) is a locally uniformly convergent series of holomorphic mappings. □

Now we move on to the infinite-dimensional case. First we state the results and then prove them both together.

Theorem 29 (Jaroslav Kurzweil, [Kur1]). *Let X be a separable Banach space that admits a separating polynomial and Y a Banach space. Let $\Omega \subset X$ be open, $f \in C(\Omega; Y)$, and $\varepsilon \in C(\Omega; \mathbb{R}^+)$. Then there exists $g \in C^\omega(\Omega; Y)$ such that $\|f(x) - g(x)\| < \varepsilon(x)$ for all $x \in \Omega$.*

Definition 30. Let X be a Banach space. We say that X has property (K) if there exists a separating real analytic function q on X and $d > 0$ such that for each $x \in X$ the radius of norm convergence of the Taylor series of q at x is at least d.

Given metric spaces P, Q we denote by $C_{\mathrm{u}}(P; Q)$ the space of all uniformly continuous mappings from P to Q.

Theorem 31 ([CH], [Fry1]). *Let X be a separable Banach space with property (K) and Y a Banach space. Let $\Omega \subset X$ be open, $f \in C_{\mathrm{u}}(\Omega; Y)$, and $\varepsilon > 0$. Then there is $g \in C^\omega(\Omega; Y)$ such that $\|f - g\|_\Omega \le \varepsilon$.*

We prove both Theorem 29 and Theorem 31 together, with the help of the next two lemmata.

Lemma 32. *Let X be a separable Banach space and $\Omega \subset X$ open. Suppose there is $\alpha > 0$ such that for any open covering $\{U(x_n, r_n)\}_{n=1}^\infty$ (or for any uniform open covering $\{U(x_n, r)\}_{n=1}^\infty$, i.e. $r_n = r$ for all $n \in \mathbb{N}$) of Ω and any sequence $\{w_n\}_{n=1}^\infty$ of positive real numbers there exists an open neighbourhood $V \subset \tilde{X}$ of Ω and a sequence of functions $\{\varphi_n\}_{n=1}^\infty \subset H(V)$ with the following properties:*

(i) *The sum $\sum_{n=1}^\infty w_n \varphi_n$ converges absolutely locally uniformly on V,*
(ii) *$\varphi_n \restriction_\Omega$ maps into $[0, +\infty)$ for every $n \in \mathbb{N}$,*
(iii) *$w_n \varphi_n(x) \le \frac{1}{4} 2^{-n}$ for every $x \in \Omega \setminus U(x_n, \alpha r_n)$, $n \in \mathbb{N}$, and*
(iv) *for every $x \in \Omega$ there is $k \in \mathbb{N}$ such that $x \in U(x_k, \alpha r_k)$ and $\varphi_k(x) \ge w_k$.*

Then for every mapping $f \in C(\Omega; C)$, where C is a closed convex subset of a Banach space Y, (resp. for every $f \in C_{\mathrm{u}}(\Omega; C)$) there is a mapping $g \in C^\omega(\Omega; C)$ satisfying $\|f - g\|_\Omega \le 1$.

Proof. Using the separability of Ω and the continuity (resp. uniform continuity) of f we find a covering (resp. uniform covering) $\{U(x_n, r_n)\}_{n=1}^\infty$ of Ω such that

$$\|f(x) - f(x_n)\| < \frac{1}{2} \quad \text{for } x \in U(x_n, \alpha r_n) \cap \Omega. \tag{11}$$

Put $w_n = 1 + \|f(x_n)\|$. Let $\{\varphi_n\}$ be the sequence of functions satisfying (i)–(iv).

The function $\varphi(z) = \sum_{n=1}^\infty \varphi_n(z)$ is well-defined for every $z \in V$ by (i) and moreover $\varphi \in H(V)$. Further, by (ii) and (iv), for every $x \in \Omega$

$$\varphi(x) \ge \varphi_k(x) \ge 1. \tag{12}$$

Hence there is an open neighbourhood W of Ω in $\tilde{X}$ such that $W \subset V$ and $\varphi \ne 0$ on W. Define $\psi_n(z) = \varphi_n(z)/\varphi(z)$ for $z \in W$. Then the functions ψ_n are holomorphic on W and have the following properties:

(a) $\{\psi_n \restriction_\Omega\}$ is a partition of unity on Ω,
(b) $\psi_n(x)\|f(x_n)\| \le \frac{1}{4}2^{-n}$ for every $x \in \Omega \setminus U(x_n, \alpha r_n)$, $n \in \mathbb{N}$, and
(c) $\psi_n(x)\|f(x)\| \le \frac{1}{4}2^{-n}$ for every $x \in \Omega \setminus U(x_n, \alpha r_n)$, $n \in \mathbb{N}$.

Indeed, property (a) follows from (ii) and the definition of ψ_n and φ, and property (b) follows from (iii) and (12). To prove (c) choose $n \in \mathbb{N}$ and $x \in \Omega \setminus U(x_n, \alpha r_n)$. Then $\varphi(x) \ge \varphi_k(x) \ge w_k = 1 + \|f(x_k)\| > \|f(x)\|$ by (iv) and (11). Thus $\psi_n(x)\|f(x)\| \le \psi_n(x)\varphi(x) = \varphi_n(x) \le w_n\varphi_n(x) \le \frac{1}{4}2^{-n}$ by (iii).

To finish the proof we put $g(z) = \sum_{n=1}^{\infty} \psi_n(z) f(x_n)$ for $z \in W$. Because $g(z) = (1/\varphi(z)) \sum_{n=1}^{\infty} \varphi_n(z) f(x_n)$ and the sum converges locally uniformly on W by (i), we obtain $g \in H(W; \tilde{Y})$. Clearly $g \restriction_\Omega \in C^\omega(\Omega; C)$. Further, choose an arbitrary $x \in \Omega$. Put $I = \{n \in \mathbb{N};\ x \in U(x_n, \alpha r_n)\}$. Then using (a), (11), (b), and (c) we obtain

$$\begin{aligned}\|f(x) - g(x)\| &= \left\| \sum_{n=1}^{\infty} \psi_n(x)\big(f(x) - f(x_n)\big) \right\| \\ &\le \sum_{n \in I} \psi_n(x)\|f(x) - f(x_n)\| + \sum_{n \in \mathbb{N} \setminus I} \psi_n(x)\big(\|f(x)\| + \|f(x_n)\|\big) < 1.\end{aligned}$$

□

Lemma 33. *Let X be a Banach space, $\Omega \subset X$ open, and $\{U(x_n, r_n)\}_{n=1}^{\infty} \subset X$ an open covering of Ω. Suppose that there are a function $q \in H(G)$ and $\alpha > 0$ such that $G = \{z \in \tilde{X};\ \|\mathrm{Im}\, z\| < \Delta \sup_{n\in\mathbb{N}} \frac{1}{r_n}\}$ for some $\Delta > 0$, $q \restriction_X$ maps into $[0, +\infty)$, $q(x) \ge 2$ for $x \in X \setminus U_X$, $\mathrm{Re}\, q(z) \le \frac{1}{4}$ for $z \in U_{\tilde{X}}(0, 1/\alpha)$, and suppose there is a sequence $\{a_n\}_{n=1}^{\infty}$ of positive real numbers such that for each $x \in \Omega$ the function*

$$z \mapsto \sum_{n=1}^{\infty} a_n \big(\mathrm{Im}\, q\big((x - x_n + z)/(\alpha r_n)\big)\big)^2 \tag{13}$$

is defined on some neighbourhood of 0 in $\tilde{X}$ and is continuous at 0.

Then for every sequence $\{w_n\}_{n=1}^{\infty}$ of positive real numbers there are an open neighbourhood $V \subset \tilde{X}$ of Ω and a sequence of functions $\{\varphi_n\}_{n=1}^{\infty} \subset H(V)$ satisfying the properties (i)–(iv) in Lemma 32.

Proof. Put $\varepsilon_n = \frac{1}{2w_n^2}\frac{1}{4}2^{-n}$ and let ζ_n be the functions and $\{\delta_n\}$ the sequence from Lemma 5. Denote $\alpha_n = 1/(\alpha r_n)$ and $G_\alpha = \{z \in \tilde{X};\ \|\mathrm{Im}\, z\| < \alpha\Delta\}$ and put

$$\varphi_n(z) = 2w_n\zeta_n\Big(q\big(\alpha_1(z - x_1)\big), \dots, q\big(\alpha_n(z - x_n)\big)\Big) \quad \text{for } z \in G_\alpha,\ n \in \mathbb{N}.$$

Then $\varphi_n \in H(G_\alpha)$ and by (Z1), $\varphi_n \restriction_X$ maps into $[0, +\infty)$.

Pick any $x \in \Omega$. Then there exists $j \in \mathbb{N}$ such that $x \in U(x_j, r_j)$ and hence $q(\alpha_j(x - x_j)) \le \frac{1}{4} < 1$. Let $k \in \mathbb{N}$ be the smallest index such that $q(\alpha_k(x - x_k)) < 1$. Then $x \in U(x_k, \alpha r_k)$ and property (Z4) implies that $\varphi_k(x) \ge w_k$.

Let $\delta_x > 0$ be such that $\|x - x_j + z\| < r_j$ and $\sum_{n=1}^{\infty} a_n \big(\mathrm{Im}\, q(\alpha_n(x - x_n + z))\big)^2 \le \delta_j$ whenever $z \in \tilde{X}$, $\|z\| \le \delta_x$. Then $\mathrm{Re}\, q(\alpha_j(x - x_j + z)) \le \frac{1}{4}$ and hence, by (Z3),

$|w_n\varphi_n(x+z)| < 2^{-n}$ for $n > j$. It follows that $\sum_{n=1}^{\infty} w_n\varphi_n$ converges absolutely uniformly on $U_{\tilde{X}}(x,\delta_x)$. We put $V = G_\alpha \cap \bigcup_{x\in\Omega} U_{\tilde{X}}(x,\delta_x)$.

Finally we show that (iii) is satisfied. Fix $n \in \mathbb{N}$. For $x \in \Omega \setminus U(x_n, \alpha r_n)$ we have $q(\alpha_n(x - x_n)) \geq 2$, hence, by (Z5), $w_n\varphi_n(x) \leq \frac{1}{4}2^{-n}$. □

The next lemma shows that in certain circumstances it is possible to pass from uniform approximations to fine approximations.

Lemma 34 ([Kur1]). *Let Ω be a topological space and Y a normed linear space. Let $S \subset C(\Omega;Y)$ and $S_1 \subset C(\Omega)$ be such that $h/\eta \in S$ for any positive function $\eta \in S_1$ and any mapping $h \in S$. Suppose that for any $f \in C(\Omega;Y)$ there is $h \in S$ such that $\|f-h\|_\Omega \leq 1$ and for any $\varphi \in C(\Omega)$ there is $\eta \in S_1$ such that $|\varphi - \eta|_\Omega \leq 1$. Then for any $f \in C(\Omega;Y)$ and any positive function $\varepsilon \in C(\Omega)$ there is $g \in S$ such that $\|f(x) - g(x)\| < \varepsilon(x)$ for every $x \in \Omega$.*

Proof. Define $\varphi \in C(\Omega)$ by $\varphi = 1 + 2/\varepsilon$. According to the assumptions there is $\eta \in S_1$ such that $|\varphi(x) - \eta(x)| \leq 1$ for every $x \in \Omega$. Since $\eta f \in C(\Omega;Y)$, there is $h \in S$ such that $\|\eta(x)f(x) - h(x)\| \leq 1$ for every $x \in \Omega$. Notice that $\eta(x) \geq \varphi(x) - 1 = 2/\varepsilon(x) > 1/\varepsilon(x) > 0$ for every $x \in \Omega$. Thus $g = h/\eta \in S$ and $\|f(x) - g(x)\| \leq 1/\eta(x) < \varepsilon(x)$ for every $x \in \Omega$. □

Proof of Theorem 29. By Fact 4.45 we may assume that there is an m-homogeneous polynomial p on X such that $p(x) \geq 2$ for $x \in X \setminus U_X$. Let $q = \tilde{p}$. Because $q(0) = 0$, from the continuity there is $\alpha > 0$ such that $\operatorname{Re} q(z) \leq \frac{1}{4}$ for $z \in U_{\tilde{X}}(0, 1/\alpha)$.

Suppose that $\{U(x_n, r_n)\}_{n=1}^{\infty}$ is an open covering of Ω. Put

$$a_n = \frac{r_n^{2m}}{2^n(1+\|x_n\|)^{2m}}$$

Then

$$\begin{aligned} a_n\big(\operatorname{Im} q((x - x_n + z)/(\alpha r_n))\big)^2 &\leq a_n\left|q((x - x_n + z)/(\alpha r_n))\right|^2 \\ &\leq a_n\|q\|^2\frac{\|x - x_n + z\|^{2m}}{(\alpha r_n)^{2m}} \leq \frac{\|q\|^2}{\alpha^{2m}}\frac{\|x - x_n + z\|^{2m}}{2^n(1+\|x_n\|)^{2m}} \\ &\leq \frac{\|q\|^2}{\alpha^{2m}}\big(1 + \|x\| + \|z\|\big)^{2m}\frac{1}{2^n} \end{aligned}$$

and hence for every $x \in X$ the sum in (13) converges absolutely locally uniformly to a continuous function on $\tilde{X}$. Thus the hypotheses of Lemma 33 are satisfied and using it together with Lemma 32 we can conclude that for any Banach space Z and any continuous mapping $f \in C(\Omega;Z)$ there is a mapping $h \in C^\omega(\Omega;Z)$ satisfying $\|f-h\|_\Omega \leq 1$. Finally Lemma 34 applied to $S = C^\omega(\Omega;Y)$ and $S_1 = C^\omega(\Omega)$ finishes the proof. □

Proof of Theorem 31. By Theorem 1.171 there are $d > 0$ and a function $q \in H(G)$, $G = \{z \in \tilde{X};\ \|\operatorname{Im} z\| < d\}$, such that $q\restriction_X : X \to [0,+\infty)$, $q(0) = 0$, $q(x) \geq 2$ for $x \in X \setminus U_X$, and the radius of norm convergence of the Taylor series of q at every point $x \in X$ is at least d. Let $\alpha > 0$ be such that $\operatorname{Re} q(z) \le \frac{1}{4}$ for $z \in U_{\tilde{X}}(0, 1/\alpha)$ and $\frac{1}{2}\alpha d > 1$.

Suppose $\{U(x_n, r)\}_{n=1}^{\infty}$ is a uniform open covering of Ω. Put

$$M_n = \sup\Big\{\big|q\big((x_j - x_n + w)/(\alpha r)\big)\big|;\ w \in \tilde{X}, \|w\| \le \tfrac{1}{2}\alpha r d, 1 \le j \le n\Big\}$$

and $a_n = 1/(2^n M_n^2)$. (Note that by the assumption on the radius of the Taylor series $M_n < +\infty$.) Fix $x \in \Omega$. There is $k \in \mathbb{N}$ such that $x \in U(x_k, r)$. For $z \in \tilde{X}$, $\|z\| \le r(\frac{1}{2}\alpha d - 1)$ we have $\|x - x_k + z\| \le \frac{1}{2}\alpha r d$ and hence for $n \ge k$

$$\begin{aligned} a_n\big(\operatorname{Im} q\big((x - x_n + z)/(\alpha r)\big)\big)^2 &\le a_n \left|q\big((x_k - x_n + x - x_k + z)/(\alpha r)\big)\right|^2 \\ &\le a_n M_n^2 = \frac{1}{2^n}. \end{aligned}$$

Therefore the sum in (13) converges absolutely uniformly on $B_{\tilde{X}}\big(0, r(\frac{1}{2}\alpha d - 1)\big)$ to a continuous function. Using Lemma 33 together with Lemma 32 and a suitable scaling finishes the proof. □

The space c_0 does not admit a separating polynomial (Proposition 5.49 or Corollary 3.59), but it has property (K) (take $P_n(x) = \big(e_n^*(x)\big)^{2n}$ in Example 1.137 and combine it with Corollary 1.165). The property (K) is inherited by subspaces and finite direct sums. In certain circumstances it can also pass to infinite direct sums: Assume that all members of a sequence of Banach spaces $\{X_n\}$ have property (K) witnessed by non-negative functions q_n with radii at least d_n and satisfying $q_n(0) = 0$ and $q_n(x) \ge 1$ whenever $\|x\| \ge 1$. Suppose that there are $0 < d \le \frac{1}{2}\inf_{n\in\mathbb{N}} d_n$ and a sequence $\{\alpha_n\} \subset \mathbb{N}$ such that $\sup_{n\in\mathbb{N}} \sup_{z\in B_{\tilde{X}_n}(0,d)} |\tilde{q}_n(z)|^{\alpha_n} < 1$, where $\tilde{q}_n$ is the analytic extension of q_n to a neighbourhood of X_n in $\tilde{X}_n$ (Theorem 1.171). Then $\big(\bigoplus_{n=1}^{\infty} X_n\big)_{c_0}$ has property (K) witnessed by $q(x_1, x_2, \dots) = \sum_{n=1}^{\infty} q_n(x_n)^{2n\alpha_n}$ with radii at least d (use Corollary 1.165). Thus for example $\big(c_0 \oplus \bigoplus_{n=1}^{\infty} \ell_{2n}\big)_{c_0}$ has property (K).

By Theorem 5.64 a space with property (K) that does not contain c_0 admits a separating polynomial. By Corollary 5.68 every space with (K) is saturated by spaces from $\{\ell_p;\ p \text{ even}\} \cup \{c_0\}$. Let us mention without proof the next result, which should be compared with Corollary 5.105.

Proposition 35 ([CH]). *Let X be a Banach space with property (K) and such that $\mathcal{P}(X) = \mathcal{P}_{\mathrm{wsc}}(X)$. Then X is isomorphic to a subspace of c_0.*

Whence all Banach spaces with the Dunford-Pettis property and (K) are isomorphic to subspaces of c_0 (Theorem 3.68). In particular, since every $C(K)$ space which is isomorphic to a subspace of c_0 is isomorphic to c_0 [LP], we have the following corollary:

Corollary 36. *If the Banach space $C(K)$ has property (K), then it is isomorphic to c_0.*

4. Infimal convolution

The infimal convolution is another global approximation technique, which similarly to the integral convolution preserves certain regularity properties of the approximated function, like for example the Lipschitzness. The undisputable advantage is that it does not need any finite-dimensional structure and it works equally well even on non-separable spaces. The drawback is that this technique is fundamentally scalar and also it usually produces only smoothness of the first order. The notion goes back to Felix Hausdorff around 1919.

Definition 37. Let X be a set, $f: X \to \mathbb{R} \cup \{+\infty\}$, and $K: X^2 \to \mathbb{R} \cup \{+\infty\}$. We define the infimal convolution of f and K by

$$(f \,\square\, K)(x) = \inf_{y \in X} \bigl(f(y) + K(x, y)\bigr), \quad x \in X.$$

The function K is called a kernel. If $(X, +)$ is a commutative group, then we associate with each $g: X \to \mathbb{R} \cup \{+\infty\}$ the kernel $K_g(x, y) = g(x - y)$. We may then define the infimal convolution of f and g as $f \,\square\, g = f \,\square\, K_g$, i.e.

$$(f \,\square\, g)(x) = \inf_{y \in X} \bigl(f(y) + g(x - y)\bigr).$$

Note that in this case $f \,\square\, g = g \,\square\, f$.

Fact 38. *Let X be a set, $f: X \to \mathbb{R} \cup \{+\infty\}$, and $K: X^2 \to \mathbb{R}$.*

(i) *If $K(x, x) = 0$ for every $x \in X$, then $f \,\square\, K \le f$.*

(ii) *If f is proper, then $f \,\square\, K < +\infty$ everywhere.*

(iii) *If X is a metric space, f is proper, and the functions $x \mapsto K(x, y)$, $y \in X$, are uniformly continuous with modulus $\omega \in \mathcal{M}$, then either $f \,\square\, K$ is identically $-\infty$, or $f \,\square\, K$ is real-valued and uniformly continuous with modulus ω.*

Proof. Both (i) and (ii) are obvious.

(iii) Let ρ be the metric on X. Suppose there is $z \in X$ such that $(f \,\square\, K)(z) = -\infty$. Then there exists a sequence $\{y_n\}$ in X satisfying $f(y_n) + K(z, y_n) < -n$ for each $n \in N$. Now for any $x \in X$ we have $(f \,\square\, K)(x) \le f(y_n) + K(x, y_n) = f(y_n) + K(z, y_n) - K(z, y_n) + K(x, y_n) < -n + \omega(\rho(x, z))$, which implies that $(f \,\square\, K)(x) = -\infty$. Therefore $f \,\square\, K$ is either identically $-\infty$, or $f \,\square\, K$ is real-valued and uniformly continuous with modulus ω, as it is an infimum of a family of uniformly continuous functions with modulus ω. □

The following extension lemma is useful for example when we deal with smooth approximations of Lipschitz functions defined on some subset of a normed linear space X: It suffices to formulate the approximation results for functions defined on the whole of X.

Lemma 39. *Let (P, ρ) be a metric space, $\emptyset \ne A \subset P$, and $f: A \to \mathbb{R}$ a uniformly continuous function with modulus $\omega \in \mathcal{M}_s$. Then there is an extension of f to the whole of P which is uniformly continuous with modulus ω.*

Proof. Define $\bar{f}: P \to \mathbb{R}$ by $\bar{f} = f$ on A and $\bar{f} = +\infty$ on $P \setminus A$. Further, put $g = \bar{f} \,\Box\, (\omega \circ \rho)$. Then

$$g(x) = \inf_{y \in A} \Big(f(y) + \omega\big(\rho(x, y)\big)\Big)$$

For any $x, y \in A$ we have $f(x) - f(y) \le \omega(\rho(x, y))$ and so $f(x) \le f(y) + \omega(\rho(x, y))$. It follows that $f \le g$ on A. This together with Fact 38(i) implies that $g = f$ on A. Consequently by Fact 38(iii) g is real-valued and uniformly continuous with modulus ω. □

The next lemma tells us that the results on uniform approximation of Lipschitz functions immediately give also approximation of uniformly continuous functions.

Lemma 40. *Let (P, ρ) be a metric space, $f: P \to \mathbb{R}$ a uniformly continuous function with modulus $\omega \in \mathcal{M}_s$, and $\varepsilon > 0$. Further, let $a \in \mathbb{R}^+$ be such that $\omega(a) \le \varepsilon$. Then there is an $\frac{\varepsilon}{a}$-Lipschitz function $g: P \to \mathbb{R}$ such that $|f - g|_P \le \varepsilon$.*

Proof. We let $g = f \,\Box\, \frac{\varepsilon}{a}\rho$. Fix $x \in P$. Clearly $g(x) \le f(x)$ (Fact 38(i)). From the sub-additivity of ω it follows that for any $y \in P$

$$\begin{aligned} f(x) - f(y) &\le \omega\big(\rho(x, y)\big) \le \omega\left(\left\lceil \frac{\rho(x, y)}{a} \right\rceil a\right) \\ &\le \left\lceil \frac{\rho(x, y)}{a} \right\rceil \omega(a) \le \left(\frac{\rho(x, y)}{a} + 1\right) \varepsilon. \end{aligned}$$

Thus $f(y) + \frac{\varepsilon}{a}\rho(x, y) \ge f(x) - \varepsilon$, which implies that $g(x) \ge f(x) - \varepsilon$. Finally, g is $\frac{\varepsilon}{a}$-Lipschitz by Fact 38(iii). □

Theorem 41 (Jean-Michel Lasry and Pierre-Louis Lions, [LL]). *Let H be a Hilbert space, $f: H \to \mathbb{R}$ an L-Lipschitz function, and $\varepsilon > 0$. Then there is an L-Lipschitz function $g \in C^{1,1}(H)$ satisfying $|f - g|_H \le \varepsilon$.*

For the proof we need a few auxiliary results concerning convex functions.

Lemma 42. *Let X be a normed linear space, $f \in C(X)$, and suppose there exist functions $\mu, \nu \in C^{1,\alpha}(X)$, $\alpha \in (0, 1]$, such that $f + \mu$ is convex and $f - \nu$ is concave. Then $f \in C^{1,\alpha}(X)$.*

Proof. It clearly suffices to show that $f + \mu \in C^{1,\alpha}(X)$ which we show using Lemma 5.20. Notice that $\mu + \nu = (f + \mu) + (-f + \nu)$ is necessarily convex. From the concavity of $f - \nu$ it follows that $(f - \nu)(x + h) + (f - \nu)(x - h) - 2(f - \nu)(x) \le 0$ for any $x, h \in X$ and hence

$$\begin{aligned} &(f + \mu)(x + h) + (f + \mu)(x - h) - 2(f + \mu)(x) \\ &\qquad \le (f + \mu)(x + h) + (f + \mu)(x - h) - 2(f + \mu)(x) \\ &\qquad\qquad - \big((f - \nu)(x + h) + (f - \nu)(x - h) - 2(f - \nu)(x)\big) \\ &\qquad = (\mu + \nu)(x + h) + (\mu + \nu)(x - h) - 2(\mu + \nu)(x) \le C\|h\|^{1+\alpha}, \end{aligned}$$

where the last inequality follows from Lemma 5.20 used on $\mu + \nu$. □

Let X be a normed linear space. For any $f: X \to \mathbb{R} \cup \{+\infty\}$ and $t > 0$ we define the Moreau-Yosida regularisation $f_t = f \,\square\, \frac{1}{2t}\|\cdot\|^2$. We note that the constant $\frac{1}{2}$ is useless (and perhaps even annoying) in our proofs but using this particular kernel is customary in convex analysis for many good reasons.

Fact 43. *Let H be a Hilbert space and $f: H \to \mathbb{R} \cup \{+\infty\}$ a proper function.*

(i) *The extended real-valued function $-f_t + \frac{1}{2t}\|\cdot\|^2$ is convex for every $t > 0$.*

(ii) *Suppose that f is real-valued and $f + \frac{1}{2t}\|\cdot\|^2$ is convex for some $t > 0$. Then $f_s + \frac{1}{2(t-s)}\|\cdot\|^2$ is convex for every $0 < s < t$.*

Proof. (i) This follows from the fact that

$$-f_t(x) + \frac{1}{2t}\|x\|^2 = -\inf_{y\in H}\left(f(y) + \frac{1}{2t}\|x-y\|^2\right) + \frac{1}{2t}\|x\|^2$$
$$= \sup_{y\in H}\left(\frac{1}{2t}(\|x\|^2 - \|x-y\|^2) - f(y)\right) = \sup_{y\in H}\left(\frac{1}{t}\langle x, y\rangle - \frac{1}{2t}\|y\|^2 - f(y)\right),$$

which is a supremum of affine functions.

(ii) We have

$$f_s(x) + \frac{1}{2(t-s)}\|x\|^2 = \inf_{y\in H}\left(f(y) + \frac{1}{2s}\|x-y\|^2 + \frac{1}{2(t-s)}\|x\|^2\right)$$
$$= \inf_{y\in H}\left(f(y) + \frac{1}{2t}\|y\|^2 + \frac{1}{2s}\|x-y\|^2 + \frac{1}{2(t-s)}\|x\|^2 - \frac{1}{2t}\|y\|^2\right)$$
$$= \inf_{y\in H}\left(f(y) + \frac{1}{2t}\|y\|^2 + \frac{t-s}{2st}\left\|\frac{t}{t-s}x - y\right\|^2\right).$$

It is easy to verify that for any convex function $\phi: X \times Y \to \mathbb{R}$, where X, Y are vector spaces, the function $x \mapsto \inf_{y\in Y}\phi(x, y)$ is also convex (in other words, "a convex body casts a convex shadow"), from which the result follows if we set $\phi(x, y) = f(y) + \frac{1}{2t}\|y\|^2 + \frac{t-s}{2st}\|\frac{t}{t-s}x - y\|^2$. □

Proof of Theorem 41. For every function $h: H \to \mathbb{R}$ that is C-Lipschitz we have $\lim_{t\to 0+} |h_t - h|_H = 0$. Indeed,

$$0 \le h(x) - h_t(x) = h(x) - \inf_{y\in H}\left(h(y) + \frac{1}{2t}\|x-y\|^2\right)$$
$$= h(x) + \sup_{y\in H}\left(-h(y) - \frac{1}{2t}\|x-y\|^2\right) = \sup_{y\in H}\left(h(x) - h(y) - \frac{1}{2t}\|x-y\|^2\right)$$
$$\le \sup_{y\in H}\left(C\|x-y\| - \frac{1}{2t}\|x-y\|^2\right) = \sup_{\delta\in[0,+\infty)}\left(C\delta - \frac{1}{2t}\delta^2\right) = \frac{C^2 t}{2}$$

for any $x \in H$, where the first inequality follows from Fact 38(i). Moreover, as $h_t = \frac{1}{2t}\|\cdot\|^2 \,\square\, h$, each h_t is C-Lipschitz by Fact 38(iii).

So choose $t > 0$ such that $|f_t - f|_H \le \frac{\varepsilon}{2}$. Then f_t is L-Lipschitz. Next, find $0 < s < t$ such that $\left|(-f_t)_s - (-f_t)\right|_H \le \frac{\varepsilon}{2}$ and put $g = -(-f_t)_s$. Then $|f - g|_H \le \varepsilon$ and the function g is L-Lipschitz.

Further, the function $-f_t + \frac{1}{2t}\|\cdot\|^2$ is convex by Fact 43(i), and hence the function $g - \frac{1}{2(t-s)}\|\cdot\|^2$ is concave by Fact 43(ii). Using Fact 43(i) again, this time on the function $-f_t$, we can conclude that $g + \frac{1}{2s}\|\cdot\|^2$ is convex. Since the function $\|\cdot\|^2$ is a 2-homogeneous polynomial, it belongs to $C^{1,1}(H)$, and so Lemma 42 finishes the proof. □

Another nice application of the infimal convolution gives the next result.

Proposition 44 ([We2]). *Let A be a closed subset of a Hilbert space H. Then there is a function $f \in C^{1,1}(H)$ such that $A = f^{-1}(\{0\})$.*

Proof. Let $\mathfrak{I}_A \colon H \to \mathbb{R} \cup \{+\infty\}$ be the indicator function of the set A, i.e. $\mathfrak{I}_A(x) = 0$ for $x \in A$ and $\mathfrak{I}_A = +\infty$ for $x \in H \setminus A$. We let $f = -(-(\mathfrak{I}_A)_1)_{\frac{1}{2}}$. Without loss of generality we may assume that A is non-empty and hence $\mathfrak{I}_A$ is proper. It is easy to see that $(\mathfrak{I}_A)_t(x) = \frac{1}{2t}\operatorname{dist}^2(x, A)$ for every $x \in H, t > 0$. Thus $(\mathfrak{I}_A)_1$ is real-valued and using Fact 43 similarly as in the proof of Theorem 41 we can conclude that $f + \|\cdot\|^2$ is convex and $f - \|\cdot\|^2$ is concave.

Next, notice the following observation: Suppose that $h \colon X \to \mathbb{R} \cup \{+\infty\}$ is a proper function on a normed linear space X and $0 < s < t$. For any $x, y, z \in X$ we have

$$\begin{aligned}
\frac{1}{2t}\|z-y\|^2 &= \frac{1}{2t}\left\|\frac{s}{t}\left(\frac{t}{s}(z-x)\right) + \left(1-\frac{s}{t}\right)\left(\frac{t}{t-s}(x-y)\right)\right\|^2 \\
&\le \frac{1}{2t}\left(\frac{s}{t}\left\|\frac{t}{s}(z-x)\right\|^2 + \left(1-\frac{s}{t}\right)\left\|\frac{t}{t-s}(x-y)\right\|^2\right) \\
&= \frac{1}{2s}\|z-x\|^2 + \frac{1}{2(t-s)}\|x-y\|^2,
\end{aligned}$$

where we used the convexity of $\|\cdot\|^2$. From this and Fact 38(i) we obtain

$$\begin{aligned}
h_t(x) \le -(-h_t)_s(x) &= \sup_{z\in X}\inf_{y\in X}\left(h(y) + \frac{1}{2t}\|z-y\|^2 - \frac{1}{2s}\|x-z\|^2\right) \\
&\le \sup_{z\in X}\inf_{y\in X}\left(h(y) + \frac{1}{2(t-s)}\|x-y\|^2\right) = h_{t-s}(x) \quad \text{for every } x \in X.
\end{aligned}$$

This gives us $\frac{1}{2}\operatorname{dist}^2(x, A) \le f(x) \le \operatorname{dist}^2(x, A)$ for every $x \in H$. It follows that $f^{-1}(\{0\}) = A$ and that $f + \|\cdot\|^2$ is locally bounded and thus convex continuous. Now Lemma 42 implies that $f \in C^{1,1}(H)$. □

5. Approximation of continuous mappings and partitions of unity

In this section we investigate smooth partitions of unity, the main tool for obtaining C^k-smooth approximations of continuous mappings in Banach spaces. We show that several rather general classes of Banach spaces admit C^k-smooth approximations provided they have a C^k-smooth bump. This applies especially to separable spaces, WCG spaces, or $C(K)$ spaces. We finish by showing that super-reflexive spaces admit partitions of unity consisting of functions with Hölder derivative.

Definition 45. Let $\mathcal{S}$ be a class of functions. We say that a topological space X admits $\mathcal{S}$-partitions of unity if for any open covering $\mathcal{U}$ of X there is a partition of unity on X subordinated to $\mathcal{U}$ such that each member of the partition belongs to $\mathcal{S}$.

Definition 46. A family of subsets of a topological space X is called

- *locally finite* if for each point $x \in X$ there is a neighbourhood of x that meets only finitely many members of this family;
- *discrete* if for each point $x \in X$ there is a neighbourhood of x that meets at most one member of this family;
- *σ-locally finite* if it can be decomposed into countably many locally finite families;
- *σ-discrete* if it can be decomposed into countably many discrete families.

A family of subsets of a metric space P is called

- *uniformly discrete* if there is $d > 0$ such that the distance of any two members of this family is at least d;
- *σ-uniformly discrete* if it can be decomposed into countably many uniformly discrete families.

A partition of unity $\{\psi_\alpha\}_{\alpha\in\Lambda}$ is called locally finite if $\{\operatorname{supp}_o \psi_\alpha\}_{\alpha\in\Lambda}$ is locally finite, it is called σ-discrete if $\{\operatorname{supp}_o \psi_\alpha\}_{\alpha\in\Lambda}$ is σ-discrete, and it is called σ-uniformly discrete if $\{\operatorname{supp}_o \psi_\alpha\}_{\alpha\in\Lambda}$ is σ-uniformly discrete.

If $\mathcal{S}$ is a class of mappings, then we use the notation $\mathcal{S}(X;Y) = \mathcal{S} \cap Y^X$, i.e. $\mathcal{S}(X;Y)$ is the set of mappings from X to Y that belong to $\mathcal{S}$. A class of C^k-smooth mappings will be denoted by $\mathcal{C}^k$ and similarly for other smoothness classes from Section 1.5.

Definition 47. Let P be a metric space and $S \subset C(P)$ a ring of functions. We say that S is a *partition ring* if it satisfies the following conditions:

(i) For each $S_0 \subset S$ with $\{\operatorname{supp}_o f;\ f \in S_0\}$ uniformly discrete in P and $\operatorname{supp}_o f$ bounded for each $f \in S_0$ there is a $g \in S$ with $\operatorname{supp}_o g = \bigcup_{f\in S_0} \operatorname{supp}_o f$.

(ii) Let $f \in S$ and $\operatorname{supp}_o f = U_1 \cup U_2$, where U_1 and U_2 are open subsets of P with $\operatorname{dist}(U_1, U_2) > 0$. Then $\chi_{U_1} \cdot f \in S$.

(iii) For each $f \in S$ bounded below and $\varepsilon > 0$ there is a $g \in S$ such that $0 \le g \le 1$, $f^{-1}\big((-\infty,\varepsilon]\big) \subset g^{-1}(\{0\})$ and $f^{-1}\big([2\varepsilon,+\infty)\big) \subset g^{-1}(\{1\})$.

Examples of partition rings: C^k-smooth functions on normed linear spaces, smooth bounded Lipschitz functions, or smooth bounded functions with bounded Hölder derivatives (see the proof of Theorem 56).

Definition 48. Let $\mathcal{S}$ be a class of mappings defined on a topological space X. We say that $\mathcal{S}$ is *determined locally* if whenever f is a mapping defined on X such that for every $x \in X$ there are a neighbourhood U of x and a mapping $g \in \mathcal{S}$ such that $f = g$ on U, then $f \in \mathcal{S}$.

Examples of classes determined locally are $\mathcal{C}^k$ classes or class of continuous Gâteaux differentiable mappings. Note that if a ring of functions on a metric space is determined locally then conditions (i) and (ii) in the definition of a partition ring are automatically satisfied.

Lemma 49. *Let P be a metric space and S a partition ring of functions on P. Consider the following statements.*

(i) *For every $A \subset W \subset P$, A closed and W open there is $\varphi \in S$ such that $\varphi = 1$ on A and $\operatorname{supp}_o \varphi \subset W$.*
(ii) *For every $V \subset W \subset P$ bounded open sets satisfying $\operatorname{dist}(V, P \setminus W) > 0$ there is $\varphi \in S$ such that $V \subset \operatorname{supp}_o \varphi \subset W$.*
(iii) *For every $V \subset W \subset P$ bounded open sets satisfying $\operatorname{dist}(V, P \setminus W) > 0$ there are $\varphi_n \in S$, $n \in \mathbb{N}$, such that $V \subset \bigcup_{n\in\mathbb{N}} \operatorname{supp}_o \varphi_n \subset W$.*
(iv) *The family $\{\operatorname{supp}_o f;\ f \in S\}$ contains a σ-uniformly discrete basis for the topology of P.*
(v) *The space P admits locally finite and σ-uniformly discrete S-partitions of unity.*
(vi) *The space P admits locally finite S-partitions of unity.*
(vii) *The family $\{\operatorname{supp}_o f;\ f \in S\}$ contains a σ-locally finite basis for the topology of P.*

Then (i)⇒(ii)⇒(iii)⇒(iv)⇔(v)⇒(vi)⇒(vii). If S is moreover determined locally, then all seven statements are equivalent.

We note that the σ-uniformly discrete partitions of unity will prove very useful in Sections 7 and 8, as they allow us to use certain separable techniques in a non-separable setting.

Proof. (i)⇒(ii)⇒(iii) is obvious.

(iii)⇒(iv) Let $\mathcal{U}_m = \{U_\alpha^m\}_{\alpha\in\Lambda_m}$ be a uniform covering of P by open balls with radius $\frac{1}{m}$. By Lemma 8 there are open refinements $\{V_{n\alpha}^m\}_{n\in\mathbb{N},\alpha\in\Lambda_m}$, $\{W_{n\alpha}^m\}_{n\in\mathbb{N},\alpha\in\Lambda_m}$ of $\mathcal{U}_m$ such that $V_{n\alpha}^m \subset W_{n\alpha}^m \subset U_\alpha^m$, $\operatorname{dist}(V_{n\alpha}^m, P \setminus W_{n\alpha}^m) \geq 2^{-n}$ and the family $\{W_{n\alpha}^m\}_{\alpha\in\Lambda_m}$ is uniformly discrete for all $n \in \mathbb{N}$. Therefore, by (iii), there exist functions $\varphi_{n\alpha k}^m \in S$ such that $V_{n\alpha}^m \subset \bigcup_{k\in\mathbb{N}} \operatorname{supp}_o \varphi_{n\alpha k}^m \subset W_{n\alpha}^m$. The family $\{\operatorname{supp}_o \varphi_{n\alpha k}^m;\ m, n, k \in \mathbb{N}, \alpha \in \Lambda_m\}$ is therefore a σ-uniformly discrete basis for the topology of P.

(iv)⇒(v) Let $\mathcal{U}$ be an open covering of P. We construct a locally finite and σ-uniformly discrete S-partition of unity subordinated to $\mathcal{U}$. Without loss of generality we may assume that $\mathcal{U}$ consists of bounded sets. By (iv) there are $S_j \subset S$, $j \in \mathbb{N}$, such that $\{\operatorname{supp_o} f;\ f \in S_j\}$ are uniformly discrete and $\{\operatorname{supp_o} f;\ f \in S_j, j \in \mathbb{N}\}$ is an open covering of P that refines $\mathcal{U}$. By property (i) of a partition ring there are functions $f_j \in S$ such that $\operatorname{supp_o} f_j = \bigcup_{f \in S_j} \operatorname{supp_o} f$. Replacing f_j by f_j^2 if necessary we may assume that $f_j \geq 0$. By property (iii) of a partition ring there are functions $g_{jk} \in S$ such that $0 \leq g_{jk} \leq 1$, $\operatorname{supp_o} g_{jk} \subset \operatorname{supp_o} f_j$, and $f_j^{-1}([\frac{1}{k}, +\infty)) \subset g_{jk}^{-1}(\{1\})$. Let $n \mapsto (j_n, k_n)$ be a bijection of $\mathbb{N}$ onto $\mathbb{N} \times \mathbb{N}$ and put $\varphi_n = g_{j_n k_n}$.

Now for $n \in \mathbb{N}$ let $\psi_n = \varphi_n \prod_{k=1}^{n-1}(1 - \varphi_k)$. Then $\psi_n \in S$ (as S is a ring) and $\{\psi_n\}_{n \in \mathbb{N}}$ is a locally finite partition of unity on P. Indeed, for any $x \in P$ there is $j \in \mathbb{N}$ such that $x \in \operatorname{supp_o} f_j$ and hence there are a neighbourhood U of x and $k \in \mathbb{N}$ such that $f_j(y) > \frac{1}{k}$ for $y \in U$. It follows that $g_{jk}(y) = 1$ for $y \in U$. Let $m \in \mathbb{N}$ be such that $j = j_m$ and $k = k_m$. Choose any $y \in U$. Then $\varphi_m(y) = g_{jk}(y) = 1$ and hence $\psi_n(y) = 0$ for $n > m$. Since

$$(1 - \varphi_1)(1 - \varphi_2) \cdots (1 - \varphi_m) = 1 - \psi_1 - \cdots - \psi_m,$$

it follows that $\sum_{n=1}^{\infty} \psi_n(y) = \sum_{n=1}^{m} \psi_n(y) = 1$.

Finally, let us define $\psi_{n,f} = \chi_{\operatorname{supp_o} f} \cdot \psi_n$ for $n \in \mathbb{N}$ and $f \in S_{j_n}$. Using the fact that $\operatorname{supp_o} \psi_n \subset \operatorname{supp_o} \varphi_n \subset \operatorname{supp_o} f_{j_n}$ and the uniform discreteness of the family $\{\operatorname{supp_o} f;\ f \in S_{j_n}\}$ it follows that $\sum_{f \in S_{j_n}} \psi_{n,f} = \psi_n$ and from property (ii) of a partition ring also that $\psi_{n,f} \in S$. As moreover $\operatorname{supp_o} \psi_{n,f} \subset \operatorname{supp_o} f$, we can conclude that $\{\psi_{n,f}\}_{n \in \mathbb{N}, f \in S_{j_n}}$ is a locally finite, σ-uniformly discrete S-partition of unity on P subordinated to $\mathcal{U}$.

(v)$\Rightarrow$(iv) Let $\mathcal{U}_m$ be a uniform covering of P by open balls with radius $\frac{1}{m}$. By (v) there exists an S-partition of unity $\{\psi_{n\alpha}^m\}_{n \in \mathbb{N}, \alpha \in \Lambda_m}$ subordinated to $\mathcal{U}_m$ such that $\{\operatorname{supp_o} \psi_{n\alpha}\}_{\alpha \in \Lambda_m}$ is uniformly discrete for each $n \in \mathbb{N}$. It follows that the family $\{\operatorname{supp_o} \psi_{n\alpha}^m;\ m \in \mathbb{N}, n \in \mathbb{N}, \alpha \in \Lambda_m\}$ is a σ-uniformly discrete basis for the topology of P.

(v)$\Rightarrow$(vi) is obvious.

(vi)$\Rightarrow$(vii) Let $\mathcal{U}_m$ be a uniform covering of P by open balls with radius $\frac{1}{m}$. By (vi) there is a locally finite S-partition of unity $\{\psi_\alpha^m\}_{\alpha \in \Lambda_m}$ subordinated to $\mathcal{U}_m$. It follows that the family $\{\operatorname{supp_o} \psi_\alpha^m;\ m \in \mathbb{N}, \alpha \in \Lambda_m\}$ is a σ-locally finite basis for the topology of P.

Now suppose that S is determined locally.

(vi)$\Rightarrow$(i) Let $\{\psi_\alpha\}_{\alpha \in \Lambda}$ be a locally finite S-partition of unity subordinated to the open covering $\{W, P \setminus A\}$ of P. Let $\Lambda_1 = \{\alpha \in \Lambda;\ \operatorname{supp_o} \psi_\alpha \subset W\}$ and put $\varphi = \sum_{\alpha \in \Lambda_1} \psi_\alpha$. As the sum is locally finite and S is determined locally, $\varphi \in S$. Obviously $\operatorname{supp_o} \varphi \subset W$. Further, $\operatorname{supp_o} \psi_\alpha \subset P \setminus A$ for $\alpha \in \Lambda \setminus \Lambda_1$ and hence $\varphi(x) = \sum_{\alpha \in \Lambda_1} \psi_\alpha(x) = \sum_{\alpha \in \Lambda} \psi_\alpha(x) = 1$ for $x \in A$.

(vii)$\Rightarrow$(iii) By (vii) there is a family $\{\varphi_{n\alpha};\ n \in \mathbb{N}, \alpha \in \Lambda_n\} \subset S$ such that $V = \bigcup_{n \in \mathbb{N}, \alpha \in \Lambda_n} \operatorname{supp_o} \varphi_{n\alpha}$ and $\{\operatorname{supp_o} \varphi_{n\alpha}\}_{\alpha \in \Lambda_n}$ is locally finite for each $n \in \mathbb{N}$. Put $\varphi_n = \sum_{\alpha \in \Lambda_n} \varphi_{n\alpha}^2$. As the sum is locally finite and S is a ring determined locally, $\varphi_n \in S$. Further, $\operatorname{supp_o} \varphi_n = \bigcup_{\alpha \in \Lambda_n} \operatorname{supp_o} \varphi_{n\alpha}$, hence (iii) follows. □

Corollary 50 ([BF]). *Let X be a separable normed linear space and S a partition ring on X such that for every $f \in S$, $a \in \mathbb{R}$, and $b \in X$ the function $g(x) = f(ax+b)$ belongs to S. Then X admits locally finite and σ-uniformly discrete S-partitions of unity if and only there is a bump function in S.*

Proof. Suppose $\varphi \in S$ is a bump function. Since S is stable under shifts and scaling, we may suppose that $\varphi(0) > 0$ and $\operatorname{supp}_o \varphi \subset U_X$. By the continuity of φ there is $0 < r < 1$ such that $\varphi > 0$ on $U(0,r)$.

We show that the statement (iii) in Lemma 49 is satisfied. So let us suppose that $V \subset W \subset X$ are bounded open sets satisfying $\operatorname{dist}(V, P \setminus W) = \delta > 0$. By the Lindelöf property of the set V there is a countable subset $\{x_n\}_{n\in\mathbb{N}}$ of V such that $V \subset \bigcup_{n\in\mathbb{N}} U(x_n,\delta r)$. We put $\varphi_n(x) = \varphi\big((x-x_n)/\delta\big)$. Then $\varphi_n \in S$ and $V \subset \bigcup_{n\in\mathbb{N}} U(x_n,\delta r) \subset \bigcup_{n\in\mathbb{N}} \operatorname{supp}_o \varphi_n \subset \bigcup_{n\in\mathbb{N}} U(x_n,\delta) \subset W$.

The reverse implication is clear for example from Lemma 49(vii). □

Definition 51. Let $\mathcal{Y}$ be a class of normed linear spaces and $\mathcal{S}$ be a class of mappings from a metric space P into spaces from $\mathcal{Y}$. We say that $\mathcal{S}$ is an *approximation class* if

- $\mathcal{S}$ is determined locally,
- $\mathcal{S}(P;\mathbb{R})$ is a partition ring,
- $f+g \in \mathcal{S}$ whenever $f, g \in \mathcal{S}$ map into the same space,
- for every $Y \in \mathcal{Y}$, every $y \in Y$, and every function $\varphi \in \mathcal{S}(P;[0,1])$ the mapping $x \mapsto \varphi(x)y$ belongs to $\mathcal{S}$.

Notice that the second property implies that the class $\mathcal{Y}$ must contain at least $\mathbb{R}$. The following theorem goes back to Kazimierz Kuratowski around 1922.

Theorem 52. *Let P be a metric space and $\mathcal{S}$ an approximation class on P. Then the following statements are equivalent:*

(i) *P admits locally finite $\mathcal{S}$-partitions of unity.*

(ii) *For any convex subset C of a normed linear space of class $\mathcal{Y}$, any $f \in C(P;C)$, and any $\varepsilon \in C(P;\mathbb{R}^+)$ there is $g \in \mathcal{S}(P;C)$ such that $\|f(x)-g(x)\| < \varepsilon(x)$ for every $x \in P$.*

(iii) *For any 1-Lipschitz $f\colon P \to [0,1]$ and any $\varepsilon > 0$ there is $g \in \mathcal{S}(P;\mathbb{R})$ such that $|f-g|_P \le \varepsilon$.*

Proof. (i)⇒(ii) For each $x \in P$ find $r(x) > 0$ such that

$$\varepsilon(y) > \frac{\varepsilon(x)}{2} \quad \text{and} \quad \|f(y)-f(x)\| < \frac{\varepsilon(x)}{2} \quad \text{for each } y \in U(x,r(x)).$$

It follows that

$$\|f(y)-f(x)\| < \varepsilon(y) \quad \text{for each } y \in U(x,r(x)). \tag{14}$$

By (i) there is a locally finite $\mathcal{S}$-partition of unity $\{\psi_\alpha\}_{\alpha\in\Lambda}$ on P subordinated to $\{U(x,r(x));\ x \in P\}$. For each $\alpha \in \Lambda$ let $U_\alpha = U(x_\alpha, r(x_\alpha))$ be such that $\operatorname{supp}_o \psi_\alpha \subset U_\alpha$. Define

$$g(x) = \sum_{\alpha\in\Lambda} \psi_\alpha(x) f(x_\alpha). \tag{15}$$

By the properties of $\mathcal{S}$ each mapping $\psi_\alpha f(x_\alpha)$ belongs to $\mathcal{S}$ as well as finite sums of these mappings. Since the sum in the definition of g is locally finite and $\mathcal{S}$ is determined locally, $g \in \mathcal{S}$. Moreover, as $g(x)$ is a convex combination of points from C for every $x \in P$, $g \in S(P;C)$. Finally, choose any $x \in P$. Then

$$\begin{aligned}\|f(x)-g(x)\| &= \Big\|\sum_{\alpha\in\Lambda}\psi_\alpha(x)\big(f(x)-f(x_\alpha)\big)\Big\| \le \sum_{\alpha\in\Lambda:\, x\in U_\alpha}\psi_\alpha(x)\|f(x)-f(x_\alpha)\| \\ &< \varepsilon(x)\sum_{\alpha\in\Lambda:\, x\in U_\alpha}\psi_\alpha(x)=\varepsilon(x),\end{aligned}$$

where the last inequality follows from (14).

(ii)$\Rightarrow$(iii) is obvious.

(iii)$\Rightarrow$(i) We show that the condition (ii) in Lemma 49 is satisfied. Let $V \subset W \subset P$ be bounded open sets satisfying $\operatorname{dist}(V, P\setminus W) > 3\delta$ for some $0 < \delta < \frac{1}{3}$. Put $f(x) = \min\{\operatorname{dist}(x, P\setminus W), 1\}$. By (iii) there is $g \in S(P;\mathbb{R})$ such that $|f-g|_P \le \delta$. Then $g \le \delta$ on $P \setminus W$ and $g > 2\delta$ on V. By property (iii) of a partition ring there is $\varphi \in S(P;\mathbb{R})$ such that $\varphi = 0$ on $P\setminus W$ and $\varphi > 0$ on V. □

Next we show how to construct smooth partitions of unity on various classes of Banach spaces. In the following theorem the mapping Φ introduces a "coordinate system" on X, while the mappings P_F serve as the "projections" associated to this "coordinate system". The requirement is that for every $x \in X$ if we take "large coordinates of x", then the associated "projection" approximates x well.

Theorem 53 ([Hay2]). *Let X be a normed linear space that admits a C^k-smooth bump function, $k \in \mathbb{N}\cup\{\infty\}$. Let Γ be a set and $\Phi\colon X \to c_0(\Gamma)$ a continuous mapping such that for every $\gamma \in \Gamma$ the function $e_\gamma^* \circ \Phi$ is C^k-smooth on the set where it is non-zero. For each finite $F \subset \Gamma$ let $P_F \in C^k(X;X)$ be such that the space* span $P_F(X)$ *admits locally finite $\mathcal{C}^k$-partitions of unity. Assume that for each $x \in X$ and each $\varepsilon > 0$ there exists $\delta > 0$ such that $\|x - P_F(x)\| < \varepsilon$ if we set $F = \{\gamma \in \Gamma;\ |\Phi(x)(\gamma)| \ge \delta\}$. Then X admits locally finite and σ-uniformly discrete $\mathcal{C}^k$-partitions of unity.*

Proof. Denote by $\mathcal{F}$ a set of all finite subsets of Γ (including an empty set). For any $q \in \mathbb{R}^+$ let $\zeta_q \in \mathbb{C}^\infty(\mathbb{R};[0,1])$ be such that $\zeta_q(t) = 0$ for $|t| \ge q$, $0 < \zeta_q(t) < 1$ for $\frac{q}{2} < |t| < q$, and $\zeta_q(t) = 1$ for $t \in [\frac{q}{2}, \frac{q}{2}]$ (Fact 2). For each $F \in \mathcal{F}$ and $q, r \in \mathbb{R}^+$ we define a function $\varphi_{F,q,r}\colon c_0(\Gamma) \to \mathbb{R}$ by

$$\varphi_{F,q,r}(x) = \prod_{\gamma\in F}\Big(1-\zeta_{2r}\big(x(\gamma)\big)\Big)\prod_{\gamma\in\Gamma\setminus F}\zeta_q\big(x(\gamma)\big).$$

For $x \in c_0(\Gamma)$ let $H = \{\gamma \in \Gamma;\ |x(\gamma)| \ge \frac{q}{4}\}$. Then $H \in \mathcal{F}$ and $|y(\gamma)| < \frac{q}{2}$ for $y \in U(x, \frac{q}{4})$, $\gamma \in \Gamma \setminus H$. Thus

$$\varphi_{F,q,r}(y) = \prod_{\gamma\in F}\Big(1-\zeta_{2r}\big(y(\gamma)\big)\Big)\prod_{\gamma\in H\setminus F}\zeta_q\big(y(\gamma)\big) \tag{16}$$

for $y \in U(x, \frac{q}{4})$, which implies that the function $\varphi_{F,q,r}$ is LFC-$\{e_\gamma^*\}_{\gamma\in\Gamma}$ and also $\varphi_{F,q,r} \in \mathbb{C}^\infty(c_0(\Gamma); [0,1])$. It is easy to check that $\operatorname{supp}_o \varphi_{F,q,r} = W_{F,q,r}$, where

$$W_{F,q,r} = \Big\{x \in c_0(\Gamma);\ \min_{\gamma\in F}|x(\gamma)| > r,\ \sup_{\gamma\in\Gamma\setminus F}|x(\gamma)| < q\Big\}.$$

Notice that $\operatorname{dist}(W_{F,q,r}, W_{H,q,r}) \geq r - q$ whenever $F, H \in \mathcal{F}$, $F \neq H$, and $r > q$. Therefore the family $\{W_{F,q,r};\ F \in \mathcal{F}, q, r \in \mathbb{Q}, r > q > 0\}$ is σ-uniformly discrete.

Further, $\varphi_{F,q,r} \circ \Phi \in C^k(X)$. This follows from the formula (16), the fact that $\zeta_s \circ e_\gamma^* \circ \Phi \in C^k(X)$ for each $\gamma \in \Gamma$ and $s \in \mathbb{R}^+$, and the continuity of Φ.

Note that the ring $C^k(Z)$, where Z is a normed linear space, is a partition ring determined locally. By the hypothesis and Lemma 49 for each $F \in \mathcal{F}$ there exists a σ-discrete basis $\mathcal{V}_F$ for the topology of span $P_F(X)$ formed by the sets in the family $\{\operatorname{supp}_o f;\ f \in C^k(\operatorname{span} P_F(X))\}$. Further, as X admits a C^k-smooth bump function, the family $\{\operatorname{supp}_o f;\ f \in C^k(X)\}$ contains a neighbourhood basis of 0, say $\{U_m\}_{m\in\mathbb{N}}$.

In X consider the family

$$\Big\{\Phi^{-1}(W_{F,q,r}) \cap P_F^{-1}(V) \cap (Id - P_F)^{-1}(U_m);\\ F \in \mathcal{F}, q, r \in \mathbb{Q}, 0 < q < r, V \in \mathcal{V}_F, m \in \mathbb{N}\Big\}.$$

Using the continuity of the mapping Φ it is easy to verify that this is a σ-discrete (and in particular σ-locally finite) subfamily of $\{\operatorname{supp}_o f;\ f \in C^k(X)\}$ (notice that $\Phi^{-1}(W_{F,q,r}) = \operatorname{supp}_o \varphi_{F,q,r} \circ \Phi$). To finish the proof using Lemma 49 we need to show that this family forms a basis for the topology of X.

To this end choose $x \in X$ and $\varepsilon > 0$. Let $m \in \mathbb{N}$ be such that $U_m \subset U(0, \frac{\varepsilon}{6})$ and further let $\delta > 0$ be such that

$$x - P_F(x) \in U_m$$

when we set $F = \{\gamma \in \Gamma;\ |\Phi(x)(\gamma)| \geq \delta\}$. Because $\Phi(x) \in c_0(\Gamma)$, there exist $q, r \in \mathbb{Q}$ with $0 < q < r < \delta$ satisfying $|\Phi(x)(\gamma)| < q$ whenever $\gamma \in \Gamma \setminus F$. Thus $x \in \Phi^{-1}(W_{F,q,r})$. Since $\mathcal{V}_F$ is a basis for the topology of span $P_F(X)$, there exists $V \in \mathcal{V}_F$ such that

$$P_F(x) \in V \subset U\Big(P_F(x), \frac{\varepsilon}{3}\Big).$$

It follows that $x \in \Phi^{-1}(W_{F,q,r}) \cap P_F^{-1}(V) \cap (Id - P_F)^{-1}(U_m)$. If y is any other member of this set, then we have $\|P_F(x) - P_F(y)\| < \frac{\varepsilon}{3}$ because $P_F(y) \in V$, while $\|P_F(y) - y\| < \frac{\varepsilon}{3}$ because $y - P_F(y) \in U_m$. Thus $\|x - y\| < \varepsilon$, which is what we wanted to prove. □

Corollary 54 ([Hay2]). *Let X be a normed linear space that admits a C^k-smooth bump, $k \in \mathbb{N} \cup \{\infty\}$. Let μ be a limit ordinal and let $\{P_\alpha \in C^k(X; X);\ \alpha < \mu\}$ be an equi-continuous family of mappings having the property that for every $x \in X$ the mapping $P_x\colon [0, \mu] \to X$ defined by $P_x(\alpha) = P_\alpha(x)$ for $\alpha < \mu$, $P_x(\mu) = x$, is continuous. If for each $\alpha < \mu$ the space* span $P_\alpha(X)$ *admits locally finite $\mathcal{C}^k$-partitions of unity, then so does X.*

Proof. Since X admits a C^k-smooth bump, there exist a function $h \in C^k(X;[0,1])$ and $\eta > 0$ such that $h(x) = 0$ for $\|x\| \le \eta$, while $h(x) = 1$ whenever $\|x\| \ge 1$. We set $\Gamma = [0,\mu) \times \mathbb{N}$ and define $\Phi\colon X \to \ell_\infty(\Gamma)$ by

$$\Phi(x)(\alpha, n) = 2^{-n} h\big(2^n (P_{\alpha+1}(x) - P_\alpha(x))\big).$$

We note that the enlargement of the index set by the factor $\mathbb{N}$ would not be necessary if we knew that h is zero only at the origin. Such function however may not exist, cf. Theorem 5.161.

Given $x \in X$ and $\varepsilon > 0$ we fix $m \in \mathbb{N}$ such that $2^{-m} < \varepsilon$ and note that the quantity $\|P_{\alpha+1}(x) - P_\alpha(x)\|$ can exceed $2^{-m}\frac{\eta}{2}$ only for α in some finite $H \subset [0,\mu)$. Indeed, otherwise there would be an increasing sequence $\{\alpha_n\}$ of ordinals such that $\|P_{\alpha_n+1}(x) - P_{\alpha_n}(x)\| > 2^{-m}\frac{\eta}{2}$, which contradicts the continuity of P_x at the ordinal $\sup \alpha_n = \lim \alpha_n$. Thus $0 \le \Phi(x)(\gamma) < \varepsilon$ except when $\gamma \in K = H \times \{0,1,2,\dots,m-1\}$. This shows that Φ maps into $c_0(\Gamma)$. Furthermore, by the equi-continuity of $\{P_\alpha\}_{\alpha<\mu}$ there is a neighbourhood U of x such that $\|P_{\alpha+1}(y) - P_\alpha(y)\| < 2^{-m}\eta$ for $y \in U$ and $\alpha \in [0,\mu) \setminus H$. It follows that $|\Phi(x)(\gamma) - \Phi(y)(\gamma)| < \varepsilon$ for $y \in U$, $\gamma \in \Gamma \setminus K$. Since K is finite, there is a neighbourhood V of x such that $\|\Phi(x) - \Phi(y)\|_\infty < \varepsilon$ whenever $y \in V$. This shows that Φ is continuous.

To define the "projections associated to Φ" we set $P_\emptyset = P_0$ and $P_F = P_{\alpha(F)+1}$, where $\alpha(F) = \max\{\alpha;\ (\alpha,n) \in F$ for some $n \in \mathbb{N}\}$ for a finite non-empty $F \subset \Gamma$. Notice that since μ is a limit ordinal, $\alpha(F) + 1 < \mu$ and the mapping P_F is thus well-defined. We shall show that the hypothesis of Theorem 53 is satisfied. Given $x \in X$ and $\varepsilon > 0$ it may be that $\|x - P_\alpha(x)\| < \varepsilon$ for all $\alpha < \mu$; in this case there is clearly no problem. Otherwise, there is a maximal ordinal $\beta < \mu$ that satisfies $\|x - P_\beta(x)\| \ge \varepsilon$. (Set $\beta = \min\{\gamma \in [0,\mu);\ \|x - P_\alpha(x)\| < \varepsilon$ for all $\gamma < \alpha < \mu\}$ and use the continuity of the mapping P_x.) It follows that $\|P_{\beta+1}(x) - P_\beta(x)\| \ge \|x - P_\beta(x)\| - \|x - P_{\beta+1}(x)\| \ge \varepsilon - \|x - P_{\beta+1}(x)\| > 0$. Now we fix $m \in \mathbb{N}$ such that $2^m\|P_{\beta+1}(x) - P_\beta(x)\| \ge 1$, noting that $\Phi(x)(\beta,m) = 2^{-m}$, and set $\delta = 2^{-m}$. If $F = \{(\alpha,n) \in \Gamma;\ |\Phi(x)(\alpha,n)| \ge \delta\}$, then $(\beta,m) \in F$ and so $\alpha(F) \ge \beta$, whence $\|x - P_F(x)\| < \varepsilon$, as required. □

We say that a class $\mathcal{X}$ of Banach spaces is a $\overline{\mathcal{P}}$-class if for every non-separable space $X \in \mathcal{X}$ there exists a projectional resolution of the identity $\{P_\alpha;\ \omega \le \alpha \le \mu\}$ on X such that $P_\alpha(X) \in \mathcal{X}$ for all $\alpha < \mu$, where μ is the first ordinal with cardinality $\operatorname{dens} X$. (We remark that this definition is slightly different from that of a $\mathcal{P}$-class in the book [HMVZ].) Examples of $\overline{\mathcal{P}}$-classes are reflexive spaces, WCG spaces, WCD spaces, WLD spaces, 1-Plichko spaces (see [KKL, Theorem 17.6]), or duals of Asplund spaces.

The following theorem was shown for $\overline{\mathcal{P}}$-class spaces by Gilles Godefroy, Stanimir Troyanski, John Whitfield, and Václav Zizler in [GTWZ], for preduals of WCG spaces by David McLaughlin in [McL], and for $C(K)$ spaces by Petr Hájek and Richard Haydon in [HáHa].

Theorem 55. *Let X be a Banach space such that*

- *X belongs to a $\overline{\mathcal{P}}$-class, or*
- *$X = C(K)$ for K compact, or*
- *X^* is a WCG space.*

The space X admits a C^k-smooth bump function, $k \in \mathbb{N} \cup \{\infty\}$, if and only if for any open $\Omega \subset X$, any convex subset C of a normed linear space, any $f \in C(\Omega; C)$, and any $\varepsilon \in C(\Omega; \mathbb{R}^+)$ there is $g \in C^k(\Omega; C)$ such that $\|f(x) - g(x)\| < \varepsilon(x)$ for every $x \in \Omega$.

Proof. $\Leftarrow$ is obvious – to construct a C^k-smooth bump function we just approximate the norm and then compose the approximation with a suitable function from $C^\infty(\mathbb{R})$.

$\Rightarrow$ Note that the class of C^k-smooth mappings from Ω into normed linear spaces is an approximation class. Further, by Lemma 49 the existence of locally finite $\mathcal{C}^k$-partitions of unity on Ω follows from the existence of locally finite $\mathcal{C}^k$-partitions of unity on the whole of X. Indeed, suppose that $\{\operatorname{supp}_o f;\ f \in C^k(X)\}$ contains a σ-locally finite basis for the topology of X. Considering only those sets in this basis which are subsets of Ω, we obtain a σ-locally finite basis for the topology of Ω consisting of sets in $\{\operatorname{supp}_o f;\ f \in C^k(\Omega)\}$. Thus it remains to prove that X admits locally finite $\mathcal{C}^k$-partitions of unity. The approximation then follows by Theorem 52.

First we consider the case that X belongs to a $\overline{\mathcal{P}}$-class $\mathcal{X}$. We use transfinite induction on the density of X. If X is separable, then by Corollary 50 the space X admits locally finite $\mathcal{C}^k$-partitions of unity. Now suppose that X is non-separable and each space in $\mathcal{X}$ with density smaller than $\operatorname{dens} X$ admits locally finite $\mathcal{C}^k$-partitions of unity. Let μ be the first ordinal with cardinality $\operatorname{dens} X$ and $\{P_\alpha;\ \omega \le \alpha \le \mu\}$ be a projectional resolution of the identity on X with $P_\alpha(X) \in \mathcal{X}$ for all $\omega \le \alpha < \mu$. Define $P_\alpha = P_\omega$ for $\alpha \in [0, \omega)$. By the inductive hypothesis all the spaces $P_\alpha(X)$ admit locally finite $\mathcal{C}^k$-partitions of unity, hence the hypotheses of Corollary 54 are satisfied and X admits locally finite $\mathcal{C}^k$-partitions of unity.

Next we consider the case $X = C(K)$, K compact. Since X admits a C^k-smooth bump function, X is an Asplund space (Corollary 5.3), which is equivalent to K being scattered (Theorem 5.125). We will exploit this fact heavily in our construction. In particular, the zero-dimensionality of K provides a rich supply of projections given by restrictions to clopen subsets of K.

First we shall examine the structure of closed subsets of K. For each $t \in K$ there is a unique ordinal $\alpha = \alpha(t)$ such that $t \in K^{(\alpha)} \setminus K^{(\alpha+1)}$. Since t is an isolated point of $K^{(\alpha)}$ and K is zero-dimensional, there is a clopen neighbourhood V of t such that $V \cap K^{(\alpha)} = \{t\}$; we choose such a V and call it V_t. Note that $V_t \cap K^{(\beta)} = \emptyset$ for $\beta > \alpha$. For finite subsets B of K we set $V_B = \bigcup_{t \in B} V_t$. We shall say that a finite subset A of K is admissible if $s \notin V_t$ whenever $s, t \in A$, $s \ne t$.

Suppose that H is a closed subset of K. We claim that there is a unique admissible set A with the property that $A \subset H \subset V_A$. If $H = \emptyset$, then this is obviously satisfied with $A = \emptyset$ and no other. For a non-empty H we construct an admissible A with the required property. Let $\alpha_0 = \max\{\alpha;\ H \cap K^{(\alpha)} \ne \emptyset\}$; thus $H \cap K^{(\alpha_0)}$ is a non-empty finite set, which we shall call A_0. If $H \subset V_{A_0}$, we set $A = A_0$ and stop. Otherwise, we set $H_1 = H \setminus V_{A_0}$, which is a closed set, $\alpha_1 = \max\{\alpha;\ H_1 \cap K^{(\alpha)} \neq \emptyset\}$, and $A_1 = H_1 \cap K^{(\alpha_1)}$, and repeat the procedure. In this way we construct a decreasing (and so necessarily finite) sequence $\alpha_0 > \alpha_1 > \cdots > \alpha_l$ of ordinals, and finite sets $A_j = \big(H \setminus (V_{A_0} \cup \cdots \cup V_{A_{j-1}})\big) \cap K^{(\alpha_j)}$, $j = 1, \ldots, l$, such that $H \subset V_{A_0} \cup \cdots \cup V_{A_l}$. By construction, the set $A = A_0 \cup \cdots \cup A_l$ is admissible and $A \subset H \subset V_A$.

Now suppose that there are two different admissible sets B and D satisfying $B \subset H \subset V_B$ and $D \subset H \subset V_D$. Let $\beta = \max\{\alpha;\ B \cap K^{(\alpha)} \neq D \cap K^{(\alpha)}\}$. Without loss of generality we may assume that there is $u \in (B \setminus D) \cap K^{(\beta)}$. Since $u \in B \subset H \subset V_D$, there is $s \in D$ such that $u \in V_s$. Because $u \in V_s \cap K^{(\beta)} \setminus D$ and $K^{(\alpha(s))} \cap V_s = \{s\}$, it must be that $\alpha(s) > \beta$. By the maximality of β we have $B \cap K^{\alpha(s)} = D \cap K^{\alpha(s)}$, whence $s \in B$, which contradicts the admissibility of B.

We now pass to the construction of partitions of unity. We shall proceed by transfinite induction on the height of K. Let μ be an ordinal satisfying $\mathrm{ht}(K) = \mu + 1$, i.e. $K^{(\mu)}$ is finite and non-empty. If $\mu = 0$, then $C(K)$ is finite-dimensional, and so has the required partitions of unity for example by Corollary 50. For $\mu > 0$ we assume inductively that if L is a compact space with $\mathrm{ht}(L) < \mu + 1$ and such that $C(L)$ has a C^k-smooth bump function, then $C(L)$ admits locally finite $\mathcal{C}^k$-partitions of unity. To show that $C(K)$ also admits locally finite $\mathcal{C}^k$-partitions of unity it will be enough to construct the partitions of unity on the finite-codimensional subspace $Z = \{f \in C(K);\ f(t) = 0 \text{ for all } t \in K^{(\mu)}\}$. (Using Lemma 49 it is not difficult to ascertain that whenever some normed linear spaces Y and Z admit locally finite $\mathcal{C}^k$-partitions of unity, then so does the space $Y \oplus Z$.) To this end we construct the mappings required by Theorem 53.

Put $\Gamma = \mathcal{Q} \times \mathcal{A}$, where $\mathcal{Q}$ is the set of all triples $(\zeta, \eta, \xi) \in \mathbb{Q}^3$ with $0 < \zeta < \eta < \xi$, and $\mathcal{A}$ consists of the admissible subsets A of K for which $A \cap K^{(\mu)} = \emptyset$. Let $n\colon \mathcal{Q} \to \mathbb{N}$ be some one-to-one mapping, $\beta_{\zeta,\eta}$ be as in Lemma 6, and $\theta \in C^\infty(\mathbb{R}; [0, 1])$ be such that $\theta^{-1}(\{0\}) = [-1, 1]$ (Fact 2). We define $\Phi\colon Z \to \ell_\infty(\Gamma)$ by

$$\Phi(f)(\zeta, \eta, \xi, A) = \frac{1}{n(\zeta, \eta, \xi)} \beta_{\zeta,\eta}(\chi_{K \setminus V_A} \cdot f) \prod_{t \in A} \theta\left(\frac{f(t)}{\xi}\right).$$

Notice that since the set V_A is clopen, $\chi_{K \setminus V_A} \in C(K)$. We shall show that Φ is actually a continuous mapping into $c_0(\Gamma)$. To do so fix a function $f \in Z$ and $\varepsilon > 0$. The quantity $n(\zeta, \eta, \xi)^{-1}$ is greater than ε only for (ζ, η, ξ) in some finite subset R of $\mathcal{Q}$. Let us put $\Delta = \min\big\{\frac{1}{2}(\xi - \eta);\ (\zeta, \eta, \xi) \in R\big\} > 0$. For a given triple $(\zeta, \eta, \xi) \in R$ we have $\Phi(g)(\zeta, \eta, \xi, A) = 0$ for each $g \in U(f, \Delta)$ unless $A \subset \big\{t \in K;\ |f(t)| \geq \frac{1}{2}(\eta + \xi)\big\} \subset V_A$, which can happen for at most one set $A \in \mathcal{A}$, as we have shown earlier. It follows easily that Φ is a continuous mapping into $c_0(\Gamma)$. Moreover, $f \mapsto \chi_{K \setminus V_A} \cdot f$ and $f \mapsto f(t)$ are bounded linear operators, whence each $e_\gamma^* \circ \Phi$, $\gamma \in \Gamma$, is C^k-smooth.

Finally we define the associated projections $P_F : Z \to Z$ as follows: if $F \subset \Gamma$ is a finite subset with elements $(\zeta_j, \eta_j, \xi_j, A_j)$, $j = 1, \ldots, m$, we set $V(F) = \bigcup_{j=1}^m V_{A_j}$, and define $P_F(f) = \chi_{V(F)} \cdot f$. Because $V(F)$ is a clopen set with $V(F) \cap K^{(\mu)} = \emptyset$, P_F is a well-defined linear projection of norm 1, the image $P_F(Z)$ is isometric to the space $C(V(F))$, and $\mathrm{ht}(V(F)) < \mu + 1$. Hence by the inductive hypothesis $P_F(Z)$ admits locally finite $\mathcal{C}^k$-partitions of unity.

It only remains to check the required relation between Φ and the projections. So let $f \in Z$ and $\varepsilon > 0$ be given. Let $H = \{t \in K;\ |f(t)| \geq \varepsilon\}$ and let A be the admissible set such that $A \subset H \subset V_A$. Then $A \in \mathcal{A}$, since $H \cap K^{(\mu)} = \emptyset$ by the definition of Z. There is $(\zeta, \eta, \xi) \in \mathcal{Q}$ satisfying $\|\chi_{K \setminus V_A} \cdot f\|_\infty \leq \zeta < \eta < \xi < \varepsilon$. It follows that $\Phi(f)(\zeta, \eta, \xi, A) > 0$. We set $\delta = \Phi(f)(\zeta, \eta, \xi, A)$, $F = \{\gamma \in \Gamma;\ |\Phi(x)(\gamma)| \geq \delta\}$, and note that $V(F) \supset V_A$. So

$$\|f - P_F(f)\|_\infty = \|\chi_{K \setminus V(F)} \cdot f\|_\infty \leq \|\chi_{K \setminus V_A} \cdot f\|_\infty < \varepsilon.$$

For the proof of the case of X^* being WCG see [McL]. □

Theorem 56 ([JTZ]). *Let X be a super-reflexive Banach space. Then X admits locally finite and σ-uniformly discrete $\mathcal{C}^{1,\alpha}$-partitions of unity for some $\alpha \in (0, 1]$.*

For the proof we need two auxiliary statements.

Lemma 57. *Let Γ be a non-empty set, $p, q \in [1, +\infty)$, and $r \in \mathbb{N}$ odd such that $rq \geq p$. Then the Mazur mapping $\Phi_r \colon \ell_p(\Gamma) \to \ell_q(\Gamma)$ defined by $\Phi_r(x)(\gamma) = x(\gamma)^r$ is a one-to-one r-homogenous polynomial with $\|\Phi_r\| = 1$.*

Proof. Define $M \colon \ell_p(\Gamma)^r \to \ell_\infty(\Gamma)$ by $M(x_1, \dots, x_r)(\gamma) = x_1(\gamma) \cdots x_r(\gamma)$. Obviously M is an r-linear mapping and $\Phi_r(x) = M(x, \dots, x)$. Hence Φ_r is an r-homogeneous polynomial. Moreover, for any $x \in \ell_p(\Gamma)$,

$$\left(\sum_{\gamma \in \Gamma} |x(\gamma)^r|^q\right)^{\frac{1}{q}} = \|x\|_p^r \left(\sum_{\gamma \in \Gamma} \left|\frac{x(\gamma)}{\|x\|_p}\right|^{rq}\right)^{\frac{1}{q}} \leq \|x\|_p^r \left(\sum_{\gamma \in \Gamma} \left|\frac{x(\gamma)}{\|x\|_p}\right|^{p}\right)^{\frac{1}{q}} = \|x\|_p^r,$$

and hence Φ_r maps into $\ell_q(\Gamma)$ and $\|\Phi_r\| \leq 1$. Further, $\Phi_r(e_\gamma) = e_\gamma$, which implies that $\|\Phi_r\| = 1$. As r is odd, Φ_r is obviously one-to-one. □

Proposition 58. *For every super-reflexive Banach space X there is $\alpha \in (0, 1]$ and a homeomorphic embedding Ψ of the space X into $\ell_2(\Gamma)$ for some set Γ such that $\Psi \in C^1(X; \ell_2(\Gamma))$ with $D\Psi$ α-Hölder on bounded sets.*

Proof. Let $\|\cdot\|$ be an equivalent norm on X which is uniformly rotund and its derivative is α-Hölder on the unit sphere for some $\alpha \in (0, 1]$, see e.g. [DGZ, Proposition IV.5.2]. It is easy to check that then the function $\|\cdot\|^2 \in C^1(X)$ with its derivative α-Hölder on bounded sets. Further, there are $p \in (1, +\infty)$ and a one-to-one bounded linear operator $T \colon X \to \ell_p(\Lambda)$ for some set Λ. This a folklore result that follows from the results of Robert Clarke James [Ja2] and Joram Lindenstrauss [AL], see also [JTZ]. Let $r \in \mathbb{N}$ be odd and satisfy $2r \geq p$. By Lemma 57 the Mazur mapping $\Phi_r \colon \ell_p(\Lambda) \to \ell_2(\Lambda)$ is a one-to-one continuous polynomial, in particular $D\Phi_r$ is Lipschitz on bounded sets and hence also α-Hölder on bounded sets. Assume that $0 \notin \Lambda$ and put $\Gamma = \Lambda \cup \{0\}$. Define

$\Psi\colon X \to \ell_2(\Gamma)$ by $\Psi(x) = \big(\Phi_r(T(x)), \|x\|^2\big)$. Then obviously Ψ is one-to-one and $\Psi \in C^1(X;\ell_2(\Gamma))$ with $D\Psi$ α-Hölder on bounded sets.

It remains to show that Ψ^{-1} is continuous. Let $x_n, x \in X$ be such that $\Psi(x_n) \to \Psi(x)$. Then $\|x_n\| \to \|x\|$ and $T(x_n)(\gamma) \to T(x)(\gamma)$ for all $\gamma \in \Lambda$. It follows that $\{x_n\}$ is bounded and hence $T(x_n) \overset{w}{\to} T(x)$. Using the weak sequential compactness of closed balls in X and the fact that T is one-to-one, we obtain that $x_n \overset{w}{\to} x$. The weak lower semi-continuity of $\|\cdot\|$ implies $\|x+x\| \le \liminf\|x_n+x\| \le \limsup\|x_n+x\| \le 2\|x\|$, hence $\lim\|x_n + x\| = 2\|x\|$, and by the uniform rotundity we finally get $x_n \to x$. □

Proof of Theorem 56. First we recall the following easy fact used several times in this proof: A bounded Lipschitz mapping is α-Hölder for every $\alpha \in (0, 1]$.

Let S be the set of functions from $C^{1,\alpha}(X)$ that are bounded and have bounded derivative. Noting that functions in S are Lipschitz, Proposition 1.129 implies that S is a ring. We show that S is a partition ring.

Property (i): Let $\{f_\gamma\}_{\gamma\in\Lambda} \subset S$ be such that $\{\operatorname{supp_o} f_\gamma\}_{\gamma\in\Lambda}$ is uniformly discrete. Let $g_\gamma = c_\gamma f_\gamma$ for some suitable constants $c_\gamma \neq 0$ chosen so that $|g_\gamma|_X \le 1$, $\|Dg_\gamma\|_X \le 1$, and Dg_γ is α-Hölder with constant 1 for all $\gamma \in \Lambda$. Put $g = \sum_{\gamma\in\Lambda} g_\gamma$. It is obvious that $g \in C^1(X)$ and both g and Dg are bounded by 1. To see that Dg is α-Hölder, pick any $x, y \in X$. Suppose there are $\gamma, \beta \in \Lambda$, $\gamma \neq \beta$, such that $x \in \operatorname{supp_o} g_\gamma$ and $y \in \operatorname{supp_o} g_\beta$. Then $\|Dg(x) - Dg(y)\| = \|Dg_\gamma(x) - Dg_\beta(y)\| = \|Dg_\gamma(x) - Dg_\gamma(y)\| + \|Dg_\beta(x) - Dg_\beta(y)\| \le 2\|x - y\|^\alpha$. The other cases are obvious. So $g \in S$ and clearly $\operatorname{supp_o} g = \bigcup_{\gamma\in\Lambda} \operatorname{supp_o} f_\gamma$.

Property (ii): Let $f \in S$ and $\operatorname{supp_o} f = U_1 \cup U_2$, where $U_1, U_2 \subset X$ are open with $d = \operatorname{dist}(U_1, U_2) > 0$. Consider the function $g = \chi_{U_1} f$. Then $g = f$ on an open set $X \setminus \overline{U_2}$ and $g = 0$ on some neighbourhood of $\overline{U_2}$, hence $g \in C^1(X)$ and both g and Dg are bounded, say by M. To see that Dg is α-Hölder, pick any $x, y \in X$. Suppose that $x \in \overline{U_1}$ and $y \in \overline{U_2}$. Then $\|Dg(x) - Dg(y)\| = \|Dg(x) - 0\| \le M \le \frac{M}{d^\alpha}\|x - y\|^\alpha$. The other cases are obvious and so $g \in S$.

Property (iii): Let $f \in S$ and $\varepsilon > 0$. Let $\theta \in \mathbb{C}^\infty(\mathbb{R}; [0, 1])$ be such that $\theta(t) = 0$ for $t \le \varepsilon$ and $\theta(t) = 1$ for $t \ge 2\varepsilon$ (Fact 2). Put $g = \theta \circ f$. Since θ' is bounded and Lipschitz and f is Lipschitz, $g \in S$ by Proposition 1.128, and has the properties required in (iii).

To finish the proof we show that (ii) of Lemma 49 is satisfied. Let Ψ be the embedding of X into $\ell_2(\Gamma)$ from Proposition 58. Let $W \subset X$ be an arbitrary bounded open set. Then $\Psi(W)$ is open in $\Psi(X)$ and so there is an open $U \subset \ell_2(\Gamma)$ such that $\Psi(W) = U \cap \Psi(X)$. By Proposition 44 there is $f \in C^{1,1}(\ell_2(\Gamma))$ such that $\operatorname{supp_o} f = U$. Put $\varphi = f \circ \Psi$.

The mapping $D\Psi$ is Hölder on bounded sets, therefore bounded on bounded sets. Consequently, Ψ is Lipschitz on bounded sets, hence bounded on bounded sets. Further, Df is Lipschitz and hence bounded on bounded sets. Therefore $D\varphi$ is bounded and α-Hölder on bounded sets by Proposition 1.128. Finally, as Ψ is one-to-one, $\operatorname{supp_o} \varphi = W$, which is a bounded set and so $D\varphi$ is globally α-Hölder and bounded. So we have found a function $\varphi \in S$ for which $\operatorname{supp_o} \varphi = W$. □

We remark that the last part of the proof, namely the fact that for any open $W \subset X$ there is $\varphi \in C^{1,\alpha}(X)$ with $\operatorname{supp}_o \varphi = W$, can be shown directly without embedding X into $\ell_2(\Gamma)$. The proof is similar in spirit to that of Proposition 44 but technically much more involved, see [Cep, Corollary 2].

6. Non-linear embeddings into $c_0(\Gamma)$

We begin by giving a characterisation of the existence of $\mathcal{C}^k$-partitions of unity on a normed linear space X by means of non-linear componentwise C^k-smooth embeddings of X into $c_0(\Gamma)$. This result is not essential in our approach to smooth partitions of unity, but it nicely completes the picture in view of the main result of this section: In our aim towards the approximation of Lipschitz functions by smooth functions preserving the Lipschitz property we introduce an important technique of supremal partitions and characterise it again by means of bi-Lipschitz componentwise C^k-smooth embeddings into $c_0(\Gamma)$. We show that every separable normed linear space with a C^k-smooth Lipschitz bump admits C^k-smooth Lipschitz sup-partitions of unity (and a bi-Lipschitz componentwise C^k-smooth embedding into c_0).

It is useful to explicitly state the following fact.

Fact 59. *For any set Γ the space $c_0(\Gamma)$ admits locally finite and σ-uniformly discrete C^∞-smooth and LFC-$\{e_\gamma^*\}_{\gamma\in\Gamma}$ partitions of unity.*

Proof. The family $\{W_{F,q,r};\ F \in \mathcal{F}, q,r \in \mathbb{Q}, r > q > 0\}$ from the proof of Theorem 53 is a σ-uniformly discrete basis for the topology of $c_0(\Gamma)$ such that $W_{F,q,r} = \operatorname{supp}_o \varphi_{F,q,r}$ and each $\varphi_{F,q,r}$ is C^∞-smooth and LFC-$\{e_\gamma^*\}_{\gamma\in\Gamma}$, so we can use Lemma 49.

It is also instructive to notice that the uniform refinements from Fact 62 for $r = \frac{1}{n}$, $n \in \mathbb{N}$, form a σ-locally finite basis for the topology of $c_0(\Gamma)$. Thus combined with the following observation it gives another proof: For any $x \in c_0(\Gamma)$ and $r > 0$ there is $\varphi \in C^\infty(c_0(\Gamma))$ which is LFC-$\{e_\gamma^*\}_{\gamma\in\Gamma}$ and such that $\operatorname{supp}_o \varphi = U(x,r)$. Indeed, it suffices to take $\varphi(y) = \prod_{\gamma\in\Gamma} \theta\big(y(\gamma) - x(\gamma)\big)$, where $\theta \in C^\infty(\mathbb{R},[0,1])$ is such that $\theta(t) = 1$ whenever $|t| \le \frac{r}{2}$ and $\theta(t) = 0$ if and only if $|t| \ge r$. □

Proposition 60 ([Tor]). *Let X be a normed linear space and $k \in \mathbb{N}_0 \cup \{\infty\}$. The space X admits locally finite $\mathcal{C}^k$-partitions of unity if and only if there are a set Γ and a homeomorphism $\Phi\colon X \to c_0(\Gamma)$ such that $e_\gamma^* \circ \Phi \in C^k(X)$ for every $\gamma \in \Gamma$.*

Proof. $\Rightarrow$ By Lemma 49 there is a basis $\mathcal{V} \subset \{\operatorname{supp}_o f : f \in C^k(X)\}$ for the topology of X such that $\mathcal{V} = \bigcup_{n\in\mathbb{N}} \mathcal{V}_n$, where each $\mathcal{V}_n$ is discrete and $\mathcal{V}_n \cap \mathcal{V}_m = \emptyset$ for $m \ne n$. For every $V \in \mathcal{V}$ we choose $\varphi_V \in C^k(X;[0,1])$ such that $V = \operatorname{supp}_o \varphi_V$. We put $\Gamma = \mathcal{V}$ and define $\Phi\colon X \to \ell_\infty(\Gamma)$ by

$$\Phi(x)(V) = \frac{1}{n}\varphi_V(x),$$

where $n \in \mathbb{N}$ is the uniquely determined number for which $V \in \mathcal{V}_n$.

The mapping Φ is one-to-one. Indeed, if $x, y \in X, x \neq y$, then there is $V \in \mathcal{V}$ such that $x \in V, y \neq V$. It follows that $\varphi_V(x) > 0$, while $\varphi_V(y) = 0$, and consequently $\Phi(x)(V) \neq \Phi(y)(V)$.

Moreover, Φ is a continuous mapping into $c_0(\Gamma)$. To see this, for a given $x \in X$ and $\varepsilon > 0$ we find $n_0 \in \mathbb{N}$ such that $\frac{1}{n_0} < \varepsilon$. Then $0 \leq \Phi(x)(V) < \varepsilon$ and also $|\Phi(x)(V) - \Phi(y)(V)| < \varepsilon$ whenever $y \in X$ and $V \in \mathcal{V}_n$ for $n \geq n_0$. Further, by the discreteness of the families $\mathcal{V}_n$, there is a neighbourhood U of x such that U meets only finitely many sets in $\mathcal{W} = \mathcal{V}_1 \cup \cdots \cup \mathcal{V}_{n_0}$, say $V_1, \ldots, V_m$, $m \in \mathbb{N}_0$. Then $\Phi(y)(V) = 0$ whenever $y \in U$ and $V \in \mathcal{W} \setminus \{V_1, \ldots, V_m\}$. It follows that Φ maps into $c_0(\Gamma)$. Finally, using the continuity of $\varphi_{V_1}, \ldots, \varphi_{V_m}$ we can find a neighbourhood $W \subset U$ of x such that $|\Phi(x)(V_n) - \Phi(y)(V_n)| < \varepsilon$ whenever $y \in W, n \in \{1, \ldots, m\}$. Thus $\|\Phi(x) - \Phi(y)\|_\infty < \varepsilon$ for each $y \in W$. This shows the continuity of Φ.

To show that Φ^{-1} continuous fix $x \in X$ and $\varepsilon > 0$. Since $\mathcal{V}$ is a basis for the topology of X, there is $V \in \mathcal{V}$ such that $x \in V \subset U(x, \varepsilon)$. Let $n \in \mathbb{N}$ be such that $V \in \mathcal{V}_n$ and choose some $0 < \delta < \frac{1}{n}\varphi_V(x)$. Suppose $y \in X$ is such that $\|\Phi(x) - \Phi(y)\|_\infty < \delta$. Then $\frac{1}{n}\varphi_V(x) - \frac{1}{n}\varphi_V(y) < \delta$ and hence $\varphi_V(y) > 0$. It follows that $y \in V \subset U(x, \varepsilon)$.

$\Leftarrow$ Denote $S = \{f \in C^\infty(c_0(\Gamma));\ f \text{ is LFC-}\{e_\gamma^*\}_{\gamma \in \Gamma}\}$. By Fact 59 and Lemma 49 there is a σ-locally finite basis $\mathcal{V}$ for the topology of $c_0(\Gamma)$ consisting of the sets $\operatorname{supp}_\circ f$ with $f \in S$. Using the homeomorphism Φ we pull this basis back onto X. Moreover, if $f \in S$, then $f \circ \Phi \in C^k(X)$ (Lemma 5.81). Lemma 49 now finishes the proof. □

In particular, when $k = 0$, the previous result together with Lemma 8 implies that any normed linear space is homeomorphic to a subset of $c_0(\Gamma)$ for some set Γ. This is no longer true for uniform homeomorphisms: the space $C([0, \omega_1])$ is not uniformly homeomorphic to a subset of any $c_0(\Gamma)$. This is a result of Jan Pelant, [PHK]. However, for any separable normed linear space X there is a bi-Lipschitz homeomorphism $\Phi\colon X \to c_0$. This result of Israel Aharoni, [Ah], can be recovered from Corollary 65 when $k = 0$.

Definition 61. Let X be a set. A collection $\{\psi_\alpha\}_{\alpha \in \Lambda}$ of functions on X is called a supremal partition (sup-partition) if

- $\psi_\alpha\colon X \to [0, 1]$ for all $\alpha \in \Lambda$,
- there is a $Q > 0$ such that $\sup_{\alpha \in \Lambda} \psi_\alpha(x) \geq Q$ for each $x \in X$,
- for each $x \in X$ and for each $\varepsilon > 0$ the set $\{\alpha \in \Lambda;\ \psi_\alpha(x) > \varepsilon\}$ is finite (or in other words $(\psi_\alpha(x))_{\alpha \in \Lambda} \in c_0(\Lambda)$).

If in the second property $Q = 1$, then $\{\psi_\alpha\}_{\alpha \in \Lambda}$ is called a sup-partition of unity.

Let $\mathcal{U}$ be a covering of X. We say that the sup-partition $\{\psi_\alpha\}_{\alpha \in \Lambda}$ is subordinated to $\mathcal{U}$ if $\{\operatorname{supp}_\circ \psi_\alpha\}_{\alpha \in \Lambda}$ refines $\mathcal{U}$. We say that $\{\psi_\alpha\}_{\alpha \in \Lambda}$ is locally finite if $\{\operatorname{supp}_\circ \psi_\alpha\}_{\alpha \in \Lambda}$ is locally finite.

Notice that in fact in the above definition for each $x \in X$ there is $\alpha \in \Lambda$ such that $\psi_\alpha(x) \geq Q$.

For a metric space P we denote by $\mathcal{U}(r) = \{U(x,r);\ x \in P\}$ the full uniform covering of P.

Fact 62. *Let Γ be an infinite set, $r > 0$, and $0 < \delta < \frac{r}{2}$. There is an open locally finite uniform refinement $\mathcal{V} = \{V_\gamma\}_{\gamma\in\Gamma}$ of the uniform covering $\mathcal{U}(r)$ of $c_0(\Gamma)$ such that $\mathcal{U}(\frac{r}{2} - \delta)$ refines $\mathcal{V}$. Moreover, $\mathcal{V}$ is formed by the translates of the open ball $U(0, r-\delta)$. Further, there is a C^∞-smooth LFC-$\{e_\gamma^*\}_{\gamma\in\Gamma}$ and $(\frac{2}{r}+\delta)$-Lipschitz locally finite sup-partition of unity $\{\psi_\gamma\}_{\gamma\in\Gamma}$ on $c_0(\Gamma)$ subordinated to $\mathcal{U}(r)$.*

Proof. Notice that by the homogeneity it suffices to prove all the statements only for $r = 1$. Let $\{a_\gamma\}_{\gamma\in\Gamma}$ be the set of all vectors in $c_0(\Gamma)$ with coordinates in $\mathbb{Z}$. (Notice that the cardinality of such set is $|\Gamma|$ and so we may index its points by Γ.) We claim that $\mathcal{V} = \{U(a_\gamma, 1-\delta)\}_{\gamma\in\Gamma}$ is the desired refinement.

Clearly, $\mathcal{V}$ is an open refinement of $\mathcal{U}(1)$. To see that it is locally finite, pick any $x \in c_0(\Gamma)$ and find a finite $F \subset \Gamma$ such that $|x(\gamma)| < \frac{\delta}{2}$ whenever $\gamma \in \Gamma \setminus F$. Suppose that $\alpha \in \Gamma$ is such that $y \in U(a_\alpha, 1-\delta)$ for some $y \in U(x, \frac{\delta}{2})$. If $\gamma \in \Gamma \setminus F$, then $|a_\alpha(\gamma)| \le |a_\alpha(\gamma) - y(\gamma)| + |y(\gamma) - x(\gamma)| + |x(\gamma)| < 1 - \delta + \frac{\delta}{2} + \frac{\delta}{2} = 1$ and so $a_\alpha(\gamma) = 0$. From $|x(\gamma) - a_\alpha(\gamma)| < 1 - \frac{\delta}{2}$ and $a_\alpha(\gamma) \in \mathbb{Z}$ it follows that there are at most two possibilities for $a_\alpha(\gamma)$ for each $\gamma \in F$. From this we can conclude that $\bigl|\{\alpha;\ U(a_\alpha, 1-\delta) \cap U(x, \frac{\delta}{2}) \neq \emptyset\}\bigr| \le 2^{|F|}$.

Finally, we show that $\mathcal{U}(\frac{1}{2} - \delta)$ refines $\mathcal{V}$. Choose any $x \in c_0(\Gamma)$ and find $\beta \in \Gamma$ such that $\|x - a_\beta\| \le \frac{1}{2}$. This is always possible, since there is a finite $F \subset \Gamma$ such that $|x(\gamma)| < \frac{1}{2}$ whenever $\gamma \notin F$, and so $a_\beta(\gamma) = 0$ for such γ. Suppose $z \in U(x, \frac{1}{2} - \delta)$. Then $\|a_\beta - z\|_\infty \le \|a_\beta - x\|_\infty + \|x - z\|_\infty < \frac{1}{2} + \frac{1}{2} - \delta = 1 - \delta$, which implies that $U(x, \frac{1}{2} - \delta) \subset U(a_\beta, 1-\delta)$.

To construct the supremal partition of unity subordinated to $\mathcal{U}(1)$ find $\varepsilon > 0$ and $0 < \eta < \frac{1}{2}$ such that $0 < 1/\bigl(1 - \eta - \frac{1+\varepsilon}{2}\bigr) < 2 + \frac{\delta}{4}$ and $(1+\varepsilon)(2 + \frac{\delta}{2}) \le 2 + \delta$. Let $\mathcal{W} = \{U(a_\gamma, 1-\eta)\}_{\gamma\in\Gamma}$ be the locally finite refinement of $\mathcal{U}(1)$ from the first part of the proof such that $\mathcal{U}(\frac{1}{2} - \eta)$ refines $\mathcal{W}$. Further, let $\|\cdot\|$ be an equivalent C^∞-smooth LFC-$\{e_\gamma^*\}_{\gamma\in\Gamma}$ norm $\|\cdot\|$ on $c_0(\Gamma)$ such that $\|x\|_\infty \le \|x\| \le (1+\varepsilon)\|x\|_\infty$ for all $x \in c_0(\Gamma)$. (To construct such a norm, take for example the Minkowski functional of the set $\{x \in c_0(\Gamma);\ \sum_{\gamma\in\Gamma} \varphi(x(\gamma)) \le 1\}$, where $\varphi \in C^\infty(\mathbb{R})$, φ is convex and even, $\varphi(1) = 1$, and $\varphi(t) = 0$ for $t \in [-\frac{1}{1+\varepsilon}, \frac{1}{1+\varepsilon}]$.) For each $\gamma \in \Gamma$ we put $\psi_\gamma(x) = q\bigl(\|x - a_\gamma\|\bigr)$, where $q \in C^\infty(\mathbb{R}; [0,1])$, q is $\bigl(2 + \frac{\delta}{2}\bigr)$-Lipschitz, $q(t) = 0$ for $t \ge 1 - \eta$, and $q(t) = 1$ for $t \le \frac{1+\varepsilon}{2}$. The collection $\{\psi_\gamma\}_{\gamma\in\Gamma}$ is a locally finite sup-partition of unity. Indeed, clearly $\operatorname{supp}_o \psi_\gamma \subset U(a_\gamma, 1-\eta)$ for each $\gamma \in \Gamma$. It also follows that the set $\{\gamma \in \Gamma;\ \psi_\gamma(x) > 0\}$ is finite for each $x \in X$. Further, fix any $x \in X$. There is an $\alpha \in \Gamma$ such that $U(x, \frac{1}{2} - \eta) \subset U(a_\alpha, 1-\eta)$, which gives $\|x - a_\alpha\|_\infty \le \frac{1}{2}$. Hence $\|x - a_\alpha\| \le (1+\varepsilon)\|x - a_\alpha\|_\infty \le \frac{1+\varepsilon}{2}$, which in turn implies that $\psi_\alpha(x) = 1$.

As the function q is $\bigl(2 + \frac{\delta}{2}\bigr)$-Lipschitz and the function $\|\cdot\|$ is $(1+\varepsilon)$-Lipschitz (with respect to the norm $\|\cdot\|_\infty$), the functions ψ_γ are $(2+\delta)$-Lipschitz according to the choice of ε. The rest of the properties of the functions ψ_γ is obvious. □

Theorem 63. *Let X be a normed linear space, Γ an infinite set, and $k \in \mathbb{N}_0 \cup \{\infty\}$. Then the following statements are equivalent:*

(i) *There is $M \in \mathbb{R}$ such that there is a C^k-smooth and M-Lipschitz sup-partition $\{\varphi_\gamma\}_{\gamma\in\Gamma}$ on X subordinated to $\mathcal{U}(1)$.*

(ii) *There is $M \in \mathbb{R}$ such that there is a C^k-smooth and M-Lipschitz locally finite sup-partition of unity $\{\varphi_\gamma\}_{\gamma\in\Gamma}$ on X subordinated to $\mathcal{U}(1)$.*

(iii) *X is uniformly homeomorphic to a subset of $c_0(\Gamma)$ and for each $\varepsilon > 0$ there is $K > 0$ such that for each 1-Lipschitz function $f: X \to [0,1]$ there is a K-Lipschitz function $g \in C^k(X)$ such that $|g - f|_X \le \varepsilon$.*

(iv) *There is a bi-Lipschitz homeomorphism $\Phi: X \to c_0(\Gamma)$ such that the component functions $e^*_\gamma \circ \Phi \in C^k(X)$ for every $\gamma \in \Gamma$.*

Proof. (ii)⇒(i) is obvious.

(i)⇒(iv) Let Q be the quantity from the definition of the sup-partition. Then there is $\beta \in \Gamma$ such that $\varphi_\beta(0) \ge Q$. By scaling and composing φ_β with a suitable function we construct a C-Lipschitz function $h \in C^k(X;[0,1])$ such that $h = 0$ on $B(0,r)$ and $h = 1$ outside $U(0,1)$ for some constants $C, r \in \mathbb{R}$, $r > 0$. (We may for example choose r such that $Q - 2Mr > 0$ and take $h(x) = q(\varphi_\beta(2x))$, where $q \in C^\infty(\mathbb{R})$, q is Lipschitz, $q([0,1]) = [0,1]$, $q(0) = 1$, and $q(s) = 0$ for $s \ge Q - 2Mr$.)

Choose $t > 1$ and for each $n \in \mathbb{Z}$ and $\gamma \in \Gamma$ define functions $\varphi^n_\gamma \in C^k(X)$ by

$$\varphi^n_\gamma(x) = t^n \varphi_\gamma\left(\frac{x}{t^n}\right) h\left(\frac{x}{t^n}\right).$$

The properties of the functions φ_γ and h guarantee that each φ^n_γ is $(M + C)$-Lipschitz. Let $d: \mathbb{Z} \times \Gamma \to \Gamma$ be some one-to-one mapping and define $\Phi: X \to \mathbb{R}^\Gamma$ by $\Phi(x)(\alpha) = \varphi^n_\gamma(x)$ if $\alpha = d(n,\gamma)$ for some $n \in \mathbb{Z}$, $\gamma \in \Gamma$; $\Phi(x)(\alpha) = 0$ otherwise.

We show that Φ actually maps into $c_0(\Gamma)$. Choose an arbitrary $x \in X$ and $\varepsilon > 0$. There is $n_0 \in \mathbb{Z}$ such that $t^n < \varepsilon$ for all $n < n_0$ and $n_1 \in \mathbb{Z}$ such that $\|x\| \le rt^n$ for all $n > n_1$. It follows that $|\varphi^n_\gamma(x)| < \varepsilon$ for all $n < n_0$ and $\gamma \in \Gamma$, and, by the properties of h, $\varphi^n_\gamma(x) = 0$ for all $n > n_1$ and $\gamma \in \Gamma$. As for each $n_0 \le n \le n_1$, $\varphi_\gamma(x/t^n) > \varepsilon/t^n$ only for finitely many $\gamma \in \Gamma$, we can conclude that $\Phi: X \to c_0(\Gamma)$. Since each φ^n_γ is $(M + C)$-Lipschitz, the mapping Φ is $(M + C)$-Lipschitz as well.

To prove that Φ is one-to-one and Φ^{-1} is Lipschitz too, choose any two points $x, y \in X$, $x \ne y$, and find $m \in \mathbb{Z}$ such that $2t^m \le \|x - y\| < 2t^{m+1}$. Without loss of generality we may assume that $\|x\| \ge t^m$. Then $h(x/t^m) = 1$ and so there is $\gamma \in \Gamma$ such that $\varphi^m_\gamma(x) \ge Qt^m$. Now suppose that $z \in X$ is such that $\varphi^m_\gamma(z) > 0$. As $\operatorname{supp}_o \varphi_\gamma \subset U(w,1)$ for some $w \in X$, $\|\frac{x}{t^m} - \frac{z}{t^m}\| < 2$ and consequently $\|x - z\| < 2t^m$. But this means that $\varphi^m_\gamma(y) = 0$ and therefore

$$\|\Phi(x) - \Phi(y)\|_\infty \ge |\varphi^m_\gamma(x) - \varphi^m_\gamma(y)| = \varphi^m_\gamma(x) \ge Qt^m > \frac{Q}{2t}\|x - y\|.$$

(iv)⇒(ii) Let $A, B > 0$ be such that $A\|x - y\| \le \|\Phi(x) - \Phi(y)\|_\infty \le B\|x - y\|$ for all $x, y \in X$. By Fact 62 there are $C > 0$ and a C^∞-smooth LFC-$\{e^*_\gamma\}_{\gamma\in\Gamma}$ and C-Lipschitz locally finite sup-partition of unity $\{\psi_\gamma\}_{\gamma\in\Gamma}$ on $c_0(\Gamma)$ subordinated to $\mathcal{U}(\frac{A}{2})$. Putting $\varphi_\gamma = \psi_\gamma \circ \Phi$, $\{\varphi_\gamma\}_{\gamma\in\Gamma}$ is a BC-Lipschitz locally finite sup-partition of unity subordinated to $\mathcal{U}(1)$. Moreover, each φ_γ is C^k-smooth by Lemma 5.81.

(ii)⇒(iii) We already know that (iv) holds and from this the first part of (iii) follows immediately. To prove the second part of (iii), let $\varepsilon > 0$. The basic idea of the proof is that Lipschitz functions are stable under the operation of pointwise supremum. To preserve the smoothness, we will use a "smoothened supremum", or an equivalent smooth norm on $c_0(\Gamma)$. Let $\|\cdot\|$ be an equivalent C^∞-smooth LFC-$\{e_\gamma^*\}_{\gamma\in\Gamma}$ norm on $c_0(\Gamma)$ and let $C > 0$ be such that $\|x\|_\infty \leq \|x\| \leq C\|x\|_\infty$ for all $x \in c_0(\Gamma)$. We will show that $K = 4C^3M/\varepsilon$ satisfies our claim.

By adding the constant 1 we may and do assume that f maps into $[1, 2]$. Put $\delta = \frac{\varepsilon}{C}$ and $\psi_\gamma(x) = \varphi_\gamma(\frac{x}{\delta})$ for all $x \in X$, $\gamma \in \Gamma$. It follows that $\{\psi_\gamma\}_{\gamma\in\Gamma}$ is a C^k-smooth and M/δ-Lipschitz sup-partition of unity subordinated to $\mathcal{U}(\delta)$. Recall that $(\psi_\gamma(x))_{\gamma\in\Gamma} \in c_0(\Gamma)$ for each $x \in X$. For each $\gamma \in \Gamma$ there is a point $x_\gamma \in X$ such that $\operatorname{supp}_o \psi_\gamma \subset U(x_\gamma, \delta)$. The boundedness of the function f guarantees that also $\bigl(f(x_\gamma)\psi_\gamma(x)\bigr)_{\gamma\in\Gamma} \in c_0(\Gamma)$ for each $x \in X$. Hence we can define $g\colon X \to \mathbb{R}$ by

$$g(x) = \frac{\bigl\|\bigl(f(x_\gamma)\psi_\gamma(x)\bigr)_{\gamma\in\Gamma}\bigr\|}{\bigl\|\bigl(\psi_\gamma(x)\bigr)_{\gamma\in\Gamma}\bigr\|}.$$

As

$$\|(\psi_\gamma(x))\| \geq \|(\psi_\gamma(x))\|_\infty = \sup_{\gamma\in\Gamma} \psi_\gamma(x) = 1 \quad \text{for each } x \in X, \tag{17}$$

the function g is well-defined on all of X.

The mapping $x \mapsto \bigl(\psi_\gamma(x)\bigr)$ and, by the boundedness of f, also the mapping $x \mapsto \bigl(f(x_\gamma)\psi_\gamma(x)\bigr)$ are Lipschitz mappings from X into $c_0(\Gamma) \setminus U(0, 1)$. (Notice that for each $x \in X$ there is $\gamma \in \Gamma$ such that $\psi_\gamma(x) = 1$ and $f(x_\gamma)\psi_\gamma(x) \geq 1$.) Since $\|\cdot\|$ is C^∞-smooth and depends locally on finitely many coordinates away from the origin, and since $\psi_\gamma \in C^k(X)$ and $f(x_\gamma)\psi_\gamma \in C^k(X)$ for each $\gamma \in \Gamma$, using Lemma 5.81 we infer that $g \in C^k(X)$.

Using the facts that f maps into $[1, 2]$, the functions ψ_γ are M/δ-Lipschitz and map into $[0, 1]$, and $\|\cdot\|$ is C-Lipschitz as a function on $(c_0(\Gamma), \|\cdot\|_\infty)$, we obtain that the function $x \mapsto \bigl\|\bigl(f(x_\gamma)\psi_\gamma(x)\bigr)\bigr\|$ is $2CM/\delta$-Lipschitz and bounded by $2C$. Similarly, the function $x \mapsto \bigl\|\bigl(\psi_\gamma(x)\bigr)\bigr\|$ is CM/δ-Lipschitz and bounded below by 1. It follows that the function g is K-Lipschitz.

Finally, to show that g approximates f, pick any $x \in X$. Applying successively the inequality (17) and the facts that $\operatorname{supp}_o \psi_\gamma \subset U(x_\gamma, \delta)$ and f is 1-Lipschitz, we obtain

$$\begin{aligned}
|g(x) - f(x)| &= \left|\frac{\bigl\|\bigl(f(x_\gamma)\psi_\gamma(x)\bigr)\bigr\|}{\bigl\|\bigl(\psi_\gamma(x)\bigr)\bigr\|} - f(x)\frac{\bigl\|\bigl(\psi_\gamma(x)\bigr)\bigr\|}{\bigl\|\bigl(\psi_\gamma(x)\bigr)\bigr\|}\right| \\
&\leq \frac{\bigl\|\bigl((f(x_\gamma) - f(x))\psi_\gamma(x)\bigr)\bigr\|}{\bigl\|\bigl(\psi_\gamma(x)\bigr)\bigr\|} \leq C\,\bigl\|\bigl((f(x_\gamma) - f(x))\psi_\gamma(x)\bigr)\bigr\|_\infty \\
&= C\sup_{\gamma\in\Gamma}\bigl\{|f(x_\gamma) - f(x)|\psi_\gamma(x)\bigr\} = C\sup_{\substack{\gamma\in\Gamma\\ x\in U(x_\gamma,\delta)}}\bigl\{|f(x_\gamma) - f(x)|\psi_\gamma(x)\bigr\} \\
&\leq C\sup_{\substack{\gamma\in\Gamma\\ x\in U(x_\gamma,\delta)}}\bigl\{\|x_\gamma - x\|\bigr\} \leq C\delta = \varepsilon.
\end{aligned}$$

(iii)⇒(ii) Let Φ be the uniform homeomorphism and let $\eta > 0$ be such that $\|\Phi^{-1}(x) - \Phi^{-1}(y)\| < 1$ whenever $x, y \in \Phi(X)$ are such that $\|x - y\| < 2\eta$. Take an open locally finite uniform refinement of the uniform covering $\mathcal{U}(\eta)$ of $c_0(\Gamma)$ from Fact 62 and pull it back onto X using Φ. We obtain an open locally finite uniform refinement $\mathcal{V} = \{V_\gamma\}_{\gamma\in\Gamma}$ of the covering $\mathcal{U}(1)$ of X. Let $0 < \delta \le 1$ be such that $\mathcal{U}(\delta)$ refines $\mathcal{V}$. For each $\gamma \in \Gamma$ we define the function $f_\gamma\colon X \to [0,1]$ by $f_\gamma(x) = \min\{\operatorname{dist}(x, X \setminus V_\gamma), \delta\}$.

Choose some $0 < \theta < \frac{\delta}{2}$. For each $\gamma \in \Gamma$ the function f_γ is 1-Lipschitz and so, by (iii), there is a K-Lipschitz function $g_\gamma \in C^k(X)$ such that $|g_\gamma - f_\gamma|_X \le \theta$. Let $q \in C^k(\mathbb{R}; [0,1])$ be a C-Lipschitz function for some $C \in \mathbb{R}$, such that $q(t) = 0$ for $t \le \theta$ and $q(t) = 1$ for $t \ge \delta - \theta$. Finally, we let $\varphi_\gamma(x) = q(g_\gamma(x))$ for each $\gamma \in \Gamma$. Clearly, each function φ_γ belongs to $C^k(X;[0,1])$ and is M-Lipschitz, where $M = CK$. Further, for any $x \in X$ there is $\alpha \in \Gamma$ such that $U(x,\delta) \subset V_\alpha$, hence $f_\alpha(x) = \delta$ and consequently $\varphi_\alpha(x) = 1$. As $\operatorname{supp}_\circ \varphi_\gamma \subset V_\gamma$ for all $\gamma \in \Gamma$ and $\mathcal{V}$ is locally finite, $\{\varphi_\gamma\}_{\gamma\in\Gamma}$ is a locally finite sup-partition of unity subordinated to $\mathcal{U}(1)$. □

We note that the proof could be made considerably shorter by proving (iv)⇒(iii) directly using Theorem 79 instead of (ii)⇒(iii) and (iv)⇒(ii). However, the reasons for our strategy of the proof were two: First, we do not need the full generality (and associated machinery) of Theorem 79 (or Theorem 74) and second, the proof of (ii)⇒(iii) shows an interesting technique for constructing smooth Lipschitz approximations (due to Robb Fry, [Fry2]), and in fact shows the reason for the definition of the notion of sup-partition of unity.

Theorem 64 (Robb Fry, [Fry2]). *Let X be a separable normed linear space that admits a C^k-smooth Lipschitz bump function, $k \in \mathbb{N}_0 \cup \{\infty\}$. Then there is $M \in \mathbb{R}$ such that there is a C^k-smooth M-Lipschitz sup-partition of unity $\{\psi_j\}_{j=1}^\infty$ on X subordinated to $\mathcal{U}(1)$.*

Proof. Using the C^k-smooth Lipschitz bump on X as a start, by shifting, scaling, and composing with a suitable real function we construct two functions $f, g \in C^k(X;[0,1])$ along with real numbers $C > 0$ and $0 < \delta < r < 1$ such that $f(x) = 0$ for all $x \in X \setminus U(0,1)$, $f(x) = 1$ for all $x \in B(0,r)$, $g(x) = 1$ for all $x \in X \setminus U(0,r)$, $g(x) = 0$ for all $x \in B(0,\delta)$, and both functions are C-Lipschitz (see also the proof of Theorem 63).

Let $\{x_j\}_{j=1}^\infty \subset X$ be a sequence such that $\{U(x_j,\delta)\}_{j=1}^\infty$ is a covering of X. We put $f_j(x) = f(x - x_j)$ and $g_j(x) = g(x - x_j)$ for each $x \in X$, $j \in \mathbb{N}$. Choose $0 < \eta < 1$ and for each $j \in \mathbb{N}$ let $\varphi_j \in C^k(\mathbb{R}^j)$ be a 1-Lipschitz function (with respect to the maximum norm) such that

$$\min\{w_1,\dots,w_j\} \le \varphi_j(w) \le \min\{w_1,\dots,w_j\} + \eta \quad \text{for each } w \in [0,1]^j$$

(use Lemma 1). We note that the functions φ_j will serve as a "smoothened minimum". Finally, to confine the sup-partition into the interval $[0,1]$, let $h \in C^k(\mathbb{R};[0,1])$ be a D-Lipschitz function such that $h(t) = 0$ for $t \le \eta$ and $h(t) = 1$ for $t \ge 1$.

For each $j \in \mathbb{N}$ we define

$$\psi_j(x) = h\Big(\varphi_j\big(g_1(x), \dots, g_{j-1}(x), f_j(x)\big)\Big) \quad \text{for each } x \in X.$$

Clearly, $\psi_j \in C^k(X; [0,1])$ and ψ_j is M-Lipschitz for each $j \in \mathbb{N}$, where $M = CD$. Moreover, $\{\psi_j\}_{j=1}^{\infty}$ is a sup-partition of unity. Indeed, choose an arbitrary $x \in X$. Let $k \in \mathbb{N}$ be the smallest index for which $x \in U(x_k, \delta)$. Then $g_k(x) = 0$, which implies that for $j > k$, $\varphi_j\big(g_1(x), \dots, g_{j-1}(x), f_j(x)\big) \le \eta$ and so $\psi_j(x) = 0$. Therefore the set $\{j \in \mathbb{N};\ \psi_j(x) > 0\}$ is finite. Further, let $n \in \mathbb{N}$ be the smallest index for which $x \in U(x_n, r)$. It follows that $g_j(x) = 1$ for each $j < n$ and $f_n(x) = 1$, hence $\psi_n(x) = 1$.

Finally, if $\|x - x_j\| \ge 1$, then $f_j(x) = 0$ and hence $\psi_j(x) = 0$, which shows that $\{\psi_j\}_{j=1}^{\infty}$ is subordinated to $\mathcal{U}(1)$. □

Corollary 65. *Let X be a separable normed linear space that admits a C^k-smooth Lipschitz bump function, $k \in \mathbb{N}_0 \cup \{\infty\}$. Then there is a bi-Lipschitz homeomorphism $\Phi\colon X \to c_0$ such that the component functions $e_j^* \circ \Phi \in C^k(X)$ for every $j \in \mathbb{N}$.*

7. Approximation of Lipschitz mappings

In this section we turn our attention to the problem of approximating Lipschitz mappings by smooth Lipschitz mappings, preferably keeping the control over the Lipschitz constant. Such approximations have applications for example in the theory of Banach manifolds. The finite-dimensional case is easy – the integral convolution respects the Lipschitz property and in fact it preserves the Lipschitz constant.

In the infinite-dimensional case, the infimal convolution preserves the Lipschitz constant too, but unfortunately it gives only the first order smoothness and works only for functions (i.e. mappings into $\mathbb{R}$). Using the local techniques (partitions of unity) alone to obtain the global property (Lipschitzness) presents some insurmountable obstacles. First, it is essentially impossible to gain any global control over the Lipschitz constant of the individual functions in the partition and regardless, there is no control over the cardinality of the (locally finite) sum in (15). Therefore we have to develop several alternative approaches to this problem.

The first one is using the integral convolution even in the infinite-dimensional setting. We show two cases when this is possible. The first one is for separable spaces, where we can use "convolution in a dense set of directions" and then exploit the Lipschitz property of the approximated mapping. The gain is however not particularly strong, as we obtain merely Gâteaux (or uniformly Gâteaux) smooth approximation. More interesting is the use of the integral convolution in the space $c_0(\Gamma)$, which is possible thanks to the very strong LFC structure in this space. The latter result has interesting corollaries when either the source or the target space have certain special properties.

The above techniques are somewhere in-between local and global – they use approximation on finite-dimensional subspaces, which are then somehow "glued together".

Another example of this approach is a technique of Nicole Moulis that uses an unconditional basis for gluing together the finite-dimensional approximations. There is also a "local" procedure: the supremal partitions, developed in the previous section, which essentially replace the sum in (15) by supremum, which preserves the Lipschitz property.

All the above results give approximations in the uniform topology. Using the σ-discrete partitions of unity we show how to proceed from uniform approximations to the approximation in fine topology. Finally, we prove an analogue of Theorem 31 (the real analytic approximation) for Lipschitz functions.

We start with a notion of a uniform Gâteaux differentiability. If f is Gâteaux differentiable and for a fixed x in the domain we require the uniformity of the limit defining $\frac{\partial f}{\partial h}(x)$ in $h \in B_X$, we obtain the notion of Fréchet differentiability. If, on the other hand, for each fixed $h \in B_X$ we require the uniformity in x, then we obtain uniform Gâteaux differentiability. Uniformity of this type will prove important later, for example in the applications of Theorem 84.

Definition 66. Let X, Y be normed linear spaces, $U \subset X$ open, and $f : U \to Y$ a Gâteaux differentiable mapping. We say that f is uniformly Gâteaux differentiable (UG for short) if for each fixed $h \in S_X$ the limit defining $\frac{\partial f}{\partial h}(x)$ is uniform for $x \in U$.

Lemma 67. *Let X, Y be normed linear spaces, $U \subset X$ open, and let $f : U \to Y$ be a Gâteaux differentiable mapping. If for each $h \in S_X$ the mapping $x \mapsto \delta f(x)[h]$ is uniformly continuous on U, then f is uniformly Gâteaux differentiable on U provided that U is convex; otherwise f is uniformly Gâteaux differentiable on any open $V \subset U$ satisfying* $\operatorname{dist}(V, X \setminus U) > 0$*. Conversely, if f is uniformly Gâteaux differentiable and uniformly continuous on U, then for each $h \in X$ the mapping $x \mapsto \delta f(x)[h]$ is uniformly continuous on any $A \subset U$ satisfying* $\operatorname{dist}(A, X \setminus U) > 0$.

Proof. Choose an arbitrary direction $h \in S_X$ and $\varepsilon > 0$, and find $\theta > 0$ such that $\|\delta f(x+th)[h] - \delta f(x)[h]\| < \varepsilon$ for all $x \in U$ and $t \in (-\theta, \theta)$ satisfying $x + th \in U$. If U is convex we set $V = U$ and $\eta = \theta$, otherwise we let $\eta = \min\{\theta, \operatorname{dist}(V, X \setminus U)\}$. Fix $x \in V$ and define a mapping $g \colon I \to Y$ by $g(t) = f(x+th) - t\delta f(x)[h]$, where $I = \{t \in (-\eta, \eta);\ x + th \in U\}$. Notice that I is an open interval containing 0 and $g'(t) = \delta f(x+th)[h] - \delta f(x)[h]$ for $t \in I$. By the assumption we have $\|g'(t)\| \le \varepsilon$ for every $t \in I$ and hence g is ε-Lipschitz on the interval I. Therefore $\left\|\frac{1}{t}\big(f(x+th) - f(x)\big) - \delta f(x)[h]\right\| = \left\|\frac{1}{t}\big(g(t) - g(0)\big)\right\| \le \varepsilon$ for all $t \in I$.

To prove the converse statement, choose $h \in X$, $h \ne 0$, a subset $A \subset U$ for which $\operatorname{dist}(A, X \setminus U) > 0$, and $\varepsilon > 0$. Find $0 < \eta < \operatorname{dist}(A, X \setminus U)/\|h\|$ such that $\left\|\frac{1}{\eta}\big(f(x+\eta h) - f(x)\big) - \delta f(x)[h]\right\| < \frac{\varepsilon}{4}$ for any $x \in A$. Let $\theta > 0$ be such that $\|f(x) - f(y)\| < \eta\frac{\varepsilon}{4}$ whenever $x, y \in A$ are such that $\|x - y\| < \theta$. Then, for such x, y, we have

$$\left\|\delta f(x)[h] - \delta f(y)[h]\right\| < \frac{\varepsilon}{2} + \frac{1}{\eta}\left\|f(x+\eta h) - f(x) - f(y+\eta h) + f(y)\right\| < \varepsilon.$$

□

We remark that if $f: U \to Y$, $U \subset X$ open, is such that for each $h \in S_X$ the mapping $x \mapsto \frac{\partial f}{\partial h}(x)$ is uniformly continuous on U, then $f\restriction_{E\cap U}$ is $C^{1,+}$-smooth for each finite-dimensional affine subspace $E \subset X$ (Theorem 1.96). In particular, if X is a Banach space and f is Baire measurable, then in view of Theorem 1.101 we do not need to assume that f is Gâteaux differentiable in Lemma 67.

Lemma 68. *Let X, Y be normed linear spaces, H a dense subset of X, $U \subset X$ open, and let $f: U \to Y$ be a Gâteaux differentiable Lipschitz mapping such that for each $h \in H$ the mapping $x \mapsto \delta f(x)[h]$ is uniformly continuous on U. Then the mapping $x \mapsto \delta f(x)[h]$ is uniformly continuous on U for every $h \in X$.*

Proof. Let L be a Lipschitz constant of f. Pick an arbitrary $h \in X$ and let $\varepsilon > 0$. Find $h_0 \in H$ such that $\|h - h_0\| < \frac{\varepsilon}{4L}$. By the uniform continuity of $x \mapsto \delta f(x)[h_0]$ there is $\eta > 0$ such that $\|\delta f(x)[h_0] - \delta f(y)[h_0]\| < \frac{\varepsilon}{2}$ whenever $x, y \in U$, $\|x - y\| < \eta$. Then

$$\begin{aligned}\big\|\delta f(x)[h] - \delta f(y)[h]\big\| &\\ &\le \big\|\delta f(x)[h_0] - \delta f(y)[h_0]\big\| + \big\|\delta f(x)[h - h_0]\big\| + \big\|\delta f(y)[h - h_0]\big\|\\ &< \frac{\varepsilon}{2} + 2L\|h - h_0\| < \varepsilon\end{aligned}$$

whenever $x, y \in U$, $\|x - y\| < \eta$. □

The following approximation lemma introduces the technique of "convolution in a dense set of directions".

Lemma 69. *Let X be a separable normed linear space, Y a Banach space, $U \subset X$ open, $f: U \to Y$ an L-Lipschitz mapping, $\varepsilon > 0$, and let $V \subset U$ be open such that $\operatorname{dist}(V, X \setminus U) > \frac{\varepsilon}{2L}$. Let $\{h_j\}$ be a dense subset of S_X and let $\varphi_j \in C^\infty(\mathbb{R})$, $j \in \mathbb{N}$, be such that $\varphi_j \ge 0$, $\int_{\mathbb{R}} \varphi_j = 1$, and $\operatorname{supp}\varphi_j \subset \big[-\frac{\varepsilon}{2L2^j}, \frac{\varepsilon}{2L2^j}\big]$. Extend f to the whole of X by setting $f(x) = 0$ for $x \in X \setminus U$ and define $g_n: V \to Y$, $n \in \mathbb{N}$, by*

$$g_n(x) = \int_{\mathbb{R}^n} f\Big(x - \sum_{j=1}^{n} t_j h_j\Big) \prod_{j=1}^{n} \varphi_j(t_j)\, \mathrm{d}\lambda_n(t), \tag{18}$$

where λ_n denotes the n-dimensional Lebesgue measure. Then $g_n \to g$ uniformly on V and the mapping $g: V \to Y$ has the following properties: It is L-Lipschitz, Gâteaux differentiable, satisfies $\|f - g\|_V < \varepsilon$, and for each $h \in X$ the mapping $x \mapsto \delta g(x)[h]$ is uniformly continuous on V. Moreover, if $Y = \mathbb{R}$ and U, V, f are convex, then so is g.

Proof. Denote $K_m = \prod_{j=1}^{m}\big[-\frac{\varepsilon}{2L2^j}, \frac{\varepsilon}{2L2^j}\big] \subset \mathbb{R}^m$. Since $x - \sum_{j=1}^{m} t_j h_j \in U$ for $x \in V$ and $(t_1, \dots, t_m) \in K_m$, using the Fubini theorem and the fact that $\int_{\mathbb{R}} \varphi_j = 1$

we obtain for $m > n$ and any $x \in V$

$$\begin{aligned}\|g_m(x) - g_n(x)\| &= \left\| \int_{\mathbb{R}^m} \left(f\left(x - \sum_{j=1}^m t_j h_j\right) - f\left(x - \sum_{j=1}^n t_j h_j\right)\right) \prod_{j=1}^m \varphi_j(t_j)\, \mathrm{d}\lambda_m \right\| \\ &\le L \int_{K_m} \left\| \sum_{j=n+1}^m t_j h_j \right\| \prod_{j=1}^m \varphi_j(t_j)\, \mathrm{d}\lambda_m \le L \int_{K_m} \left(\sum_{j=n+1}^m |t_j| \right) \prod_{j=1}^m \varphi_j(t_j)\, \mathrm{d}\lambda_m \\ &\le L \left(\sum_{j=n+1}^m \frac{\varepsilon}{2L2^j} \right) \int_{K_m} \prod_{j=1}^m \varphi_j(t_j)\, \mathrm{d}\lambda_m < \frac{\varepsilon}{2 \cdot 2^n}.\end{aligned}$$

It follows that there is $g \colon V \to Y$ such that $g_n \to g$ uniformly on V.

The mappings g_n are L-Lipschitz on V. Indeed, for any $x, y \in V$ we have

$$\begin{aligned}\|g_n(x) - g_n(y)\| &\le \int_{K_n} \left\| f\left(x - \sum_{j=1}^n t_j h_j\right) - f\left(y - \sum_{j=1}^n t_j h_j\right) \right\| \prod_{j=1}^n \varphi_j(t_j)\, \mathrm{d}\lambda_n \\ &\le L\|x - y\| \int_{K_n} \prod_{j=1}^n \varphi_j(t_j)\, \mathrm{d}\lambda_n = L\|x - y\|.\end{aligned}$$

Therefore g is also L-Lipschitz. Similarly we can check that the functions g_n and g are convex under the additional convexity assumptions.

Moreover, $\|f - g\|_V < \varepsilon$. Indeed, pick $n \in \mathbb{N}$ such that $\|g_n - g\|_V < \frac{\varepsilon}{2}$. Then

$$\begin{aligned}\|f(x) - g(x)\| &\le \|f(x) - g_n(x)\| + \|g_n(x) - g(x)\| \\ &< \int_{\mathbb{R}^n} \left\| f(x) - f\left(x - \sum_{j=1}^n t_j h_j\right) \right\| \prod_{j=1}^n \varphi_j(t_j)\, \mathrm{d}\lambda_n + \frac{\varepsilon}{2} \\ &\le L \int_{K_n} \left(\sum_{j=1}^n |t_j| \right) \prod_{j=1}^n \varphi_j(t_j)\, \mathrm{d}\lambda_n + \frac{\varepsilon}{2} < \varepsilon\end{aligned}$$

for any $x \in V$.

Next we show that g is Gâteaux differentiable on V. Fix $n \in \mathbb{N}$, $x \in V$ and define $T \colon \mathbb{R}^n \to X$ by $T(t) = x + \sum_{j=1}^n t_j h_j$. Let $\Delta > 0$ be such that $T\big((-\Delta, \Delta)^n\big) \subset V$. Using substitution $t \to s - t$ we obtain

$$g_n \circ T(s) = \int_{R^n} f \circ T(s - t) \prod_{j=1}^n \varphi_j(t_j)\, \mathrm{d}\lambda_n(t) = \int_K f \circ T(t) \prod_{j=1}^n \varphi_j(s_j - t_j)\, \mathrm{d}\lambda_n(t)$$

for any $s \in (-\Delta, \Delta)^n$, where $K = \prod_{j=1}^n \big[-\frac{\varepsilon}{2L2^j} - \Delta, \frac{\varepsilon}{2L2^j} + \Delta\big]$. It follows from Corollary 1.91 that the mapping $g_n \circ T$ is C^1-smooth on $(-\Delta, \Delta)^n$. Since by the definition $\frac{\partial g_n}{\partial h}(x) = D(g_n \circ T)(0)[(s_1, \dots, s_n)]$ for all $h = s_1 h_1 + \dots + s_n h_n$, it follows that $h \mapsto \frac{\partial g_n}{\partial h}(x)$ is linear on $\operatorname{span}\{h_1, \dots, h_n\}$ and

$$\frac{\partial g_n}{\partial h_i}(x) = \int_K f \circ T(t)\varphi_i'(-t_i) \prod_{\substack{j=1\\ j\neq i}}^{n} \varphi_j(-t_j)\,\mathrm{d}\lambda_n(t)$$
$$= \int_{\mathbb{R}^n} f\Big(x - \sum_{j=1}^{n} t_j h_j\Big)\varphi_i'(t_i) \prod_{\substack{j=1\\ j\neq i}}^{n} \varphi_j(t_j)\,\mathrm{d}\lambda_n(t)$$

for each $i \in \{1, \dots, n\}$.

Further, $\{\frac{\partial g_n}{\partial h_i}\}_{n=i}^{\infty}$ converges uniformly for $x \in V$. Indeed, using the Fubini theorem and the fact that $\int_{\mathbb{R}} \varphi_j = 1$ we have for $m > n \geq i$ and any $x \in V$

$$\Big\|\frac{\partial g_m}{\partial h_i}(x) - \frac{\partial g_n}{\partial h_i}(x)\Big\|$$
$$= \Big\|\int_{\mathbb{R}^m}\Big(f\Big(x - \sum_{j=1}^{m} t_j h_j\Big) - f\Big(x - \sum_{j=1}^{n} t_j h_j\Big)\Big)\varphi_i'(t_i) \prod_{\substack{j=1\\ j\neq i}}^{m} \varphi_j(t_j)\,\mathrm{d}\lambda_m\Big\|$$
$$\leq L\int_{K_m}\Big(\sum_{j=n+1}^{m} |t_j|\Big)|\varphi_i'(t_i)| \prod_{\substack{j=1\\ j\neq i}}^{m} \varphi_j(t_j)\,\mathrm{d}\lambda_m \leq \frac{\varepsilon}{2\cdot 2^n}\int_{\mathbb{R}} |\varphi_i'(t)|\,\mathrm{d}t.$$

By Theorem 1.85 (used on the restrictions to $x+\operatorname{span}\{h_i\}$) we obtain that the directional derivative $\frac{\partial g}{\partial h_i}(x)$ exists for all $x \in V$, $i \in \mathbb{N}$. From the above it also follows that $\frac{\partial g}{\partial (h_i + h_j)}(x) = \frac{\partial g}{\partial h_i}(x) + \frac{\partial g}{\partial h_j}(x)$ for all $x \in V$, $i, j \in \mathbb{N}$.

To see that for given $x \in V$ the derivative $\frac{\partial g}{\partial h}(x)$ exists for all $h \in S_X$ choose $\eta > 0$ and let $i \in \mathbb{N}$ be such that $\|h - h_i\| < \frac{\eta}{3L}$. Then for any $\tau \in \mathbb{R} \setminus \{0\}$ small enough so that $x + \tau h \in V$, $x + \tau h_i \in V$ we have

$$\Big\|\frac{1}{\tau}\big(g(x+\tau h) - g(x)\big) - \frac{1}{\tau}\big(g(x+\tau h_i) - g(x)\big)\Big\| \leq \frac{L}{|\tau|}\|\tau(h - h_i)\| < \frac{\eta}{3}.$$

Thus there is $\theta > 0$ such that

$$\Big\|\frac{1}{\tau_1}\big(g(x+\tau_1 h) - g(x)\big) - \frac{1}{\tau_2}\big(g(x+\tau_2 h) - g(x)\big)\Big\|$$
$$< \frac{2}{3}\eta + \Big\|\frac{1}{\tau_1}\big(g(x+\tau_1 h_i) - g(x)\big) - \frac{1}{\tau_2}\big(g(x+\tau_2 h_i) - g(x)\big)\Big\| < \eta$$

for $0 < |\tau_1| < \theta$, $0 < |\tau_2| < \theta$.

Next we show that the mapping $h \mapsto \frac{\partial g}{\partial h}(x)$ is L-Lipschitz. For arbitrary $u, v \in X$ and $\eta > 0$ we have

$$\begin{aligned}\left\|\frac{\partial g}{\partial u}(x) - \frac{\partial g}{\partial v}(x)\right\| &\le \left\|\frac{\partial g}{\partial u}(x) - \frac{1}{\tau}\big(g(x+\tau u) - g(x)\big)\right\| \\ &\quad + \left\|\frac{1}{\tau}\big(g(x+\tau u) - g(x+\tau v)\big)\right\| \\ &\quad + \left\|\frac{\partial g}{\partial v}(x) - \frac{1}{\tau}\big(g(x+\tau v) - g(x)\big)\right\| \\ &\le \eta + L\|u - v\|\end{aligned}$$

for τ small enough. Thus $\left\|\frac{\partial g}{\partial u}(x) - \frac{\partial g}{\partial v}(x)\right\| \le L\|u - v\|$. It follows that the mapping $h \mapsto \frac{\partial g}{\partial h}(x)$ belongs to $\mathcal{L}(X; Y)$, since it is a Lipschitz mapping that is linear on a dense subset of X. Therefore g is Gâteaux differentiable on V.

It remains to prove that $x \mapsto \delta g(x)[h]$ is uniformly continuous on V for any $h \in X$. To this end, first note that the mapping $x \mapsto \frac{\partial g_n}{\partial h_i}(x)$ is L_i-Lipschitz for any $n \ge i$, where $L_i = L\int_{\mathbb{R}} |\varphi_i'(t)|\,\mathrm{d}t$:

$$\begin{aligned}&\left\|\frac{\partial g_n}{\partial h_i}(x) - \frac{\partial g_n}{\partial h_i}(y)\right\| \\ &\quad\le \left\|\int_{\mathbb{R}^n}\left(f\Big(x - \sum_{j=1}^n t_j h_j\Big) - f\Big(y - \sum_{j=1}^n t_j h_j\Big)\right)\varphi_i'(t_i)\prod_{\substack{j=1\\ j\ne i}}^n \varphi_j(t_j)\,\mathrm{d}\lambda_n\right\| \\ &\quad\le L\|x - y\|\int_{\mathbb{R}} |\varphi_i'(t)|\,\mathrm{d}t = L_i\|x - y\|.\end{aligned}$$

Thus the mapping $x \mapsto \delta g(x)[h_i]$ is L_i-Lipschitz for each $i \in \mathbb{N}$. It follows from Lemma 68 that $x \mapsto \delta g(x)[h]$ is uniformly continuous on V for any $h \in X$. □

Corollary 70. *Let X be a separable normed linear space, Y a Banach space, $U \subset X$ open, $k \in \mathbb{N}_0$, $f \in C^k(U; Y)$ such that $d^j f$ is L_j-Lipschitz for $j = 0, \dots, k$, $\varepsilon > 0$, and let $V \subset U$ be open such that* $\operatorname{dist}(V, X \setminus U) > 0$. *Then there is $g \in C^k(V; Y)$ such that $d^j g$ is L_j-Lipschitz for $j = 0, \dots, k$, $d^k g$ is uniformly Gâteaux differentiable (in particular, g is G^{k+1}-smooth), and $\|d^j g - d^j f\|_V < \varepsilon$ for all $j \in \{0, \dots, k\}$.*

Proof. Let $W \subset U$ be open such that $\operatorname{dist}(W, X \setminus U) > 0$ and $\operatorname{dist}(V, X \setminus W) > 0$. We define mappings $g_n : W \to Y$ by formula (18). By Corollary 1.91 we have $g_n \in C^k(W; Y)$ and

$$d^j g_n(x) = \int_{\mathbb{R}^n} d^j f\Big(x - \sum_{l=1}^n t_l h_l\Big)\prod_{l=1}^n \varphi_l(t_l)\,\mathrm{d}\lambda_n(t) \tag{19}$$

for $x \in W$, $j = 0, \dots, k$. Since each $d^j f$ is L_j-Lipschitz, by Lemma 69 used on (19) we obtain that there are L_j-Lipschitz mappings $q_j : W \to \mathcal{P}(^j X; Y)$ such that $d^j g_n \to q_j$ uniformly on W and $\|d^j f - q_j\|_W < \varepsilon$. Moreover, q_k is Gâteaux differentiable on W and $x \mapsto \delta q_k[h]$ is uniformly continuous on W for each $h \in X$. Therefore q_k is uniformly Gâteaux differentiable on V by Lemma 67. Theorem 1.85 implies that $g_n \to g \in C^k(W;Y)$ uniformly on W and $d^j g = q_j$, $j = 0, \dots, k$. □

The following version of Lemma 69 is for mappings that are only locally Lipschitz.

Lemma 71. *Let X be a separable Banach space, Y a Banach space, $U \subset X$ an open set, $f: U \to Y$ a locally Lipschitz mapping, and let $V \subset U$ be open such that $\delta = \operatorname{dist}(V, X \setminus U) > 0$. Let $\{h_j\}$ be a dense subset of S_X and let $\varphi_j \in C^\infty(\mathbb{R})$, $j \in \mathbb{N}$, be such that $\varphi_j \geq 0$, $\int_{\mathbb{R}} \varphi_j = 1$, and $\operatorname{supp} \varphi_j \subset \left[-\frac{\delta}{2^j}, \frac{\delta}{2^j}\right]$. Extend f to the whole of X by setting $f(x) = 0$ for $x \in X \setminus U$ and define $g_n: V \to Y$, $n \in \mathbb{N}$, by formula* (18)*. Then $g_n \to g$ locally uniformly on V and the mapping $g: V \to Y$ is locally Lipschitz and Gâteaux differentiable. Moreover, if $Y = \mathbb{R}$ and U, V, f are convex, then so is g.*

Proof. Let $K = \left\{\sum_{j=1}^\infty t_j h_j;\ |t_j| \leq \frac{\delta}{2^j}\right\}$. Then K is a compact subset of X and so it is easy to show that for each $x \in V$ there is a neighbourhood $V_x \subset V$ of x such that f is Lipschitz on $V_x - K$. Note that for $y \in V_x$ each $g_n(y)$ is defined using values of f on $V_x - K$ only. So we may repeat the proof of Lemma 69 with the following differences:

- $g_n \to g$ only locally uniformly on V.
- g_n are only locally Lipschitz and so is g.
- $\left\{\frac{\partial g_n}{\partial h_i}\right\}_{n=i}^\infty$ converges only locally uniformly on V.
- In the proof of the Gâteaux differentiability of g we use the fact that g is locally Lipschitz. □

Corollary 72. *Let X be a separable Banach space that admits a $C^{k,1}_{\mathrm{loc}}$-smooth bump for some $k \in \mathbb{N}_0$. Then X admits a $C^{k,1}_{\mathrm{loc}}$-smooth bump with Gâteaux differentiable kth derivative (in particular it admits G^{k+1}-smooth bump).*

Proof. Let $f \in C^{k,1}_{\mathrm{loc}}$ be a non-negative bump function with $\operatorname{supp} f \subset B(0,1)$. Let $\varphi_j \in C^\infty(\mathbb{R})$, $j \in \mathbb{N}$, be such that $\varphi_j \geq 0$, $\int_{\mathbb{R}} \varphi_j = 1$, and $\operatorname{supp} \varphi_j \subset \left[-\frac{1}{2^j}, \frac{1}{2^j}\right]$. Define mappings $g_n: X \to Y$ by formula (18). By Corollary 1.91 we have $g_n \in C^k(X)$ and (19) holds for $x \in X$, $j = 0, \dots, k$. Since each $d^j f$ is locally Lipschitz, by Lemma 71 used on (19) we obtain that there are locally Lipschitz mappings $q_j: X \to \mathcal{P}(^jX)$ such that $d^j g_n \to q_j$ locally uniformly on X. Moreover, q_k is Gâteaux differentiable on X. Theorem 1.85 implies that $g_n \to g \in C^k(X)$ locally uniformly on X and $d^j g = q_j$, $j = 0, \dots, k$. Finally, since by the definition each g_n is zero outside $B(0,2)$, g is a bump. □

To proceed to integral convolutions in $c_0(\Gamma)$ we need an auxiliary notion. Let X be a topological vector space, $\Omega \subset X$ an open subset, E an arbitrary set, $M \subset X^*$, and $g: \Omega \to E$. Let U be a neighbourhood of zero in X. We say that g depends U-uniformly locally on finitely many coordinates from M (U-ULFC-M for short) if for each $x \in \Omega$ there is a finite subset $F \subset M$ such that g depends only on F on $(x + U) \cap \Omega$ (cf. Definition 5.78).

For $F \subset \Gamma$ we denote the associated coordinate projection in $c_0(\Gamma)$ by P_F, i.e. $P_F(x) = \sum_{\gamma \in F} e^*_\gamma(x)e_\gamma$ for $x \in c_0(\Gamma)$. By $c_{00}(\Gamma)$ we denote the linear subspace of $c_0(\Gamma)$ consisting of finitely supported vectors.

Lemma 73. *Let Γ be an arbitrary set, Y a Banach space, and let $f: c_0(\Gamma) \to Y$ be a mapping that is $U(0,r)$-ULFC-$\{e^*_\gamma\}_{\gamma\in\Gamma}$ for some $r > 0$. Further, let $\Omega \subset c_0(\Gamma)$ be open, let f be uniformly continuous on Ω with modulus ω, and suppose that $f = 0$ on $c_0(\Gamma) \setminus \Omega$. Then for every $V \subset \Omega$ with $\operatorname{dist}(V, c_0(\Gamma) \setminus \Omega) > 0$ and for every $\varepsilon > 0$ there is a $U(0, \frac{r}{2})$-ULFC-$\{e^*_\gamma\}_{\gamma\in\Gamma}$ mapping $g \in C^\infty(c_0(\Gamma); Y)$ such that $\|f - g\|_V \le \varepsilon$, g is uniformly continuous on V with modulus ω, and the mapping $x \mapsto Dg(x)[h]$ is uniformly continuous on V for any $h \in c_{00}(\Gamma)$. If f is even, then so is g. If moreover $Y = \mathbb{R}$ and f is convex, then so is g.*

Proof. Let $\eta = \operatorname{dist}(V, c_0(\Gamma) \setminus \Omega)$ and find $0 < \delta < \min\{\eta, \frac{r}{2}\}$ such that $\omega(\delta) < \varepsilon$. Choose an even C^∞-smooth non-negative function φ on $\mathbb{R}$ such that $\operatorname{supp}\varphi \subset [-\delta, \delta]$ and $\int_{\mathbb{R}} \varphi = 1$. We denote $C = \int_{\mathbb{R}} |\varphi'(t)| \,\mathrm{d}\lambda$. Let $\mathcal{F} \subset 2^\Gamma$ be a partially ordered set of non-empty finite subsets of Γ ordered by inclusion. For any $F \in \mathcal{F}$ we define the mapping $g_F : c_0(\Gamma) \to Y$ by

$$g_F(x) = \int_{\mathbb{R}^{|F|}} f\Big(x - \sum_{\gamma\in F} t_\gamma e_\gamma\Big) \prod_{\gamma\in F} \varphi(t_\gamma) \,\mathrm{d}\lambda_{|F|}(t).$$

Notice that the integral is well-defined, since $f = 0$ on the closed set $c_0(\Gamma) \setminus \Omega$ and f is uniformly continuous on Ω and so it is bounded on totally bounded sets.

The net $\{g_F\}_{\mathcal{F}}$ converges on $c_0(\Gamma)$ to a mapping $g: c_0(\Gamma) \to Y$. In fact, we claim that for any $x \in c_0(\Gamma)$ there is an $F \in \mathcal{F}$ such that $g_F(y) = g_H(y)$ for any $F \subset H \in \mathcal{F}$ and any $y \in U(x, \frac{r}{2})$. Indeed, for a fixed $x \in c_0(\Gamma)$ let $F \in \mathcal{F}$ be such that the mapping f depends only on the functionals $\{e^*_\gamma\}_{\gamma\in F}$ on $U(x, r)$ and $\|x - P_F(x)\| < \frac{r}{2}$. Choose any $y \in U(x, \frac{r}{2})$ and $H \in \mathcal{F}$ satisfying $H \supset F$. Suppose that $t_\gamma \in [-\frac{r}{2}, \frac{r}{2}]$ for all $\gamma \in H$. Then $\big\|x - \big(y - \sum_{\gamma\in H} t_\gamma e_\gamma\big)\big\| < r$ and consequently $f\big(y - \sum_{\gamma\in H} t_\gamma e_\gamma\big) = f\big(y - \sum_{\gamma\in F} t_\gamma e_\gamma\big)$. Thus, by the Fubini theorem,

$$\begin{aligned} g_H(y) &= \int_{[-\delta,\delta]^{|H|}} f\Big(y - \sum_{\gamma\in H} t_\gamma e_\gamma\Big) \prod_{\gamma\in H} \varphi(t_\gamma) \,\mathrm{d}\lambda_{|H|}(t) \\ &= \int_{[-\delta,\delta]^{|F|}} f\Big(y - \sum_{\gamma\in F} t_\gamma e_\gamma\Big) \prod_{\gamma\in F} \varphi(t_\gamma) \,\mathrm{d}\lambda_{|F|}(t) \prod_{\gamma\in H\setminus F} \int_{[-\delta,\delta]} \varphi(t_\gamma) \,\mathrm{d}\lambda = g_F(y). \end{aligned}$$

Moreover, $\|x - P_F(y)\| \le \|x - P_F(x)\| + \|P_F\|\|x - y\| < r$ and so we can easily see that $g_F(y) = g_F(P_F(y))$. The mapping $g_F \restriction_{P_F(c_0(\Gamma))}$ is in fact a finite-dimensional convolution with a smooth kernel on $R^{|F|}$, and so g_F is a C^∞-smooth mapping on $U(x, \frac{r}{2})$ (Corollary 1.91; recall that a uniformly continuous mapping is bounded on totally bounded sets). The mapping g is therefore $U(0, \frac{r}{2})$-ULFC-$\{e^*_\gamma\}_{\gamma\in\Gamma}$ and $g \in C^\infty(c_0(\Gamma); Y)$, as for any $x \in c_0(\Gamma)$, $g = g_F \circ P_F$ on $U(x, \frac{r}{2})$ for some $F \in \mathcal{F}$.

To show that $\|f-g\|_V \le \varepsilon$ choose an arbitrary $x \in V$. Let $F \in \mathcal{F}$ be such that $g(x) = g_F(x)$. Notice that $\|x - (x - \sum_{\gamma\in F} t_\gamma e_\gamma)\| = \|\sum_{\gamma\in F} t_\gamma e_\gamma\| \le \delta < \eta$ whenever $t_\gamma \in [-\delta,\delta]$ for all $\gamma \in F$. Hence $x - \sum_{\gamma\in F} t_\gamma e_\gamma \in \Omega$ and

$$
\begin{aligned}
\|f(x)-g(x)\| &= \|f(x)-g_F(x)\| \\
&= \left\| \int_{\mathbb{R}^{|F|}} f(x)\prod_{\gamma\in F}\varphi(t_\gamma)\,\mathrm{d}\lambda_{|F|}(t) - \int_{\mathbb{R}^{|F|}} f\Big(x-\sum_{\gamma\in F} t_\gamma e_\gamma\Big)\prod_{\gamma\in F}\varphi(t_\gamma)\,\mathrm{d}\lambda_{|F|}(t)\right\| \\
&\le \int_{[-\delta,\delta]^{|F|}} \left\| f(x) - f\Big(x-\sum_{\gamma\in F} t_\gamma e_\gamma\Big)\right\| \prod_{\gamma\in F}\varphi(t_\gamma)\,\mathrm{d}\lambda_{|F|}(t) \\
&\le \int_{[-\delta,\delta]^{|F|}} \omega(\delta)\prod_{\gamma\in F}\varphi(t_\gamma)\,\mathrm{d}\lambda_{|F|}(t) = \omega(\delta) < \varepsilon.
\end{aligned}
$$

To see that the mapping g is uniformly continuous on V with modulus ω, choose $x, y \in V$ and find $F, H \in \mathcal{F}$ such that $g(x) = g_F(x)$ and $g(y) = g_H(y)$. Then for $K = F \cup H$ we have $g(x) = g_K(x)$ and $g(y) = g_K(y)$. As $x - \sum_{\gamma\in K} t_\gamma e_\gamma \in \Omega$ and $y - \sum_{\gamma\in K} t_\gamma e_\gamma \in \Omega$ whenever $t_\gamma \in (-\eta,\eta)$ for all $\gamma \in K$,

$$
\begin{aligned}
\|g(x)-g(y)\| &= \|g_K(x)-g_K(y)\| \\
&\le \int_{[-\delta,\delta]^{|K|}} \left\| f\Big(x-\sum_{\gamma\in K} t_\gamma e_\gamma\Big) - f\Big(y-\sum_{\gamma\in K} t_\gamma e_\gamma\Big)\right\| \prod_{\gamma\in K}\varphi(t_\gamma)\,\mathrm{d}\lambda_{|K|}(t) \\
&\le \omega\big(\|x-y\|\big).
\end{aligned}
$$

Similarly we can check that g is even if f is even and g is convex under the additional assumptions that $Y = \mathbb{R}$ and f is convex.

We finish the proof by showing that the directional derivatives of g in the directions of $c_{00}(\Gamma)$ are uniformly continuous on V. So first, choose any $\alpha \in \Gamma$. For $x, y \in V$ find $F, H \in \mathcal{F}$ such that $g(x) = g_F(x)$ on $U(x, \frac{r}{2})$ and $g(y) = g_H(y)$ on $U(y, \frac{r}{2})$. Put $K = F \cup H \cup \{\alpha\}$. Using Corollary 1.91 and substitution we obtain

$$
Dg_K(x)[e_\alpha] = \int_{\mathbb{R}^{|K|}} f\Big(x-\sum_{\gamma\in K} t_\gamma e_\gamma\Big)\varphi'(t_\alpha)\prod_{\gamma\in K\setminus\{\alpha\}}\varphi(t_\gamma)\,\mathrm{d}\lambda_{|K|}(t).
$$

Hence, similarly as above,

$$
\begin{aligned}
&\big\|Dg(x)[e_\alpha] - Dg(y)[e_\alpha]\big\| \\
&\quad\le \int_{\mathbb{R}^{|K|}} \left\| f\Big(x-\sum_{\gamma\in K} t_\gamma e_\gamma\Big) - f\Big(y-\sum_{\gamma\in K} t_\gamma e_\gamma\Big)\right\| |\varphi'(t_\alpha)| \prod_{\gamma\in K\setminus\{\alpha\}}\varphi(t_\gamma)\,\mathrm{d}\lambda_{|K|}(t) \\
&\quad\le \omega\big(\|x-y\|\big)\int_{\mathbb{R}} |\varphi'(t)|\,\mathrm{d}\lambda = C\omega\big(\|x-y\|\big).
\end{aligned}
$$

Finally, choose any $h \in c_{00}(\Gamma)$ and $x, y \in V$. It follows from the above estimate that

$$\begin{aligned}\|Dg(x)[h] - Dg(y)[h]\| &\le \sum_{\gamma \in \operatorname{supp} h} \|Dg(x)[e_\gamma^*(h)e_\gamma] - Dg(y)[e_\gamma^*(h)e_\gamma]\| \\ &\le C\omega(\|x-y\|) \sum_{\gamma \in \operatorname{supp} h} |e_\gamma^*(h)| = C\|h\|_{\ell_1}\omega(\|x-y\|).\end{aligned}$$

□

Theorem 74. *Let Γ be an arbitrary set, Y a Banach space, $U \subset c_0(\Gamma)$ an open set, and let $f : U \to Y$ be a uniformly continuous mapping with modulus ω. Then for every $V \subset U$ with* $\operatorname{dist}(V, c_0(\Gamma) \setminus U) > 0$ *and every $\varepsilon > 0$ there is a mapping $g \in C^\infty(c_0(\Gamma); Y)$ which uniformly locally depends on finitely many coordinates $\{e_\gamma^*\}_{\gamma\in\Gamma}$, such that $\|f - g\|_V \le \varepsilon$, and g is uniformly continuous on V with modulus ω. If f is moreover L-Lipschitz, then g is L-Lipschitz on V and uniformly Gâteaux differentiable on* $\operatorname{Int} V$.

Proof. Let $r = \operatorname{dist}(V, c_0(\Gamma) \setminus U)$ and find $0 < \eta \le \frac{r}{2}$ such that $\omega(\eta) < \frac{\varepsilon}{2}$. Define $\varphi\colon \mathbb{R} \to \mathbb{R}$ by $\varphi(t) = \max\{0, t-\eta\} + \min\{0, t+\eta\}$. Then φ is 1-Lipschitz and $|\varphi(t) - t| \le \eta$ for all $t \in \mathbb{R}$. Further, define a mapping $\Phi\colon c_0(\Gamma) \to c_0(\Gamma)$ by $\Phi(x) = \sum_{\gamma\in\Gamma} \varphi(e_\gamma^*(x))e_\gamma$. (Notice that in fact Φ maps into $c_{00}(\Gamma)$.) Then Φ is 1-Lipschitz and $\|\Phi(x) - x\| \le \eta$ for all $x \in c_0(\Gamma)$. Moreover, we claim that Φ is $U(0, \frac{\eta}{2})$-ULFC-$\{e_\gamma^*\}_{\gamma\in\Gamma}$.

Indeed, fix $x \in c_0(\Gamma)$ and find a finite $F \subset \Gamma$ such that $\|x - P_F(x)\| < \frac{\eta}{2}$. Then for any $y \in U(x, \frac{\eta}{2})$ we have $\|y - P_F(y)\| < \eta$. This means that if $y, z \in U(x, \frac{\eta}{2})$ are such that $e_\gamma^*(y) = e_\gamma^*(z)$ for all $\gamma \in F$, then $\varphi(e_\gamma^*(y)) = 0 = \varphi(e_\gamma^*(z))$ for all $\gamma \in \Gamma \setminus F$ and of course $\varphi(e_\gamma^*(y)) = \varphi(e_\gamma^*(z))$ for all $\gamma \in F$. Hence $\Phi(y) = \Phi(z)$, and so Φ depends only on $\{e_\gamma^*\}_{\gamma\in F}$ on $U(x, \frac{\eta}{2})$.

We extend f to the whole of $c_0(\Gamma)$ by $f(x) = 0$ for $x \in c_0(\Gamma) \setminus U$ and put $h = f \circ \Phi$. The mapping $h\colon c_0(\Gamma) \to Y$ is clearly $U(0, \frac{\eta}{2})$-ULFC-$\{e_\gamma^*\}_{\gamma\in\Gamma}$. Put $\Omega = \Phi^{-1}(U)$. Then the set Ω is open and $\operatorname{dist}(V, c_0(\Gamma) \setminus \Omega) \ge \frac{r}{2}$ (in particular $V \subset \Omega$). Indeed, choose $x \in V$ and $y \in c_0(\Gamma)$ such that $\|x - y\| < \frac{r}{2}$. Then $\|\Phi(y) - x\| \le \|\Phi(y) - y\| + \|y - x\| < \eta + \frac{r}{2} \le r$, which means that $\Phi(y) \in U$ and therefore $y \in \Omega$. Moreover, the mapping h is uniformly continuous on Ω with modulus ω. To see this, choose any $x, y \in \Omega$. Then $\Phi(x), \Phi(y) \in U$ and consequently $\|h(x) - h(y)\| \le \omega(\|\Phi(x) - \Phi(y)\|) \le \omega(\|x - y\|)$.

Finally, $\|f - h\|_V \le \sup_{x\in V} \omega(\|x - \Phi(x)\|) \le \omega(\eta) < \frac{\varepsilon}{2}$, and Lemma 73 used on h together with Lemma 68 and Lemma 67 finishes the proof. □

Let (X, ν) be a normed linear space. The norm ν is said to be UG-smooth (or just UG) if it is Gâteaux differentiable on $X \setminus \{0\}$ and for each fixed $h \in S_X$ the limit defining $\frac{\partial\nu}{\partial h}(x)$ is uniform for $x \in S_X$.

Proof of Theorem 5.126. Let us define a function $f : c_0(\Gamma) \to \mathbb{R}$ by the formula $f(x) = \max\{0, \|x\| - 1\}$. Then f is a 1-Lipschitz convex even function which is $U(0, \frac{1}{2})$-ULFC-$\{e_\gamma^*\}_{\gamma\in\Gamma}$. (Notice that $f = \|\cdot\| \circ \Phi$ as in the proof of Theorem 74 for $\eta = 1$.)

Let $g \in C^\infty(c_0(\Gamma))$ be a 1-Lipschitz convex even function with uniformly continuous directional derivatives produced by Lemma 73 combined with Lemma 68, such that $|g(x) - f(x)| \le 1$ for all $x \in c_0(\Gamma)$. Then g is separating, as $g(0) \le 1$ and $g(x) \ge 2$ on $4S_{c_0(\Gamma)}$. The function g is also UG by Lemma 67, and so we can finish by using the next lemma. □

Lemma 75. *Let X be a normed linear space, $k \in \mathbb{N} \cup \{\infty\}$, and let $g\colon X \to \mathbb{R}$ be a C^k-smooth, UG, Lipschitz, even, and convex separating function. Then X admits an equivalent C^k-smooth UG norm.*

Proof. As shown in [HJ5], the Minkowski functional of a sub-level set of a convex separating UG function need not be UG. To be able to use the Minkowski functional we need to gain more control over $Dg(x)[x]$. To this end we introduce an additional transformation. Basically, we construct a function that is "directionally primitive" to g in a sense, so that its directional derivative is g back again (more or less), hence Lipschitz. So, define $f\colon X \to \mathbb{R}$ by

$$f(x) = \int_{[0,1]} g(tx)\,\mathrm{d}\lambda(t).$$

Let L be the Lipschitz constant of g. It is easy to check that f is $\frac{L}{2}$-Lipschitz, even, and convex.

Without loss of generality we may assume that $g(0) = 0$. By the convexity of g and the fact that g is even, $g(x) \ge 0$ for $x \in X$. Since g is separating, there are $r > 0$ and $a > 0$ such that $g(x) \ge a$ for all $x \in rS_X$. Hence $g(tx) \ge a - Lr(1-t)$ whenever $t \in [0,1]$ and $\|x\| = r$. It follows that

$$f(x) \ge \int_{1-a/(Lr)}^{1} \big(a - Lr(1-t)\big)\,\mathrm{d}\lambda(t) = \frac{a^2}{2Lr} = b \quad \text{for any } x \in rS_X.$$

By Corollary 1.91 the function f is C^k-smooth and

$$Df(x)[h] = \int_{[0,1]} Dg(tx)[th]\,\mathrm{d}\lambda(t). \tag{20}$$

By the proof of Lemma 5.23 there is an equivalent C^k-smooth norm ν on X satisfying $\nu(x) = 1$ if and only if $f(x) = b$, and

$$D\nu(x) = \frac{1}{Df\big(\frac{x}{\nu(x)}\big)\big[\frac{x}{\nu(x)}\big]} Df\Big(\frac{x}{\nu(x)}\Big).$$

Using Lemma 67 and (20) we can see that the function $x \mapsto Df(x)[h]$ is uniformly continuous on X for any $h \in X$. Moreover, we claim that the function $x \mapsto Df(x)[x]$ is also Lipschitz on X. Indeed, to see this we use the substitution $t(1+\tau) = s$ to obtain $f(x + \tau x) = \int_{[0,1]} g(tx + t\tau x)\,\mathrm{d}\lambda(t) = \frac{1}{1+\tau} \int_{[0,1+\tau]} g(sx)\,\mathrm{d}\lambda(s)$. Thus, using the continuity of g along the way,

$$\begin{aligned}
Df(x)[x] &= \lim_{\tau\to 0}\frac{1}{\tau}\big(f(x+\tau x)-f(x)\big)\\
&= \lim_{\tau\to 0}\frac{1}{\tau}\left(\left(\frac{1}{1+\tau}-1\right)\int_0^{1+\tau} g(tx)\,\mathrm{d}\lambda(t)+\int_1^{1+\tau} g(tx)\,\mathrm{d}\lambda(t)\right)\\
&= \lim_{\tau\to 0}\frac{-1}{1+\tau}\int_0^{1+\tau} g(tx)\,\mathrm{d}\lambda(t)+\lim_{\tau\to 0}\frac{1}{\tau}\int_1^{1+\tau} g(tx)\,\mathrm{d}\lambda(t)\\
&= g(x)-\int_0^1 g(tx)\,\mathrm{d}\lambda(t)=g(x)-f(x).
\end{aligned}$$

Since both f and g are L-Lipschitz, the function $x\mapsto Df(x)[x]$ is $2L$-Lipschitz. Clearly, $f(0)=0$. So, the convexity of f implies that $Df(x)[x]\geq f(x)$ for any $x\in X$, and in particular $Df(x)[x]\geq b$ for $x\in X$, $\nu(x)=1$.

Finally, we claim that the function $x\mapsto D\nu(x)[h]$ is uniformly continuous on $A_R=X\setminus B(0,R)$ for any $h\in X$ and any $R>0$, which according to Lemma 67 means that the norm ν is UG. Fix any $R>0$ and $h\in X$. Denote $S=\{x\in X;\ \nu(x)=1\}$. As the mapping $\psi\colon A_R\to S$, $\psi(x)=x/\nu(x)$ is Lipschitz and $D\nu(x)=D\nu(\psi(x))$, it is enough to show that $x\mapsto D\nu(x)[h]$ is uniformly continuous on S. Let $\varepsilon>0$. Find $0<\delta<\varepsilon$ such that $|Df(x)[h]-Df(y)[h]|<\varepsilon$ whenever $\|x-y\|<\delta$. Then for any $x,y\in S$, $\|x-y\|<\delta$ we have

$$\begin{aligned}
\big|D\nu(x)[h]-D\nu(y)[h]\big| &= \left|\frac{Df(x)[h]}{Df(x)[x]}-\frac{Df(y)[h]}{Df(y)[y]}\right|\\
&\leq \frac{1}{|Df(x)[x]|}\big|Df(x)[h]-Df(y)[h]\big|+|Df(y)[h]|\left|\frac{1}{Df(x)[x]}-\frac{1}{Df(y)[y]}\right|\\
&\leq \frac{\varepsilon}{b}+\frac{L}{2}\|h\|\frac{\big|Df(x)[x]-Df(y)[y]\big|}{|Df(x)[x]|\cdot|Df(y)[y]|}\leq\frac{\varepsilon}{b}+\frac{L}{2}\|h\|\frac{2L\|x-y\|}{b^2}\\
&< \varepsilon\left(\frac{1}{b}+\frac{L^2}{b^2}\|h\|\right).
\end{aligned}$$

□

Assume that a normed linear space X can be embedded in $c_0(\Gamma)$ by a bi-Lipschitz homeomorphism Φ. Then we can use Theorem 74 to approximate mappings on X provided that we can extend the approximated mapping from $\Phi(X)$ onto some uniform neighbourhood of $\Phi(X)$. Since the extensions are intimately tied with retractions (see e.g. [BenLi]), we recall the following notions.

A retraction of a set A onto $B\subset A$ is a mapping $r\colon A\to B$ such that $r\restriction_B=Id$. A metric space P is called an absolute Lipschitz (resp. uniform) retract if for every metric space Q containing P as a subspace there is a Lipschitz (resp. uniformly continuous) retraction of Q onto P. The space P is called an absolute Lipschitz (resp. uniform) uniform neighbourhood retract if for every metric space Q containing P as a subspace there is a uniform neighbourhood U of P in Q (i.e. $\operatorname{dist}(P, Q \setminus U) > 0$) and a Lipschitz (resp. uniformly continuous) retraction of U onto P.

Fact 76. *Let Γ be any set, (P, ρ) a metric space, $Q \subset P$, and $f: Q \to \ell_\infty(\Gamma)$ a uniformly continuous mapping with modulus $\omega \in \mathcal{M}_s$. Then f can be extended to a uniformly continuous mapping $g: P \to \ell_\infty(\Gamma)$ with modulus ω. In particular, an L-Lipschitz mapping can be extended to an L-Lipschitz mapping.*

Proof. For each $\gamma \in \Gamma$ the function $f_\gamma = e^*_\gamma \circ f$ is a uniformly continuous function on Q with modulus ω. By Lemma 39 there exist extensions $g_\gamma: P \to \mathbb{R}$ of f_γ which are uniformly continuous with modulus ω. Fix an arbitrary $a \in Q$. For any $x \in P$ we have $|g_\gamma(x)| \le |g_\gamma(a)| + |g_\gamma(x) - g_\gamma(a)| \le |g_\gamma(a)| + \omega(\rho(x,a)) = |e^*_\gamma(f(a))| + \omega(\rho(x,a)) \le \|f(a)\| + \omega(\rho(x,a))$. It follows that $(g_\gamma(x))_{\gamma\in\Gamma} \in \ell_\infty(\Gamma)$ and the mapping g can be defined as $g(x) = (g_\gamma(x))_{\gamma\in\Gamma}$. Further, $\omega_g \le \omega$, as $\|g(x) - g(y)\| = \sup_{\gamma\in\Gamma}|g_\gamma(x) - g_\gamma(y)| \le \omega(\rho(x,y))$. □

Fact 77. *Every metric space X is isometric to a subset of $\ell_\infty(X)$.*

Proof. Let ρ be the metric on X. Fix $a \in X$ and define the isometric embedding $\Phi: X \to \ell_\infty(X)$ by $\Phi(x)(y) = \rho(x,y) - \rho(a,y)$ for $x, y \in X$. □

Proposition 78. *Let X be a metric space. Then X is an absolute Lipschitz uniform neighbourhood retract if and only if there are $K > 0$ and $\delta > 0$ such that for any two metric spaces $Q \subset P$ and every L-Lipschitz mapping $f: Q \to X$ there are a uniform neighbourhood $U \subset P$ of Q with* $\operatorname{dist}(Q, P \setminus U) \ge \frac{\delta}{L}$ *and a KL-Lipschitz mapping $g: U \to X$ which extends f.*

Similarly, X is an absolute uniform uniform neighbourhood retract if and only if there are $\omega_0 \in \mathcal{M}$ and $\delta > 0$ such that for any two metric spaces $Q \subset P$ and every uniformly continuous mapping $f: Q \to X$ with modulus $\omega \in \mathcal{M}_s$ there are a uniform neighbourhood $U \subset P$ of Q with $\operatorname{dist}(Q, P \setminus U) \ge \eta$*, where η is any number satisfying $\omega(\eta) < \delta$, and a uniformly continuous mapping $g: U \to X$ with modulus $\omega_0 \circ \omega$ which extends f.*

Proof. $\Rightarrow$ By Fact 77 we can embed X isometrically into $\ell_\infty(\Gamma)$. Let $V \subset \ell_\infty(\Gamma)$ be a uniform open neighbourhood of X and let $r: V \to X$ be a K-Lipschitz retraction. Let $\delta = \operatorname{dist}(X, \ell_\infty(\Gamma) \setminus V)$. By Fact 76 there is an L-Lipschitz extension $h: P \to \ell_\infty(\Gamma)$ of the mapping $f: Q \to X \subset \ell_\infty(\Gamma)$. Put $U = h^{-1}(V)$. Then U is open in P and $\operatorname{dist}(Q, P \setminus U) \ge \frac{\delta}{L}$. Indeed, if $y \in U(z, \delta/L)$ for some $z \in Q$, then we have $h(y) \in U(h(z), \delta)$, where $h(z) \in X$; hence $h(y) \in V$. Finally, put $g(x) = r(h(x))$ for any $x \in U$.

$\Leftarrow$ Let X be a subspace of a metric space P and put $Q = X$. The Lipschitz extension of the identity mapping $Id: X \to X$ to a uniform neighbourhood of X in P serves as the desired retraction.

The proof for the uniformly continuous version is analogous. □

Now we are ready to prove the approximation theorem.

Theorem 79. *Let Y be a Banach space, $k \in \mathbb{N} \cup \{\infty\}$, and let X be a normed linear space such that there are a set Γ and a bi-Lipschitz homeomorphism $\Phi\colon X \to c_0(\Gamma)$ such that the component functions $e_\gamma^* \circ \Phi \in C^k(X)$ for every $\gamma \in \Gamma$. Assume further that X or Y is an absolute Lipschitz uniform neighbourhood retract. There is a constant $C \in \mathbb{R}$ such that if $f : X \to Y$ is L-Lipschitz and $\varepsilon > 0$, then there is a CL-Lipschitz mapping $g \in C^k(X; Y)$ such that $\|f - g\|_X \leq \varepsilon$.*

Moreover, if $C_1, C_2 \in \mathbb{R}$ are such that Φ is C_1-Lipschitz and Φ^{-1} is C_2-Lipschitz, and if K is the constant from Proposition 78, then $C = C_1C_2K$.

This theorem immediately follows from the following more general version.

Theorem 80. *Let Y be a Banach space, $k \in \mathbb{N} \cup \{\infty\}$, and let X be a normed linear space such that there are a set Γ and a uniform homeomorphism $\Phi\colon X \to c_0(\Gamma)$ such that $\omega_{\Phi^{-1}} \leq \omega_1 \in \mathcal{M}_s$ and the component functions $e_\gamma^* \circ \Phi \in C^k(X)$ for every $\gamma \in \Gamma$. Assume further that X or Y is an absolute uniform uniform neighbourhood retract. If $f : X \to Y$ is uniformly continuous and $\varepsilon > 0$, then there is a uniformly continuous mapping $g \in C^k(X; Y)$ such that $\|f - g\|_X \leq \varepsilon$.*

Moreover, if $\omega_0 \in \mathcal{M}$ is the modulus from Proposition 78 for the space X, then $\omega_g \leq \omega_f \circ \omega_0 \circ \omega_1 \circ \omega_\Phi$. If $\omega_0 \in \mathcal{M}$ is the modulus from Proposition 78 for the space Y, then $\omega_g \leq \omega_0 \circ \omega_f \circ \omega_1 \circ \omega_\Phi$.

Proof. Define $\hat{f}\colon \Phi(X) \to Y$ by $\hat{f}(z) = f(\Phi^{-1}(z))$ for any $z \in \Phi(X)$. The mapping $\hat{f}$ is uniformly continuous with modulus $\omega_f \circ \omega_1 \in \mathcal{M}_s$. If Y is an absolute uniform uniform neighbourhood retract, then by Proposition 78 there are a uniform open neighbourhood U of $\Phi(X)$ in $c_0(\Gamma)$ and an extension $\bar{f}\colon U \to Y$ of $\hat{f}$ such that $\omega_{\bar{f}} \leq \omega_2 = \omega_0 \circ \omega_f \circ \omega_1$.

In case that X is an absolute uniform uniform neighbourhood retract, we use the Proposition 78 to a mapping Φ^{-1} to obtain a uniform open neighbourhood U of $\Phi(X)$ in $c_0(\Gamma)$ and an extension $q\colon U \to X$ of Φ^{-1} such that $\omega_q \leq \omega_0 \circ \omega_1$. Now put $\bar{f} = f \circ q$. Then $\omega_{\bar{f}} \leq \omega_2 = \omega_f \circ \omega_0 \circ \omega_1$ and $\bar{f}$ is an extension of $\hat{f}$.

By Theorem 74 there is a mapping $\bar{g} \in C^\infty(c_0(\Gamma); Y)$ locally dependent on finitely many coordinates and such that $\|\bar{g}(z) - \bar{f}(z)\| \leq \varepsilon$ for any $z \in \Phi(X)$ and $\bar{g}$ is uniformly continuous on $\Phi(X)$ with modulus ω_2. We define the mapping $g\colon X \to Y$ by $g = \bar{g} \circ \Phi$. By Lemma 5.81, $g \in C^k(X; Y)$. Clearly, $\omega_g \leq \omega_2 \circ \omega_\Phi$. To see that g approximates f, choose any $x \in X$. Then

$$\|g(x) - f(x)\| = \|\bar{g}(\Phi(x)) - \hat{f}(\Phi(x))\| = \|\bar{g}(\Phi(x)) - \bar{f}(\Phi(x))\| \leq \varepsilon. \qquad \square$$

Let V be a topological space and $v_0 \in V$. By $B_0(V)$ we denote the space of all bounded real-valued functions f on V for which $f(v) \to 0$ whenever $v \to v_0$, considered with the supremum norm. Given a metric space P we denote by $C_{\mathrm{ub}}(P)$ the space of all bounded, uniformly continuous real-valued functions on P with the supremum norm. By the result of Joram Lindenstrauss [Li3, Theorem 6] (see also [BenLi]), both $B_0(V)$ and $C_{\mathrm{ub}}(P)$ are absolute Lipschitz retracts. Therefore using Corollary 65 and Theorem 79 we obtain the following result:

Corollary 81. *Let X be a separable normed linear space that admits a C^k-smooth Lipschitz bump function, $k \in \mathbb{N} \cup \{\infty\}$. Let Y be a Banach space. If at least one of the spaces X or Y is equal to either $B_0(V)$ for some topological space V, or $C_{\mathrm{ub}}(P)$ for some metric space P, then there is a constant $C \in \mathbb{R}$ such that for any L-Lipschitz mapping $f: X \to Y$ and any $\varepsilon > 0$ there is a CL-Lipschitz mapping $g \in C^k(X;Y)$ for which $\|f - g\|_X \leq \varepsilon$.*

Further, by another result of J. Lindenstrauss, [Li3, Theorem 8] (see also [BenLi, Corollary 1.26]), super-reflexive Banach spaces are absolute uniform uniform neighbourhood retracts. Hence using Corollary 65 and Theorem 80 we obtain the following result:

Corollary 82. *Let X be a separable normed linear space that admits a C^k-smooth Lipschitz bump function, $k \in \mathbb{N} \cup \{\infty\}$. Let Y be a Banach space. If X or Y is a super-reflexive Banach space, then there are a constant $C \in \mathbb{R}$ and a modulus $\omega_0 \in \mathcal{M}$ such that for any uniformly continuous mapping $f: X \to Y$ and any $\varepsilon > 0$ there is a uniformly continuous mapping $g \in C^k(X;Y)$ for which $\|f - g\|_X \leq \varepsilon$ and $\omega_g(\delta) \leq \omega_f(\omega_0(C\delta))$ (if X is super-reflexive) or $\omega_g(\delta) \leq \omega_0(\omega_f(C\delta))$ (if Y is super-reflexive) for $\delta \in [0, +\infty)$.*

Next, we show another way how to glue together the finite-dimensional approximations. In what follows, the unconditional basis is instrumental so that we can arbitrarily perturb the coordinates of a given vector with a control over the norm.

Lemma 83 ([Mo2]). *Let X be a Banach space with an unconditional Schauder basis $\{e_j\}_{j=1}^{\infty}$ that admits a C^k-smooth Lipschitz bump function. Denote $X_n = \operatorname{span}\{e_j\}_{j=1}^{n}$, $X_\infty = \operatorname{span}\{e_j\}_{j=1}^{\infty} = \bigcup_{n=1}^{\infty} X_n$. Then there is a constant $K > 0$ such that for any $\varepsilon > 0$ there is a K-Lipschitz mapping $\Psi \in C^k(X; X_\infty)$ such that for each $x \in X$ there are a neighbourhood U of x and $n \in \mathbb{N}$ such that $\Psi(U) \subset X_n$ and $\|x - \Psi(x)\| < \varepsilon$.*

Proof. Let $\{f_n\}$ be the biorthogonal functionals to $\{e_n\}$. Let A be the unconditional basis constant of $\{e_n\}$ and B the basis constant of $\{e_n\}$. Let $\varphi \in C^k(X; [0,1])$ be a Lipschitz function such that $\varphi(x) = 1$ whenever $\|x\| \geq 1/A$ and $\varphi(x) = 0$ whenever $\|x\| \leq r$ for some $r > 0$. Such a function can be constructed from the C^k-smooth Lipschitz bump function, which we have at our disposal, by translating, scaling, and composing with a suitable real function. Let M be the Lipschitz constant of φ and $K = A + M(1 + B)$. Denote $R_n = Id - P_n$ for $n \in \mathbb{N}$, where P_n are the projections associated with the basis, $P_0 = 0$, and $R_0 = Id$.

Let $\varepsilon > 0$. Define $\Psi: X \to X_\infty$ by $\Psi(x) = \sum_{n=1}^{\infty} \varphi(R_{n-1}(x)/\varepsilon) f_n(x) e_n$. Suppose that $x \in X$. Then there is an $N \in \mathbb{N}$ such that $\|R_n(x)\| < \varepsilon \frac{r}{2}$ for all $n \geq N$ and thus there is a neighbourhood U of x such that $\|R_n(y)\| < \varepsilon r$ whenever $n \geq N$ and $y \in U$. It follows that $\Psi(y) = \sum_{n=1}^{N} \varphi(R_{n-1}(y)/\varepsilon) f_n(y) e_n$ and hence $\Psi(U) \subset X_N$. This fact also implies that $\Psi \in C^k(X; X_\infty)$, as it is locally a finite sum of C^k-smooth mappings.

To see that $\|x - \Psi(x)\| < \varepsilon$ for all $x \in X$, fix an arbitrary $x \in X$ and find $n_0 \in \mathbb{N}_0$ such that $\|R_{n_0}(x)\| < \varepsilon/A$ and $\|R_n(x)\| \geq \varepsilon/A$ for all $0 \leq n < n_0$. Then

$$\begin{aligned}
\|x - \Psi(x)\| &= \left\| \sum_{n=1}^{\infty} \big(1 - \varphi(R_{n-1}(x)/\varepsilon)\big) f_n(x) e_n \right\| \\
&= \left\| \sum_{n>n_0} \big(1 - \varphi(R_{n-1}(x)/\varepsilon)\big) f_n(x) e_n \right\| \\
&\leq A \left\| \sum_{n>n_0} f_n(x) e_n \right\| = A\|R_{n_0}(x)\| \\
&< A\frac{\varepsilon}{A} = \varepsilon.
\end{aligned}$$

To show that Ψ is Lipschitz we estimate the norm of the derivative of Ψ. Let $x \in X$. Find $n_0 \in \mathbb{N}_0$ such that $\|R_{n_0}(x)\| \leq \varepsilon/A$ and $\|R_n(x)\| > \varepsilon/A$ for all $0 \leq n < n_0$. Let $h \in B_X$. Notice that all the sums in the following computation are in fact finite and that $D\varphi(R_{n-1}(x)/\varepsilon) = 0$ for $1 \leq n \leq n_0$.

$$\begin{aligned}
&\|D\Psi(x)[h]\| = \left\| \sum_{n=1}^{\infty} D\big(\varphi(R_{n-1}(\cdot)/\varepsilon) f_n e_n\big)(x)[h] \right\| \\
&= \left\| \sum_{n=1}^{\infty} \varphi(R_{n-1}(x)/\varepsilon) f_n(h) e_n + f_n(x) e_n \cdot D\varphi(R_{n-1}(x)/\varepsilon)[R_{n-1}(h)/\varepsilon] \right\| \\
&\leq \left\| \sum_{n=1}^{\infty} \varphi(R_{n-1}(x)/\varepsilon) f_n(h) e_n \right\| + \left\| \sum_{n=1}^{\infty} f_n(x) e_n \cdot D\varphi(R_{n-1}(x)/\varepsilon)[R_{n-1}(h)/\varepsilon] \right\| \\
&\leq A\|h\| + \left\| \sum_{n>n_0} f_n(x) e_n \cdot D\varphi(R_{n-1}(x)/\varepsilon)[R_{n-1}(h)/\varepsilon] \right\| \\
&\leq A + A \sup_{n>n_0} \big| D\varphi(R_{n-1}(x)/\varepsilon)[R_{n-1}(h)/\varepsilon] \big| \left\| \sum_{n>n_0} f_n(x) e_n \right\| \\
&\leq A + A \sup_{n>n_0} \big\| D\varphi(R_{n-1}(x)/\varepsilon) \big\| (1 + B) \|h/\varepsilon\| \cdot \|R_{n_0}(x)\| \\
&\leq A + AM(1 + B)\frac{1}{\varepsilon}\frac{\varepsilon}{A} = A + M(1 + B) = K.
\end{aligned}$$

□

Theorem 84 (Nicole Moulis, [Mo2]). *Let X be a Banach space with a monotone unconditional Schauder basis $\{e_j\}_{j=1}^{\infty}$ that admits a C^k-smooth Lipschitz bump function. Denote $X_n = \operatorname{span}\{e_j\}_{j=1}^{n}$. There is a constant $C > 0$ such that if Y is a Banach space, $M \subset X$ is such that $P_n(M) \subset M$ for all $n \in \mathbb{N}$, Ω is a uniform open neighbourhood of M, $f : \Omega \to Y$ is an L-Lipschitz mapping such that $f\restriction_{\Omega \cap X_n}$ is $C^{1,+}$-smooth for each $n \in \mathbb{N}$, $V \subset M$ is open such that* $\operatorname{dist}(V, X \setminus M) > 0$, *and $\varepsilon > 0$, then there is $g \in C^k(X; Y)$ such that $\|Dg\|_V \leq C(1 + \varepsilon)L$ and $\|f - g\|_V \leq \varepsilon$.*

Proof. Without loss of generality we may assume that $\{e_j\}$ is normalised. Denote $X_\infty = \operatorname{span}\{e_j\}_{j=1}^\infty = \bigcup_{n=1}^\infty X_n$. Let us extend the mapping f to the whole of X by $f(x) = 0$ for $x \in X \setminus \Omega$ and denote $f_n = f\restriction_{X_n}$. For each $n \in \mathbb{N}$ denote by T_n the isomorphism $T_n\colon \mathbb{R}^n \to X_n$, $T_n(y) = \sum_{j=1}^n y_j e_j$, and define a mapping $g_n\colon X_n \to Y$ by the Bochner integral

$$g_n(x) = \int_{\mathbb{R}^n} f_n\bigl(x - T_n(y)\bigr)\varphi_n(y)\,\mathrm{d}\lambda_n(y),$$

where $\varphi_n \in C^\infty(\mathbb{R}^n)$ are smooth functions with sufficiently small compact supports chosen so that

$$\|g_n(x) - f_n(x)\| < \frac{\varepsilon}{4}\frac{1}{2^n} \quad \text{for every } x \in M \cap X_n, \tag{21}$$

$$\|Dg_n(x) - Df_n(x)\| < L\frac{\varepsilon}{2}\frac{1}{2^n} \quad \text{for every } x \in M \cap X_n. \tag{22}$$

This is possible, since the mappings f_n and Df_n are uniformly continuous on $\Omega \cap X_n$ (see the proof of Lemma 1). Using Corollary 1.91 and substitution (recall that a uniformly continuous mapping is bounded on totally bounded sets) it is easy to see that $g_n \in C^\infty(X_n; Y)$.

Now let us define inductively a sequence of mappings $\bar{g}_n\colon X_n \to Y$. Let $\bar{g}_1 = g_1$. Suppose that $n \in \mathbb{N}$, $n > 1$, and the mapping $\bar{g}_{n-1}$ is already defined. Then we put

$$\bar{g}_n(x) = g_n(x) + \bar{g}_{n-1}(P_{n-1}(x)) - g_n(P_{n-1}(x)) \quad \text{for all } x \in X_n.$$

Notice that $\bar{g}_n\restriction_{X_{n-1}} = \bar{g}_{n-1}$, that is $\bar{g}_n$ is an extension of $\bar{g}_{n-1}$. Furthermore, by induction we can show that

$$\bar{g}_n \in C^\infty(X_n; Y) \quad \text{for each } n \in \mathbb{N}, \tag{23}$$

$$\|\bar{g}_n(x) - f_n(x)\| < \frac{\varepsilon}{2}\left(1 - \frac{1}{2^n}\right) \quad \text{for every } x \in M \cap X_n,\ n \in \mathbb{N}, \tag{24}$$

$$\|D\bar{g}_n(x) - Df_n(x)\| < L\varepsilon\left(1 - \frac{1}{2^n}\right) \quad \text{for every } x \in M \cap X_n,\ n \in \mathbb{N}. \tag{25}$$

Indeed, (23) is obvious. For $n = 1$ the inequality (24) follows from (21). Let $n \in \mathbb{N}$, $n > 1$, and suppose the inequality (24) holds for $n - 1$. Then, using (21),

$$\begin{aligned}
\|\bar{g}_n(x) - f_n(x)\| &\le \|g_n(x) - f_n(x)\| + \bigl\|\bar{g}_{n-1}(P_{n-1}(x)) - g_n(P_{n-1}(x))\bigr\| \\
&< \frac{\varepsilon}{4}\frac{1}{2^n} + \bigl\|\bar{g}_{n-1}(P_{n-1}(x)) - f_{n-1}(P_{n-1}(x))\bigr\| + \bigl\|f_n(P_{n-1}(x)) - g_n(P_{n-1}(x))\bigr\| \\
&< \frac{\varepsilon}{4}\frac{1}{2^n} + \frac{\varepsilon}{2}\left(1 - \frac{1}{2^{n-1}}\right) + \frac{\varepsilon}{4}\frac{1}{2^n} = \frac{\varepsilon}{2}\left(1 - \frac{1}{2^n}\right)
\end{aligned}$$

for any $x \in M \cap X_n$. Notice, that here we used the fact that $P_{n-1}(M) \subset M$.

The inequality (25) for $n = 1$ follows from (22). Let $n \in \mathbb{N}$, $n > 1$, and suppose the inequality (25) holds for $n - 1$. Then, using (22),

$$\begin{aligned}
&\|D\bar{g}_n(x) - Df_n(x)\| \\
&\quad \le \|Dg_n(x) - Df_n(x)\| + \big\|D(\bar{g}_{n-1} \circ P_{n-1})(x) - D(g_n \circ P_{n-1})(x)\big\| \\
&\quad < L\frac{\varepsilon}{2}\frac{1}{2^n} + \big\|D(\bar{g}_{n-1} \circ P_{n-1})(x) - D(f_n \circ P_{n-1})(x)\big\| \\
&\qquad\qquad + \big\|D(f_n \circ P_{n-1})(x) - D(g_n \circ P_{n-1})(x)\big\| \\
&\quad = L\frac{\varepsilon}{2}\frac{1}{2^n} + \big\|D\bar{g}_{n-1}(P_{n-1}(x)) \circ P_{n-1} - Df_{n-1}(P_{n-1}(x)) \circ P_{n-1}\big\| \\
&\qquad\qquad + \big\|Df_n(P_{n-1}(x)) \circ P_{n-1} - Dg_n(P_{n-1}(x)) \circ P_{n-1}\big\| \\
&\quad < L\frac{\varepsilon}{2}\frac{1}{2^n} + L\varepsilon\left(1 - \frac{1}{2^{n-1}}\right) + L\frac{\varepsilon}{2}\frac{1}{2^n} = L\varepsilon\left(1 - \frac{1}{2^n}\right)
\end{aligned}$$

for any $x \in M \cap X_n$. Here we used the fact that $P_{n-1}(M) \subset M$ and also the fact that $\|P_{n-1}\| = 1$.

Next, we define the mapping $\bar{g}\colon X_\infty \to Y$ by $\bar{g}(x) = \lim_{n\to\infty} \bar{g}_n(x)$ for all $x \in X_\infty$. Recall that $\bar{g}_n(x) = \bar{g}_m(x)$ for all $n \ge m$ whenever $x \in X_m$ and thus the mapping $\bar{g}$ is well-defined. From (24) it readily follows that

$$\|\bar{g}(x) - f(x)\| < \frac{\varepsilon}{2} \quad \text{for every } x \in M \cap X_\infty. \tag{26}$$

Now, let $\Psi \in C^k(X; X_\infty)$ be the mapping supplied by Lemma 83 that satisfies $\|\Psi(x) - x\| < \min\{\frac{\varepsilon}{2L}, \varepsilon, \operatorname{dist}(V, X \setminus M)\}$ for all $x \in X$. Let $C > 0$ be the Lipschitz constant of Ψ. We define the mapping $g\colon X \to Y$ by $g = \bar{g} \circ \Psi$. For each $x \in X$ there are a neighbourhood U of x and $n \in \mathbb{N}$ such that $\Psi(U) \subset X_n$. Hence $g = \bar{g}_n \circ \Psi$ on U which together with (23) implies that $g \in C^k(X; Y)$.

To see that g approximates f, choose an arbitrary $x \in V$. Then $\Psi(x) \in M \cap X_\infty$, which used together with (26) gives

$$\|g(x) - f(x)\| \le \|\bar{g}(\Psi(x)) - f(\Psi(x))\| + \|f(\Psi(x)) - f(x)\| < \frac{\varepsilon}{2} + L\frac{\varepsilon}{2L} = \varepsilon.$$

Finally, we estimate the derivative of g on V. Fix any $x \in V$. There are a neighbourhood U of x and $n \in \mathbb{N}$ such that $\Psi(U) \subset X_n$ and so $g = \bar{g}_n \circ \Psi$ on U. Also, $\Psi(x) \in M \cap X_n$ and therefore we can use (25) to obtain

$$\begin{aligned}
\|Dg(x)\| &= \|D(\bar{g}_n \circ \Psi)(x)\| = \|D\bar{g}_n(\Psi(x)) \circ D\Psi(x)\| \le \|D\bar{g}_n(\Psi(x))\|\|D\Psi(x)\| \\
&\le C\Big(\big\|D\bar{g}_n(\Psi(x)) - Df_n(\Psi(x))\big\| + \|Df_n(\Psi(x))\|\Big) < C(L\varepsilon + L) = C(1+\varepsilon)L.
\end{aligned}$$

□

Combining Lemma 69 and Theorem 84 we would obtain a uniform approximation result on spaces with unconditional basis. However, we postpone the precise formulation until Corollary 87, where we obtain even stronger statement.

Next, we prove a result that allows us to pass from uniform approximations to fine approximations. We start with the existence of smooth and Lipschitz σ-discrete partitions of unity.

Lemma 85. *Let X, Y be normed linear spaces and $k \in \mathbb{N} \cup \{\infty\}$. Suppose that for each 1-Lipschitz mapping $f\colon 2U_X \to Y$ and $\varepsilon > 0$ there is a Lipschitz mapping $g \in C^k(U_X; Y)$ satisfying $\|f - g\|_{U_X} \leq \varepsilon$. Let $\Omega \subset X$ be open. Then for any open covering $\mathcal{U}$ of Ω there is a Lipschitz and C^k-smooth locally finite and σ-uniformly discrete partition of unity on Ω subordinated to $\mathcal{U}$.*

Proof. Let $S \subset C^k(\Omega)$ be the subset consisting of bounded Lipschitz functions. Analogously as in the proof of Theorem 56 it can be shown that S is a partition ring. Further, notice that approximation of mappings into Y gives us also approximations of functions. Indeed, if $f\colon 2U_X \to \mathbb{R}$ is 1-Lipschitz, then choose some $y \in S_Y$ and consider the mapping $\bar{f}\colon 2U_X \to Y$, $\bar{f}(x) = f(x) \cdot y$. Let $\bar{g} \in C^k(U_X; Y)$ be an approximation of $\bar{f}$ provided by our assumption and $F \in Y^*$ be a Hahn-Banach extension of the norm-one functional $ty \mapsto t$ defined on $\operatorname{span}\{y\}$. Then $g = F \circ \bar{g}$ is the desired approximation of the function f. Thus by approximating the function $x \mapsto \operatorname{dist}(x, \Omega \setminus W)$ we can show that (ii) in Lemma 49 is satisfied, which finishes the proof. □

Theorem 86. *Let X, Y be normed linear spaces and $k \in \mathbb{N} \cup \{\infty\}$. Suppose that there is a $C \geq 1$ such that for each L-Lipschitz mapping $f\colon 2U_X \to Y$ and $\varepsilon > 0$ there is a CL-Lipschitz mapping $g \in C^k(U_X; Y)$ satisfying $\|f - g\|_{U_X} \leq \varepsilon$. Let $\Omega \subset X$ be open. Then for any L-Lipschitz mapping $f\colon \Omega \to Y$, any continuous function $\varepsilon\colon \Omega \to \mathbb{R}^+$, and any $\eta > 1$ there is an ηCL-Lipschitz mapping $g \in C^k(\Omega; Y)$ such that $\|f(x) - g(x)\| < \varepsilon(x)$ for all $x \in \Omega$.*

Proof. First notice that from approximations on U_X by translating and scaling we immediately obtain approximations on any open ball in X. For each $x \in \Omega$ find $r(x) > 0$ such that $U(x, 4r(x)) \subset \Omega$ and

$$\varepsilon(y) > \frac{\varepsilon(x)}{3} \quad \text{for each } y \in U(x, r(x)). \tag{27}$$

By Lemma 85 there is a locally finite and σ-discrete C^k-smooth Lipschitz partition of unity on Ω subordinated to $\{U(x, r(x));\ x \in \Omega\}$. We may assume that the partition of unity is of the form $\{\psi_{n\alpha}\}_{n\in\mathbb{N},\alpha\in\Lambda}$, where for each $n \in \mathbb{N}$ the family $\{\operatorname{supp}_o \psi_{n\alpha}\}_{\alpha\in\Lambda}$ is discrete in Ω. For each $n \in \mathbb{N}$ and $\alpha \in \Lambda$ let $U_{n\alpha} = U(x_{n\alpha}, r(x_{n\alpha}))$ be such that $\operatorname{supp}_o \psi_{n\alpha} \subset U_{n\alpha}$. Let $L_{n\alpha}$ be the Lipschitz constant of $\psi_{n\alpha}$, and without loss of generality assume that $L_{n\alpha} \geq 1$. Further, denote $V_{n\alpha} = U(x_{n\alpha}, 2r(x_{n\alpha}))$.

For each $n \in \mathbb{N}$ and $\alpha \in \Lambda$ we approximate f on $V_{n\alpha}$ by CL-Lipschitz mapping $g_{n\alpha} \in C^k(V_{n\alpha}; Y)$ such that

$$\|f(x) - g_{n\alpha}(x)\| \leq \min\left\{\frac{(\eta - 1)CL}{2^n L_{n\alpha}}, \frac{\varepsilon(x_{n\alpha})}{3}\right\} < \varepsilon(x) \quad \text{for each } x \in U_{n\alpha}. \tag{28}$$

(The second inequality follows from (27).) Define the mapping $\bar{g}_{n\alpha}\colon \Omega \to Y$ by $\bar{g}_{n\alpha}(x) = g_{n\alpha}(x)$ for $x \in V_{n\alpha}$, $\bar{g}_{n\alpha}(x) = 0$ otherwise.

Finally, we define the mapping $g\colon \Omega \to Y$ by

$$g(x) = \sum_{n\in\mathbb{N},\alpha\in\Lambda} \psi_{n\alpha}(x)\bar{g}_{n\alpha}(x).$$

Since $\operatorname{supp}_{\mathrm{o}} \psi_{n\alpha} \subset U_{n\alpha}$, $g_{n\alpha} \in C^k(V_{n\alpha}; Y)$, and the sum is locally finite, the mapping g is well-defined and moreover $g \in C^k(\Omega; Y)$.

Choose $x \in \Omega$ and let us compute how far $g(x)$ is from $f(x)$:

$$\begin{aligned}\|f(x) - g(x)\| &= \Bigl\| \sum_{n\in\mathbb{N},\alpha\in\Lambda} \psi_{n\alpha}(x)\bigl(f(x) - \bar{g}_{n\alpha}(x)\bigr)\Bigr\| \\ &\le \sum_{\substack{n\in\mathbb{N}\\ \alpha\in\Lambda\colon x\in U_{n\alpha}}} \psi_{n\alpha}(x)\|f(x) - g_{n\alpha}(x)\| < \varepsilon(x) \sum_{\substack{n\in\mathbb{N}\\ \alpha\in\Lambda\colon x\in U_{n\alpha}}} \psi_{n\alpha}(x) = \varepsilon(x),\end{aligned}$$

where the last inequality follows from (28).

To estimate the derivative of g at some fixed $x \in \Omega$, notice that by the discreteness of $\{\operatorname{supp}_{\mathrm{o}} \psi_{n\alpha}\}_{\alpha\in\Lambda}$, for each $n \in \mathbb{N}$ there is at most one $\alpha \in \Lambda$ such that $D\psi_{n\alpha}(x) \neq 0$. Put $M = \{n \in \mathbb{N};\ \exists\alpha \in \Lambda\colon D\psi_{n\alpha}(x) \neq 0\}$. Then there is a mapping $\beta\colon M \to \Lambda$ such that for each $n \in M$, $D\psi_{n\alpha}(x) = 0$ whenever $\alpha \neq \beta(n)$ and moreover $x \in U_{n\beta(n)}$. (Notice that if $D\psi_{n\alpha}(x) \neq 0$, then necessarily $x \in U_{n\alpha}$.) Further, since $\sum \psi_{n\alpha} = 1$, it follows that $\sum D\psi_{n\alpha} = 0$. Hence

$$\begin{aligned}\|Dg(x)\| &= \Bigl\| \sum_{n\in\mathbb{N},\alpha\in\Lambda} D(\psi_{n\alpha}\bar{g}_{n\alpha})(x)\Bigr\| = \Bigl\| \sum_{\substack{n\in\mathbb{N}\\ \alpha\in\Lambda\colon x\in U_{n\alpha}}} D(\psi_{n\alpha}\bar{g}_{n\alpha})(x)\Bigr\| \\ &= \Bigl\| \sum_{\substack{n\in\mathbb{N}\\ \alpha\in\Lambda\colon x\in U_{n\alpha}}} \psi_{n\alpha}(x)Dg_{n\alpha}(x) + \sum_{\substack{n\in\mathbb{N}\\ \alpha\in\Lambda\colon x\in U_{n\alpha}}} D\psi_{n\alpha}(x)\cdot g_{n\alpha}(x)\Bigr\| \\ &= \Bigl\| \sum_{\substack{n\in\mathbb{N}\\ \alpha\in\Lambda\colon x\in U_{n\alpha}}} \psi_{n\alpha}(x)Dg_{n\alpha}(x) + \sum_{\substack{n\in\mathbb{N}\\ \alpha\in\Lambda\colon x\in U_{n\alpha}}} D\psi_{n\alpha}(x)\cdot \bigl(g_{n\alpha}(x) - f(x)\bigr)\Bigr\| \\ &\le \sum_{\substack{n\in\mathbb{N}\\ \alpha\in\Lambda\colon x\in U_{n\alpha}}} \psi_{n\alpha}(x)\|Dg_{n\alpha}(x)\| + \sum_{n\in M} \|D\psi_{n\beta(n)}(x)\|\|g_{n\beta(n)}(x) - f(x)\| \\ &\le \sum_{\substack{n\subset\mathbb{N}\\ \alpha\in\Lambda\colon x\in U_{n\alpha}}} CL\psi_{n\alpha}(x) + \sum_{n\in M} L_{n\beta(n)}\|g_{n\beta(n)}(x) - f(x)\| \\ &\le CL + \sum_{n\in M} L_{n\beta(n)}\frac{(\eta-1)CL}{2^n L_{n\beta(n)}} \le \eta CL,\end{aligned}$$

where the last but one inequality follows from (28).

To finish the proof we show that g is ηCL-Lipschitz on the set Ω. Without loss of generality we may assume that $\varepsilon(x) \le (\eta C - 1)L \operatorname{dist}(x, X \setminus \Omega)$ for every $x \in \Omega$. Now fix $x, y \in \Omega$. If the line segment l with end points x and y lies in Ω, then $\|g(x)-g(y)\| \le \eta CL\|x-y\|$ by Proposition 1.71. Otherwise there is $z \in l \cap (X \setminus \Omega)$. Then

$$\begin{aligned}\|g(x) - g(y)\| &\le \|g(x) - f(x)\| + \|f(x) - f(y)\| + \|f(y) - g(y)\| \\ &< \varepsilon(x) + L\|x - y\| + \varepsilon(y) \\ &\le (\eta C - 1)L\|x - z\| + L\|x - y\| + (\eta C - 1)L\|y - z\| = \eta CL\|x - y\|.\end{aligned}$$

□

Corollary 87. *Let X be a separable normed linear space that admits a C^k-smooth Lipschitz bump function, $k \in \mathbb{N} \cup \{\infty\}$. Let Y be a Banach space. Suppose further that one of the following conditions is satisfied:*

- *X is a Banach space with an unconditional Schauder basis, or*
- *at least one of the spaces X or Y equals $B_0(V)$ for some topological space V, or*
- *at least one of the spaces X or Y equals $C_{\mathrm{ub}}(P)$ for some metric space P.*

Then there is a constant $C \in \mathbb{R}$ such that for any open $\Omega \subset X$, any L-Lipschitz mapping $f: \Omega \to Y$, and any continuous function $\varepsilon: \Omega \to \mathbb{R}^+$ there is a CL-Lipschitz mapping $g \in C^k(\Omega; Y)$ for which $\|f(x) - g(x)\| < \varepsilon(x)$ for all $x \in \Omega$.

Proof. It suffices to notice that under our assumptions the hypothesis of Theorem 86 is satisfied. Indeed, since B_X is a 2-Lipschitz retract of X, every L-Lipschitz mapping defined on B_X can be extended to a $2L$-Lipschitz mapping defined on X. Thus we may apply either Corollary 81, or (in the case of the unconditional basis) we combine Lemma 69 and Theorem 84. □

Further, Theorem 86 together with Theorem 41 gives us the next corollary.

Corollary 88. *Let X be a Hilbert space and $\Omega \subset X$ an open set. Then for any L-Lipschitz function $f: \Omega \to \mathbb{R}$, any continuous function $\varepsilon: \Omega \to \mathbb{R}^+$, and any $\eta > 1$ there is an ηL-Lipschitz function $g \in C^1(\Omega)$ such that $|f(x) - g(x)| < \varepsilon(x)$ for all $x \in \Omega$.*

Similarly, Theorem 86 together with Theorem 74 produces the following corollary.

Corollary 89. *Let Γ be an arbitrary set, $\Omega \subset c_0(\Gamma)$ an open set, and Y a Banach space. Then for any L-Lipschitz mapping $f: \Omega \to Y$, any continuous function $\varepsilon: \Omega \to \mathbb{R}^+$, and any $\eta > 1$ there is an ηL-Lipschitz mapping $g \in C^\infty(\Omega; Y)$ such that $\|f(x) - g(x)\| < \varepsilon(x)$ for all $x \in \Omega$.*

In the rest of the section we deal with the approximation of Lipschitz functions by real analytic Lipschitz functions.

Theorem 90. *Let X be a separable Banach space that admits a Lipschitz separating real-analytic function with uniform radii of convergence as in property (K). Then there is a constant $K \in \mathbb{R}$ such that for each $\varepsilon > 0$ and any L-Lipschitz function $f: X \to \mathbb{R}$ there is a KL-Lipschitz function $g \in C^\omega(X)$ satisfying $|f - g|_X \le \varepsilon$.*

We remark that the assumption is in particular satisfied if X admits a separating polynomial. Indeed, by Fact 4.45 there is $P \in \mathcal{P}(^nX)$ for some $n \in \mathbb{N}$ even such that $P(x) \geq 1$ for $x \in S_X$. Let $0 < r \leq 1$ be such that $r^2\|\check{P}\|(1+t)^n < 1+t^n$ for each $t \in [0,+\infty)$. Consider the function $q(z) = \big(1+\tilde{P}(z)\big)^{\frac{1}{n}}$. By Proposition 1.61 we have $\operatorname{Re}\tilde{P}(x+iy) \geq P(x) - \|\check{P}\|\sum_{k=1}^{[\frac{n}{2}]}\binom{n}{2k}\|x\|^{n-2k}\|y\|^{2k} \geq \|x\|^n - \|\check{P}\|r^2(1+\|x\|)^n > -1$ for all $x, y \in X$, $\|y\| \leq r$. Thus $q \in H(G)$, where $G = \{z \in \tilde{X};\ \|\operatorname{Im} z\| < r\}$, and by Corollary 1.165 the radius of norm convergence of the Taylor series of q at each $x \in X$ is at least r. Further, $q\restriction_X$ is clearly a separating function. Finally, $\|Dq(x)\| = \frac{1}{n}(1+P(x))^{\frac{1}{n}-1}\|DP(x)\| \leq \frac{1}{n}(1+\|P\|\|x\|^n)^{\frac{1-n}{n}}\|DP\|\|x\|^{n-1} = \frac{\|DP\|}{n}\big(\frac{\|x\|^n}{1+\|P\|\|x\|^n}\big)^{1-\frac{1}{n}}$ which is bounded on X and so $q\restriction_X$ is Lipschitz.

The proof of Theorem 90 is divided into a few steps (Proposition 91, Proposition 92, and Lemma 93). We begin by introducing an auxiliary notion. Let X be a normed linear space and $\mathcal{U} = \{U_x;\ x \in U_x \subset \tilde{X}, x \in X\}$ a collection of open neighbourhoods in $\tilde{X}$. Let $A \subset X$. We say that a function $h\colon \bigcup\mathcal{U} \to \mathbb{C}$ *separates A with respect to $\mathcal{U}$* if

(S1) $h\restriction_X$ maps into $\mathbb{R}$,
(S2) $h(x) \geq 1$ whenever $x \in A$,
(S3) $|h(z)| \leq \frac{1}{4}$ whenever $z \in U_x$, $x \in X$, $\operatorname{dist}(x, A) \geq 1$.

Proposition 91. *Let X be a Banach space. Assume that there are a constant $C > 0$ and $\mathcal{U} = \{U_x;\ x \in U_x \subset \tilde{X}, x \in X\}$ a collection of open neighbourhoods in $\tilde{X}$ such that for each $A \subset X$ there is a function $h_A \in H\big(\bigcup\mathcal{U}\big)$ which separates A with respect to $\mathcal{U}$ and such that $h_A\restriction_X$ is C-Lipschitz. Then for every $\varepsilon > 0$ and every L-Lipschitz function $f\colon X \to \mathbb{R}$ there is a $10CL$-Lipschitz function $g \in C^\omega(X)$ satisfying $|f-g|_X \leq \varepsilon$.*

Proof. Let us define a function $\bar{f}\colon X \to \mathbb{R}$ by $\bar{f}(x) = \frac{4}{\varepsilon}f\big(\frac{\varepsilon}{4L}x\big)$. This function is obviously 1-Lipschitz. Denote $\bar{f}^+ = \max\{\bar{f}, 0\}$ and $\bar{f}^- = \max\{-\bar{f}, 0\}$ and notice that both functions are again 1-Lipschitz. Next, we set $A_n = \{x \in X;\ \bar{f}^+(x) \geq n\}$ for $n \in \mathbb{N}_0$. Clearly, $A_n \subset A_{n-1}$ for all $n \in \mathbb{N}$, and using the 1-Lipschitz property of $\bar{f}^+$ it is easy to check that

$$\operatorname{dist}(X \setminus A_n, A_{n+1}) \geq 1 \quad \text{for all } n \in \mathbb{N}. \tag{29}$$

Denote $h_n = \theta_n \circ h_{A_n}$ for $n \in \mathbb{N}$, where the functions θ_n come from Lemma 4. For each $n \in \mathbb{N}$, $h_n \in H\big(\bigcup\mathcal{U}\big)$ and $h_n\restriction_X$ is $4C$-Lipschitz. Put $h^+ = \sum_{n=1}^\infty h_n$.

Fix an arbitrary $x \in X$. Then there is $m \in \mathbb{N}$ such that $x \in A_{m-1} \setminus A_m$. Hence

$$x \in A_n \text{ for } n < m \quad \text{and} \quad x \in X \setminus A_{n-1} \text{ for } n > m. \tag{30}$$

From this, (29), (S3), and (T3) it follows that $|h_n(z)| \leq 2^{-n}$ for all $n > m$ and $z \in U_x$. Hence the sum in the definition of h^+ converges absolutely uniformly on U_x and so $h^+ \in H\big(\bigcup\mathcal{U}\big)$. This together with (S1) and (T1) implies that $h^+\restriction_X \in C^\omega(X)$.

Using (30), (S2) and (T4), (29), (S3) and (T3), and finally $m-1+h_m(x) \in [m-1,m]$ and $\bar{f}^+(x) \in [m-1,m)$, we obtain

$$\begin{aligned}
\left|h^+(x) - \bar{f}^+(x)\right| &= \left|\sum_{n=1}^{m-1} h_n(x) + h_m(x) + \sum_{n=m+1}^{\infty} h_n(x) - \bar{f}^+(x)\right| \\
&\le \sum_{n=1}^{m-1} |h_n(x) - 1| + \sum_{n=m+1}^{\infty} |h_n(x)| + \left|m - 1 + h_m(x) - \bar{f}^+(x)\right| \\
&< \sum_{n=1}^{m-1} 2^{-n} + \sum_{n=m+1}^{\infty} 2^{-n} + 1 < 2.
\end{aligned}$$

Further, (30) and (29) together with the property (T5) imply $\|D(h_n\restriction_X)(x)\| = |(\theta_n\restriction_{\mathbb{R}})'(h_{A_n}(x))|\|D(h_{A_n}\restriction_X)(x)\| \le 2^{-n}C$ for $n \in \mathbb{N} \setminus \{m\}$. Hence by Corollary 1.166

$$\begin{aligned}
\|D(h^+\restriction_X)(x)\| &\le \sum_{n=1}^{\infty} \|D(h_n\restriction_X)(x)\| \\
&\le \sum_{n\neq m} 2^{-n}C + \|D(h_m\restriction_X)(x)\| < C + 4C = 5C.
\end{aligned}$$

Similarly we obtain an approximation of the function $\bar{f}^-$ denoted by h^-. We put $h = h^+ - h^-$. Then $h\restriction_X \in C^\omega(X)$, $\left|h(x) - \bar{f}(x)\right| < 4$ for every $x \in X$, and $\|D(h\restriction_X)(x)\| \le \|D(h^+\restriction_X)(x)\| + \|D(h^-\restriction_X)(x)\| < 10C$ for every $x \in X$.

Finally, let $g(x) = \frac{\varepsilon}{4}h\left(\frac{4L}{\varepsilon}x\right)$ for $x \in X$. It is straightforward to check that g satisfies the conclusion of our proposition. □

Proposition 92. *Let X be a Banach space. Suppose that there are an open neighbourhood G of X in $\tilde{X}$ and a collection $\{\psi_n\}_{n\in\mathbb{N}}$ of functions on G with the following properties:*

(P1) *$\{\psi_n\restriction_X\}_{n\in\mathbb{N}}$ is a sup-partition on X,*

(P2) *the mapping $z \mapsto (b_n\psi_n(z))_{n\in\mathbb{N}}$ is a holomorphic mapping from G into $\tilde{c_0}$ for any $(b_n) \in \ell_\infty$,*

(P3) *there is $M > 0$ such that each $\psi_n\restriction_X$ is M-Lipschitz,*

(P4) *for each $n \in \mathbb{N}$ there is $x_n \in X$ such that $\psi_n(x) \le \frac{Q}{8}$ for $x \in X$, $\|x - x_n\| \ge \frac{1}{2}$, where Q is the quantity from the definition of a sup-partition.*

Then there is a collection $\mathcal{U}$ of open neighbourhoods in $\tilde{X}$ such that for each $A \subset X$ there is a function $h_A \in H\left(\bigcup\mathcal{U}\right)$ which separates A with respect to $\mathcal{U}$ and such that $h_A\restriction_X$ is C-Lipschitz, where $C = \sqrt{2}M/Q$.

Proof. Let W, μ, and Δ_w be as in Proposition 7 for $q = \frac{8}{Q}$. Further, let us denote $w(z) = (\psi_n(z))_{n\in\mathbb{N}}$ for $z \in G$. By the continuity of the mapping w (which follows from (P2)), for each $x \in X$ there exists an open neighbourhood U_x of x in $\tilde{X}$ such that $U_x \subset G$ and $\|w(z) - w(x)\| < \Delta_{w(x)}/q$ whenever $z \in U_x$. (Notice that $w(x) \in c_0 \setminus \{0\}$.) Put $\mathcal{U} = \{U_x;\ x \in X\}$.

Let $A \subset X$. For each $n \in \mathbb{N}$ put $b_n = q$ if $\operatorname{dist}(x_n, A) \le \frac{1}{2}$ and $b_n = 1$ otherwise. Choose $z \in \bigcup \mathcal{U}$ and let $x \in X$ be such that $z \in U_x$. Then

$$\begin{aligned}\big\|(b_n\psi_n(z)) - (b_n\psi_n(x))\big\| &= \sup_{n\in\mathbb{N}} \big|b_n(\psi_n(z) - \psi_n(x))\big| \\ &\le q \sup_{n\in\mathbb{N}} |\psi_n(z) - \psi_n(x)| = q\|w(z) - w(x)\| < \Delta_{w(x)}\end{aligned} \tag{31}$$

and since $0 \le w(x) \le (b_n\psi_n(x)) \le qw(x)$ in the lattice sense, from (M1) it follows that $(b_n\psi_n(z)) \in W$. Therefore we may define $h_A(z) = \frac{1}{8}\mu\big((b_n\psi_n(z))\big)$ for $z \in \bigcup \mathcal{U}$ and (P2) implies that $h_A \in H\big(\bigcup \mathcal{U}\big)$. Further, $h_A \restriction_X$ is obviously C-Lipschitz.

Next we show that the function h_A separates A with respect to $\mathcal{U}$. Clearly h_A has property (S1). Pick any $x \in A$. From the properties (P1) and (P4) it follows that $\sup\{\psi_n(x);\ n \in \mathbb{N}, \operatorname{dist}(x_n, A) \le \frac{1}{2}\} \ge Q$. Therefore $\|(b_n\psi_n(x))\| \ge qQ = 8$ and consequently (M2) gives property (S2). Finally, to show (S3) let $x \in X$ be such that $\operatorname{dist}(x, A) \ge 1$. Then, by (P4), $\psi_n(x) \le \frac{Q}{8}$ for those $n \in \mathbb{N}$ for which $\operatorname{dist}(x_n, A) \le \frac{1}{2}$. Thus $\|(b_n\psi_n(x))\| \le \max\{q\frac{Q}{8}, 1\} = 1$. Now (31) together with (M3) implies $|h_A(z)| \le \frac{1}{4}$ for $z \in U_x$. □

Lemma 93. *Let X be a separable Banach space and further let $\{x_n\}_{n=1}^\infty$ be a dense sequence in X. Suppose that there are $\Delta > 0$ and a function $q \in H(\Omega)$ where $\Omega = \{z \in \tilde{X};\ \|\operatorname{Im} z\| < \Delta\}$, such that $q \restriction_X$ is Lipschitz and maps into $[0, +\infty)$, $q(0) = 0$, $q(x) \ge 2$ for $x \in X$, $\|x\| \ge \frac{1}{2}$, and suppose there is a sequence $\{a_n\}_{n=1}^\infty$ of positive real numbers such that for each $x \in X$ the function*

$$z \mapsto \sum_{n=1}^{\infty} a_n\big(\operatorname{Im} q(x - x_n + z)\big)^2 \tag{32}$$

is defined on some neighbourhood of 0 in $\tilde{X}$ and is continuous at 0. Then there are an open neighbourhood G of X in $\tilde{X}$ and a collection of functions $\{\psi_n\}_{n\in\mathbb{N}}$ satisfying the properties (P1)–(P4) in Proposition 92.

Proof. Put $\varepsilon_n = \min\{2^{-n}, \frac{1}{16}\}$ and let ζ_n be the functions and $\{\delta_n\}$ the sequence from Lemma 5. Put

$$\psi_n(z) = \zeta_n\big(q(z - x_1), \dots, q(z - x_n)\big) \quad \text{for } z \in \Omega, n \in \mathbb{N}.$$

Then $\psi_n \in H(\Omega)$ and by (Z1) $\psi_n \restriction_X$ maps into $[0, 1]$.

Pick any $x \in X$. From the density of $\{x_n\}$ and the fact that $q(0) = 0$ it follows that there is $l \in \mathbb{N}$ such that $q(x - x_l) < 1$. Let $k \in \mathbb{N}$ be the smallest such number. Then property (Z4) implies that $\psi_k(x) \ge \frac{1}{2}$. Thus $\sup_{n\in\mathbb{N}} \psi_n(x) \ge Q$ for each $x \in X$, where $Q = \frac{1}{2}$.

By the continuity of q there is $\rho > 0$ such that $\operatorname{Re} q(z) \le \frac{1}{4}$ whenever $z \in \tilde{X}$, $\|z\| < \rho$. Now fix $x \in X$ and find an index $j \in \mathbb{N}$ such that $\|x_j - x\| < \rho$. Let $\delta_x > 0$ be such that $\|x - x_j + w\| < \rho$ and $\sum_{n=1}^\infty a_n\big(\operatorname{Im} q(x - x_n + w)\big)^2 \le \delta_j$ whenever $w \in \tilde{X}$, $\|w\| \le \delta_x$. Then $\operatorname{Re} q(x - x_j + w) \le \frac{1}{4}$ and hence, by (Z3), $|\psi_n(z)| \le 2^{-n}$ for $n > j$ and $z \in U_x = U_{\tilde{X}}(x, \delta_x)$. It follows that for any $(b_n) \in \ell_\infty$,

$(b_n\psi_n(z))_{n\in\mathbb{N}} = \sum_{n=1}^{\infty} b_n\psi_n(z)e_n \in \tilde{c_0}$ and the sum converges absolutely uniformly on U_x. As the mappings $z \mapsto b_n\psi_n(z)e_n$ are holomorphic as mappings from Ω into $\tilde{c_0}$, we can conclude that $(b_n\psi_n)$ is a holomorphic mapping from $G = \bigcup_{x\in X} U_x$ into $\tilde{c_0}$, which gives (P2). (P1) then immediately follows.

Property (P3) obviously holds by (Z2). Finally we show that (P4) is satisfied. Indeed, fix $n \in \mathbb{N}$. For $x \in X$, $\|x - x_n\| \geq \frac{1}{2}$ we have $q(x - x_n) \geq 2$, hence, by (Z5), $\psi_n(x) \leq \frac{1}{16} = \frac{Q}{8}$. □

Proof of Theorem 90. There are $d > 0$ and $q \in H(G)$, $G = \{z \in \tilde{X};\ \|\mathrm{Im}\, z\| < d\}$, such that $q\restriction_X$ is Lipschitz and maps into $[0, +\infty)$, $q(0) = 0$, $q(x) \geq 2$ for $x \in X$, $\|x\| \geq \frac{1}{2}$, and the radius of convergence of the Taylor expansion of q at every point $x \in X$ is at least d (Theorem 1.171).

Let $\{x_n\}_{n=1}^{\infty}$ be a dense sequence in X. Put

$$M_n = \sup\Big\{\big|q(x_j - x_n + w)\big|;\ w \in \tilde{X}, \|w\| \leq \tfrac{d}{2}, 1 \leq j \leq n\Big\}$$

and $a_n = 1/(2^n M_n^2)$. (Note that by the assumption on the radius of the Taylor series $M_n < +\infty$.) Fix $x \in X$. There is $k \in \mathbb{N}$ such that $x \in U\big(x_k, \frac{d}{4}\big)$. For $z \in \tilde{X}$ satisfying $\|z\| \leq \frac{d}{4}$ we have $\|x - x_k + z\| \leq \frac{d}{2}$ and hence for $n \geq k$

$$a_n\big(\mathrm{Im}\, q(x - x_n + z)\big)^2 \leq a_n\big|q(x_k - x_n + x - x_k + z)\big|^2 \leq a_n M_n^2 = \frac{1}{2^n}.$$

Therefore the sum in (32) converges absolutely uniformly on $B_{\tilde{X}}\big(0, \frac{d}{4}\big)$ to a continuous function. Using Lemma 93 together with Proposition 92 and Proposition 91 finishes the proof. □

8. Approximation of C^1-smooth mappings

In this section we prove our most general result on an approximation of a C^1-smooth mapping together with its first derivative by a C^k-smooth mapping in the fine topology. A concise formulation is in Corollary 96.

In order to avoid repeating the same argument in various contexts, we prove the following somewhat technical proposition. One of the main ideas is based on the same argument as the proof of Theorem 86.

Proposition 94. *Let X, Y be normed linear spaces, $k \in \mathbb{N} \cup \{\infty\}$, and $\Omega \subset X$ open. Suppose that for any open covering $\mathcal{U}$ of Ω there is a C^k-smooth Lipschitz locally finite and σ-discrete partition of unity on Ω subordinated to $\mathcal{U}$. Suppose further that $\{Y_\gamma\}_{\gamma\in\Gamma}$ is a collection of closed subspaces of Y such that for each $\gamma \in \Gamma$ there is a constant $C_\gamma \in \mathbb{R}$ such that for any L-Lipschitz mapping $f \in C^1(2U_X; Y_\gamma)$ and any $\varepsilon > 0$ there is a $C_\gamma L$-Lipschitz mapping $g \in C^k(U_X; Y)$ satisfying $\|f - g\|_{U_X} \leq \varepsilon$. Let $f \in C^1(\Omega; Y)$ be such that it is locally a mapping into some Y_γ, $\gamma \in \Gamma$. Then for any continuous function $\varepsilon\colon \Omega \to \mathbb{R}^+$ there is a mapping $g \in C^k(\Omega; Y)$ such that $\|f(x) - g(x)\| < \varepsilon(x)$ and $\|Df(x) - Dg(x)\| < \varepsilon(x)$ for all $x \in \Omega$.*

Proof. First notice that from approximations on U_X by translating and scaling we immediately obtain approximations on any open ball in X. For each $x \in \Omega$ find $r(x) > 0$ and $\gamma(x) \in \Gamma$ such that $U(x, 4r(x)) \subset \Omega$, $f\big(U(x, 4r(x))\big) \subset Y_{\gamma(x)}$,

$$\varepsilon(y) > \frac{\varepsilon(x)}{3} \quad \text{for each } y \in U(x, 4r(x)), \text{ and} \tag{33}$$

$$\|Df(x) - Df(y)\| < \frac{\varepsilon(x)}{9C_{\gamma(x)}} \quad \text{for each } y \in U(x, 4r(x)). \tag{34}$$

By our assumption there is a locally finite and σ-discrete C^k-smooth Lipschitz partition of unity on Ω subordinated to $\{U(x, r(x));\ x \in \Omega\}$. We may assume that the partition of unity is of the form $\{\psi_{n\alpha}\}_{n\in\mathbb{N},\alpha\in\Lambda}$, where for each $n \in \mathbb{N}$ the family $\{\operatorname{supp}_o \psi_{n\alpha}\}_{\alpha\in\Lambda}$ is discrete in Ω. For each $n \in \mathbb{N}$ and $\alpha \in \Lambda$ let $U_{n\alpha} = U(x_{n\alpha}, r(x_{n\alpha}))$ be such that $\operatorname{supp}_o \psi_{n\alpha} \subset U_{n\alpha}$. Let $L_{n\alpha}$ be the Lipschitz constant of $\psi_{n\alpha}$. Further, denote $C_{n\alpha} = C_{\gamma(x_{n\alpha})}$ and $V_{n\alpha} = U(x_{n\alpha}, 2r(x_{n\alpha}))$. Without loss of generality assume that $L_{n\alpha} \geq 1$ and $C_{n\alpha} \geq 1$.

For each $n \in \mathbb{N}$ and $\alpha \in \Lambda$ we define the mapping $f_{n\alpha}\colon U(x_{n\alpha}, 4r(x_{n\alpha})) \to Y_{\gamma(x_{n\alpha})}$ by $f_{n\alpha}(x) = f(x) - Df(x_{n\alpha})[x]$. Then, by (34) and (33),

$$\begin{aligned} \|Df_{n\alpha}(x)\| &= \|Df(x) - Df(x_{n\alpha})\| \\ &< \frac{\varepsilon(x_{n\alpha})}{9C_{n\alpha}} < \frac{\varepsilon(x)}{3C_{n\alpha}} \leq \frac{\varepsilon(x)}{3} \quad \text{for each } x \in U(x_{n\alpha}, 4r(x_{n\alpha})). \end{aligned} \tag{35}$$

According to our assumption, for each $n \in \mathbb{N}$ and $\alpha \in \Lambda$ we can approximate $f_{n\alpha}$ on $V_{n\alpha}$ by $g_{n\alpha} \in C^k(V_{n\alpha}; Y)$ such that

$$\|Dg_{n\alpha}(x)\| \leq \frac{\varepsilon(x_{n\alpha})}{9} < \frac{\varepsilon(x)}{3} \quad \text{for each } x \in V_{n\alpha}, \tag{36}$$

$$\|f_{n\alpha}(x) - g_{n\alpha}(x)\| \leq \frac{\varepsilon(x_{n\alpha})}{9 \cdot 2^n L_{n\alpha}} < \frac{\varepsilon(x)}{3 \cdot 2^n L_{n\alpha}} < \varepsilon(x) \quad \text{for each } x \in V_{n\alpha}. \tag{37}$$

(The second inequalities follow from (33).) Define the mapping $\bar{g}_{n\alpha}\colon \Omega \to Y$ by $\bar{g}_{n\alpha}(x) = g_{n\alpha}(x)$ for $x \in V_{n\alpha}$, $\bar{g}_{n\alpha}(x) = 0$ otherwise. Finally, we define the mapping $g\colon \Omega \to Y$ by

$$g(x) = \sum_{n\in\mathbb{N},\alpha\in\Lambda} \psi_{n\alpha}(x)\big(\bar{g}_{n\alpha}(x) + Df(x_{n\alpha})[x]\big).$$

Since $\operatorname{supp}_o \psi_{n\alpha} \subset U_{n\alpha}$, $g_{n\alpha} \in C^k(V_{n\alpha}; Y)$, and the sum is locally finite, the mapping g is well-defined and moreover $g \in C^k(\Omega; Y)$.

Choose $x \in \Omega$ and let us compute how far $g(x)$ is from $f(x)$:

$$\begin{aligned}\|f(x)-g(x)\| &= \Big\| \sum_{n\in\mathbb{N},\alpha\in\Lambda} \psi_{n\alpha}(x)\big(f(x)-\bar{g}_{n\alpha}(x)-Df(x_{n\alpha})[x]\big)\Big\| \\ &= \Big\| \sum_{\substack{n\in\mathbb{N}\\ \alpha\in\Lambda:\, x\in U_{n\alpha}}} \psi_{n\alpha}(x)\big(f_{n\alpha}(x)-g_{n\alpha}(x)\big)\Big\| \\ &\le \sum_{\substack{n\in\mathbb{N}\\ \alpha\in\Lambda:\, x\in U_{n\alpha}}} \psi_{n\alpha}(x)\|f_{n\alpha}(x)-g_{n\alpha}(x)\| < \varepsilon(x) \sum_{\substack{n\in\mathbb{N}\\ \alpha\in\Lambda:\, x\in U_{n\alpha}}} \psi_{n\alpha}(x) = \varepsilon(x),\end{aligned}$$

where the last inequality follows from (37).

To estimate the distance between the derivatives at a fixed $x \in \Omega$, note that by the discreteness of $\{\operatorname{supp}_o \psi_{n\alpha}\}_{\alpha\in\Lambda}$, for each $n \in \mathbb{N}$ there is at most one $\alpha \in \Lambda$ such that $D\psi_{n\alpha}(x) \neq 0$. Put $M = \{n \in \mathbb{N};\ \exists\alpha \in \Lambda\colon D\psi_{n\alpha}(x) \neq 0\}$. Then there is a mapping $\beta\colon M \to \Lambda$ such that for each $n \in M$, $D\psi_{n\alpha}(x) = 0$ whenever $\alpha \neq \beta(n)$, and moreover $x \in U_{n\beta(n)}$. (Notice that if $D\psi_{n\alpha}(x) \neq 0$, then necessarily $x \in U_{n\alpha}$.) Hence

$$\begin{aligned}&\|Df(x)-Dg(x)\| = \|D(f-g)(x)\| \\ &= \Big\| \sum_{n\in\mathbb{N},\alpha\in\Lambda} D\Big(\psi_{n\alpha}\cdot\big(f-\bar{g}_{n\alpha}-Df(x_{n\alpha})\big)\Big)(x)\Big\| \\ &= \Big\| \sum_{\substack{n\in\mathbb{N}\\ \alpha\in\Lambda:\, x\in U_{n\alpha}}} D\Big(\psi_{n\alpha}\cdot\big(f-g_{n\alpha}-Df(x_{n\alpha})\big)\Big)(x)\Big\| \\ &= \Big\| \sum_{\substack{n\in\mathbb{N}\\ \alpha\in\Lambda:\, x\in U_{n\alpha}}} D\big(\psi_{n\alpha}\cdot(f_{n\alpha}-g_{n\alpha})\big)(x)\Big\| \\ &= \Big\| \sum_{\substack{n\in\mathbb{N}\\ \alpha\in\Lambda:\, x\in U_{n\alpha}}} \psi_{n\alpha}(x)D(f_{n\alpha}-g_{n\alpha})(x) + \sum_{\substack{n\in\mathbb{N}\\ \alpha\in\Lambda:\, x\in U_{n\alpha}}} D\psi_{n\alpha}(x)\cdot\big(f_{n\alpha}(x)-g_{n\alpha}(x)\big)\Big\| \\ &\le \sum_{\substack{n\in\mathbb{N}\\ \alpha\in\Lambda:\, x\in U_{n\alpha}}} \psi_{n\alpha}(x)\big\|Df_{n\alpha}(x)-Dg_{n\alpha}(x)\big\| \\ &\qquad + \sum_{n\in M} \|D\psi_{n\beta(n)}(x)\|\,\big\|f_{n\beta(n)}(x)-g_{n\beta(n)}(x)\big\| \\ &\le \sum_{\substack{n\in\mathbb{N}\\ \alpha\in\Lambda:\, x\in U_{n\alpha}}} \psi_{n\alpha}(x)\Big(\|Df_{n\alpha}(x)\| + \|Dg_{n\alpha}(x)\|\Big) \\ &\qquad + \sum_{n\in M} L_{n\beta(n)}\,\big\|f_{n\beta(n)}(x)-g_{n\beta(n)}(x)\big\| \\ &< \Big(\frac{\varepsilon(x)}{3}+\frac{\varepsilon(x)}{3}\Big) \sum_{\substack{n\in\mathbb{N}\\ \alpha\in\Lambda:\, x\in U_{n\alpha}}} \psi_{n\alpha}(x) + \sum_{n\in M} L_{n\beta(n)}\frac{\varepsilon(x)}{3\cdot 2^n L_{n\beta(n)}} \le \varepsilon(x),\end{aligned}$$

where the last but one inequality follows from (35), (36), and (37). □

Theorem 95. *Let X, Y be normed linear spaces, $k \in \mathbb{N} \cup \{\infty\}$. Consider the following statements:*

(i) *There is $C \in \mathbb{R}$ such that for any L-Lipschitz mapping $f: 2U_X \to Y$ and any $\varepsilon > 0$ there is a CL-Lipschitz mapping $g \in C^k(U_X; Y)$ such that $\|f - g\|_{U_X} \leq \varepsilon$.*

(ii) *For any open $\Omega \subset X$ and any open covering $\mathcal{U}$ of Ω there is a C^k-smooth Lipschitz locally finite and σ-discrete partition of unity on Ω subordinated to $\mathcal{U}$. There is $C \in \mathbb{R}$ such that for any L-Lipschitz mapping $f \in C^1(2U_X; Y)$ and any $\varepsilon > 0$ there is a CL-Lipschitz mapping $g \in C^k(U_X; Y)$ such that $\|f - g\|_{U_X} \leq \varepsilon$.*

(iii) *For any open $\Omega \subset X$, any mapping $f \in C^1(\Omega; Y)$, and any continuous function $\varepsilon: \Omega \to \mathbb{R}^+$ there is $g \in C^k(\Omega; Y)$ such that $\|f(x) - g(x)\| < \varepsilon(x)$ and $\|Df(x) - Dg(x)\| < \varepsilon(x)$ for all $x \in \Omega$.*

(iv) *For any open $\Omega \subset X$, any L-Lipschitz mapping $f \in C^1(\Omega; Y)$, any continuous function $\varepsilon: \Omega \to \mathbb{R}^+$, and any $\eta > 1$ there exists an ηL-Lipschitz mapping $g \in C^k(\Omega; Y)$ such that $\|f(x) - g(x)\| < \varepsilon(x)$ for all $x \in \Omega$.*

Then (i)$\Rightarrow$(ii)$\Rightarrow$(iii)$\Rightarrow$(iv).

Proof. (i)$\Rightarrow$(ii) follows from Lemma 85, (ii)$\Rightarrow$(iii) follows from Proposition 94 (consider the collection of subspaces of Y consisting only of the space Y itself), and for (iii)$\Rightarrow$(iv) see the end of the proof of Theorem 86. □

Corollary 96. *Let X be a separable normed linear space that admits a C^k-smooth Lipschitz bump function, $k \in \mathbb{N} \cup \{\infty\}$. Let Y be a Banach space. Suppose further that one of the following conditions is satisfied:*

- *at least one of the spaces X or Y equals $B_0(V)$ for some topological space V, or*
- *at least one of the spaces X or Y equals $C_{ub}(P)$ for some metric space P, or*
- *X is a Banach space with an unconditional Schauder basis, or*
- *Y is a Banach space with an unconditional Schauder basis and with a separable dual.*

Then for any open $\Omega \subset X$, any mapping $f \in C^1(\Omega; Y)$, and any continuous function $\varepsilon: \Omega \to \mathbb{R}^+$ there exists $g \in C^k(\Omega; Y)$ such that $\|f(x) - g(x)\| < \varepsilon(x)$ and $\|Df(x) - Dg(x)\| < \varepsilon(x)$ for all $x \in \Omega$.

Proof. Suppose that one of the first three conditions is satisfied. Then our corollary follows from Theorem 95. It suffices to notice that under our assumptions the statement (i) of Theorem 95 holds. We may either apply Corollary 87, or the less involved Corollary 81 together with the observation that since B_X is a 2-Lipschitz retract of X, every L-Lipschitz mapping defined on B_X can be extended to a $2L$-Lipschitz mapping defined on X.

It remains to prove the case that Y has an unconditional Schauder basis $\{e_n\}$ and has a separable dual (which means that Y admits a C^1-smooth Lipschitz bump function, Theorem 5.2). We will show that statement (ii) in Theorem 95 is satisfied, which will prove our claim. For the first part we use Lemma 85 together with the approximation of Lipschitz functions given by either Corollary 81, or the less involved combination of

Theorem 63 and Theorem 64. (Although, since X is separable, it is not overly difficult to construct the required partitions of unity directly.)

To prove the second assertion in statement (ii) of Theorem 95 let K be the constant from Lemma 83 used on the space Y. Put $C = 2K$. Let $f \in C^1(U_X; Y)$ be L-Lipschitz and $\varepsilon > 0$. Denote $Y_n = \operatorname{span}\{e_j\}_{j=1}^n$. By Lemma 83 there is a K-Lipschitz mapping $\Psi \in C^1(Y; Y)$ which locally maps into some Y_n and such that $\|y - \Psi(y)\| < \frac{\varepsilon}{2}$ for every $y \in Y$. Put $h = \Psi \circ f$. Then $h \in C^1(U_X; Y)$ is a KL-Lipschitz mapping which locally maps into some Y_n and such that $\|f - h\|_{U_X} \leq \frac{\varepsilon}{2}$. Since the spaces Y_n, $n \in \mathbb{N}$, are finite-dimensional, by Corollary 81 there are constants C_n such that any M-Lipschitz mapping from U_X into Y_n can be approximated by C^k-smooth $C_n M$-Lipschitz mapping. Therefore we can use Proposition 94 to find a CL-Lipschitz mapping $g \in C^k(U_X; Y)$ such that $\|g - h\|_{U_X} \leq \frac{\varepsilon}{2}$. As $\|f - g\|_{U_X} \leq \varepsilon$, we have just shown that the statement (ii) in Theorem 95 holds. □

Finally, combining Theorem 74 and Theorem 95 we obtain the following corollary.

Corollary 97. *Let Γ be an arbitrary set, Y a Banach space, $\Omega \subset c_0(\Gamma)$ open, $f \in C^1(\Omega; Y)$, and $\varepsilon\colon \Omega \to \mathbb{R}^+$ a continuous function. Then there is $g \in C^\infty(\Omega; Y)$ such that $\|f(x) - g(x)\| < \varepsilon(x)$ and $\|Df(x) - Dg(x)\| < \varepsilon(x)$ for all $x \in \Omega$.*

9. Approximation of norms

In this section we consider the approximation problem under the additional convexity assumption. Suppose that U is an open, bounded, and convex subset of a normed linear space X, and $f\colon U \to \mathbb{R}$ is a bounded convex function. Is there a C^k-smooth and convex function g on U such that $|f - g|_U < \varepsilon$? If this is true for every f, U, and $\varepsilon > 0$, we will say that convex functions on X are approximable by C^k-smooth convex functions. It is clear that this property implies a formally weaker condition, stating that every equivalent norm on X can be approximated by C^k-smooth norms. The opposite implication, in a slightly weaker sense, is also easy to prove. It suffices to make the graph of f part of a unit sphere of some equivalent norm on $X \oplus \mathbb{R}$ by considering two shifted symmetric copies of it and taking the convex hull. In this way we can see that C^k-smooth approximations of norms in $X \oplus \mathbb{R}$ together with the use of the Implicit function theorem imply the C^k-smooth approximations for convex functions in X. A necessary condition for this result is clearly the existence of at least some equivalent C^k-smooth norm on X (similarly to the existence of a C^k-smooth bump in the general case). This condition is satisfied for X if and only if it is satisfied for $X \oplus \mathbb{R}$, which implies that the problem of approximating norms is essentially equivalent to the problem of approximating convex functions.

In this section we will give a general solution to the norm (and hence the general convex) case for separable Banach spaces. The norm setting has some technical advantages, namely a convenient representation of the norm as a supremum (in fact a maximum) of the values of a sequence of C^k-smooth convex functions. This technique bears strong resemblance to the sup-partitions of unity used in the previous sections, which can in fact be considered its generalisation.

Theorem 98. *Let X be a normed linear space, $U \subset X$ an open convex set, and let $f\colon U \to \mathbb{R}$ be G^2-smooth. Then f is convex if and only if $\widehat{\delta^2 f(x)}$ is positive semi-definite for each $x \in U$, i.e. $\widehat{\delta^2 f(x)}[h] \geq 0$ for every $h \in X$.*

Proof. This is a one-dimensional result. Indeed, for $x \in U$ and $h \in X$ we put $f_{x,h}(t) = f(x+th)$, which is defined on an open interval containing zero. Then f is convex if and only if $f_{x,h}$ is convex for each $x \in U$, $h \in X$. Directly from the definition we obtain $f''_{x,h}(t) = \widehat{\delta^2 f(x+th)}[h]$ and so the statement follows. □

Fact 99. *Let X be a normed linear space and let Y, Z be closed subspaces of X such that $X = Y \oplus Z$. Let $G\colon X \to \mathbb{R}$ be an even convex continuous function such that the Minkowski functional $|||\cdot|||$ of the set $\{x \in X;\ G(x) \leq 1\}$ is a norm on X. Put $g(z) = \inf_{y\in Y} G(y,z)$ for $z \in Z$. Then the Minkowski functional ν of the set $\{z \in Z;\ g(z) \leq 1\}$ is a norm on Z such that the canonical quotient mapping $q\colon (Z,\nu) \to (X/Y, |||\cdot|||_{X/Y})$ is an isometry.*

Proof. Clearly, g is convex on Z ("a convex body casts a convex shadow"), even, and continuous (as it is locally bounded above). We have $\nu(z) \geq 1$ if and only if $g(z) \geq 1$ if and only if $G(y,z) \geq 1$ for all $y \in Y$ if and only if $|||(y,z)||| \geq 1$ for all $y \in Y$ if and only if $\inf_{y\in Y} |||(y,z)||| \geq 1$ if and only if $|||q(z)|||_{X/Y} \geq 1$. □

Lemma 100. *Let X be a normed linear space such that its norm $\|\cdot\|$ is C^k-smooth, $k \in \mathbb{N} \cup \{\infty\}$, and let $Y \subset X$ be a finite-dimensional subspace. Then $\|\cdot\|$ can be approximated uniformly on bounded sets by a C^k-smooth norm $|||\cdot|||$ such that the quotient norm $|||\cdot|||_{X/Y}$ is also C^k-smooth.*

Proof. Let $\{e_j\}_{j=1}^n$ be a normalised basis of Y and let $\{f_j\}_{j=1}^n$ be the corresponding biorthogonal functionals. Let Z be a (topological) complement of Y in X and K be the norm of the projection $(z,y) \mapsto z$ in X. Define an isomorphism $T \in \mathcal{L}(\mathbb{R}^n; Y)$ by $T(t) = \sum_{j=1}^n t_j e_j$. Further, let $\varepsilon > 0$ and $\varphi \in \mathbb{C}^\infty(\mathbb{R}^n)$ be a non-negative function with $\operatorname{supp}\varphi \subset B(0,\delta)$ and $\int_{\mathbb{R}^n} \varphi = 1$, where $\delta > 0$ is chosen so that $Kn\delta < 1$. Put

$$\nu(z,y) = \int_{\mathbb{R}^n} \|z+y-T(t)\|\varphi(t)\,\mathrm{d}\lambda(t)$$

for $z \in Z$, $y \in Y$. Note that ν is a convex function on $X = Z \oplus Y$ and if δ is small enough, then $\left|\nu - \|\cdot\|\right|_X \leq \varepsilon$ (see the proof of Lemma 69). Corollary 1.91 implies that ν is C^k-smooth on $X \setminus B(0,n\delta)$. Using substitution and Corollary 1.91 (similarly as in the proof of Lemma 69) it is easy to see that for each $z \in Z$ the function $y \mapsto \nu(z,y)$ is C^∞-smooth on Y and

$$D_2\nu(z,y) = \int_{\mathbb{R}^n} \|z+y-T(t)\| D\varphi(t) \circ T^{-1}\,\mathrm{d}\lambda(t).$$

Using Corollary 1.91 again, we see that $D_2\nu$ is C^k-smooth on $X \setminus B(0,n\delta)$.

Put $P(y) = \sum_{j=1}^{n} \frac{1}{2^j \|f\|_j^2} f_j(y)^2$. Then $P \in \mathcal{P}(^2Y)$, P is strictly convex, and $\|P\| \le 1$. Let $\varepsilon > 0$ and define $G\colon Z \oplus Y \to \mathbb{R}$ by $G(z, y) = \nu(z, y) + \varepsilon P(y)$. Then G is a convex function on X. Further, set $g(z) = \inf_{y \in Y} G(z, y)$. Note that since Y is finite-dimensional and $y \mapsto G(z, y)$ is linearly coercive and strictly convex, for each $z \in Z$ the infimum is attained at a unique point $y = \rho(z) \in Y$. Thus $g(z) = G(z, \rho(z))$. We claim that g is C^k-smooth on $Z \setminus B(0, Kn\delta)$, for which it suffices to show this for ρ.

Put $F = D_2 G$. Then $F\colon Z \oplus Y \to Y^*$ and F is C^k-smooth on $X \setminus B(0, n\delta)$. Note that $F(z, y) = 0$ if and only if $y = \rho(z)$. Indeed, due to the convexity $F(z, y) = 0$ if and only if $y \mapsto G(z, y)$ attains its minimum at y.

We claim that for each $z \in Z$ and $y \in Y$ the derivative $D_2 F(z, y) \in \mathcal{L}(Y; Y^*)$ is an isomorphism. Since $\dim Y = \dim Y^* = n$, it suffices to prove that $\ker D_2 F(z, y) = \{0\}$. To this end we show that $(D_2 F(z, y)[u])[u] \ne 0$ for each $u \in Y$, $u \ne 0$. Fix $z \in Z$. Observe that $D_2 F(z, y) = D^2 h(y)$, where $h(y) = G(z, y)$. Thus it suffices to show that $d^2 h(y)$ is positive definite. Denote $\nu_z(y) = \nu(z, y)$. We have $d^2 h(y) = d^2 \nu_z(y) + \varepsilon d^2 P(y) = d^2 \nu_z(y) + 2\varepsilon P$ (Lemma 1.99). The differential $d^2 \nu_z(y)$ is positive semi-definite for each $y \in Y$ by Theorem 98, while P is clearly positive definite and so the claim follows.

Therefore by the Implicit function theorem (Theorem 1.87) the mapping $z \mapsto \rho(z)$, which is a unique solution to $F(z, y) = 0$, is C^k-smooth on $X \setminus B(0, Kn\delta)$.

Let $|||\cdot|||$ be the Minkowski functional of the set $\{(z, y) \in X;\ G(z, y) \le 1\}$. Then the norm $|||\cdot|||$ satisfies the requirements (use Lemma 5.23 on G and on g and then Fact 99). Notice also that through the choice of ε this norm can be made arbitrarily close to the norm $\|\cdot\|$. □

Fact 101. *Let X be a normed linear space, $C \subset X$ a closed convex set, $W \subset X$, $r > 0$, and $0 < \eta < 1$ such that $B(0, r) \subset W \subset C + B\big(0, r(1 - \eta)\big)$. Then $\eta W \subset C$.*

Proof. Assume the contrary. By the separation theorem there are $x \in \eta W$, $f \in S_{X^*}$, and $a \in \mathbb{R}$ such that $\sup_C f \le a < f(x)$. Given any $\varepsilon > 0$ there is $z \in S_X$ such that $f(z) > 1 - \varepsilon$. Thus $rz \in W$ and so there is $u \in C$ such that $\|rz - u\| \le r(1 - \eta)$. It follows that $(1 - \varepsilon)r < f(rz) = f(u) + f(rz - u) \le a + r(1 - \eta)$. Thus $a \ge \eta r$.

Because $\frac{1}{\eta} x \in W$, there is $v \in C$ such that $\|\frac{1}{\eta} x - v\| \le r(1 - \eta)$. It follows that $\frac{1}{\eta} a < f(\frac{1}{\eta} x) = f(v) + f(\frac{1}{\eta} x - v) \le a + r(1 - \eta)$. Thus $r(1 - \eta) \le a(\frac{1}{\eta} - 1) < r(1 - \eta)$, a contradiction. □

The following lemma comes from [DFH2] and is similar to the results in [Zip2].

Lemma 102. *Let X be a Banach space that has a shrinking Markushevich basis $\{(x_j; f_j)\}_{j \in \mathbb{N}}$. Let $W \subset X^*$ be a w^*-compact convex set such that $0 \in \operatorname{Int} W$ and $0 < \varepsilon < 1$. Then there exist a w^*-compact set $F \subset W$, a sequence $\{g_k\}_{k \in \mathbb{N}} \subset F$, and a sequence $\{n_k\}_{k \in \mathbb{N}} \subset \mathbb{N}$ such that $n_k \to \infty$,*

(i) $(1 - \varepsilon)W \subset \overline{\operatorname{conv}}\, F$, *and*

(ii) $F \subset \bigcup_{k \in \mathbb{N}} \big(g_k + (\varepsilon B_{X^*} \cap Z_{n_k})\big)$,

where $Z_n = \{x_1, \dots, x_n\}^{\perp}$.

Proof. Let $r > 0$ be such that $B(0,r) \subset W$. Note also that by the assumptions $qW \subset W$ for any $q \in [0,1]$. Each set $x_j(W) \subset \mathbb{R}$ is bounded and thus there exists a finite $\frac{\varepsilon r}{4\cdot 2^j \|f_j\|}$-dense set $C_j \subset x_j(W)$. Put

$$A = \left\{ \sum_{j=1}^{n} a_j f_j \in W;\ n \in \mathbb{N}, a_j \in C_j \right\}$$

and $F = \overline{A}^{w^*}$. Clearly $x_j(F) = x_j(A) \subset C_j \cup \{0\}$, $j \in \mathbb{N}$.

To show the property (i), in view of Fact 101 it suffices to prove that for any $f \in (1-\frac{\varepsilon}{2})W$ there is $h \in F$ such that $\|f-h\| \le r\frac{\varepsilon}{2}$. So choose any $f \in (1-\frac{\varepsilon}{2})W$. Since $\operatorname{span}\{f_j\}$ is dense in X^* and W is convex with non-empty interior, there is $g = \sum_{j=1}^{n} b_j f_j \in W$ such that $\|f-g\| < r\frac{\varepsilon}{4}$. Further, there are $a_j \in C_j$ such that $|b_j - a_j| < \frac{\varepsilon r}{4\cdot 2^j \|f_j\|}$, $j = 1,\dots,n$. Put $h = \sum_{j=1}^{n} a_j f_j$. Then $\|g-h\| < r\frac{\varepsilon}{4}$ and so $\|f-h\| < r\frac{\varepsilon}{2}$. It remains to show that $h \in W$, which in turn implies that $h \in A \subset F$. This is of course true if $h = f$. Otherwise denote $s = (1-\frac{\varepsilon}{2})r$, $u = s\frac{h-f}{\|h-f\|}$, and

$$v = \frac{s}{s+\|h-f\|} f + \frac{\|h-f\|}{s+\|h-f\|} u = \frac{s}{s+\|h-f\|} h.$$

As $\|u\| = s$, we have $u \in (1-\frac{\varepsilon}{2})W$, and hence by the convexity also $v \in (1-\frac{\varepsilon}{2})W$. Since $\frac{s+\|h-f\|}{s}(1-\frac{\varepsilon}{2}) < 1$, we conclude that $h \in W$.

Next, we construct the sequences $\{g_k\}$ and $\{n_k\}$. Put $\mathcal{Z}_n = \{g + Z_n;\ g \in F\}$ and $\mathcal{Z} = \bigcup_{n\in\mathbb{N}} \mathcal{Z}_n$. Note that for each $n \in \mathbb{N}$ the set $\mathcal{Z}_n$ is finite and hence $\mathcal{Z}$ is countable. Indeed, since for each $j \in \mathbb{N}$ the set $x_j(F)$ is finite, there is $Q \subset F$ finite such that $\{(f(x_1),\dots,f(x_n)) \in \mathbb{R}^n;\ f \in Q\} = \{(f(x_1),\dots,f(x_n)) \in \mathbb{R}^n;\ f \in F\}$. Thus given $g \in F$ there is $f \in Q$ such that $g(x_j) = f(x_j)$, $j = 1,\dots,n$, and hence $g \in f + Z_n$. This means that $\mathcal{Z}_n = \{h + Z_n;\ h \in Q\}$.

For each ordinal α we define sets $F_\alpha \subset F$ and $h_\alpha + Z_{n(\alpha)} \in \mathcal{Z}$ such that $F_\alpha \subsetneq F_\beta$ and

$$h_\alpha + Z_{n(\alpha)} \neq h_\beta + Z_{n(\beta)} \quad \text{whenever } \beta < \alpha \text{ and } F_\alpha \neq \emptyset, \text{ and} \tag{38}$$

$$(h_\alpha + Z_{n(\alpha)}) \cap F_\alpha = \big(h_\alpha + (\varepsilon B_{X^*} \cap Z_{n(\alpha)})\big) \cap F_\alpha. \tag{39}$$

This will be done by transfinite induction: we put $F_0 = F$, $F_{\alpha+1} = F_\alpha \setminus (h_\alpha + Z_{n(\alpha)})$, and $F_\alpha = \bigcap_{\beta<\alpha} F_\beta$ if α is a limit ordinal. Further, if $F_\alpha \neq \emptyset$, then since X^* is w^*-dentable [FHHMZ, Theorem 11.8], there are $x \in X$ and $a \in \mathbb{R}$ such that the slice $S = \{f \in F_\alpha;\ f(x) > a\}$ is non-empty and has diameter less than ε. Since F_α is bounded, there is $h_\alpha + Z_{n(\alpha)} \in \mathcal{Z}$ such that $h_\alpha \in S$ and $(h_\alpha + Z_{n(\alpha)}) \cap F_\alpha \subset S$. Thus (39) holds. Clearly, $F_\alpha \cap (h_\beta + Z_{n(\beta)}) = \emptyset$ for all $\beta < \alpha$, and so (38) holds.

By (38) and the countability of $\mathcal{Z}$ there is a countable ordinal η such that $F_\eta = \emptyset$ and $F_\alpha \neq \emptyset$ for $\alpha < \eta$. We re-index the family $\{h_\alpha + Z_{n(\alpha)}\}_{\alpha<\eta}$ as $\{g_k + Z_{n_k}\}_{k\in\mathbb{N}}$. In this sequence each member of $\mathcal{Z}$ appears at most once, and since each $\mathcal{Z}_n$ is finite, it follows that $n_k \to \infty$. Since for each $x \in F$ there is an ordinal α such that $x \in F_\alpha \setminus F_{\alpha+1}$, i.e. $x \in (h_\alpha + Z_{n(\alpha)}) \cap F_\alpha$, by (39) the property (ii) clearly holds. □

Theorem 103 ([DFH1], [HáTa]). *Let X be a separable Banach space with a C^k-smooth norm, $k \in \mathbb{N} \cup \{\infty\}$. Then every equivalent norm on X can be approximated uniformly on bounded sets by C^k-smooth equivalent norms.*

Proof. Let $\|\cdot\|$ be the equivalent norm on X that we want to approximate and let B_{X^*} be its dual unit ball. Since X admits a C^k-smooth norm, the dual X^* is separable and therefore X admits a shrinking Markushevich basis $\{(x_j; f_j)\}_{j\in\mathbb{N}}$. Choose an arbitrary $0 < \varepsilon < 1$. Let $F \subset B_{X^*}$, $\{g_j\}_{j\in\mathbb{N}} \subset F$, and $\{n_j\}_{j\in\mathbb{N}} \subset \mathbb{N}$ be the w^*-compact set and the sequences from Lemma 102 used on $W = B_{X^*}$. Denote $Y_j = \operatorname{span}\{x_1, \dots, x_{n_j}\}$ and let $P_j : X \to Y_j$ be the canonical projections given by $P_j(x) = \sum_{l=1}^{n_j} f_l(x)x_l$. Let $q_j : X \to X/Y_j$ denote the canonical quotient mapping. By Lemma 100 there are $M > 0$ and equivalent C^k-smooth norms $|||\cdot|||_j$ on X such that the semi-norms $|||q_j(\cdot)|||_j$ are C^k-smooth on X away from Y_j, where $|||\cdot|||_j$ denotes also the canonical quotient norm on $(X, |||\cdot|||_j)/Y_j$, and $|||x|||_j \le \|x\| \le M|||x|||_j$ for each $x \in X$, $j \in \mathbb{N}$.

We will construct a norm on X with a smooth generalised James boundary. Let $\varphi \in C^\infty(\mathbb{R})$ be an even convex function such that $\varphi(t) = 0$ whenever $t \in [-\frac{1}{2}, \frac{1}{2}]$ and $\varphi(1) = 1$. Let ψ be the Minkowski functional of the set $\{(x, y) \in \mathbb{R}^2;\ \varphi(x)+\varphi(y) \le 1\}$. Then ψ is a C^∞-smooth norm on $\mathbb{R}^2$ that is strongly lattice (i.e. $\psi(s,t) \le \psi(x, y)$ whenever $|s| \le |x|$, $|t| \le |y|$) and $\max\{|x|, |y|\} \le \psi(x, y) \le |x| + |y|$ for every $(x, y) \in \mathbb{R}^2$. Observe also that $\psi(x, y) = |x|$ on the set $\{(x, y) \in \mathbb{R}^2;\ |y| \le \frac{1}{2}|x|\}$ and $\psi(x, y) = |y|$ on the set $\{(x, y) \in \mathbb{R}^2;\ |x| \le \frac{1}{2}|y|\}$.

Further, we define norms ν_j on X by

$$\nu_j(x) = \psi\left(\frac{1}{j|||P_j|||_j}|||P_j(x)|||_j, |||q_j(x)|||_j\right).$$

It is easy to see that ν_j is C^k-smooth. This follows from the fact that if $x \in Y_j \setminus \{0\}$, then $\nu_j(y) = \frac{1}{j|||P_j|||_j}|||P_j(y)|||_j$ for y from a neighbourhood of x, while if $x \in \ker P_j \setminus \{0\}$, then $\nu_j(y) = |||q_j(y)|||_j$ for y from a neighbourhood of x. For any $x \in X$ we have $\max\{\frac{1}{j|||P_j|||_j}|||P_j(x)|||_j, |||q_j(x)|||_j\} \le \nu_j(x) \le \frac{1}{j|||P_j|||_j}|||P_j(x)|||_j + |||q_j(x)|||_j \le (\frac{1}{j}+1)|||x|||_j \le 2\|x\|$. Since $q_j \restriction_{\ker P_j}$ is an isomorphism, the norm ν_j is an equivalent norm on X. Denote by $B_j \subset X^*$ the dual unit ball of ν_j. The preceding inequality implies that $\|f\| \le 2\nu_j(f)$ for any $f \in X^*$, i.e.

$$B_j \subset 2B_{X^*}. \tag{40}$$

Further, denote by $T_j : Y_j^\perp \to (X/Y_j)^*$ the canonical isometry. As $\nu_j(x) \ge |||q_j(x)|||_j$ for each $x \in X$, we have $\nu_j(f) = \sup_{\nu_j(x)\le 1} f(x) \le \sup_{|||q_j(x)|||_j \le 1} f(x) = \sup_{|||q_j(x)|||_j \le 1} T_j(f)(q_j(x)) = |||f|||_j$ for each $f \in Y_j^\perp$. Therefore

$$\frac{1}{M}B_{X^*} \cap Y_j^\perp \subset B_j. \tag{41}$$

Put

$$K = \bigcup_{j \in \mathbb{N}} (g_j + \varepsilon M B_j).$$

By (41) and the property (ii) from Lemma 102 we get $F \subset K$. Hence by (40)

$$F \subset K \subset (1 + 2\varepsilon M) B_{X^*}. \tag{42}$$

Next, for any $f \in B_j$ there is $g \in 3B_{X^*} \cap Y_j^{\perp}$ such that $|||f - g|||_j \leq \frac{1}{j}$. Indeed, we put $g(x) = f\big(x - P_j(x)\big)$. Then clearly $g \in Y_j^{\perp}$ and

$$|||f - g|||_j = \sup_{|||x|||_j \leq 1} f\big(P_j(x)\big) \leq \sup_{|||P_j(x)|||_j \leq |||P_j|||_j} f\big(P_j(x)\big) = \sup_{\substack{|||P_j(x)|||_j \leq |||P_j|||_j \\ x \in Y_j}} f(x)$$

$$= \sup_{\substack{\nu_j(x) \leq \frac{1}{j} \\ x \in Y_j}} f(x) \leq \sup_{\nu_j(x) \leq \frac{1}{j}} f(x) = \frac{1}{j} \nu_j(f) \leq \frac{1}{j}.$$

Also, $\|g\| \leq \|f\| + \frac{1}{j} \leq 3$ by (40).

Now we show that K is a w^*-compact set. Since X is separable, $((1+2M)B_{X^*}, w^*)$ is metrisable. Let $\{h_m\}_{m \in \mathbb{N}} \subset K$ be such that $w^*\text{-}\lim h_m = h_0 \in (1 + 2M) B_{X^*}$. If one of the sets $g_j + \varepsilon M B_j$ contains infinitely many of h_ms, then of course $h_0 \in K$. Otherwise by passing to a subsequence we may assume that $h_m \in g_{j_m} + \varepsilon M B_{j_m}$, where $j_m \to \infty$, and $w^*\text{-}\lim g_{j_m} = g \in F \subset K$. Then by the above paragraph we may write $h_m = g_{j_m} + u_m + v_m$, where $u_m \in 3\varepsilon M B_{X^*} \cap Y_{j_m}^{\perp}$ and $\|v_m\| \leq \frac{1}{j_m}$. Since $j_m \to \infty$, we have $n_{j_m} \to \infty$ and therefore $w^*\text{-}\lim u_m = 0$. Thus $h_0 = g \in K$, which means that K is w^*-closed and consequently w^*-compact.

Finally, put $K_s = K \cup -K$ and define the new norm on X by

$$\nu(x) = \sup_{f \in K_s} f(x).$$

The property (i) from Lemma 102 gives $(1 - \varepsilon) B_{X^*} \subset \overline{\operatorname{conv}}\, K$, which combined with the relation (42) implies that $(1 - \varepsilon)\|x\| \leq \nu(x) \leq (1 + 2\varepsilon M)\|x\|$ for every $x \in X$ and so the norm ν approximates the norm $\|\cdot\|$ uniformly on bounded sets. Let $F_j^{\theta}(x) = \sup\{f(x);\ f \in \theta g_j + \varepsilon M B_j\}$ for $j \in \mathbb{N}$ and $\theta = \pm 1$. Clearly $F_j^{\theta}(x) = \theta g_j(x) + \varepsilon M \nu_j(x)$. Thus each F_j^{θ} is C^k-smooth away from the origin and the collection $\{F_j^{\theta};\ j \in \mathbb{N}, \theta = \pm 1\}$ is equi-continuous. Given $x \in X$, by the w^*-compactness of K_s there is $f \in K_s$ such that $f(x) = \max_{f \in K_s} f(x) = \nu(x)$. This means that $\nu(x) = \max_{j,\theta} F_j^{\theta}(x)$, i.e. $\{F_j^{\theta};\ j \in \mathbb{N}, \theta = \pm 1\}$ is a generalised James boundary of X. Now it suffices to apply Theorem 5.99. □

We remark that one of the difficulties in the proof was that we needed both (40) and (41) to hold. Since the projections P_j are not uniformly bounded, we could not use the semi-norms $x \mapsto |||x - P_j(x)|||_j$ in the definition of ν_j and we had to employ the quotient norms instead.

The following result can be proved by a similar argument.

Theorem 104 ([DFH2]). *Let X be a separable polyhedral Banach space. Then every equivalent norm on X can be approximated uniformly on bounded sets by C^∞-smooth and LFC (resp. real analytic) equivalent norms.*

10. Notes and remarks

In [ArSc] approximations in the spaces of holomorphic functions under various topologies are studied.

Section 2. Theorem 9 was proved by Karl Weierstraß in 1885 for functions in $C(\mathbb{R})$, by Émile Picard in 1891 for functions in $C(\mathbb{R}^n)$, and by Charles de la Vallée Poussin in 1908–12 for functions in $C^k(\mathbb{R}^n)$. Theorem 12 is due to Stanisław Mazur, Theorem 16 is due to Maurice Fréchet. Theorem 26 was proved by Arkadij Semenovich Nemirovskij.

The origin of some of the polynomial approximation results in infinite-dimensional spaces can be traced to Guillermo Restrepo's work [Res], and also Georgiy Evgenievich Shilov's paper [Shi] where the relevant problem of characterising the closure of the space of continuous polynomials was posed. For results regarding approximations on compact sets together with higher derivatives see e.g. [Pr], [AP], [ArSc].

In connection with Theorem 21 there is the next problem.

Problem 105 ([AP]). Does $\mathcal{P}_{\mathrm{w}}(^nX) = \overline{\mathcal{P}_{\mathrm{f}}(^nX)}$ for all $n \in \mathbb{N}$ imply that X^* has the approximation property?

Section 3. The real-analytic part of Theorem 28 was proved by Hassler Whitney in [Wh]. Torsten Carleman ([Carle]) proved the approximation of functions in $C(\mathbb{R})$ by entire functions and Stephen Scheinberg generalised it in [Schei] for functions in $C(\mathbb{R}^n)$.

Theorem 29 is a pioneering result of Jaroslav Kurzweil, whose influence on this area cannot be exaggerated. Various versions of Theorem 31 have been proved independently by Robb Fry [Fry1], Manuel Cepedello-Boiso, and Petr Hájek. The last two authors decided to publish a joint paper [CH].

The following is one of the main open problems in this area.

Problem 106. Is every continuous function on c_0 uniformly approximable by real analytic functions?

The method of proof of Theorem 31 is not strong enough to make this conclusion. If the answer is negative, then Theorem 29 would have a converse, via Deville's Theorem 5.67, namely the existence of analytic approximations for all continuous functions on a separable space X would imply that X has a separating polynomial. This fact was also noted, for super-reflexive spaces, in [Kur2].

Section 4. Lemma 39 is usually attributed to Edward James McShane [McS], but it was known at least two years earlier to H. Whitney [Wh]. For more information on the infimal convolution see e.g. the survey [Str]. Extending further the infimal convolution technique M. Cepedello-Boiso obtained the following result.

Theorem 107 ([Cep]). *Let X be a super-reflexive Banach space. Then there is $\alpha \in (0, 1]$ such that for any Lipschitz function $f: X \to \mathbb{R}$ and any $\varepsilon > 0$ there is a function $g \in C^1(X)$ with its derivative α-Hölder on bounded sets (and so g is Lipschitz on bounded sets) such that $|f - g|_X \leq \varepsilon$.*

Many versions of Proposition 44 exist in the literature, e.g. for compact K it is true for analytic functions on ℓ_p, [Do2]. This is related to the negligibility theory of subsets of Banach spaces, initiated by Czesław Bessaga in [Bes], and studied in detail e.g. in [Do1], [Do2], [Do3], [Az], [AzDo], [Dev3], [DeH1].

Section 5. Most of Lemma 49 was obtained in a seminal work by Henryk Toruńczyk, [Tor]. The main open problem in this area is the following one.

Problem 108 ([DGZ]). Suppose that a Banach space X admits a C^k-smooth bump function. Does X have C^k-smooth partitions of unity, or equivalently are continuous mappings on X approximable by C^k-smooth mappings?

As we have seen, the answer is positive for some classes of Banach spaces, including separable, WCG, and $C(K)$ spaces. Some earlier results on smooth partitions of unity on $C(K)$ spaces were obtained in [DGZ1]. In [JohnZiz] it is shown that WCG spaces always admit Gâteaux smooth partitions of unity. It is shown in [Fro] that if a Banach space X has an LUR norm and every Lipschitz convex function on X can be approximated by C^k-smooth functions, then X has C^k-smooth partitions of unity.

Recently, several papers focused on the problem of C^k-smooth approximations by functions that lack critical points, or more generally their derivative avoids a prescribed set of values, e.g. [AC], [HJ1], [AJ2], [Ji]. This interest was spurred by the result on the existence of bump functions without critical points we referred to earlier.

Section 6. The important technique of approximation of Lipschitz functions using supremal partitions of unity is due to R. Fry [Fry2]. It is closely related, in spirit and technique, to the method of boundaries or generalised boundaries used for obtaining C^k-smooth renormings. It is however a purely scalar method, and so the results concerning C^k-smooth Lipschitz approximations are less satisfactory than the continuous case.

The first part of Fact 62 was shown in [Pel, Proposition 2.3]. Many of the results in Sections 6–8 come from [HJ7].

A natural question is the following.

Problem 109. Let X be a Banach space (e.g. WCG) with a Lipschitz C^k-smooth bump function. Does X admit C^k-smooth Lipschitz sup-partitions of unity?

Section 7. The method of Lemma 69 originated from [FWZ]. Similar results are for example in [Joha], [FZ3]. Corollary 72 for $k = 1$ is proved in [FZ2] and its older norm variant (for $k = 1$) is proved in [FWZ]. Theorem 90 is proved in [AFK2] under slightly stronger assumptions.

The results given in this section are partial cases of the following general open problem.

Problem 110. Suppose that a Banach space X admits a C^k-smooth and Lipschitz bump function. Are Lipschitz mappings into another Banach space Y approximable by C^k-smooth and Lipschitz mappings?

This problem is open even for a pair of general separable X and Y, or for a general X and $Y = \mathbb{R}$.

Section 8. The implication (i)$\Rightarrow$(iii) in Theorem 95 in the separable case was proved by Nicole Moulis [Mo2]. The case of Corollary 96 when X has an unconditional Schauder basis was proved by N. Moulis [Mo2], although the result is stated only for c_0 and ℓ_p spaces. This seminal paper has essentially started the line of research into approximations together with derivatives. The motivation for N. Moulis's results apparently comes from the investigation of Banach manifolds, where they find important applications. The only result available for approximations together with higher derivatives comes again from [Mo2] and applies only to the Hilbert space. The statement of this result, unlike that of the first order, is nevertheless not very satisfactory, as it requires C^{2k-1}-smoothness for approximation of the kth order derivatives.

The most general form of the approximation problem can be stated as follows.

Problem 111. Let X be a (separable) Banach space admitting a C^k-smooth bump function. Is it true that every C^n-smooth mapping from X into a Banach space Y (or just into $\mathbb{R}$) can be approximated, together with all derivatives of order up to $n < k$, by C^k-smooth mappings?

Section 9. This section is based on the results from [DFH1], [DFH2], where the results were proved under the assumption of Schauder basis. This was removed in [HáTa].

Problem 112. Suppose that a separable Banach space X admits a convex homogeneous separating polynomial. Are analytic norms dense among all equivalent renormings?

The above problem is open even for $X = \ell_2$. The positive solution to this problem given in [DFH1] is unfortunately not correct. Of course, we do not know the following:

Problem 113. Suppose that a separable Banach space X has a real analytic norm. Are analytic norms dense among all equivalent renormings?

In full generality, the problem is the following:

Problem 114. Let X be a Banach space with a C^k-smooth norm, $k \in \mathbb{N} \cup \{\infty, \omega\}$. Is every equivalent norm on X approximable by C^k-smooth norms?

We have shown that the above problem has a positive solution for every separable X, but it is open for any non-separable Banach space. Some very partial results can be found in [FHZ] (the approximation of strongly lattice norms on $c_0(\Gamma)$ follows also from the proof of Theorem 74) and [HP]. The existence of C^k-smooth norms and bumps in the non-separable setting has been studied in seminal papers of Richard Haydon [Hay2], [Hay3].

Bibliography

[Ah] Israel Aharoni, *Every separable metric space is Lipschitz equivalent to a subset of c_0*, Israel J. Math. **19** (1974), no. 3, 284–291, DOI: 10.1007/BF02757727.

[Ai] Lev Aizenberg, *Multidimensional analogues of Bohr's theorem on power series*, Proc. Amer. Math. Soc. **128** (2000), no. 4, 1147–1155, DOI: 10.1090/S0002-9939-99-05084-4.

[AM] Lev A. Aizenberg and Boris Samuilovich Mityagin, *Prostranstva funktsiĭ, analiticheskikh v kratno-krugovykh oblastyakh* (Russian), Sibirsk. Mat. Zh. **1** (1960), no. 2, 153–170.

[AK] Fernando Albiac and Nigel J. Kalton, *Topics in Banach space theory*, Grad. Texts in Math. 233, Springer, New York, 2006.

[AlDia] Felix Albrecht and Harold G. Diamond, *A converse of Taylor's theorem*, Indiana Univ. Math. J. **21** (1971), no. 4, 347–350, DOI: 10.1512/iumj.1972.21.21025.

[AAD] Raymundo Alencar, Richard Martin Aron, and Seán Dineen, *A reflexive space of holomorphic functions in infinitely many variables*, Proc. Amer. Math. Soc. **90** (1984), no. 3, 407–411, DOI: 10.1090/S0002-9939-1984-0728358-5.

[AAF] Raymundo Alencar, Richard Martin Aron, and Gerd Fricke, *Tensor products of Tsirelson's space*, Illinois J. Math. **31** (1987), no. 1, 17–23.

[AO] Andrzej Alexiewicz and Władysław Orlicz, *Analytic operations in real Banach spaces*, Studia Math. **14** (1953), no. 1, 57–78.

[Al] Dale Edward Alspach, *A quotient of $C(\omega^\omega)$ which is not isomorphic to a subspace of $C(\alpha)$, $\alpha < \omega_1$*, Israel J. Math. **35** (1980), no. 1–2, 49–60, DOI: 10.1007/BF02760938.

[AlBen] Dale Edward Alspach and Yoav Benyamini, *Primariness of spaces of continuous functions on ordinals*, Israel J. Math. **27** (1977), no. 1, 64–92, DOI: 10.1007/BF02761606.

[AL] Dan Amir and Joram Lindenstrauss, *The structure of weakly compact sets in Banach spaces*, Ann. of Math. **88** (1968), no. 1, 35–46.

[AnFl] José María Martínez Ansemil and Klaus Floret, *The symmetric tensor product of a direct sum of locally convex spaces*, Studia Math. **129** (1998), no. 3, 285–295.

[Are] Richard Arens, *The adjoint of a bilinear operation*, Proc. Amer. Math. Soc. **2** (1951), no. 6, 839–848, DOI: 10.1090/S0002-9939-1951-0045941-1.

[ArgFe] Spiros A. Argyros and Vaggelis Felouzis, *Interpolating hereditarily indecomposable Banach spaces*, J. Amer. Math. Soc. **13** (2000), no. 2, 243–294, DOI: 10.1090/S0894-0347-00-00325-8.

[ArgRa] Spiros A. Argyros and Theocharis Raikoftsalis, *The cofinal property of the reflexive indecomposable Banach spaces*, Ann. Inst. Fourier (Grenoble) **62** (2012), no. 1, 1–45, DOI: 10.5802/aif.2697.

[AriFa] Alvaro Arias and Jeffrey Darrell Farmer, *On the structure of tensor products of ℓ_p-spaces*, Pacific J. Math. **175** (1996), no. 1, 13–37.

[Aro1] Richard Martin Aron, *Polynomial approximation and a question of G. E. Shilov*, Approximation theory and functional analysis. Proceedings of the International symposium on approximation theory, ed. by João Bosco Prolla, Mathematics studies 35, North-Holland, Amsterdam, 1979, pp. 1–12, DOI: 10.1016/S0304-0208(08)72462-9.

[Aro2] Richard Martin Aron, *Weakly uniformly continuous and weakly sequentially continuous entire functions*, Advances in holomorphy. Proceedings of the Seminario de holomorfia, ed. by Jorge Alberto Barroso, Mathematics studies 34, North-Holland, Amsterdam, 1979, pp. 47–66, DOI: 10.1016/S0304-0208(08)70753-9.

[Aro3] Richard Martin Aron, *An introduction to polynomials on Banach spaces*, Extracta Math. **17** (2002), no. 3, 303–329.

[ABE] Richard Martin Aron, Bernard Beauzamy, and Per Enflo, *Polynomials in many variables: real vs complex norms*, J. Approx. Theory **74** (1993), no. 2, 181–198, DOI: 10.1006/jath.1993.1060.

[ArBer] Richard Martin Aron and Paul D. Berner, *A Hahn-Banach extension theorem for analytic mappings*, Bull. Soc. Math. France **106** (1978), 3–24.

[ACL] Richard Martin Aron, Yun Sung Choi, and José Luís González Llavona, *Estimates by polynomials*, Bull. Aust. Math. Soc. **52** (1995), no. 3, 475–486, DOI: 10.1017/S0004972700014957.

[ArDim] Richard Martin Aron and Verónica Dimant, *Sets of weak sequential continuity for polynomials*, Indag. Math. (N. S.) **13** (2002), no. 3, 287–299, DOI: 10.1016/S0019-3577(02)80012-X.

[AG] Richard Martin Aron and Josip Globevnik, *Analytic functions on c_0*, Rev. Mat. Complut. **2** (1989), no. extra, 27–33.

[AGZ] Richard Martin Aron, Raquel Gonzalo, and Andriy Zagorodnyuk, *Zeros of real polynomials*, Linear Multilinear Algebra **48** (2000), no. 2, 107–115, DOI: 10.1080/03081080008818662.

[ArHá1] Richard Martin Aron and Petr Hájek, *Zero sets of polynomials in several variables*, Arch. Math. **86** (2006), no. 6, 561–568, DOI: 10.1007/s00013-006-1314-9.

[ArHá2] Richard Martin Aron and Petr Hájek, *Odd degree polynomials on real Banach spaces*, Positivity **11** (2007), no. 1, 143–153, DOI: 10.1007/s11117-006-2035-9.

[AHV] Richard Martin Aron, Carlos Hervés-Beloso, and Manuel Valdivia Ureña, *Weakly continuous mappings on Banach spaces*, J. Funct. Anal. **52** (1983), no. 2, 189–204, DOI: 10.1016/0022-1236(83)90081-2.

[ALRT] Richard Martin Aron, Miguel Lacruz, Raymond A. Ryan, and Andrew Morton Tonge, *The generalized Rademacher functions*, Note Mat. **12** (1992), 15–25, DOI: 10.1285/i15900932v12p15.

[AP] Richard Martin Aron and João Bosco Prolla, *Polynomial approximation of differentiable functions on Banach spaces*, J. Reine Angew. Math. **313** (1980), 195–216, DOI: 10.1515/crll.1980.313.195.

[AroRu] Richard Martin Aron and María Pilar Rueda Segado, *A problem concerning zero subspaces of polynomials*, Linear topological spaces and complex analysis **3** (1997), 20–23.

[ArSc] Richard Martin Aron and Martin Schottenloher, *Compact holomorphic mappings on Banach spaces and the approximation property*, J. Funct. Anal. **21** (1976), no. 1, 7–30, DOI: 10.1016/0022-1236(76)90026-4.

[AvSm1] Vladimir Iosifovich Averbuch and Oleg Georgievich Smolyanov, *Theory of differentiation in linear topological spaces*, Russian Math. Surveys **22** (1967), no. 6, 201–258, DOI: 10.1070/RM1967v022n06ABEH003761.

[AvSm2] Vladimir Iosifovich Averbuch and Oleg Georgievich Smolyanov, *Various definitions of the derivative in linear topological spaces*, Russian Math. Surveys **23** (1968), no. 4, 67–113, DOI: 10.1070/RM1968v023n04ABEH003770.

[AT] Antonio Avilés and Stevo Todorčević, *Zero subspaces of polynomials on $\ell_1(\Gamma)$*, J. Math. Anal. Appl. **350** (2009), no. 2, 427–435, DOI: 10.1016/j.jmaa.2007.08.020.

[Az] Daniel Azagra, *Diffeomorphisms between spheres and hyperplanes in infinite-dimensional Banach spaces*, Studia Math. **125** (1997), no. 2, 179–186.

[AC] Daniel Azagra and Manuel Cepedello-Boiso, *Uniform approximation of continuous mappings by smooth mappings with no critical points on Hilbert manifolds*, Duke Math. J. **124** (2004), no. 1, 47–66, DOI: 10.1215/S0012-7094-04-12412-1.

[AzDe] Daniel Azagra and Robert Deville, *James' theorem fails for starlike bodies*, J. Funct. Anal. **180** (2001), no. 2, 328–346, DOI: 10.1006/jfan.2000.3696.

[ADJ] Daniel Azagra, Robert Deville, and Mar Jiménez-Sevilla, *On the range of the derivatives of a smooth function between Banach spaces*, Math. Proc. Cambridge Philos. Soc. **134** (2003), no. 1, 163–185, DOI: 10.1017/S0305004102006278.

[AzDo] Daniel Azagra and Tadeusz Dobrowolski, *Smooth negligibility of compact sets in infinite-dimensional Banach spaces, with applications*, Math. Ann. **312** (1998), no. 3, 445–463, DOI: 10.1007/s002080050231.

[AFJ] Daniel Azagra, Marián Fabian, and Mar Jiménez-Sevilla, *Exact filling of figures with the derivatives of smooth mappings between Banach spaces*, Canad. Math. Bull **48** (2005), no. 4, 481–499, DOI: 10.4153/CMB-2005-045-9.

[AFK1] Daniel Azagra, Robb Fry, and Lee Keener, *Erratum to: Smooth extensions of functions on separable Banach spaces*, Math. Ann. **350** (2011), no. 2, 497–500, DOI: 10.1007/s00208-010-0576-5.

[AFK2] Daniel Azagra, Robb Fry, and Lee Keener, *Real analytic approximation of Lipschitz functions on Hilbert space and other Banach spaces*, J. Funct. Anal. **262** (2012), no. 1, 124–166, DOI: 10.1016/j.jfa.2011.09.009.

[AJ1] Daniel Azagra and Mar Jiménez-Sevilla, *On the size of the sets of gradients of bump functions and starlike bodies on the Hilbert space*, Bull. Soc. Math. France **130** (2002), no. 3, 337–347.

[AJ2] Daniel Azagra and Mar Jiménez-Sevilla, *Approximation by smooth functions with no critical points on separable Banach spaces*, J. Funct. Anal. **242** (2007), no. 1, 1–36, DOI: 10.1016/j.jfa.2006.08.009.

[AzHa] Parviz Azimi and James N. Hagler, *Examples of hereditarily ℓ_1 Banach spaces failing the Schur property*, Pacific J. Math. **122** (1986), no. 2, 287–197.

[Bae] Albert Baernstein, *On reflexivity and summability*, Studia Math. **42** (1972), no. 1, 91–94.

[Bana] Stefan Banach, *Über homogene Polynome in* (L^2) (German), Studia Math. **7** (1938), no. 1, 36–44.

[BaSa] Stefan Banach and Stanisław Saks, *Sur la convergence forte dans les champs L_p* (French), Studia Math. **2** (1930), no. 1, 51–57.

[Bann] Eiichi Bannai, *Spherical designs and group representations*, Combinatorics and algebra. Proceedings of the AMS-NSF joint summer research conference on combinatorics and algebra, ed. by Curtis Greene, Contemp. Math. 34, American Mathematical Society, Providence, RI, 1984, pp. 95–107, DOI: 10.1090/conm/034/777697.

[BGV] Michael O. Bartlett, John R. Giles, and Jon Vanderwerff, *Directional moduli of rotundity and smoothness*, Comment. Math. Univ. Carolin. **40** (1999), no. 1, 39–51.

[Bat] Sean Michael Bates, *On smooth, nonlinear surjections of Banach spaces*, Israel J. Math. **100** (1997), no. 1, 209–220, DOI: 10.1007/BF02773641.

[Bea] Bernard Beauzamy, *Banach-Saks properties and spreading models*, Math. Scand. **44** (1979), 357–384.

[BeaLa] Bernard Beauzamy and Jean-Thierry Lapresté, *Modèles étalés des espaces de Banach* (French), Travaux en Cours 4, Hermann, Paris, 1984.

[Ben] Grahame Bennett, *Inclusion mappings between ℓ_p spaces*, J. Funct. Anal. **13** (1973), no. 1, 20–27, DOI: 10.1016/0022-1236(73)90064-5.

[BenLi] Yoav Benyamini and Joram Lindenstrauss, *Geometric nonlinear functional analysis*, Amer. Math. Soc. Colloq. Publ. 48, American Mathematical Society, Providence, RI, 2000.

[BerLö] Jöran Bergh and Jörgen Löfström, *Interpolation spaces. An introduction*, Grundlehren Math. Wiss. 223, Springer-Verlag, Berlin, Heidelberg, 1976.

[Bes] Czesław Bessaga, *Every infinite-dimensional Hilbert space is diffeomorphic with its unit sphere*, Bull. Acad. Polon. Sci., Sér. Sci. Math. Astronom. Phys. **14** (1966), no. 1, 27–31.

[Bie] Otto Biermann, *Über näherungsweise Cubaturen* (German), Monatsh. Math. Phys. **14** (1903), no. 1, 211–225, DOI: 10.1007/BF01706869.

[Bir] Bryan John Birch, *Homogeneous forms of odd degree in a large number of variables*, Mathematika **4** (1957), no. 2, 102–105, DOI: 10.1112/S0025579300001145.

[Bl] Fernando Blasco, *Complementation in spaces of symmetric tensor products and polynomials*, Studia Math. **123** (1997), no. 2, 165–173.

[Boa] Harold Philip Boas, *Majorant series*, J. Korean Math. Soc. **37** (2000), no. 2, 321–337.

[BoKh] Harold Philip Boas and Dmitry Khavinson, *Bohr's power series theorem in several variables*, Proc. Amer. Math. Soc. **125** (1997), no. 10, 2975–2979, DOI: 10.1090/S0002-9939-97-04270-6.

[Boc] Jacek Bochnak, *Analytic functions in Banach spaces*, Studia Math. **35** (1970), no. 3, 273–292.

[BoSi] Jacek Bochnak and Józef Siciak, *Polynomials and multilinear mappings in topological vector spaces*, Studia Math. **39** (1971), no. 1, 59–76.

[Bog] Witold M. Bogdanowicz, *Slabaya nepreryvnost' polinomnykh funktsionalov na prostranstve* c_0 (Russian), Bull. Acad. Polon. Sci., Cl. III **5** (1957), no. 3, 243–246.

[BH] Henri Frédéric Bohnenblust and Einar Hille, *On the absolute convergence of Dirichlet series*, Ann. of Math. **32** (1931), no. 3, 600–622.

[Boj] Borislav Bojanov, *Optimal cubature formulas*, Russian Math. Surveys **60** (2005), no. 6, 1035–1055, DOI: 10.1070/RM2005v060n06ABEH004280.

[BojPe] Borislav Bojanov and Guergana Petrova, *On minimal cubature formulae for product weight functions*, J. Comput. Appl. Math. **85** (1997), no. 1, 113–121, DOI: 10.1016/S0377-0427(97)00133-7.

[BPV] Fernando Bombal, David Pérez-García, and Ignacio Villanueva, *Multilinear extensions of Grothendieck's theorem*, Q. J. Math. **55** (2004), no. 4, 441–450, DOI: 10.1093/qmath/hah017.

[BomVi] Fernando Bombal and Ignacio Villanueva, *On the Dunford-Pettis property of the tensor product of $C(K)$ spaces*, Proc. Amer. Math. Soc. **129** (2001), no. 5, 1359–1363, DOI: 10.1090/S0002-9939-00-05662-8.

[BF] Robert Bonic and John Frampton, *Smooth functions on Banach manifolds*, J. Mathematics and Mechanics **15** (1966), no. 5, 877–898.

[BFKL] Jonathan M. Borwein, Marián Fabian, Ivaylo S. Kortezov, and Philip Daniel Loewen, *The range of the gradient of a continuously differentiable bump*, J. Nonlinear Convex Anal. **2** (2001), no. 1, 1–19.

[BFL] Jonathan M. Borwein, Marián Fabian, and Philip Daniel Loewen, *The range of the gradient of a Lipschitz C^1-smooth bump in infinite dimensions*, Israel J. Math. **132** (2002), no. 1, 239–251, DOI: 10.1007/BF02784514.

[BN] Jonathan M. Borwein and Dominikus Noll, *Second order differentiability of convex functions in Banach spaces*, Trans. Amer. Math. Soc. **342** (1994), no. 1, 43–81, DOI: 10.1090/S0002-9947-1994-1145959-4.

[BorPr] Jonathan M. Borwein and David Preiss, *A smooth variational principle with applications to subdifferentiability and to differentiability of convex functions*, Trans. Amer. Math. Soc. **303** (1987), no. 2, 517–527, DOI: 10.1090/S0002-9947-1987-0902782-7.

[BorVa] Jonathan M. Borwein and Jon Vanderwerff, *Convex functions. Constructions, characterizations and counterexamples*, Encyclopedia Math. Appl. 109, Cambridge University Press, Cambridge, 2010.

[BE] Peter Borwein and Tamás Erdélyi, *Polynomials and polynomial inequalities*, Grad. Texts in Math. 161, Springer-Verlag, New York, 1995.

[BDVX] Len Bos, Stefano De Marchi, Marco Vianello, and Yuan Xu, *Bivariate Lagrange interpolation at the Padua points: the ideal theory approach*, Numer. Math. **108** (2007), no. 1, 43–57, DOI: 10.1007/s00211-007-0112-z.

[BourgJ1] Jean Bourgain, *The Szlenk index and operators on $C(K)$-spaces*, Bull. Soc. Math. Belg., Sér. B **31** (1979), 87–117.

[BourgJ2] Jean Bourgain, *New classes of $\mathcal{L}_p$ spaces*, Lecture Notes in Math. 889, Springer-Verlag, Berlin, Heidelberg, 1981.

[BD] Jean Bourgain and Freddy Delbaen, *A class of special $\mathcal{L}_\infty$ spaces*, Acta Math. **145** (1980), no. 1, 155–176, DOI: 10.1007/BF02414188.

[BFT] Jean Bourgain, David H. Fremlin, and Michel Talagrand, *Pointwise compact sets of Baire-measurable functions*, Amer. J. Math. **100** (1978), no. 4, 845–886, DOI: 10.2307/2373913.

[BourgR] Richard David Bourgin, *Geometric aspects of convex sets with the Radon-Nikodým property*, Lecture Notes in Math. 993, Springer-Verlag, Berlin, Heidelberg, 1983.

[BR] Christopher Boyd and Raymond A. Ryan, *Bounded weak continuity of homogeneous polynomials at the origin*, Arch. Math. **71** (1998), no. 3, 211–218, DOI: 10.1007/s000130050254.

[BrSu1] Antoine Brunel and Louis Sucheston, *On B-convex Banach spaces*, Math. Systems Theory **7** (1973), no. 4, 294–299, DOI: 10.1007/BF01795947.

[BrSu2] Antoine Brunel and Louis Sucheston, *On J-convexity and some ergodic super-properties of Banach spaces*, Trans. Amer. Math. Soc. **204** (1975), 79–90, DOI: 10.1090/S0002-9947-1975-0380361-2.

[BuKu] Dan Burghelea and Nicolaas Hendrik Kuiper, *Hilbert manifolds*, Ann. of Math. **90** (1969), no. 3, 379–417.

[Carl] Bernd Carl, *Absolut-$(p, 1)$-summierende identische Operatoren von ℓ_u in ℓ_v* (German), Math. Nachr. **63** (1974), no. 1, 353–360, DOI: 10.1002/mana.3210630132.

[Carle] Torsten Carleman, *Sur un théorème de Weierstrass* (French), Ark. Mat. Astr. Fys. **20B** (1927), no. 4, 1–5.

[CCG] Thomas Keith Carne, Brian James Cole, and Theodore William Gamelin, *A uniform algebra of analytic functions on a Banach space*, Trans. Amer. Math. Soc. **314** (1989), no. 2, 639–659, DOI: 10.1090/S0002-9947-1989-0986022-0.

[Cas] Peter George Casazza, *Approximation properties*, Handbook of the geometry of Banach spaces, ed. by William Buhmann Johnson and Joram Lindenstrauss, vol. 1, North-Holland, Amsterdam, 2001, chap. 7, pp. 271–316, DOI: 10.1016/S1874-5849(01)80009-7.

[CKT] Peter George Casazza, Nigel J. Kalton, and Lior Tzafriri, *Decompositions of Banach lattices into direct sums*, Trans. Amer. Math. Soc. **304** (1987), no. 2, 771–800, DOI: 10.1090/S0002-9947-1987-0911095-9.

[CasaSh] Peter George Casazza and Thaddeus J. Shura, *Tsirelson's space*, Lecture Notes in Math. 1363, Springer-Verlag, Berlin, Heidelberg, 1989.

[CGG] Jesús M. F. Castillo, Ricardo García, and Raquel Gonzalo, *Banach spaces in which all multilinear forms are weakly sequentially continuous*, Studia Math. **136** (1999), no. 2, 121–145.

[CaGo] Jesús M. F. Castillo and Manuel González, *An approach to Schreier's space*, Extracta Math. **6** (1991), no. 3, 166–169.

[CastSá1] Jesús M. F. Castillo and Fernando Sánchez, *Upper ℓ_p-estimates in vector sequence spaces, with some applications*, Math. Proc. Cambridge Philos. Soc. **113** (1993), no. 2, 329–334, DOI: 10.1017/S030500410007599X.

[CastSá2] Jesús M. F. Castillo and Fernando Sánchez, *Weakly p-compact, p-Banach-Saks, and super-reflexive Banach spaces*, J. Math. Anal. Appl. **185** (1994), no. 2, 256–261, DOI: 10.1006/jmaa.1994.1246.

[Cem] Pilar Cembranos, *The hereditary Dunford-Pettis property on $C(K, E)$*, Illinois J. Math. **31** (1987), no. 3, 365–373.

[Cep] Manuel Cepedello-Boiso, *On regularization in superreflexive Banach spaces by infimal convolution formulas*, Studia Math. **129** (1998), no. 3, 265–284.

[CH] Manuel Cepedello-Boiso and Petr Hájek, *Analytic approximations of uniformly continuous functions in real Banach spaces*, J. Math. Anal. Appl. **256** (2001), no. 1, 80–98, DOI: 10.1006/jmaa.2000.7291.

[Cha] Vladimir L. Chakalov, *Formules de cubatures mécaniques a coefficients non négatifs* (French), Bull. Sci. Math. **81** (1957), 123–134.

[CHL] Yun Sung Choi, Petr Hájek, and Han Ju Lee, *Extensions of smooth mappings into biduals and weak continuity*, Adv. Math. **234** (2013), 453–487, DOI: 10.1016/j.aim.2012.11.001.

[Cho] Gustave Choquet, *Lectures on analysis*, 3 vols., Mathematics lecture note, W. A. Benjamin, Reading, MA, 1969.

[CiGu1] Raffaella Cilia and Joaquín M. Gutiérrez, *Factorization of weakly continuous differentiable mappings*, Bull. Braz. Math. Soc. (N. S.) **40** (2009), no. 3, 371–380, DOI: 10.1007/s00574-009-0016-x.

[CiGu2] Raffaella Cilia and Joaquín M. Gutiérrez, *Weakly sequentially continuous differentiable mappings*, J. Math. Anal. Appl. **360** (2009), no. 2, 609–623, DOI: 10.1016/j.jmaa.2009.07.002.

[CGS] Raffaella Cilia, Joaquín M. Gutiérrez, and Giuseppe Saluzzo, *Compact factorization of differentiable mappings*, Proc. Amer. Math. Soc. **137** (2009), no. 5, 1743–1752, DOI: 10.1090/S0002-9939-08-09716-5.

[Col] James B. Collier, *The dual of a space with the Radon-Nikodým property*, Pacific J. Math. **64** (1976), no. 1, 103–106.

[Con] John B. Conway, *A course in functional analysis*, Grad. Texts in Math. 96, Springer, New York, 1985.

[CF] Raúl Enrico Curto and Lawrence Arthur Fialkow, *A duality proof of Tchakaloff's theorem*, J. Math. Anal. Appl. **269** (2002), no. 2, 519–532, DOI: 10.1016/S0022-247X(02)00034-3.

[DaKr] Didier Dacunha-Castelle and Jean-Louis Krivine, *Applications des ultraproduits à l'étude des espaces et des algèbres de Banach* (French), Studia Math. **41** (1972), no. 3, 315–334.

[D'H] Stefania D'Alessandro and Petr Hájek, *Polynomial algebras and smooth functions in Banach spaces*, J. Funct. Anal. **266** (2014), no. 3, 1627–1646, DOI: 10.1016/j.jfa.2013.11.017.

[DHJ] Stefania D'Alessandro, Petr Hájek, and Michal Johanis, *Corrigendum to the paper "Polynomial algebras on classical Banach spaces", Israel J. Math. 106 (1998), 209–220*, Israel J. Math. (to appear).

[Da] Alexander Munro Davie, *Quotient algebras of uniform algebras*, J. Lond. Math. Soc. **7** (1973), no. 1, 31–40, DOI: 10.1112/jlms/s2-7.1.31.

[DefFl] Andreas Defant and Klaus Floret, *Tensor norms and operator ideals*, Mathematics studies 176, North-Holland, Amsterdam, 1993.

[DefFr] Andreas Defant and Leonhard Frerick, *A logarithmic lower bound for multi-dimensional Bohr radii*, Israel J. Math. **152** (2006), no. 1, 17–28, DOI: 10.1007/BF02771973.

[DFOOS] Andreas Defant, Leonhard Frerick, Joaquim Ortega-Cerdà, Myriam Ounaïes, and Kristian Seip, *The Bohnenblust-Hille inequality for homogeneous polynomials is hypercontractive*, Ann. of Math. **174** (2011), no. 1, 485–497, DOI: 10.4007/annals.2011.174.1.13.

[DGM] Andreas Defant, Domingo García, and Manuel Maestre, *Bohr's power series theorem and local Banach space theory*, J. Reine Angew. Math. **557** (2003), 173–197, DOI: 10.1515/crll.2003.030.

[DeKal] Andreas Defant and Nigel J. Kalton, *Unconditionality in spaces of m-homogeneous polynomials*, Q. J. Math. **56** (2005), no. 1, 53–64, DOI: 10.1093/qmath/hah022.

[DMP] Andreas Defant, Manuel Maestre, and Christopher Prengel, *The arithmetic Bohr radius*, Q. J. Math. **59** (2008), no. 2, 189–205, DOI: 10.1093/qmath/ham028.

[DPS] Andreas Defant, Dumitru Popa, and Ursula Schwarting, *Coordinatewise multiple summing operators in Banach spaces*, J. Funct. Anal. **259** (2010), no. 1, 220–242, DOI: 10.1016/j.jfa.2010.01.008.

[DePr] Andreas Defant and Christopher Prengel, *Harald Bohr meets Stefan Banach*, Methods in Banach space theory. Proceedings of the V conference on Banach spaces, Cáceres, 2004, ed. by Jesús M. F. Castillo and William Buhmann Johnson, London Math. Soc. Lecture Note Ser. 337, Cambridge University Press, Cambridge, 2006, pp. 317–339.

[DeSe] Andreas Defant and Pablo Sevilla-Peris, *A new multilinear insight on Littlewood's* 4/3-*inequality*, J. Funct. Anal. **256** (2009), no. 5, 1642–1664, DOI: 10.1016/j.jfa.2008.07.005.

[Dei] Klaus Deimling, *Ordinary differential equations in Banach spaces*, Lecture Notes in Math. 596, Springer-Verlag, Berlin, Heidelberg, 1977.

[DJP] Freddy Delbaen, Hans Jarchow, and Aleksander Pełczyński, *Subspaces of L_p isometric to subspaces of ℓ_p*, Positivity **2** (1998), no. 4, 339–367, DOI: 10.1023/A:1009764511096.

[Dev1] Robert Deville, *Geometrical implications of the existence of very smooth bump functions in Banach spaces*, Israel J. Math. **67** (1989), no. 1, 1–22, DOI: 10.1007/BF02764895.

[Dev2] Robert Deville, *A characterization of C^∞-smooth Banach spaces*, Bull. Lond. Math. Soc. **22** (1990), no. 1, 13–17, DOI: 10.1112/blms/22.1.13.

[Dev3] Robert Deville, *On the range of the derivative of a smooth function and applications*, Rev. R. Acad. Cienc. Exactas Fís. Nat., Ser. A Mat., RACSAM **100** (2006), no. 1–2, 63–74.

[DevFa] Robert Deville and Marián Fabian, *Principes variationnels et différentiabilité d'applications définies sur un espace de Banach* (French), Publ. Math. Fac. Sci. Besançon Anal. Non Linéaire **10** (1989), 79–102.

[DFH1] Robert Deville, Vladimir P. Fonf, and Petr Hájek, *Analytic and C^k approximations of norms in separable Banach spaces*, Studia Math. **120** (1996), no. 1, 61–74.

[DFH2] Robert Deville, Vladimir P. Fonf, and Petr Hájek, *Analytic and polyhedral approximation of convex bodies in separable polyhedral Banach spaces*, Israel J. Math. **105** (1998), no. 1, 139–154, DOI: 10.1007/BF02780326.

[DeGh] Robert Deville and Nassif Ghoussoub, *Perturbed minimization principles and applications*, Handbook of the geometry of Banach spaces, ed. by William Buhmann Johnson and Joram Lindenstrauss, vol. 1, North-Holland, Amsterdam, 2001, chap. 10, pp. 393–435, DOI: 10.1016/S1874-5849(01)80012-7.

[DGZ1] Robert Deville, Gilles Godefroy, and Václav Zizler, *The three space problem for smooth partitions of unity and $C(K)$ spaces*, Math. Ann. **288** (1990), no. 1, 613–625, DOI: 10.1007/BF01444554.

[DGZ2] Robert Deville, Gilles Godefroy, and Václav Zizler, *Smooth bump functions and geometry of Banach spaces*, Mathematika **40** (1993), no. 2, 305–321, DOI: 10.1112/S0025579300007075.

[DGZ3] Robert Deville, Gilles Godefroy, and Václav Zizler, *A smooth variational principle with applications to Hamilton-Jacobi equations in infinite dimensions*, J. Funct. Anal. **111** (1993), no. 1, 197–212, DOI: 10.1006/jfan.1993.1009.

[DGZ] Robert Deville, Gilles Godefroy, and Václav Zizler, *Smoothness and renormings in Banach spaces*, Pitman Monographs and Surveys in Pure and Applied Mathematics 64, Longman Scientific & Technical, Harlow, 1993.

[DGJ] Robert Deville, Raquel Gonzalo, and Jesús Angel Jaramillo, *Renormings of $L_p(L_q)$*, Math. Proc. Cambridge Philos. Soc. **126** (1999), no. 1, 155–169, DOI: 10.1017/S0305004198003119.

[DeH1] Robert Deville and Petr Hájek, *On the range of the derivative of Gâteaux-smooth functions on separable Banach spaces*, Israel J. Math. **145** (2005), no. 1, 257–269, DOI: 10.1007/BF02786693.

[DeH2] Robert Deville and Petr Hájek, *Smooth noncompact operators from $C(K)$, K scattered*, Israel J. Math. **162** (2007), no. 1, 29–56, DOI: 10.1007/s11856-007-0086-7.

[DM] Robert Deville and Étienne Matheron, *Pyramidal vectors and smooth functions on Banach spaces*, Proc. Amer. Math. Soc. **128** (2000), no. 12, 3601–3608, DOI: 10.1090/S0002-9939-00-05519-2.

[DeZi] Robert Deville and Václav Zizler, *A note on cotype of smooth spaces*, Manuscripta Math. **62** (1988), no. 3, 375–382, DOI: 10.1007/BF01246841.

[DíD] Juan Carlos Díaz and Seán Dineen, *Polynomials on stable spaces*, Ark. Mat. **36** (1998), no. 1, 87–96, DOI: 10.1007/BF02385668.

[Dies1] Joseph Diestel, *A survey of results related to the Dunford-Pettis property*, Proceedings of the Conference on integration, topology, and geometry in linear spaces, ed. by William Howard Graves, Contemp. Math. 2, American Mathematical Society, Providence, RI, 1980, pp. 15–60, DOI: 10.1090/conm/002/621850.

[Dies2] Joseph Diestel, *Sequences and series in Banach spaces*, Grad. Texts in Math. 92, Springer-Verlag, New York, 1984.

[DFS] Joseph Diestel, Jan Hendrik Fourie, and Johan Swart, *The metric theory of tensor products. Grothendieck's Résumé revisited*, American Mathematical Society, Providence, RI, 2008.

[DJT] Joseph Diestel, Hans Jarchow, and Andrew Morton Tonge, *Absolutely summing operators*, Cambridge Stud. Adv. Math. 43, Cambridge University Press, Cambridge, 1995.

[DU] Joseph Diestel and John Jerry Uhl Jr, *Vector measures*, Mathematical surveys 15, American Mathematical Society, Providence, RI, 1977.

[Dieu] Jean Dieudonné, *Foundations of modern analysis*, Pure and Applied Mathematics 10-I, Academic Press, New York and London, 1969.

[DiD] Verónica Dimant and Seán Dineen, *Banach subspaces of spaces of holomorphic functions and related topics*, Math. Scand. **83** (1998), no. 1, 142–160.

[DiGo] Verónica Dimant and Raquel Gonzalo, *Block diagonal polynomials*, Trans. Amer. Math. Soc. **353** (2001), no. 2, 733–747, DOI: 10.1090/S0002-9947-00-02735-5.

[DiZa] Verónica Dimant and Ignacio Zalduendo, *Bases in spaces of multilinear forms over Banach spaces*, J. Math. Anal. Appl. **200** (1996), no. 3, 548–566, DOI: 10.1006/jmaa.1996.0224.

[Din] Seán Dineen, *Complex analysis on infinite dimensional spaces*, Springer monogr. math. Springer-Verlag, London, 1999.

[DL] Seán Dineen and Mikael Lindström, *Spaces of homogeneous polynomials containing c_0 or ℓ_∞*, Functional analysis. Proceedings of the First international workshop held at Trier university, 1994, ed. by Susanne Dierolf, Seán Dineen, and Paweł Domański, Walter de Gruyter, Berlin, 1996, pp. 119–127.

[DT1] Seán Dineen and Richard M. Timoney, *Absolute bases, tensor products and a theorem of Bohr*, Studia Math. **94** (1989), no. 3, 227–234.

[DT2] Seán Dineen and Richard M. Timoney, *Complex geodesics on convex domains*, Progress in functional analysis. Proceedings of the International functional analysis meeting, Peñíscola, 1990, ed. by Klaus D. Bierstedt, José Bonet, John Horváth, and Manuel Maestre, Mathematics studies 170, North-Holland, Amsterdam, 1992, pp. 333–365, DOI: 10.1016/S0304-0208(08)70330-X.

[Do1] Tadeusz Dobrowolski, *Extension of Bessaga's negligibility technique to certain infinite-dimensional groups*, Bull. Acad. Polon. Sci., Sér. Sci. Math. Astronom. Phys. **26** (1978), no. 6, 535–545.

[Do2] Tadeusz Dobrowolski, *Smooth and R-analytic neglibility of subsets and extension of homeomorphisms in Banach spaces*, Studia Math. **65** (1979), no. 2, 115–139.

[Do3] Tadeusz Dobrowolski, *Every infinite-dimensional Hilbert space is real-analytically isomorphic with its unit sphere*, J. Funct. Anal. **134** (1995), no. 2, 350–362, DOI: 10.1006/jfan.1995.1149.

[DoKar] Vladimir Leonidovich Dol'nikov and Roman Nikolaevich Karasev, *Dvoretzky type theorems for multivariate polynomials and sections of convex bodies*, Geom. Funct. Anal. **21** (2011), no. 2, 301–318, DOI: 10.1007/s00039-011-0109-8.

[DMNZ] Roberto Dragoni, Jack W. Macki, Paolo Nistri, and Pietro Zecca, *Solution sets of differential equations in abstract spaces*, Pitman Research Notes in Mathematics Series 342, Longman, Harlow, 1996.

[DuSc] Richard James Duffin and Albert Charles Schaeffer, *A refinement of an inequality of the brothers Markoff*, Trans. Amer. Math. Soc. **50** (1941), no. 3, 517–528, DOI: 10.1090/S0002-9947-1941-0005942-4.

[Dun] Nelson Dunford, *Uniformity in linear spaces*, Trans. Amer. Math. Soc. **44** (1938), no. 2, 305–356, DOI: 10.1090/S0002-9947-1938-1501971-X.

[DuPe] Nelson Dunford and Billy James Pettis, *Linear operations on summable functions*, Trans. Amer. Math. Soc. **47** (1940), no. 3, 323–392, DOI: 10.1090/S0002-9947-1940-0002020-4.

[Dur] Alain Durand, *Quelques aspects de la théorie analytique des polynômes I, II* (French), Cinquante ans de polynômes – Fifty years of polynomials, ed. by Michel Langevin and Michel Waldschmidt, Lecture Notes in Math. 1415, Springer-Verlag, Berlin, Heidelberg, 1990, pp. 1–85, DOI: 10.1007/BFb0084878.

[Dv] Aryeh Dvoretzky, *Some results on convex bodies and Banach spaces*, Proceedings of the International symposium on linear spaces, Jerusalem, 1960, Jerusalem academic press, Jerusalem, 1961, pp. 123–160.

[EW] Gerald Arthur Edgar and Robert Francis Wheeler, *Topological properties of Banach spaces*, Pacific J. Math. **115** (1984), no. 2, 317–350.

[EE] James Eells and Kenneth David Elworthy, *Open embeddings of certain Banach manifolds*, Ann. of Math. **91** (1970), no. 3, 465–485.

[El1] William John Ellison, *A 'Waring's problem' for homogeneous forms*, Proc. Cambridge Philos. Soc. **65** (1969), no. 3, 663–672, DOI: 10.1017/S0305004100003455.

[El2] William John Ellison, *Waring's problem*, Amer. Math. Monthly **78** (1971), no. 1, 10–36.

[EJ] Giovanni Emmanuele and Kamil John, *The space of compact operators contains c_0 when a noncompact operator is suitably factorized*, Czechoslovak Math. J. **50** (2000), no. 1, 75–82, DOI: 10.1023/A:1022489103715.

[Enf] Per Enflo, *Banach spaces which can be given an equivalent uniformly convex norm*, Israel J. Math. **13** (1972), no. 3–4, 281–288, DOI: 10.1007/BF02762802.

[Eng] Ryszard Engelking, *General topology*, Sigma ser. pure math. 6, Heldermann Verlag, Berlin, 1989.

[Er] Pál Erdős, *Some remarks on polynomials*, Bull. Amer. Math. Soc. **53** (1947), no. 12, 1169–1176, DOI: 10.1090/S0002-9904-1947-08938-2.

[Fab1] Marián Fabian, *On minimum principles*, Acta Polytechnica **20** (1983), 109–117.

[Fab2] Marián Fabian, *Lipschitz smooth points of convex functions and isomorphic characterizations of Hilbert spaces*, Proc. Lond. Math. Soc. **51** (1985), no. 1, 113–126, DOI: 10.1112/plms/s3-51.1.113.

[Fab3] Marián Fabian, *Gâteaux differentiability of convex functions and topology. Weak Asplund spaces*, CMS series of monographs and advanced texts, John Wiley & Sons, New York, 1997.

[FGZ] Marián Fabian, Gilles Godefroy, and Václav Zizler, *The structure of uniformly Gâteaux smooth Banach spaces*, Israel J. Math. **124** (2001), no. 1, 243–252, DOI: 10.1007/BF02772620.

[FHHMZ] Marián Fabian, Petr Habala, Petr Hájek, Vicente Montesinos, and Václav Zizler, *Banach space theory. The basis for linear and nonlinear analysis*, CMS Books in Mathematics, Springer, New York, 2011, DOI: 10.1007/978-1-4419-7515-7.

[FHV] Marián Fabian, Petr Hájek, and Jon Vanderwerff, *On smooth variational principles in Banach spaces*, J. Math. Anal. Appl. **197** (1996), no. 1, 153–172, DOI: 10.1006/jmaa.1996.0013.

[FHZ] Marián Fabian, Petr Hájek, and Václav Zizler, *A note on lattice renormings*, Comment. Math. Univ. Carolin. **38** (1997), no. 2, 263–272.

[FKK] Marián Fabian, Ondřej F. K. Kalenda, and Jan Kolář, *Filling analytic sets by the derivatives of C^1-smooth bumps*, Proc. Amer. Math. Soc. **133** (2005), no. 1, 295–303, DOI: 10.1090/S0002-9939-04-07730-5.

[FMZ] Marián Fabian, Vicente Montesinos, and Václav Zizler, *Smoothness in Banach spaces. Selected problems*, Rev. R. Acad. Cienc. Exactas Fís. Nat., Ser. A Mat., RACSAM **100** (2006), no. 1–2, 101–125.

[FPWZ] Marián Fabian, David Preiss, John H. M. Whitfield, and Václav Zizler, *Separating polynomials on Banach spaces*, Q. J. Math. **40** (1989), no. 4, 409–422, DOI: 10.1093/qmath/40.4.409.

[FWZ] Marián Fabian, John H. M. Whitfield, and Václav Zizler, *Norms with locally Lipschitzian derivatives*, Israel J. Math. **44** (1983), no. 3, 262–276, DOI: 10.1007/BF02760975.

[FZ1] Marián Fabian and Václav Zizler, *A note on bump functions that locally depend on finitely many coordinates*, Bull. Aust. Math. Soc. **56** (1997), no. 3, 447–451, DOI: 10.1017/S0004972700031233.

[FZ2] Marián Fabian and Václav Zizler, *An elementary approach to some questions in higher order smoothness in Banach spaces*, Extracta Math. **14** (1999), no. 3, 295–327.

[FZ3] Marián Fabian and Václav Zizler, *On uniformly Gâteaux smooth $C^{(n)}$-smooth norms on separable Banach spaces*, Czechoslovak Math. J. **49** (1999), no. 3, 657–672, DOI: 10.1023/A:1022487822852.

[Far] Jeffrey Darrell Farmer, *Polynomial reflexivity in Banach spaces*, Israel J. Math. **87** (1994), no. 1–3, 257–273, DOI: 10.1007/BF02772998.

[FaJ] Jeffrey Darrell Farmer and William Buhmann Johnson, *Polynomial Schur and polynomial Dunford-Pettis properties*, Banach spaces. Proceedings of An international workshop on Banach space theory, ed. by Bor-Luh Lin and William Buhmann Johnson, Contemp. Math. 144, American Mathematical Society, Providence, RI, 1993, pp. 95–105, DOI: 10.1090/conm/144/1209450.

[Fed] Herbert Federer, *Geometric measure theory*, Grundlehren Math. Wiss. 153, Springer-Verlag, New York, 1969.

[Fern] Maite Fernández-Unzueta, *Zeroes of polynomials on ℓ_∞*, J. Math. Anal. Appl. **324** (2006), no. 2, 1115–1124, DOI: 10.1016/j.jmaa.2006.01.021.

[Ferr] Jesús Ferrer, *Zeroes of real polynomials on $C(K)$ spaces*, J. Math. Anal. Appl. **336** (2007), no. 2, 788–796, DOI: 10.1016/j.jmaa.2007.02.083.

[FGL] Juan Ferrera, Javier Gómez Gil, and José Luís González Llavona, *On completion of spaces of weakly continuous functions*, Bull. Lond. Math. Soc. **15** (1983), no. 3, 260–264, DOI: 10.1112/blms/15.3.260.

[Fi] Tadeusz Figiel, *On the moduli of convexity and smoothness*, Studia Math. **56** (1976), no. 2, 121–155.

[FiJ] Tadeusz Figiel and William Buhmann Johnson, *A uniformly convex Banach space which contains no ℓ_p*, Compos. Math. **29** (1974), no. 2, 179–190.

[Fl] Klaus Floret, *Natural norms on symmetric tensor products of normed spaces*, Note Mat. **17** (1997), 153–188, DOI: 10.1285/i15900932v17p153.

[Fo1] Vladimir P. Fonf, *Massiveness of the set of extreme points of the dual ball of a Banach space. Polyhedral spaces*, Funct. Anal. Appl. **12** (1978), no. 3, 237–239, DOI: 10.1007/BF01681448.

[Fo2] Vladimir P. Fonf, *Some properties of polyhedral Banach spaces*, Funct. Anal. Appl. **14** (1980), no. 4, 323–324, DOI: 10.1007/BF01078320.

[Fo3] Vladimir P. Fonf, *Polyhedral Banach spaces*, Math. Notes **30** (1981), no. 4, 809–813, DOI: 10.1007/bf01137813.

[Fo4] Vladimir P. Fonf, *Three characterizations of polyhedral Banach spaces*, Ukrainian Math. J. **42** (1990), no. 9, 1145–1148, DOI: 10.1007/BF01056615.

[FPST] Vladimir P. Fonf, Antonio José Pallarés, Richard James Smith, and Stanimir L. Troyanski, *Polyhedral norms on non-separable Banach spaces*, J. Funct. Anal. **255** (2008), no. 2, 449–470, DOI: 10.1016/j.jfa.2008.03.001.

[Fré1] Maurice Fréchet, *Toute fonctionnelle continue est développable en série de fonctionnelles d'ordres entiers* (French), C. R. Acad. Sci. Paris **148** (1909), 155–156.

[Fré2] Maurice Fréchet, *Une définition fonctionnelle des polynomes* (French), Nouv. Ann. Math., Sér. 4 **9** (1909), 145–162.

[FOS] Daniel Becker Freeman, Edward Wilfred Odell, and Thomas Schlumprecht, *The universality of ℓ_1 as a dual space*, Math. Ann. **351** (2011), no. 1, 149–186, DOI: 10.1007/s00208-010-0601-8.

[Fri] Yaakov Friedman, *Subspaces of $LC(H)$ and c_p*, Proc. Amer. Math. Soc. **53** (1975), no. 1, 117–122, DOI: 10.1090/S0002-9939-1975-0377592-X.

[Fro] Julien Frontisi, *Smooth partitions of unity in Banach spaces*, Rocky Mountain J. Math. **25** (1995), no. 4, 1295–1304, DOI: 10.1216/rmjm/1181072147.

[Fry1] Robb Fry, *Analytic approximation on c_0*, J. Funct. Anal. **158** (1998), no. 2, 509–520, DOI: 10.1006/jfan.1998.3302.

[Fry2] Robb Fry, *Approximation by functions with bounded derivative on Banach spaces*, Bull. Aust. Math. Soc. **69** (2004), no. 1, 125–131, DOI: 10.1017/S0004972700034316.

[GGMM] Pablo Galindo, Domingo García, Manuel Maestre, and Jorge Mujica, *Extension of multilinear mappings on Banach spaces*, Studia Math. **108** (1994), no. 1, 55–76.

[GaJL] María Isabel Garrido, Jesús Angel Jaramillo, and José Luís González Llavona, *Polynomial topologies on Banach spaces*, Topology Appl. **153** (2005), no. 5–6, 854–867, DOI: 10.1016/j.topol.2005.01.015.

[GaSa] Mariano Gasca and Thomas Sauer, *Polynomial interpolation in several variables*, Adv. Comput. Math. **12** (2000), no. 4, 377–410, DOI: 10.1023/A:1018981505752.

[GaspaT1] Thierry Gaspari, *On the range of the derivative of a real-valued function with bounded support*, Studia Math. **153** (2002), no. 1, 81–99, DOI: 10.4064/sm153-1-6.

[GaspaT2] Thierry Gaspari, *Bump functions with Hölder derivatives*, Canad. J. Math. **56** (2004), no. 4, 699–715, DOI: 10.4153/CJM-2004-032-3.

[GaspaI1] Ioannis Gasparis, *New examples of c_0-saturated Banach spaces*, Math. Ann. **344** (2009), no. 2, 491–500, DOI: 10.1007/s00208-008-0319-z.

[GaspaI2] Ioannis Gasparis, *New examples of c_0-saturated Banach spaces II*, J. Funct. Anal. **256** (2009), no. 11, 3830–3840, DOI: 10.1016/j.jfa.2008.11.021.

[GhMa] Nassif Ghoussoub and Bernard Maurey, *G_δ-embeddings in Hilbert space*, J. Funct. Anal. **61** (1985), no. 1, 72–97, DOI: 10.1016/0022-1236(85)90039-4.

[Gi] John R. Giles, *Classes of semi-inner-product spaces*, Trans. Amer. Math. Soc. **129** (1967), no. 3, 436–446, DOI: 10.1090/S0002-9947-1967-0217574-1.

[GlMc] Alan Gleit and Robert McGuigan, *A note on polyhedral Banach spaces*, Proc. Amer. Math. Soc. **33** (1972), no. 2, 398–404, DOI: 10.1090/S0002-9939-1972-0295055-4.

[God] Gilles Godefroy, *Renormings of Banach spaces*, Handbook of the geometry of Banach spaces, ed. by William Buhmann Johnson and Joram Lindenstrauss, vol. 1, North-Holland, Amsterdam, 2001, chap. 18, pp. 781–835, DOI: 10.1016/S1874-5849(01)80020-6.

[GTWZ] Gilles Godefroy, Stanimir L. Troyanski, John H. M. Whitfield, and Václav Zizler, *Smoothness in weakly compactly generated Banach spaces*, J. Funct. Anal. **52** (1983), no. 3, 344–352, DOI: 10.1016/0022-1236(83)90073-3.

[GR] Boris V. Godun and Sergey Anatolyovich Rakov, *Banach-Saks property and the problem of three spaces*, Math. Notes **31** (1982), no. 1, 32–39, DOI: 10.1007/BF01146265.

[GoSe] Jean-Marie Goethals and Johan Jacob Seidel, *Cubature formulae, polytopes, and spherical designs*, The geometric vein. The Coxeter Festschrift, ed. by Chandler Davis, Branko Grünbaum, and F. A. Sherk, Springer-Verlag, New York, 1981, pp. 203–218, DOI: 10.1007/978-1-4612-5648-9_13.

[GGo] Manuel González and Raquel Gonzalo, *Banach spaces admitting a separating polynomial and L_p spaces*, Monatsh. Math. **135** (2002), no. 2, 97–113, DOI: 10.1007/s006050200009.

[GGJ] Manuel González, Raquel Gonzalo, and Jesús Angel Jaramillo, *Symmetric polynomials on rearrangement-invariant function spaces*, J. Lond. Math. Soc. **59** (1999), no. 2, 681–697, DOI: 10.1112/S0024610799007164.

[GGu1] Manuel González and Joaquín M. Gutiérrez, *When every polynomial is unconditionally converging*, Arch. Math. **63** (1994), no. 2, 145–151, DOI: 10.1007/BF01189888.

[GGu2] Manuel González and Joaquín M. Gutiérrez, *Polynomial Grothendieck properties*, Glasg. Math. J. **37** (1995), no. 2, 211–219, DOI: 10.1017/S0017089500031116.

[GGu3] Manuel González and Joaquín M. Gutiérrez, *Unconditionally converging polynomials on Banach spaces*, Math. Proc. Cambridge Philos. Soc. **117** (1995), no. 2, 321–331, DOI: 10.1017/S030500410007314X.

[GGu4] Manuel González and Joaquín M. Gutiérrez, *Orlicz-Pettis polynomials on Banach spaces*, Monatsh. Math. **129** (2000), no. 4, 341–350, DOI: 10.1007/s006050050080.

[GGu5] Manuel González and Joaquín M. Gutiérrez, *Polynomials on Schreier's space*, Rocky Mountain J. Math. **30** (2000), no. 2, 571–585, DOI: 10.1216/rmjm/1022009282.

[Gon1] Raquel Gonzalo, *Upper and lower estimates in Banach sequence spaces*, Comment. Math. Univ. Carolin. **36** (1995), no. 4, 641–653.

[Gon2] Raquel Gonzalo, *Multilinear forms, subsymmetric polynomials, and spreading models on Banach spaces*, J. Math. Anal. Appl. **202** (1996), no. 2, 379–397, DOI: 10.1006/jmaa.1996.0322.

[GJ1] Raquel Gonzalo and Jesús Angel Jaramillo, *Compact polynomials between Banach spaces*, Proc. R. Ir. Acad., Sec. A **95A** (1995), no. 2, 213–226.

[GJ2] Raquel Gonzalo and Jesús Angel Jaramillo, *Smoothness and estimates of sequences in Banach spaces*, Israel J. Math. **89** (1995), no. 1–3, 321–341, DOI: 10.1007/BF02808207.

[GJ3] Raquel Gonzalo and Jesús Angel Jaramillo, *Higher order smoothness in sequence spaces and spreading models*, Function spaces. Proceedings of the Third conference on function spaces, ed. by Krzysztof Jarosz, Contemp. Math. 232, American Mathematical Society, Providence, RI, 1999, pp. 151–160, DOI: 10.1090/conm/232.

[GJT] Raquel Gonzalo, Jesús Angel Jaramillo, and Stanimir L. Troyanski, *High order smoothness and asymptotic structure in Banach spaces*, J. Convex Anal. **14** (2007), no. 2, 249–269.

[Gr] Alexander Grothendieck, *Sur les applications linéaires faiblement compactes d'espaces du type $C(K)$* (French), Canad. J. Math. **5** (1953), 129–173, DOI: 10.4153/CJM-1953-017-4.

[Gue] Sylvie Guerre-Delabrière, *Classical sequences in Banach spaces*, Pure and applied mathematics 166, Marcel Dekker, New York, 1992.

[GHM] Antonio José Guirao, Petr Hájek, and Vicente Montesinos, *Ranges of operators and derivatives*, J. Math. Anal. Appl. **367** (2010), no. 1, 29–33, DOI: 10.1016/j.jmaa.2009.12.010.

[Gut] Joaquín M. Gutiérrez, *Weakly continuous functions on Banach spaces not containing ℓ_1*, Proc. Amer. Math. Soc. **119** (1993), no. 1, 147–152, DOI: 10.1090/S0002-9939-1993-1158000-9.

[GuJL] Joaquín M. Gutiérrez, Jesús Angel Jaramillo, and José Luís González Llavona, *Polynomials and geometry of Banach spaces*, Extracta Math. **10** (1995), no. 2, 79–114.

[GL] Joaquín M. Gutiérrez and José Luís González Llavona, *Polynomially continuous operators*, Israel J. Math. **102** (1997), no. 1, 179–187, DOI: 10.1007/BF02773798.

[HaHá] Petr Habala and Petr Hájek, *Stabilization of polynomials*, C. R. Acad. Sci. Paris, Sér. I **320** (1995), no. 7, 821–825.

[Hag1] James N. Hagler, *Some more Banach spaces which contain ℓ_1*, Studia Math. **46** (1973), no. 1, 35–42.

[Hag2] James N. Hagler, *A counterexample to several questions about Banach spaces*, Studia Math. **60** (1977), no. 3, 289–308.

[Há1] Petr Hájek, *Polynomials and injections of Banach spaces into superreflexive spaces*, Arch. Math. **63** (1994), no. 1, 39–44, DOI: 10.1007/BF01196297.

[Há2] Petr Hájek, *Smooth norms that depend locally on finitely many coordinates*, Proc. Amer. Math. Soc. **123** (1995), no. 12, 3817–3821, DOI: 10.1090/S0002-9939-1995-1285993-3.

[Há3] Petr Hájek, *Analytic renormings of $C(K)$ spaces*, Serdica Math. J. **22** (1996), no. 1, 25–28.

[Há4] Petr Hájek, *Polynomial algebras on classical Banach spaces*, Israel J. Math. **106** (1998), no. 1, 209–220, DOI: 10.1007/BF02773469.

[Há5] Petr Hájek, *Smooth functions on c_0*, Israel J. Math. **104** (1998), no. 1, 17–27, DOI: 10.1007/BF02897057.

[Há6] Petr Hájek, *Smooth functions on $C(K)$*, Israel J. Math. **107** (1998), no. 1, 237–252, DOI: 10.1007/BF02764011.

[Há7] Petr Hájek, *A remark on S. M. Bates' theorem*, Serdica Math. J. **25** (1999), no. 3, 257–258.

[Há8] Petr Hájek, *Smooth norms on certain $C(K)$ spaces*, Proc. Amer. Math. Soc. **131** (2003), no. 7, 2049–2051, DOI: 10.1090/S0002-9939-03-06819-9.

[HáHa] Petr Hájek and Richard Haydon, *Smooth norms and approximation in Banach spaces of the type $C(K)$*, Q. J. Math. **58** (2007), no. 2, 221–228, DOI: 10.1093/qmath/ham010.

[HJ1] Petr Hájek and Michal Johanis, *Smooth approximations without critical points*, Cent. Eur. J. Math. **1** (2003), no. 3, 284–291, DOI: 10.2478/BF02475210.

[HJ2] Petr Hájek and Michal Johanis, *Isomorphic embeddings and harmonic behaviour of smooth operators*, Israel J. Math. **143** (2004), no. 1, 299–315, DOI: 10.1007/BF02803504.

[HJ3] Petr Hájek and Michal Johanis, *Polyhedrality in Orlicz spaces*, Israel J. Math. **168** (2008), no. 1, 167–188, DOI: 10.1007/s11856-008-1062-6.

[HJ4] Petr Hájek and Michal Johanis, *Smoothing of bump functions*, J. Math. Anal. Appl. **338** (2008), no. 2, 1131–1139, DOI: 10.1016/j.jmaa.2007.06.006.

[HJ5] Petr Hájek and Michal Johanis, *Uniformly Gâteaux smooth approximations on $c_0(\Gamma)$*, J. Math. Anal. Appl. **350** (2009), no. 2, 623–629, DOI: 10.1016/j.jmaa.2008.05.008.

[HJ6] Petr Hájek and Michal Johanis, *On Peano's theorem in Banach spaces*, J. Differential Equations **249** (2010), no. 12, 3342–3351, DOI: 10.1016/j.jde.2010.09.013.

[HJ7] Petr Hájek and Michal Johanis, *Smooth approximations*, J. Funct. Anal. **259** (2010), no. 3, 561–582, DOI: 10.1016/j.jfa.2010.04.020.

[HK] Petr Hájek and Michal Kraus, *Polynomials and identities on real Banach spaces*, J. Math. Anal. Appl. **385** (2012), no. 2, 1015–1026, DOI: 10.1016/j.jmaa.2011.07.028.

[HL] Petr Hájek and José Luís González Llavona, *P-continuity on classical Banach spaces*, Proc. Amer. Math. Soc. **128** (2000), no. 3, 827–830, DOI: 10.1090/S0002-9939-99-05056-X.

[HMVZ] Petr Hájek, Vicente Montesinos, Jon Vanderwerff, and Václav Zizler, *Biorthogonal systems in Banach spaces*, CMS Books in Mathematics 26, Springer, New York, 2008.

[HP] Petr Hájek and Antonín Procházka, *C^k-smooth approximations of LUR norms*, Trans. Amer. Math. Soc. **366** (2014), no. 4, 1973–1992, DOI: 10.1090/S0002-9947-2013-05899-0.

[HáTa] Petr Hájek and Jarno Talponen, *Smooth approximations of norms in separable Banach spaces*, Q. J. Math. **Advance access** (2013), DOI: 10.1093/qmath/hat053.

[HáTr] Petr Hájek and Stanimir L. Troyanski, *Analytic norms in Orlicz spaces*, Proc. Amer. Math. Soc. **129** (2001), no. 3, 713–717, DOI: 10.1090/S0002-9939-00-05773-7.

[HV1] Petr Hájek and Paola Vivi, *On ω-limit sets of ordinary differential equations in Banach spaces*, J. Math. Anal. Appl. **371** (2010), no. 2, 793–812, DOI: 10.1016/j.jmaa.2010.05.059.

[HV2] Petr Hájek and Paola Vivi, *Some problems on ordinary differential equations in Banach spaces*, Rev. R. Acad. Cienc. Exactas Fís. Nat., Ser. A Mat., RACSAM **104** (2010), no. 2, 245–255, DOI: 10.5052/RACSAM.2010.16.

[HV3] Petr Hájek and Paola Vivi, *Cross-sections of solution funnels*, J. Math. Anal. Appl. (submitted).

[HLP] Godfrey Harold Hardy, John Edensor Littlewood, and George Pólya, *Inequalities*, Cambridge University Press, London, 1934.

[Harr1] Lawrence Albert Harris, *Bounds on the derivatives of holomorphic functions of vectors*, Analyse fonctionnelle et applications. C. r. du Colloque d'analyse instituto de matemática Universidade federal do Rio de Janeiro, ed. by Leopoldo Nachbin, Actualités scientifiques et industrielles 1367, Hermann, Paris, 1975, pp. 145–163.

[Harr2] Lawrence Albert Harris, *A Bernstein-Markov theorem for normed spaces*, J. Math. Anal. Appl. **208** (1997), no. 2, 476–486, DOI: 10.1006/jmaa.1997.5339.

[Harr3] Lawrence Albert Harris, *Multivariate Markov polynomial inequalities and Chebyshev nodes*, J. Math. Anal. Appl. **338** (2008), no. 1, 350–357, DOI: 10.1016/j.jmaa.2007.05.044.

[Harr4] Lawrence Albert Harris, *A proof of Markov's theorem for polynomials on Banach spaces*, J. Math. Anal. Appl. **368** (2010), no. 1, 374–381, DOI: 10.1016/j.jmaa.2010.02.050.

[Hart] Fritz Hartogs, *Zur Theorie der analytischen Funktionen mehrerer unabhängiger Veränderlichen, insbesondere über die Darstellung derselben durch Reihen, welche nach Potenzen einer Veränderlichen fortschreiten* (German), Math. Ann. **62** (1906), no. 1, 1–88, DOI: 10.1007/BF01448415.

[Hay1] Richard Haydon, *Normes infiniment différentiables sur certains espaces de Banach* (French), C. R. Acad. Sci. Paris, Sér. I **315** (1992), 1175–1178.

[Hay2] Richard Haydon, *Smooth functions and partitions of unity on certain Banach spaces*, Q. J. Math. **47** (1996), no. 4, 455–468, DOI: 10.1093/qmath/47.4.455.

[Hay3] Richard Haydon, *Trees in renorming theory*, Proc. Lond. Math. Soc. **78** (1999), no. 3, 541–584, DOI: 10.1112/S0024611599001768.

[Hay4] Richard Haydon, *Subspaces of the Bourgain-Delbaen space*, Studia Math. **139** (2000), no. 3, 275–293.

[He] Stefan Heinrich, *Ultraproducts in Banach space theory*, J. Reine Angew. Math. **313** (1980), 72–104.

[HeTr] Francisco L. Hernández and Stanimir L. Troyanski, *On Gâteaux differentiable bump functions*, Studia Math. **118** (1996), no. 2, 135–143.

[Hilb] David Hilbert, *Beweis für die Darstellbarkeit der ganzen Zahlen durch eine feste Anzahl n-ter Pontenzen (Waringsches Problem)* (German), Math. Ann. **67** (1909), no. 3, 281–300, DOI: 10.1007/BF01450405.

[Hill] Einar Hille, *Functional analysis and semi-groups*, Amer. Math. Soc. Colloq. Publ. 31, American Mathematical Society, New York, 1948.

[Ho] Jørgen Hoffmann-Jørgensen, *The strong law of large numbers and the central limit theorem in Banach spaces*, Proceedings of the Seminar on random series, convex sets, and geometry of Banach space, Various publications series 24, Mat. Inst., Århus Univ., Århus, 1975, pp. 74–99.

[HŠZ] Petr Holický, Michal Šmídek, and Luděk Zajíček, *Convex functions with non-Borel set of Gâteaux differentiability points*, Comment. Math. Univ. Carolin. **39** (1998), no. 3, 469–482.

[IT] Milen Ivanov and Stanimir L. Troyanski, *Uniformly smooth renorming of Banach spaces with modulus of convexity of power type* 2, J. Funct. Anal. **237** (2006), no. 2, 373–390, DOI: 10.1016/j.jfa.2006.03.024.

[Ja1] Robert Clarke James, *Super-reflexive Banach spaces*, Canad. J. Math. **24** (1972), no. 5, 896–904, DOI: 10.4153/CJM-1972-089-7.

[Ja2] Robert Clarke James, *Super-reflexive spaces with bases*, Pacific J. Math. **41** (1972), no. 2, 409–419.

[JP] Jesús Angel Jaramillo and Ángeles Prieto, *Weak-polynomial convergence on a Banach space*, Proc. Amer. Math. Soc. **118** (1993), no. 2, 463–468, DOI: 10.1090/S0002-9939-1993-1126196-0.

[Ji] Mar Jiménez-Sevilla, *A note on the range of the derivatives of analytic approximations of uniformly continuous functions on c_0*, J. Math. Anal. Appl. **348** (2008), no. 2, 573–580, DOI: 10.1016/j.jmaa.2008.07.050.

[JS] Mar Jiménez-Sevilla and Luis Sánchez-González, *LFC bumps on separable Banach spaces*, J. Math. Anal. Appl. **365** (2010), no. 1, 315–319, DOI: 10.1016/j.jmaa.2009.10.049.

[Joha] Michal Johanis, *Approximation of Lipschitz mappings*, Serdica Math. J. **29** (2003), no. 2, 141–148.

[JohaZa] Michal Johanis and Luděk Zajíček, *Smoothness via directional smoothness and Marchaud's theorem in Banach spaces*, preprint (2014).

[John] Kamil John, *On the uncomplemented subspace $K(X,Y)$*, Czechoslovak Math. J. **42** (1992), no. 1, 167–173.

[JTZ] Kamil John, Henryk Toruńczyk, and Václav Zizler, *Uniformly smooth partitions of unity on superreflexive Banach spaces*, Studia Math. **70** (1981), no. 2, 129–137.

[JohnZiz] Kamil John and Václav Zizler, *Gâteaux smooth partitions of unity on weakly compactly generated Banach spaces*, Studia Math. **60** (1977), no. 2, 131–135.

[Johns1] William Buhmann Johnson, *A reflexive Banach space which is not sufficiently Euclidean*, Studia Math. **55** (1976), no. 2, 201–205.

[Johns2] William Buhmann Johnson, *On quotients of L_p which are quotients of ℓ_p*, Compos. Math. **34** (1977), no. 1, 69–89.

[JL] William Buhmann Johnson and Joram Lindenstrauss, eds., *Handbook of the geometry of Banach spaces*, 2 vols., North-Holland, Amsterdam, 2001, 2003.

[JO] William Buhmann Johnson and Edward Wilfred Odell, *Subspaces of L_p which embed into ℓ_p*, Compos. Math. **28** (1974), no. 1, 37–49.

[JRZ] William Buhmann Johnson, Haskell Paul Rosenthal, and Mordecay Zippin, *On bases, finite dimensional decompositions and weaker structures in Banach spaces*, Israel J. Math. **9** (1971), no. 4, 488–506, DOI: 10.1007/BF02771464.

[JohnsZip1] William Buhmann Johnson and Mordecay Zippin, *On subspaces of quotients of $(\sum G_n)_{\ell_p}$ and $(\sum G_n)_{c_0}$*, Israel J. Math. **13** (1972), no. 3–4, 311–316, DOI: 10.1007/BF02762805.

[JohnsZip2] William Buhmann Johnson and Mordecay Zippin, *Separable L_1 preduals are quotients of $C(\Delta)$*, Israel J. Math. **16** (1973), no. 2, 198–202, DOI: 10.1007/BF02757870.

[JN] Pascual Jordan and John von Neumann, *On inner products in linear, metric spaces*, Ann. of Math. **36** (1935), no. 3, 719–723.

[Kai] Sten Kaijser, *Some results in the metric theory of tensor products*, Studia Math. **63** (1978), no. 2, 157–170.

[KKL] Jerzy Kąkol, Wiesław Kubiś, and Manuel López-Pellicer, *Descriptive topology in selected topics of functional analysis*, Developments in Mathematics 24, Springer, New York, 2011, DOI: 10.1007/978-1-4614-0529-0.

[Kal] Nigel J. Kalton, *Spaces of compact operators*, Math. Ann. **208** (1974), no. 4, 267–278, DOI: 10.1007/BF01432152.

[KS] Boris Sergeevich Kashin and Stanisław J. Szarek, *The Knaster problem and the geometry of high-dimensional cubes*, C. R. Math. Acad. Sci. Paris **336** (2003), no. 11, 931–936, DOI: 10.1016/S1631-073X(03)00226-7.

[Kec] Alexander S. Kechris, *Classical descriptive set theory*, Grad. Texts in Math. 156, Springer-Verlag, New York, 1995.

[Kel] Oliver Dimon Kellogg, *On bounded polynomials in several variables*, Math. Z. **27** (1928), no. 1, 55–64, DOI: 10.1007/BF01171085.

[Kl1] Maciej Klimek, *Pluripotential theory*, London Math. Soc. Monogr. New Ser. 6, Oxford University Press, Oxford, 1991.

[Kl2] Maciej Klimek, *Metrics associated with extremal plurisubharmonic functions*, Proc. Amer. Math. Soc. **123** (1995), no. 9, 2763–2770, DOI: 10.1090/S0002-9939-1995-1307539-3.

[Kn] Helmut Knaust, *Orlicz sequence spaces of Banach-Saks type*, Arch. Math. **59** (1992), no. 6, 562–565, DOI: 10.1007/BF01194848.

[KO1] Helmut Knaust and Edward Wilfred Odell, *On c_0-sequences in Banach spaces*, Israel J. Math. **67** (1989), no. 2, 153–169, DOI: 10.1007/BF02937292.

[KO2] Helmut Knaust and Edward Wilfred Odell, *Weakly null sequences with upper ℓ_p-estimates*, Functional analysis. Proceedings of the seminar at the University of Texas at Austin 1987–89, ed. by Edward Wilfred Odell and Haskell Paul Rosenthal, Lecture Notes in Math. 1470, Springer-Verlag, Berlin, Heidelberg, 1991, pp. 85–107, DOI: 10.1007/BFb0090216.

[Koe] Donald Otto Koehler, *The analytic properties of G_{2n} spaces*, Proc. Amer. Math. Soc. **35** (1972), no. 1, 201–206, DOI: 10.1090/S0002-9939-1972-0301486-6.

[Kor] Pavel Petrovich Korovkin, *Linear operators and approximation theory*, International monographs on advanced mathematics & physics, Hindustan Publishing Corp., Delhi, 1960.

[Kra] Steven G. Krantz, *Function theory of several complex variables*, AMS Chelsea Publishing, Providence, RI, 2001.

[Kri] Jean-Louis Krivine, *Plongement des espaces normés dans les L_p pour $p > 2$* (French), C. R. Acad. Sci. Paris **261** (1965), 4307–4310.

[Kuc] Marek Kuczma, *An introduction to the theory of functional equations and inequalities. Cauchy's equation and Jensen's inequality*, Prace naukowe Uniwersytetu Śląskiego w Katowicach 489, Państwowe wydawnictwo naukowe, Kraków, 1985.

[Kur1] Jaroslav Kurzweil, *On approximation in real Banach spaces*, Studia Math. **14** (1954), no. 2, 214–231.

[Kur2] Jaroslav Kurzweil, *On approximation in real Banach spaces by analytic operations*, Studia Math. **16** (1957), no. 2, 124–129.

[Kw1] Stanisław Kwapień, *Some remarks on (p, q)-absolutely summing operators in ℓ_p-spaces*, Studia Math. **29** (1968), no. 3, 327–337.

[Kw2] Stanisław Kwapień, *Isomorphic characterizations of inner product spaces by orthogonal series with vector valued coefficients*, Studia Math. **44** (1972), no. 6, 583–595.

[La] Miguel Lacruz, *Norms of polynomials and capacities on Banach spaces*, Integral Equations Operator Theory **34** (1999), no. 4, 494–499, DOI: 10.1007/BF01272887.

[LL] Jean-Michel Lasry and Pierre-Louis Lions, *A remark on regularization in Hilbert spaces*, Israel J. Math. **55** (1986), no. 3, 257–266, DOI: 10.1007/BF02765025.

[Lej1] Franciszek Leja, *Sur les suites de polynômes bornées presque partout sur la frontière d'un domaine* (French), Math. Ann. **108** (1933), no. 1, 517–524, DOI: 10.1007/BF01452850.

[Lej2] Franciszek Leja, *Teoria funkcji analitycznych* (polish), Biblioteka matematyczna 14, Państwowe wydawnictwo naukowe, Warszawa, 1957.

[LeSu] Isaac Edward Leonard and Kondagunta Sundaresan, *Geometry of Lebesgue-Bochner function spaces – smoothness*, Trans. Amer. Math. Soc. **198** (1974), 229–251, DOI: 10.1090/S0002-9947-1974-0367652-5.

[Leu1] Denny Ho-Hon Leung, *Some isomorphically polyhedral Orlicz sequence spaces*, Israel J. Math. **87** (1994), no. 1–3, 117–128, DOI: 10.1007/BF02772988.

[Leu2] Denny Ho-Hon Leung, *On c_0-saturated Banach spaces*, Illinois J. Math. **39** (1995), no. 1, 15–29.

[Leu3] Denny Ho-Hon Leung, *Symmetric sequence subspaces of $C(\alpha)$*, J. Lond. Math. Soc. **59** (1999), no. 3, 1049–1063, DOI: 10.1112/S0024610799007218.

[Li1] Joram Lindenstraus, *Extension of compact operators*, Mem. Amer. Math. Soc. (1964), no. 48.

[Li2] Joram Lindenstrauss, *On the modulus of smoothness and divergent series in Banach spaces*, Michigan Math. J. **10** (1963), no. 3, 241–252, DOI: 10.1307/mmj/1028998906.

[Li3] Joram Lindenstrauss, *On nonlinear projections in Banach spaces*, Michigan Math. J. **11** (1964), no. 3, 263–287, DOI: 10.1307/mmj/1028999141.

[LP] Joram Lindenstrauss and Aleksander Pełczyński, *Absolutely summing operators in $\mathcal{L}_p$-spaces and their applications*, Studia Math. **29** (1968), no. 3, 275–326.

[LPT] Joram Lindenstrauss, David Preiss, and Jaroslav Tišer, *Fréchet differentiability of Lipschitz functions and porous sets in Banach spaces*, Annals of Mathematics Studies 179, Princeton University Press, Princeton, NJ, 2012.

[LindeRo] Joram Lindenstrauss and Haskell Paul Rosenthal, *The $\mathcal{L}_p$-spaces*, Israel J. Math. **7** (1969), no. 4, 325–349, DOI: 10.1007/BF02788865.

[LiTz1] Joram Lindenstrauss and Lior Tzafriri, *Classical Banach spaces*, Lecture Notes in Math. 338, Springer-Verlag, Berlin, Heidelberg, 1973.

[LiTz2] Joram Lindenstrauss and Lior Tzafriri, *Classical Banach spaces I. Sequence spaces*, Springer-Verlag, Berlin, Heidelberg, 1996.

[LiTz3] Joram Lindenstrauss and Lior Tzafriri, *Classical Banach spaces II. Function spaces*, Springer-Verlag, Berlin, Heidelberg, 1996.

[LindsRy] Mikael Lindström and Raymond A. Ryan, *Applications of ultraproducts to infinite dimensional holomorphy*, Math. Scand. **71** (1992), 229–242.

[LjSo] Lazar Aronovich Ljusternik and Vladimir Ivanovich Sobolev, *Elements of functional analysis*, Frederick Ungar Publishing Co., New York, 1961.

[Loh] Robert Henry Lohman, *A note on Banach spaces containing ℓ_1*, Canad. Math. Bull **19** (1976), no. 3, 365–367, DOI: 10.4153/CMB-1976-056-x.

[LóTo] Jordi López-Abad and Stevo Todorčević, *Partial unconditionality of weakly null sequences*, Rev. R. Acad. Cienc. Exactas Fís. Nat., Ser. A Mat., RACSAM **100** (2010), no. 1–2, 237–277.

[Lor] Rudolph Alexander Lorentz, *Multivariate Birkhoff interpolation*, Lecture Notes in Math. 1516, Springer-Verlag, Berlin, Heidelberg, 1992.

[Lov] László Lovász, *Combinatorial problems and exercises*, 2nd ed., North-Holland, Amsterdam, 1993.

[LV] Yuri Illich Lyubich and Leonid N. Vaserstein, *Isometric embeddings between classical Banach spaces, cubature formulas, and spherical designs*, Geom. Dedicata **47** (1993), no. 3, 327–362, DOI: 10.1007/BF01263664.

[Mal1] Rumen Petrov Maleev, *Norms of best smoothness in Orlicz spaces*, Z. Anal. Anwend. **12** (1993), no. 1, 123–135.

[Mal2] Rumen Petrov Maleev, *Higher order uniformly Gâteaux differentiable norms on Orlicz spaces*, Rocky Mountain J. Math. **25** (1995), no. 3, 1117–1136, DOI: 10.1216/rmjm/1181072209.

[MaT1] Rumen Petrov Maleev and Stanimir L. Troyanski, *On the moduli of convexity and smoothness in Orlicz spaces*, Studia Math. **54** (1975), no. 2, 131–141.

[MaT2] Rumen Petrov Maleev and Stanimir L. Troyanski, *Smooth functions in Orlicz spaces*, Banach space theory. Proceedings of A research workshop held July 5–25, 1987, ed. by Bor-Luh Lin, Contemp. Math. 85, American Mathematical Society, Providence, RI, 1989, pp. 355–370, DOI: 10.1090/conm/085/983394.

[MaT3] Rumen Petrov Maleev and Stanimir L. Troyanski, *Smooth norms in Orlicz spaces*, Canad. Math. Bull. **34** (1991), no. 1, 74–82, DOI: 10.4153/CMB-1991-012-7.

[MZ] Rumen Petrov Maleev and Boyan Zlatanov, *Smoothness in Musielak-Orlicz sequence spaces*, C. R. Acad. Bulgare Sci. **55** (2002), no. 6, 11–16.

[MarkoA] Andrei Andreevich Markov, *Ob odnom voprose D. I. Mendeleeva* (Russian), Zapiski Imperatorskoi akademii nauk **62** (1890), 1–24.

[MarkoV] Vladimir Andreevich Markov, *O funktsiyakh, naimenee uklonyayushchikhsya ot nulya v dannom promezhutke* (Russian), Tipografiya Imperatorskoi akademii nauk, Sankt-Peterburg, 1892.

[Mas1] John Charles Mason, *Near-best multivariate approximation by Fourier series, Chebyshev series and Chebyshev interpolation*, J. Approx. Theory **28** (1980), no. 4, 349–358, DOI: 10.1016/0021-9045(80)90069-6.

[Mas2] John Charles Mason, *Minimal projections and near-best approximations by multivariate polynomial expansion and interpolation*, Multivariate approximation theory II. Proceedings of the conference held at Oberwolfach, 1982, ed. by Walter Schempp and Karl Zeller, Internat. Ser. Numer. Math. 61, Birkhäuser, Basel, 1982, pp. 241–254.

[MH] John Charles Mason and David Christopher Handscomb, *Chebyshev polynomials*, Chapman & Hall/CRC, Boca Raton, FL, 2003.

[Mat] Jiří Matoušek, *Using the Borsuk-Ulam theorem. Lectures on topological methods in combinatorics and geometry*, Universitext, Springer-Verlag, Berlin, Heidelberg, 2003.

[MaPi1] Bernard Maurey and Gilles Pisier, *Caractérisation d'une classe d'espaces de Banach par des propriétés de séries aléatoires vectorielles* (French), C. R. Acad. Sci. Paris, Sér. A **277** (1973), no. 14, 687–690.

[MaPi2] Bernard Maurey and Gilles Pisier, *Séries de variables aléatoires vectorielles indépendantes et propriétés géométriques des espaces de Banach* (French), Studia Math. **58** (1976), no. 1, 45–90.

[MO1] Stanisław Mazur and Władysław Orlicz, *Grundlegende Eigenschaften der polynomischen Operationen. Erste Mitteilung* (German), Studia Math. **5** (1934), 50–68.

[MO2] Stanisław Mazur and Władysław Orlicz, *Grundlegende Eigenschaften der polynomischen Operationen. Zweite Mitteilung* (German), Studia Math. **5** (1934), 179–189.

[McC] Charles Alan McCarthy, c_p, Israel J. Math. **5** (1967), no. 4, 249–271, DOI: 10.1007/BF02771613.

[McL] David P. McLaughlin, *Smooth partitions of unity in preduals of WCG spaces*, Math. Z. **211** (1992), 189–194, DOI: 10.1007/BF02571426.

[MPVZ] David P. McLaughlin, René A. Poliquin, Jon Vanderwerff, and Václav Zizler, *Second-order Gâteaux differentiable bump functions and approximations in Banach spaces*, Canad. J. Math. **45** (1993), no. 3, 612–625, DOI: 10.4153/CJM-1993-032-9.

[MV] David P. McLaughlin and Jon Vanderwerff, *Higher order Gâteaux smooth bump functions on Banach spaces*, Bull. Aust. Math. Soc. **51** (1995), no. 2, 291–300, DOI: 10.1017/S000497270001412X.

[McS] Edward James McShane, *Extension of range of functions*, Bull. Amer. Math. Soc. **40** (1934), no. 12, 837–842, DOI: 10.1090/S0002-9904-1934-05978-0.

[Me] Victor Zakharovich Meshkov, *Smoothness properties in Banach spaces*, Studia Math. **63** (1978), no. 2, 111–123.

[MW] Aristotle D. Michal and Max Wyman, *Characterization of complex couple spaces*, Ann. of Math. **42** (1941), no. 1, 247–250.

[Mi] Vitali Davidovich Milman, *A few observations on the connections between local theory and some other fields*, Geometric aspects of functional analysis. Israel seminar (GAFA) 1986–87, ed. by Joram Lindenstrauss and Vitali Davidovich Milman, Lecture Notes in Math. 1317, Springer-Verlag, Berlin, Heidelberg, 1988, pp. 283–289, DOI: 10.1007/BFb0081748.

[MS] Vitali Davidovich Milman and Gideon Schechtman, *Asymptotic theory of finite dimensional normed spaces*, Lecture Notes in Math. 1200, Springer-Verlag, Berlin, Heidelberg, 1986.

[MMR] Gradimir V. Milovanović, Dragoslav S. Mitrinović, and Themistocles M. Rassias, *Topics in polynomials: extremal problems, inequalitites, zeros*, World Scientific Publishing, River Edge, NJ, 1994.

[MiPe] Boris Samuilovich Mityagin and Aleksander Pełczyński, *Nuclear operators and approximative dimensions*, Proc. internat. congr. math. (Moscow, 1966), ed. by Ivan Georgievich Petrovskii, Mir, Moscow, 1968, pp. 366–372.

[Mö] Hans Michael Möller, *Kubaturformeln mit minimaler Knotenzahl* (German), Numer. Math. **25** (1976), no. 2, 185–200, DOI: 10.1007/BF01462272.

[MoT] Aníbal Moltó Martínez and Stanimir L. Troyanski, *On uniformly Gâteaux differentiable norms in $C(K)$*, Mathematika **41** (1994), no. 2, 233–238, DOI: 10.1112/S0025579300007348.

[MoPa] Corbett R. Morrow and Thomas N. L. Patterson, *Construction of algebraic cubature rules using polynomial ideal theory*, SIAM J. Numer. Anal. **15** (1978), no. 5, 953–976, DOI: 10.1137/0715062.

[Mo1] Nicole Moulis, *Sur les variétés Hilbertiennes et les fonctions non dégénérées* (French), Indag. Math. **30** (1968), no. 5, 497–511.

[Mo2] Nicole Moulis, *Approximation de fonctions différentiables sur certains espaces de Banach* (French), Ann. Inst. Fourier (Grenoble) **21** (1971), no. 4, 293–345.

[Mu] Jorge Mujica, *Complex analysis in Banach spaces. Holomorphic functions and domains of holomorphy in finite and infinite dimensions*, Mathematics studies 120, North-Holland, Amsterdam, 1986.

[MST] Gustavo A. Muñoz, Yannis Sarantopoulos, and Andrew Morton Tonge, *Complexifications of real Banach spaces, polynomials and multilinear maps*, Studia Math. **134** (1999), no. 1, 1–33.

[Na] Leopoldo Nachbin, *Topology on spaces of holomorphic mappings*, Ergeb. Math. Grenzgeb. 47, Springer-Verlag, Berlin, Heidelberg, 1969.

[NS] Arkadij Semenovich Nemirovskij and S. M. Semenov, *On polynomial approximation of functions on Hilbert spaces*, Math. USSR Sb. **21** (1973), no. 2, 255–277, DOI: 10.1070/SM1973v021n02ABEH002016.

[Ne] Abraham Neyman, *Representation of L_p-norms and isometric embedding in L_p-spaces*, Israel J. Math. **48** (1984), no. 2–3, 129–138, DOI: 10.1007/BF02761158.

[O] Edward Wilfred Odell, *Applications of Ramsey theorems to Banach space theory*, Notes in Banach spaces, ed. by Howard Elton Lacey, University of Texas Press, Austin, 1980, pp. 379–404.

[OR] Edward Wilfred Odell and Haskell Paul Rosenthal, *A double-dual characterization of separable Banach spaces containing ℓ_1*, Israel J. Math. **20** (1975), no. 3–4, 375–384, DOI: 10.1007/BF02760341.

[OS1] Edward Wilfred Odell and Thomas Schlumprecht, *The distortion problem*, Acta Math. **173** (1994), no. 2, 259–281, DOI: 10.1007/BF02398436.

[OS2] Edward Wilfred Odell and Thomas Schlumprecht, *Distortion and asymptotic structure*, Handbook of the geometry of Banach spaces, ed. by William Buhmann Johnson and Joram Lindenstrauss, vol. 2, North-Holland, Amsterdam, 2003, chap. 31, pp. 1333–1360, DOI: 10.1016/S1874-5849(03)80038-4.

[PWZ] Jaroslav Pechanec, John H. M. Whitfield, and Václav Zizler, *Norms locally dependent on finitely many coordinates*, An. Acad. Brasil. Ciênc. **53** (1981), no. 3, 415–417.

[Pel] Jan Pelant, *Embeddings into c_0*, Topology Appl. **57** (1994), no. 2–3, 259–269, DOI: 10.1016/0166-8641(94)90053-1.

[PHK] Jan Pelant, Petr Holický, and Ondřej F. K. Kalenda, *$C(K)$ spaces which cannot be uniformly embedded into $c_0(\Gamma)$*, Fund. Math. **192** (2006), no. 3, 245–254, DOI: 10.4064/fm192-3-4.

[Peł1] Aleksander Pełczyński, *A property of multilinear operations*, Studia Math. **16** (1957), no. 2, 173–182.

[Peł2] Aleksander Pełczyński, *Projections in certain Banach spaces*, Studia Math. **19** (1960), no. 2, 209–228.

[Peł3] Aleksander Pełczyński, *A theorem of Dunford-Pettis type for polynomial operators*, Bull. Acad. Polon. Sci., Sér. Sci. Math. Astronom. Phys. **11** (1963), no. 6, 379–386.

[Peł4] Aleksander Pełczyński, *On weakly compact polynomial operators on B-spaces with Dunford-Pettis property*, Bull. Acad. Polon. Sci., Sér. Sci. Math. Astronom. Phys. **11** (1963), no. 6, 371–378.

[Peł5] Aleksander Pełczyński, *On Banach spaces containing $L_1(\mu)$*, Studia Math. **30** (1968), no. 2, 231–246.

[PS] Aleksander Pełczyński and Zbigniew Semadeni, *Spaces of continuous functions (III)*, Studia Math. **18** (1959), no. 2, 211–222.

[Pie] Albrecht Pietsch, *Operator ideals*, North-Holland mathematical library 20, North-Holland, Amsterdam, 1980.

[Pis1] Gilles Pisier, *Sur les espaces de Banach qui ne contiennent pas uniformément de ℓ_n^1* (French), C. R. Acad. Sci. Paris, Sér. A **277** (1973), no. 20, 991–994.

[Pis2] Gilles Pisier, *Un exemple concernant la super-réflexivité* (French), Séminaire d'analyse fonctionnelle (Polytechnique) (1974–1975), no. A.2.

[Pis3] Gilles Pisier, *Martingales with values in uniformly convex spaces*, Israel J. Math. **20** (1975), no. 3–4, 326–350, DOI: 10.1007/BF02760337.

[Pis4] Gilles Pisier, *Factorization of linear operators and geometry of Banach spaces*, CBMS Reg. Conf. Ser. Math. 60, American Mathematical Society, Providence, RI, 1986.

[Pis5] Gilles Pisier, *The volume of convex bodies and Banach space geometry*, Cambridge Tracts in Math. 94, Cambridge University Press, Cambridge, 1989, DOI: 10.1017/CBO97805116 62454.

[PZ] Anatolij Plichko and Andriy Zagorodnyuk, *On automatic continuity and three problems of "The Scottish Book" concerning the boundedness of polynomial functions*, J. Math. Anal. Appl. **220** (1998), no. 2, 477–494, DOI: 10.1006/jmaa.1997.5826.

[Pr] João Bosco Prolla, *On the Weierstrass-Stone theorem*, J. Approx. Theory **78** (1994), no. 3, 299–313, DOI: 10.1006/jath.1994.1080.

[RS] Qazi Ibadur Rahman and Gerhard Schmeisser, *Analytic theory of polynomials*, London Math. Soc. Monogr. New Ser. 26, Oxford University Press, Oxford, 2002.

[Raj] Matías Raja, *On dual locally uniformly rotund norms*, Israel J. Math. **129** (2002), no. 1, 77–91, DOI: 10.1007/BF02773154.

[Rak] Sergey Anatolyovich Rakov, *Banach-Saks property of a Banach space*, Math. Notes **26** (1979), no. 6, 909–916, DOI: 10.1007/BF01142075.

[Res] Guillermo Restrepo, *An infinite dimensional version of a theorem of Bernstein*, Proc. Amer. Math. Soc. **23** (1969), no. 1, 193–198, DOI: 10.1090/S0002-9939-1969-0246092-7.

[Rez1] Bruce Reznick, *Banach spaces which satisfy linear identities*, Pacific J. Math. **74** (1978), no. 1, 221–233.

[Rez2] Bruce Reznick, *Banach spaces with polynomial norms*, Pacific J. Math. **82** (1979), no. 1, 223–235.

[Rez3] Bruce Reznick, *Sums of even powers of real linear forms*, Mem. Amer. Math. Soc. **96** (1992), no. 463.

[Ri] Theodore Joseph Rivlin, *The Chebyshev polynomials*, John Wiley & Sons, New York, 1974.

[Rog] Werner Wolfgang Rogosinski, *Some elementary inequalities for polynomials*, Math. Gaz. **39** (1955), no. 327, 7–12.

[Ros1] Haskell Paul Rosenthal, *On factors of $C([0,1])$ with non-separable dual*, Israel J. Math. **13** (1972), no. 3–4, 361–378, DOI: 10.1007/BF02762811.

[Ros2] Haskell Paul Rosenthal, *A characterization of Banach spaces containing ℓ_1*, Proc. Natl. Acad. Sci. USA **71** (1974), no. 6, 2411–2413.

[Ros3] Haskell Paul Rosenthal, *The banach spaces $C(K)$*, Handbook of the geometry of Banach spaces, ed. by William Buhmann Johnson and Joram Lindenstrauss, vol. 2, North-Holland, Amsterdam, 2003, chap. 36, pp. 1547–1602, DOI: 10.1016/S1874-5849(03)80043-8.

[Ros4] Haskell Paul Rosenthal, *Several new characterizations of Banach spaces containing ℓ_1*, Israel J. Math. (submitted).

[RudinM] Mary Ellen Rudin, *A new proof that metric spaces are paracompact*, Proc. Amer. Math. Soc. **20** (1969), no. 2, 603, DOI: 10.1090/S0002-9939-1969-0236876-3.

[RudinW] Walter Rudin, *Real and complex analysis*, 3rd ed., McGraw-Hill Book Co., New York, 1987.

[Ry1] Raymond A. Ryan, *Dunford-Pettis properties*, Bull. Acad. Polon. Sci., Sér. Sci. Math. **27** (1979), no. 5, 373–379.

[Ry2] Raymond A. Ryan, *Applications of topological tensor products to infinite dimensional holomorphy*, PhD thesis, Trinity College Dublin, 1980.

[Ry3] Raymond A. Ryan, *Introduction to tensor products of Banach spaces*, Springer monogr. math. Springer, New York, 2002.

[Sa] Yannis Sarantopoulos, *Bounds on the derivatives of polynomials on Banach spaces*, Math. Proc. Cambridge Philos. Soc. **110** (1991), no. 2, 307–312, DOI: 10.1017/S0305004100070389.

[Schec] Eric Schechter, *Handbook of analysis and its foundations*, Academic Press, San Diego, 1997.

[Schei] Stephen Scheinberg, *Uniform approximation by entire functions*, J. Anal. Math. **29** (1976), 16–18, DOI: 10.1007/BF02789974.

[Schr] Józef Schreier, *Ein Gegenbeispiel zur Theorie der schwachen Konvergenz* (German), Studia Math. **2** (1930), no. 1, 58–62.

[Schu] Issai Schur, *Über das Maximum des absoluten Betrages eines Polynoms in einem gegebenen Intervall* (German), Math. Z. **4** (1919), no. 3–4, 271–287, DOI: 10.1007/BF01203015.

[Shi] Georgiy Evgenievich Shilov, *Certain solved and unsolved problems in the theory of functions in Hilbert space*, Moscow Univ. Math. Bull. **25** (1970), no. 2, 87–89.

[Shk] Stanislav A. Shkarin, *On Rolle's theorem in infinite-dimensional Banach spaces*, Math. Notes **51** (1992), no. 3, 311–317, DOI: 10.1007/BF01206400.

[Sic] Józef Siciak, *A polynomial lemma and analytic mappings in topological vector spaces*, Séminaire Pierre Lelong (analyse) année 1970/71, ed. by Pierre Lelong, Lecture Notes in Math. 275, Springer-Verlag, Berlin, Heidelberg, 1972, pp. 131–142, DOI: 10.1007/BFb0058496.

[Sin] Ivan Singer, *Bases in Banach spaces I*, Grundlehren Math. Wiss. 154, Springer-Verlag, Berlin, Heidelberg, 1970.

[Sk] Valentin Ivanovich Skalyga, *Bounds for the derivatives of polynomials on centrally symmetric convex bodies*, Izv. Math. **69** (2005), no. 3, 607–621, DOI: 10.1070/IM2005v069n03ABEH000539.

[ST] Richard James Smith and Stanimir L. Troyanski, *Renormings of $C(K)$ spaces*, Rev. R. Acad. Cienc. Exactas Fís. Nat., Ser. A Mat., RACSAM **104** (2010), no. 2, 375–412, DOI: 10.5052/RACSAM.2010.24.

[Sta] Richard Peter Stanley, *Enumerative combinatorics. Volume 2*, Cambridge Stud. Adv. Math. 62, Cambridge University Press, Cambridge, 1999.

[Steg1] Charles Stegall, *Banach spaces whose duals contain $\ell_1(\Gamma)$ with applications to the study of dual $L_1(\mu)$ spaces*, Trans. Amer. Math. Soc. **176** (1973), 463–477, DOI: 10.1090/S0002-9947-1973-0315404-3.

[Steg2] Charles Stegall, *Optimization of functions on certain subsets of Banach spaces*, Math. Ann. **236** (1978), no. 2, 171–176, DOI: 10.1007/BF01351389.

[Steg3] Charles Stegall, *Optimization and differentiation in Banach spaces*, Linear Algebra Appl. **84** (1986), 191–211, DOI: 10.1016/0024-3795(86)90314-9.

[Ster] Jacques Stern, *Some applications of model theory in Banach space theory*, Ann. Math. Logic **9** (1976), no. 1–2, 49–121, DOI: 10.1016/0003-4843(76)90006-1.

[Str] Thomas Strömberg, *The operation of infimal convolution*, Dissertationes Math. (Rozprawy Mat.) **352** (1996).

[Tal1] Michel Talagrand, *Deux exemples de fonctions convexes*, C. R. Acad. Sci. Paris, Sér. A **288** (1979), no. 8, 461–464.

[Tal2] Michel Talagrand, *Cotype and $(q,1)$-summing norm in a Banach space*, Invent. Math. **110** (1992), no. 1, 545–556, DOI: 10.1007/BF01231344.

[Tal3] Michel Talagrand, *Cotype of operators from $C(K)$*, Invent. Math. **107** (1992), no. 1, 1–40, DOI: 10.1007/BF01231879.

[Tay1] Angus Ellis Taylor, *Additions to the theory of polynomials in normed linear spaces*, Tohoku Math. J. **44** (1938), 302–318.

[Tay2] Angus Ellis Taylor, *Linear operations which depend analytically on a parameter*, Ann. of Math. **39** (1938), no. 3, 574–593.

[Tay3] Angus Ellis Taylor, *Analysis in complex Banach spaces*, Bull. Amer. Math. Soc. **49** (1943), no. 9, 652–669, DOI: 10.1090/S0002-9904-1943-07968-2.

[Tay4] Angus Ellis Taylor, *Historical notes on analyticity as a concept in functional analysis*, Problems in analysis. A symposium in honor of Salomon Bochner, ed. by Robert C. Gunning, Princeton Math. Ser. 31, Princeton University Press, Princeton, NJ, 1970, pp. 325–343.

[Tay5] Angus Ellis Taylor, *Notes on the history of the uses of analyticity in operator theory*, Amer. Math. Monthly **78** (1971), no. 4, 331–342.

[Th] Edward Oakley Thorp, *Projections onto the subspace of compact operators*, Pacific J. Math. **10** (1960), no. 2, 693–696.

[Tom1] Nicole Tomczak-Jaegermann, *On the differentiability of the norm in trace classes S_p*, Séminaire d'analyse fonctionnelle (Polytechnique) (1974–1975), no. 22.

[Tom2] Nicole Tomczak-Jaegermann, *Banach-Mazur distances and finite-dimensional operator ideals*, Pitman Monographs and Surveys in Pure and Applied Mathematics 38, Longman Scientific & Technical, Harlow, 1989.

[Ton] Alfred Esperanza Tong, *On the existence of noncompact bounded linear operators between certain Banach spaces*, Israel J. Math. **10** (1971), no. 4, 451–456, DOI: 10.1007/BF02771732.

[TW] Alfred Esperanza Tong and Donald R. Wilken, *The uncomplemented subspace $K(E, F)$*, Studia Math. **37** (1971), no. 3, 227–236.

[Tor] Henryk Toruńczyk, *Smooth partitions of unity on some non-separable Banach spaces*, Studia Math. **46** (1973), no. 1, 43–51.

[Tr] Stanimir L. Troyanski, *Gâteaux differentiable norms in L_p*, Math. Ann. **287** (1990), no. 1, 221–227, DOI: 10.1007/BF01446889.

[Va1] Jon Vanderwerff, *Smooth approximations in Banach spaces*, Proc. Amer. Math. Soc. **115** (1992), no. 1, 113–120, DOI: 10.1090/S0002-9939-1992-1081100-8.

[Va2] Jon Vanderwerff, *Second-order Gâteaux differentiability and an isomorphic characterization of Hilbert spaces*, Q. J. Math. **44** (1993), no. 2, 249–255, DOI: 10.1093/qmath/44.2.249.

[Ve] Libor Veselý, *Boundary of polyhedral spaces: an alternative proof*, Extracta Math. **15** (2000), no. 1, 213–217.

[Vi] Ignacio Villanueva, *Completely continuous multilinear operators on $C(K)$ spaces*, Proc. Amer. Math. Soc. **128** (2000), no. 3, 793–801, DOI: 10.1090/S0002-9939-99-05396-4.

[We1] John Wells, *Differentiable functions on c_0*, Bull. Amer. Math. Soc. **75** (1969), no. 1, 117–118, DOI: 10.1090/S0002-9904-1969-12167-1.

[We2] John Wells, *C^1 partitions of unity on nonseparable Hilbert space*, Bull. Amer. Math. Soc. **77** (1971), no. 5, 804–807, DOI: 10.1090/S0002-9904-1971-12813-6.

[Wh] Hassler Whitney, *Analytic extensions of differentiable functions defined in closed sets*, Trans. Amer. Math. Soc. **36** (1934), no. 1, 63–89, DOI: 10.1090/S0002-9947-1934-1501735-3.

[Wi] William Harold Wilson, *On a certain general class of functional equations*, Amer. J. Math. **40** (1918), no. 3, 263–282, DOI: 10.2307/2370484.

[Woj] Przemysław Wojtaszczyk, *Banach spaces for analysts*, Cambridge Stud. Adv. Math. 25, Cambridge University Press, Cambridge, 1991.

[Wol] John Wolfe, *$C(\alpha)$ preserving operators on $C(K)$ spaces*, Trans. Amer. Math. Soc. **273** (1982), no. 2, 705–719, DOI: 10.1090/S0002-9947-1982-0667169-9.

[X1] Yuan Xu, *Gaussian cubature and bivariate polynomial interpolation*, Math. Comp. **59** (1992), no. 200, 547–555, DOI: 10.1090/S0025-5718-1992-1140649-8.

[X2] Yuan Xu, *Lagrange interpolation on Chebyshev points of two variables*, J. Approx. Theory **87** (1996), no. 2, 220–238, DOI: 10.1006/jath.1996.0102.

[Za1] Ignacio Zalduendo, *A canonical extension for analytic functions on Banach spaces*, Trans. Amer. Math. Soc. **320** (1990), no. 2, 747–763, DOI: 10.1090/S0002-9947-1990-1001952-X.

[Za2] Ignacio Zalduendo, *An estimate for multilinear forms on ℓ_p spaces*, Proc. R. Ir. Acad., Sec. A **93A** (1993), no. 1, 137–142.

[Zie] Günter Matthias Ziegler, *Lectures on polytopes*, Grad. Texts in Math. 152, Springer-Verlag, New York, 1998.

[Zip1] Mordecay Zippin, *On perfectly homogeneous bases in Banach spaces*, Israel J. Math. **4** (1966), no. 4, 265–272, DOI: 10.1007/BF02771642.

[Zip2] Mordecay Zippin, *The separable extension problem*, Israel J. Math. **26** (1977), no. 3–4, 372–387, DOI: 10.1007/BF03007653.

[Ziz] Václav Zizler, *Smooth extension of norms and complementability of subspaces*, Arch. Math. **53** (1989), no. 6, 585–589, DOI: 10.1007/BF01199818.

[Zo] Vladimir Antonovich Zorich, *Mathematical analysis I*, Universitext, Springer-Verlag, Berlin, Heidelberg, 2004.

Notation

Index

De Gruyter Series in Nonlinear Analysis and Applications

Volume 22
Convex Analysis and Optimization in Hadamard Spaces
Miroslav Bácak, 2014
ISBN 978-3-11-036103-2, e-ISBN 978-3-11-036162-9, Set-ISBN 978-3-11-036163-6

Volume 21
Nonlinear Second Order Elliptic Equations Involving Measures
Moshe Marcus, Laurent Véron, 2013
ISBN 978-3-11-030515-9, e-ISBN 978-3-11-030531-9, Set-ISBN 978-3-11-030532-6

Volume 20
Impulsive Differential Inclusions: A Fixed Point Approach
John R. Graef, Johnny Henderson, Abdelghani Ouahab, 2013
ISBN 978-3-11-029361-6, e-ISBN 978-3-11-029531-3, Set-ISBN 978-3-11-029532-0

Volume 19
Smooth Analysis in Banach Spaces
Petr Hájek, Michal Johanis, 2014
ISBN 978-3-11-025898-1, e-ISBN 978-3-11-025899-8, Set-ISBN 978-3-11-220385-9

Volume 18
Solution Sets for Differential Equations and Inclusions
Smaïl Djebali, Lech Górniewicz, Abdelghani Ouahab, 2012
ISBN 978-3-11-029344-9, e-ISBN 978-3-11-029356-2, Set-ISBN 978-3-11-029357-9

Volume 17
Bounded Variation and Around
Jürgen Appell, Józef Banas, Nelson José Merentes Díaz, 2013
ISBN 978-3-11-026507-1, e-ISBN 978-3-11-026511-8, Set-ISBN 978-3-11-026624-5

Bei Fragen zur Produktsicherheit wenden Sie sich bitte an:
If you have any questions regarding product safety,
please contact:

Walter de Gruyter GmbH
Genthiner Straße 13
10785 Berlin
productsafety@degruyterbrill.com